COATINGS TRIBOLOGY

TRIBOLOGY AND INTERFACE ENGINEERING SERIES

Vol. 27 Dissipative Processes in Tribology (Dowson et al., Editors)
Vol. 28 Coatings Tribology – Properties, Techniques and Applications in Surface Engineering (Holmberg and Matthews)
Vol. 29 Friction Surface Phenomena (Shpenkov)
Vol. 30 Lubricants and Lubrication (Dowson et al., Editors)
Vol. 31 The Third Body Concept: Interpretation of Tribological Phenomena (Dowson et al., Editors)
Vol. 32 Elastohydrodynamics – '96: Fundamentals and Applications in Lubrication and Traction (Dowson et al., Editors)
Vol. 33 Hydrodynamic Lubrication – Bearings and Thrust Bearings (Frêne et al.)
Vol. 34 Tribology for Energy Conservation (Dowson et al., Editors)
Vol. 35 Molybdenum Disulphide Lubrication (Lansdown)
Vol. 36 Lubrication at the Frontier – The Role of the Interface and Surface Layers in the Thin Film and Boundary Regime (Dowson et al., Editors)
Vol. 37 Multilevel Methods in Lubrication (Venner and Lubrecht)
Vol. 38 Thinning Films and Tribological Interfaces (Dowson et al., Editors)
Vol. 39 Tribological Research: From Model Experiment to Industrial Problem (Dalmaz et al., Editors)
Vol. 40 Boundary and Mixed Lubrication: Science and Applications (Dowson et al., Editors)
Vol. 41 Tribological Research and Design for Engineering Systems (Dowson et al., Editors)
Vol. 42 Lubricated Wear – Science and Technology (Sethuramiah)
Vol. 43 Transient Processes in Tribology (Lubrecht, Editor)
Vol. 44 Experimental Methods in Tribology (Stachowiak and Batchelor)
Vol. 45 Tribochemistry of Lubricating Oils (Pawlak)
Vol. 46 An Intelligent System For Tribological Design In Engines (Zhang and Gui)
Vol. 47 Tribology of Elastomers (Si-Wei Zhang)
Vol. 48 Life Cycle Tribology (Dowson et al., Editors)
Vol. 49 Tribology in Electrical Environments (Briscoe, Editor)
Vol. 50 Tribology & Biophysics of Artificial Joints (Pinchuk)
Vol. 52 Tribology of Metal Cutting (Astakhov)
Vol. 53 Acoustic Emission in Friction (Baranov et al.)
Vol. 54 High Pressure Rheology for Quantitative Elastohydrodynamics (Bair)
Vol. 55 Tribology of Polymeric Nanocomposites: Friction and Wear of Bulk Materials and Coatings (Friedrich and Schlarb)

Aims & Scope

The Tribology Book Series is well established as a major and seminal archival source for definitive books on the subject of classical tribology. The scope of the Series has been widened to include other facets of the now-recognised and expanding topic of Interface Engineering.

The expanded content will now include:
• colloid and multiphase systems; • rheology; • colloids; • tribology and erosion; • processing systems; • machining; • interfaces and adhesion; as well as the classical tribology content which will continue to include • friction; contact damage; • lubrication; and • wear at all length scales.

TRIBOLOGY AND INTERFACE ENGINEERING SERIES, 56
EDITOR: B.J. BRISCOE

COATINGS TRIBOLOGY

Properties, Mechanisms, Techniques and Applications in Surface Engineering

Second Edition

Kenneth Holmberg

VTT – Technical Research Centre of Finland

Allan Matthews

The University of Sheffield, UK

ELSEVIER

Amsterdam • Boston • Heidelberg • London • New York • Oxford
Paris • San Diego • San Francisco • Singapore • Sydney • Tokyo

Elsevier
Radarweg 29, PO Box 211, 1000 AE Amsterdam, The Netherlands
The Boulevard, Langford Lane, Kidlington, Oxford OX5 IGB, UK

First edition **1994**

British Library Cataloguing in Publication Data
Holmberg, Kenneth.
Coatings tribology: contact mechanisms, deposition techniques and application. – 2nd ed. – (Tribology and interface engineering series; v. 56)
1. Tribology. 2. Coatings.
I. Title II. Series III. Matthews, A. (Allan)
621.8′9-dc22

Library of Congress Cataloging-in-Publication Data
A catalog record for this book is available from the Library of Congress
Library of Congress Control Number: 2009920710

ISBN: 978-0-444-52750-9
ISSN: 1572-3364

For information on all **Elsevier** publications
visit our web site at books.elsevier.com

Printed and bound in the United Kingdom

Transferred to Digital Print on Demand 2009

To
Ville, Maya and Hanna
and
Dave and Lynda

CONTENTS

PREFACE

This book gives a comprehensive description of thin surface coatings, their behaviour and potential uses in tribological applications. The deposition techniques, tribological mechanisms, properties of coatings, characterization, evaluation and selection methodology in addition to application examples are described and discussed. One aim of the book has been to bring together, systematically in a single volume, the state-of-the-art knowledge on tribological coatings.

The book is the result of very fruitful research cooperation between the tribology research group at the VTT Technical Research Centre of Finland in Helsinki and the Surface Engineering research group at the University of Sheffield in the UK. It brings together accumulated knowledge both on coating processes and on the tribology of coatings.

The authors published the first edition of the book in 1994. The contents in this second edition are thoroughly revised and expanded by more than 50%. Completely new sections have been written related to deposition of composite and nanostructured coatings, nanotribological contact mechanisms, modelling and simulation of stresses in coated surfaces, fracture mechanics analysis, wear product generation, lubricated contact mechanisms, diamond-like carbon and molybdenum disulphide coatings, coating selection and application examples. In the earlier text, we tried to take into account the most important published works at that time which covered over 1500 in all and we cited over 800 of these. Since then, the field has seen a remarkable growth in published output, and in the present book more than 3000 publications were scrutinized, of which about 1700 are cited herein.

As in the earlier book, we have tried to present the contents in a logical, easily assimilated, manner. We have retained the systematic structure of the earlier book, which has remained appropriate and robust, despite the many developments over the ensuing years.

Kenneth Holmberg
Allan Matthews
Helsinki and Sheffield
September 2008

ACKNOWLEDGEMENTS

A great number of people, including colleagues, friends and family, have through the years supported and encouraged the authors with their work in the field of coatings and tribology. We are most grateful to all of them and express our sincere thanks.

The authors also extend their thanks to Mr Erkki Makkonen for meticulous preparation of the illustrations.

The financial support of the Swedish Academy of Engineering Sciences in Finland and the VTT Technical Research Centre of Finland is gratefully acknowledged.

NOTATION

Symbols

a length of crack or deformation, or after
A area, e.g. contact area or projected area of indentation
A_a contact area contributed by deformed asperities
A_c load support area of the coating
A_d debris contact area
A_s load support area of the substrate
b half contact length or before
c constant or crack length
C content
d diameter or indentation diagonal length or crack depth
E Young's modulus of elasticity or elastic modulus
E' reduced modulus of elasticity, $1/E' = 0.5((1 - \nu_1^2)/E_1 + (1 - \nu_2^2)/E_2)$
F friction force
G material parameter in elastohydrodynamic lubrication, $G = \alpha \cdot E'$, or toughness
$\mathbf{G}$ strain–energy release rate
H hardness
H' film thickness parameter, $H' = h/R'$
h film thickness or indentation displacement
k constant
K wear rate, $K = V/w \cdot s$, or fracture toughness or stress intensity factor
K' coefficient of wear, $K' = V \cdot H/w \cdot s$
l length of contact
L_c critical load
m indentation size effect (ISE) index
M torque
n number of cycles
p pressure
r radius
R radius

R' radius of conformity, $1/R' = (1/R_1) + (1/R_2)$
R_a centreline average value (CLA or R_a) for the roughness of the surface
Rc Rockwell hardness (c-scale)
RH relative humidity
R_{max} maximum height of irregularities
R_t peak-to-valley height value for the roughness of the surface
R_z ten points height value for the roughness of the surface
s distance
S shear strength or slope of unloading curve
t' dummy variable
T temperature or transition zone
u velocity
u' displacement in x-direction
U speed parameter, $U = (\eta_o \cdot u')/(E' \cdot R')$ or elastic energy stored
v velocity
v' displacement in y-direction
V volume of worn or deformed material
w normal load
w' load per unit length
W load parameter, $W = w/(E' \cdot R' \cdot l)$, or work
x distance in x-direction or length of distance
y distance in y-direction or length of distance
Y yield stress or correction factor
z distance in z-direction
Z crack shape factor

Greek symbols

α pressure exponent of viscosity or coefficient of thermal expansion or empirical contact
β percentage of asperity contact area under compression
γ Boussinesq function or surface energy
Γ deviator stress or unit area
ε strain or indenter geometry constant
ζ standard deviation of asperity height distribution
ρ crack spacing
η absolute dynamic viscosity
Θ angle
χ empirical interface parameter
λ specific film thickness, $\lambda = h_{min}/R'_a$
μ coefficient of friction, $\mu = F/w$
v Poisson's ratio
σ stress or strength in tension
τ stress or strength in shear
τ_a shear strength within asperity adhesion area

Subscripts

a adhesion or asperity or abrasive
c coating or critical or contact

comp	composite
d	deformation or debris
e	effective
f	film or final
fat	fatigue
H	hydrogen
Hz	Hertzian
i	counter, $i = 1, 2, \ldots, N$
int	interface
I	interface
j	junction
k	Knoop
m	melting or hydrostatic
max	maximum
min	minimum
o	ambient
p	ploughing or peak
r	radial or rough or reduced or residual
res	residual
s	substrate
th	thermal
u	upper surface
v	Vickers
vM	von Mises
y	yield

CHAPTER 1

Introduction

Tribology is the field of science and technology dealing with contacting surfaces in relative motion – which means that it deals with phenomena related to friction, wear and lubrication.

Tribology played a central role in early technological evolution, even in ancient times. Reducing friction by using wheels made it possible for humans to move further and faster, and the lubrication of sleds made it possible to transport building blocks and raise large structures. Together with good tribological engineering knowledge, metal as a construction material and oil as a lubricant eventually smoothed the path for the modern industrial revolution, and allowed new inventions like high strength and low friction bearings and gears that were key components in high power machinery, as described by Dowson (1998), Ludema (2001) and Holmberg (2004).

In modern industrialized societies there is a growing need to reduce or control friction and wear for several reasons, for example to extend the lifetime of machinery and bio-systems, to make engines and devices more efficient, to develop new advanced products, to conserve scarce material resources, to save energy, and to improve safety.

Historically these aims have been achieved by design changes, selecting improved bulk materials, or by utilizing lubrication techniques. Bulk material changes might involve applications with ceramics or polymers. The lubrication techniques would include the use of liquid lubricants such as mineral or synthetic oils or solid lubricants such as molybdenum disulphide.

Recently, tribologists have made increasing use of another approach to friction and wear control – that is to utilize surface treatments and coatings. This has led to, and to some extent been fuelled by, the growth of a new discipline called surface engineering. This growth has been encouraged by two main factors. The first has been the development of new coating and treatment methods, which provide coating characteristics and tribochemical properties that were previously unachievable. The second reason for the growth in this subject area has been the recognition by engineers and materials scientists that the surface is the most important part in many engineering components. It is at the surface that most failures originate, either by wear, fatigue or corrosion. The surface has a dominant influence on lifetime cost and performance, including machinery maintainability.

The surface may also have other functionally important attributes, not confined to mechanical or chemical properties, such as thermal, electronic, magnetic and optical characteristics that influence the choice of surface material. The retention of these physical surface properties is clearly essential throughout the life of the product. It is a further reason why surface durability enhancement by appropriate coating selection is critical to any product's effectiveness, and therefore its saleability in the marketplace.

Like all products, mechanical components and tools are today facing higher performance requirements. The use of surface coatings opens up the possibility for a material design in which the specific properties are located where they are most needed. The substrate material can be designed for strength and toughness while the coating is responsible for the resistance to wear, corrosion and thermal loads, and the achievement of the required frictional characteristics.

Tribologists developed an understanding of the behaviour of surfaces in contact, providing a theoretical basis for the prediction of the desirable attributes of surfaces, even before fully optimized coatings were available.

It is against this background that new coating and treatment methods are being developed, and are already having a significant impact. Devices and bearing systems which operate under near-vacuum conditions, as in space mechanisms or satellites, or engine components operating under hot corrosive and erosive conditions, as in aero gas turbines, could not function without advanced tribological coatings.

At present, hard coatings such as titanium nitride, titanium carbide and aluminium oxide are commonly used on cutting tools in the manufacturing industry. Chromium nitride and molybdenum disulphide coatings are used on metal forming tools. Very hard but also low friction diamond-like carbon coatings are deposited for wear protection on magnetic storage devices produced for computers. Optical lenses are produced with hard erosion-resistive thin transparent coatings. Various carbon-based coatings are used on components in the automotive industry to reduce energy consumption. The coatings in some applications are deposited as multicomponent coatings, multilayers, gradient layers, superlattice structures and duplex-treated surfaces with various material combinations as shown by Hogmark *et al.* (2001).

While coatings for the applications cited are in commercial use, there are many others that are still at a developmental stage. In this book we set out the background to present developments in tribological coatings, emphasizing the newer processes. We place this in the context of current thinking on the theories and experimental findings relating to the use of surface coatings in tribology.

We shall take as *our definition of tribological coatings those which are sufficiently thin that the substrate material plays a role in the friction and wear performance*. Thus, we exclude coatings which are so thick that there is little or no substrate influence on the tribological behaviour – the coating in effect acts as a bulk material. Weld deposits are a typical example of such thick coatings that have been excluded from in-depth study.

With this definition of tribological coatings we focus on solid surface films that are typically in a thickness range of 0.1–10 μm. New deposition techniques and improved nanotribological understanding have made it possible to produce tribologically effective solid surface films even as thin as 1–5 nm for magnetic storage device applications (Bhushan, 1999a; Wang *et al.*, 2005).

The coating processes can conveniently be divided into four generic groups – gaseous, solution, molten and solid – depending on the state of the depositing phase (Rickerby and Matthews, 1991a). Our definition of tribological coatings means that we shall concentrate mainly on the gaseous state processes, which are attracting considerable scientific and commercial interest. In particular, the main coatings that we shall consider will be those deposited by plasma-assisted techniques, since those can provide excellent adhesion to the substrate and the dense coating structural morphologies which are needed for tribological applications.

A general design appraisal of the tribological requirements on contact with coated surfaces can be formulated as follows:

1. The initial coefficient of friction, the steady-state coefficient of friction and the friction instability must not exceed certain design values.
2. The wear of the coated surface and that of the counterface must not exceed certain design values.
3. The lifetime of the system must, with a specified probability, be longer than the required lifetime. The lifetime limit of the system may be defined as occurring when at least one of the earlier requirements is not maintained.

To meet the tribological requirements, the coated surface must possess a suitable combination of properties, for example in terms of hardness, elasticity, shear strength, fracture toughness, thermal expansion and adhesion. As shown in Fig. 1.1 we can distinguish between four different zones, each with different properties which must be considered.

The properties required by the substrate and by the coating involve material strength and thermal attributes determined by their composition and microstructure as well as the porosity and homogeneity of the material. At the interface between them, the adhesion and shear strength of the junction is important. At the surface of the coating the chemical reactivity and the roughness must be considered in addition to the shear strength.

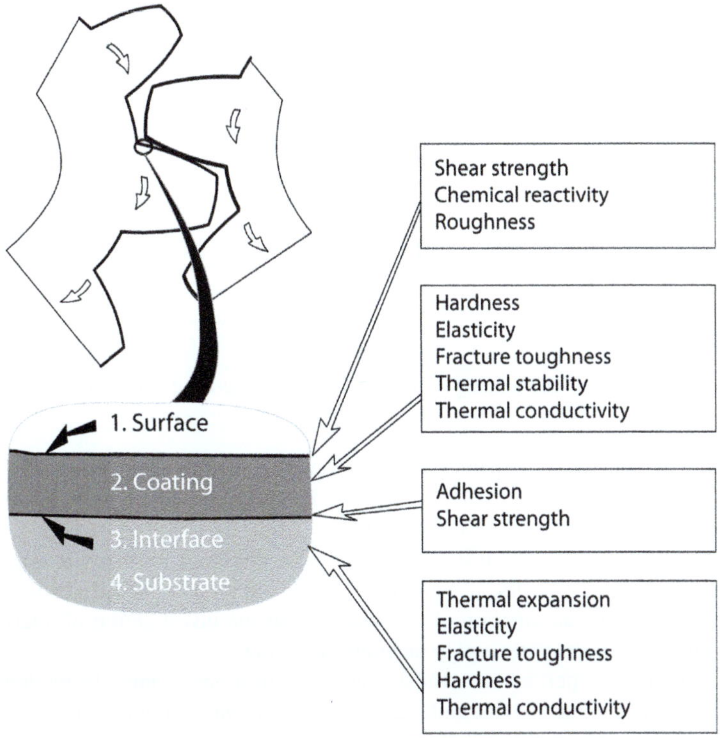

Fig. 1.1. Tribologically important properties in different zones of the coated surface.

A primary problem in surface design is that many desired properties, such as good adhesion at the coating/substrate interface and no surface interactions with the counterface, or high hardness and high toughness of the coating, cannot easily be obtained simultaneously. Increased hardness and strength is often concomitant with decreasing toughness and adherence. For this reason the final coating design is always a compromise between many different technical requirements on the properties of the coating system and the economical requirements on the deposition of the coating on to products.

The factors that determine the coating material properties are the constitution of the material system and the fabrication parameters, such as the coating process and the thickness, as shown in Fig. 1.2. Both of these determine the microstructure of the coating, including, for example, its density, grain size, grain boundaries and grain orientation.

The whole coated surface system with its properties and functional parameters can be considered as a composite system to be optimized to gain maximum benefit (Rickerby and Matthews, 1991a, b). This, in essence, specifies the fundamental philosophy of surface engineering, which has been defined by Melford (1991) as 'the design of surface and substrate together, as a system, to give a cost-effective performance enhancement of which neither is capable on its own'.

The selection criteria for choosing one surface engineering technique over another are complex. It is often not realized that the properties of the bulk material may be impaired by the coating or surface treatment. A second important point is that, to get maximum benefit from a coating, a redesign procedure may be necessary. This further complicates an already complex task for the design engineer faced with the prospect of selecting from the many coatings available.

There have been various attempts to devise a selection procedure for surface coatings. Originally coatings were seen as a last resort solution to problems which had their roots in poor design or poor

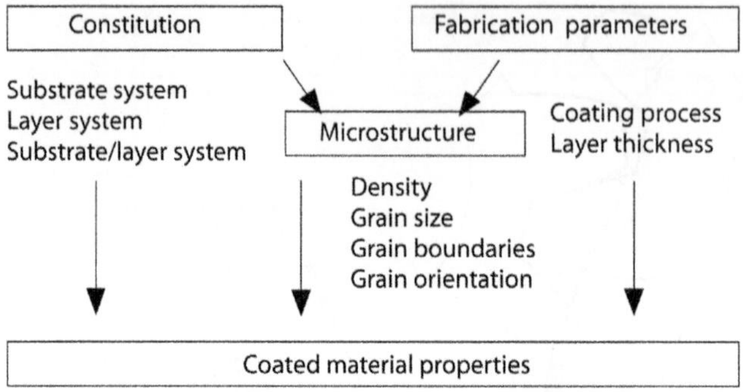

Fig. 1.2. Examples of factors influencing material properties of coated surfaces (after Holleck, 1986).

material selection. Frequently, companies would adopt what James (1978) describes as a positive selection approach; that is, a coating will be tried, and if it works it will be used. However, this often results in better solutions being rejected by default.

Progressive elimination of all feasible solutions is to be preferred for many reasons, even though it is more time consuming. The discipline inherent in this approach encourages the acquisition of information, rather than the assumption of knowledge, and the investigation of alternative ways of solving minor problems obstructing highly desirable solutions.

It is important to make expert advice available at the earliest opportunity. If freedom to contribute to the original design and to feed advice and suggestions before the design is finalized is permitted, the surface may play a significant role in producing a part with superior performance at a reasonable cost. Thus tribological coatings need not necessarily be viewed as an additional cost feature; they often allow the use of a lower-grade, cheaper substrate material while providing improved quality and performance.

The potential offered by computer-based modelling, simulation and coating selection tools coupled with the new coating processes, makes possible the solution of problems to which tribologists previously had only theoretical solutions. The holistic modelling of a coated tribological contact is a very complex task. However, the rapid increase in computer processing capacity in the last few decades together with improved software tools for modelling has brought us to the point where realistic model-based optimization of a coated surface for tribological use is a reality, as shown by Holmberg *et al.* (2003).

It has been known for many years that tribochemical reactions at the contact interface are critical in many tribological systems. Often these lead to the formation of stable compounds, after a period of rubbing, which then control the ensuing friction and wear behaviour. This occurs in applications as diverse as cutting and lubricated sliding. In the latter case steps are often taken to encourage the formation of specific compounds at the interface, for example by the use of extreme pressure additives. Advanced surface engineering techniques allow the design and production of such compounds as surface layers, prior to in-service use, thereby ensuring effective control of the tribological system.

Furthermore, not only is the sliding mechanism improved by such possibilities, it can also be possible to modify the thermal behaviour, altering both thermal and chemical diffusion effects. This in turn leads to the possibility of improving the efficiency of machines, with the consequential economic advantages of good tribological design.

The economical impact of friction and wear on society is huge. It has been estimated that the economic cost of wear and friction in USA exceeds US$100 billion per annum (Blau, 2008). Improved coatings and selection procedures lead to decreased friction and wear. This will enable improved efficiency over a longer lifetime and provide a considerable reduction in overall energy consumption, not to mention reductions in costs caused by in-service failure or maintenance down-time. An important objective for tribological coatings should be the achievement of extended and predictable lives, and ideally non-catastrophic failure modes.

The potential of tribological coatings is extensive, and it is hoped that this book will assist designers and others who specify coatings to do so more effectively and more widely. The book is structured in such a way that individual sections and chapters can be read, free-standing, in their own right. The main aim, however, is to provide a text that covers the important aspects of coatings tribology in a progressive manner, first covering the coating methods, then the fundamentals of tribology and how it can be applied to coatings and then the tribological properties of specific coatings. Following this there is a chapter dealing with coating characterization and evaluation, and one on coating selection. Finally, application examples are described and discussed. An Appendix gives data sheets which summarize tribological test results for a wide range of coatings.

CHAPTER 2

Deposition processes and coating structures

Contents

2.1 Introduction

The rapid development of tribological coatings in recent years is largely due to the availability of new coating methods which can provide enhancements in properties such as morphology, composition, structure, cohesion and adhesion, which were previously unachievable. The deposition techniques that have mostly caused the increasing interest in this field are the plasma- and ion-based methods. This chapter will therefore concentrate primarily on processes within this category.

Although the main principles of most of these processes have been known for over 70 years, the requirements for large-scale commercial exploitation have only been fulfilled during the last few decades. Their introduction was delayed due to difficulties with the advanced technologies involved, including: (1) high-current and high-voltage power supply technology, (2) process control and related electronic technologies, (3) plasma physics and chemistry, and (4) vacuum technology. Plasma-based techniques now offer the solution to many of these problems and considerable benefits to many industrial sectors.

Other coating methods are not to be neglected, however, and indeed it must be realized that researchers investigating more traditional processes have responded to the challenges of the newer methods by introducing significant improvements. In line with our objectives in this book, we shall concentrate on the methods that are used to produce relatively thin surface layers, such that the substrate and coating combine to provide a tribological performance which neither can achieve on its own.

We shall use the general classification system for the deposition processes presented earlier by Rickerby and Matthews (1991a). This divides the processes into the following four categories:

- gaseous state processes,
- solution state processes,
- molten or semi-molten state processes, and
- solid-state processes.

The solid-state processes will not be discussed in detail since they tend to be used to produce the thicker coatings which are outside our remit. The surface engineering techniques included in this general classification are shown in Fig. 2.1.

Two important characteristic parameters for the coating processes are the thickness of the coatings that can be achieved and the deposition temperature. The range of the typical coating thicknesses varying from 0.1 μm to 10 mm is shown in Fig. 2.2a, and the range of deposition temperatures varying

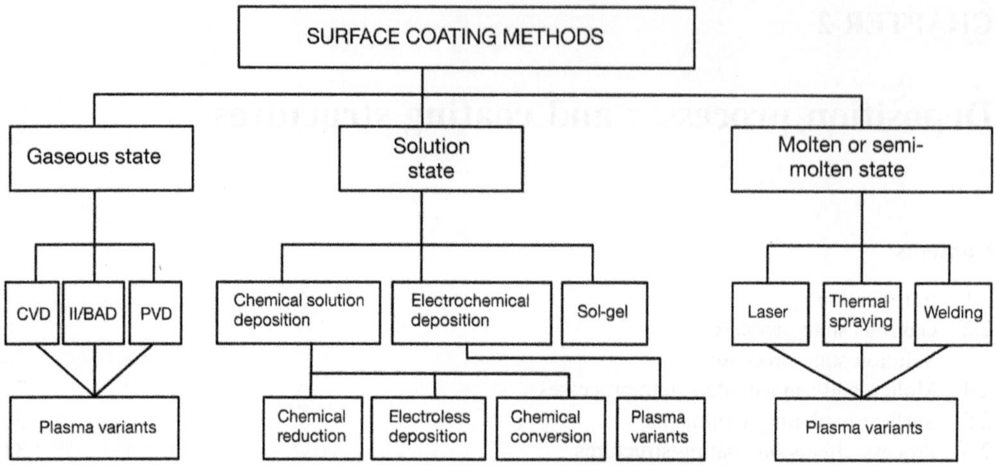

Fig. 2.1. A general classification of surface engineering techniques.

from room temperature up to 1000°C is shown in Fig. 2.2b. Recent research is investigating possibilities of using even thinner coatings, down to thicknesses of 1–3 nm, for tribological purposes (e.g. Wang *et al.*, 2005).

It is a feature of the newer technologies that acronyms are rife – indeed any one process may be known by a variety of such acronyms. In the figures and in the text the following abbreviations are used:

PA	Plasma assisted
CVD	Chemical vapour deposition
PVD	Physical vapour deposition
IBAD	Ion beam assisted deposition
IAC	Ion assisted coating
II	Ion implantation

More detailed presentations on the surface coating methods are found in Matthews (1990), Rickerby and Matthews (1991a, b), Gissler and Jehn (1992), *ASM Handbook* (1994), Kennedy and Gibson (1997), Cartier (2003) and Mellor (2006).

Some characteristics of coating methods lying within the different processing categories are presented in Table 2.1. As can be seen, there are wide differences in the range of deposition rates, temperatures, pre- and post-treatments and thicknesses achievable. What is not evident from this table are the enormous differences in the types and character of the deposits produced, for example in compositions and morphologies.

2.2 Gaseous State Processes

2.2.1 General

Gaseous state processes cover surface engineering techniques in which the coating or surface treatment material passes through a gaseous or vapour phase prior to depositing on to or modifying the surface. The main generic coating subgroups are chemical vapour deposition (CVD) and physical vapour deposition (PVD). The former utilizes gaseous reagents as the source of coating species, whereas in the latter at least one of the coating species is evaporated or otherwise atomized from solid within the coating chamber.

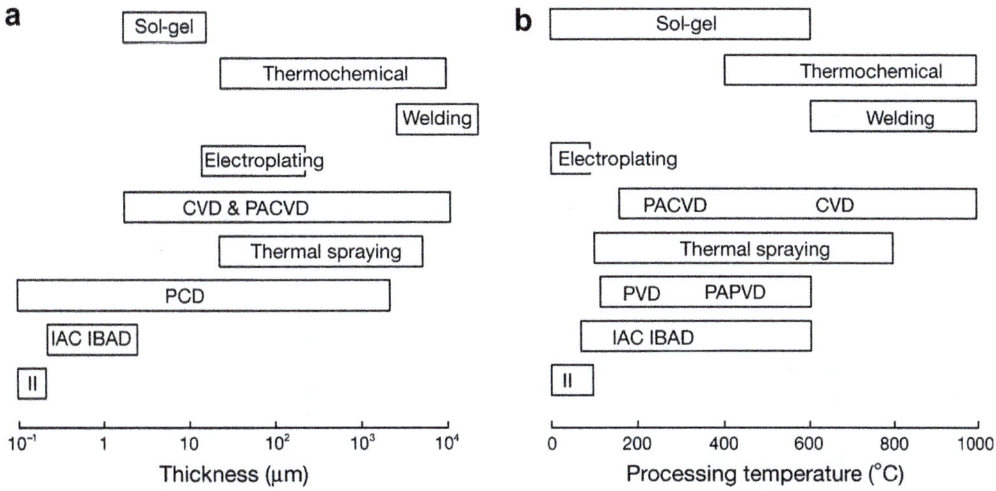

Fig. 2.2. Typical ranges for (a) depths of surface modifications and thicknesses of coatings, and (b) processing temperatures for coating technologies.

As shown in Table 2.1, these methods are of considerable interest for a number of reasons, not least because they permit the deposition of pure ceramic films. In the case of PVD this requires a means of increasing the energy of the coating species, most typically by ionizing them and accelerating the ions towards the growing film. This can be achieved by utilizing an ion beam source, or by initiating plasma around the substrate, from which ions can be accelerated. Below we overview the CVD and PVD process technologies.

2.2.2 Chemical vapour deposition

In the basic chemical vapour deposition (CVD) process, gases containing material to be deposited are introduced into a reaction chamber, and condense on to the substrate to form a coating. The process normally requires deposition temperatures in the range of 800–1200°C. Figure 2.3 shows a hot-wall CVD layout typically used for tool coating with titanium nitride and titanium carbide.

Figure 2.4 shows the many derivatives of the basic CVD process. These have arisen in response to a need to achieve specific coating characteristics, such as epitaxial growth, improved hole penetration or lower deposition temperatures. Furthermore, certain coating types (e.g. diamond) are only achievable using particular process technologies and parameters. In this sense CVD is now almost as diverse in the variants available as PVD.

The deposition pressure in CVD can range from atmospheric down to 1 Pa or less. There are various means of assisting the process, such as through the use of laser or electron beams, or by ion bombardment of the growing films. These methods promise the benefit of a reduction in coating temperatures necessary to obtain dense and well-adhered films.

Notwithstanding these developments, there is still considerable scope for the wider use of conventional non-enhanced thermally activated CVD coating methods for tribological applications. The strength of the technique lies in its ability to produce well-adhered, uniform and dense surface layers. The grain orientation and size, coating composition and its properties can be varied by the selection of appropriate process parameters. The technique can be used to deposit a large number of wear-resistant coatings such as borides, carbides, nitrides, oxides, carbo-nitrides and oxy-nitrides of almost all the transition metals (Sudarshan, 1992). In particular the use of such coatings on substrates which are not

Table 2.1. Comparative typical characteristics of some of the main coating methods.

	Gaseous State Processes					Solution Processes		Molten or Semi-Molten State Processes		
	PVD	PAPVD	CVD	PACVD	Ion Implantation	Sol-Gel	Electro-Plating	Laser	Thermal Spraying	Welding
Deposition rate (kg/h)	Up to 0.5 per source	Up to 0.2	Up to 1	Up to 0.5		0.1–0.5	0.1–0.5	0.1–1	0.1–10	3.0–50
Coating thickness or treatment depth (μm)	0.1–1000	0.1–100	0.5–2000	1–20	0.01–0.5	1–10	10–500	50–2000	50–1000	1000–10,000
Component size		Limited by chamber size				Limited by solution bath		May be limited by chamber size		
Substrate deposition or treatment temperature (°C)	50–500	25–500	150–12,000	150–700	50–200	25–1000	25–100	200–2000	100–800	500–1200
Substrate material	Metals, ceramics, polymers	Metals, ceramics	Metals, ceramics	Metals, ceramics	Metals, ceramics, polymers	Metals, ceramics, polymers	Metals, ceramics, polymers	Metals		
Pretreatment	Mechanical/ chemical	Mechanical/ chemical plus ion bombardment	Mechanical/ chemical	Mechanical/ chemical plus ion bombardment	Chemical plus ion bombardment	Grit blast and/or chemical clean	Chemical cleaning and etching	Mechanical and chemical cleaning		
Post-treatment	None	None	Substrate stress relief	None	None	High temperature	None/thermal treatment	None/substrate stress relief		None
Uniformity of coating	Good	Good	Very good	Good	Line of sight	Fair/good	Fair/good	Fair	Variable	Variable
Bonding mechanism	Atomic	Atomic plus diffusion	Atomic	Atomic plus diffusion	Integral	Surface forces		Mechanical/ chemical/ metallurgical		

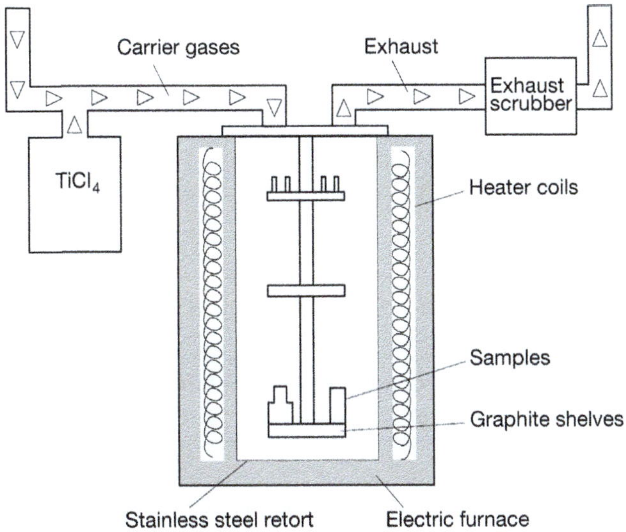

Fig. 2.3. A typical CVD process layout.

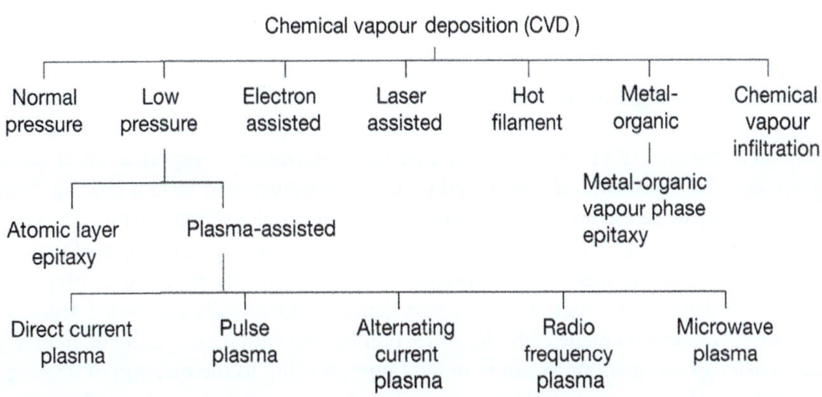

Fig. 2.4. Some chemical vapour deposition-based coating processes and ion-based derivatives.

temperature sensitive, such as cemented carbides, will continue, especially when the excellent coating uniformity and recess penetration capabilities of this technique are required.

Carlsson (1991) has described schematically the reaction zones present in thermally activated CVD, as shown in Fig. 2.5. A feature of this model is that a boundary layer is identified across which the reactants and reaction products are transported. This controls the deposition rate, while the heterogeneous reactions in zone 2 usually define the microstructure of the coating. Typical CVD-deposited films for tribological applications, such as titanium nitride, have large-angle grain boundaries and grain sizes in the range 0.5–5 μm.

The microstructure depends critically on the deposition temperature, which for TiN is typically 950°C, but for CVD in general can vary from room temperature up to over 1200°C. At higher temperatures various solid-state reactions, such as phase transformations, precipitation, recrystallization

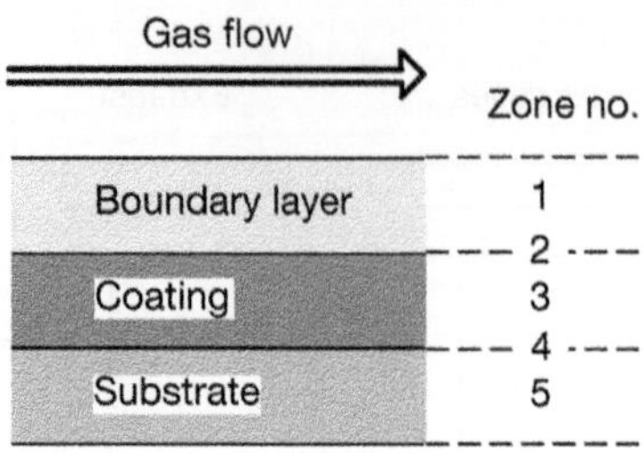

Fig. 2.5. Reaction zones in thermally activated CVD.

and grain growth, can occur in zones 3 to 5. In zone 4 interdiffusion between coating and substrate can lead to the formation of intermediate phases. Such effects are critical to the effective adhesion of the coating.

The above gives a generalized overview of the thermally activated CVD process. In practice, as indicated by Fig. 2.4 there are many variations on this and the aim of these modifications can be to allow lower deposition temperatures, to control deposition rates or to permit selective deposition. For plasma-assisted or -enhanced CVD many of the beneficial effects are analogous with those observed for plasma-assisted PVD. Since in that case plasma assistance is now almost mandatory, more details will be given on this in the following section. Plasma-assisted CVD is discussed in more detail by d'Agostino *et al.* (1992), Bachmann *et al.* (1988), Wheeler (2006) and in the *ASM Handbook* (1994).

2.2.3 Physical vapour deposition

Physical vapour deposition (PVD) processes involve the atomization or vaporization of material from a solid source and the deposition of that material on to the substrate to form a coating. They can be categorized according to the means of atomizing the source material, as shown in Fig. 2.6. Materials can be removed from the source or target either by transfer of kinetic energy or by input of thermal energy.

PVD methods are often divided into evaporation and sputtering. Evaporation involves the thermal vaporization of the deposition material source. Sputtering is a kinetically controlled process in which the source material or target is made cathodic and is bombarded with ions, usually of an inert gas. This results in a transfer of momentum to atoms in the target, leading to the ejection of coating atoms.

Both sputtering and evaporation techniques originated at about the same time. The first sputtering experiments were reported by Grove (1852), while the first reports of evaporation deposition were made by Faraday (1857). Recently, a number of new PVD techniques have been developed that are hybrids of thermal and kinetic methods, as discussed later.

Although PVD was originally used as a means of depositing elemental metallic coatings, it has been increasingly used for alloy and ceramic deposition. In the latter case this can be achieved either by using a ceramic source or by using a metal source in combination with a reactive gas such as oxygen, nitrogen or methane, to produce, e.g., oxides, nitrides or carbides. In addition, means now exist to deposit multi-element alloy or even composite metal/ceramic films. Some techniques are better suited for these applications than others, and this will be addressed when discussing each method.

Thermal and arc evaporation processes have been transformed by the addition of a plasma within the deposition chamber, which provides control of film nucleation and growth kinetics, allowing the production of coatings with previously unachievable properties. We shall thus describe how ion-assisted processes are incorporated into the most common PVD techniques and also outline some aspects of the influence of energetic particle bombardment on coating morphologies and properties.

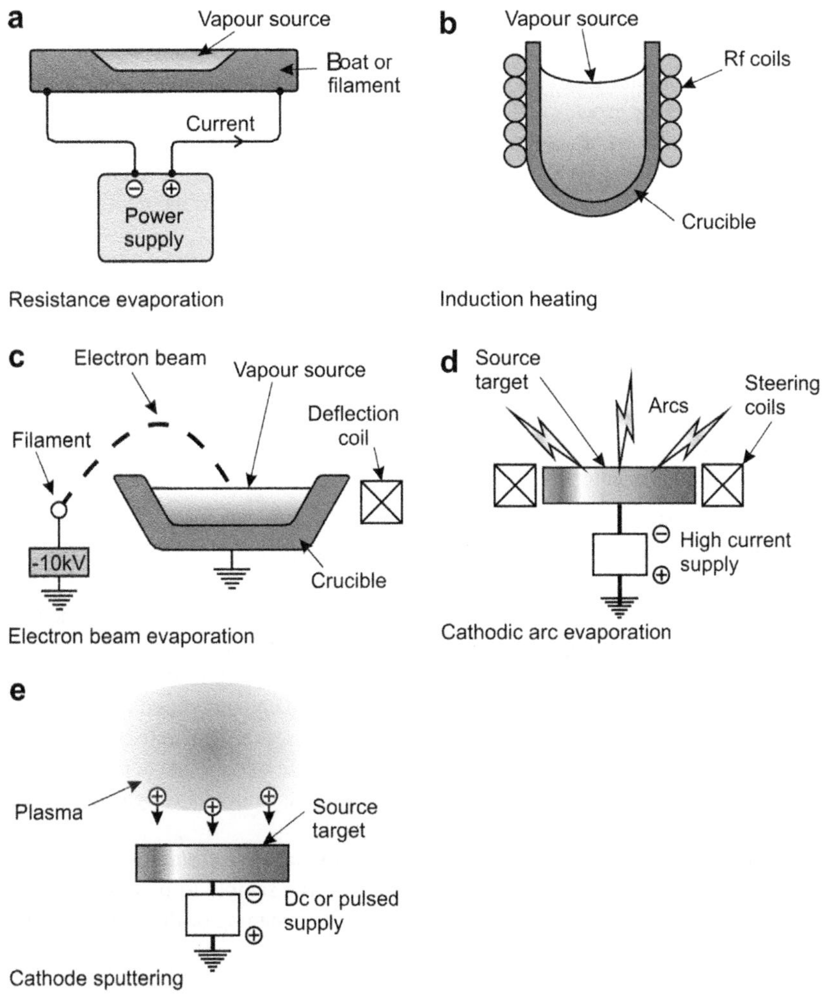

Fig. 2.6. Techniques used to atomize the source materials in PVD.

2.2.3.1 Plasma-assisted PVD

Although the basic PVD process has been known for over 150 years and the process which we now call plasma-assisted physical vapour deposition (PAPVD) was patented more than 70 years ago (Berghaus, 1938), it is only in the last few decades that PAPVD has begun to make a major mark as a tribological coating method. This is because the process has only recently been more fully understood at a fundamental level and the necessary modifications introduced to provide benefits such as those in the list below. These are the main advantages for such plasma-assisted processes, which principally relate to PVD but in several cases are also pertinent to plasma-assisted CVD:

1. Improved coating adhesion, due to the ability to clean and pre-heat substrates by energetic ion and neutral bombardment of the substrate surface. This mechanism is sometimes called sputter cleaning.
2. Uniform coating thicknesses, through gas-scattering effects and the ability to rotate or displace samples relative to the vapour source during deposition.

3. Avoidance of a final machining or polishing stage after coating, as in most cases the coating replicates the original surface finish.
4. Controlled coating structures, due to the effect of bombardment in encouraging adatom mobility and obviating columnar growth.
5. Deposition of a wide range of coating and substrate materials, including dielectrics, usually by the use of radio frequency or pulsed substrate biasing.
6. Controllable deposition rates, using a wide variety of vapour sources including resistance heated, electron beam, arc, induction and sputter magnetron.
7. Usually no effluents or pollutants are produced, as in most cases there are no harmful by-products or toxic chemical solutions used.
8. High purity deposits through the use of a controlled vacuum environment and pure source materials.
9. Lower deposition temperatures, as direct energization of the coating species provides many of the benefits previously only achievable on hot substrates. Thus lower bulk substrate temperatures can be used for ceramic deposition compared to non-plasma-assisted vapour deposition.
10. Avoidance of the hydrogen embrittlement problems sometimes experienced in electroplating.

The flexibility of the PVD coating process can be seen from Fig. 2.7, which shows how the coating structure can be varied as a function of deposition pressure and temperature according to Thornton (1974). Morphology is discussed in detail later in this chapter and in Chapter 5.

The advantages of the PAPVD processes extend beyond those listed. They include the possibility to deposit alloy compounds, multilayer compositions and structures, and the ability to vary coating characteristics continuously throughout the film, giving the concept of a functionally graded coating. The developments leading to these improvements have been many and varied and this has produced a proliferation of techniques and acronyms for various processes, as illustrated in Fig. 2.8.

This figure gives only a few of the many variations. In the field of diamond deposition, for example, there are probably almost as many processes or process variants being investigated as laboratories in the field.

Consider first the basic PAPVD process, which is also called *ion plating*. This was originally envisaged as being carried out at a reduced pressure, with the sample to be coated as the negative electrode in a glow discharge of argon. This led to the sputter-cleaning of the surface with argon ions

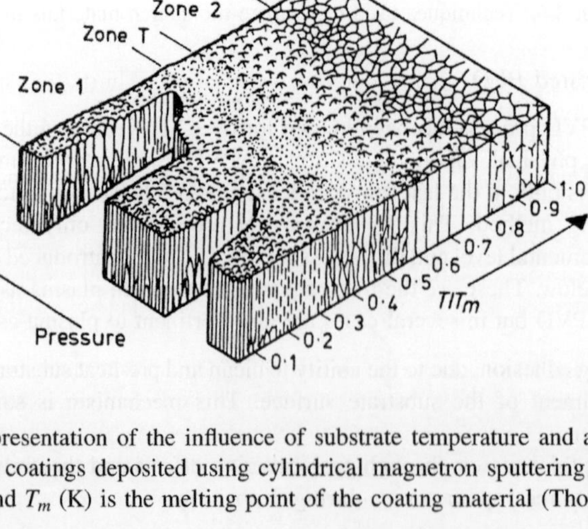

Fig. 2.7. Schematic representation of the influence of substrate temperature and argon pressure on the microstructure of metal coatings deposited using cylindrical magnetron sputtering sources. T (K) is the substrate temperature and T_m (K) is the melting point of the coating material (Thornton, 1974).

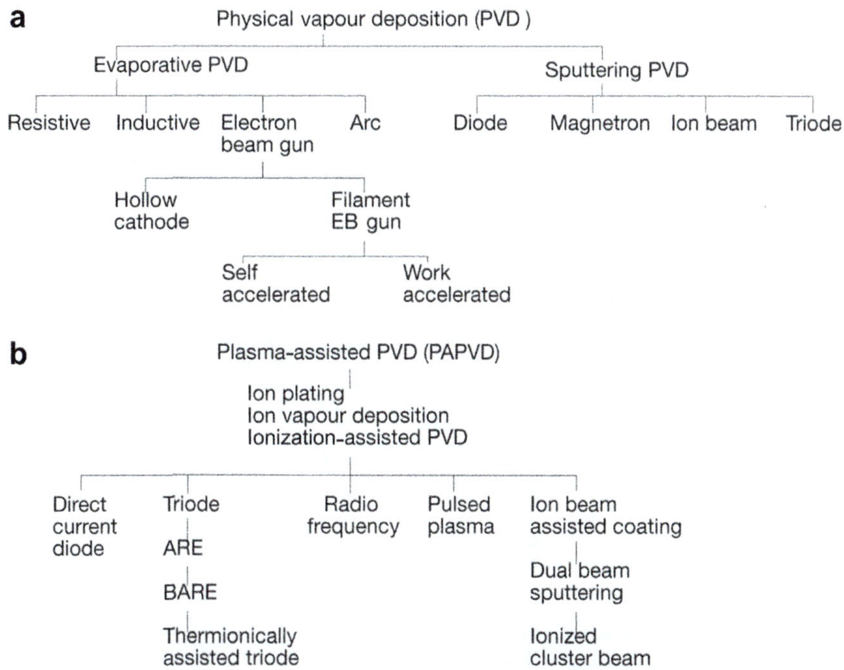

Fig. 2.8. (a) Some physical vapour deposition-based coating processes and (b) plasma derivatives.

and accelerated neutrals. Subsequently, the metal to be deposited was evaporated into the glow discharge or plasma, itself being ionized to an extent and arriving at the substrate surface with increased energy. This simple concept still lies at the heart of all PAPVD methods, though increasingly the level of ionization achieved is enhanced by some means in order to maximize or control the effects occurring.

The main mechanisms arising due to accelerated ion or neutral bombardment of a surface are illustrated by Fig. 2.9. One of the most important is the *sputtering mechanism*. This is generally considered to involve a momentum transfer process, in which the arriving atoms cause other atoms or clusters of atoms to be knocked free from the surface. This is important for three main reasons. First, it allows the surface to be cleaned of unwanted species prior to coating. Second, it allows the formation of a pseudo-diffusion layer between the substrate and coating, due to the forced intermixing of substrate and coating atoms during interface formation. Third, the sputtering mechanism is responsible for the continuous redistribution of surface coating atoms during film growth – which can lead to effects such as film densification.

Many of the effects created by accelerated ion or neutral bombardment can be enhanced by increasing the amount of ionization occurring, i.e. by ensuring that a larger proportion of the species arriving at the surface are in fact ionized. It has been shown that, provided some critical minimum energy is exceeded, it is more advantageous to increase the amount of ionization, i.e. the ionization efficiency, than to increase the energy, i.e. the acceleration voltage of the individual ions (Matthews, 1980).

One of the first systems to achieve ionization enhancement was the activated reactive evaporation (ARE) system (Bunshah and Raghuram, 1972), which was then modified to the biased ARE (BARE) system (Koboyashi and Doi, 1978) shown in Fig. 2.10a. These systems utilized an additional electrode in the deposition chamber, to increase the potential of the plasma, and ensure more effective ionization, essentially by making use of electrons emitted from the vapour source or cathodic surfaces

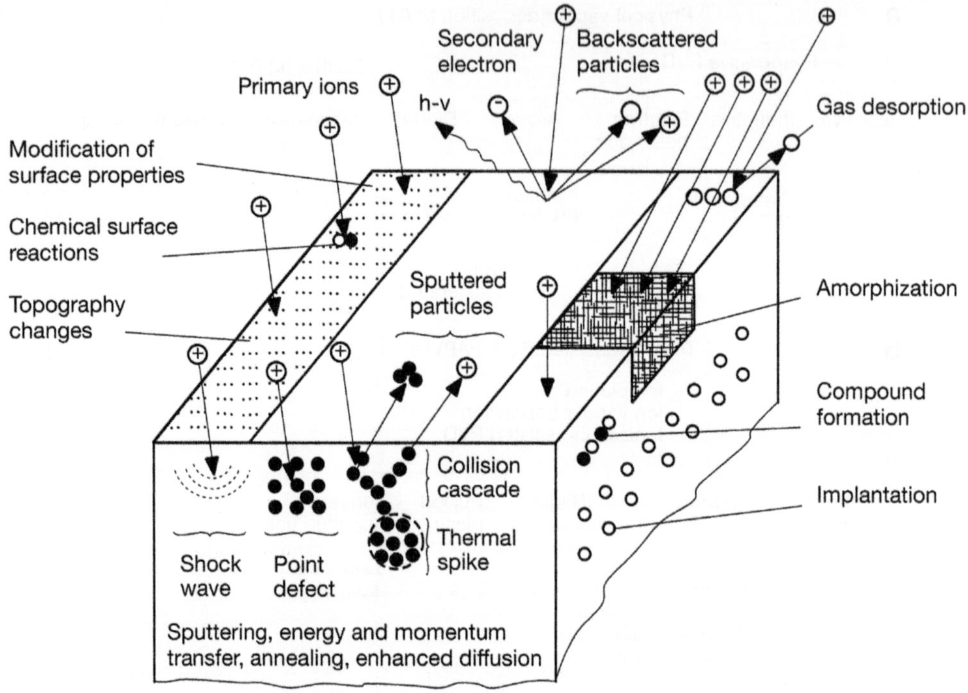

Fig. 2.9. Surface effects occurring during ion bombardment (after Weissmantel, 1983).

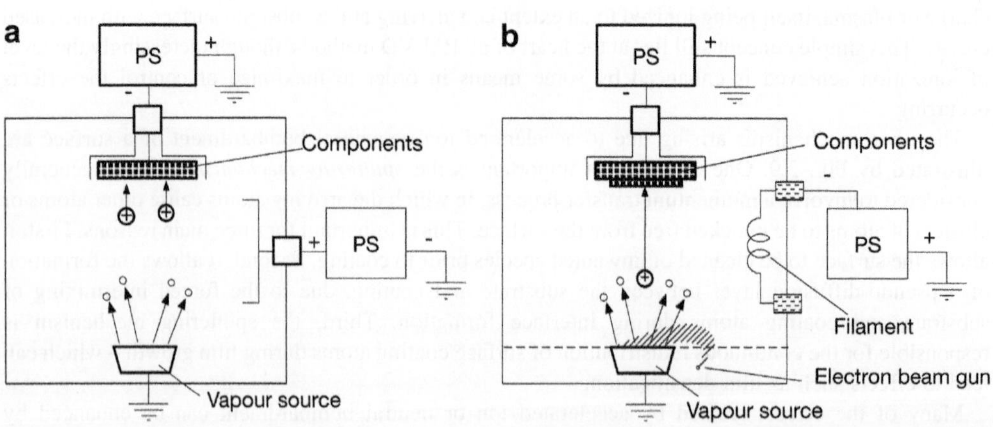

Fig. 2.10. (a) Layout of the biased activated reactive evaporation system and (b) the thermionically enhanced triode system.

within the chamber or both. An alternative is to insert an electron emitter into the system, in the form of a heated filament. This is used in the thermionically enhanced triode system (Matthews, 1980) shown in Fig. 2.10b.

As originally conceived the above systems utilized an electron beam gun to evaporate the source material. However, a variety of other methods can be used to achieve this. Some of the vapour sources themselves increase ionization.

As described in Fig. 2.6, the main techniques available to vaporize the coating material are: resistance-heated sources, electron beam guns, induction-heated sources, arc sources and sputter sources.

1. Resistance heating involves holding the evaporant in a holder, sometimes a filament or a so-called 'boat', made from a refractory material such as tungsten or molybdenum, or an intermetallic compound such as titanium diboride/boron nitride. The container is heated by passing a current through it. This method has been used for evaporating low-melting point materials such as aluminium, copper, silver and lead.
2. Electron beam guns have become popular for depositing thin ceramic coatings such as TiN and also for thicker bond coats and thermal barriers for gas turbine components. There is effectively no limit on the melting point of materials that can be evaporated, and the deposition rates achievable tend to be higher than other PVD methods. The most popular types of gun used are the bent beam self-accelerated gun (Schiller *et al.*, 1986; Pulker, 1984), the work accelerated gun (Moll and Daxinger, 1980; Hill, 1986), and a variant of the latter, known as the hollow cathode discharge (HCD) electron beam gun (Morley and Smith, 1972).

   The last two systems, shown in Fig. 2.11, are very effective in increasing the ionization when used in ion plating systems. In the work accelerated electron beam gun system the electron beam (shown dotted) passes through the deposition chamber, thereby enhancing ionization of the coating species, which deposit on the sample. In the hollow cathode discharge gun system the electron beam is generated by confining plasma inside the tube. This leads to the emission of the beam.
3. Like electron beam heating, induction heating has the advantage that the heat can be directed to the melt material, rather than to the boat or crucible. This can reduce evaporant crucible reactions and is particularly attractive for materials such as titanium and aluminium.
4. Arc evaporation was used for many years as a means of vaporizing carbon – simply by striking an arc between two carbon electrodes. Considerable interest has surrounded the use of the arc technique to evaporate various metals (Sanders *et al.*, 1990; Boxman *et al.*, 1996; Sanders and Anders, 2000). A typical layout is shown in Fig. 2.12. This method also enhances the ionization of the depositing species. A further benefit is that, because no molten pool is formed, evaporators can be fitted in any orientation. However, a reported disadvantage for this technique is that droplets are ejected from the source producing macroparticles in the coatings.

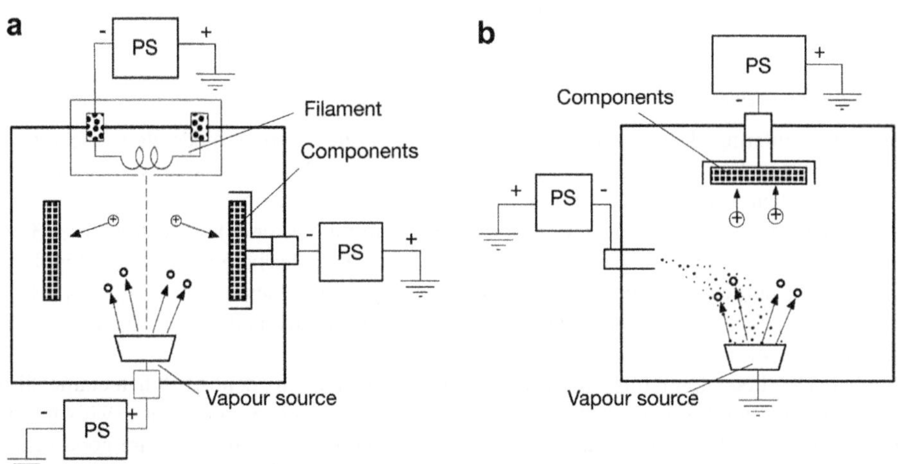

Fig. 2.11. (a) Layout of the work accelerated electron beam gun system and (b) the hollow cathode discharge gun system.

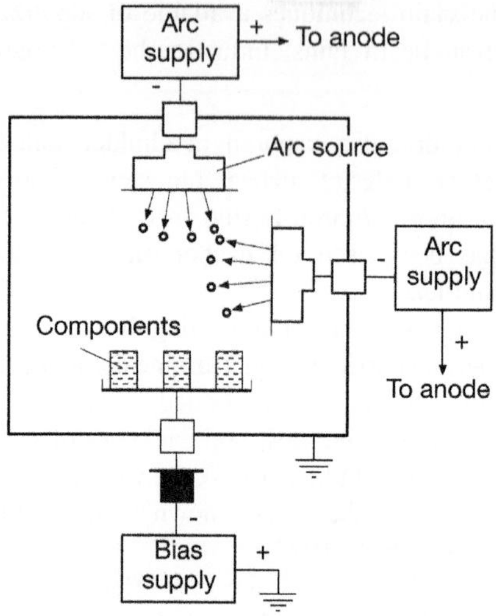

Fig. 2.12. Layout of an arc source system.

5. Sputtering sources utilize the sputter mechanism described earlier, to produce the atomistic coating species through the bombardment of a source or target by ions and accelerated neutrals, usually of argon. A major advantage of this approach is that alloys can be sputter deposited to retain their composition. New magnetrons have been developed which unbalance the magnetic field, such that the plasma associated with the target is released into the deposition volume, to provide a higher level of ionization of the depositing species (Sproul *et al.*, 1990; Rohde *et al.*, 1990; Rohde and Munz, 1991; Matthews *et al.*, 1993; Matthews, 2003).

Since it is probably true to say that the majority of new coatings reported in this book are produced using sputter deposition technologies, we shall outline below some of the key technological developments pertinent to sputter deposition.

2.2.3.2 Sputter deposition

The simplest sputtering plasma deposition layout is that of direct current diode glow discharge, since there are only two electrodes, a positive anode and a negatively biased cathode. Very simply, sputter deposition can then be carried out with the substrate earthed. If the potential applied between these two electrodes is constant over time it is termed a direct current (DC or dc) diode discharge. However, if this potential changes with time it may be either a pulsed-diode discharge (unipolar or bipolar), or if the frequency is sufficiently high and reverses in polarity, it becomes a radio frequency (RF or rf) diode discharge. Each of these has relative advantages and disadvantages, particularly in terms of ceramic deposition.

Although DC-diode sputtering was used in the 19th century to deposit thin metallic films on mirrors, it was not until reliable radio frequency sputtering was developed in the 1970s that it was possible to sputter dielectric materials. Since many ceramic thin film materials are good insulators, most early applications of sputtering were limited to the deposition of metallic coatings. Direct current diode sputtering would probably not have been widely used in the deposition of ceramic thin films if the technology necessary for high-rate reactive sputtering from a metallic target had not been developed in the early 1980s (Sproul, 1983; Sproul and Tomashek, 1984).

Provided that the target material is sufficiently conductive, the target can be effectively sputtered using simple DC-diode geometry. Typically in DC-diode sputtering a potential of several hundred volts is applied between the cathode and anode, and gas pressures in the range from about 0.5 Pa up to tens of Pa are used to generate the plasma. The substrates can be heated or cooled, and held at ground potential, electrically isolated (i.e. floated), or biased relative to the plasma.

The use of an oscillating power source to generate the sputtering plasma offers several advantages over direct current methods. The main one is that when the frequency of oscillation is greater than about 50 kHz, it is no longer necessary for both electrodes to be conductive because the electrode can be coupled through an impedance (Hill, 1986) and will take up a negative DC offset voltage due to the greater mobility of electrons compared to ions in the reversing field. The coupled electrode must be much smaller than the direct electrode in order to effectively sputter only the insulating (coupled) electrode. This is usually accomplished by utilizing the grounded chamber walls as the other electrode. An impedance matching network is integrated into the circuit between the RF generator and the load to introduce the inductance necessary to form a resonant circuit.

An additional benefit of using frequencies above 50 kHz is that the electrons have a longer residence in the negative glow region and also have sufficient energy to directly ionize the gas atoms; hence, the number of electrons required to sustain the discharge is substantially reduced (Vossen and Cuomo, 1978; Chapman, 1980; Thorton, 1982). This, in turn, means that lower sputtering pressures can be used, reducing the risk of film contamination. The most commonly used frequency for rf sputtering is 13.56 MHz.

Unfortunately, the deposition rates in RF sputtering of ceramic target are often limited by the low sputter yield and thermal conductivity of the insulating target materials. This low thermal conductivity of the materials can lead to the formation of 'hot spots' on the target. The hot spots generate stresses that may cause fracture of the brittle target materials. Thus, it may be preferable to deposit insulating films reactively from a metal source.

The availability of sophisticated electrical supplies, capable of providing controlled pulsed power waveforms in tandem with excellent arc suppression, has opened up new opportunities in sputter deposition (e.g. Schiller *et al.*, 1993, 1995). In particular, pulse power has proved to be highly effective in the reactive sputtering of oxide coatings. Sproul (1996) has described the use of bipolar pulse power for sputtering of metal targets in oxygen as follows. The polarity of the target power is switched from negative to positive, and during the positive pulse any charging of oxide layers formed on the target surface is discharged when electrons are attracted to the positive surface. During the negative pulse, ions are attracted to the target surface and sputtering takes place initially from all surfaces on the target, even those that have formed a compound, since the charge on that surface has been neutralized during the positive pulse.

Bipolar pulse power can be either symmetric or asymmetric, depending on the relative positive and negative maximum voltages (Scholl, 1996). Pulsed DC power is said to be symmetric if these voltages are equal. With modern power supplies it is possible to control the on and off time for the pulses.

The successful use of symmetric bipolar pulsed DC power has been reported for the deposition of oxide coatings from two magnetron targets. These targets, mounted side by side, are both connected to the same bipolar pulsed DC power supply. One DC power lead is connected to one target, and the other to the second target. With this system, one of the sputter targets is briefly the anode, while the other is the cathode, and this continuously swaps, with reversals in polarity. Sputtering from the cathode surface during the negative pulse keeps the target surface clean, and when it switches to act as an anode it is not covered by an oxide. As pointed out by Sproul, this procedure avoids the 'disappearing anode' problem, which can occur in pulsed DC sputtering of oxides when all surfaces in the chamber become covered with an insulating oxide.

Asymmetric bipolar pulsed DC has unequal pulse heights and will be applied to one target. Usually, the negative pulse voltage is greater than the positive one and in this case there is usually no off time between pulses. The width of the positive pulse is typically 10–20% of the negative one, thus a high proportion of the cycle is spent in sputtering mode, providing deposition rates which are near to those for non-reactive DC sputtering of metals provided that pulsed DC power is combined with effective partial pressure control.

2.2.3.3 Magnetron sputter deposition

Usually, the beneficial biasing arrangements described above are applied to magnetron sputtering targets, in which a magnetic field is used to trap electrons in the vicinity of the target and to spiral them round a 'racetrack', thereby increasing the degree of ionization occurring and therefore the sputtering rate. Probably the greatest research advances in PVD tribological coatings have occurred around the magnetron sputtering process.

The benefits of magnetrons in sputtering have been known for several decades, i.e. in terms of reducing the required pressure and increasing the deposition rate. However, until recently magnetron sputtering was found to be less effective as a means of producing hard ceramic coatings on tools and components than electron beam and arc evaporation. This was because the levels of ionization achieved were considerably less than those possible with the enhanced plasmas which are characteristic of these other methods.

The most important step in improving the competitiveness of magnetron sputtering was reported by Window and Savvides (1986) and Savvides and Window (1986), when they described their studies on the unbalanced magnetron (UBM) effect. They found that the ion flux to the substrate could be considerably increased by running magnetron cathodes in unbalanced mode (UM). They identified three main magnetic arrangements shown in Fig. 2.13. In the Type I configuration all the field lines originate from the central magnet, with some not passing into the outer magnet. In the intermediate case all of the field lines starting on the central pole go to the outer pole. In the Type II configuration, all of the field lines originate in the outer magnet, with some not passing to the central pole. Following the work of Window and Savvides, a number of researchers began building upon these concepts.

In the late 1980s Sproul and his co-workers researched multi-cathode high-rate reactive sputtering systems (Sproul *et al.*, 1990; Rohde *et al.*, 1990). They studied different magnetic configurations, principally in the dual-cathode arrangement shown schematically in Fig. 2.14. They found that by strengthening the outer magnets of a magnetron cathode with NdFeB and arranging two of these magnetrons in an opposed closed field configuration, i.e. with opposite poles facing each other, the substrate bias current could be increased to well over 5 mA/cm^2 at a pressure of 0.7 Pa. Under these conditions they were able to deposit hard well-adhered titanium nitride coatings.

Another person utilizing the opposed magnetron configuration at that time was Tominaga (1990). He presented an arrangement whereby the permanent magnetic field could be increased by external electromagnetic coils. He stated that this allowed the trapping of electrons by the closed field arrangement to increase ionization or to allow lower working gas pressure with the same level of ionization as achievable with conventional magnetron sputtering.

Another group which has researched confined and unbalanced magnetron sputtering is that led by Kadlec (Kadlec *et al.*, 1990a, b), who in early collaboration with Munz investigated a circular planer unbalanced magnetron arrangement surrounded by two magnetic coils and a set of permanent magnets. They described this as a multipolar magnetic plasma confinement (MMPC), and cite high ion currents at substrates, even when located at large distances from the magnetron.

Howson and co-workers have also studied unbalanced magnetrons, especially for high-power large area applications (Howson *et al.*, 1990, 1992) and demonstrated how additional anodes or electromagnets can be used to control the ion current to the substrate. This may be seen as an extension of early work by Morrison (1982) and Morrison and Welty (1982) who demonstrated that a 'magnetically hidden' anode placed in a magnetron sputtering system could double the plasma density and provide a more uniform plasma throughout the chamber. Several commercial companies have now adopted variations on closed field and anodically enhanced systems, as discussed by Matthews (2003).

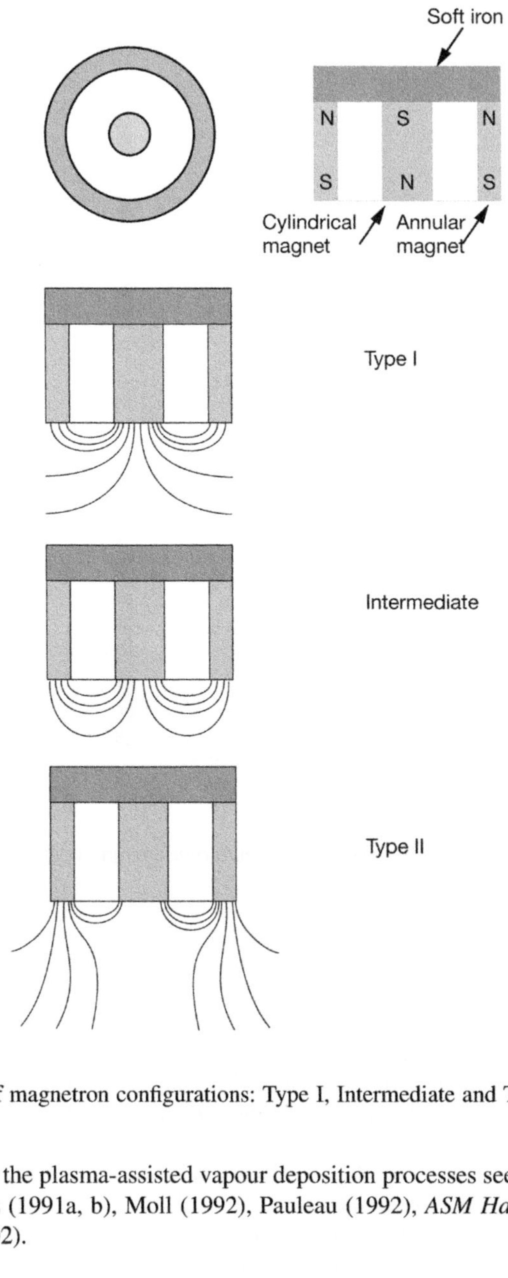

Fig. 2.13. Three types of magnetron configurations: Type I, Intermediate and Type II (after Window and Savvides, 1986).

For further details on the plasma-assisted vapour deposition processes see Matthews (1985, 2003), Rickerby and Matthews (1991a, b), Moll (1992), Pauleau (1992), *ASM Handbook* (1994), Bunshah (1994) and Ohring (2002).

2.2.4 Ion and laser beam-assisted deposition and surface treatment

The ion–surface interaction effects illustrated in Fig. 2.9 include the possibility that the ions will have sufficient energy to become embedded within the surface. This can provide a means of enhancing the tribological properties, although not producing a coating. Typically, this is achieved by generating a high-energy ion beam in a separate source as shown in Fig. 2.15.

The introduction of elemental species such as nitrogen, titanium, carbon or yttrium into the surface of many metals, ceramics, cermets or even polymers has been shown to improve wear, friction, fatigue

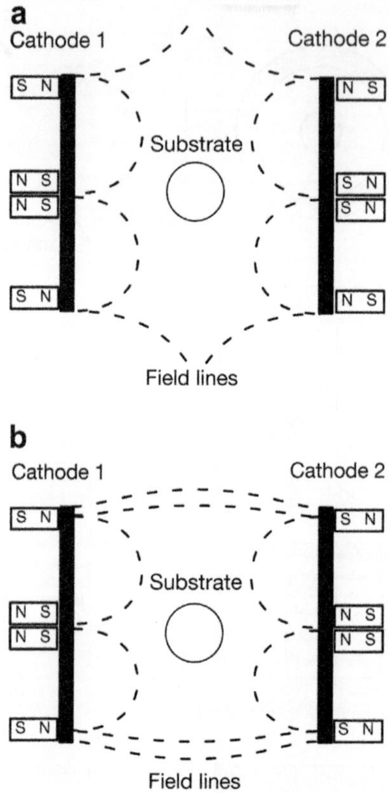

Fig. 2.14. The (a) mirrored and (b) closed field sputtering configuration (after Sproul *et al.*, 1990).

and corrosion properties (Legg *et al.*, 1985). Dearnaley (1991) has explained that there are several mechanisms which could account for this. In ferrous alloys mobile interstitial additives such as nitrogen can decorate dislocations and thus impede dislocation movement. Also, in certain alloy steels and in other metals, such as titanium, it is possible to form stable compound phases, such as metal nitrides, which can harden the surface. Such a mechanism can also enhance corrosion performance, for example by forming a protective oxide layer. The implantation of additional species such as yttrium and rare-earth metals has been shown to further control corrosion mechanisms.

Developments have moved towards improving the economics of the process, for example by scaling up ion beam systems, or by moving to plasma source ion implantation (Conrad *et al.*, 1987) or plasma immersion ion implantation (Tendys *et al.*, 1988; Anders, 2000). Such systems utilize a high-voltage pulsed plasma, analogous to a PAPVD or PACVD system, to achieve implantation. In the authors' view these systems have similarities to plasma or ion-based thermochemical processing systems, such as plasma nitriding or plasma carburizing and there is clearly significant potential for using such systems in two-stage or duplex processes for pre- or post-treatment during plasma-assisted PVD or CVD processing (Matthews, 1985).

The concept of combining ion-implantation technology with PVD or CVD has been quite widely studied. For example, a beam source can be utilized to achieve so-called ion beam mixing. This involves the deposition of a thin layer of coating material, typically up to 50 nm, and impinging an ion beam, e.g. Ar^+, on the surface. The recoil of ions from nuclear stopping collisions mixes the two materials together. Typical ion dose levels are 2×10^{17} ions/cm^{-2}.

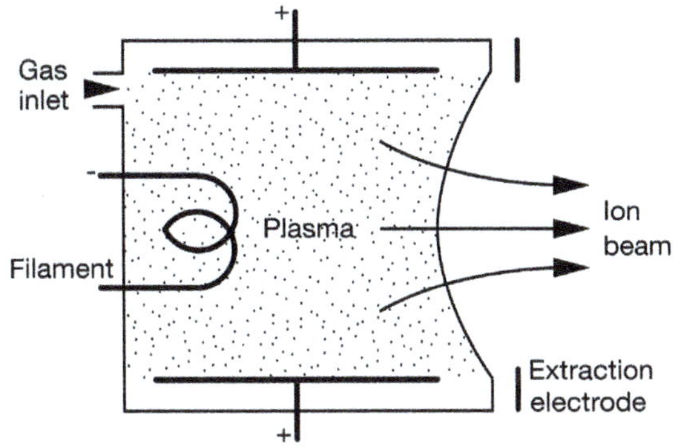

Fig. 2.15. Schematic of a typical ion source.

It is not always necessary to physically ion beam mix the interface atoms to improve the film-to-substrate bond, especially on ceramic and polymer surfaces (Parker, 1974). The bond of a thin coating to a substrate can often be improved by lightly implanting through the interface. This process is sometimes called ion beam stitching. Low doses can be used for this technique such as 10^{15} to 10^{16} ions/cm^2.

Ion beam-assisted deposition (IBAD) in which deposition and ion irradiation are simultaneously combined is useful in forming compound films. Both structural and chemical properties can be modified. Sometimes the technique goes under the name ion-assisted coating (IAC). If the bombarding ions are reactive species then the process is known as reactive ion beam-assisted deposition (RIBAD). Sometimes a combination of techniques is used. For example, a very thin layer may first be deposited, e.g. by sputtering or electron beam evaporation, and then intermixed with the substrate using an ion beam. Further depositions are then combined with ion implantations to produce a final coating over a highly intermixed interface. This type of ion-assisted coating, usually carried out with reactive ion beams such as nitrogen, carbon or oxygen, has been termed reactive ion-assisted coating (RIAC).

The four main ion beam techniques for surface modification or treatment are shown in Fig. 2.16 (Legg *et al.*, 1985) and an ion-assisted coating layout is schematically shown in Fig. 2.17.

Over many years there have been many publications dealing with ion beam and related plasma-based techniques to deposit diamond and diamond-like carbon (DLC) (Aisenberg and Chabot, 1971; Angus *et al.*, 1986; Dehbi-Alaoui *et al.*, 1991; Voevodin *et al.*, 1995). A further variation on this theme is the filtered-arc source (McKenzie *et al.*, 1991; Martin *et al.*, 1992; Zhitomirsky *et al.*, 1994) which utilizes a deflection coil to eliminate macrodroplets and produce a beam of ionized atoms, which can be carbon, for DLC films, or a metal such as titanium, for reactive deposition of a ceramic.

Another important beam technique in PVD is pulsed laser deposition. According to Voevodin and Donley (1996) this is especially appropriate for DLC coatings, as the controlled energy range achievable for the arriving species is wider than other techniques. Also flux control and species control are said to be excellent, for example non-hydrogenated DLC can readily be produced. Several ceramics have also been deposited by this method (Prasad *et al.*, 1995; Voevodin *et al.*, 1996). Laser surface treatments are further discussed in section 2.4.1.

2.3 Solution State Processes

The main coating techniques in this category are electroplating and electroless plating. The solutions used are usually aqueous, and deposits can be produced on metallic or non-metallic substrates. Methods may be divided into the categories of chemical and electrochemical, but this division may not be straightforward since some reactions which appear to be purely chemical may in fact be electrochemical (Lowenheim, 1978; Cartier, 2003).

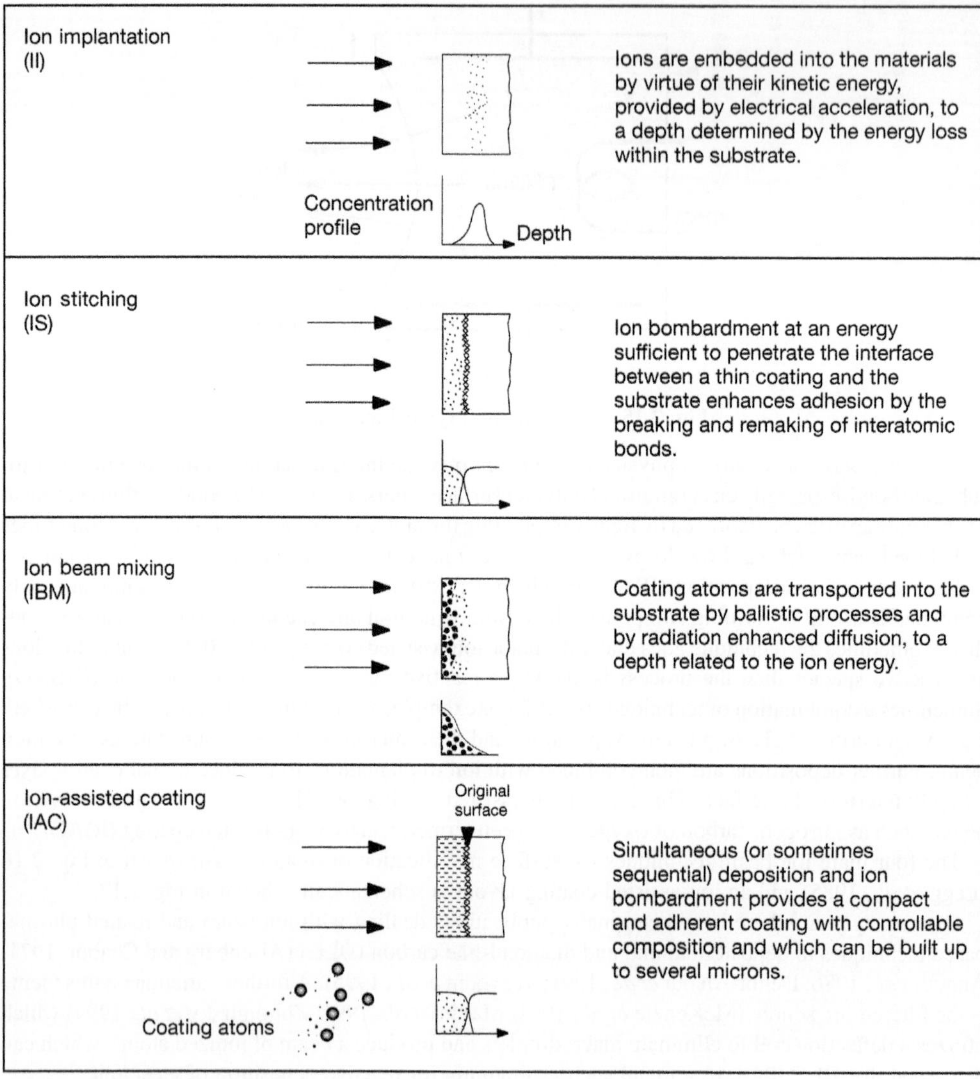

Fig. 2.16. The four main ion beam-based coating techniques (after Legg *et al.*, 1985).

One of the benefits of the solution processes is that they have no upper limit on thickness. In this sense they differ from vapour deposition methods, which, because of stress build-up leading to debonding, are typically used for films less than 10 μm thick, when hard ceramics are deposited.

2.3.1 Chemical solution deposition

Within chemical solution deposition methods we find:

- homogeneous chemical reactions, e.g. reduction of a metal ion in solution by a reducing agent,
- electroless or autocatalytic deposition, which is similar to the previous, except that the reaction takes place on surfaces which are catalytic, rather than throughout the solution, and
- conversion coatings, where a reagent in solution reacts with the substrate, to form a compound.

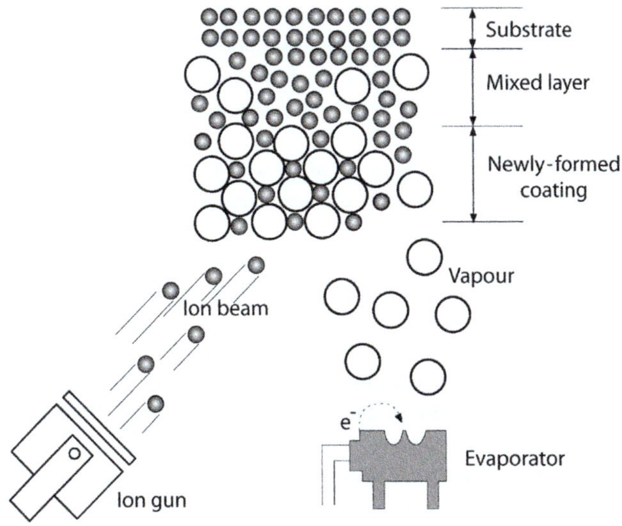

Fig. 2.17. An ion-assisted coating arrangement (original copyright AEA Technology, Rickerby and Matthews, 1991a).

The chemical reduction process is used, for example, by spraying a solution of metal ions and reducing agents on to a surface, such that the metal coating builds up. The technique can work on metals and plastics, but the adhesion levels can be poor. Examples of coating materials which can be deposited in this way are silver, gold and copper.

Electroless deposition of nickel, normally with phosphorus or boron additions, is an increasingly important coating technology. The additions derive from the reducing agents used, e.g. sodium hypophosphite or aminoboranes, typically with nickel sulphate. A significant advantage of the electroless nickel process is that it can be used in conjunction with finely divided particles, in the Ni–P bath for example, to produce composite coatings. Examples of materials used to reinforce the coatings in this way are SiC, WC, CrC, Al_2O_3, or diamond (Parker, 1974; Celis *et al.*, 1988). Alternatively, solid lubricant particles, such as graphite, PTFE or calcium fluoride, may be added (Ebdon, 1987). Electroless plating has been reviewed by Ponce de León *et al.* (2006).

Chemical conversion coating methods include phosphating and chromating. The main use of the former is to enhance the corrosion barrier properties of steels or aluminium prior to painting. The layer, produced by reaction with phosphoric acid, can be either an amorphous or a crystalline phosphate. Sometimes this layer is applied as an aid to the lubrication of moving parts by providing a degree of protection against scuffing and to facilitate running-in. Chromate conversion coatings are applied by immersing or spraying a surface with an aqueous solution of chromic acid, chromic salts, phosphoric acid or other mineral acids. The surfaces develop an oxide film which is sealed by a metallic chromate. The process is typically applied to steel, aluminium, magnesium, cadmium and zinc (Celis *et al.*, 1988). It improves the corrosion properties and can enhance paint adhesion.

2.3.2 Electrochemical deposition

Electrochemical deposition, also commonly termed *electroplating*, involves the deposition of a metallic coating on an electrode by a process of electrolysis, whereby chemical changes are produced by the passage of a current. Michael Faraday first put forward the laws of electrolysis in 1833, and they still form the basis of the technology, such as (1) the amount of chemical change produced is proportional to the quantity of electricity which passes, and (2) the amounts of substances liberated by a given quantity of electricity are proportional to their equivalent weights. A schematic electroplating layout is shown in Fig. 2.18. The electrochemical deposition techniques are described, e.g., by Wills and Walsh (2006).

Without going into detail about the chemical aspects of the process, we can state that the metals depositable from aqueous solutions generally occupy a region in the middle of the periodic table as shown in Fig. 2.19.

Among the metals listed, many are deposited as alloys, e.g. brass (Cu–Zn), bronze (Cu–Sn), Co–Ni, Fe–Ni, Sn–Ni, Sn–Zn and Sn–Pb (Lowenheim, 1978). The most electronegative metal commonly plated is Zn, which when deposited on steel provides galvanic protection, as does cadmium.

One of the most common electroplated tribological coatings is chromium plate, which is often Cu/Ni/Cr or Ni/Cr, the Cr being only a very thin overlayer to prevent tarnishing of the Ni (Lowenheim, 1978). Hard chrome is a term often used for thicker chromium deposits which have good abrasion and lubricant-retention properties.

A few electroplating processes use non-aqueous solutions, and some metals have been deposited from solvents such as liquid ammonia. The two main alternatives to aqueous baths are organic solvents and fused salts, though both present problems, e.g. in terms of toxicity hazards, disposal and cost. However, there are certain metals, such as aluminium, whose deposition by these routes is actively carried out (Ross, 1988).

2.3.3 Sol-gel processing

A sol is a colloidal dispersion of particles in a liquid, usually aqueous but sometimes an organosol. Sol-gel processing may be carried out by dipping, spinning, spraying or roller bar draw coating. It involves applying the sol to a substrate, whereupon it undergoes hydrolysis and condensation reactions to form an aggregate gel. Final drying leads to further condensation and the formation of a dense surface film.

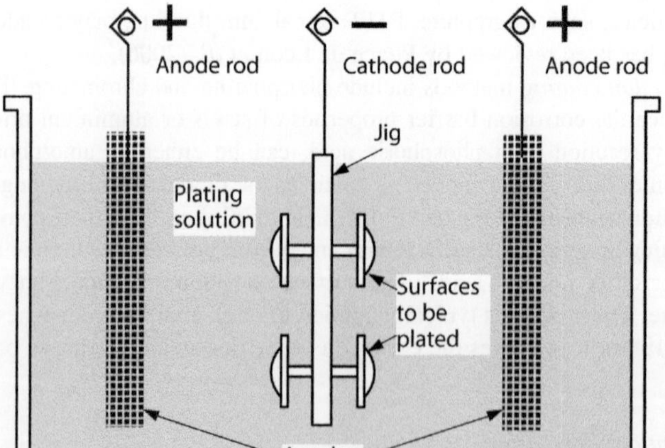

Fig. 2.18. Typical electroplating layout.

H																	He
Li	Be											B	C	N	O	F	Ne
Na	Mg											Al	Si	P	S	Cl	Ar
K	Ca	Sc	Ti	V	Cr	Mn	Fe	Co	Ni	Cu	Zn	Ga	Ge	As	Se	Br	Kr
Rb	Sr	Y	Zr	Nb	Mo	Tc	Ru	Rh	Pd	Ag	Cd	In	Sn	Sb	Te	I	Xe
Cs	Ba	La	Hf	Ta	W	Re	Os	Ir	Pt	Au	Hg	Ti	Pb	Bi	Pa	At	Rn
Fr	Ra	Ac															

Fig. 2.19. The long-form periodic table, showing the elements depositable from aqueous solution within the shaded area.

Typically, the technique is used for the production of oxide ceramic films, in which case the gel is fired at above 150°C to leave the ceramic, but also much higher temperatures of up to 900°C can be used (Bhushan and Gupta, 1991; Schmidt, 1994). There is less knowledge about the tribological properties of sol-gel films than those produced by other techniques (Zhang *et al.*, 2002). The main published use of sol-gel-derived coatings in tribology seems to be for friction or corrosion control (Wang and Akid, 2007), and then, typically, a hybrid polymer/metal oxide film is applied (Lai *et al.*, 2007; Voevodin *et al.*, 2003). There are various reports of such films being used to improve the abrasion and scratch resistance of polymers (Li *et al.*, 2000).

Where sol-gel-derived coatings have been used to enhance the abrasion resistance of metallic substrates, a high-temperature post-treatment is almost always used and variously described as annealing (Zhang *et al.*, 2006a) or sintering (Liu *et al.*, 2003). It should be noted though that where the polymeric constituent in the hybrid composites are beneficial, as in erosive impact conditions, such a very high-temperature treatment is avoided (Gründwurmer *et al.*, 2007).

Thus one can summarize sol-gel methods as being useful for producing composite coatings, often with controlled structural characteristics such as porosity, but they should be used as part of an extended processing route, usually requiring additional heat treatment in order to obtain the desired structure and adhesion. The technique also has potential as a means of producing spherical particles of controlled size, which can then be used in plasma spraying processes (Bhushan and Gupta, 1991; Schmidt, 1994; Liu *et al.*, 2003).

2.3.4 Plasma electrolysis

This is an electrolytic plasma-based process, which is carried out in a liquid electrolyte. It is especially suitable for producing hard ceramic surface layers on lightweight metals such as aluminium, magnesium and titanium (Yerokhin *et al.*, 1999). It is probably best known as a means of forming a hard oxide surface on aluminium and magnesium, although variants of this technique can be used to produce carbonitrides, and also to create surfaces which incorporate other metals, ceramics and even organic compounds. The plasma electrolytic processes are therefore gaining increasing importance in tribological applications, especially for lightweight components, and as a supporting layer for subsequent deposits or impregnation treatments (Nie *et al.*, 2000).

2.4 Molten and Semi-Molten State Processes

Within this category may be included laser surface treatments, and the hard-facing techniques of thermal spraying and welding. Each of these subjects covers a very wide field, but since this book is intended primarily to emphasize thin tribological coatings, only a brief mention will be made of each.

2.4.1 Laser surface treatments

A review of this subject has been compiled by Fellowes and Steen (1991), who list the following tribological uses for lasers in surface treatment and coating:

- surface heating for transformation hardening or annealing,
- surface melting for homogenization, microstructure refinement, generation of rapid solidification structures and surface sealing,
- surface alloying for improvement of corrosion, wear or aesthetic properties,
- surface cladding for similar reasons as well as changing thermal properties such as melting point or thermal conductivity, and
- plating by laser chemical vapour deposition (LCVD) or laser physical vapour deposition (LPVD). The latter can use laser heating of the evaporant in PVD, usually using a continuous wave laser, or a pulsed laser to ablate the source material in which case the term laser ablation deposition (LAD) is sometimes used.

In the context of this book the coating techniques which lasers facilitate are of great interest. One benefit of lasers is that they can provide a directable deposition in defined areas (e.g. in LCVD). When used as a means of producing the vapour in PVD, where the laser ablates the source material, the technique is said to allow the composition of an alloy target to be retained in the coating. For further details of laser surface treatments, see Frenk and Kurz (1992) and Man (2006), and for a wider perspective on pulsed laser deposition, see Eason (2006).

2.4.2 Thermal spraying

Thermal spraying covers a wide range of techniques in which material is heated rapidly in a hot gaseous medium and simultaneously projected at high velocity onto a surface to produce a coating (Scott and Kingswell, 1991; Guilemany and Nin, 2006). According to Grainger (1989), processes for thermal spraying can be grouped into two categories. First, there are lower energy processes often referred to as metallizing, which include arc and flame spraying. These are frequently used for spraying metals for corrosion resistance, such as zinc and aluminium. Second, there are the higher energy processes such as plasma spraying, the detonation gun and high-velocity combustion spraying. A technique attracting considerable interest is the high-velocity oxy-fuel (HVOF) process, as shown in Fig. 2.20.

In all cases the types of structures differ considerably from those produced by either gaseous or solution-state processes. Further details on plasma spraying are given by Brossa and Lang (1992), Cartier (2003) and Pawlowski (2008).

2.4.3 Welding

Various welding processes can be used to deposit a wide range of metals and metal/ceramic composites. The two most widely used methods are the oxyacetylene torch and arc welding. These techniques can be applied to most metals, but they are not generally recommended for coating non-ferrous ones having melting points below 1100°C. The thickness range is generally 1.5 to 6.0 mm. Coating materials are usually applied to resist abrasive wear, and typically contain carbides of chromium, tungsten or boron dispersed in a matrix of iron, cobalt or nickel. Those used to resist impact generally have austenitic structures which work harden, e.g. austenitic manganese steel. This, as deposited, is relatively soft (approx. 170 Hv), but hardens to about 550 Hv when subjected to impact conditions. This material gives an example of how the processing route can affect the coating properties. The austenitic structure of manganese steel is normally produced by a water quench from

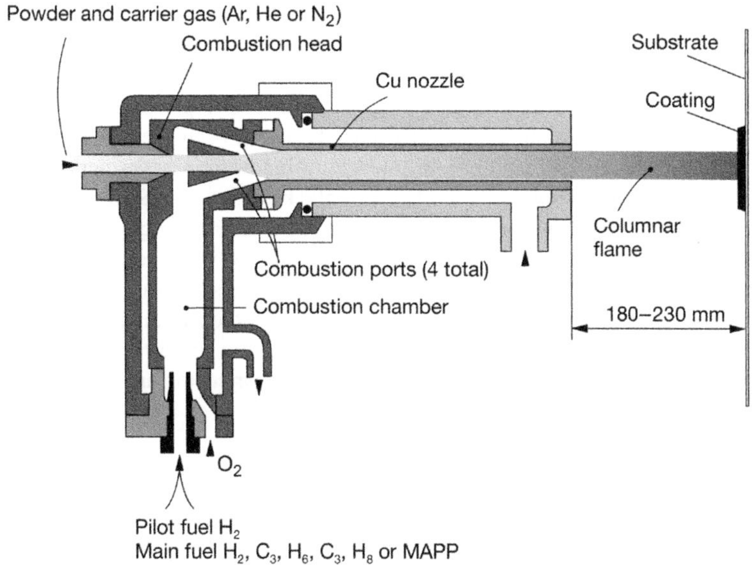

Fig. 2.20. Typical high-velocity oxy-fuel process layout.

about 980°C. The alloys used for weld deposition are usually formulated to ensure that the austenitic structure is obtained on air cooling from the welding temperature (HMSO, 1986; Mellor, 2006).

2.4.4 Other developments

Although molten and semi-molten state processes are regarded by many as mature technologies, there are several developments being made of which two are described below.

2.4.4.1 Electro-spark deposition

This is a pulsed-arc micro-welding process using short-duration, high-current electrical pulses to deposit an electrode material on a metallic substrate. It is one of the few methods available by which a fused, metallurgically bonded coating can be applied with such a low total heat input that the substrate material remains at or near ambient temperature (Johnson, 1988). This is achieved by controlling the spark duration to a few microseconds and the spark frequency to around 1000 Hz. The substrate is cathodic and the source material is the anode. The sparking frequency can be controlled by vibrating one of these electrodes to vary the gap distance or, alternatively, triggering circuits can be used.

One of the earliest reported examples of electro-spark deposition (ESD) was the sparking of a titanium surface under oil, to produce a titanium carbide layer (Welsh, 1958). Nowadays, ESD coatings are applied in air, argon, nitrogen, helium, CO_2 and various mixtures of gases commonly used in other arc welding processes. While gloveboxes may be required for control of environments for highly reactive materials, an appropriately controlled flow of argon can usually effectively shield the surface.

The transfer of material from the anode has been described as globular or spray, depending on the voltage and current conditions and the gas used. Anode materials can be virtually any electrically conducting metal or cermet capable of being melted in an electric arc. The most commonly deposited ones are the refractory metal carbides and the cobalt- and nickel-based hard-facing alloys (Levashov *et al.*, 2004, 2007). The next most important are materials to improve corrosion or oxidation resistance of surfaces, such as chromium and aluminium alloys, and noble metals such as iridium, platinum and gold. Carbon, which does not have a molten phase at atmospheric pressure, does not appear to be a suitable material for ESD. Graphite electrodes have, however, been used to thinly carburize some strong carbide formers such as titanium. An interesting innovation is the deposition of Fe-based amorphous alloys with Vickers hardness levels over 1500 kgf/mm^2 by ESD (Liu *et al.*, 2007).

Rough surfaces are difficult to coat by ESD, as are holes or internal angles. Likewise, sharp knife-edges tend to blunt when subjected to ESD. Coating rates vary with the coating material and the thickness required. For example, to produce dense smooth coatings 25 μm thick, the rate of coverage might be as low as 1 cm^2/min. This low rate may be mitigated by the automation of the process and the use of multiple applicator heads. Typical thicknesses from 3 μm to 250 μm have been quoted, although some deposits greater than 1 mm thick have been reported for nickel-based alloys. Further information on the process is given in Lesnjak and Tusek (2002).

2.4.4.2 Friction surfacing

This method achieves many of the benefits of welding but without the torch or flame. The coating material, usually in rod form, is rotated at high speed and forced against the substrate. Frictional heating generates a hot plasticized zone at the interface. Movement of the substrate across the face of the rotating rod transfers a layer of coating material onto the substrate, giving a build-up of 0.2 to 2.5 mm. The solid phase joining of the coating to the substrate avoids many of the problems associated with fusion welding or thermal spray processes. The result is a well-bonded dense coating with a fine microstructure. The process is capable of depositing many of the hard wear-resistant alloys, from low alloy steels through tool steels to the cobalt-based alloys. Likewise, many of the alloys traditionally used for corrosion resistance such as stainless steels, nickel-based alloys and aluminium bronzes can also be deposited (Doyle and Jewsbury, 1986; Bedford and Richards, 1986). Further details on the process are given in Mellor (2006).

2.5 Surface Hardening Treatments

Engineering surfaces are often hardened, usually by thermal or thermochemical means, to improve their tribological and corrosion properties. While not within the definition of tribological coatings we have adopted within this book, these treatments are mentioned here, especially since they are sometimes used as precursors to provide a load supporting surface layer beneath the tribological coating. This aspect is discussed further in sections 3.4.2 and 4.6.4.

Thermal treatments involve the application of a specified thermal cycle to a material, in order to obtain desired mechanical and sometimes chemical properties. Common examples are the heating and quench hardening of steel, which are typically applied to change the iron–carbon distribution and modify the grain structure. Such treatments can be applied selectively by the local application of heat by induction, flame, electron beam or laser. Typically, surface layers hardened in this way can be up to several mm in depth.

Like thermal treatments, thermochemical treatments rely on modifying the existing surface rather than adding a coating on top of it. However, whereas a thermal treatment will utilize the existing carbon in steel, in a thermochemical treatment the composition of the surface is altered by adding one or more additional elements. This means that a much wider range of steels and other non-ferrous alloys can be treated.

The two main thermochemical treatment methods are carburizing and nitriding. The former is used, for example, to raise the carbon content in the near-surface regions of low carbon steels, and subsequent quenching can cause a martensitic transformation to occur, creating a hard surface with beneficial compressive stresses which can improve the fatigue strength. Typical case depths are up to about 4 mm. The main processes used are gas, vacuum and plasma carburizing. The last two are coming increasingly to the fore due to the possibility of limiting the availability of oxygen at the surface. Carburizing is usually followed by a direct quench from the processing temperature, typically around 900°C. The addition of about 0.5% nitrogen in the process can allow lower process temperatures and reduces the case thickness, creating a carbonitrided surface.

Nitriding is a lower-temperature process typically carried out at around 500°C for steels. Gas nitriding and plasma nitriding are the most popular methods. The latter provides a shorter processing time and is claimed to give improved control over the nature of the surface layers. Case depths are typically 0.4–0.6 mm. Note that nitriding elements, such as chromium, must be present in the steel for nitriding to be effective. Nitrocarburizing is a thermochemical process that can be applied to plain carbon, i.e. non-alloyed, steels. It is usually carried out at about 570°C in ammonia with a carbon-bearing gas and a carbonitride phase about 20 μm thick is formed on the surface in about 2 hours. Some porosity may occur, which can be beneficial for oil retention, or even for impregnating with a sealing and/or lubricating medium. Table 2.2 gives some details of the main surface hardening treatments based on carbon and nitrogen (HMSO, 1986).

2.6 Process Effects on Coating Structures

In this chapter we have described several coating and treatment processes, emphasizing the emerging technologies. The nature of the films produced by each method can be very different, for example a chromium coating produced by electroplating would be different to a chromium coating produced by PVD. The differences will include properties such as internal stress, morphology, hardness and toughness which produce considerably different tribological behaviours.

These broad differences in the nature of coatings produced by different methods provide an initial filtering route when selecting a coating, as will be discussed in Chapter 6. However, the authors feel that more work is needed to quantify more fully the influences of the coating process on the tribological properties. This can most effectively be achieved by considering the influence on the intrinsic film and substrate properties, including their anisotropic nature.

Intrinsic properties are determined by interatomic forces and influence the elastic modulus, the yield and the shear strengths and fracture behaviour of the coating. In practice, the differences we see in different coatings are largely determined by structural differences such as in atomic structure, grain sizes, grain boundary effects and internal stresses. These effects largely dictate the current patterns of usage. Examples of this are the use of micro-cracked electroplated chrome coatings which can retain lubricant, segmented-structure plasma-sprayed thermal barrier coatings for aero-engines and dense equiaxed PVD coatings for cutting tools. All of these rely on both the intrinsic and morphological properties imparted by the different coating methods. Also, designed structural changes such as multilayering can be used to provide further control over properties such as hardness and toughness.

2.6.1 Morphological growth structures

In section 2.2.3 we presented the Thornton structure zone diagram (Fig. 2.7), which describes the influence of temperature and pressure on the microstructure of PVD coatings. Implicit within this diagram is the concept that the greater the energy input to a growing coating, e.g. by means of increased substrate temperature, the greater will be the densification, and the lower will be the porosity. In modern plasma-assisted processes, the energy can be input directly by ionizing the deposition species and accelerating them to the substrate surface. By such means the dense equiaxed Zone 3-type structure can be obtained at a homologous temperature even lower than 0.2, compared to the 0.8 or so predicted by the Thornton diagram for a non-biased substrate. Many researchers have

Table 2.2. Some thermomechanical treatments based on carbon and nitrogen.

Process	Description	State	Typical treatment temp. (°C)	Typical case depth (mm)	Typical surface hardness (H_y)
Carburizing	A process in which a steel surface is enriched with carbon, at a temperature above the ferrite/austenite transformation. On subsequent quenching, an essentially martensitic case formed	Solid Liquid Gaseous Plasma	850–950	0.25–4.0	700–900*
Carbonitriding	Similar to carburizing, but involving nitrogen as well as carbon enrichment	Liquid Gaseous Plasma	700–900	0.05–0.75	600–850*
Nitrocarburizing	A process in which a steel or cast iron surface is enriched with nitrogen, carbon and possibly sulphur at a temperature below the ferrite/austenite transformation	Liquid Gaseous Plasma	570	0.02 max† 1.0 max‡	500–650†
Nitriding	A process in which a steel surface is enriched with nitrogen, at a temperature below the ferrite/austenite transformation	Gaseous Liquid Plasma	500–525	0.4–0.6	800–1050

*Depending on temperature treatment. The upper figure represents typical as-quenched hardness.

†Thickness and microhardness of compound layer on mild steel. Values are dependent on alloy content of material.

‡Total depth of diffusion.

examined the influence of process parameters on the coating structure in PVD (e.g. Messier *et al.*, 1984; Müller, 1986; Petrov *et al.*, 2003).

In practice, all studies concluded that arriving atoms should have sufficient energy to allow adatom diffusion to occur on the surface, and as many arriving atoms as possible should possess that required energy. Hence concepts such as the ion-to-neutral ratio (Martin *et al.*, 1987; Müller, 1987; Petrov *et al.*, 1992) and the ionization efficiency (Matthews, 1985) are introduced to quantify the extent to which the arriving species are ionized and therefore the number of arriving atoms that have the required energy to ensure effective coating densification.

In plasma- and ion-assisted PVD systems, the influence of process parameters on coating growth morphology is now well understood, and it is possible to create dense equiaxed structures without having high substrate temperatures. Also, these systems can produce amorphous structures with certain deposition species, due to the high cooling rates achieved during film formation. PVD systems differ from most thermal spraying systems in that the depositing species can be regarded as atomic during the nucleation and growth stages. Even if they were to arrive as clusters, they can be expected to break up into atoms with high mobility in most modern PVD systems.

The same cannot be said of thermal spray systems, in which typically the depositing species arrive as molten globules which create splats on the surface. The film structure in that case is thus very different from that seen in PVD films. For example, whereas a PVD film will often have an underlying columnar grain growth structure with grain boundaries normal to the surface, in thermal spraying many boundaries between splats will be parallel to the surface. This means that the wear mode in thermally or plasma sprayed coatings can be one of fracture at poorly bonded splat boundaries, whereas a PVD film will be less likely to undergo large-scale particle detachment in the same way.

The structure of electrodeposited coatings, as for plasma-assisted PVD, depends on the current characteristics at the substrate, with the growth following some analogous steps to those seen in PVD films; i.e. an adatom diffuses across the surface to find a site of minimum surface or interfacial energy. Thus, electrodeposited coatings can also be dense and pore free. However, the stress state of the growing film can strongly influence its functionality. Hard chrome coatings typically exhibit a microcracked structure which can actually be beneficial in retaining lubricants. Cracked structures actually contain more chromium oxide, which can enhance wear resistance.

2.6.2 Composite structures

In the above we have briefly outlined the intrinsic structural characteristics of the main coating methods used to produce tribological coatings. Methods exist to design structures to further enhance the tribological characteristics of coatings. In this category we include layered and other composite structures which typically comprise a combination of phases and elemental compositions.

Composite bulk materials can be created to improve mechanical properties such as stiffness, strength, toughness and resistance to fatigue. Composite coatings can be designed specifically also to improve tribological functionality and also electrical, optical, electronic and magnetic properties (Hogmark *et al.*, 2001). It is the first of these that concerns us most here, and there can be little doubt that it has been the creation of advanced composite, including multilayered, nanocomposite and functionality graded coating structures that has allowed the most significant progress in the enhancement of tribological surface properties in recent years.

Holleck was one of the first to write about the benefits of composite multilayered and multiphased coatings (Holleck, 1986). He also laid out a comprehensive basis for considering how knowledge of the dominant bonding mechanism in each of the phases could be utilized to design the appropriate mixed-phase structure (Holleck, 1986; Holleck *et al.*, 1990). He presented the information shown in Fig. 2.21 to illustrate how the type of bonding present is related to the crystal structure. He also presented the schematic view shown in Fig. 2.22 to illustrate how a combination of structural

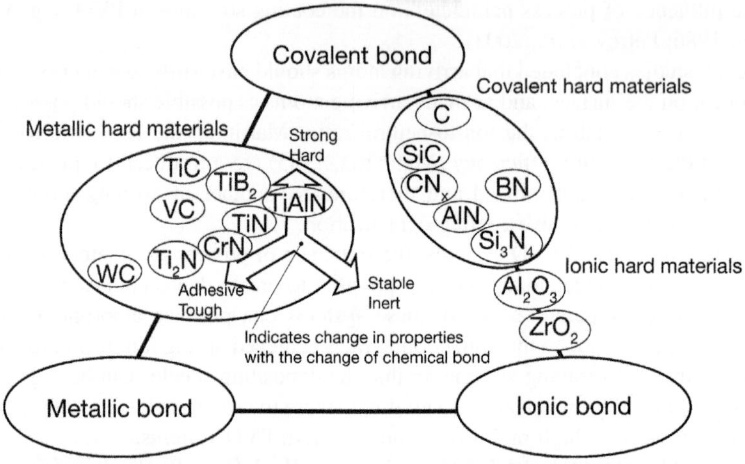

Fig. 2.21. Different groups of hard materials with characteristic crystal structures (after Holleck and Schier, 1995).

characteristics could enhance wear resistance and crack-arresting capabilities. Writers such as Subramanian and Strafford (1993) have endorsed the thinking behind Holleck's work.

The interfaces in a multilayer ceramic coating can toughen a coating in different ways, as shown in Fig. 2.23 (Holleck and Schier, 1995). They considered the interface not as a plane surface, but as an interfacial volume which can be regarded as a softer binder phase which can dissipate energy and effectively reduce the brittleness of the coating, i.e. by inhibiting crack growth, deflective cracks, allowing crack opening to reduce stress concentrations, or imparting nanoplasticity.

Holleck's early concepts now have a clear resonance with the more recent moves to combine adequate hardness with a degree of elasticity, or even ductility, when these attributes are required as, for example, in impact conditions where toughness and the ability to absorb energy without failure is an important goal. One of the key recent developments which has allowed the fulfilment of Holleck's early objectives has been the production of nanocomposite and nanolayered structures. Due to the nanometre-scale size of the features, they can respond in a truly composite manner to applied stresses. This means that they are less likely to exhibit the individual characteristics of the phases present, which may inhibit the optimized performance. Many researchers have taken these ideas forward (Leyland and Matthews, 2000, 2004; Voevodin and Zabinski, 1998a; Musil and Jirout, 2007).

The creation of nanostructured coatings that are either nanolayered or nanocomposites is a significant topic in tribological coating development. As discussed by Jehn (2006) the term nanoscale structure is commonly used to describe coatings with features from nanometre to almost micron range, but strictly speaking it should only be used when individual structures are below about 20 nm (Cavaleiro and De Hosson, 2006). According to Jehn, coatings having grain sizes of that dimension, or less, can be achieved by controlling deposition parameters or by subsequent heat treatment (Huang *et al.*, 2006; Mayrhofer *et al.*, 2006). Such heating-induced changes in grain size can be achieved in both single- and multiphased materials (Jehn, 2006), with remarkable improvements in mechanical and chemical properties (Veprek and Reiprich, 1995; Veprek, 1999; Veprek and Argon, 2002; Veprek *et al.*, 2006; Fischer-Cripps *et al.*, 2006; Cavaleiro and De Hosson, 2006).

2.6.2.1 Chemical reactivity considerations

It is known that thin ceramic films, despite their chemical inertness, do not usually provide adequate corrosion resistance. They tend to suffer defects and porosity through which the coating/substrate

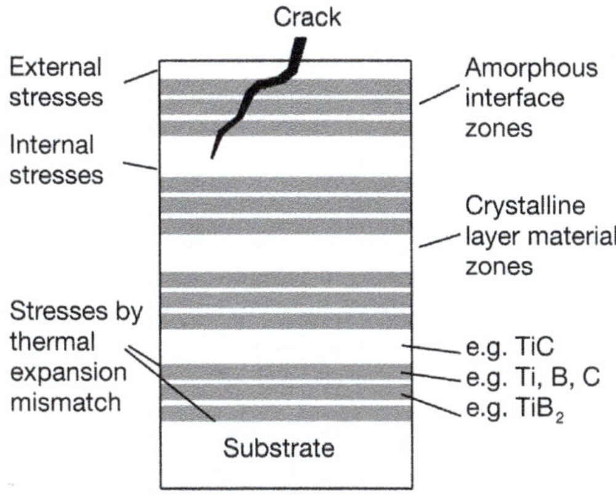

Fig. 2.22. Possible mechanism of crack deflection and stress relaxation in a multilayer coating system (after Holleck *et al.*, 1990).

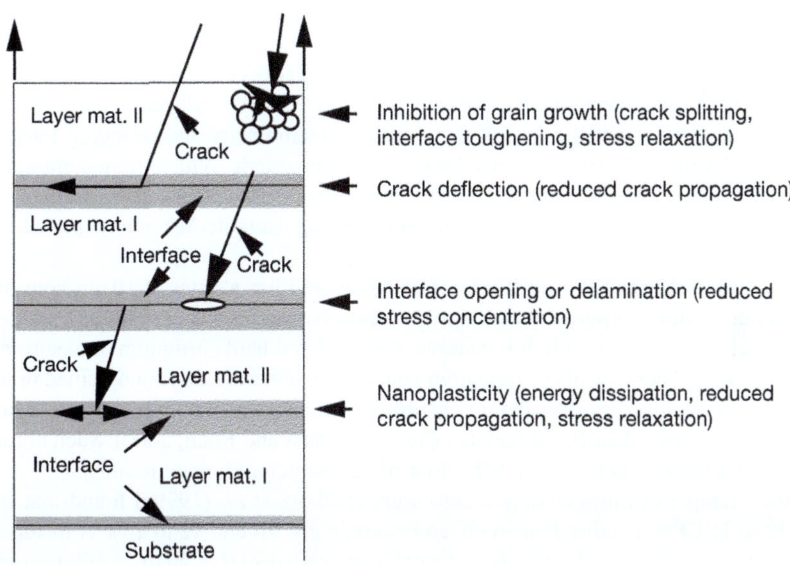

Fig. 2.23. Toughening mechanism in ceramic multilayer surface (after Hollek and Schier, 1995).

interface can be attacked, leading to spallation. Such effects may be worsened when a thin metallic interlayer is deposited between coating and substrate. The latter is already highly activated by the sputter pre-cleaning process to improve interfacial adhesion and accommodate strain mismatch. The titanium interlayer deposited between a steel substrate and a PVD TiN coating is known to be susceptible to galvanic attack (Liu *et al.*, 2001).

2.6.2.2 Mechanical property considerations

The group of Matthews has investigated the nanocomposite coating concept to produce coatings that have a high hardness, combined with a low modulus (Leyland and Matthews, 2000, 2004). This is in keeping with the reasoning behind the need to reduce the ratio of hardness (H) to elastic modulus (E). In practice, there are applications in which a high stiffness is beneficial, for example on cutting tools, while other applications require the toughness and long elastic strain to failure which a low H/E coating can provide.

One coating system which has been shown to have great potential in the latter regard is the Ti–Al–B–N system (Rebholz *et al.*, 1998, 1999a, b). The property improvement achievements using this system have now been extended to other systems. These include other ternary and quaternary systems based on the substitution of chromium for titanium or aluminium and silicon for aluminium or boron. There are a number of commercial applications for such nanostructured films now emerging; particularly for CrAlN, TiSiN, TiCrAlN and TiAlSiN – as well as the now widely used TiAlN (Veprek *et al.*, 2008).

Following on from the work on Ti–Al–B–N, Leyland and Matthews and co-workers have proposed an approach to tribological coatings development based on predominantly metallic rather than ceramic coatings (Leyland and Matthews, 2004). The advantages of vapour deposition techniques, particularly low-temperature plasma-assisted PVD methods, can be harnessed to provide the required structure, adhesion and high hardness – together with improved toughness and resilience, compared to ceramic coatings. The inherent ability of such techniques to provide a high quench rate in the depositing film and generate unusual levels of supersaturation of both interstitial and low-miscibility substitutional alloying elements is a key factor in promoting nanocrystalline or even glassy phase formation in vapour-deposited coatings.

Metallic films can generally be vapour deposited with lower residual compressive stress than their ceramic counterparts, allowing thicker films to be deployed with reduced risk of spallation. The maximum rate at which a metallic film of the required stoichiometry and structural integrity can be deposited is usually many times higher than for ceramic compounds. With current commercial vapour deposition technology, growth rates of only a few μm/hour are realistically attainable for dense ceramic films at the 200–400°C deposition temperatures suitable for many metallic substrate materials.

There is realistic scope to deposit nanostructured metallic films at rates of 10 μm/hour or higher at these temperatures with low porosity and high abrasion/erosion resistance. This creates prospects to compete on thickness and cost with, for example, electroplated hard chromium or electroless nickel–phosphorus coatings. There are also strong arguments in the literature that nanograined or amorphous metals might provide more uniform and predictable sacrificial corrosion protection, due primarily to their small and uniformly distributed features (Suryanarayana and Koch, 2000). Such arguments can be expected to translate directly to the behaviour of nanostructured metallic films.

Taking the example of nitrogen-doped chromium Rebholz *et al.* (1999c) found that moderately hard ($H = 12$ to 15 GPa) metallic films with, correspondingly, around 12 to 15% N in supersaturated solid solution gave results superior to those of significantly harder ($H = 20$ to 25 GPa) ceramic nitride films, when subjected to severe impact wear. Experiments were performed using a ball-on-plate configuration, whereby a 10 mm diameter ball of either chromium steel or WC–Co was repeatedly driven against a coated plate at a load of several hundred newtons. This was sufficient to cause considerable plastic deformation of the underlying substrate material. For both ball materials, the impact crater wear volume after 50,000 cycles was high for low-nitrogen chromium films, but similarly low for both high-nitrogen chromium and ceramic nitride films. Increased hardness beyond 15 GPa had little further effect in reducing the impact crater volume. The overall performance of the high-nitrogen metallic films was superior, in that, despite substantial substrate plastic deformation, little or no cracking or debonding of tough, lower modulus metallic films occurred. The brittle, higher-modulus ceramic films exhibited extensive circumferential cracking around the crater rim, with substantial fracture and debonding in the central zone of the crater itself.

Analogous observations can be made regarding the superior abrasion and erosion behaviour of PAPVD Ti/TiN multilayer films (Leyland and Matthews, 1994) or CrN/electroless-nickel duplex coatings (Bin-Sudin *et al.*, 1996), where the relatively ductile metallic constituents facilitate plastic strain accommodation and thicker coating deposits, to the benefit of both wear and corrosion behaviour.

In the light of the above results with nitrogen-doped chromium film, a logical step is to modify a nitrogen-doped chromium film by the introduction of a secondary, ideally low-modulus, immiscible-metal constituent. This can either be captured substitutionally in the chromium with the aim of increasing lattice friction or, with suitable choice of deposition parameters, form a thin intergranular phase and promote a nanograined composite structure, with similarly increased resistance to plastic deformation (Leyland and Matthews, 2000). The result can be a further substantial increase in yield strength, and hardness, with E remaining similar or, with appropriate alloying, substantially reduced.

This leads to a significant increase in the H/E ratio, to the benefit of coating resilience and toughness. This generates prospects to tailor both composition and structure to engineer coating resilience, toughness and modulus to match closely the mechanical properties of almost any, but particularly any metallic, substrate material that might be chosen. Initial work in this area was directed towards addition of copper to the Cr(N) metallic film system. Copper exhibits almost complete immiscibility with chromium under most near-equilibrium conditions. It has in bulk, microcrystalline form roughly half the elastic modulus (130 GPa compared to 280 GPa) and, from a practical perspective, is relatively inexpensive and widely available at suitably high purity. Work on this coating system, with additions of both nitrogen and boron, has demonstrated a number of encouraging results (Baker *et al.*, 2003, 2005; Tsotsos *et al.*, 2006). There are many other candidate coating systems which might be considered for the synthesis of wear-resistant metallic films based on the above-mentioned concepts and on other materials-related design considerations (Leyland and Matthews, 2004, 2006).

Other authors have achieved success with ceramic–ceramic nanocrystalline/amorphous films, such as nc-TiN/a-SiNx (Veprek, 1999; Veprek and Argon, 2002; Veprek *et al.*, 2008), at similar phase ratios and very small grain sizes, less than 5 nm. The key to good performance in this case appears to be the avoidance of a fully percolated structure, with nanograins completely separated from each other by the matrix (Patscheider *et al.*, 2001), since interpenetrating networks of crystalline and amorphous regions may control possible grain boundary rotation. There is a lack of tribological data for such superhard coatings. According to Ma *et al.* (2005), they have a lower friction coefficient than TiN films against a hardened G Cr15 steel ball at high temperature (550°C), i.e. 0.4–0.5 compared to 0.7.

One of the main challenges in developing PAPVD metallic nanocomposite coatings will be to determine which possible structural features outlined above can give the most appropriate tribological behaviour in individual applications. The chromium–copper–nitrogen system appears to be an obvious materials choice for further investigation of the potential benefits of metallic nanocomposite films - over and above the benefits of, for example, PAPVD nitrogen-doped hard chromium metal films. A number of other candidate systems exist (Leyland and Matthews, 2004), with the potential to develop a diverse spectrum of unusual and extreme mechanical and tribological properties, which might not be attainable through many of the conventional materials processing or vapour deposition routes currently favoured for scientific investigation.

Leyland and Matthews (2006) discuss the issues involved in selecting a suitable nanocomposite tribological coating. They first consider the nitride-forming elements which are generally selected for the production of refractory ceramic nitride, carbide, boride or oxide materials used in tribology: the Group IVb/VIb elements (Ti/Zr/Hf; V/Nb/Ta; Cr/Mo/W) and the Group IIIa/IVa elements Al and Si. Whether bulk materials or coatings, these are the 11 elements which, in various combinations, tend almost exclusively to be used – together with B, C, N and/or O – to produce wear-resistant engineering ceramics, although other light and rare-earth elements (e.g. Be, Mg, Sc, Y) might arguably also qualify. Excepting Al, the remaining 10 elements all have melting temperatures (T_m) of $\sim$1700 K or above (up to $\sim$3700 K in the case of W) and are therefore inherently quite refractory.

Considering a homologous temperature (T/T_m) of 1/3 as a point at which diffusion mechanisms could start to significantly alter the microstructure of a bulk material over a fairly short timeframe, of the order of a few hours to a few tens of hours, one might infer that the 10 elementary materials in question would tend to exhibit some inherent thermal stability up to a few hundred degrees Celsius or more. They could therefore be considered quite suitable as candidate tribological materials in their own right – assuming, for instance, that a non-equilibrium nanocrystalline microstructure could be generated to maximize their hardness and load-bearing capability (Leyland and Matthews, 2004).

For the purposes of creating such a small grain size, the addition of a second, low-miscibility element has been shown to be very effective – particularly in vapour deposition of metallic thin films such as Al–Y (Liu *et al.*, 1997) and Cr–Cu (Baker *et al.*, 2003, 2005). There are a high number of metallic elements which, when introduced to create a binary alloy with the candidate elements identified above, exhibit a wide miscibility gap in the solid state. Examples of elements for which there is evidence of such behaviour when mixed with some or all of the IIIa, IVa, IVb, Vb and VIb nitride-forming elements identified are, in order of atomic number: Mg, Ca, Sc, Ni, Cu, Y, Ag, In, Sn, La, Au, Pb. Of these 12, only five, Sc, Ni, Y – and perhaps also Cu and Au – might be considered sufficiently refractory ($T_m > 1000°C$) that they could be expected to demonstrate some inherent thermal stability above ambient. Such considerations may be relevant in selecting material combinations in applications such as dry or marginally lubricated sliding, where local flash temperatures may reach several hundred degrees Celsius.

Low-melting point metals such as In, Sn and Pb with little inherent thermal stability also exhibit very low elastic moduli (11, 40 and 16 GPa, respectively), which may be attractive in adjusting the overall modulus of a tribological film to meet a specific substrate requirement such as protective coatings for magnesium alloys, where the substrate elastic modulus is only ~45 GPa.

Conversely, the more refractory, IVb transition metal elements (e.g. Zr, Hf), and their low-miscibility rare-earth neighbours (Sc, Y), have surprisingly low elastic moduli in the 60–80 GPa range, making them potentially very suitable candidates for coating aluminium alloys with similar elastic properties. Refractory combinations such as Ta–Ni ($T_m = 2996°C$ and 1455°C, $E = 186$ GPa and 200 GPa, respectively) might be considered for protection of steel substrates with similar moduli. So, it can be seen that many possibilities now exist to develop nanocomposite tribological coatings, based on metallic constituents (Leyland and Matthews, 2006).

2.6.2.3 Glassy metal films structures

Glassy metal films are said to exhibit superelastic deformation behaviour (Peker and Johnson, 1993). Numerous other physical benefits have also been observed for metallic glasses due largely to the absence of grain boundaries, such as better corrosion resistance (Pang *et al.*, 2002). They have more uniform sacrificial behaviour, reduced pitting and improved fatigue behaviour. The potential benefits of glassy phase formation in vapour-deposited thin films are significant, but are as yet under-researched, with only a few authors (Sanchette *et al.*, 1995, 1998; Sanchette and Billard, 2001) having explored the topic in any systematic way.

The unusual capability to post-coat amorphize a nanocrystalline film through plasma-diffusion treatment (Matthews *et al.*, 1995) or, conversely, to post-coat anneal a glassy film – promoting partial devitrification and a controlled nanocomposite structure (Tsotsos *et al.*, 2006; Inoue, 2001; Louzguine and Inoue, 2002) – are two examples of topics in the vapour-deposited thin film field which merit further investigation for tribological applications.

Low-temperature (<350°C) plasma-assisted PVD techniques in particular raise the exciting prospect of investigating new metallic glass compositions which may currently be difficult to synthesize via conventional bulk casting methods. Still they might yield novel physical behaviour particularly when devitrified – whether that be mechanical, tribological, chemical, magnetic, or other functional properties (Leyland and Matthews, 2004).

Several of the recently developed bulk metallic glass systems are based substantially on mixtures of low-miscibility nitride-forming and non-nitride-forming transition metals such as those mentioned earlier. It seems likely that further developments of metal–metal and doped-metal nanocomposite coatings of a nanocomposite or glassy structure will allow surfaces with tailored and optimized tribological properties to be produced.

CHAPTER 3

Tribology of coatings

Contents

3.1 Friction, Wear and Lubrication

3.1.1 General

The controlled movement of one surface in relation to another, or in engineering terms of one component in relation to another, is typically achieved by machine elements such as bearings, joints, gears, sliders and cam and tappet mechanisms. In manufacturing processes such as metal cutting or extrusion the tool surface is exposed to severe conditions by a counterface which is in a fast-moving, high-temperature plastic or liquid state. In mining, on the other hand, the moving counterfaces can be hard and sharp edged rocks, often imposing impact loads.

A high degree of precision is required from surfaces such as those of bearings and joints in space instruments, precision robotics, microdevices such as MEMS (micro electro mechanical systems) as well as the fast-moving reader heads in magnetic storage systems. Effects from the biological environment are present in the moving contacts involving surgical knives and dental drills or human and animal joints. The wheel on the road and the moving shoe on the pavement represent flexible tribological contacts; in skiing the moving counterface is soft snow. A degree of flexibility is also involved in the moving surfaces of human joints.

Interacting surfaces in relative motion thus occur extensively in human society. Many of its key functions, especially in our industrialized community, depend on them. The variety of conditions where the surfaces are expected to fulfil demanding requirements is thus extremely wide. Some of the examples mentioned are illustrated in Fig. 3.1. The tribophysical and tribochemical processes involved are complex because of the many different parameters involved, as shown in Fig. 3.2.

3.1.2 Surface characteristics

A typical metallic surface for engineering purposes has a surface profile with hills and valleys, as shown in Fig. 3.3, where the slopes are around 5° and rarely steeper than 30°. The height difference from the bottom of a valley to the top of a hill is typically 0.1 to 3 μm.

The most commonly used parameter to quantify the roughness of a surface in engineering is the R_a or centre-line-average (CLA) value. It is simply the arithmetic average deviation of the profile from the mean line. In many occasions it interrelates with other surface roughness parameters but its limitation is a lack of sensitivity to the shape of the surface profile. There are a number of other roughness parameters, like the peak-to-valley height value R_t and the 10 points height R_z that have been used. They are described in detail by Dunaevsky *et al.* (1997). Different aspects of their

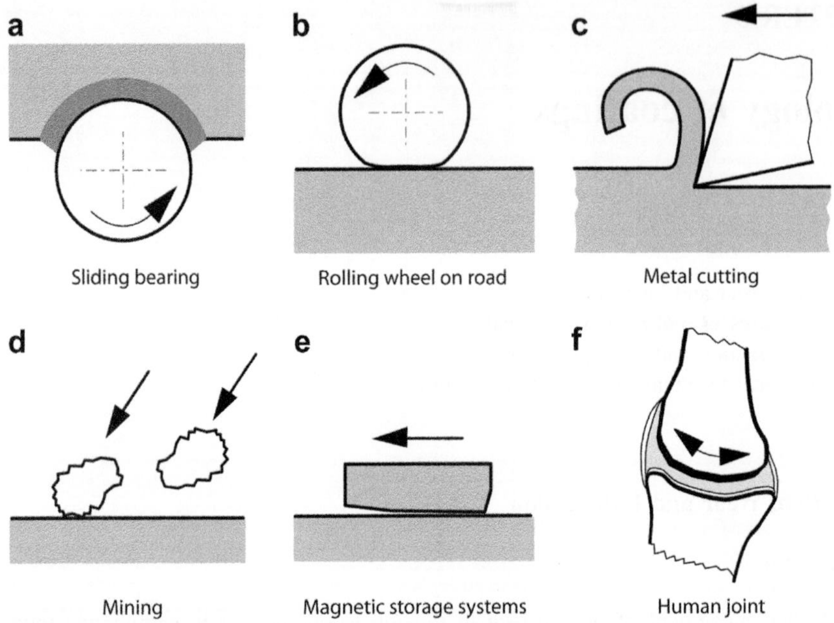

Fig. 3.1. There are a wide variety of different kinds of conditions where interacting surfaces in relative motion are expected to fulfil their requirements, as in (a) sliding bearings, (b) rolling wheel on road, (c) metal cutting, (d) mining, (e) magnetic storage systems and (f) human joints.

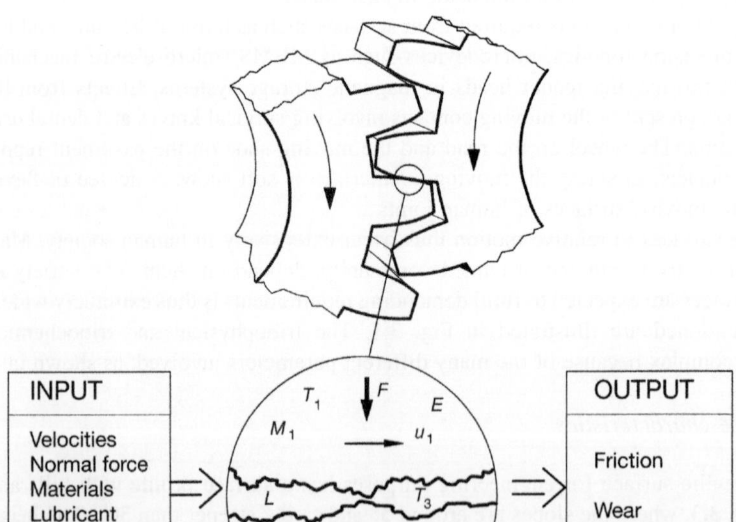

Fig. 3.2. Several material, contact condition and environmental input parameters influence the dynamic tribophysical and tribochemical contact processes which control friction and wear.

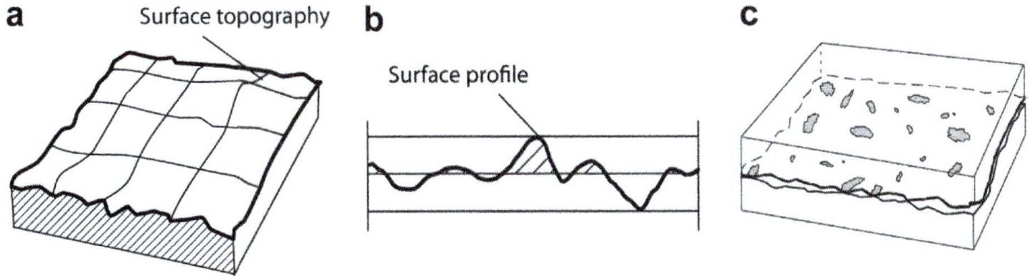

Fig. 3.3. The shape of a surface is characterized by (a) the surface topography and (b) the surface profile. For two surfaces in contact, the real area of contact, shown in (c), is considerably smaller than the apparent area of contact.

usefulness in tribology have been discussed by Whitehouse (1974), Czichos (1978), Williamson (1984), O'Connor *et al.* (1991), Bhushan (1999b), and Cai and Bhushan (2007).

It has, however, been difficult to find a suitable parameter that would uniquely satisfy the need to characterize the surface roughness with respect to its influence on friction and wear. In lubricated contacts especially, there is also a need to take into account the topographical pattern, such as the direction of grooves in the surface (Holmberg, 1984, 1989).

Tribological surfaces normally exhibit a random roughness profile across different length scales and this is changing during the process of sliding contact. It is difficult to represent this by numerical methods, as those in use today normally work well with surface data exhibiting a fixed random profile and provide surface descriptors closely related to a scale at which surface data were acquired. A comparison of different methods such as Fourier transform, windowed Fourier transform, Cohen's class distributions, wavelet transform, fractal methods and hybrid-wavelet method was carried out by Podsiadlo and Stachowiak (2002). They concluded that the most suitable method for tribological surfaces appears to be a hybrid fractal-wavelet method, i.e. a method in which a partition iterated function system and a wavelet transform are combined.

When one surface is placed upon another, the first contact between the surfaces will take place at the top points of the highest peaks of the surfaces. Two metallic surfaces that are pressed together have a real contact area which may be only about 10% of the surface area, as shown in Fig. 3.3c. Analysis of gap-maps with a lightly bead-blasted glass surface suggests that normal contact is almost entirely confined to the highest 25% of each surface and mainly occurs in the top 10% (Williamson, 1984).

The chemical and physical nature of a surface depends on the composition of the material and the surrounding environment. A schematic representation of a solid surface is shown in Fig. 3.4 (Bhushan, 1999b). Finished mechanical components have a surface layer with properties different from the bulk material depending on the machining process. Energy that goes to the near surface region during machining can result in deformation, strain hardening, recrystallization and texturing. In polishing, grinding, lapping and machining, the surface layers are plastically deformed with or without a temperature gradient. They become highly strained and residual stresses may be released or created. A deformed layer may also be produced during the friction process. The typical thickness of the deformed layer is 10–100 μm. The Beilby layer in metals and alloys is produced by melting and surface flow during machining. The layer typically has an amorphous or microcrystalline structure and is 1–100 nm thick.

Metals and alloys react with oxygen in air and form oxide layers. In other environments they may form nitrides, sulphides and chlorides. Polymers generally do not form an oxide layer. Iron can form the following oxides: Fe_2O_3, Fe_3O_4 or FeO. Oxides present on an alloy surface depend upon the concentration of alloying elements, the affinity of alloying elements for oxygen, the ability of oxygen

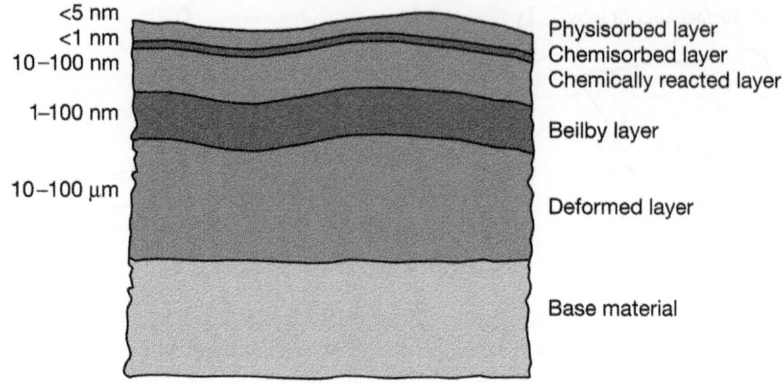

Fig. 3.4. A solid surface consists typically of a deformed layer, the Beilby layer, a chemically reacted layer such as typically the oxide layer, a chemisorbed layer and a physisorbed layer.

to diffuse into surface layers and the segregation of alloy constituents at the surface (Buckley, 1984). The thickness of the chemically reacted layer is typically 10–100 nm and depends on the reactivity of the material to the environment, reaction temperature and reaction time.

On the chemically reacted layer, adsorbed contaminant layers can be formed from the environment by chemical or physical adsorption. The most common constituents of physisorbed layers are molecules of water vapour, oxygen or hydrocarbons that may be condensed and physically adsorbed on the surface. The layer can be monomolecular (about 0.3 nm thick) or polymolecular and is then less than 5 nm thick. Oil droplets in the environment may form a physically adsorbed oil layer on the surface, which may be only 3 nm thick. A monolayer of molecules may be formed by chemical adsorption. This layer forms strong chemical bonds with the layer below and retains its own individual identity; it does not change as in the chemically reacted layer. The physical and chemical nature of surfaces has been described by Buckley (1981, 1984), Suh (1986), Bhushan (1999b) and Abel (2001).

It is important to keep in mind the dimensions of the different surface elements when problems connected with surfaces in relative motion are considered. Size is one parameter influencing the interaction of different surface elements such as surface films, debris, asperities, junctions and gaps. Figure 3.5 shows the comparative sizes of surface elements and related phenomena.

3.1.3 Friction

Friction is the resistance to motion which is experienced when one body moves tangentially over another with which it is in contact. Thus friction is not a material property; it is a system response in the form of a reaction force. The coefficient of friction, μ, is the tangential frictional force F divided by the normal load w on the contact:

$$\mu = \frac{F}{w} \tag{3.1}$$

Basically friction can be divided in two components, an adhesion force, F_a, due to adhesion between the two surfaces, and a deformation force, F_d, due to deformation of the surfaces so that

$$F = F_a + F_d \tag{3.2}$$

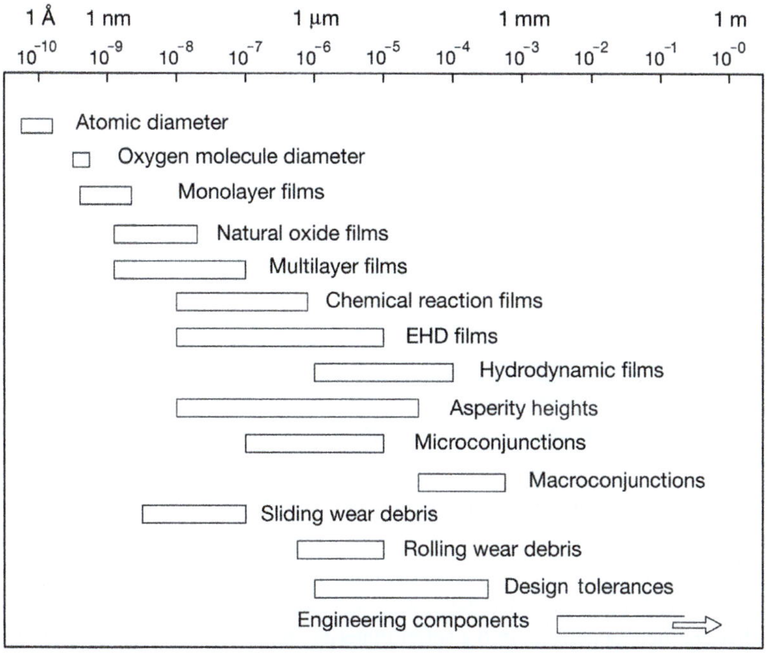

Fig. 3.5. Sizes of surface elements and surface-related phenomena.

The deformation force is called 'ploughing friction' when it takes place on a macroscale and 'asperity deformation' when it takes place on a microscale. In spite of its importance and after many centuries of investigations and research there is still no complete physical explanation for friction.

3.1.3.1 Early friction theories

An excellent overview of the history of friction investigations has been presented by Dowson (1998). In the 20th century a new approach to friction was introduced by Bowden and Tabor (1950) and Kragelsky (1957). They shifted the focus of models explaining friction from component and macro-contact scales – such as apparent contact area and Hertz contact pressure – to phenomena of micrometre size. Surface topography, asperity collisions and molecular boundary layers now became of particular interest.

Bowden and Tabor explained that when two asperities are forced into contact they will weld together due to the adhesion between the two materials as shown in Fig. 3.6a. When one of the solids is moved in a tangential direction relative to the other the micro-welded junctions will break but their shear strength causes a resistance to motion. During sliding, new micro-welded junctions are formed and broken continuously.

The adhesion model of friction was, however, criticized on the basis of several objections, such as:

- the agreement between the theoretically calculated and the experimentally obtained values of the coefficient of friction is not particularly good,
- the model does not take into account the surface roughness effect on friction,
- there is a lack of evidence that the junctions produced will be of necessary strength, and
- when the normal force pressing the surfaces together is removed the adhesion cannot be detected.

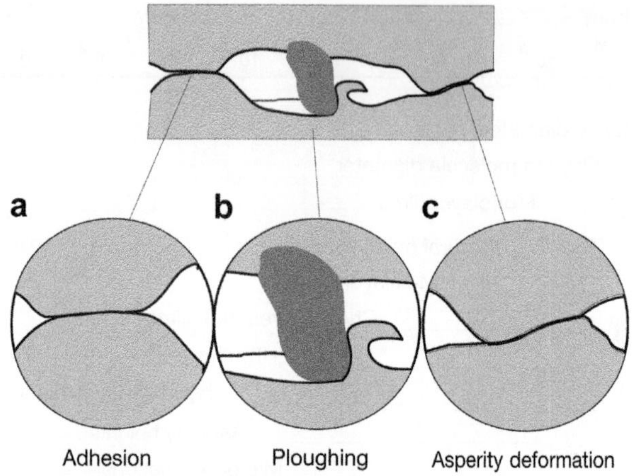

Fig. 3.6. The three components of sliding friction are (a) adhesion, (b) ploughing and (c) asperity deformation.

Bowden and Tabor (1950) also included the ploughing effect in their concept of friction. When a hard asperity or a hard particle penetrates into a softer material and ploughs out a groove by plastic flow in the softer material this action creates a resistance to motion as shown in Fig. 3.6b. To overcome the shortcomings of the adhesion theory of friction Rabinowicz (1965) offered a theory according to which the surface energy of adhesion as well as the roughness angle of the surface and the radius of the junctions is taken into account.

A molecular–mechanical theory of friction was developed by Kragelsky (1957, 1965). Here the word 'molecular', according to the Russian terminology, stands for the adhesion component and the word 'mechanical' stands for the material deformation component. Kragelsky includes in the adhesion component a so-called piezo coefficient, which characterizes the increase of shear strength with compressive force, according to a binomial law introduced by Deryagin in 1934 (Kragelsky *et al.*, 1982).

Some workers (reviewed by Rigney and Hirth, 1979) have concentrated more on the effect of plastic deformation losses in friction. The influence of plastic deformation of asperities was considered by Green (1955) and extended by Edwards and Halling (1968). They showed that resistance to motion arises from the work done when asperities are plastically deformed as they slide over each other, as shown in Fig. 3.6c. On the basis of studies concentrating on plastic deformation losses in the material, Heilmann and Rigney (1981) suggest an energy-based friction model where metallurgical effects are included. However, their results have been disputed by Tangena *et al.* (1984). In an analysis of the adhesion, deformation and surface roughness effects on friction Tangena (1987) concludes that if surface asperities are plastically deformed the influence of surface roughness on the contact parameters will be small, because the asperities will deform in such a way that a continuous surface is obtained. This is supported by Rabinowicz (1992) who found only a rather modest influence of surface roughness on the coefficient of friction in static friction experiments with noble metals. A review of the classical friction theories can be found in Tabor (1981).

3.1.3.2 Genesis of friction at the microscale

A new concept of friction, called the genesis of friction, was presented by Suh and Sin (1981). They focused on microscale friction mechanisms and showed that the mechanical properties affect the frictional behaviour to a greater extent than chemical properties when sliding occurs without a significant interfacial temperature rise. The effect of friction can be divided into three basic mechanisms, one due to asperity deformation, one due to adhesion and one due to particle ploughing, as shown in Fig. 3.6. The adhesion mechanism in many practical cases may not be the most dominant of the three. They also stress the influence of the history of sliding on the friction force. On the basis of crossed-cylinder experiments carried out with specimens made of engineering materials they give the following typical values for the three components of the coefficient of friction (with maximum values in brackets):

1. Friction due to asperity deformation, $\mu_d = 0$–0.43 (0.75). It appears that asperity deformation is responsible for the static coefficient of friction. Once the original asperities have been deformed, asperity interlocking cannot take place. This component can contribute to steady-state friction if new asperities are continuously generated as a result of the wear process.
2. Friction due to adhesion, $\mu_a = 0$–0.4. The low value is for a well-lubricated surface. The high value is for identical metals sliding against each other without any surface contamination or oxide layers.
3. Friction due to ploughing, $\mu_p = 0$–0.4 (1.0). The low value is obtained either when wear particles are totally absent from the interface or when a soft surface is slid against a hard surface with a mirror finish. The high value is associated with two identical metals sliding against each other with deep penetration by wear particles.

In a more detailed study of the microscopic mechanisms of friction Kim and Suh (1991) claim that the ploughing component of the surfaces by wear particles is the most important in most sliding situations. This is supported by a study by Miyoshi (2001) showing that there is no clear relationship between the coefficient of friction and the theoretical shear strength. However, for most material combinations, with different metals sliding on themselves and on non-metals, friction increased as the theoretical shear strength decreased. The relation between friction and the theoretical tensile strength had a similar behaviour.

In contradiction to this it can be argued that the lubrication of initially dry surfaces usually reduces friction drastically. In a review on adhesion, friction and wear, Landheer and de Gee (1991) point out that a boundary lubricant film only a few molecules thick, in the nanometre range, can effectively reduce friction, e.g. from 1.0 to 0.1, without affecting asperity interaction and thus the ploughing effect. The influence of surface roughness on both the initial and the steady-state friction has been found to be considerable for sliding steel and ceramic pairs by Hwang and Zum Ghar (2003). A change of the surface roughness from polished, $R_a = 0.02$ μm, steel plate to a ground one, $R_a = 0.57$ μm, sliding against a polished, $R_a = 0.02$ μm, steel ball resulted in a drop in the steady-state friction of about 30%. When they introduced oil lubrication the initial friction dropped moderately, by 0–10%, but the steady-state friction dropped considerably, by 10–60%.

Based on knowledge available on the mechanisms of friction in sliding contacts it seems that both the particle ploughing effect and the adhesion effect are important and either of them may dominate the frictional behaviour, depending on the tribological contact conditions. This certainly means that the effect of wear particle ploughing in friction should be given more consideration than previously the case.

3.1.3.3 Development of friction with time

The chronological stage of a sliding contact influences considerably its frictional behaviour. Essentially, a contact starts with a running-in period followed by a steady-state period and ends by a breakdown or destruction period. From a practical point of view it is important to achieve a controlled running-in period resulting in optimal frictional behaviour during the steady-state period. The steady-state period represents the major part of the lifetime of the contact.

A sliding contact goes through several stages of different friction mechanisms in the running-in period before the steady-state period is reached. Suh and Sin (1981) identified five different friction

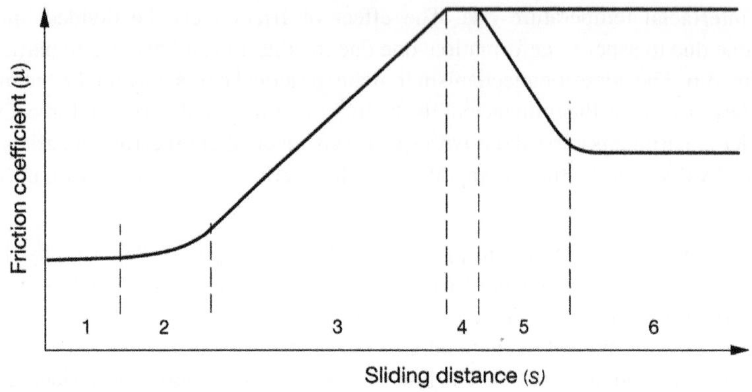

Fig. 3.7. Six stages of friction mechanisms occurring in sliding steel contacts in the initial period of sliding (after Suh and Sin, 1981).

stages that occurred in steel contacts before steady-state friction. These stages are specific for the materials tested, the experimental arrangement and the environmental conditions. However, as an example of the mechanisms they are very illuminating. The different stages are shown in Fig. 3.7 and are as follows:

Stage 1. Initially, the friction force is largely a result of ploughing of the surface by asperities. Adhesion does not play any significant role due to surface contamination. Asperity deformation takes place and affects the static coefficient of friction and the surface is easily polished. Consequently, the coefficient of friction in the initial stage is largely independent of material combinations, surface conditions and environmental conditions.

Stage 2. The polishing wear process at stage 1 has removed surface contamination and elements of bare surface will appear, resulting in a slow increase in the coefficient of friction due to increased adhesion.

Stage 3. The coefficient of friction increases due to a rapid increase in the number of wear particles entrapped between the sliding surfaces as a consequence of higher wear rates. The deformation of asperities continues and the adhesion effect increases due to larger clean interfacial areas. Some of the wear particles are trapped between the surfaces, causing ploughing. If the wear particles are entrapped between metals of equal hardness they will penetrate into both surfaces, preventing any slippage between the particle and the surface and resulting in maximum ploughing friction.

Stage 4. The number of wear particles entrapped between the surfaces remains constant because the number of entrapped particles entering is the same as the number leaving the interface. The adhesion contribution also remains constant and the asperity deformation continues to contribute, since the wear by delamination creates new rough surfaces with asperities. Stage 4 represents the steady-state friction when two identical materials slide against each other or when the mechanisms of stage 5 are not significant.

Stage 5. In some cases, such as when a very hard stationary slider is slid against a soft specimen, the asperities of the hard surface are gradually removed, creating a mirror-like smooth surface. The frictional force decreases, due to the decrease in asperity deformation and ploughing, because wear particles cannot anchor so easily to a polished surface.

Stage 6. The coefficient of friction slowly levels off and reaches a steady-state value as the hard surface becomes mirror smooth to a maximum extent and the softer surface also acquires a mirror finish.

3.1.3.4 Influence of wear debris and layers

The above stages of initial friction are representative for the combinations of steel materials studied by Suh and Sin (1981) and the experimental conditions used. They will be different in other cases, for example in contacts where transfer layers are formed, and the interfacial friction is then caused primarily by the combined effect of the three basic mechanisms of friction (Blau, 1989). The influence of metal transfer in sliding friction was studied by Landheer and Zaat (1974) and they showed that the metal transfer layers always occur according to the same fundamental process; there is only a difference in scale, at which the process presents itself.

The important but often neglected influence of wear particles in a sliding contact has been shown for steel contacts in crossed-cylinder experiments with 10 N normal load, 0.02 m/s sliding speed and 6.35 mm sample diameter. After a running-in period the experiment was stopped and the wear particles were removed, which resulted in a decrease in the coefficient of friction from 0.71 to 0.4 as shown in Fig. 3.8. The initial coefficient of friction was 0.13. New direct evidence of the wear debris and wear layer influence on friction in sliding contacts was demonstrated by Singer *et al.* (2003). The mechanisms of wear debris generation, agglomeration, mechanical mixing and formation of reaction and transfer layers is discussed in sections 3.3.7–3.3.9.

3.1.3.5 Friction equations

During recent years there have been numerous attempts to formulate mathematical expressions for the friction force or the coefficient of friction (Dunayevsky, 1997; Bhushan, 1999b; Andersson, 2004). All these equations suffer from a lack of generality. They are limited to certain frictional mechanisms, material combinations or contact conditions and thus have not been very suitable for use in practical applications.

Two basic laws of friction are generally valid over a wide range of applications. They are often referred to as Amonton's laws and have recently been shown to work surprisingly well at the macroscale, microscale and even on the nanoscale (Gao *et al.*, 2004; Tysoe and Spencer, 2004). The first states that there is a linear relationship between the friction force and the normal load. This means that the coefficient of friction, μ, is a constant independent of the normal load (see equation 3.1).

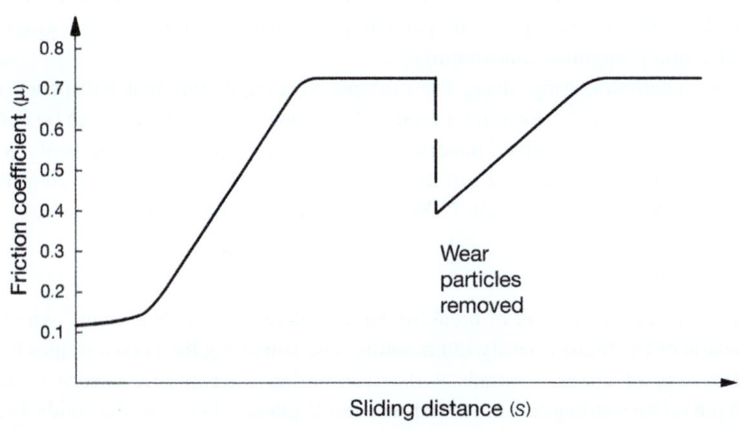

Fig. 3.8. Effect of removing wear particles from a crossed-cylinder contact with steel sliding on steel (after Suh and Sin, 1981).

The second law states that the friction force or the coefficient of friction is independent of the apparent area of contact between the contacting bodies (see Fig. 3.3c). Thus the physical size does not influence the coefficient of friction.

A third law is sometimes linked to these two. It states that the friction force is independent of the sliding velocity. The applicability of this third law is much more limited. These three laws are entirely based on empirical observations.

Based on recent understanding of friction mechanisms there has been, however, important work published in the direction of a more general friction equation. One such example is the model for friction of two elastoplastic surfaces in quasi-steady-state sliding by Zhang *et al.* (1991a). The model considers the separate effects of asperity interaction and deformation, debris interaction and ploughing, and adhesion, with partitioning of normal load and contact areas between different contacting elements. It illustrates well the complexity of describing the contacting system and its relation to friction. The analysis is restricted to nominally steady-state sliding where the contact parameters are statistically invariant. The authors use the word 'quasi' to emphasize that in tribodynamic systems there is no true equilibrium condition in which the frictional effects are constant.

The coefficient of friction, μ, can, according to the Zhang *et al.* (1991a) model, be formulated as:

$$\mu = \frac{S_1}{w}\left[(1-\beta)\,A_a \cdot f\left(\tau_{a1}\frac{1}{S_1}\right) + \beta \cdot A_a + A_{d1}\right] + \frac{S_2}{w}\left[(1-\beta)\,A_a \cdot f\left(\tau_{a2}\frac{1}{S_2}\right) + \beta\,A_a \cdot f\left(\frac{S_1 \cdot H_1}{S_2 \cdot H_2}\right) + A_{d2}\right] \tag{3.3}$$

and

$$f(t) = 1 - \frac{2\ln(1+t) - t}{\ln(1-t^2)} \tag{3.4}$$

where S_i is the shear strength of the contacting material (Pa), w is the applied normal load (N), β is the percentage of asperity contact area under compression, A_a is the contact area contributed by deformed asperities (m^2), τ_{ai} is the shear stress within the asperity adhesion area of the material, A_{di} is the total debris contact area (m^2), H_i is the hardness of the contacting material (Pa) and t' is a dummy variable.

The model has been compared with experimental measurements (Zhang *et al.*, 1991b) with the conclusion that it can predict the coefficient of friction with reasonable accuracy. According to the model, adhesive effects are negligible. To enable the use of the formula, estimations must be made for average debris diameter, debris density and percentage of debris ploughing in the softer material and asperity adhesive and ploughing contributions.

In the case of macroploughing, using the example of a rigid spherical ball penetrating a softer surface and moving tangentially forming a groove, the ploughing component can be derived directly by calculating the volume of deformed material. The adhesive component consists of a material strain constant and the geometrically derived surface of sliding. The equations are derived in Komvopoulos (1991b), Bhushan (1999b), Lafaye *et al.* (2005) and Felder and Bucaille (2006).

3.1.3.6 Atomic scale friction

Friction at the microscale is a result of breaking bonds between individual atoms, whether instigated by strong adhesion or by micro-asperity interlocking, and slip along the crystal planes in the presence of defects. Breakage of cohesive bonds will occur during microplastic deformation or fracture. Depending on the relative strengths of the interfacial and cohesive bonds of the solids, breakage of the bonds will occur at locations that offer the least resistance to sliding (Kim and Suh, 1991).

Emerging techniques such as atomic force microscopy and other surface analysis methods (Israelachvili and Tabor, 1972; Bhushan, 1999b) have opened a new possibility to study friction and wear phenomena on a molecular scale and to measure friction forces between contacting molecules at the nano-newton level. Increased computation power has made it possible to study friction and associated phenomena by molecular dynamic simulations of sliding surfaces and to investigate the atomic scale contact mechanisms (Landman *et al.*, 1992; Harrison *et al.*, 1993, 1995; Zhang and Tanaka, 1997; Gao *et al.*, 2003; Yang and Komvopoulos, 2005).

Only some aspects of these complex nanophysical phenomena have so far been investigated such as the friction that arises from slippage between solid to solid interfaces (Thompson and Robbins, 1989) and between closely packed films in sliding contact (McClelland and Glosli, 1992) and that atomic friction increases logarithmically with sliding speed (Gnecco *et al.*, 2000). The atomic scale mechanisms of friction when two hydrogen-terminated diamond surfaces are in sliding contact has been studied and the dependence of the coefficient of friction on load, crystallographic sliding direction, roughness and methane molecules as third bodies in the contact have been investigated by Harrison *et al.* (1992, 1993, 1995, 1998) and Perry and Harrison (1996).

The improved understanding of the origin of friction at the atomic scale and even why friction exists has resulted in an examination of the relationship between the commonly used laws of friction at the macroscale and the molecular frictional behaviour on a nanoscale. There have been suggestions that friction arises from phonons, quantized modes of vibration, when atoms close to one surface are set into motion by the sliding action of atoms in the opposing surface (Celis, 1987). The lattice vibrations are produced when mechanical energy needed to slide one surface over the other is converted to sound energy, which is eventually transformed into heat.

Information on adsorbate vibration modes is needed to estimate the magnitude of surface corrugation, that is, how much the binding energy to a surface varies as an atom moves from one position to another along the surface. The phononic component of friction depends on the corrugation in the potential. The phononic friction is hypersensitive to commensurability and interfacial orientation effects. Liquid surface layers are more flexible than solid ones and therefore slightly more commensurate with the underlying surface, exhibiting higher friction than solid counter parts. A transition

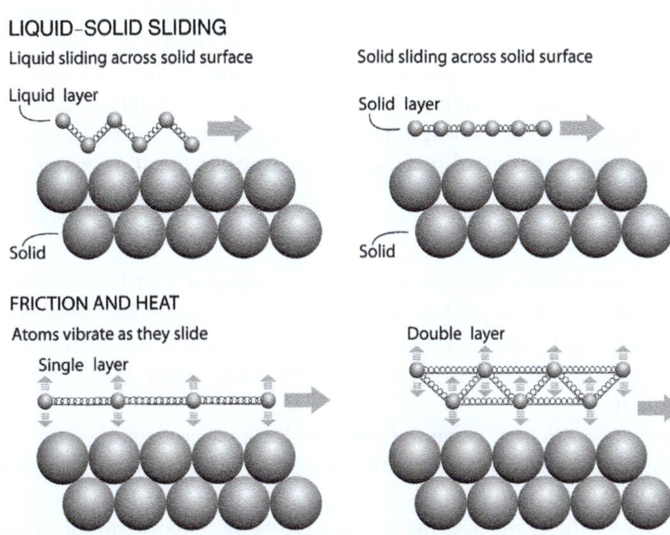

Fig. 3.9. Schematic illustration of phonon friction when liquid and solid layers, single and double, slide across a solid surface (after Krim, 2002b).

from commensurate to incommensurate sliding conditions can theoretically reduce the sliding friction levels by many orders of magnitude (Krim, 2002b).

A schematic illustration of phonon friction is shown in Fig. 3.9. Unlike matter in the visible world, a solid layer that is one atom thick slides more easily over a solid surface than an equivalent liquid layer. That is because atoms in solids are more tightly bound together than atoms in liquids. Liquid atoms tend to fall between atoms of the surface beneath them, hindering their sideways movement. Solid atoms slip across the surface as a cohesive sheet. At the atomic level, a double layer of atoms is harder to slide over a solid surface than a single layer because the two layers of atoms jostle each other, producing extra heat. The extra heat creates more friction, and that means it takes more effort to slide the double layer of atoms. More mechanical energy must be added – one has to push harder – to maintain the sliding motion (Krim, 1996, 2002a, b; Robbins and Krim, 1998).

In a molecular level study of Amonton's law Berman *et al.* (1998) found that the projected area is not necessarily proportional to the load and that the shear strength is not constant. Despite this, Amonton's laws are obeyed and the friction force is still proportional to load on a macro level. They offer a physical model, based on intermolecular forces and thermodynamic considerations, which explains why the friction force is proportional to the applied load, and why the case of adhering surfaces – where the friction force is found to be proportional to the molecular contact area – is quite different from that of non-adhering surfaces. Gao *et al.* (2004) state that the 'real' area of contact is not a fundamental quantity, whether on nano-, micro- or macroscale. However, it may serve as a convenient scaling parameter for describing the really fundamental parameters, which are the number or density of atoms, molecules and bonds involved in an adhesive or frictional interaction.

New and interesting approaches to understanding friction on a nanolevel are represented by the quantum chemical study on the atomic scale tribology of diamond and boron nitride surfaces (Koskilinna, 2007) and the movable cellular automata method that allows simulation of the friction force as a function of material and loading parameters, development of surface topography and wear (Popov and Psakhie, 2007). The experimental and theoretical aspects of nanotribology have been reviewed by Dedkov (2000).

Investigations of friction at the atomic scale have taught us that the many assumptions we make about friction mechanisms are, in the best cases, oversimplified and, in the worst cases, wrong. It has also been shown that a statistical approach, such as using the Weibull distribution on the local friction forces, may be a useful way to approach the friction phenomena (Gao *et al.*, 2004; Tysoe and Spencer, 2004). Today, we are only at the very beginning of understanding the nanomechanical tribological contact phenomena that explain the origin of friction and there is no doubt that in the near future many new theories and explanations for the origin of tribological phenomena will become available. The nanotribological mechanisms of coated surfaces are discussed in detail in section 3.4.6.

3.1.3.7 Ultra-low sliding friction

The development of new surface deposition techniques, especially PVD and CVD and their combinations, has made it possible to produce surface structures and contact systems with extremely low friction. The measured values of the coefficient of friction for carbon and molybdenum disulphide-based coated surfaces have been about 50 times lower than what has earlier been achieved in dry sliding.

Very low coefficients of friction, down to 0.001 and below, were measured for a sputtered, about 350 nm thick, MoS_2 coated surface sliding against a steel hemispherical pin in high vacuum (Martin *et al.*, 1993, 1994; Le Mogne *et al.*, 1994). The very low friction is thought to depend on the appropriate combination of the grain size, the crystal orientation effects and the absence of contaminants. The MoS_2 crystal structure had a friction-induced orientation of very easily sheared basal planes parallel to the sliding direction. They report that even a zero friction state can be theoretically predicted in these conditions.

Low coefficients of friction of the same level, down to 0.001, have been measured by Erdemir *et al.* (2000, 2001) for a 1 μm thick diamond-like carbon coating deposited on a flat sapphire surface sliding against a similarly coated sapphire ball in inert-gas environments such as dry nitrogen and argon. The surfaces were extremely smooth so asperity interlocking effects were eliminated; the surfaces were hard so ploughing effects were eliminated; and the dangling bonds of the carbon structure were attached to hydrogen atoms creating inert hydrophobic surfaces that exhibit almost no frictional resistance (Harrison *et al.*, 1995; Erdemir, 2002). Actually, no internal material shear is taking place at all, the hydrogen planes just 'fly' over each other.

The above ultra-low friction experiments were carried out in special environments, in high vacuum or in an inert gas. However, low friction values have also been measured in normal air recently. Abreu *et al.* (2005) measured a coefficient of friction of 0.03 for a CVD diamond coated surface sliding against a similarly coated ball in normal air. Vander Wal *et al.* (2005) have reported extremely low coefficients of friction, between 0.002 and 0.07, for a quartz disk that was carbon nanotube surface-fluorinated and sliding in contact with a sapphire ball in air of about 40% relative humidity and room temperature.

The property of two surfaces sliding over each other with extremely low friction is also referred to as 'superlubricity'. Erdemir and Martin (2007) present studies of superlubricity related to the theoretical aspects, various sliding systems at nanoscale, lamellar solids over a broad range of scales, carbon-based thin coatings at macroscale and superlubricity in elastohydrodynamic and boundary lubrication regimes.

3.1.3.8 Rolling friction

We have so far considered friction mechanisms only in relation to the sliding of one surface over another. In the case of rolling, the coefficient of friction is normally smaller. The general considerations in relation to contact mechanisms are also valid for rolling contacts, but, because of the difference in contact kinematics, other contact effects dominate the frictional behaviour.

The main contributions to friction in rolling contacts are:

- micro-slip effects within the contact area,
- elastic hysteresis of the contacting materials,
- plastic deformation of the materials, and
- adhesion effects in the contact.

The contact mechanics for rolling are certainly complex but fairly well understood on a macroscopic level (Halling, 1975; Czichos, 1978; Johnson, 1985; Bhushan, 1999b). On the microscopic level they are less well understood, but on this level the mechanisms of contact physics and chemistry dominate and they are mainly similar to the mechanisms in sliding contacts discussed in more detail earlier.

3.1.3.9 Typical values for the coefficient of friction

The coefficient of friction depends on the contact system. It can vary a great deal with regard to the material pair used. Some typical values for the coefficient of friction in sliding contacts are given in Table 3.1.

3.1.4 Wear

Wear is commonly defined as the removal of material from solid surfaces as a result of one contacting surface moving over another. Thus both friction and wear are simultaneously the results of the same tribological contact process that takes place between two moving surfaces. However, their interrelationship is not well understood. It is common to find that low friction corresponds to low wear

Table 3.1. Values for the coefficient of friction in dry sliding contacts. Some indicative values for the coefficient of friction in different dry sliding systems are given below (data from Dowson, 1998; Avallone and Baumeister, 1987; Hutchings, 1992; Bhushan, 1999b).

Sliding materials coefficient of friction	
Gold on gold	1–2
Iron on iron	0.8–1.5
Aluminium on aluminium	0.8–1.2
Copper on copper	0.8–1.2
Silver on silver	0.8–1
Grey cast iron on grey cast iron	0.8–1
Mild steel on mild steel	0.7–0.9
Copper on mild steel	0.6–0.7
Titanium on titanium	0.5–0.6
Aluminium on mild steel	0.5–0.6
Silver on steel	0.5
Brake materials	0.5
Titanium on mild steel	0.4–0.6
Gold on mild steel	0.4–0.5
Sandstone on sandstone	0.35
Titanium carbide on titanium carbide	0.3–0.7
Aluminium oxide on aluminium oxide	0.3–0.6
Granite on granite	0.3
Rubber on steel	0.3–3
Titanium nitride on titanium nitride	0.25–0.5
Leather on iron	0.25
Polyimide on hard surface	0.2–0.4
Polyamide (nylon) on hard surface	0.15–0.3
Hardwood on hardwood	0.15
Diamond on diamond	0.1–0.2
Dry sliding in general	0.1 to 1.0
Graphite solid lubricant on hard surface	0.05–0.15
PTFE (teflon) on hard surface	0.05–0.1
MoS_2 solid lubricant on hard surface	0.05–0.1
Fullerenes (c_{60}) on hard surface	0.05–0.1
Rolling friction (balls or rollers)	0.001–0.01
DLC on DLC in dry nitrogen or argon	0.001–0.01
MoS_2 coated surface on steel in high vacuum	0.001 and below

and high friction to high wear, as shown in the set of experimental results in Fig. 3.10 by Saka (1980). However, this is not a general rule and there are several examples (e.g. Franklin, 1991) of contradictory behaviour. The figure also illustrates another typical feature. The value for the wear coefficient may well change, in different conditions, by several orders of magnitude. This can be contrasted with the coefficient of friction, which rarely changes by more than a single order of magnitude, unless a lubricant is introduced or removed from a contact – and in that case the wear would show even greater changes.

A *wear mechanism* is a classification of the process by which material is removed from the contact surface. Typical wear mechanisms are adhesive, abrasive, fatigue and chemical wear. It is very common that in a real contact more than one wear mechanism is acting at the same time (Holmberg, 1991). The precise combination of wear mechanisms depends on the contact conditions and will result in a specific type of wear.

A *wear mode* is a classification of the type of contact, which is characterized by a specific kind of movement, geometry or environment. Examples of wear modes are sliding wear, rolling wear,

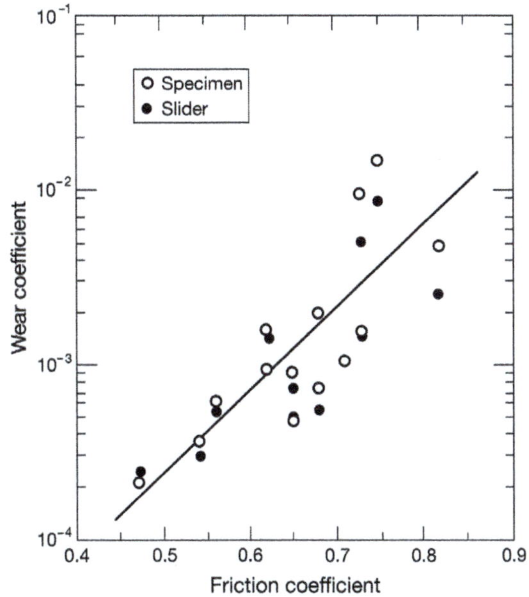

Fig. 3.10. The coefficient of wear of different steels as a function of the coefficient of friction (data and wear coefficient definition according to Saka, 1980).

fretting, erosion, impact wear and cavitation. A *wear failure mode* is typically referred to on the basis of how the surface appears after the contact, e.g. scuffing, pitting, scoring, spalling, seizure and delamination. The wear mode defines a certain set of conditions where the interrelationship between the influencing parameters is similar and thus the different effects can more easily and accurately be described and mathematically formulated. One example of this is the suite of wear-rate equations used in the wear maps devised by Lim and Ashby (1987), and shown in Fig. 3.16.

There is, however, still no general agreement about the classification of wear mechanisms and modes. From the examples of different proposed wear classification schemes reviewed by Blau (1989) it can be seen that the one we use in this book seems to be the most common. The basic wear mechanisms are shown in Fig. 3.11:

- adhesive wear,
- abrasive wear,
- fatigue wear, and
- chemical wear.

3.1.4.1 Adhesive wear

When asperities of one surface come into contact with asperities of the counter surface they may adhere strongly to each other and form asperity junctions as shown in Fig. 3.11a. Relative tangential motion of the surfaces causes separation in the bulk of the softer asperities and material is removed. In adhesive wear the surface material properties, as well as possible protecting surface films or contaminants, play important roles.

The adhesive wear mechanism was described by Bowden and Tabor (1950) and it has been a dominating wear theory for sliding contacts ever since. The adhesive wear mechanism has been described in several works (Rabinowicz, 1965; Merchant, 1968; Tabor, 1977; Rabinowicz, 1980;

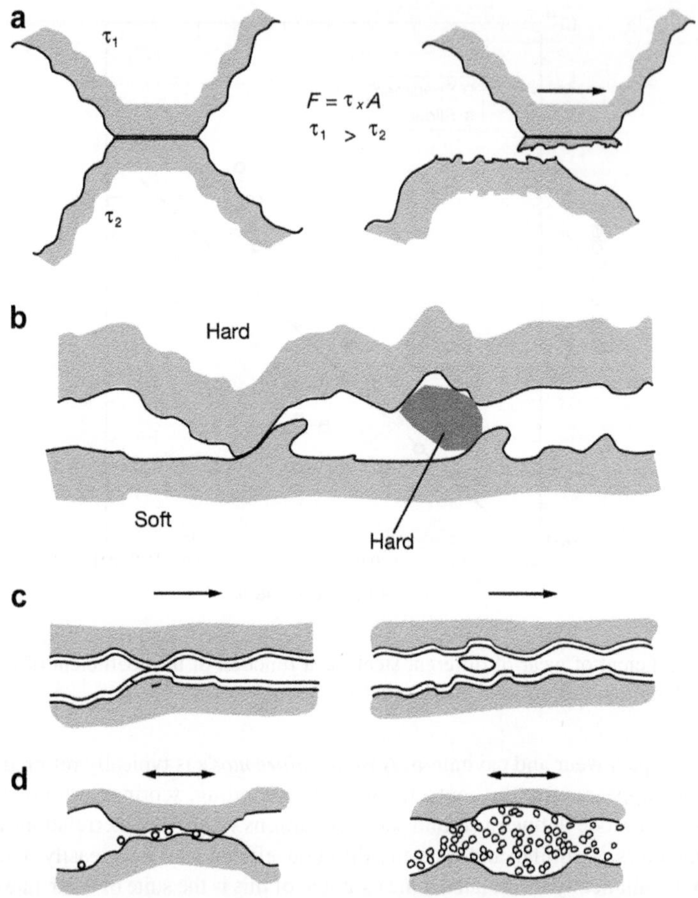

Fig. 3.11. The basic mechanisms of wear are (a) adhesive, (b) abrasive, (c) fatigue and (d) chemical wear.

Buckley, 1981; Kragelskii, 1982; Landheer and de Gee, 1991; Bayer, 1994; Bhushan, 1999b; Kato and Adachi, 2001b; Stachowiak and Batchelor, 2005). The adhesion wear theory has, however, been criticized (e.g. Jahanmir, 1980) on the following grounds:

- because of impurities it seems unlikely that the interface will be stronger than either of the bulk materials,
- it cannot explain the wear of the harder surface which is frequently observed,
- it cannot explain observed effects of surface roughness,
- it cannot explain the creation of loose wear particles,
- the maximum work required to generate wear particles is two to three orders of magnitude smaller than the external work done,
- it cannot explain the relation between friction and wear, and
- it does not take into account microstructure and metallurgical parameters.

There is, though, enough experimental evidence in the references given above to prove that adhesive wear does occur in many sliding situations. However, based on analyses by Suh (1973), Jahanmir (1980) and Suh (1986) of friction and wear mechanisms, it seems that the adhesion wear mechanism does not necessarily play such a dominating role in real contacts as had traditionally been considered. There are, however, conditions like sliding wear in vacuum or inert atmospheres, or when an asperity flash temperature is large enough to promote asperity adhesion, in which the effect of adhesive wear is likely to be the dominating mechanism. The influence of humidity has been observed to be considerable in dry sliding of steel against steel in normal air while the friction behaviour is less affected (Klafke, 2004a; Klafke *et al.*, 2004b). In failure modes like scuffing and seizure the effect of adhesive wear is very important.

3.1.4.2 Abrasive wear and asperity deformation

Wear by plastic deformation takes place both in abrasive wear and during asperity deformation as shown in Fig. 3.11b.

Abrasive wear occurs in contacts where one of the surfaces is considerably harder than the other or where hard particles are introduced into the contact. The harder surface asperities are pressed into the softer surface which results in plastic flow of the softer material around the hard one. When the harder surface moves tangentially, ploughing and removal of softer material takes place resulting in grooves or scratches in the surface.

Depending on the geometry of the harder surface and the degree of penetration, the removal of material can take different forms, such as ploughing, wedge formation or cutting, as shown by Hokkirigawa and Kato (1988a, b). Ploughing abrasive wear is present to a large extent in mining and earth moving equipments. Wear modes such as gouging abrasion, grinding abrasion and erosion abrasion can be distinguished (Czichos, 1978). The action of metal cutting has been described as abrasive wear on a large scale.

A distinction is often made between two-body and three-body abrasive wear, where the latter refers to situations in which hard particles are introduced between the moving surfaces. The mechanism of abrasive wear has been extensively studied, and described by several authors (Kruschov, 1974; Moore, 1974; Sin *et al.*, 1979; Zum Gahr, 1989; Hutchings, 1992; Kato, 1992; Bayer, 1994; Bhushan, 1999b; Kato and Adachi, 2001b; Adachi and Hutchings, 2003; Stachowiak and Batchelor, 2005; Atkins and Liu, 2007; Siniawski *et al.*, 2007).

Asperity deformation is due to the roughness and waviness of the contacting surfaces. A collision of two asperities results in plastic deformation of one or both of the asperities leading to material removal from the asperities. The asperity deformation mechanisms have been studied in detail by Suh (1986) and it seems possible that a large proportion of the wear effects earlier considered as adhesive have in fact been processes of asperity deformation.

Sliding contacts deforming the surface plastically cause large plastic strains and strain hardening in the surface. The plastic deformation changes the near-surface microstructure in ways which make the material resistant to local shear. This leads to the transfer of pieces of material and elements from the environment to produce an ultra-fine-grained transfer film. The very fine microstructure in this transfer material is stabilized by the mixing in, actually a form of mechanical alloying, of a second phase (Rigney *et al.*, 1984).

Mechanical intermixing as a result of asperity deformation can also change the chemical composition of the near-surface volume. The depth profile in Fig 3.12 shows the change in chemical composition of a grey cast iron surface before and after oil lubricated sliding against a steel pin at a pressure of less than 100 MPa (Scherge *et al.*, 2003a). The modified zone here exhibits an extension of more than 150 nm. The distributions of calcium, oxygen and sodium have their maxima underneath the surface, clearly indicating mechanical mixing.

3.1.4.3 Fatigue and delamination wear

Fatigue crack growth is now a well-documented phenomenon which results from loading and unloading of a surface, at a stress level in the material that it can sustain once but not if repeated many

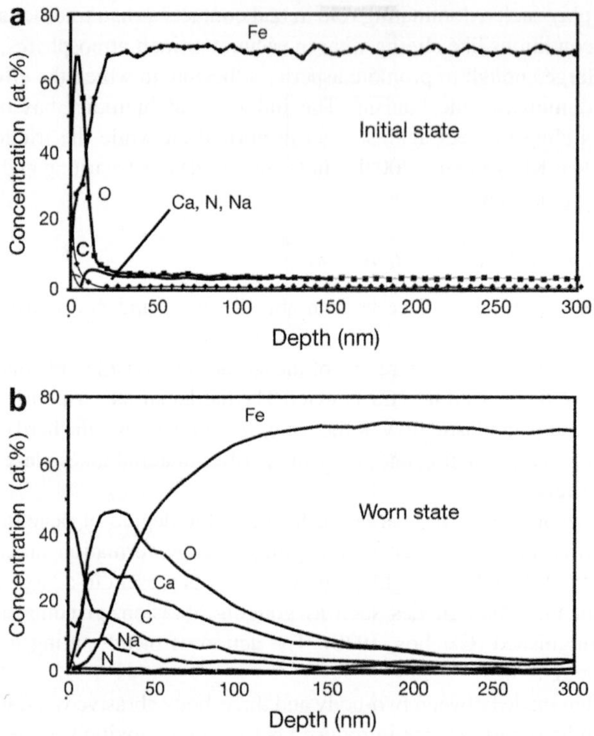

Fig. 3.12. Depth profile by Auger electron spectroscopy of grey cast iron: (a) the initial state and (b) worn state after 200 hours of sliding (after Scherge *et al.*, 2003a).

times. Fatigue can form the origin for large-scale cracking and liberation of surface material to form wear debris as shown in Fig. 3.11c.

Classic surface fatigue wear is a lifetime limiting failure mechanism in ball and roller bearings and in gear contacts. In these concentrated contacts the Hertzian pressure at the surface creates a stress field beneath the contact zone. The maximum shear stress occurs about one-third of the contact length beneath the surface in the case of pure rolling and moves to the surface with increasing traction. For coefficients of friction exceeding 0.32 the maximum shear stress will be found on the surface, as shown in Fig. 3.13 (Schouten, 1973a, b).

In rolling, where the stress field moves repeatedly over the surface, fatigue of the near-surface material takes place. Material voids or dislocation pile-ups may form the nuclei for the first crack to occur. After that the crack will often propagate quite rapidly, unite with other cracks and liberate surface material. This often results in large craters in the surface with diameters of 1 to 2 millimetres and rapid destruction of the whole surface. The mechanism of rolling contact fatigue is not fully understood even though it has been extensively studied (Lundberg and Palmgren, 1947, 1952; Tallian, 1967–68, 1982; Tallian and McCool, 1971; Scott, 1979; Bamberger and Clark, 1982; Sundquist, 1981; Spikes *et al.*, 1986; Longching *et al.*, 1989; Bayer, 1994; Ding *et al.*, 1996; Zaretsky, 1997; Murakami *et al.*, 1997; Ioannides *et al.*, 1999). Effects of lubricant film thickness and chemistry on fatigue life and crack propagation have been observed but are also not fully understood (Fowles *et al.*, 1981; Cantley, 1983; Sullivan and Middleton, 1989). Pitting and spalling are wear failure modes in which the fatigue wear mechanism dominates.

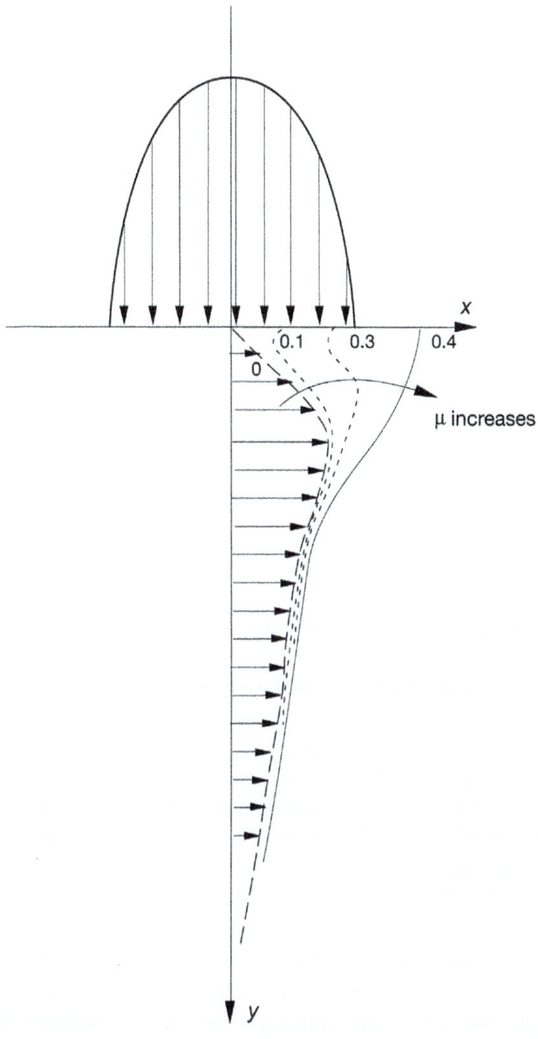

Fig. 3.13. The position of maximum shear stress moves towards the surface with increasing values of the coefficient of friction (after Schouten, 1973a).

Delamination wear is a kind of fatigue wear that occurs on a more microscopic scale in sliding contacts when surface asperities repeatedly slide over each other. Small cracks are nucleated below the surface. Crack nucleation very near the surface cannot occur, due to the triaxial state of compressive loading which exists just below the contact region. Further loading and deformation causes cracks to extend, propagate and join with neighbouring cracks. The cracks tend to propagate parallel to the surface, resulting in the delamination of long and thin wear sheets. Figure 3.14 shows the formation of parallel subsurface cracks and plate-like wear debris in a reciprocating fretting wear contact where delamination is one important wear mechanism.

The delamination wear theory was developed and has been described in detail by Suh (1973, 1986) and Suh *et al.* (1974). Some of the effects in sliding wear that have been traditionally referred to as adhesive wear are probably more correctly described by the delamination wear mechanism (Ko *et al.*, 1992; Johnson, 1995).

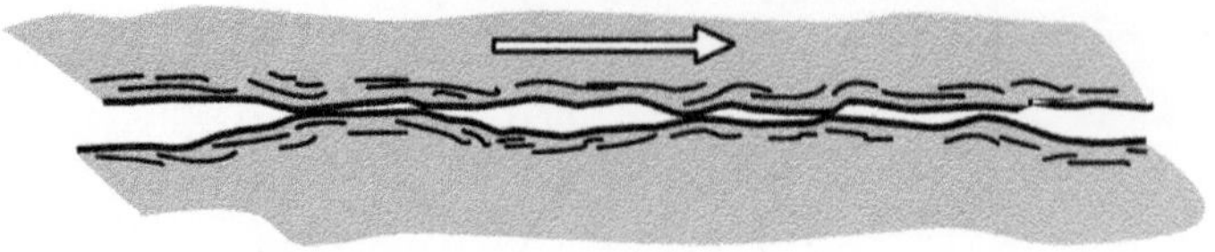

Fig. 3.14. Parallel subsurface cracks and plate-like wear debris are formed as a result of repeated loading in delamination wear.

3.1.4.4 Chemical wear

In chemical wear the wear process is dominated by detrimental chemical reactions in the contact, initiated by the influence of the environment, in combination with mechanical contact mechanisms as shown in Fig. 3.11d.

Rubbing, in combination with the chemical reactions in the contact, results in removal of material and wear debris formation. The rubbing action results in increased temperatures at the surface and creates surface cracks which are favourable for more chemical reactions to take place. On the other hand, the chemical reactions at the surfaces may make them softer and weaker and thus decrease their resistance to new crack formations and liberation of surface material to produce wear products. The mechanisms of chemical wear have been investigated by Czichos (1978), Buckley (1981), Celis *et al.* (2006) and Jiang and Stack (2006).

Oxidational wear is the most common chemical wear process. A thin layer of oxides will be formed on the top of metal surfaces. This is an important protecting layer because without it both the friction and wear in metal contacts would be extremely high. If this layer is continuously removed by a rubbing action, and the formation of new layers is speeded up by a high humidity environment that can easily reach the contact, the result is typical oxidational wear. The oxidational wear process has been reviewed by Quinn (1983a, b), Sullivan (1987a, b) and Bayer (1994) and the thermal aspects treated by Quinn and Winer (1985).

In certain cases wear can be due to chemical instability of the materials. Examples of this are found in the metal cutting process where the chip moves rapidly along the tool surface (Fig. 3.1c). Temperatures at the interface can be over 700°C and the surface of the chip that slides against the tool is virgin and may be partly molten. This process can result in two kinds of wear due to chemical instability, as shown by Suh (1986). One is the solution wear due to dissolution of the tool material in the chip. The other is the diffusion of elements from the tool material into the chip, leading to weakening and ultimate failure of the cutting tool material. These wear mechanisms are discussed in more detail in section 7.5.

It is arguable whether the chemical wear should really be classified as a separate wear mechanism because the chemical actions are always combined and interacting with some of the other wear mechanisms. If no other mechanical wear mechanism like adhesive, abrasive or fatigue wear is involved it is no longer a tribological contact but pure surface corrosion. It would perhaps be more correct to talk only about three mechanical wear mechanisms, and consider the chemical effects as an additional influencing parameter in the tribological system which changes the material properties of the surface layer as a function of time.

3.1.4.5 Velocity accommodation concept

A different approach to the question of how wear mechanisms should be classified is the velocity accommodation approach developed by Berthier *et al.* (1988, 1989), Godet (1984, 1989), Berthier (2005) and Fillot *et al.* (2007). They identify four basic mechanisms with which the velocity

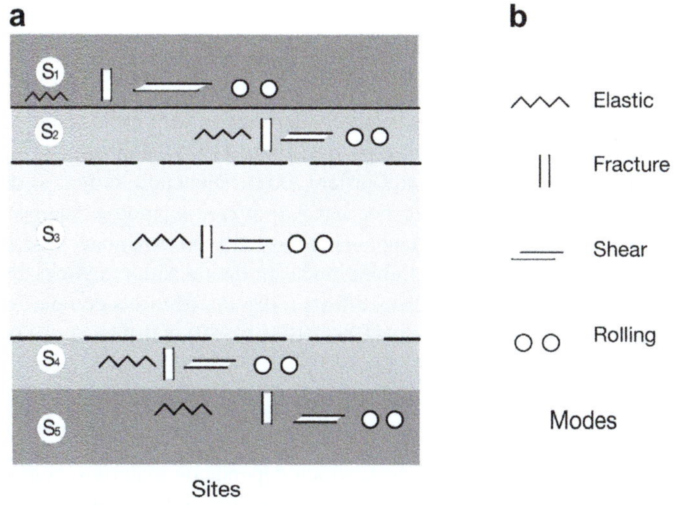

Fig. 3.15. The mechanisms of velocity accommodation of two surfaces in relative motion (after Godet, 1989).

difference between two surfaces in relative motion can be accommodated; elastic deformation, fracture, shear and rolling. These basic mechanisms may occur in either of the two surfaces (first bodies) or in the layer between them (third body), as shown in Fig. 3.15. The major advantage with this approach is that the fundamental tribological mechanisms are simplified to well-known mechanisms of material behaviour that can be characterized by properties such as Young's modulus, fracture toughness and shear strength. It is, however, clear that much work has to be done before these basic properties can be combined in an adequate and useful way to form a model that represents friction and wear behaviour in real contacts. New modelling techniques as well as increased computation capacity offer an improved possibility to determine the tribological behaviour in a contact based on the basic material performance specifications of the contacting surfaces (Holmberg *et al.*, 2003, 2006a, b; Laukkanen *et al.*, 2006).

3.1.4.6 Wear failure modes or types of wear

The wear failure mode or type of wear describes a certain combination of wear mechanism with a well-described regular process of material removal that corresponds to a particular range of contact conditions. The most common wear failure modes are:

- fretting,
- pitting,
- scuffing,
- gouging,
- electrical wear,
- solution wear,
- melt wear,
- spalling,
- scoring,
- galling,
- cavitation,

- diffusive wear,
- mild wear and
- severe wear.

The published work on wear failure modes is largely descriptive with little analytical basis. Different modes have been described and reviewed, e.g., by Moore (1975), Jahanmir (1980), Peterson and Winer (1980), Blau (1989), Kato (2002) and Cartier (2003). The whole subject is still in a somewhat confused state because much work is published but with non-comparable data because of the large variations in test conditions and equipment. There is still no adequate description of the interrelationship between the different wear failure modes or the transition regimes and mechanisms of transfer from one mode to another. Pioneering efforts to try to cover this complex and most difficult field are found in the works of Lim and Ashby (1987), Blau (1989), Kitsunai *et al.* (1990), Kato (1992) and Kato and Adachi (2001b).

3.1.4.7 Wear equations

Because of the complexity of the wear process, which depends on materials, contact conditions and environmental parameters in a number of different combinations, it has not been possible to formulate a universal equation of wear. Numerous authors (reviewed by Bahadur, 1978; Meng and Ludema, 1995; Dunaevsky, 1997; Bhushan, 1999b) have developed wear equations for the different wear modes but all of them are quite limited in their range of validity. Meng and Ludema (1995) listed 100 different variables and constants they found in the equations for general sliding wear, which illustrates well the complexity of the wear process. The Russian school originating from Kragelsky has the longest tradition and experience in formulating wear equations for different kinds of contact conditions and machine components (Kragelsky, 1965; Kragelsky *et al.*, 1982; and Kragelskii and Marchenko, 1982).

A more general approach is taken by Lim and Ashby (1987) in which they show the validity ranges of six different wear modes in a wear map with normalized contact pressure and normalized velocity as coordinates and also give wear equations for each wear mode. The identified wear modes are ultra-mild wear, delamination wear, mild oxidational wear, severe oxidational wear, melt wear and seizure. The representation is, however, limited to a steel sliding pair in a pin-on-disk machine as shown in Fig. 3.16.

It is possible to observe some common features when studying the proposed wear equations for the different wear modes. Often the worn volume is directly proportional to the normal load and the distance of movement and inversely proportional to the hardness of the material. This relationship was observed by Holm (1946) and Archard (1953) and can be formulated as

$$V = K' \cdot \frac{w \cdot s}{H} \tag{3.5}$$

where V is the worn volume, w is the normal load, s is the distance moved, H is the hardness and K' is a constant, frequently called coefficient of wear. Even if this relationship is valid in many cases, it should not be considered as a universal wear equation because there are several examples of tribological contacts where it is invalid.

3.1.4.8 Wear rate

For design and material development purposes it is necessary to have some universal quantitative parameter for wear. To use the volume of removed material is not useful or illustrative because of the very different test conditions in use. The so-called Holm or Archard wear relationship (Holm, 1946; Archard, 1953) is often used for formulating the wear rate (also called specific wear rate) which is a practical and more general value for the amount of wear. We will use for the wear rate, K, the value

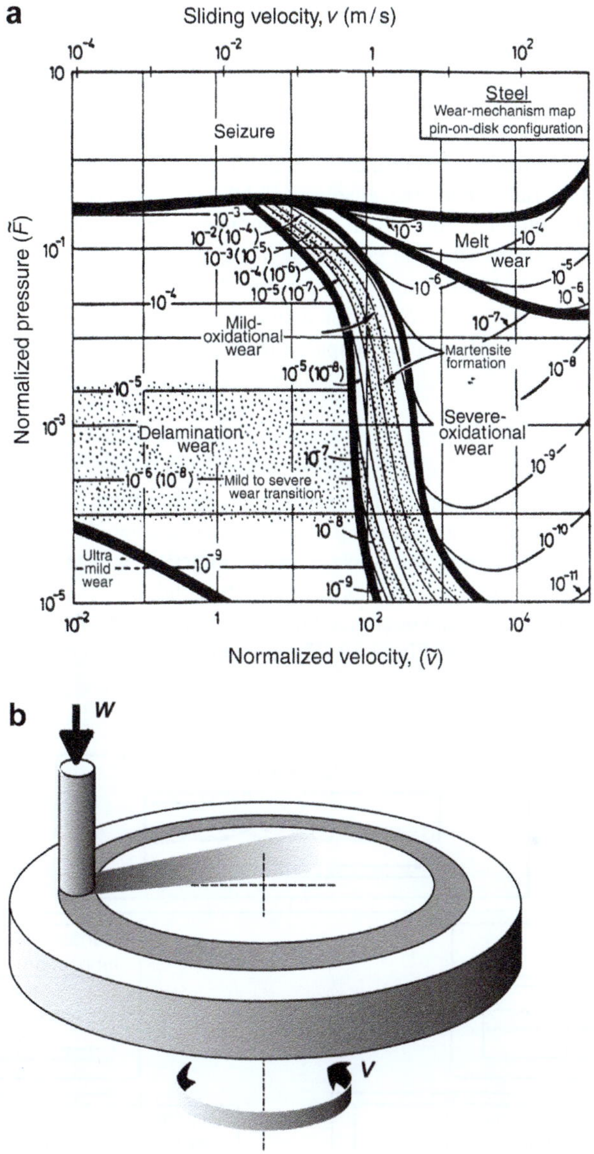

Fig. 3.16. (a) The Lim and Ashby (1987) wear map for a sliding steel pair. (b) The pin-on-disk test device used.

of the constant K' divided by the hardness H, in accordance with common practice. Thus the wear rate is

$$K = \frac{V}{w \cdot s} \tag{3.6}$$

which is normally given with the dimensions (10^{-6} mm^3/Nm). Even if this is not from all aspects the ideal way of expressing wear it is to our understanding the most suitable, and it has wide support.

There is also a clear physical argument for using the wear rate as defined above because it is the worn volume divided by the mechanical energy input into the contact. The contact energy input can be described as the product of the normal load, the velocity and the applied time which again is the same as the product of load and distance which we find in the equation. It is strongly recommended that this way of expressing wear should be taken into use as widely as possible, to make it possible to compare and utilize wear data produced in different contact conditions. The International Research Group on Wear of Advanced Materials under the auspices of OECD (IRG OECD) has recently discussed a suggestion to give the unit '10^{-6} mm^3/Nm' a specific name for more convenient data comparison.

In this book the different tribological coatings will be quantitatively compared to each other by comparing the coefficient of friction and the wear rate. In the literature it is, however, more often possible to find data for the coefficient of friction than for the wear rate. This is unfortunate, because in many cases it would be more relevant to compare the wear properties than the friction properties.

3.1.5 Lubrication

The most common way of reducing friction and wear is to introduce a lubricant between the two moving surfaces. If the lubricant manages to prevent contact between the solids, the shearing action, which creates resistance to motion and stresses in the material, will take place either within the lubricant or in the boundary between the lubricant and the solid.

The mechanisms of lubrication can be divided into two main groups, fluid pressure lubrication and surface film lubrication as shown in Fig. 3.17. These expressions are considered more suitable than

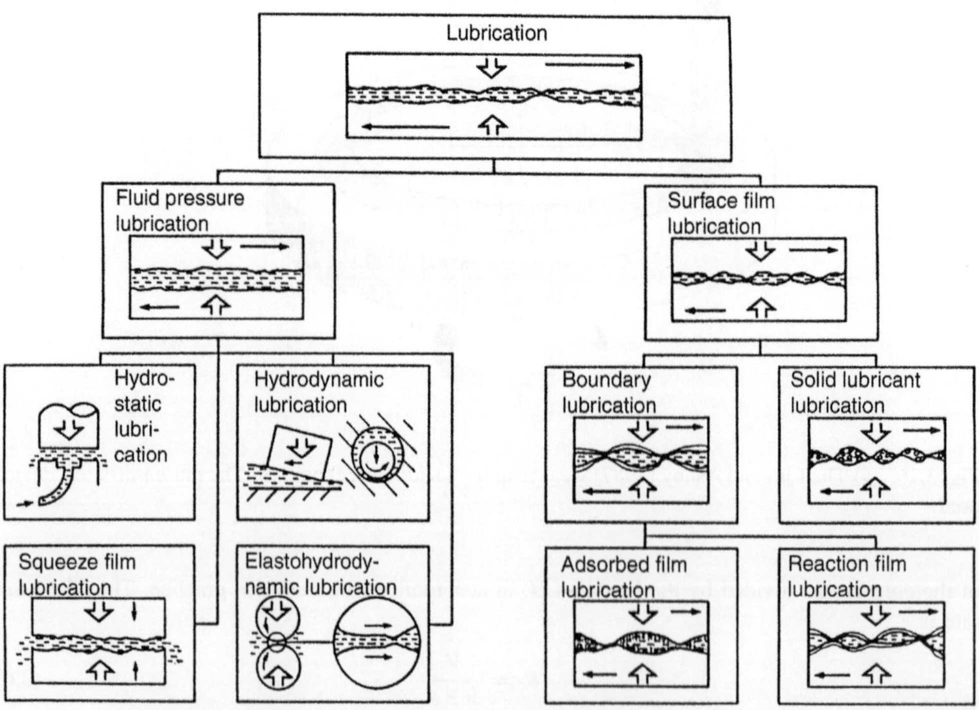

Fig. 3.17. Classification of the mechanisms of lubrication.

those used previously: thin film and thick film lubrication. As the latter expressions are not related to physical effects, the distinction between the two groups becomes unclear.

Fluid pressure lubrication can be defined as the group of lubrication mechanisms where the surfaces are kept apart from each other by fluid pressure created in the lubricant within the contact. Fluid pressure lubrication is mainly governed by fluid dynamics and rheology. This is a relatively well-known area and it can be mathematically formulated to a large extent.

Surface film lubrication can be defined as the group of lubrication mechanisms where the surfaces are partly kept apart from each other by a protecting film attached to the surfaces. The film may be formed chemically or physically and can be attached to the surfaces by chemical or physical bonds. This area is dominated by lubricant chemistry and materials knowledge which is generally expressed in a fairly unsystematic way and the behaviour cannot generally be formulated mathematically.

These two classifications can be subdivided as shown in Fig. 3.17. In particular hydrodynamic, elastohydrodynamic and boundary lubrication will be discussed below.

3.1.5.1 Hydrodynamic lubrication

In hydrodynamic lubrication a pressure is created within the lubricant due to the relative motion of the opposing surfaces. The pressure is high enough to keep the two solid surfaces entirely separated. The mechanism of hydrodynamic lubrication is illustrated in Fig. 3.18 by the flow of fluid between two surfaces in relative motion with a convergent shape in the direction of motion.

The fluid adhering to the moving surfaces will be dragged into the narrowing gap. Due to the law of continuity the amount of fluid entering must be equal to the amount leaving. Thus the velocity distribution must be concave at the inlet and convex at the outlet, as shown in Fig. 3.18. Forcing the fluid into the converging region creates a pressure between the surfaces. The high pressure in the centre slows down the fluid at the inlet and makes it go faster at the outlet.

The three basic requirements for the generation of hydrodynamic pressure are:

- the surfaces must be in relative motion,
- the surfaces must be converging in the direction of motion, and
- there must be a viscous fluid between the surfaces.

For lubricated machine parts the hydrodynamic films are normally 10 to 100 times thicker than the height of the surface irregularities. The minimum film thickness normally exceeds 1 micrometre. The coefficient of friction is typically of the order 0.001 to 0.01 (Rabinowicz, 1968).

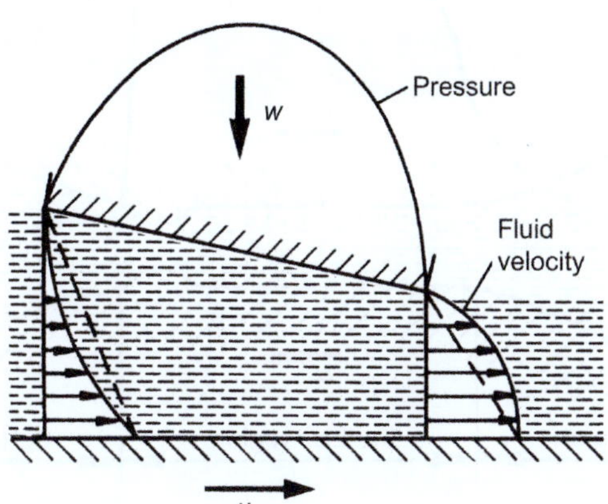

Fig. 3.18. Hydrodynamic pressure generation between two converging flat surfaces.

The hydrodynamic pressure is low compared to the strength properties of the solids and will not cause appreciable local deformation. The condition of hydrodynamic lubrication is governed by the bulk physical properties of the lubricant, mainly by the viscosity, and surface effects are negligible.

The basic equation for hydrodynamics was developed by Reynolds (1886). A detailed treatment of hydrodynamic lubrication is found in numerous books on this subject, e.g. Cameron (1966), Lang and Steinhilpher (1978), Szeri (2001) and Stachowiak and Batchelor (2005).

3.1.5.2 Elastohydrodynamic lubrication

In elastohydrodynamic lubrication (EHD or EHL) the local elastic deformation of the solids provides a coherent hydrodynamic film, which prevents asperity interaction. In non-conformal contacts of metal surfaces the changes in the rheological properties of the lubricant, typically a mineral oil, are essential for the formation of an elastohydrodynamic film.

When a viscous lubricant is introduced into the contact between two rotating metal discs, a hydrodynamic pressure will be generated in the inlet region as shown in Fig. 3.19. This pressure tries to force the surfaces apart at the same time as the lubricant is dragged into the contact. The deformed contact region is called the Hertzian region because here the pressure distribution is very close to the Hertzian pressure distribution, except for the pressure peak at the rear of the contact.

Even if the maximum pressure in steel contacts is as high as 2 to 3 GPa, it does not squeeze out the oil film for the following reasons:

- a typical elastohydrodynamic film is extremely thin ($h = 0.05$ to 2 μm) and the length and width of the contact region is often more than 1000 times greater,
- the viscosity of oils increases almost exponentially with pressure so that the lubricant in the centre of the contact region has a very thick, solid-like consistency, and
- the lubricant passes very quickly through the contact region, typically in the order of milliseconds.

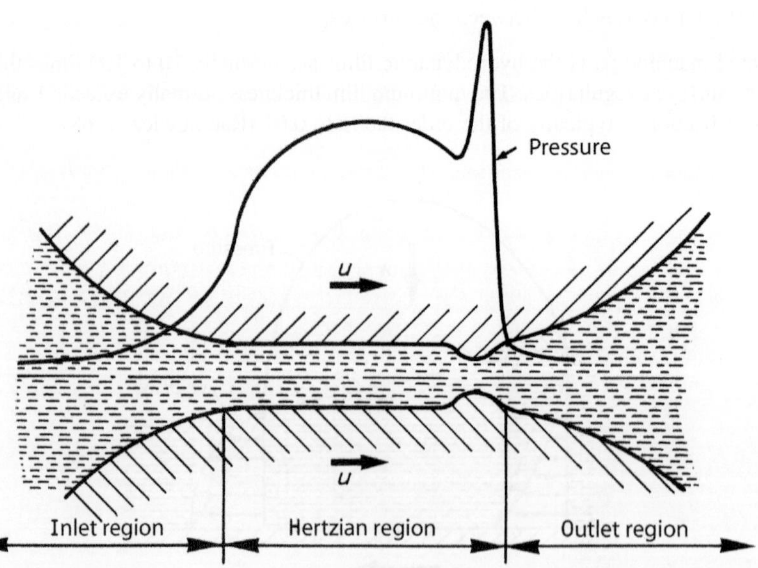

Fig. 3.19. Pressure distribution and lubricant film shape in the elastohydrodynamically lubricated contact of two rotating discs.

There will be no time for the high Hertzian pressure to squeeze out this poorly flowing thick fluid from the long and extremely narrow gap. The coefficient of friction is typically of the order of 0.01 to 0.1.

The first satisfactory analysis in which the effect of elastic deformation and the influence of pressure on viscosity were taken into account was worked out by Erthel and published by Grubin (1949). More complete numerical solutions were developed by Dowson and Higginson (1966) for line contacts and Hamrock and Dowson (1981) for elliptical contacts.

The most important practical result of elastohydrodynamic lubrication theory is the formula for the calculation of the minimum film thickness in the contact. For line contacts the minimum film thickness formula can be written as suggested by Dowson and Higginson (1961) and later corrected by Dowson (1968) in the following form:

$$H'_{\min} = 2.65 \cdot G^{0.54} \cdot U^{0.70} \cdot W^{0.13} \tag{3.7}$$

where

$$H'_{\min} = h_{\min} / R' \tag{3.8}$$

$$G = \alpha \cdot E' \tag{3.9}$$

$$U = \frac{\eta_0 \cdot u'}{E' \cdot R'} \tag{3.10}$$

$$W = \frac{w}{E' \cdot R' \cdot l} \tag{3.11}$$

$$u' = 0.5(u_1 + u_2) \tag{3.12}$$

$$\frac{1}{E'} = \frac{1}{2} \cdot \left(\frac{1 - \nu_1^2}{E_1} + \frac{1 - \nu_1^2}{E_2} \right) \tag{3.13}$$

$$\frac{1}{R'} = \frac{1}{R_1} + \frac{1}{R_2} \tag{3.14}$$

and $h_{\min}$ is the minimum film thickness, α is the pressure exponent of viscosity, η_0 is the lubricant viscosity at ambient pressure, w is the total load, l is the axial length of the contact, u_1 and u_2 are the velocities of the surfaces, R_1 and R_2 are the radii of the curvatures of the surfaces, ν_1 and ν_2 are the Poisson's ratios and E_1 and E_2 the Young's moduli of elasticity for the solid materials in contact.

The absolute value of the thickness of the film is not a good measure of the effectiveness of lubrication when very thin films, such as elastohydrodynamic ones, are considered. A better picture is obtained by considering the film thickness in relation to the surface roughness, which is often of the same order of magnitude. This relationship is called the specific film thickness, λ, and it is defined as

$$\lambda = \frac{h_{\min}}{R'_a} \tag{3.15}$$

where

$$R'_a = \sqrt{R_{a1}^2 + R_{a2}^2} \tag{3.16}$$

and R_{a1} and R_{a2} are the centreline average values for the roughness of the two surfaces. Typical values for the lubricant film properties are shown in Table 3.2 and a comparison of lubrication regimes at different contact severities is shown in Fig. 3.20.

Table 3.2. Typical lubricant film properties.

Mechanisms of lubrication	Lubricant film properties		
	Lubricant film thickness	Specific film thickness	Coefficient of friction
	h_{min}	λ	μ
	µm	–	–
Boundary lubrication	0.005–0.1	<1	0.03–0.2
Mixed lubrication	0.01–1	1–4	0.02–0.15
Elastohydrodynamic lubrication	0.01–10	3–10	0.01–0.1
Hydrodynamic lubrication	1–100	10–100	0.001–0.01

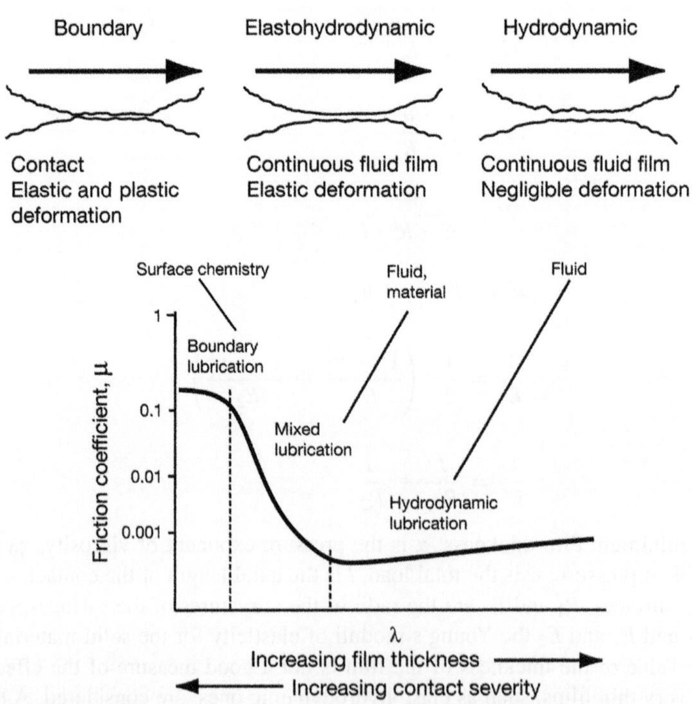

Fig. 3.20. Comparison of contact severities and dominating contact interactions in the different contact regimes (after Hsu and Gates, 2001).

It is a very complex task to explain the frictional behaviour in elastohydrodynamically lubricated contacts because Newtonian fluid analysis is not appropriate for predicting the observed traction behaviour. A more accurate model incorporating the three classical forms of material shear behaviour, Newtonian viscous, Hookean elastic and plastic yield, was developed by Bair and Winer (1979). The three primary physical properties required to use the model are low shear stress viscosity, the limiting elastic shear modulus and the limiting yield shear stress. The last one determines the maximum traction that can be transmitted in an elastohydrodynamic contact.

The concept of micro-elastohydrodynamic lubrication consists of the theory that the approach between two individual asperities will be delayed by a local microscopic elastohydrodynamic action. This action will contribute to the load-carrying capacity and the separation of the surfaces (Cheng, 1978).

Reviews of the state of the art in the field of elastohydrodynamic lubrication have been published by Cheng (1980), Holmberg (1980, 1982), Spikes (1999), Szeri (2001), Larsson and Höglund (2004) and Stachowiak and Batchelor (2005).

3.1.5.3 Boundary lubrication

In boundary lubrication the solid surfaces move so close to each other that there is considerable asperity interaction. A very thin lubricant film adsorbs to the solid surfaces and significantly inhibits asperity welding, as shown in Fig. 3.17. The coefficient of friction is reduced because the lubricant film has lower shear strength than the welded asperity junctions and the wear is reduced because of a smaller number of welded asperity contacts.

Surface interaction phenomena between thin layers of boundary lubricants and the solids dominate the condition of contact. Thus the tribological characteristics are determined primarily by the physical and chemical properties of the lubricant and the solids. The rheological properties of the bulk lubricant, such as viscosity, play little or no part in the friction and wear behaviour. In boundary lubrication the coefficient of friction is typically of the order 0.03 to 0.2 (Godfrey, 1968).

The following fundamental mechanisms of interaction between the lubricant and the solid are involved to produce a protective lubricant film:

- physical adsorption (physisorption),
- chemical adsorption (chemisorption), and
- chemical reaction.

Physical adsorption occurs when the molecules of the lubricant are held to the surface only by bonds such as van der Waal's forces as shown in Fig. 3.21a. No exchange of electrons takes place between the molecules of the adsorbate and the adsorbent. The molecules are weakly bound, the boundary lubricant film has a mono- or multilayer structure and its formation process is reversible. Polar molecules, i.e. molecules that have a changing electrical charge along their length, such as long chain alcohols, aldehydes or acids, adsorb on to the surface with vertical orientation. The films can be of either monomolecular (typically <3 nm) or polymolecular thickness. Many molecules pack in as closely as possible and strengthen the film with lateral cohesive forces. The film is temperature sensitive and provides lubrication only at low loads and low sliding velocities.

Chemical adsorption occurs when the lubricant molecules are held to the surface by chemical bonds as shown in Fig. 3.21b. Electron exchange between the molecules of the adsorbate and the adsorbent takes place. The film might be only a monolayer thick, and the formation is characterized by irreversibility and strong bonding energies. Lubricant films formed by polar molecules of, e.g., fatty acids, soaps and esters lubricate effectively up to their melting point. At temperatures typically in the range of about 120 to 180°C these lubricants fail as a result of reorientation, softening and melting. Films formed by chemical adsorption provide good lubrication at moderate loads, temperatures and sliding velocities.

Chemical reactions between the lubricant molecules and the solid surface occur when there is an exchange of valence electrons and a new chemical compound is formed as shown in Fig. 3.21c. The lubricating films are unlimited in thickness and characterized by high activation, high bonding energies and irreversibility. The chemically reactive boundary lubricants usually contain sulphur, chlorine or phosphorus atoms in the molecule. They may additionally even contain oxygen, lead, zinc, boron, selenium, etc. (Fein, 1970). The films behave more like solids than viscous liquids but their shear strength is considerably lower than that of the substrates (Cavdar and Ludema, 1991). The films

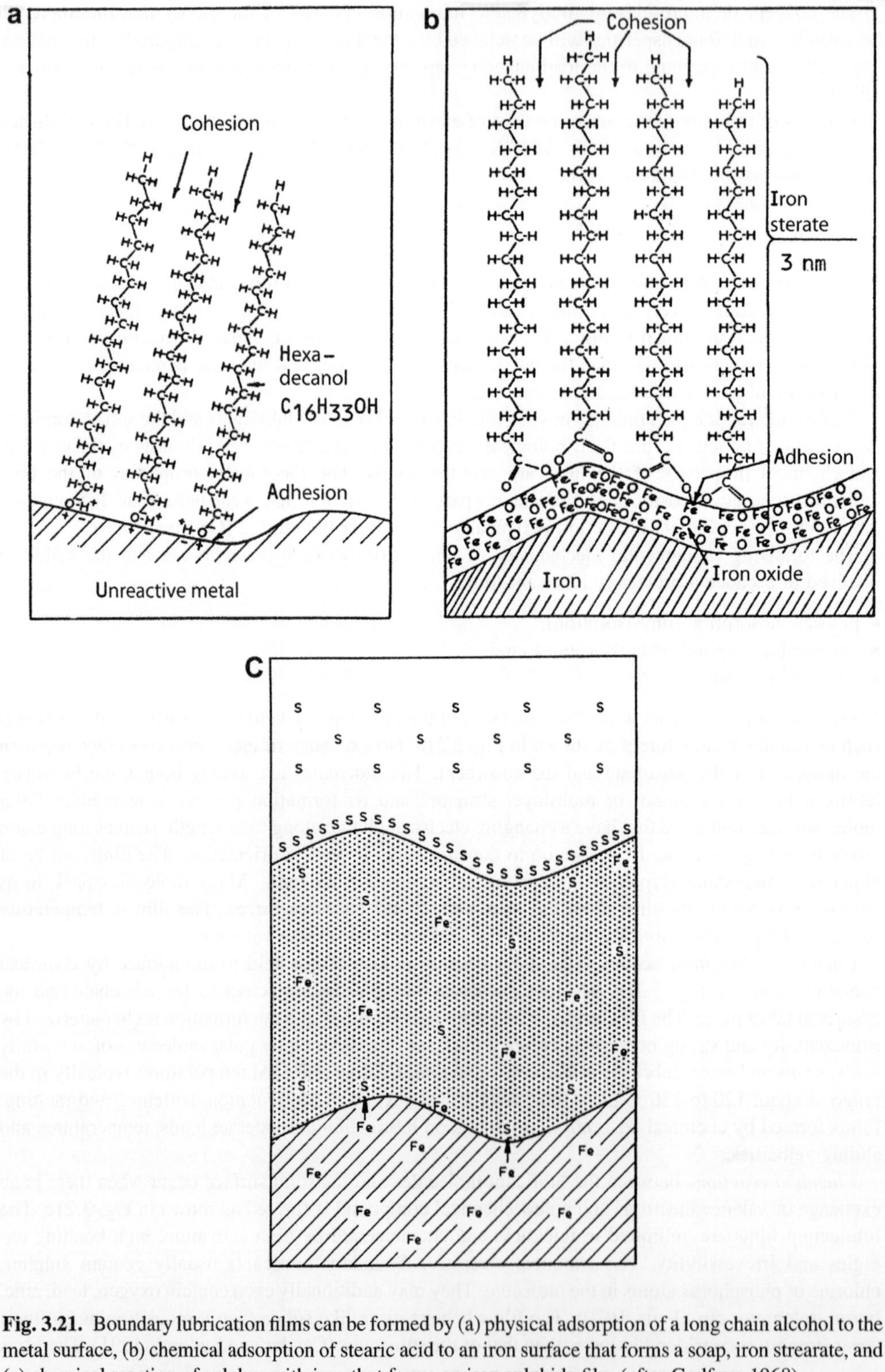

Fig. 3.21. Boundary lubrication films can be formed by (a) physical adsorption of a long chain alcohol to the metal surface, (b) chemical adsorption of stearic acid to an iron surface that forms a soap, iron strearate, and (c) chemical reaction of sulphur with iron that forms an iron sulphide film (after Godfrey, 1968).

are more stable than physically and chemically adsorbed films. The chemical reaction and the film formation is strongly activated at temperatures above 150 to 180°C. The maximum temperature limit for the reactions varies in the range 300 to 800°C. The lubricant films formed by chemical reaction provide good lubrication for high loads, high temperatures and high sliding speeds but are limited to reactive materials. These conditions are often referred to as extreme pressure (EP) conditions (Bowden and Tabor, 1950; Fein, 1970; Godfrey, 1968; Cameron, 1980; Bhushan, 1999b; Kajdas, 2001; Gellman and Spenser, 2002, 2005; Hsu and Gates, 2001, 2005; Hsu, 2004a, b; Vizintin, 2004; Neville and Morina, 2005; Furlong *et al.*, 2007).

Lubricants are often classified based on the behaviour of their additives: non-reactive, low friction (lubricity), anti-wear and extreme pressure. Non-reactive lubricants are mineral oils without additives and esters. Low-friction or high-lubricity lubricants are defined as lubricants that have the ability to reduce friction below that of the base oil. The additives are adsorbed on, or they react with, the metal surface or its oxide layer and form monolayers with low shear strength. Anti-wear additives react with the surface and form a relatively thick surface film that is not easily removed by load or shear. They form organic, metallo-organic or metal salt films on the surface. Zinc dialkyldithiophosphate (ZDDP) and tricresyl phosphate (TCP) are the most widely used anti-wear additives. Extreme pressure additives are chemically corrosive additives that have strong affinity to the surface. At high loads they form thick films of high melting point metal salts on the surface which prevent metal to metal contact. By frictional heating they may be removed but new additive molecules rapidly react and build new protective films. Sulphurized olephines, ZDDP and TCP are common EP additives (Bhushan, 1999b; Hsu and Gates, 2001; Gellman and Spenser, 2005; Morina and Neville, 2007).

The tribological mechanisms of some lubricant additives, such as ZDDP, molybdenum dialkyldhitiocarbamate (MoDTC) and polymers, have recently been the subject of detailed studies. ZDDP forms pad-like structures on the rubbed surfaces. The film formation takes place during rubbing even at room temperature. The morphology of the reaction film formed depends strongly on the nature of sliding. Primary and secondary forms of the additive form films of different morphology and properties. The ZDDP film formation has been found to be proportional to the product of sliding distance and rubbing time (Fujita and Spikes, 2004; Topolovec and Spikes, 2005). As the severity of the loading increases, the resistive forces within the film also increase as the sheared region penetrates down to the deeper layers of the film (Bec *et al.*, 1999). Canter (2005) showed that in rubbing the ZDDP film the zinc changes coordination at a pressure of 17 GPa to form a hexacoordinate, highly crosslinked structure resulting finally in a fully crosslinked network. The films are formed due to an increase in pressure.

The MoDTC friction modifier additive forms tiny domains of low friction on the high spots of rubbed surfaces. Raman analysis indicates that these domains represent crystallites of MoS_2. Functionalized viscosity modifier polymer additives can form viscous boundary films on rubbed surfaces which produce much lower friction than corresponding non-functionalized polymers. It has been shown that dispersant olefin co-polymer films adsorb to surfaces to form viscous films and that these result in reduced friction in thin film conditions (Topolovec and Spikes, 2005).

In boundary lubrication the solid surface should have a high surface energy and high chemical reactivity, so that there will be a strong tendency for additive molecules to adsorb on the surface and react with it. Thus, metals tend to be the easiest surfaces to be lubricated. The solid surface should also have high wetting property so that the liquid lubricant wets the surface easily (Bhushan, 1999b). A surface coating may radically change the surface properties and also the boundary lubrication mechanisms (Kalin and Vizintin, 2006a; Barros Bouchet *et al.*, 2005; Kano *et al.*, 2005; Barros Bouchet and Martin, 2008). These aspects are discussed in detail in sections 3.4.7 and 6.4.8.

In practice boundary lubrication is a complex phenomenon where the above-mentioned mechanisms of surface interaction are mixed with effects of metallurgy, surface roughness, corrosion, catalysis, temperature, pressure and time of reaction. Other environmental components such as oxygen, water and competing surfactants affect the film formation. The load-carrying capacity and the breakthrough conditions for a boundary lubricated coated surface were analysed by Komvopoulos (1991a) and he showed that the elastic or plastic deformation of the surface depends on coating thickness, surface roughness, microstructure of the coating, normal and friction forces and the mechanical properties of the coating and the substrate.

The work of Akhmatov (1966) offers an early insight into the molecular physics involved in boundary lubrication. More recently, Hsu (2004a, b) has reviewed the molecular basis for boundary lubrication. Israelachvili (1992) showed on a molecular level that the static and dynamic properties of ordinary liquids, when confined within a very thin film between two surfaces, can be very different from those of the bulk liquid. Films thinner than 10 molecular diameters can become very much more viscous. Films below 4 molecular diameters in thickness can undergo a phase transition into a liquid-crystalline or solid-like phase. Such surface films can support a limited load and shear stress. These modified properties depend on the geometry of the liquid molecules, and how these molecules can fit in between the two atomically corrugated surfaces.

Different molecular configurations in shearing of boundary molecular thin films are possible, as shown by Fig. 3.22 (Israelachvili, 1992). With no external load, the two surfaces are at rest and stuck to each other through the film (Fig. 3.22a). When a progressively increasing lateral shear stress is applied (Fig. 3.22b) there is a small increase in the lateral displacement and thickness but only by a small fraction of the lattice spacing or molecular dimension. The film retains its solid-like 'frozen' state and all the strains are elastic and reversible and the surfaces are still stuck to each other. When the applied shear stress has reached a certain critical value the film suddenly melts (known as shear melting) and the two surfaces begin to slip rapidly over each other (Figs 3.22c, c′ and c″). Depending on the system, a number of different molecular

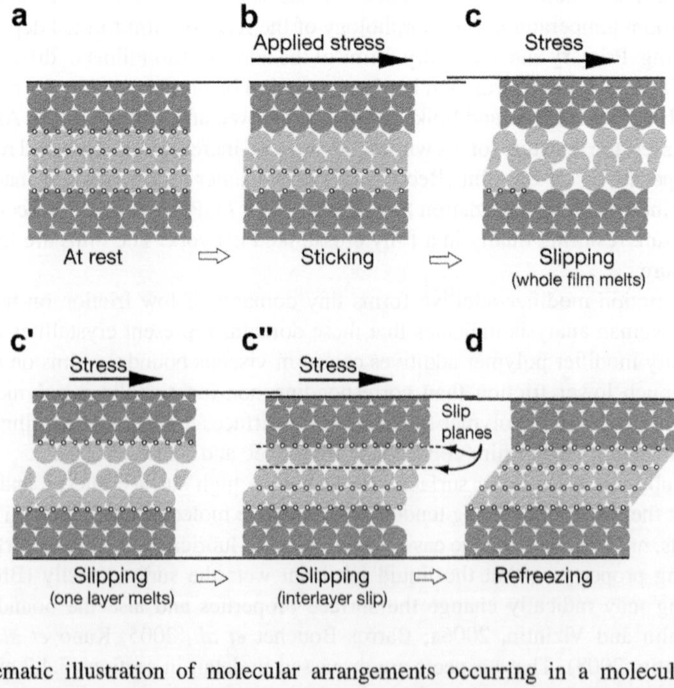

Fig. 3.22. Schematic illustration of molecular arrangements occurring in a molecularly thin film of spherical or simple chain molecules between two solid surfaces in shear. (a) The surfaces are at rest, (b) a small shear stress is applied but the surfaces are still stuck to each other, (c, c′ and c″) slipping occurs with three different molecular configurations within the film and (d) refreezing, with the surfaces again stuck to each other (after Israelachvili, 1992).

configurations within the film are possible during slipping, shown as (c) total disorder, (c′) partial disorder and (c″) full order. The slip-plane may occur either within the film or at the film–surface interface. The configuration of branched-chained molecules is much less ordered and remains amorphous during sliding, leading to smoother sliding with reduced friction and little or no stick-slip. The slipping ends once the applied shear falls below a critical value, when the film freezes and the surfaces become stuck once again (Fig. 3.22d). If the slip mechanism is slow the surfaces will continue to slide smoothly and there will be no stick-slip. However, unless the liquid molecules are highly entangled or irregular in shape, there will always be a single stick-slip spike on starting.

The traditional concept of boundary lubrication where adhesion plays a dominating role was criticized by Komvopoulos *et al.* (1986b). They claim that adhesion has only a minor effect and that ploughing due to wear particles entrapped at the interface is the predominant factor in controlling friction and wear. For evidence they refer to wear grooves formed in surfaces in pin-on-disk experiments with aluminium, copper and chromium and a relatively inert additive-free mineral oil. It should, however, be noted that their measured coefficients of friction of 0.15 to 0.2 represent very high values for boundary lubrication and that the lubricant and material choice is not good for the formation of stronger boundary lubrication surface films. This indicates that only a poor boundary lubrication film was formed in their experiments. In those conditions, very probably the wear debris ploughing mechanism is dominating, but there are strong indications that this is not the situation in sliding conditions when a good boundary lubrication film has been formed resulting in low friction and wear.

The nanomechanical properties of an anti-wear tribofilm produced from phosphorus-containing additives on boundary-lubricated steel surfaces was measured by Komvopoulos *et al.* (2004). They used a Berkovich diamond tip with 100 nm curvature diameter and normal load between 100 and 600 μN to perform indentations on and off a wear track. They found that the phosphate tribofilm exhibited an average hardness of 6.0 GPa and reduced elastic modulus of 122.7 GPa compared to 12.5 GPa and 217.6 GPa of the steel substrate, respectively. In nanoscratch test friction measurements with a 20 μm diamond tip the coefficient of friction of the phosphate tribofilm was 0.23–0.29 while it was 0.16–0.17 for the steel surface. They attribute the higher friction obtained for the phosphate tribofilm to the contaminant effects of the relatively high surface roughness of the wear track and the greater plastic flow of the tribofilm compared to the steel substrate. The tribofilm surface contained a large number of ploughing grooves produced by the sliding process, whereas the steel surface only had a small number of surface ridges from polishing.

To support the concept of adhesion dominating the tribology in boundary lubrication, Landheer and de Gee (1991) correctly point out that a boundary lubricant film only a few molecules thick, i.e. in the nanometre range, can effectively reduce friction even from 1.0 to 0.1, without affecting asperity interaction and the ploughing effect. So the friction-reducing effect must be due to a reduction in the interfacial shear strength. The change in shear strength will naturally also affect how the ploughing phenomenon develops, but that is a secondary effect. However, it is obvious that especially in more severe boundary lubricated conditions with higher friction and wear the effect of wear debris ploughing should be taken into account to a larger extent than is generally done. A number of different equations representing special cases of boundary lubrication have been published. Even though there are no serious inconsistencies in the models, each of them is so limited in its coverage that considerable gaps exist.

It can be deduced from the above that the term ‘boundary lubrication’ is not always well defined. In the literature it is often ill understood and overused. However, the recent more systematic studies on a nano- and microlevel are a good starting point for a more holistic understanding of the tribological mechanisms involved in the complex contacts of boundary lubrication. The complexity of boundary lubrication is illustrated by the seven characteristic features that Ludema (2005) suggests for more detailed investigations:

1. Boundary lubrication films are present as patches or islands, ‘washed’ to the front slopes of asperity clusters. They are neither continuous nor stable.

2. There is a gradation of composition and also of properties through the thickness of the film.
3. The composition and properties vary with applied load, time of operation, stationary time and changes of atmosphere.
4. New surfaces, though unprotected by a boundary film, survive for a time though they are incapable of carrying an overload. After some time of sliding the boundary film forms, which prepares the surface for overload, and even then some time at overload is required to cause failure.
5. Surface material makes a difference: boundary lubricated grey cast iron will carry twice the load before scuffing failure than will tempered martensite of the same hardness.
6. The shape of sliding contact regions and pathway of sliding are important, probably relating to the availability of protecting chemical elements or compounds.
7. The usual failure criteria for lubricated sliding are not applicable, such as the lambda ratio, the plasticity indices or the maximum surface temperature limit criterion.

3.1.5.4 Mixed lubrication

Mixed lubrication is often referred to as a separate lubrication regime. In this regime the contact behaviour is governed by a mixture of elastohydrodynamic (or hydrodynamic) lubrication and boundary lubrication. The surfaces are largely separated by a thin lubricant film but asperity contact still takes place. The total applied load is thought to be partly carried by asperity contacts and partly by hydrodynamic action.

Mixed lubrication has been defined for conditions where the specific film thickness, λ, is in the range of about 1 to 4. In a comprehensive study of mixed lubrication phenomena Tallian (1972) found that boundary lubrication effects dominate the load-carrying mechanism when $\lambda < 1.5$, while elastohydrodynamic effects dominate when $\lambda > 1.5$. According to this theoretical treatment the entire load is carried by asperity interactions when $\lambda < 0.4$, which was considered a minimum value for the specific film thickness. More recent experience indicates that the elastohydrodynamic lubrication mechanism may play an active role and separate the moving surfaces in some cases even at values as low as 0.1 and below. Theoretical approaches to the contact behaviour in mixed lubrication have been presented by Tallian *et al.* (1966), Tallian (1972), Christensen (1971, 1972), Johnson *et al.* (1972), Patir and Cheng (1978), Andersson *et al.* (2004) and Geike and Popov (2008).

Our knowledge of mixed lubrication is still poor and the theories suggested need experimental confirmation. This will not be an easy task for the future. When concentrated contacts are considered, some authors prefer to use the term partial elastohydrodynamic lubrication instead of mixed lubrication.

3.1.5.5 Lubrication with solid lubricants

In solid lubrication the lubricating material is not a liquid but a solid material, often in powder or flake form. The solid material acts in the same way as a liquid lubricant. It inhibits direct contact between the two sliding surfaces and allows the shear to take place within the solid lubricant. For that reason materials with a lamellar structure are especially suitable as solid lubricants, such as molybdenum disulphide, graphite, hexagonal boron nitride and boric acid. Other materials that are used as solid lubricants are polytetrafluoroethylene (PTFE), polyimide, soft metals, some oxides and rare-earth fluorides, diamond, diamond-like amorphous carbon and fullerenes. The coefficient of friction in sliding contacts with solid lubricants is typically in the range of $\mu = 0.05$–0.25 (Savan *et al.*, 2000; Lansdown, 1999; Erdemir, 2001a; Stachowiak and Batchelor, 2005).

Solid lubricants are often used in very severe conditions where liquid lubrication has limitations. Such conditions are vacuum, very high and very low temperatures, extremely high contact pressure, radiation, etc. However, the limitations of solid lubrication are typically low capability to conduct heat transfer from the contact, limited wear life because the material replenishment is more difficult compared to liquid lubrication, unstable frictional behaviour and chemical changes which affect their tribological behaviour.

Solid lubricants may also be dispersed in water, oil or grease resulting in a combined effect of solid and liquid lubrication. There has been a growing interest in the use of fullerenes as solid additives in liquid lubrication (Rapoport *et al.*, 2005). The mechanisms related to solid lubrication by different coating materials such as graphite, polytetrafluoroethylene, molybdenum disulphide and diamond-like carbon surface films are discussed in Chapter 4.

3.2 Surface Stresses and Response to Loading

3.2.1 Response of materials to loading

In tribological contacts there are at least two surfaces in relative motion interacting with each other. The phenomena that take place in the contact are considerably influenced by the force pressing the two surfaces together. The normal force and the shape of the surfaces determine the contact pressure that produces a load on both surfaces. The frictional force that resists motion is an additional tangential load on the area of contact.

A distinction between conforming and non-conforming contacts is made based on the shapes of the surfaces. A contact is said to be conforming if the surfaces fit close together without appreciable deformation as in Fig. 3.23a. Non-conforming contacts have dissimilar surface profiles, as illustrated in Fig. 3.23b. When non-conforming contacts are brought together they will first touch at a point or along a line. According to this they are called point contacts or line contacts. For conforming contacts the contact area is large and the contact pressure is low while for non-conforming contacts the contact area is small and the contact pressure high.

The contact pressure will create stresses and deformation in the material close to the contact. We know that different materials respond differently when loaded. Based on their behaviour we distinguish between elastic, plastic and viscous materials. Materials which combine these characteristics are common, for example steel exhibits elastoplastic properties and polymers are generally viscoelastic.

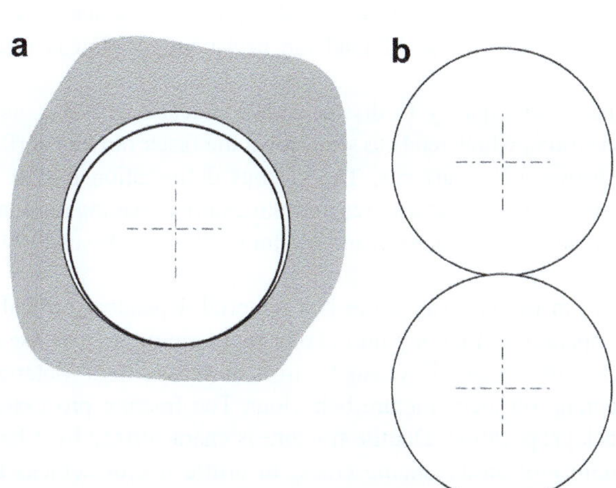

Fig. 3.23. (a) The conforming contact between a shaft and bushing and (b) the non-conforming contact between two cylinders.

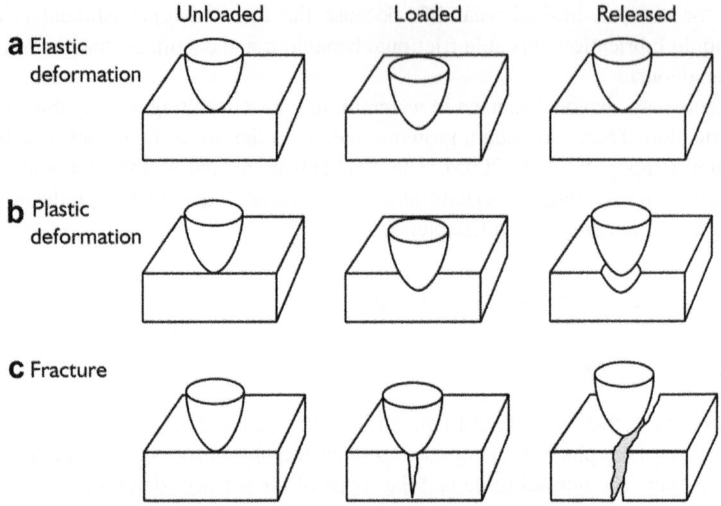

Fig. 3.24. Three characteristic ways for a material to respond to external load: (a) elastic deformation, (b) plastic deformation and (c) material fracture.

Elastic materials are characterized by a reversible deformation when loaded. They first deform but when the load is removed they nearly instantaneously return to their original shape, as shown in Fig. 3.24a. In deformation the atoms are not permanently displaced. The work done by external load is stored in the body as distortion of the interatomic bonds. The relation between stress and strain below the yield point is essentially linear.

Plastically deforming materials represent a more complex stress–strain relationship and the deformation persists even after removal of the load. When the load-induced stresses increase above a yield point the material can no longer accommodate the high stresses and it retains time-independent deformation and will no longer return to its original shape, Fig. 3.24b. The bonds between initially neighbouring atoms are broken and a new set of bonds is established which are as stable as the original ones. On release of load the new configuration is maintained and the deformation is permanent.

Viscous materials have the capacity to dissipate mechanical energy but not store it. In viscous deformation the viscous force, which tends to slow down the faster moving surface, is proportional to the velocity gradient between the surfaces. The viscous deformation is time dependent. Detailed treatments of the different forms of material response to external loading are found in textbooks such as Timoshenko and Goodier (1970), Johnson and Mellor (1983), Jacobsson (1991), Ashby (1999) and Bhushan (1999b).

When the stress level in the material exceeds a material-dependent critical value fracture takes place. Fracture is the separation of a body into two or more pieces in response to an imposed stress, Fig. 3.24c. The fracture can be caused by tensile stress or shear stress. Materials are categorized as *ductile* or *brittle* depending on their fracture behaviour. The fracture processes involves first crack initiation and then crack propagation. Ductile fracture is characterized by a large amount of plastic deformation in the vicinity of an advancing crack. In brittle fracture cracks are spread extremely rapidly, with very little accompanying plastic deformation, typically less than 5%, and result in a sudden separation of a stressed body into two or more parts.

Propagation of the crack to a larger area of fracture results in material liberation and the creation of wear particles. The process from loading to material removal is shown schematically in Fig. 3.25.

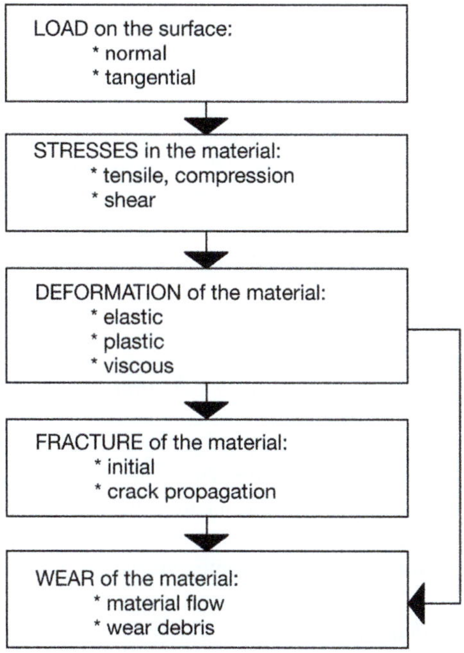

Fig. 3.25. Schematic diagram of the process from loading of the surface to wear.

A more detailed treatment of the mechanisms of fracture and crack propagation near the surface is found in Buckley (1981), Suh (1986), Ko *et al.* (1992), Anderson (1995) and Shukla (2005).

3.2.2 Material parameters E, σ_y, H, G *and* K_c

The tribological properties of surfaces are related to their response to surface loading and their deformation behaviour. The most important material parameters controlling this are the elastic modulus describing elastic deformation; the material strength, often given as hardness, describing plastic deformation; and the toughness and fracture toughness describing fracture behaviour (Pharr, 1998; Holmberg, 2000).

Elastic modulus or *Young's modulus E* (GPa) is defined as the slope of the linear–elastic part of the stress–strain curve. A low E value corresponds to very elastic materials such as rubber while a high value is related to non-elastic materials such as ceramics.

Strength (MPa) of a solid is more complex. For metals the strength is often defined by the yield strength, σ_y, or as the 0.2% offset yield strength, that is, the stress required to produce 0.2% of strain when the stress is removed. Sometimes the *ultimate tensile strength* (UTS) is defined, but since this occurs in a heavily plastically deformed state it is not as common to utilize UTS as a design parameter unless a high safety factor is applied. As the load is increased beyond the elastic limit, the strain increases at a greater rate.

Hardness H (MPa) is a crude measure of material strength. The hardness is defined by the indenter normal force divided by the projected area of the indent as a diamond or hardened steel or cermet tip is pressed into the surface of a material. It is related to the yield strength by $H \approx 3\,\sigma_y$. Hardness is often measured in other units, the most common of which is Vickers hardness H_v (kg/mm^2). It is related to H with the units used above as $H \approx 10\,H_v$. However, increasingly GPa is being used as a universal hardness unit – sometimes inappropriately, since a hardness value is test method dependent.

Toughness G (kJ/m^2) is the ability of a material to absorb energy during deformation without fracture. The ability to absorb energy elastically, without yielding, is generally termed resilience. *Fracture toughness* K_c (MPa · m$^{1/2}$) is the ability of a material to resist the growth of a pre-existing crack. According to this, toughness contains both the energy required to create the crack and to enable the crack to propagate until fracture. Fracture toughness, on the other hand, contains only the energy required to facilitate the crack propagation to fracture. The crack propagates when the *stress intensity factor* (*K* or SIF) is greater than the *fracture toughness* of the material, that is $K_I > K_{Ic}$ (Ashby, 1999). The subscript '*Ic*' denotes to mode *I* crack opening, see Fig. 3.29.

Detailed definitions and descriptions of material parameters are given in books such as those by Ashby (1999) and Bhushan (1999b).

3.2.3 Analytical solutions of contact stresses and deformations at surfaces of solid materials

It is a very complex task to calculate mathematically the stress field and the deformations in a typical tribological contact. However, useful information about the stresses to which the materials are exposed can be achieved even when the mathematical calculations are based on certain simplifications. We will start by looking at a simple case before going on to more complex situations. First, we make the assumption that the contacting surfaces are homogeneous and isotropic solids.

3.2.3.1 Line load on semi-infinite solid

For simplicity we consider the situation when a line load w' per unit length is distributed along the *y* axis on a semi-infinite solid as shown in Fig. 3.26. The stress function defining this is the Boussinesq function (Johnson, 1985) which in polar coordinates is

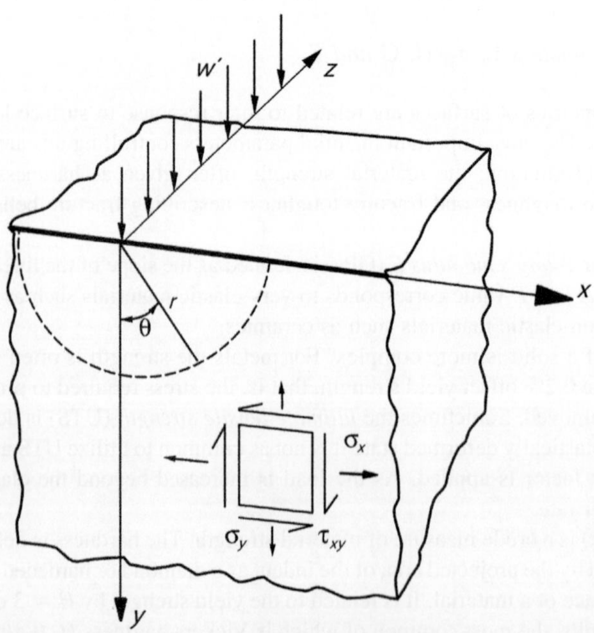

Fig. 3.26. Stress coordinates in the material when a line load is acting on a semi-infinite body.

$$\gamma(r,\Theta) = \frac{w'}{\pi} \cdot r \cdot \Theta \cdot \sin\Theta \tag{3.17}$$

where r is the radius and Θ is the angle according to Fig. 3.26. By changing the radial stress distribution to rectangular coordinates the following stress components are obtained

$$\sigma_x = \sigma_r \cdot \sin^2\Theta = \frac{2w'}{\pi} \cdot \frac{x^2 \cdot y}{(x^2 + y^2)^2} \tag{3.18}$$

$$\sigma_y = \sigma_r \cdot \cos^2\Theta = \frac{2w'}{\pi} \cdot \frac{y^3}{(x^2 + y^2)^2} \tag{3.19}$$

$$\tau_{xy} = \sigma_r \cdot \sin\theta \cdot \cos\theta = \frac{2w'}{\pi} \cdot \frac{x \cdot y^2}{(x^2 + y^2)^2} \tag{3.20}$$

where σ_x, σ_y and τ_{xy} are the stress components according to the coordinates in the figure.

The displacements u' in the x-direction and v' in the y-direction can now be calculated by using the definition of strain and Hooke's law (Johnson, 1985). In the case of plane strain

$$\varepsilon_x = \frac{du'}{dx} = \frac{1}{E}[\sigma_x - \nu(\sigma_y + \sigma_z)] \tag{3.21}$$

$$\varepsilon_y = \frac{dv'}{dx} = \frac{1}{E}[\sigma_y - \nu(\sigma_z + \sigma_x)] \tag{3.22}$$

$$\varepsilon_{xy} = \frac{\partial u'}{\partial y} + \frac{\partial v'}{\partial x} = \frac{2(1 + \sigma)}{E} \cdot \tau_{xy} \tag{3.23}$$

where ε_x, ε_y and ε_{xy} are strain components in the solid, ν is Poisson's ratio and E is Young's modulus of elasticity. From this the displacement can be calculated by substituting the expressions for the stress components and integrating.

3.2.3.2 Hertzian contact

Contacts between two non-conforming bodies with a circular shape are of particular interest in engineering because they represent the contact condition we find in common machine components like rolling bearings, gears, cam-and-tappets and wheel-on-rail contacts. These problems were first solved theoretically by Hertz (1882) for elastic contacts and are generally referred to as Hertzian contacts. Hertz assumed that the surfaces are continuous, smooth and non-conforming; the strains are small; each solid can be considered as an elastic half-space in the proximity of the contact region; and the surfaces are frictionless.

Consider two identical cylinders, with the radii $R_1 = R_2 = R$ and the same elastic modulus $E_1 = E_2 = E$, which are pressed together by a normal load w' per unit axial length as shown in Fig. 3.27. According to the Hertz theory this will result in a flat rectangular contact zone with a width of $2b$. Since the normal deformation at the centre of the contact zone is greater than at the extremities, the pressure distribution, $p(x)$, will be (Halling, 1975)

$$p(x) = \frac{2 \cdot w'}{\pi \cdot b}\sqrt{1 - \frac{x^2}{b^2}} \tag{3.24}$$

The solution for the half contact length, b, is

$$b^2 = \frac{4\,w\,R\,(1 - \nu^2)}{\pi \cdot E} \tag{3.25}$$

The solution defined by equation (3.24) gives reasonably good results for cylinders with different elastic moduli and different geometries, if the angle subtended by the contact width at the centre of the cylinder is less than 30° and by using

$$b^2 = \frac{4 \cdot w \cdot R'}{\pi \cdot E'} \tag{3.26}$$

which can be derived from the Hertzian equations.

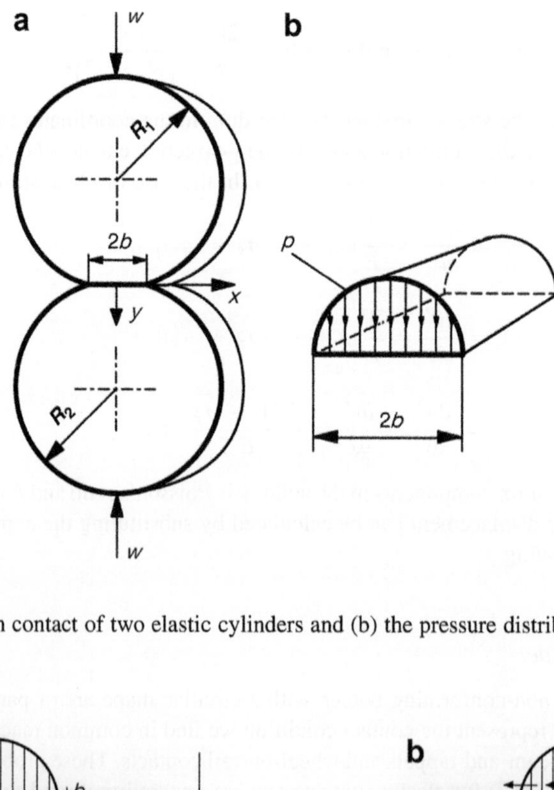

Fig. 3.27. (a) Hertzian contact of two elastic cylinders and (b) the pressure distribution in the flat contact area.

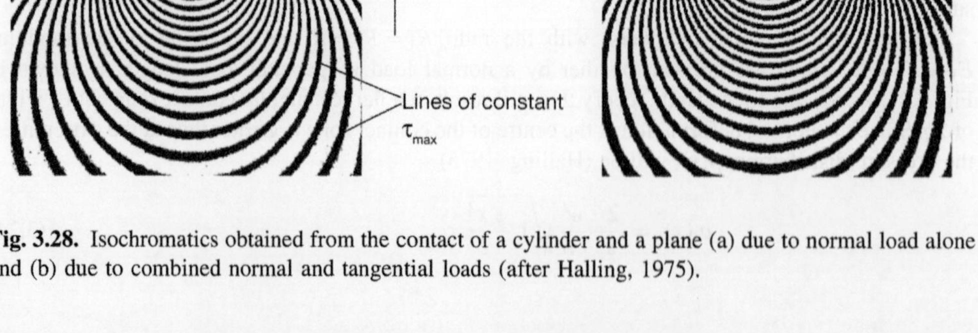

Fig. 3.28. Isochromatics obtained from the contact of a cylinder and a plane (a) due to normal load alone and (b) due to combined normal and tangential loads (after Halling, 1975).

The maximum contact pressure can be derived to be (Bhushan, 1999b)

$$p_0 = \frac{3 \cdot w}{2 \cdot \pi \cdot b^2} = \sqrt[3]{\frac{6 \cdot w \cdot E'^2}{\pi^3 \cdot R'^2}} \tag{3.27}$$

The stress field created by a spherical sliding contact on a flat surface was derived by Hamilton and Goodman (1966) and Hamilton (1983). Complete mathematical analysis of the stress distribution in a Hertzian contact is found in Johnson (1985) and Suh (1986).

The patterns of lines of constant maximum shear stress (isochromatics) in a Hertzian contact of a cylinder and a plane due to normal load alone are shown in Fig. 3.28a. The greatest value of maximum shear stress occurs below the surface at a distance of $0.67b$ which corresponds to about one-third of the total contact length.

When the contact is due to combined normal and tangential loads, as shown in Fig. 3.28b, the position of maximum shear stress moves towards the surface with increasing tangential load. It can be shown (Schouten, 1973a, b) that for coefficients of friction exceeding 0.32 the maximum shear stress will be found on the surface as shown in Fig. 3.13. This is extremely important when stresses in a coated surface are considered because it determines whether the maximum shear stress will occur in the coating, in the substrate or at the interface.

3.2.4 Criteria for plastic yield

The limit for the elastic deformation can be described as a criterion for plastic yield. The conditions at which the stresses in the material are higher than that which the material can take elastically define the beginning of plastic flow or plastic yield. This is related to the yield point of the material in simple tension or pure shear through an appropriate yield criterion. Two commonly used criteria are Tresca's maximum shear stress criterion and von Mises shear strain energy criterion.

In *Tresca's maximum shear stress criterion* the yielding will occur when the maximum shear stress, which is half of the difference between the maximum and minimum principal stress, reaches the yield stress in pure shear or half of the yield stress in simple tension. This can be expressed as

$$\sigma_{max} - \sigma_{min} = 2\sigma_y \tag{3.28}$$

where σ_{max} and σ_{min} are the maximum and minimum principal stresses, and σ_y is the yield strength of the material in shear.

In the *von Mises shear strain energy criterion*, yielding will occur when the distortion energy equals the distortion energy at yield in simple tension or pure shear. Shear stresses in the material result in dislocation movements and deformations. The deformations are, however, not influenced by the hydrostatic stress which only results in changes of volume. The yield criteria can therefore be expressed in terms of deviator stresses, Γ_i, instead of actual principal stresses, σ_i, as

$$\Gamma_i = \sigma_i - \sigma_m \tag{3.29}$$

where σ_m is the hydrostatic stress.

The von Mises maximum shear strain energy relation is a deviatoric stress relation for yield that is in wide use and has been found to be in good agreement with experiments (Tangena, 1987; Sainsot *et al.*, 1990). The criterion states that yield is dependent on all three values of deviatoric stresses and that deformation occurs when the distortion energy reaches a critical value. This occurs at a yield stress, σ', when

$$\sigma' = \frac{1}{\sqrt{2}}\sqrt{(\Gamma_1 - \Gamma_2)^2 + (\Gamma_2 - \Gamma_3)^2 + (\Gamma_1 - \Gamma_3)^2} \tag{3.30}$$

The von Mises criterion typically predicts a pure shear yield stress which is about 15% higher than predicted by the Tresca criterion. It is generally considered that von Mises criterion usually fits experimental data for metallic specimens better than other theories. The difference is not large but the von Mises criterion tends to be more commonly used due to its more convenient continuous mathematical formulation (Diao and Kato, 1994; Lee *et al.*, 2001; Yang and Komvopoulos, 2004). Tangena (1987) found a remarkably good correlation between the wear behaviour of coated surfaces and the von Mises stress averaged over the contact area. Different criteria for material yield were developed by Timoshenko (1941) and are reviewed in Anderson (1995) and Bhushan (1999b).

3.2.5 Criteria for material fracture

Under excessive loading, ductile and brittle materials fail ultimately by fracture. The fracture process starts with crack initiation and is followed by crack propagation and finally separation of material into two or more parts. This process is normally classified as being either ductile or brittle. Analytical and numerical techniques can be used to determine the elastic energy needed to propagate a crack. This quantity, the strain–energy release rate, **G**, is defined as:

$$\mathbf{G} = -\frac{\delta(U - W)}{\delta A} \tag{3.31}$$

where U is the elastic energy stored in the system, W is the work done by external loads and A is the crack area. A crack can only advance if the total energy available from external loads relieved by stored strain energy exceeds that required for crack propagation. Thus **G** must exceed a critical value $\mathbf{G_c}$ for the crack to propagate. $\mathbf{G_c}$ includes the term involved in creating new crack surfaces and also the work absorbed by any non-linear deformation of the crack tip. $\mathbf{G_c}$ is a measure for the toughness of the material.

Fracture toughness, K, is the expression for the ability of a material to resist the growth of an existing crack. Tensile loading or normal separation is called mode I and for this K_I is given as:

$$K_I = Y \cdot \sigma \cdot \sqrt{\pi \cdot c} \tag{3.32}$$

where the term Y is a geometric correction factor, σ is the nominal tensile stress and c is the crack length. Mode II is the sliding mode in which the shear forces are parallel to the crack surface and perpendicular to the crack front, while mode III is the tearing mode in which the tearing forces are parallel to the crack surface as well as to the crack front, as shown in Fig. 3.29.

The former parameter, **G**, quantifies the net change in potential energy that accompanies an increment of crack extension; the latter parameter, K, characterizes the stresses, strains and displacements near the crack tip. For linear–elastic materials the two parameters can be related for plane stress as

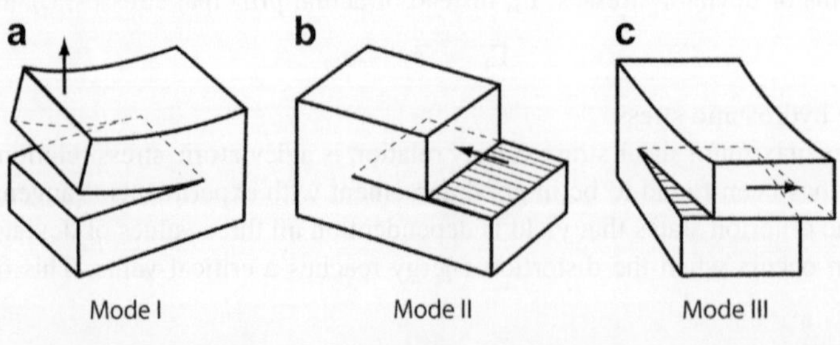

Fig. 3.29. The three principal modes of crack opening: (a) mode I: opening mode, (b) mode II: sliding mode and (c) mode III: tearing mode.

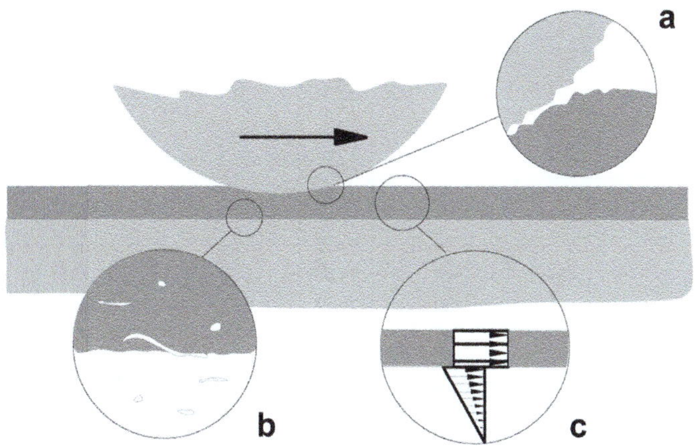

Fig. 3.30. The stress field and deformations in the ideal case of a slider moving tangentially on a coated surface is disturbed in practical situations by (a) surface roughness, (b) cracks and flaws, and (c) residual internal stresses such as thermal and intrinsic stresses.

$$G = K_I^2/E \tag{3.33}$$

For plane strain conditions E must be replaced by $E/(1 - \nu^2)$ where E is Young's modulus and ν is Poisson's ratio (Thouless, 1988; Anderson, 1995; Bhushan, 1999b).

3.2.6 Analytical solutions of stresses and deformations in normally loaded coated surfaces

The correct determination of stresses and deformations in surfaces in tribological contacts is important as it forms the basis for the surface mechanisms resulting in friction and wear. In the following paragraphs we will discuss early analytical and numerical solutions for coated surfaces in loaded contacts. Today, a more powerful tool for solving stresses and deformations in solids is the finite element method (FEM) for modelling and simulation, which we will discuss in detail in section 3.2.8.

We will first consider the stress field and deformations generated in a layered surface when it is loaded normally by an indenter. We assume that the material is homogeneous, the surfaces are smooth and no residual stresses are present. This will be followed by separate considerations of how the stress field and the deformations are influenced by tangential movement of the indenter, surface roughness, cracks and flaws near the surface and residual stresses such as thermal and intrinsic stresses (Fig. 3.30).

A number of analytical solutions have been developed for the contact situation when one smooth non-conformal surface is pressed against a layered counterface without sliding. The layers are considered elastic and the substrates either elastic or rigid. In some of the solutions the friction between the contacting surfaces is taken into account. Most solutions are for the contact of a ball or a cylinder against a coated plane substrate. The analytical solutions give the possibility to study effects of parameters such as coating thickness, geometry, load, shear strength, Young's modulus and Poisson's ratio on the contact stresses, contact area, deformations, subsurface stresses and subsurface strains. A summary of published analytical treatments is presented in Table 3.3.

Most of the analytical solutions are based on the classical elasticity theory starting from the Boussinesq stress function. This makes the mathematical treatment fairly complex. Because the

Table 3.3. Analytical solutions for the contact problem with layered surfaces.

Reference	Elastic coating	Elastic substrate	Rigid substrate	Elastic counterface	Rigid counterface	Ball on plane	Cylinder on plane	Cone on plane	Cylinder on cylinder	Multilayers
Indentation										
Meijers (1968)	X		X		X		X			
Chen and Engel (1972)	X	X		X	X	X				X
Gupta *et al.* (1973)	X	X			X		X			
Gupta and Walowit (1974)	X	X	X	X	X		X			
El-Sherbiney and Halling (1976)	X		X	X		X				
Matthewson (1981)	X		X		X	X		X		
Chiu and Harnett (1983)	X	X		X			X			
Solecki and Ohgushi (1984)	X	X	X	X	X				X	
Ramsey *et al.* (1991)	X	X			X			X		
Djabella and Arnell (1992)	X	X			X	X	X			
Bragallini *et al.* (2003)	X	X		X		X				
Sliding										
Chang *et al.* (1983)	X		X		X	X				
King and O'Sullivan (1987)	X	X			X		X			
O'Sullivan and King (1988)	X	X			X	X				
Ahn and Roylance (1990)	X	X		X			X			

Reference	Contact pressure	Contact deformation	Contact area	Subsurface stresses	Subsurface strain	Three dimensional	Analytical	Numerical computations	Experimental
Indentation									
Meijers (1968)	X	X					X	X	
Chen and Engel (1972)	X	X					X	X	X
Gupta *et al.* (1973)				X			X	X	
Gupta and Walowit (1974)	X						X	X	
El-Sherbiney and Halling (1976)		X					X	X	
Matthewson (1981)			X	X	X		X	X	X
Chiu and Harnett (1983)	X	X	X	X		X	X	X	
Solecki and Ohgushi (1984)	X	X					X	X	
Ramsey *et al.* (1991)		X			X		X	X	
Djabella and Arnell (1992)	X			X			X	X	
Bragallini *et al.* (2003)				X	X		X		
Sliding									
Chang *et al.* (1983)				X	X		X	X	
King and O'Sullivan (1987)	X			X			X		
O'Sullivan and King (1988)	X			X			X		
Ahn and Roylance (1990)				X				X	

solutions are mathematically quite intensive they will only be commented on briefly below. The full mathematical treatment is found in the references given.

The agreement between the theoretical models and real contacts is often not easy to estimate because most of the reported models do not include experimental comparisons.

A deviation from experimental results will naturally occur as long as the models do not take into account influencing parameters like surface roughness, material inhomogeneity and voids and the changing material properties in the mixed layer at the interface between the coating and substrate surface.

3.2.6.1 Contact pressure and stresses

The pressure distribution for a rigid cylinder indenting an elastic layer on a rigid substrate has been solved analytically by Meijers (1968), on an elastic substrate by Chiu and Harnett (1983) and for both elastic and rigid cylinders with coatings on elastic substrates by Gupta and Walowit (1974). The situation of one or several elastic layers indented by a parabolic counterface has been solved by Chen and Engel (1972). They also performed ball dropping experiments with a steel ball on to a nylon layer on a granite surface. By measuring the contact times they obtained good correlation with analytical predictions.

The cylinder against cylinder situation with either one or both coated and with either elastic or rigid substrates under the elastic coating has been solved by Solecki and Ohgushi (1984). A strain approach has been proposed by Bragallini *et al.* (2003) as a more simple method, where the stress analysis is carried out in terms of strains of the coating/substrate system. This method is used to evaluate the risk of coating failure for both hard and soft coatings, once the elastic modulus of the coating and the substrate and the coating thickness are known.

3.2.6.2 Normal approach and contact area

Solutions for the normal approach are given in the above situations by Meijers (1968), Chen and Engel (1972), Gupta and Walowit (1974), Chiu and Harnett (1983) and Solecki and Ohgushi (1984). Calculations for both the normal approach and the size of the contact area with an elastic spherical indenter against an elastic coating on a rigid substrate are presented by El-Sherbiney and Halling (1976). The size of the contact area is solved for a similar situation but with an elastic substrate by Chiu and Harnett (1983).

3.2.6.3 Subsurface stresses and strains

An analytical solution including a simplifying approximation enabled Matthewson (1981) to determine the stresses and strains within an elastic coating on a rigid substrate, averaged through the thickness of the coating and with particular profiles of rigid indenters. The stress distribution within the coating was found to be extremely sensitive to the Poisson's ratio of the coating. It was also found that the interfacial shear stress between the coating and the substrate is largest for almost incompressible materials, and therefore these materials are more likely to debond than others. The analysis can be used to predict properties and minimum useful thicknesses for coatings required in situations like a rubber coating on a hard substrate. Good agreement of the model with experimental results from contact radius and indentation measurements with spherical and conical indenters against silicone rubber on glass was achieved.

Subsurface stresses have been calculated by Chiu and Harnett (1983) for arbitrary surface shapes with an elastic layer on an elastic substrate. The numerical calculations show that when Young's modulus of the substrate is chosen to be one-third of that of the layer, as in steel layers on aluminium substrates, some degree of 'contact softening' occurs and is characterized by increased deformation and decreased contact pressure. These kinds of situations are present in transmission systems where a thin steel bearing is supported by a housing made of softer material such as aluminium or magnesium. The calculation of the equivalent stress in the aluminium substrate can be used to predict whether the housing will yield plastically, causing a progressively looser fit between the bearing and the housing, which is not acceptable in most applications.

An analytical solution for the stress distributions in plane strain layered elastic solids subjected to arbitrary boundary loading has been presented by Gupta *et al.* (1973). They show that extremely high tensile stresses may be developed in the layer when it is rigid compared to the substrate. The shear stress on the surface of the layer may also induce a high tensile stress in it.

The stress–strain relationship of elastic layers on elastic substrates loaded by rigid spheres has been investigated experimentally by Tangena and Hurkx (1986). They showed that a Brinell indentation test can be used on thin layers to obtain the plastic part of the stress–strain relationship and that finite element methods are suitable for checking the accuracy of the stress–strain curves obtained.

3.2.7 Analytical solutions of stresses and deformations in normally and tangentially loaded coated surfaces

The stress distributions and deformations in a layered surface under a sliding non-conformal counterface are complex and have not been easy to approach by analytical solutions. They have been investigated more in depth and understood only from the mid-1990s by the FEM techniques discussed later. It is, however, useful first to recall what effects the addition of tangential traction on the surface originating from sliding has on the stresses in the contact of a sphere against an uncoated solid (Hamilton and Goodman, 1966; Hamilton, 1983; Arnell, 1990):

- The point of first probable yield at or below the surface due to maximum von Mises stresses moves forwards and towards the surface as the tangential traction is increased, and reaches the surface when the coefficient of friction exceeds 0.32, as shown in Figs 3.13 and 3.28.

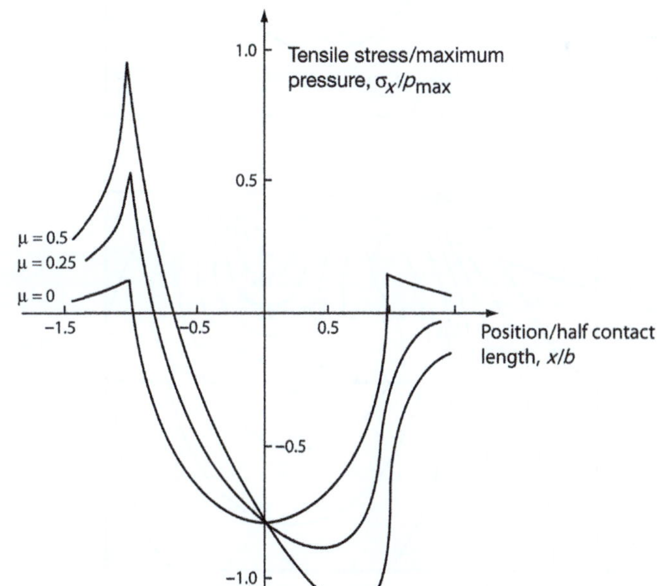

Fig. 3.31. The influence of the coefficient of friction on the tensile stress, σ_x, at the surface in the direction of sliding when loaded against a sliding sphere (after Hamilton, 1983).

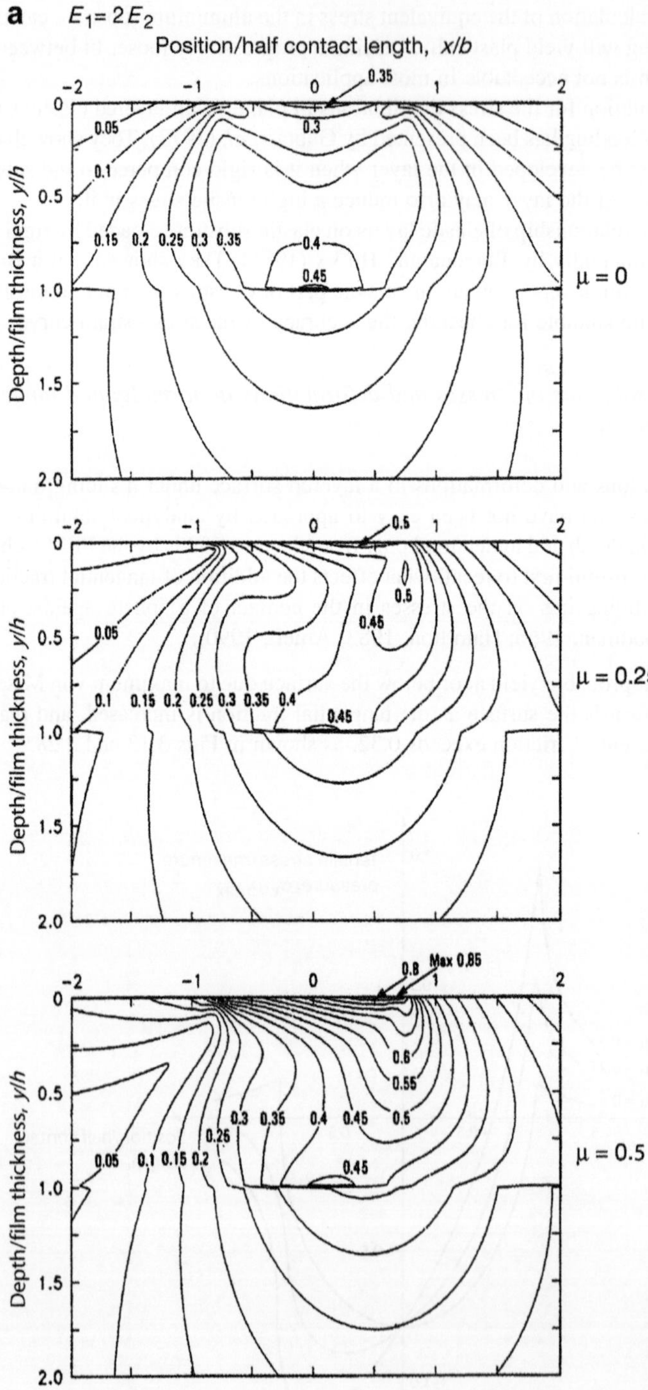
a
E1= 2E2
Position/half contact length, x/b
Depth/film thickness, y/h
μ = 0
μ = 0.25
μ = 0.5
Max 0.85

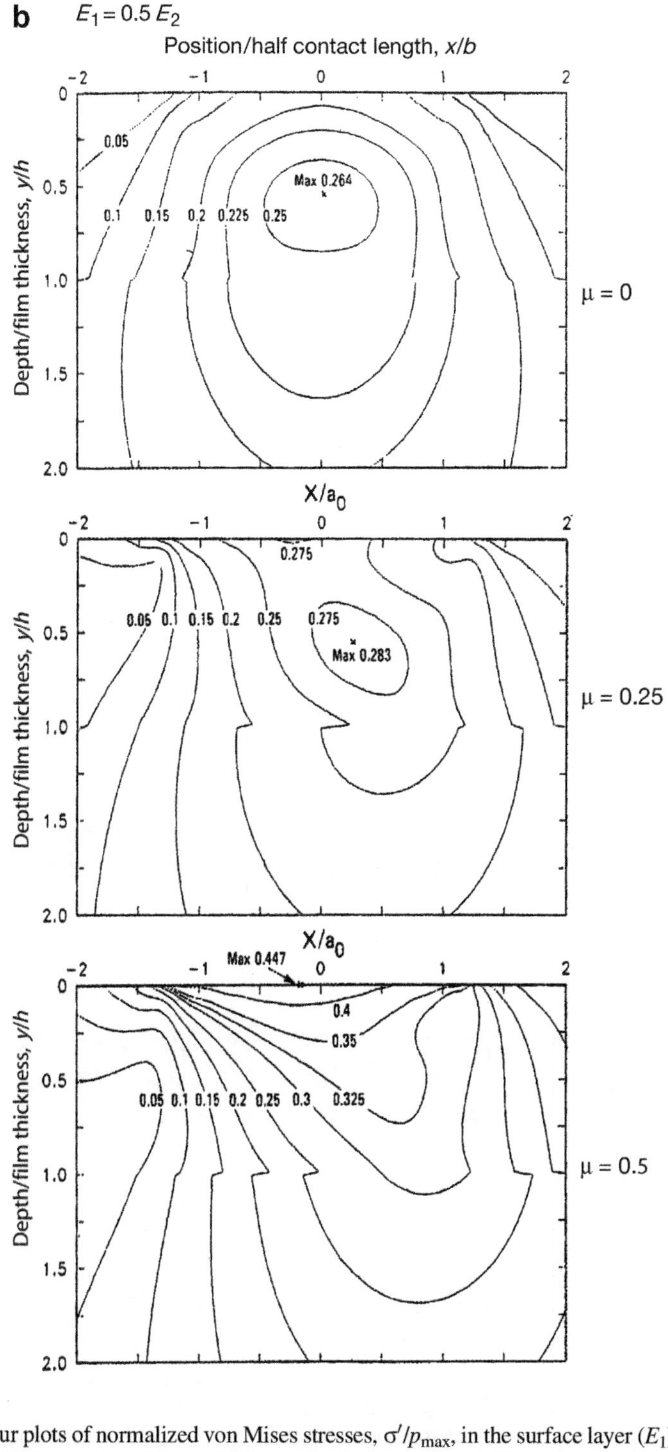

Fig. 3.32. Contour plots of normalized von Mises stresses, σ'/p_{max}, in the surface layer (E_1) and the substrate (E_2) with (a) a stiff layer, $E_1 = 2E_2$, and (b) an elastic layer, $E_1 = 0.5E_2$, when loaded by a sliding rigid *cylindrical* countersurface with coefficients of friction of $\mu = 0, 0.25$ and 0.5 (after King and O'Sullivan, 1987).

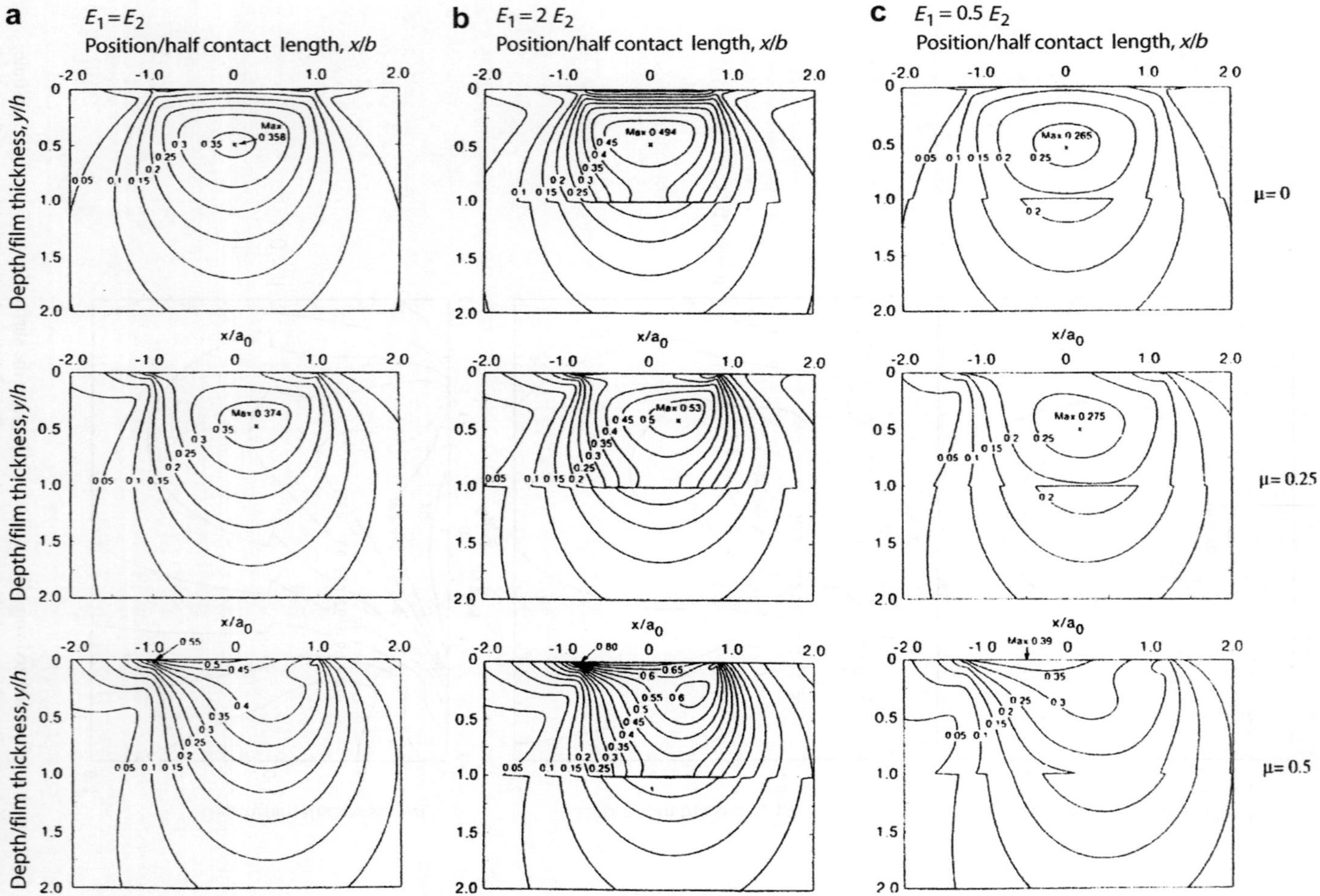

Fig. 3.33. Contour plots of normalized von Mises stresses, σ'/p_{max}, in the surface layer (E_1) and the substrate (E_2) with (a) no surface layer, $E_1 = E_2$, (b) a stiff layer, $E_1 = 2E_2$, and (c) an elastic layer, $E_1 = 0.5E_2$, when loaded by a sliding rigid *spherical* countersurface with coefficients of friction of $\mu = 0$, 0.25 and 0.5 (after O'Sullivan and King, 1988).

- A new high-stress region develops at the back edge of the contact, and this is the most likely region for yield for friction coefficients greater than 0.25.
- The frictional traction adds a compressive stress at the front edge of the contact and intensifies the tensile stress at the back edge. The intensification of the tensile stress is considerable, as shown in Fig. 3.31. Thus the region near the back edge of the contact is the most likely place for failure due to yield in tension.

The situation changes as a thin layer, like a surface coating, is deposited on top of the surface. The slip between an elastic layer and a substrate uniformly pressed together, caused by a normal force moving steadily over the surface, has been analysed and solved numerically by Chang *et al.* (1983). They show that two asymmetric slip zones are generated if the force exceeds a friction-dependent critical value. The slip is opposite in direction in the two zones, but due to the lack of symmetry, a net tangential shift of the layer is left behind by the passage of the force. This net shift is influenced by the coefficient of friction. For high coefficients of friction, above 0.5, the layer creeps in the direction of motion, while for low coefficients, below 0.25, the creep direction is backwards. The latter result has been observed experimentally.

The contact stress conditions in a layered elastic flat surface at and below the surface were analytically solved with a rigid cylindrical sliding countersurface by King and O'Sullivan (1987) and with a rigid spherical sliding countersurface by O'Sullivan and King (1988). They relate the stresses to the risk of yield in the coating system, and the calculated von Mises stresses for a stiff layer are shown in Fig. 3.32 and for an elastic layer in Fig. 3.33 at different values of the coefficient of friction. For $\mu = 0$ the presence of the stiff layer, $E_1 = 0.5E_2$, brings the point of maximum von Mises stress close to the interface, and the maximum is 40% higher than in the non-layered case. For $\mu = 0.25$ the maximum stress moves closer to the surface compared with the non-layered case, and its value is also approximately 40% higher than in the non-layered case. For $\mu = 0.5$ the von Mises stress has a maximum at the surface for both the layered and the non-layered cases but its value is 52% higher in the non-layered case. The presence of an elastic layer, $E_1 = 0.5E_2$, results in correspondingly lower von Mises stresses as shown in Fig. 3.33.

The results show that the value of Young's modulus of the layer relative to the substrate has a strong effect on the potential for yielding in both the layer and the substrate as well as on the adhesion stresses at the layer/substrate interface. The maximum tensile stress on the surface depends strongly on both the friction coefficient and the value of E_1 relative to E_2. In general, for a fixed value of the substrate Young's modulus, the stresses increase for higher values of the layer's Young's modulus. The strong influence of the coefficient of friction reported is partly due to the high values, $\mu = 0.25$ and 0.5, used in the calculations. Corresponding subsurface stress distributions for the contact with a rigid sphere sliding against uncoated and coated surfaces are shown in Fig. 3.33.

A thermomechanical wear transition model for coated surfaces was used by Cowan and Winer (1993) to study how sliding surfaces in contact produce a stress field and frictional heat source that may induce severe wear from material yielding. Based on isothermal and thermal factors, film thickness and material combinations are evaluated so as to minimize the probability of thermomechanical wear. A procedure for determining the combined thermal and mechanical stresses in concentrated sliding contacts has been utilized to predict the failure condition of layered solids by Ahn and Roylance (1990, 1992) and Ahn (1993).

A displacement formulation method was used by Ramalingam and Zhen (1995) to calculate layer/substrate interface stresses for line and extended distributed surface loadings. The layer/substrate interface stress calculations show that a hard coating is a viable solution for wear protection in extended contacts even when the substrate Young's modulus is as low as 0.2 of that of the coating modulus.

The stress distribution of a coated surface with a range of film thickness and elastic properties was analysed by Fujino *et al.* (2007) in elastohydrodynamic lubricated conditions. They used a three-dimensional analysis of the maximum shear stress applied into the coated film and substrate and concluded that an optimal surface design solution is one where the coated film has a large thickness and a smaller modulus of elasticity than that of the substrate is preferable.

3.2.8 Finite element method modelling and simulation of stresses and deformations

The analytical solutions of the stresses and deformations give important information about the general behaviour of a sliding contact. Still, due to the complexity of a real tribological contact and due to the great number of contact parameters influencing both the friction and wear process, the analytical solutions describe a contact idealized with regard to, e.g., geometry, surface roughness, material homogeneity, etc. The rapid development in finite element method (FEM) modelling and simulation techniques as well as developments in computing capacity offers a possibility to calculate the stresses and deformations in tribological contacts with a completely new level of accuracy and better coverage of the whole influencing parameter space.

3.2.8.1 Surfaces of bulk materials

Various tribological mechanisms of solid surfaces have been studied by the FEM technique. The stress analyses are often carried out to show the evolution of the von Mises equivalent stresses at various loadings at the surface and in the subsurface. Examples of such studies are elastic–plastic analysis including the influence of strain hardening on the loading and residual stresses that simulate material behaviour correlating with metallic contacts (Kral *et al.*, 1993). Subsurface deformation and the damage accumulation process in aluminium–silicon alloys subjected to repeated sliding contacts has been modelled and analysed (Akarca *et al.*, 2007), as well as the ploughing deformation, the effect of work hardening and friction behaviour when a conical tip in a scratch test is sliding over an aluminium alloy surface (Ben Tkaya *et al.*, 2007) and the residual groove formation on an amorphous polymeric surface (Pelletier *et al.*, 2008). The influence of several surface parameters on the stress distribution and deformation in a multi-asperity contact was investigated by Sabelkin and Mall (2007).

Advanced FEM modelling and stress and strain simulations including the use of a polycrystal plasticity model, cyclic plasticity, strain ratchetting, surface hardening and pad size effect have been used for analysing the failure process and for fatigue prediction in fretting wear by Bernardo *et al.* (2006), Dick and Cailletaud (2006b) and Dick *et al.* (2006). Nanotribological interactions related to the instability of crystalline material structure have been investigated by adopting an atomic-scale FEM model and using the quasi-static method to reduce the computation time (Jeng and Tan, 2004). Viscoplastic analysis was used to simulate the friction and wear mechanisms of polymeric materials (Lee *et al.*, 2001). Computational methods used in contact mechanics are found in, e.g., Wriggers (2002).

3.2.8.2 Coated surfaces in two dimensions

The early FEM models to calculate stresses and deformations in coated surfaces were in two dimensions and gave much useful information even if the number of elements was not more than 178 (Komvopoulos, 1988). The rapid development both in modelling techniques and computing capacity makes it possible to use advanced three-dimensional FEM with more than 200,000 elements (Leopold *et al.*, 2001; Laukkanen *et al.*, 2006).

A comprehensive analysis of stresses and deformations in coated surfaces exposed to frictional sliding was performed by Tangena (1987). He analysed the sliding contact when a rigid indenter moves over a soft elastic coating on an elastic substrate. The stresses and strains were calculated for both elastic and elastic–plastic material behaviour by FEM including coulomb friction behaviour at the interface between the indenter and the coating. The influence of surface roughness was also taken into account. If the asperities are plastically deformed the influence of surface roughness on contact parameters was found to be small.

The calculations show that the contact area increases continuously during indentation, independent of the value for the coefficient of friction, and that the influence of the coefficient of friction on the contact parameters is very small. In pure indentation the occurrence of plastic deformation has a major influence on the contact area and the contact stresses while the influence of the coefficient of friction is minor. During sliding the coefficient of friction has a large influence on the subsurface stresses for friction coefficients above 0.3. At coefficient of friction values below 0.3 the subsurface stress field for sliding resembles that of a pure indentation.

The radial distribution of von Mises stresses was calculated by Tangena by the FEM method for gold layers of different thicknesses on a glass substrate. The stresses are similarly high, independent of the thickness of the layer at the edge of the contact zone but considerably higher for the thinner layer in the middle of the contact zone as shown in Fig. 3.34.

The group led by Komvopoulos in Berkeley, USA, has had a leading role in developing the FEM modelling and simulation technique for investigating different aspects of material response to loading in coated tribological contacts. They have studied both single layer and multilayer systems, elastic and elastic–plastic contacts, stress fields, yield criteria, contact pressure, deformations, strains, residual stresses, surface patterns, thermal effects, crack propagation and fracture toughness (Komvopoulos *et al.*, 1987; Komvopoulos, 1988, 1989; Kral *et al.*, 1995a, b; Gong and Komvopoulos, 2003, 2004a, b). The information from the FEM analysis of a coated surface has been transferred into surface yield maps by Diao, Kato and co-workers (Diao *et al.*, 1993, 1994a; Diao and Kato, 1994; Diao, 1999). The different aspects studied by two-dimensional FEM analysis are summarized in Table 3.4.

The two-dimensional FEM studies include a great deal of interesting new information on the coating behaviour. Some of the main conclusions are:

- maximum tensile stresses are important when estimating the risk of coating failure by fracture (Mansour *et al.*, 1987),
- interfacial shear stresses and tensile stresses normal to the interface are important when estimating the risk of spalling-type surface failure (Mansour *et al.*, 1987),

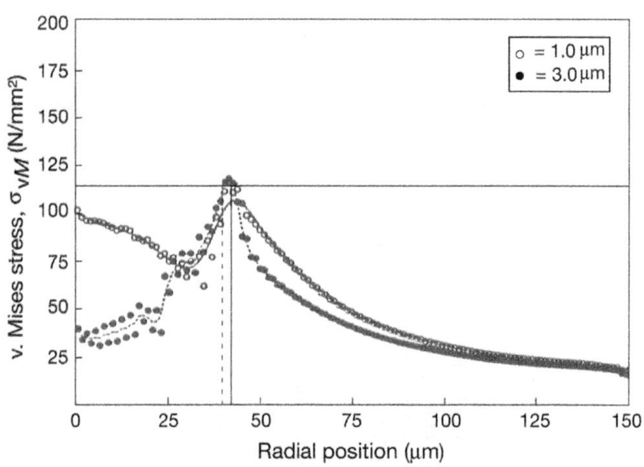

Fig. 3.34. The von Mises stresses averaged over the depth of the gold layer as a function of the radial position for a layered system with 1.0 and 3.0 μm thick gold films on a glass substrate. The indenter radius was 2.5 mm and the normal load 2.5 N (data from Tangena, 1987).

Table 3.4. Finite element method (FEM) studies of the contact problem with layered surfaces (E = elastic, P = plastic, R = rigid)

Reference	2 D FEM	3 D FEM	Coating	Sub-strate	Counterface	Single layer	Multi layers	Multi-contacts	Contact pressure	Stress distrib.	Modelled coating/ substrate materials
Mansour *et al.* (1987)	X		E	E	E	X				X	
Tangena (1987); Tangena and Wijnhoven (1988)	X		EP	EP	R	X	X		X	X	Gold/glass CoAu/Ni/Cu
Komvopoulos *et al.* (1987)	X		E	E	R	X					TiN/Ti and steel
Komvopoulos (1988, 1989)	X		E and EP	E and EP	R	X			X	X	
Djabella and Arnell (1992, 1993a, b, c)	X		E	E	E		X			X	
Diao *et al.* (1993, 1994); Diao and Kato (1994); Diao (1999)	X		EP	EP	E	X	X		X	X	
Kral *et al.* (1995a)	X		EP	EP	R	X		X	X	X	
Kral *et al.* (1995b)	X		EP	EP	R	X		X		X	
Kral and Komvopoulos (1996a)		X	EP	EP	R	X		X		X	
Kral and Komvopoulos (1996b)		X	EP	EP	R	X		X	X	X	
Rabinovich and Sarin (1996)		X	EP	EP	R	X			X		TiC/WC–Co
Wong and Kapoor (1996)	X		EP	EP	R	X		X			
Kral and Komvopoulos (1997)		X	EP	EP	R	X				X	
Bell *et al.* (1998)	X		EP	EP	R	X					TiN/HSS
Stephens *et al.* (2000)	X		E	EP	E	X			X	X	DLC/Ti–6Al–4V alloy
Souza *et al.* (2001)	X		E	EP	R	X				X	
Xiaoyu *et al.* (2001)	X		E	EP	R	X				X	TiN/HSS
Xiaoyu *et al.* (2002)		X	E	EP	R					X	TiN, Al, polymer/steel, Al, polymer, glass
Gong and Komvopoulos (2003)	X		EP	EP	R	X			X	X	
Holmberg *et al.* (2003)		X	E	EP	R	X				X	TiN/high-speed steel
Ye and Komvopoulos (2003a)		X	EP	EP	R	X				X	
Ye and Komvopoulos (2003c)		X	EP	EP	E	X			X	X	Carbon/magnetic layer
Gong and Komvopoulos (2004a)		X	E	EP	R		X	X	X		

Gong and Komvopoulos (2004b)	X		E	EP	R		X	X	X		Pyrolitic carbon/ graphite
Lindholm *et al.* (2006)		X	E	E	E	X				X	
Holmberg *et al.* (2006a, b)		X	E	EP	R	X				X	TiN/high-speed steel
Li and Beres (2006)		X	EP	EP	R	X				X	TiN/Ti–6Al–4V
Laukkanen *et al.* (2006)		X	E	EP	R	X				X	TiN/high-speed steel
Panich and Sun (2006)		X	EP	EP	R	X				X	TiB_2/HSS
Dick *et al.* (2006a)	X		EP	E	E		X		X		Composite layer
Fukumasu *et al.* (2006)	X		E	EP	R	X				X	2-component substrate
Tilbrook *et al.* (2007)	X		EP	EP	R					X	TiN/metal
Jungk *et al.* (2008)	X		EP	EP	R	X					DLC/metal
Holmberg *et al.* (2008c)		X	E	EP	R	X				X	TiN and DLC/HSS

Reference	Mises yield stress	Deformations	Strain	Residual stress	Thermal analysis	Surface pattern	Defects	Crack growth	K/SIF	μ	Comment
Mansour *et al.* (1987)											
Tangena (1987); Tangena and Wijnhoven (1988)	X	X	X			X				0–1	Wear modelling
Komvopoulos *et al.* (1987)	X	X								0–0.5	Lubricated
Komvopoulos (1988, 1989)	X	X	X							0–0.5	
Djabella and Arnell (1992, 1993a, b, c)										0–0.5	
Diao *et al.* (1993, 1994); Diao and Kato (1994); Diao (1999)	X									0–0.7	Yield map
Kral *et al.* (1995a)				X						0	Indentation
Kral *et al.* (1996b)	X			X						0	Indentation
Kral and Komvopoulos (1996a)	X		X	X						0.1	Shakedown effect
Kral and Komvopoulos (1996b)	X	X		X						0.1/0.25	
Rabinovich and Sarin (1996)								X			Scratch test
Wong and Kapoor (1996)				X						0.1–0.5	Shakedown limit
Kral and Komvopoulos (1997)	X		X	X						0.1/0.25	
Bell *et al.* (1998)	X	X									
Stephens *et al.* (2000)	X									0–0.5	Graded substrate
Souza *et al.* (2001)							X	X			Indentation

(Continued)

Table 3.4. Continued

Reference	Mises yield stress	Deformations	Strain	Residual stress	Thermal analysis	Surface pattern	Defects	Crack growth	K/SIF	μ	Comment
Xiaoyu *et al.* (2001)		X						X		0–0.8	Coating detachment, Scratch test
Xiaoyu *et al.* (2002)										0.2	Scratch test
Gong and Komvopoulos (2003)	X		X			X				0.1/0.5	
Holmberg *et al.* (2003)								X	X	0.08	Scratch test
Ye and Komvopoulos (2003a)	X		X	X						0–0.5	
Ye and Komvopoulos (2003c)	X		X		X					0.5	Frictional heating
Gong and Komvopoulos (2004a)	X		X		X	X				0.5	Frictional heating
Gong and Komvopoulos (2004b)	X		X					X	X	0.1/0.5	
Lindholm *et al.* (2006)						X					
Holmberg *et al.* (2006a, b)		X	X							0.08	Scratch test
Li and Beres (2006)		X								0.26	Scratch test
Laukkanen *et al.* (2006)				X				X	X	0.08	Scratch test
Panich and Sun (2006)		X									Scratch test
Dick *et al.* (2006a)			X			X				0.1–0.5	Fretting
Fukumasu *et al.* (2006)		X	X								
Tilbrook *et al.* (2007)	X	X	X	X							
Jungk *et al.* (2008)	X		X								
Holmberg *et al.* (2008c)										0.05 and 0.07	

- the stress distributions are strongly dependent on the thickness of the coating (or thickness/contact length ratio), the coefficient of friction and the Young's moduli of the coating and the substrate (Mansour *et al.*, 1987; Komvopoulos, 1988, 1989),
- for soft plastically deformable coatings the wear will increase rapidly as the average von Mises stresses increase above the yield stress of the coating, indicating that plastic deformation dominates the wear particle formation process (Tangena, 1987; Tangena and Wijnhoven, 1988),
- in sliding on soft coatings the friction force has a large influence on the subsurface stresses at coefficients of friction above 0.3; at friction coefficients below 0.3 the subsurface stress field resembles that of pure indentation (Tangena, 1987),
- in pure indentation, yielding in the coated surface always initiates at the coating/substrate interface below the centre of the contact; converse to the uncoated surface, the plastic zone does not grow toward the surface, it is restricted to the bottom of the hard layer and the substrate; the layer thickness plays a key role in the size and location of the plastic zone; thinner and harder coatings promote large tensile residual stresses at the coating/substrate interface (Komvopoulos *et al.*, 1989; Kral *et al.*, 1995a, b),
- multilayer coatings generally have less severe stress distributions than monolayers having the same total thickness and the same top layer/substrate modulus ratio; at the surface the maximum shear stress is higher for a coated surface compared to an uncoated one; conversely, in the substrate the maximum shear stress is lower than that of an uncoated surface (Djabella and Arnell, 1993a, b, c),
- a microcrack in a hard coating initiates usually from the local yield position; the local yield position can be identified by using a local yield map, which shows that yield at the coating/substrate interface is the most common case in a wide range of contact conditions and the yield at the surface of the coating is the second common case at higher coefficients of friction (Diao *et al.*, 1993),
- a pattern of residual stresses will develop after repeated sliding cycles and is essentially protective in that it makes subsequent yielding less likely; the load limit for the elastic steady state, known as the shakedown limit, provides a method to estimate the load that can be carried by the surface after repeated sliding contacts without plastic deformation; it depends on the hardness and stiffness of

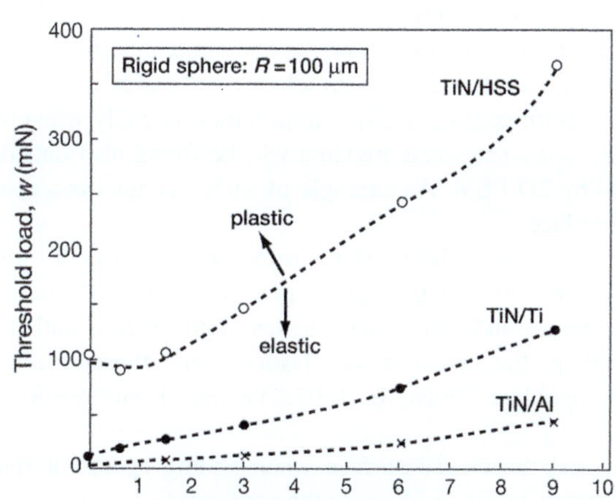

Fig. 3.35. Deformation maps for various elastic–plastic coating/substrate systems with TiN coating on HSS, Ti and Al substrates, given as threshold load of plasticity as a function of coating thickness (data from Bell *et al.*, 1998b).

the coating and the substrate, the coating thickness and the coefficient of friction (Wong and Kapoor, 1996),

- deformation maps, with the threshold load for plasticity as a function of coating thickness, can be developed for designers to assist in selection of coating thickness and substrates to meet specific loading conditions, as shown in Fig. 3.35 (Bell *et al.*, 1998b),
- a coating substrate system with an appropriate gradient in substrate yield strength can withstand significantly higher applied contact stresses, shift the location of the initial yield point deeper into the substrate, and coatings of greater thickness can be used; an appropriate gradient in Young's modulus results in a dramatic reduction in equivalent stress on the contact surface and the interface as compared to a non-graded one; the initial yielding in a graded substrate is largely independent of the coating thickness, friction coefficient and Young's modulus; the initial yield point remains well below the coating/substrate interface, decreasing the possibility of coating detachment due to excessive strain at the interface (Stephens *et al.*, 2000),
- maximum contact pressure is a strong function of surface topographical pattern; patterned surfaces produce higher tensile stresses but lower plastic strains and smaller plastic zones than flat surfaces as the surface can store significant strain energy without undergoing plastic deformation (Gong and Komvopoulos, 2003),
- surface cracking in a multilayer surface containing a crack perpendicular to the free surface due to repetitive sliding of a rigid asperity is controlled by the tensile fracture mode as the stress intensity factor K_I is considerably higher than K_{II}; higher friction at the sliding contact region increases both K_I and K_{II} significantly; the increase of friction at the crack interface promotes stress relaxation; initial crack growth was found to occur at an angle of ~10° from the original crack plane, independent of the initial crack length; after the first few crack growth increments the crack growth paths are almost parallel to each other, exhibiting a common deviation from the original crack plane by 57° (Gong and Komvopoulos, 2004b).

3.2.8.3 Coated surfaces in three dimensions

FEM modelling can be carried out with a very complex finite element mesh; the more complex it is, the more detailed information can be achieved. The drawback with complex meshes is that they result in very long computing times. So the optimal solution is always a balance between the complexity of the mesh and minimizing of the computing time. Nonetheless, the rapid development in computing capacity and speed offers a possibility to run even relatively complex three-dimensional FEM meshes in a reasonable time. Three-dimensional FEM is important especially when the stress fields of other than line contacts are studied since stress maxima will be found also outside the vertical plane of symmetry that is studied by 2D FEM. An example of such a contact is a coated flat contact sliding against a spherical counterface.

The group of Komvopoulos used 3D FEM to study the contact of a rigid sphere sliding on an elastic–plastic coated surface in both a single contact and repeated sliding. They described the shakedown effect, real stresses and von Mises stresses, deformations and strains, residual stresses, reyielding during unloading, the evolution of plasticity, and thermomechanical and topography effects (Kral and Komvopoulos, 1996a, b, 1997; Ye and Komvopoulos, 2003a, c; Gong and Komvopoulos, 2004a).

It is interesting to model the coated surface when sliding against a spherical diamond, as in a scratch test device, as shown in Fig. 3.36, since the configuration with an increasing load shows the evolution of the stress and strain fields at different loads. Three effects are important to consider with regards to the fracture behaviour of the coating. They are the ploughing of the spherical stylus resulting in plastic deformation, the sliding of the stylus on the coating resulting in friction and from this the pulling force on the coating behind the contact resulting in tensile stresses and fracture (Holmberg *et al.*, 2003).

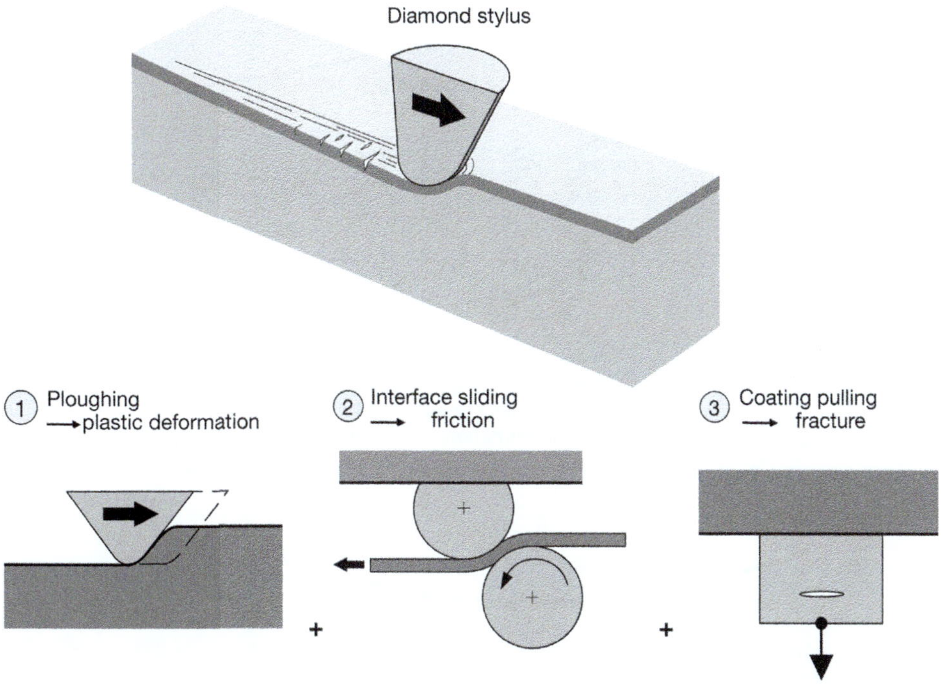

Fig. 3.36. Schematic illustration of the scratch tester spherical stylus drawn along the coated surface. The material loading and response can be divided into three phases: ploughing, interface sliding and pulling the coating.

A detailed analysis of the contact conditions is needed for accurate FEM stress measurements. Holmberg *et al.* (2006a) investigated the contact of a rigid sphere sliding on an elastic–plastic substrate coated with a hard elastic coating, correlating with the conditions of a thin ceramic coating on a steel substrate. The complex stress field in the surface is formed as a result of the following four effects, see Fig. 3.37:

1. Friction force. The friction force between the sliding stylus and the surface generates compressional stresses in front of the stylus originating from the pushing action, and tensional stresses behind the tip originated from the pulling.
2. Geometry changes. The elastic and plastic deformations are spherical indent, groove and torus shaped. They result in bending, stretching and compressing of the coating. The stresses arising are both compressive and tensile.
3. Bulk plasticity concentration. The spherical indentation causes the hard coating to deform elastically in a circular wave-like shape and the substrate under the coating to deform plastically, reaching its peak value at an angle of about 45° from the plane of symmetry in the plane of the coating. This can be identified with local minima and maxima of deformation between the tensile stress peaks around the indenter (at locations of 0 and 90°, respectively).
4. Residual stresses. It is very common that especially thin ceramic coatings, due to the deposition process, contain even very considerable compressive residual stresses. These are typically of the order of 0.5–4 GPa but values even as high as 10 GPa may appear, see section 3.2.10.

The bulk plasticity effect can be illustrated by the similarity with the situation experienced when a piece of paper is placed over a soft substrate, e.g. the hand, and pressed down at one point with

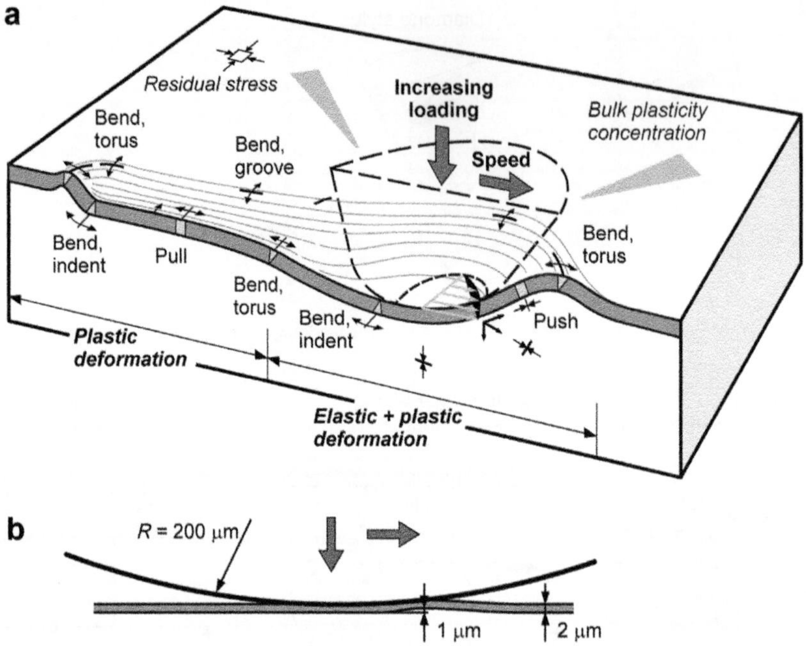

Fig. 3.37. The stress field in the coated surface resulting from a sliding sphere is a result of four loading effects: friction force, geometrical deformations, bulk plasticity concentration and residual stresses. Illustration (a) shows the loading effects with exaggerated dimensions and deformations and (b) with correct dimension interrelationships.

a finger. The paper will respond with a circular wave-like shape around the finger. If the paper was glued to the hand it would generate stresses and strains on the skin accordingly.

The contact conditions of a 200 μm radius diamond stylus sliding 10 mm with a load increasing from 5 N to 50 N and indentation depth increasing from 0.5 μm to 3 μm, over a 2 μm thick TiN coating attached with complete adhesion to a high-speed steel substrate was modelled by 3D FEM by Holmberg *et al.* (2003, 2006a). For the adhesive component of the coefficient of friction they used the measured value 0.08. The friction originating from ploughing is integrated as plastic deformation in the FEM model. The stylus was modelled as linear elastic but with Young's modulus of a diamond it will behave practically as rigid. The ceramic TiN coating was assumed not to deform plastically and thus modelled as linear–elastic. The high-speed steel substrate was modelled as elastic–plastic including the strain hardening effect. A schematic illustration of the FEM mesh is shown in Fig. 3.38.

Holmberg and co-workers were interested in investigating the fracture behaviour of the hard coating and thus they primarily analyse and infer the FEM stress calculations with respect to *first principal stress*, which is defined as the largest stress in tension in any direction. The frequently used von Mises stress is not usually associated with brittle failure and has its greatest potential in understanding and modelling deformation-related events such as plasticity of metals. Topographical stress field maps showing the first principal stresses on the top of the coating and at the subsurface symmetry plane in the direction of sliding at four different positions are shown in Figs 3.39a–d.

The topographical stress field maps illustrate the following effects. In these loading conditions there are:

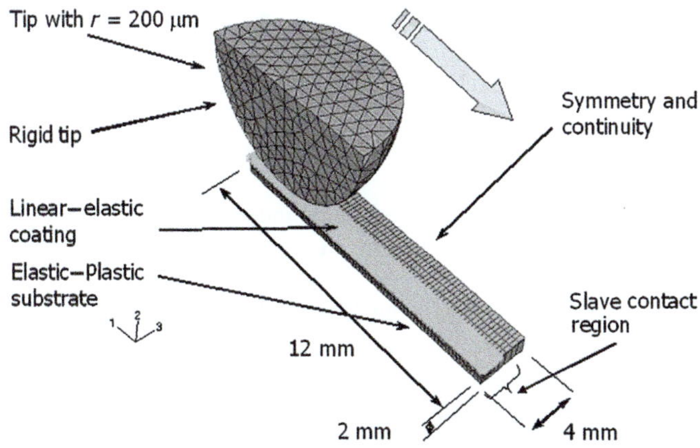

Fig. 3.38. Schematic illustration of the three-dimensional finite element mesh. The mesh sizing is in the range of 0.25–100 μm and the number of mesh nodes is about 190,000.

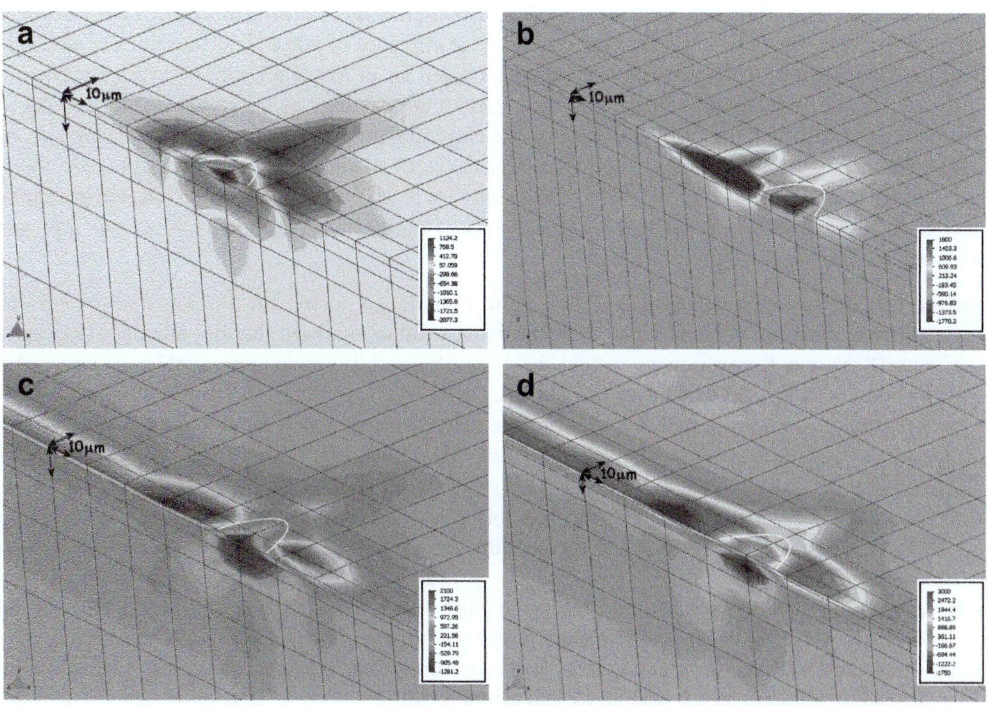

Fig. 3.39. Topographical stress field maps showing first principal stresses on the top of the coating and at the symmetry plane intersection of a high-speed steel sample coated with a 2 μm thick TiN coating and loaded by a sliding spherical diamond stylus. Sliding direction is from left to right, the sliding stylus is invisible in the figure and the contact zone is indicated by the white curve. The values on the colour scale are given in MPa. Stress fields shown are at (a) 5 N pre-load and 0.5 μm indentation depth without sliding, (b) 5.3 N load and 0.06 mm sliding, (c) 10 N load and 1.2 mm sliding and (d) 20 N load and 3.3 mm sliding. See Appendix B for coloured figure.

- fairly high tensional stresses, about 1150 MPa, at the top of the surface around the stylus (Fig. 3.39a),
- stress minima down to zero MPa at about 45° and 135° from the symmetry plane originating from the bulk plasticity concentration (Fig. 3.39a),
- high, about −2100 MPa, compressive stresses right under the stylus, and tensile stresses, about 400 MPa, some 20 μm under the stylus (Fig. 3.39a),
- residual stresses, about 2000 MPa, on the remaining indent after some sliding (Fig. 3.39b),
- high residual stresses, about 2000 MPa, in the groove formed behind the sliding stylus (Fig.3.39c),
- tensile stress maxima, about 2100 MPa, about 15 μm behind the back edge of the contact area and about 10 μm on the left side of the symmetry plane (Fig. 3.39c); this is especially interesting since the first angular cracks observed empirically in similar conditions appear exactly at this location,
- movement of the compressive stress maxima from the top of the coating right under the sliding stylus (Fig. 3.39a) to a depth of about 15 μm below the surface (Fig. 3.39d); this illustrates the compensation of the compressive stresses in the coating by tensile stresses from the coating stretching due to substrate deformation, and
- high tensile stresses, about 3000 MPa, behind the sliding stylus (Fig. 3.39d); these are, however, overestimated since in the real conditions they will partly be released by the formed crack field.

The finite element method can be used to analyse tribological mechanisms in surfaces even down to the nanolevel. Holmberg *et al.* (2008b) have studied the influence of 100 and 500 nm thick bond layers between a 2 μm thick coating and the substrate and used an FEM model with nod density down to 25 nm. They showed the rapid changes in the interface region stresses under the sliding contact in a scratch test and how these are influenced by the elasticity and thickness of the bond layer.

The three-dimensional FEM stress and strain analyses of coated surfaces have also bought much new understanding and some of the main conclusions are listed below. The conclusions are presented in a more generalized form and reflect typically the situation when a sphere slides over a plastic substrate covered with a harder coating:

- main mechanisms generating stresses and strain in a loaded coated surface are: friction, surface geometrical changes, bulk plasticity concentration, residual stresses, thermal effects, topography effects (Kral and Komvopoulos, 1996a, b, 1997; Ye and Komvopoulos, 2003a, c; Gong and Komvopoulos, 2004a; Holmberg *et al.*, 2006a),
- bulk plasticity concentration effects result in a tetra arm-shaped tensile stress pattern on the coating around the counterface when a sphere is sliding on a substrate covered with a harder coating (Holmberg, 2003; Li and Beres, 2006),
- the first coating crack in the total stress field volume is generated at the top surface of the coating behind the sliding counterface at the groove edge and grows downward (Holmberg *et al.*, 2003, 2006a; Laukkanen *et al.*, 2006),
- a hard coating on a softer substrate is carrying part of the load and decreasing the stresses in the substrate (Kral and Komvopoulos, 1996b; Holmberg *et al.*, 2006a),
- high residual tensile stresses are formed in the coating on the groove behind the slider (Xiaoyu *et al.*, 2002; Holmberg *et al.*, 2006a; Li and Beres, 2006),
- there is a pile up effect of material generated from compression in front of a slider that results in high tensile stresses on the surface; the maximum forward flow at the contact surface increase primarily with friction and decreasing stiffness and hardness of the coating and secondarily with normal load (Kral and Komvopoulos, 1996a, b; Xiaoyu *et al.*, 2002; Holmberg *et al.*, 2006b; Li and Beres, 2006),
- Young's modulus and coating thickness are principal parameters affecting the magnitude of stresses in a loaded soft substrate covered with a hard coating (Holmberg *et al.*, 2006b),

- a thicker hard coating has a better load-carrying capacity than a thinner one but generates higher stresses at the coating/substrate interface (Holmberg *et al.*, 2006b),
- a stiffer coating (high E modulus) has a better load-carrying capacity than a more compliant one (low E modulus) but generates higher tensile stresses in the coating and lower tensile stresses in the substrate with the same indentation depth (Kral and Komvopoulos, 1996b, 1997; Holmberg *et al.*, 2006b),
- maximum first principal stress in the coating due to loaded sliding increases, respectively decreases, with increasing tensile, respectively compressive, residual stress in the coating; in high-friction sliding, plasticity in the substrate is affected predominantly by the coefficient of friction and secondarily by the residual stress (Ye and Komvopoulos, 2003a),
- residual compressional stresses in a hard coating increase both the maximum tensile stresses at the top of the coating and compressional stresses in the coating right under the sliding counterface (Laukkanen *et al.*, 2006),
- residual stresses existing in the coating are relaxed by the deformations taking place under a loaded sliding counterface and will have little influence on crack formation in the groove behind the slider (Laukkanen *et al.*, 2006),
- reyielding in the substrate may occur at unloading; it may also occur after a second sliding cycle for a highly loaded sliding contact and result in a larger plastic zone than is produced in the first load cycle (Kral and Komvopoulos, 1996a, 1997),
- the maximum plastic strain arises in the subsurface region slightly behind the centre of the contact area at low friction while at high friction it occurs on the surface within the contact area (Kral and Komvopoulos, 1997),
- high-friction sliding promotes plasticity and intensifies the von Mises and first principal stresses in both the coating and the substrate, thus increasing the tendency for yielding and cracking in each of them (Ye and Komvopoulos, 2003a),
- coating thickness exhibits a more pronounced effect on the temperature rise from frictional heating at the coating surface and the coating/substrate interface than the layer thermal conductivity; frictional heating and shear surface traction may intensify the stress field significantly; the likelihood of yielding and cracking in the coated surface increases with decreasing coating thickness (Ye and Komvopoulos, 2003c),
- a patterned coated surface results in reduced plastic deformation in the substrate compared to a smooth surface (Gong and Komvopoulos, 2004a), and
- when a bond layer is used, e.g. for coating/substrate adhesion improvement, should the bond layer be less stiff than the coating then high and critical tensile stresses will not be generated; the thickness of the bond layer may vary and is normally not critical with respect to generated stresses in the surface (Holmberg *et al.*, 2008b).

3.2.8.4 The E/H relationship

The FEM analyses aim to consider the whole parameter space influencing friction and wear in a coated contact. Another approach is to focus on some dominating parameters to easily obtain some information on the tribological behaviour of the coated surface. One such approach is to calculate the E/H or H/E relationship for a contact. It is commonly known that the hardness, H, of a material is often strongly related to wear and also to friction. However, the elastic property represented by Young's modulus, E, is often neglected even if it also has an important influence in many contact conditions. Thus is it very relevant to study these two parameters together.

It was pointed out many years ago by Oberle (1951) that in a sliding contact, wear of an engineering material is closely related to the elastic limit of strain of the material. A practical measure of the elastic limit of strain is the H/E relationship. Halling and Arnell (1984) used this relationship to evaluate the elastic contact of a rough surface. They used the ratio E'/H, were E' combines Young's modulus values of both mating surfaces and found that this ratio is related to the wear resistance of the two surfaces.

This approach can be applied to a coated surface as discussed by Matthews and Leyland (2000, 2002) and Leyland and Matthews (2000). They reiterate that the ratio H_C/E_C, where H_C and E_C are the hardness and Young's modulus of the coating, defines the material's elastic strain limit, that is the amount by which the coating can be stretched before permanent deformation occurs. This gives an indication of the ability of the film to deform with the substrate under load without yielding. Minimization of E_C/H_C will ensure that plastic deformation and wear can be avoided and the ploughing component in friction is reduced, as shown by Gupta and Meletis (2004) and Ni *et al.* (2004a). A comparison of the E/H values of the coating and the substrate gives an indication of the amount of plastic deformation occurring in the coating and in the substrate, respectively (Korunsky *et al.*, 1998; Komvopoulos *et al.*, 2004).

The ratio H/E is also part of the plasticity index (Greenwood and Williamson, 1966). The importance of the H/E ratio has been known by bearing designers for many years. They routinely use the plasticity index when determining the degree of polish required by a bearing surface to ensure that surface asperity deformation under load is avoided.

The ratio $H_C{}^3/E_C{}^2$, called the plastic resistance parameter, has been theoretically derived from the material behaviour and used as an indicator of the resistance to plastic flow of the coating (Tsui *et al.*, 1995; Mayrhofer *et al.*, 2003; Musil, 2005; Musil and Jirout, 2007).

3.2.9 Influence of surface roughness

All surfaces we deal with in our daily life are rough and the roughness influences the contact and subsurface stresses in a loaded contact. This influence depends very much on whether the asperities are deformed elastically or plastically. The mode of asperity deformation also influences the real contact area and thus the friction and wear (Halling, 1975, 1986). When the asperities are deformed plastically the load is linearly related to the real area of contact for any distribution of asperity heights. When the asperities deform elastically, linearity between load and real area of contact will occur only when the asperity distribution is exponential, which is common for practical surfaces.

The influence of surface roughness on surface and subsurface stresses and other contact parameters is small if the asperities are deformed plastically, as is often the case for soft coatings (Tangena, 1987). Ogilvy (1992, 1993) showed by numerical simulation how the average contact pressure across all contact

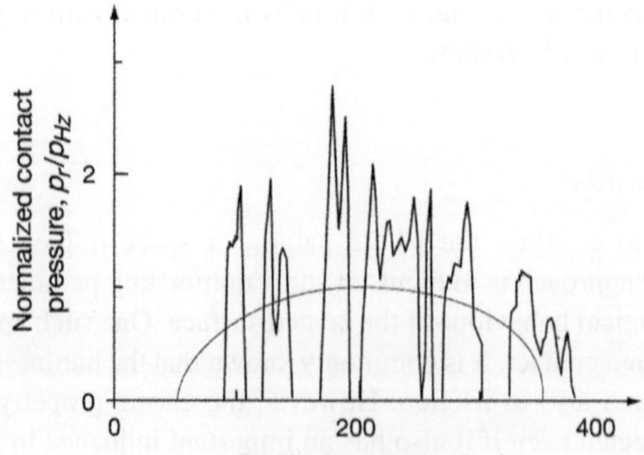

Fig. 3.40. Contact pressure distribution for a smooth uncoated surface loaded by a smooth cylinder (Hertzian distribution) and by a rough cylinder with $R_a = 0.44$ µm and $p_{Hz} = 2.2$ GPa (after Sainsot *et al.*, 1990).

junctions increases as a function of surface roughness until, for very rough surfaces, it reaches a value close to the asperity yield stress and the load is supported by only a few plastically deformed contact points.

The influence of surface roughness on contact and subsurface stresses is, however, considerable when the surfaces behave elastically (Sainsot *et al.*, 1990; Lee and Ren, 1994, 1996; Yu and Bhushan, 1996; Bhushan, 1998a; Mihailidis *et al.*, 2000; Lin and Ovaert, 2004). This is typically the situation for steel surfaces uncoated or coated with hard layers. Computer simulation models have been used to calculate the contact pressure and the von Mises stresses in rough contacts with coated surfaces of different hardness, film thickness and coefficients of friction values (Sainsot *et al.*, 1990; Mao *et al.*, 1996; Bell *et al.*, 1998). The contact between a rough elastic cylinder and a smooth coated and uncoated elastic body was studied by Sainsot *et al.* (1990). The Hertzian pressure distribution was considerably disturbed by the roughness of the cylinder, which was $R_a = 0.44$ µm, as shown in Fig. 3.40. The microgeometry generated surface pressures were almost three times higher than the maximum Hertzian pressure.

The real interrupted pressure distribution will smooth out for soft coatings and the maximum pressure drops with increasing thickness as shown in Fig. 3.41a and b. Hard coatings do not modify the pressure distribution significantly, the maximum pressure might slightly increase and the influence of thickness is minor, as shown in Fig. 3.41c and d.

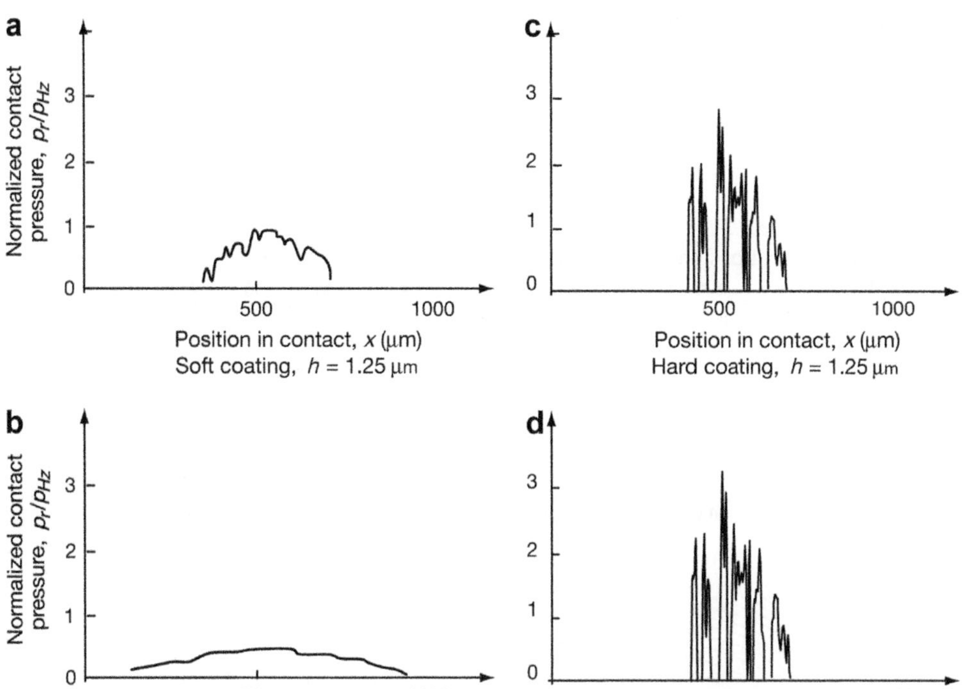

Fig. 3.41. The normalized contact pressure distribution for (a) soft and thin, (b) soft and thick, (c) hard and thin and (d) hard and thick smooth coating loaded by a rough cylinder with $R_a = 0.44$ µm and $p_{Hz} = 2.2$ GPa (after Sainsot *et al.*, 1990).

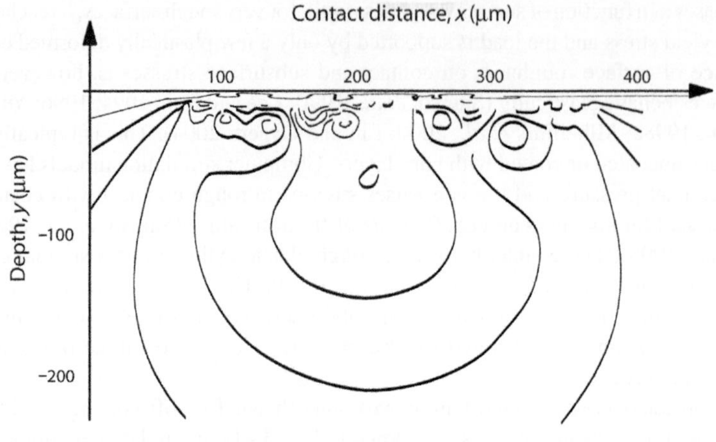

Fig. 3.42. Von Mises stresses in a smooth uncoated surface loaded by a rough cylinder with $R_a = 0.44$ µm and $p_{Hz} = 2.2$ GPa (after Sainsot *et al.*, 1990).

When the load distribution is known, the subsurface stresses can be calculated. The von Mises stress distribution of the uncoated rough contact referred to earlier (Fig. 3.40) is shown in Fig. 3.42. The roughness of the surface generates some stress perturbations near the surface of the loaded contact. These stress maxima perturbations are often at about the same depth from the surface as where the coating/substrate interface occurs in coated surfaces and thus their influence on interface stresses is important.

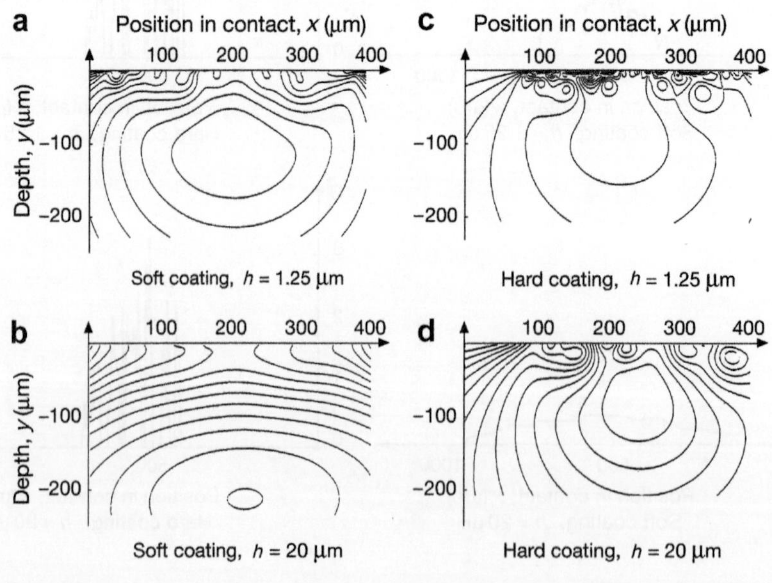

Fig. 3.43. Von Mises stresses in (a) soft and thin, (b) soft and thick, (c) hard and thin and (d) hard and thick smooth surface loaded by a rough cylinder with $R_a = 0.44$ µm and $p_{Hz} = 2.2$ GPa (after Sainsot *et al.*, 1990).

A soft coating will decrease the stress perturbations under the surface and this effect is stronger with thicker coatings as shown in Fig. 3.43a and b. Thin hard coatings have almost no influence on the perturbations under the surface but with thicker hard coatings they are slightly decreased as shown in Fig. 3.43c and d.

The calculations by Sainsot *et al.* (1990) show that for hard coatings with a coating thickness of 15 μm or less the maximum von Mises stresses in the coating are located at the interface between the coating and the substrate as shown in Fig. 3.44a. For both soft and hard coatings maximum von Mises stresses decrease with increasing coating thickness. With hard coatings of thickness less than 20 μm the maximum von Mises stress in the substrate is found at the interface between the coating and the substrate, as shown in Fig. 3.44b.

The development of advanced coating deposition techniques has opened the possibility to produce designed patterned surfaces with controlled surface topographies. Regular surface patterns have been modelled and related contact pressures, subsurface stresses and thermo-mechanical effects calculated (Gong and Komvopoulos, 2003, 2004a; Lindholm *et al.*, 2006; Aoki and Ohtake, 2004). A segment structured surface is effective for stress relaxation in coatings in the case of substrate deformation.

The apparent contact area of a coated patterned surface in contact with a rigid cylindrical asperity consists of several microcontacts exhibiting a trend to merge with each other with increasing indentation depth (Gong and Komvopoulos, 2003). During sliding, high peak pressures occur at the trailing edges of microcontacts in the contact zone. The maximum contact pressure strongly depends

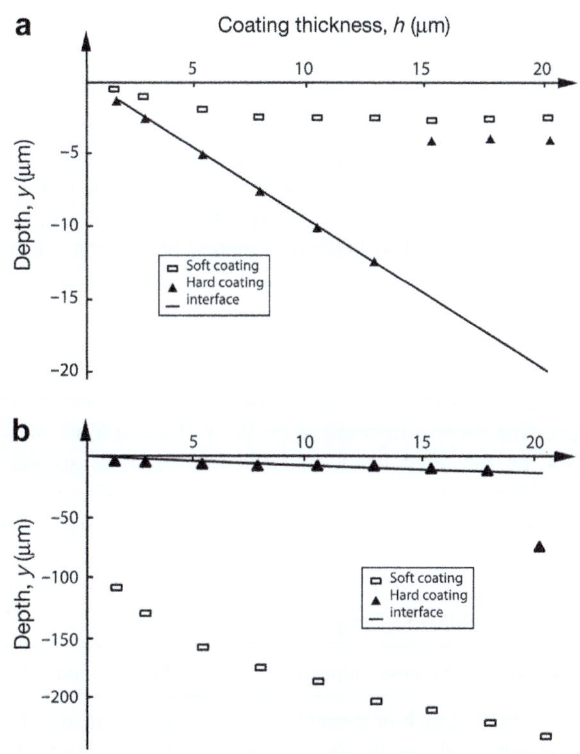

Fig. 3.44. Location of maximum von Mises stresses (a) in the coatings and (b) in the substrate for soft and hard coatings on a smooth substrate loaded by a rough cylinder with $R_a = 0.44$ μm and $p_{Hz} = 2.2$ GPa (after Sainsot *et al.*, 1990).

on the pattern geometry. The maximum tensile residual stress on a coating with a sinusoidal surface pattern occurs at the trailing edge of the contact region. This stress is much higher than that obtained with a flat coated surface and depends on the pattern geometry and the coefficient of friction. In a patterned coated surface with a high amplitude-to-wavelength ratio and high friction, a significant tensile residual stress is generated in the groove behind the contact. Patterned surfaces produce lower plastic strains and smaller plastic zones than flat surfaces due to the lower stresses resulting from the increased compliance of the coating that can store strain energy without undergoing plastic deformation. The decreased surface stiffness of a hard patterned coated surface reduces the maximum plastic strain and size of the plastic zone in the substrate. However, this results in a higher surface tensile stress at the trailing edge of the contact interface, creating a greater risk of surface crack initiation in the patterned surface.

Contact fatigue of a coated surface with an elastic surface coating on the top of three elastic–plastic underlying coatings in sliding contact with a rigid and rough surface was analysed by FEM by Gong and Komvopoulos (2005). The rough surface was characterized by scale-invariant fractal geometry in order to include multiscale roughness effects and self-affine surface features. The critical segment of the rough surface was identified and used for contact fatigue simulations. They showed that a transition from tensile to shear dominant mode of fatigue crack growth occurs as the crack tip approaches the interface. This results in further crack growth almost parallel to the layer interface. The results show the importance of surface roughness in the contact fatigue of coated surfaces.

Detailed analysis of tribology-related effects in contacts with rough surfaces has been reported by Bhushan and co-workers (Yu and Bhushan, 1996; Bhushan, 1998a, 1999b).

3.2.10 Residual stresses

A residual stress is a stress in the surface that for some reason cannot be relaxed and thus remains in the surface even when no external load is present. A coated surface generally contains residual stresses that are either a result of intrinsic changes in the structure due to the deposition process or deformations, or a result of mismatch in thermal expansion at temperature changes. The residual stresses are normally in the range of 0.1 to 4 GPa in compression, but can sometimes be very high, up to 10 GPa in compression and up to a few GPa in tension. The stresses due to temperature changes are called thermal stresses and those due to structural changes are called intrinsic stresses. Both are discussed in more detail below. The residual stresses can be divided into four categories, as shown in Table 3.5 (Holmberg *et al.*, 2008d).

Most ceramic coatings are improved tribologically by high compressive residual stresses parallel to the surface. The compressive stresses in coatings on flat surfaces inhibit crack initiation and crack growth. Tensile stresses, on the other hand, tend to promote crack growth and are not desired (Bower and Fleck, 1994; Eberhardt and Peri, 1995; Bromark, 1996; Gunnars and Alahelisten, 1996; Nordin, 2000; Zoestbergen and De Hosson, 2002; Cammarata, 2004). High compressive stresses may

Table 3.5. Categories of residual stresses.

Residual stresses	Deposition process induced	Sliding induced
Thermal stress	due to mismatch in thermal expansion of coating and substrate after temperature change from deposition to ambient.	due to mismatch in thermal expansion of coating and substrate from temperature change originating from frictional heating.
Intrinsic stress	due to defects built into coating during deposition.	due to strains created by deformation.

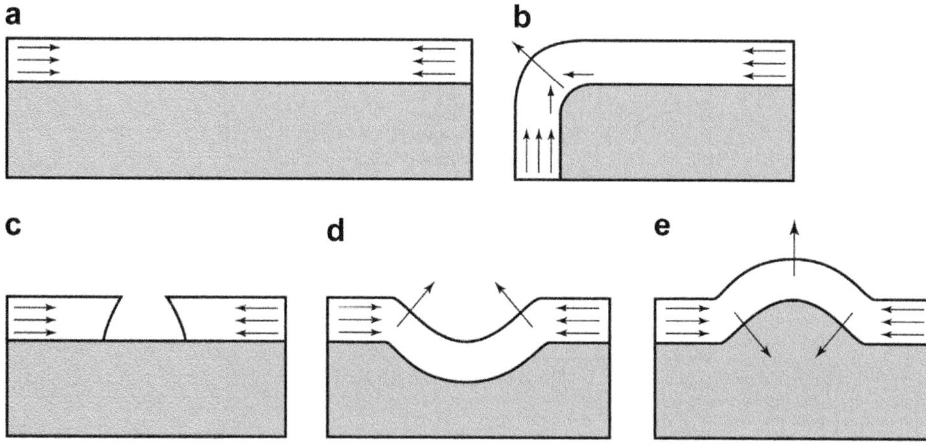

Fig. 3.45. Illustration of stresses induced in a coated surface having compressive residual stresses when (a) coated on a flat and smooth surface, (b) coated over an edge, (c) including a pore, (d) in a groove and (e) at a ridge (after Wiklund *et al.*, 1999b).

result in decohesion and delamination at edges or corners or when the coating is deposited on rough surfaces. This is illustrated in Fig. 3.45.

Generally, coatings, which are thin with respect to the interface topography, are less sensitive to residual stress-induced failure. The critical interface stress state becomes independent of the coating thickness if the coating is thicker than about three times the amplitude of the interface roughness. Coating/substrate interfacial stresses of about half the magnitude of the residual stresses in the coating are induced at corners, edges and topographical irregularities. The highest value of the interfacial stress is obtained for a coating thickness equal to the minimum radius of the surface contour (Wiklund *et al.*, 1999b).

A mechanism explaining the level of residual compressive stresses needed to inhibit crack propagation in a diamond coating has been proposed by Gunnars and Alahelisten (1996). At the beginning of the wear process the protruding grains will take up the externally applied stresses, as shown in Fig. 4.53. The result is that high local stress concentrations and fragmentation occur. The residual stresses are low in the protruding grains since the geometry permits relaxation. A crack nucleated in a protruding grain will propagate preferentially along easy cleavage planes. However, a crack that propagates deeper into the coating will encounter higher and higher compressive residual stresses which will deflect the crack into a path, irrespective of cleavage planes. The large residual stresses will prevent cracks initiated at the surface from propagating towards the interface and will thus promote crack paths parallel to the surface.

An idealized representation of the total stress generated in a thin film as a function of the parameter T/T_m, where T (K) is the substrate temperature and T_m is the coating material melting temperature, is shown in Fig. 3.46. The intrinsic stress is assumed to be tensile by Thornton and Hoffman (1989), although it must be remembered that this is not typical for ceramic tribological coatings on metal substrates produced by PVD. The deposition temperature is assumed to be greater than the final measurement temperature and the film thermal expansion coefficient is assumed to be greater than that of the substrate so that a tensile stress is generated when the substrate is cooled. At low T/T_m the intrinsic stress dominates the thermal stress, whereas at high values the thermal stresses dominate. When the T/T_m exceeds about 0.25, the recovery processes are operative so that the intrinsic component of the stress is reduced (Thornton and Hoffman, 1989).

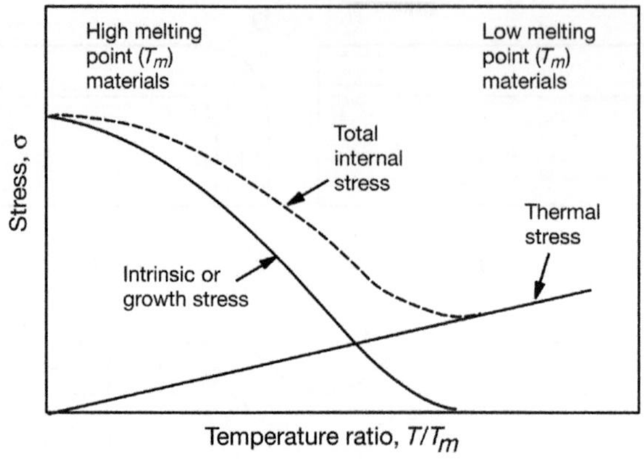

Fig. 3.46. The thermal and the intrinsic stress contributions as functions of the T/T_m ratio (after Thornton and Hoffman, 1989).

The most common methods for measuring the residual stresses in coated surfaces are those based on direct measurement of the elastic strains in the film by X-ray diffraction and those based on measuring the associated curvature or deflection of a thin substrate (Sue, 1992; Larsson *et al.*, 1996; Borland, 1997; Malzbender *et al.*, 2002; Zoestbergen and De Hosson, 2002; Cammarata, 2004; and Pauleau, 2008). Some characteristic residual stress values are given in Table 3.6. Stress relaxation of coatings with high residual stresses on flexible substrates can be achieved by using a segmented structure where the film is intentionally separated into small segments as if it was full of small regular cracks (Aoki and Ohtake, 2004).

Finite element analysis has been used to investigate the influence of residual stresses on the stress and strain field in coated surfaces (Ye and Komvopoulos, 2003a; Laukkanen *et al.*, 2006). The stress simulations show that the maximum first principal stress in the coating due to sliding decreases with increasing compressive residual stress. The residual stress effect on the first principal stress is relatively less pronounced in high-friction sliding. The tendency for yielding and cracking at the surface and at the coating/substrate interface is strongly affected by the magnitudes of the residual stress and the coefficient of friction. An optimum residual stress can be determined, depending on the type of the loading, the dominant deformation mode in the coating (i.e. plastic deformation or cracking) and the coefficient of friction.

Residual compressive stresses of 1 GPa in a 2 μm TiN coating on a steel substrate were found to increase the maximum tensile stresses on the top of the coating by 10–50% and also increase the compressive stresses on the surface under the tip to about double in the contact with a spherical diamond tip sliding with increasing load from 5 to 20 N over the coated surface (Laukkanen *et al.*, 2006). During sliding of the spherical tip on the surface, the residual stresses in the coating are relaxed under the tip due to the plastic deformation of the substrate. At the location of appearance of the first cracks, at the groove edge behind the sliding tip, plastic deformation has already taken place and the compressive residual stresses have been relaxed. Thus they are not active in hindering the process of crack opening due to the concentration of tensile stresses behind the tip.

The stresses and strains that a point on top of the coated surface at the plane of symmetry experience when a sliding spherical tip in a scratch test moves over it are shown in Fig. 3.47 (Holmberg *et al.*, 2008d). Starting from left and moving to the right there is first a 1 GPa compressive lateral residual stress in the coating. Coming closer to the contact zone the compressive stresses increase to about 5000 MPa right under the tip and then decrease after the contact to zero and even go to close to 1000 MPa in tension. After that the stresses level off to a residual level due to inelasticity within the

Table 3.6. Example of measured residual stresses in coatings.

Coating/substrate	Deposition process	Coating thickness (μm)	Characteristic residual stress (GPa)	Reference
TiN/Ti–6Al–4V	PVD	8.8	−3.7	Sue (1992)
TiN/Inconel 718	PVD	8.2	−3.4	Sue (1992)
TiN/AISI 304	PVD	8.4	−4.1	Sue (1992)
TiN/tool steel	PVD	4.1	−3.5 to −4.1	Bromark *et al.* (1994)
TiN/HSS	PVD	4–5	−4	Larsson *et al.* (1996); Wiklund *et al.* (1999b)
TiN/steel			−0.2	Komvopoulos (1991a)
TiN/Ti			0	Komvopoulos (1991a)
TiN/stainless steel	PVD	5–7	−6.1	Bull (1997)
TiN/CC	PVD	3.8	−2.5	Nordin *et al.* (1999)
TiN/CC	PVD	4	−4	Wiklund *et al.* (1999b)
TiN/AISI tool steel	PVD	2.5–4.4	−4 to −4.5	Zoestbergen *et al.* (2002)
TiAlN/tool steel	PVD	3.3	−5 to −6.1	Bromark *et al.* (1994)
Ti(C,N)/AISI tool steel	PVD	3.9–4.2	−6.2 to −6.8	Zoestbergen *et al.* (2002)
TiCN/tool steel	PVD	3.2	−3.7 to −4.5	Bromark *et al.* (1994)
TiC–C/HSS	PVD	1.8–2.6	−0.5 to 7	Wiklund *et al.* (1999c)
CrN/tool steel	PVD	4.1	−2.1 to −2.6	Bromark *et al.* (1994)
CrN/tool steel	PVD	4	−1	Wiklund *et al.* (1999b)
CrN/CC	PVD	4.4	−0.3	Nordin *et al.* (1999)
Cr_2O_3/Cr/steel			−1	Komvopoulos (1991a)
Al_2O_3/Al			−3	Komvopoulos (1991a)
Al_2O_3/alloy MA956	Oxide surface	5	−3.7 to −3.9	Bull(1997)
TiB_2/stainless steel	PVD	2	−10	Wiklund *et al.* (1999b)
TiB_2/?			−6.8	Berger *et al.* (2000)
TiC/CC	CVD	4	+0.2	Wiklund *et al.* (1999b)

(Continued)

Table 3.6. Continued

Coating/substrate	Deposition process	Coating thickness (μm)	Characteristic residual stress (GPa)	Reference
Ti/			0	Berger *et al.* (2000)
W–N/			−1.6	Gachon *et al.* (1999)
DLC/Si			−0.5 to 3.6	Kodali *et al.* (1997)
DLC/Si			−1.7 to 3.6	Nastasi *et al.* (1999)
DLC a-C/Si	PVD	0.4	−12.5	Bhushan (1999a)
DLC a-C:H/Si	PVD	0.4	−0.6 to −2.0	Bhushan (1999a)
DLC/			−1.8	Blug *et al.* (2001)
DLC/	CVD		−0.4 to −2	Gupta and Meletis (2004)
DLC a-C	PVD		−0.9 to −2.8	Mounier *et al.* (1997)
DLC a-C	PVD		−0.1 to −0.9	Mounier *et al.* (1996)
DLC–Si	PVD		−1 to −3.2	Damasceno *et al.* (2000)
DL(C:H:Si:O)/			−0.5 to −1.4	Scharf *et al.* (2003b)
Diamond/CC	CVD	5 − 10	−2	Larsson *et al.* (1996); Wiklund *et al.* (1999b)
Diamond/	CVD		−1.7 to −5.8	Silva *et al.* (2003)
Organic–inorganic film/ float glass	Sol-gel spin	2–20	+0.1 to 0	Malzbender *et al.* (2000a)
Multilayers				
TiN/CrN/CC	PVD	3.3	−2.1	Nordin *et al.* (1998)
TiN/CrN/CC	PVD	2.5–5	−1.3 to −3.2	Nordin *et al.* (1999)
TiN/MoN/CC	PVD	3.5	−6.1	Nordin *et al.* (1998)
TiN/NbN/CC	PVD	2.9	−1.1	Nordin *et al.* (1998)
TiN/TaN/CC	PVD	3.5	−3.1	Nordin *et al.* (1998)
TiN/TaNx	PVD	1.0–1.4	+3.5 to −1.5	Nordin (2000)
TiN/TiAlN/AISI	PVD	2.8–3.2	−2.5 to −9.7	Zoestbergen *et al.* (2002)

CC = cemented carbide; HSS = high-speed steel; − = compressive stresses

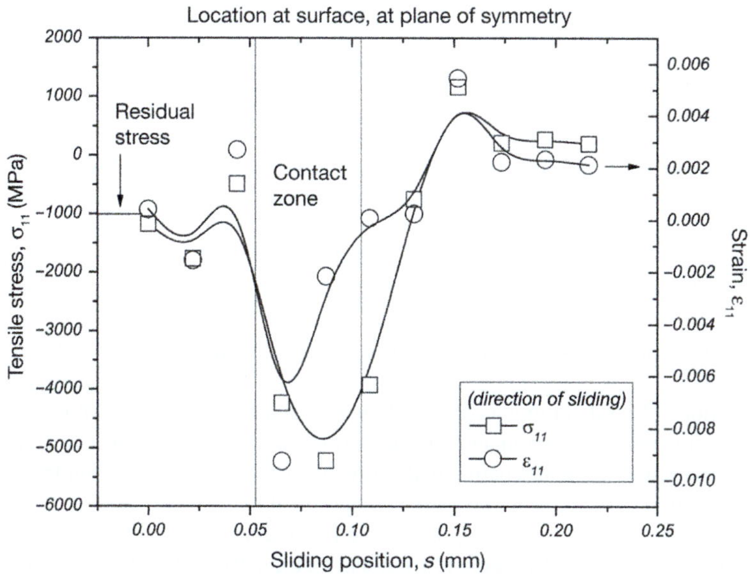

Fig. 3.47. Stresses in sliding direction at the symmetry line along the groove on the top of a TiN coated steel surface in a scratch test with spherical diamond tip sliding from left to right. The contact zone extends from about $s = 0.05$ to 0.09 mm.

substrate. The strains go through similar large changes as shown in the figure. The figure illustrates well the very radical changes both in stress level and strain that every single point that travels under the sliding tip experiences. It also shows that due to these changes the 1 GPa compressive residual stresses before the contact will be relaxed and not appear any longer after the contact.

3.2.10.1 Thermal stresses

Thermal stresses are residual stresses that originate from a change in temperature when a coated surface is used at a different temperature to that of coating deposition. These are a result of the differences in the thermal expansion coefficients of the coating and the substrate. If the coating is thin compared to the substrate, the thermal expansion of the substrate will be relatively unaffected by the presence of the coating and essentially all mismatch will be accumulated as thermal stresses in the coating. The difference in temperature can originate from the environment or from the heat generated in a rubbing contact. The latter is important from a tribological point of view and may even result in temperature differences between the coating and the substrate. The change in stress due to temperature change in a coated surface (the thermal stress $\Delta\sigma_{th}$) can be calculated from the equation

$$\Delta\sigma_{th} = [\Delta T \cdot (\alpha_s - \alpha_c) \cdot E_c]/(1 - \nu_c) \qquad (3.34)$$

where ΔT is the temperature change, α_s and α_c are the thermal expansion coefficients of the substrate and the coating, E_c is Young's modulus for the coating and ν_c is Poisson's ratio of the coating (Kramer, 1983; Musil, 2005).

The thermal and mechanical consequences of sliding, including heating, friction, deformation and wear, originate within the real area of contact. The contact spots are subject to mechanical loading by shear and normal pressure and to thermal loading resulting from temperature gradients due to frictional heating or temperature differences between the contacting bodies. The thermo-mechanical stress and strain problem for coated surfaces can be studied by three different approaches. An analytical solution has been presented by Ju and Chen (1984) and Ju and Liu (1988), a half analytical and half numerical approach by Leroy *et al.* (1989, 1990) and a finite element method (FEM) approach by Kennedy and Hussain (1987), Gunnars and Alahelisten (1996) and Ye and Komvopoulos (2003c). The thermomechanics of single and multilayer films has been analysed by Cammarata (2004) and Dunn and Cunningham (2004).

Based on a rapidly traversing friction source model Ju and Chen (1984) show that the thermal effect on the stress field is considerable. Substantial reductions in the thermal stress can be achieved by decreasing the thermal diffusivity, by decreasing the modulus of elasticity of the surface layer, or by designing the layer thickness such that the interface is away from the depth of the maximum temperature gradient.

FEM calculations of hard ceramic coatings (Kennedy and Hussain, 1987) showed that the stress field around a sliding coating is dominated by the thermal contribution from frictional heating, which tends to cause tensile stresses in the coating in the direction of sliding, which might lead to a low-cycle fatigue crack initiation by increasing the thermo-mechanical tensile stress in the coating.

The thermal stresses in the coating may be reduced by reducing the normal load or velocity, thus achieving a lower coefficient of friction and frictional heating effect. On the other hand, the heating during sliding may also actually reduce the thermal stresses if the deposition has been carried out at higher temperatures. A reduction in coating thermal stress may be achieved by using a thicker coating or a coating with lower modulus of elasticity or higher thermal conductivity (Kennedy and Hussain, 1987). An increase in the thermal expansion coefficient of the coating would result in significantly lower tensile stress in the coating but increased plastic deformation. The properties of the substrate are of broadly equal importance and thus a reduction in stresses can be achieved in both coating and substrate by having a decreased coefficient of thermal expansion, decreased modulus of elasticity and increased thermal conductivity of the substrate material.

Three-dimensional finite element analysis of an elastic–plastic coated surface under thermo-mechanical surface loading has been studied both for flat and patterned surfaces by Ye and Komvopoulos (2003c) and Gong and Komvopoulos (2004a). The coating thickness exhibited a more pronounced effect on the temperature rise at the coating surface and the coating/substrate interfaces than the coating thermal conductivity. Frictional heating and surface traction had a significant intensifying effect on the stress field. The likelihood of yielding and cracking in the coated surface increased with decreasing coating thickness. The location of the maximum stresses and strain is controlled by the thermal conductivity of the coating, the coefficient of friction, the contact interface compliance and the applied load. Sliding intensifies the temperature field causing the maximum temperature to appear at the trailing edge of the contact region. The resulting temperature gradients lead to the development of a high thermal tensile stress slightly below the trailing edge of the contact region, which is considered to be responsible for the thermal cracking in the wake of the sliding microcontacts.

Leroy *et al.* (1989) modelled a moving heat source over a coated surface and showed that the substrate temperature and stress field were hardly influenced by the coating but that even a slight difference in the coefficients of thermal expansion α_c and α_s for coating and substrate, respectively, generated tension in a thin coating if $\alpha_c > \alpha_s$ and the temperature increased. The von Mises stress in the coating was minimum for a given α_c/α_s ratio which varied with the properties and the thickness of the coating. High von Mises stresses were obtained when the coating was rigid ($E_2/E_1 > 1$) or the coefficient of thermal expansion of the coating was higher than that of the substrate. In this case stresses rapidly increased with the coating thickness.

The influence of coating thickness on the von Mises stresses at the surface of a hard titanium nitride (TiN) layer on a steel substrate subject to a heat flux was shown to be insignificant, as illustrated in

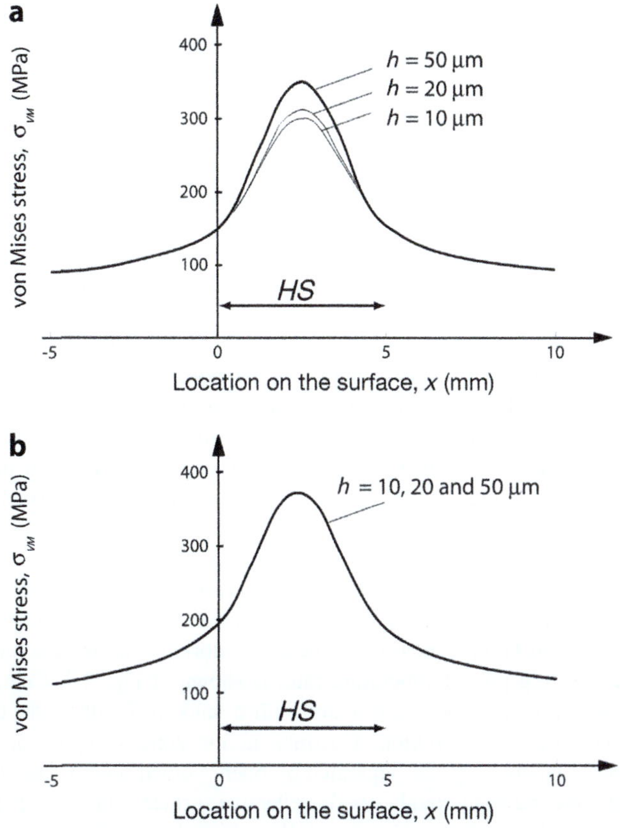

Fig. 3.48. The effect of coating thickness, h, on the distribution of von Mises thermal stresses at different locations along the surface, (a) at the interface between a TiN layer and a steel substrate and (b) at the surface of the TiN layer. *HS* shows the width of the constant heat source on the surface (after Leroy *et al.*, 1989).

Fig. 3.48b. The heat source was 5 mm in width and had a sinusoidal power distribution with a maximum of 10^6 W/m^2. At the interface between the TiN coating and the substrate, higher von Mises stresses were obtained for thicker coatings as shown in Fig. 3.48a.

The influence of coating thickness on maximum surface temperature depends on the thermal conductivity of the coating material. TiN, for example, has a relatively low conductivity and thus the maximum surface temperature increases with increasing coating thickness while the effect of thickness is small for tungsten carbide (WC) with a higher thermal conductivity (Leroy *et al.*, 1989).

Thermal stresses for 2 to 15 μm thick diamond coatings on curved cemented carbide substrates were calculated by FEM by Gunnars and Alahelisten (1996). The stresses were normally very high, with tensile stresses up to 1 GPa and compressive stresses up to 2 GPa, and their influence on cracking of the coating was considered to be substantial.

3.2.10.2 Intrinsic stresses

Intrinsic stresses are residual stresses that originate from the manner of growth of the film or from structural changes due to sliding. The former are for vacuum-deposited coatings typically due to the accumulated effect of the crystallographic defects that are built into the coating during deposition. In soft, low melting point materials like aluminium, bulk diffusion tends to relax the internal stresses and to prevent their accumulation. However, the diffusion processes can also cause defects to form. In hard, high melting point materials like chromium deposited at low temperatures, intrinsic stresses accumulate and tend to dominate thermal stresses. Stress cracking and buckling are commonly observed (Thornton and Hoffman, 1989; Thompson and Carel, 1996; Cammarata, 2004; Musil, 2004; Pauleau, 2008).

The growth-originated stresses are similar to the intrinsic stresses that are formed in the bulk material during cold working, with strains associated with various lattice defects created by the deformation. The density of defects that are formed in the film during deposition can be two orders of magnitude higher than those produced by the severest cold-work treatment of a bulk material. The intrinsic stresses can be in the range of 0.1 to 3 GPa and are therefore comparable to the yield strength of most materials (Thornton and Hoffman, 1989).

The intrinsic stresses in hard PVD coatings are generally compressive and strongly dependent on the process parameters, especially on the energy of ion bombardment on the growing film. The creation of internal stress has been explained by Knotek *et al.* (1991) by a phenomenological model describing the interaction of bombarding ions from the plasma and coating ions, whose mobility is restricted by their position on the substrate or previously deposited layers in the coating. Ions or atoms of the vaporized or sputtered material will occupy sites that would normally be too small for them, due to their high kinetic energy or as a result of impacting with other high-velocity particles (Fig. 2.9).

The intrinsic stresses in vacuum-deposited coatings are related to the microstructure and thus sensitive to the deposition conditions. A schematic representation of the dependence of microstructure in sputtered coatings on the pressure of the argon working gas and the substrate temperature and coating material melting point temperature ratio is shown in Fig. 2.7. Coatings in the zone T (transition) region have a dense fibrous structure with a smooth, highly reflective surface. Large intrinsic stresses can form in this region. Coatings in the zone 1 region are characterized by a structure consisting of tapered crystals separated by open, voided boundaries. The structure is too porous to support stresses and has a rough, poorly reflective surface. The zone 2 structure consists of columnar grains separated by distinct, intercrystalline boundaries. Recovery limits the intrinsic stresses. Zone 3 is characterized by a range of conditions where bulk diffusion has a dominant influence on the final structure of the coating. Recrystallization occurs if sufficient strain is built into the coating during deposition. Recovery and recrystallization limit the intrinsic stresses (Thornton and Hoffman, 1989).

For hard TiN coatings in the thickness range of 4.6 to 28 μm, which exhibit compressive stresses, Kennedy and Tang (1990) found that the residual stresses were beneficial to the stability of coatings because the tensile stresses that occurred in the coatings during sliding were reduced. The tendency for spalling and delamination of coatings tends to increase with increased coating thickness as stresses build up.

For even harder CVD diamond films deposited on polycrystalline tungsten substrates Guo and Alam (1991) found that the residual strain in the film was non-uniform and tensile in nature. The magnitude of strain increased with substrate temperature and was less than the thermal mismatch between diamond and tungsten between room and deposition temperature.

The modulus of elasticity influences the direction of crack propagation in coated surfaces with residual stresses. This was shown by Drory *et al.* (1988) in an analytical and experimental study. For films having a moderately high Young's modulus, cracks close to the interface divert into the substrate; this was observed, for example, for chromium films on silicon and on silicon oxide. For low Young's modulus films, such as polymers, brittle substrate cracks tend to be attracted toward the interface and the film cohesion is more likely to be affected by the interface. When the substrate is relatively tough, such as a metal, the crack would be constrained to propagate along the interface or in the film close to the interface, and decohesion is determined by a mixed mode fracture resistance of the interfacial zone. Drory *et al.* (1988) also showed that the existence of an asymptotic limit for plane cracks allows the definition of a critical film thickness, below which complete decohesion of films is inhibited.

3.2.11 Influence of interface cracks

Fracture near the surface often starts from imperfections in the material that form the nucleus for fracture initiation and propagation. Common imperfections are broken bonds on the interfaces, and voids, inclusions and dislocations in the layer. These imperfections have a disturbing effect on the stress field and the strains generated in the material.

Analytical solutions for coated surfaces with a flaw or a crack have been developed by a number of authors starting from Erdogan and Gupta (1971a, b) and reviewed by Hills *et al.* (1990). Many of the solutions are limited in use, because of the small number of cracks analysed, usually only one, and assumptions on the location of displacement zones along the crack faces.

An attempt to solve the problem of a spall more typical to those observed experimentally has been presented by Breton and Dubourg (1991a, b). Using a half analytical and half numerical model based on the dislocation theory they simultaneously analysed the combination of an interfacial crack that propagates at the interface and a surface breaking crack that propagates normally to the interface. The energy release rate was determined at the crack tips. With the model they could show that the crack interaction favours propagation and that the surface breaking crack propagates down to the interface and the interfacial crack propagates at the interface in a direction opposite to that of the load displacement.

A theoretical model for multiple fatigue cracks situated close to a loading zone has been developed by Dubourg and Villechaise (1992) and Dubourg *et al.* (1992). They considered straight arbitrarily oriented cracks and tested the method for up to five cracks without any indication of a limitation in crack number. For the analysis they identified five interaction mechanisms which depend on the distance between cracks, their relative position with respect to the loading zone, the interfacial crack coefficient of friction, and loading mode conditions: i.e. the step effect, the plug effect, the tilt effect, the stretch effect and the stretch–tilt effect as shown in Fig. 3.49.

The surface cracking in a multilayer coated surface containing a crack perpendicular to the surface due to repeated sliding of a rigid counterbody was analysed using linear elastic fracture mechanics and finite element method by Gong and Komvopoulos (2004b). They conclude that the surface cracking is controlled by the tensile fracture mode, longer cracks produce significantly higher stress intensity factor, K_I, values, higher friction increases significantly both K_I and K_{II} due to the strong effect of surface shear traction on the crack tip stresses and an increase in friction at the crack interface promotes stress relaxation. The initial crack growth was found to occur at an angle of about 10° from the original crack plane, independent of the initial crack length. At the first few crack growth increments the crack growth paths exhibited a common deviation from the original crack plane by about 57°.

3.3 Surface Fracture and Wear Products

Load-induced deformations of a coated surface generate surface stresses that may be high enough to result in surface fracture. Fracture is the separation of a body into two or more pieces in response to an imposed stress. The wear process is related to fracture since material is separated from the surface. Wear always starts with breaking of cohesive or adhesive material bonds. This is called surface cracking when it is observed on a macro- or microlevel. However, the crack starts with the breaking of atomic bonds and material separation taking place already on a molecular level. Cracking and

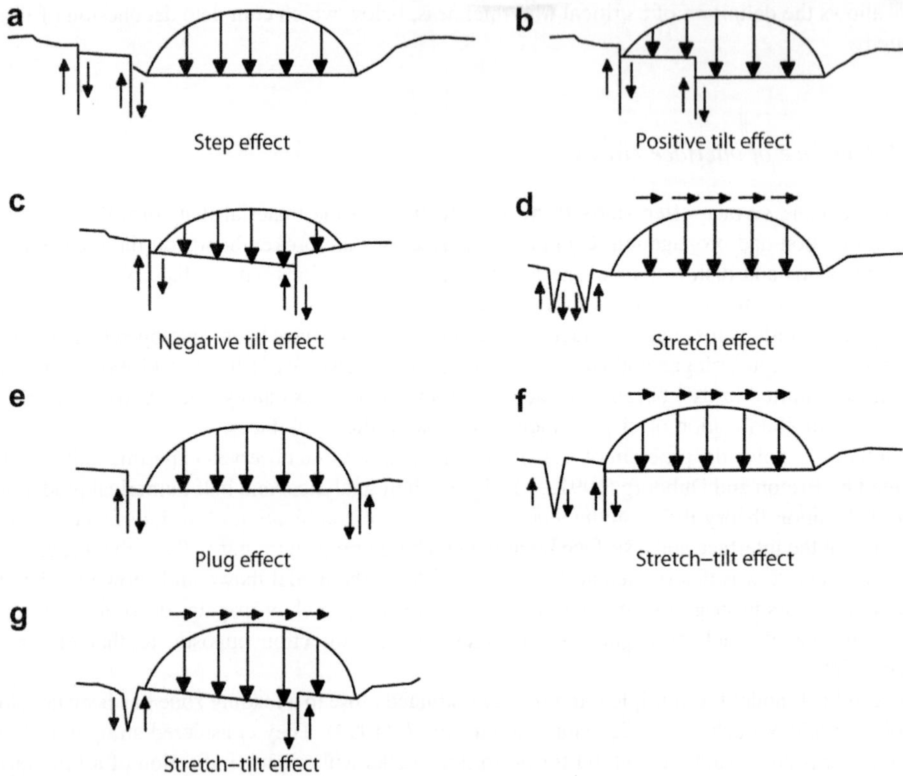

Fig. 3.49. Different interaction mechanisms for a multiply cracked surface (after Dubourg *et al.*, 1992).

material fracture is typical for brittle materials like ceramic coatings. Any fracture process involves three steps: crack nucleation, crack propagation and failure by material separation. Fracture is classified as being ductile or brittle. In ductile fracture the body breaks after extensive plastic deformation has occurred in a relatively large material volume. Pure brittle fracture happens without significant permanent deformations.

3.3.1 Crack nucleation

The structures of engineering materials are, with few exceptions, never perfect. They include defects like dislocations, cavities, pores, voids, contamination, flaws and submicroscopic cracks, as shown in Fig. 3.50. The crack nucleation and initiation may take place according to several different mechanisms depending on the material structure, geometrical features and state of applied and residual stress. One possible mechanism for crack nucleation is that dislocations in a work-hardened region are piled up by shear stresses, resulting in a crack. Cracks can nucleate at sites where deformation is hindered, e.g. by inclusions, second-phase particles or fine oxide particles. In high purity metals voids can form at grain boundary triple points. With very ductile materials the cracks grow essentially by a process of void unification, where the voids are located along slip planes (Dieter, 1986; Jahanmir, 1980; Tangena, 1987).

The nucleation of cavities around non-deforming particles smaller than approximately 1 μm appears to be controlled by the build-up of internal stresses as a result of incomplete plastic relaxation

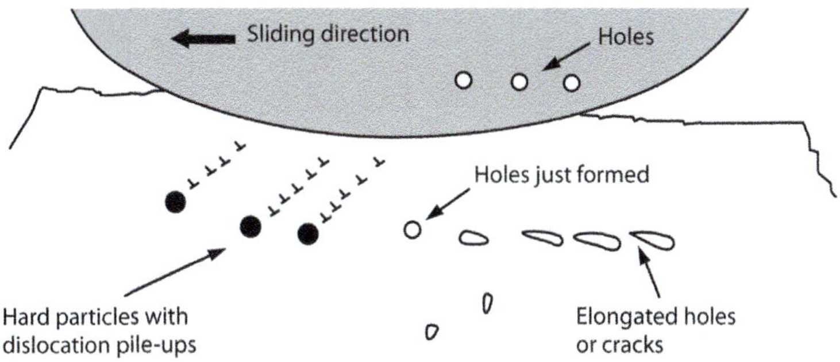

Fig. 3.50. The initial stage of the process of wear particle formation by shear deformation of voids in the material near the surface (after Suh, 1973).

due to local dislocation storage and hence local work hardening around the particles (Goods and Brown, 1979). In ceramic surface coatings with a columnar structure the crack nucleation is expected to take place between the columnar grains.

Based on macroscale experimental observations in rolling contact tests with bearing steel materials, Nélias *et al.* (1999) report that the cracks are either initiated from subsurface inclusions or surface microcracks. They claim that the initiation and propagation of inclusion-initiated cracks is the main mode of subsurface originated cracks in rolling bearings. Under the load a stress concentration is built up around inhomogenities like inclusions and primary carbides. This is due to the elastoplastic mismatch between the inhomogenities and the martensitic matrix. Once the yield stress is reached, plastic strain is induced in a small volume surrounding the inclusion. Under repeated action the dislocations will move back and forth and accumulate. This process leads to localized changes in the steel microstructure and a crack is nucleated in this domain once a critical density is reached.

Surface microcracks are typically generated in the discontinuities of the surface topography such as at grinding marks or short wavelength surface roughness in operating conditions like rolling contact fatigue (Nélias *et al.*, 1999). The process of initiation of surface microcracks and formation of microspalls along grinding marks is shown schematically in Fig. 3.51a. Microcracks may also propagate parallel to the surface at a few micrometres depth along the machining direction, leading to the formation of microspalls as shown in Fig. 3.51c. These processes have mainly been observed in pure rolling. Sliding may play a significant role, as it produces transverse microcracks at an angle of 30 to 40° to the direction of the friction force. The cracks are located mainly on the top of large roughness peaks as shown in Fig. 3.51b. There are indications that the local friction coefficient at the top of asperities may be large enough to initiate transverse microcracks. Such microcracks are at the origin of microspalls, usually wider and deeper than those observed under pure rolling conditions. Depending on the combination of load movement, friction direction and crack orientation a lubricant may help to propagate surface cracks by hydraulic pressure effects.

The crack nucleation conditions in partial slip fretting wear contact conditions has been analysed by Fouvry *et al.* (2004a) and they found only a very minor effect of normal force on the nucleation condition. The crack nucleation started at a constant threshold value of the tangential force.

3.3.2 Surface crack propagation

The modern fracture mechanics approach is based on linear elastic fracture mechanics (LEFM) and is widely used to explain the process of crack propagation and to predict crack growth (Anderson, 1995; Shukla, 2005) typically for systems with cracks longer than about 1 mm. However,

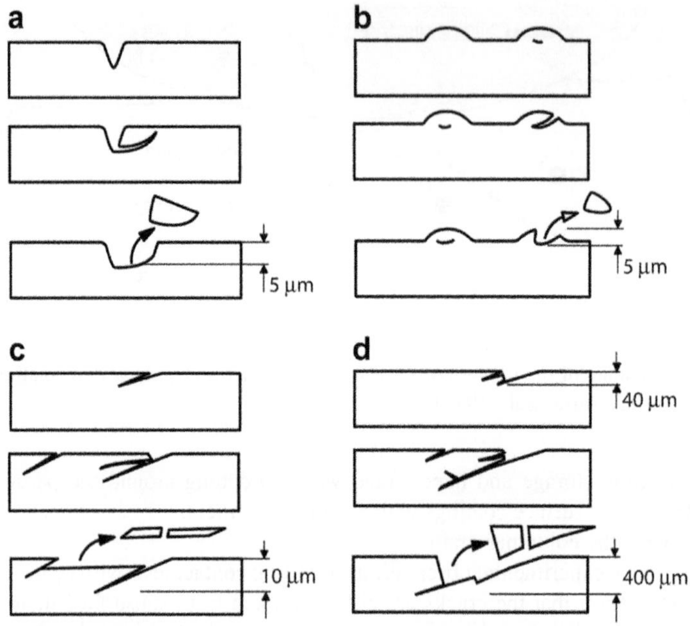

Fig. 3.51. Schematic stages of (a) microspall formation along grinding marks (transverse cut), (b) microspall formation below longitudinal roughness of large wavelength (transverse cut), (c) microcrack and microspall formation (longitudinal cut) and (d) surface-initiated deep spall formation (longitudinal cut). Direction of friction force is from left to right in c and d (after Nélias *et al.*, 1999).

for the case of a loaded very thin coated surface there exists no general theory or model for crack initiation or crack propagation.

Three regions of crack growth, (1) the LEFM region, (2) the physically short crack region and (3) the microstructurally short crack region, are illustrated in a diagram with logarithmic stress range, $\Delta\sigma$, as a function of logarithmic crack length, a, as first suggested by Kitagawa and Takahashi (1976) and shown in Fig. 3.52. In contrast to long fatigue crack propagation, there exists no adequate definition of the growth mechanism, the crack tip stress–strain fields, or the interaction with microstructural features for short fatigue crack propagation, which would be applicable to the dimensions in very thin coated surfaces subject to loading.

The models and measurement methods developed in fracture mechanics, such as the LEFM techniques for expressing the fracture toughness of a material, are in most cases applicable for large cracks longer than 1 mm, or with special modifications for physically short cracks in the range of 100 μm to 1 mm.

Cracks less than 100 μm are considered as microstructurally short cracks because their dimensions are in the order of the grain size in metals. At these scales the material no longer behaves as a homogeneous continuum because the crack growth is strongly influenced by microstructural features. In this region crack growth is influenced by grain size and grain boundaries, inclusion spacing, precipitate spacing, micro-asperities, macro-asperities, etc. The crack growth is often very sporadic, such that it may grow rapidly in certain intervals and then be followed by crack arrest when it meets barriers such as grain boundaries and second phase particles (El Haddad *et al.*, 1979; Meguid, 1989; Anderson, 1995; Pippan, 1999). Microstructurally short cracks in a plain specimen may not affect

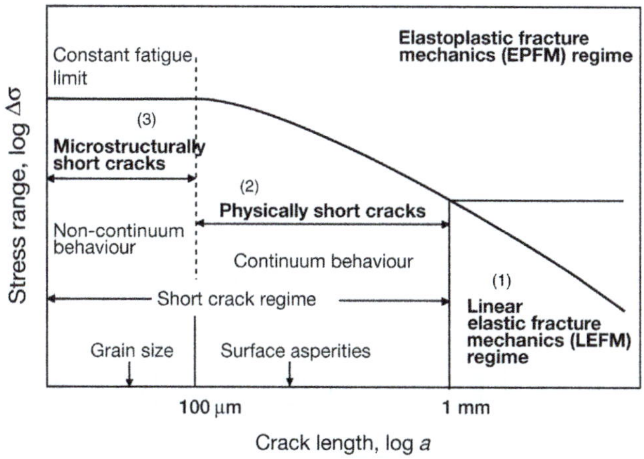

Fig. 3.52. The Kitagawa–Takahashi curve, illustrating the three regions of different crack propagation mechanisms: (1) the LEFM regime, (2) the physically short crack regime and (3) the microstructurally short crack regime (from Meguid, 1989).

its fatigue limit. Fracture mechanical aspects on short cracks and their behaviour have been reported by, e.g., Miller (1982), Miller and de los Rios (1986) and Tanaka (1989).

There is a wide range of conditions, sources and features that may influence the process of short crack initiation and propagation. Many of them have been collected to a schematic classification applied for metallic components, shown in Fig. 3.53. The process is influenced by material parameters, the geometrical features and the state of stress, both applied and residual.

The tribological loading conditions will influence the crack propagation mechanism and the type of fracture failure of the surface. In a sliding contact when a slider moves over a brittle solid surface, short cracks will form near the surface in its wake behind the contact zone as shown in Fig. 3.54. A periodic array of cracks appears in the wake of the slider, with a critical spacing between the cracks (Bower and Fleck, 1994).

For a rolling contact it has been argued that the cracking of the surface either originates from the surface or from somewhere below the surface. Both hypotheses have supporters and it seems likely that either of them can apply depending on the contact conditions. When a first crack has been formed at the surface of a lubricated rolling contact, such as in a rolling bearing, the lubricant will be pressed into the crack and creates a hydraulic pressure that enhances the crack growth (Murakami *et al.*, 1997). A similar crack growth process takes place beneath the pitch line of a gear tooth in a gear transmission. In addition to surface cracks, vertical material ligament collapse may have a crucial role in pitting-type wear particle formation when debris is formed by the surface material above a lateral crack close to the surface (Ding *et al.*, 1996). The vertical surface material ligament collapse is actually a crack initiation process due to compression from a surface normal load and not to tangential tension at the surface as more frequently observed.

3.3.3 Crack growth in coated surfaces

It follows from the above that mechanisms of crack initiation and crack growth in surfaces covered by thin PVD and CVD coatings and used in tribological contacts are not well understood. The problem is especially challenging for two reasons. The surface is a compound of a substrate and one or several layers having different material properties, and the dimensions of the layers, typically 0.5 to 5 μm, is four to five

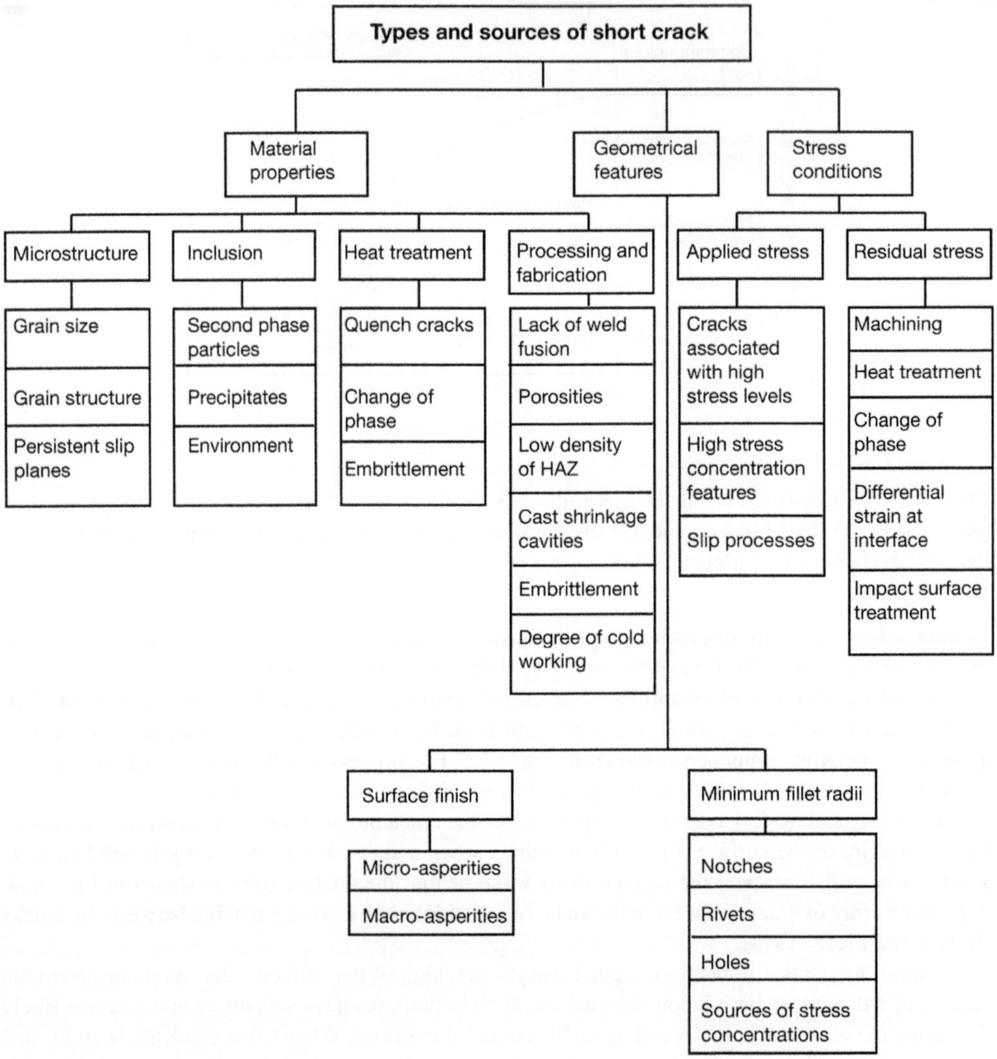

Fig. 3.53. Classification of different sources and features for short crack initiation and propagation in metallic components (from Meguid, 1989).

orders of magnitude smaller than the size range of the well-established linear elastic fracture mechanics concepts. Actually, the dimension of the coating overlaps the size range, 0.1–1 μm, where material scientists normally define crack nucleation and initiation to take place (Miller, 1982).

There are some special features in the cracking of a surface covered by a thin coating. As a crack is initiated in a thin coating it normally goes all the way down to the coating/substrate interface. It may be only one single crack but often a whole system of cracks is generated. When reaching the interface the crack may change direction and propagate along the coating/substrate interface or it may penetrate deeper down into the substrate (y-direction in Fig. 3.55). The crack growth in the lateral direction (z-direction in Fig. 3.55) is important since it influences the conditions for coating delamination.

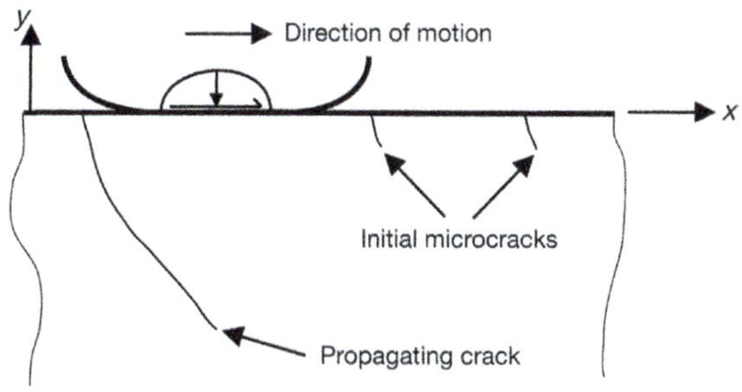

Fig. 3.54. A propagating crack is formed in the wake behind a slider moving over a brittle flat surface.

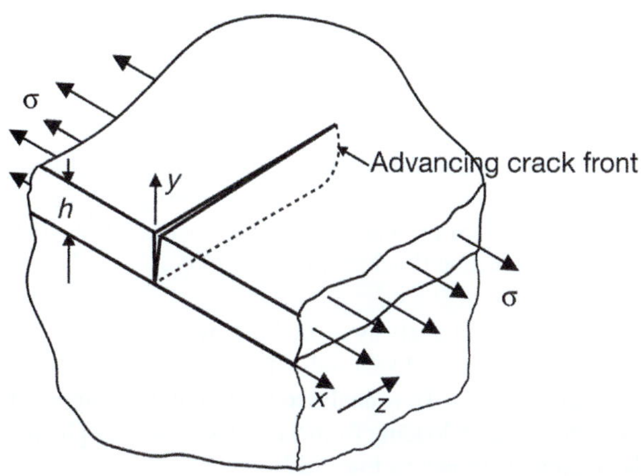

Fig. 3.55. Crack propagation in a thin coating loaded by stress field in tension.

Analytical and numerical solutions of fully and partially cracked thin coatings in a stress field in tension and unstressed have been presented by Beuth (1992), Menčik (1996) and Oliveira and Bower (1996). Still, the use of these solutions in practical applications is not easy and straightforward due to the mathematical complexity.

Surface cracking of a multilayered surface due to repeated sliding of a rigid asperity was analysed by linear elastic fracture mechanics and the finite element method by Gong and Komvopoulos (2004b, 2005). They studied the propagation of a partial perpendicular crack in the first coating layer and came up with several very interesting conclusions. The significantly higher values, by an order of magnitude, of the tensile stress intensity factor, K_I, than those of the shear intensity factor, K_{II}, indicate that surface cracking in a multilayered surface due to sliding is controlled by the tensile fracture mode. Longer surface cracks produce significantly higher K_I values while higher friction increases both K_I and K_{II} significantly due to the strong effect of the surface shear traction on the crack-tip stresses. The increase of friction at the crack interface promotes stress relaxation but the effect on K_I is negligible. The initial crack growth was found to occur at an angle of ~10° from the original crack plane, independent of the initial crack length (see propagating crack in Fig. 3.54). After the first one to three crack growth increments, the crack growth paths are almost parallel to each other, exhibiting a common deviation from the original crack by ~57° and this was reported to be in fair agreement with experimental observations.

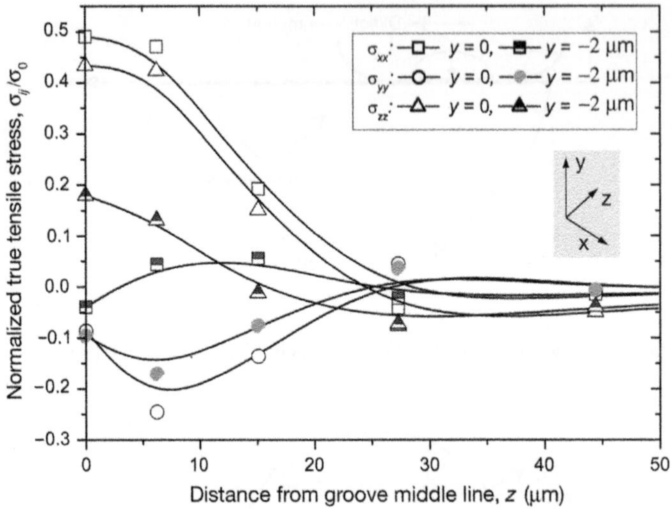

Fig. 3.56. True *tensile stress* components normalized with the substrate yield strength on a coated surface ($y=0$) and at the coating/substrate interface ($y=-2\,\mu m$) in three coordinate axis directions for a flat steel surface coated with a 2 μm TiN coating and loaded at 11 N with a 200 μm radius rigid sliding sphere. The sliding takes place in the *x*-direction (Laukkanen *et al.*, 2006).

Based on the stress and strain FEM analysis carried out by Holmberg *et al.* (2006a, b) referred to earlier section 3.2.8, Laukkanen *et al.* (2006) did a fracture mechanics evaluation of thin coated surfaces by boundary element analysis. They analysed the fracture conditions in the contact of a rigid diamond sphere sliding with increased loading from 5 to 50 N on a flat high-speed steel surface coated with a 2 μm thick TiN coating, as shown in Fig. 3.37.

In addition to calculating the first principal stresses on the top of the surface and at the coating/substrate interface, Laukkanen *et al.* (2006) also calculated the true stresses in tension and in shear. Figure 3.56 shows the true stress tensile components normalized with the substrate yield strength on a coated surface and at the coating/substrate interface in sliding. It is interesting to see that the highest tensile stress levels are at the surface in the *x*- and *z*-directions. This explains why the first cracks observed in these conditions are typically angular cracks at the groove edge behind the slider. It also clearly indicates that the surface cracking starts from the top of the coating and advances down to the interface and into the substrate. The true shear stresses in Fig. 3.57 show that the shear stress level is lower than the tensile stresses and that the highest shear stresses are generated on top of the coating in the surface plane (*zx*-plane).

Laukkanen *et al.* (2006) further studied the influence of different crack parameters on crack growth by calculating the stress intensity factor, *K*, along the crack front, characterized by a parametric coordinate, *s*, for the different cases shown in Fig. 3.58. The tension mode I dominates the crack growth in lateral directions even if the shear mode II seems to be important for crack growth in the vertical direction, as shown in Fig. 3.59. The influence from mode III is negligible in the analyses. From the summary of the crack parameter analysis, shown in Fig. 3.60, the following can be concluded for the case studied:

- loading modes: mode I dominates the crack growth in the lateral direction while mode II is important for crack growth in the vertical direction,
- crack orientation: cracks vertical to the sliding direction are more prone to grow compared with cracks 20° and 45° from the vertical direction,

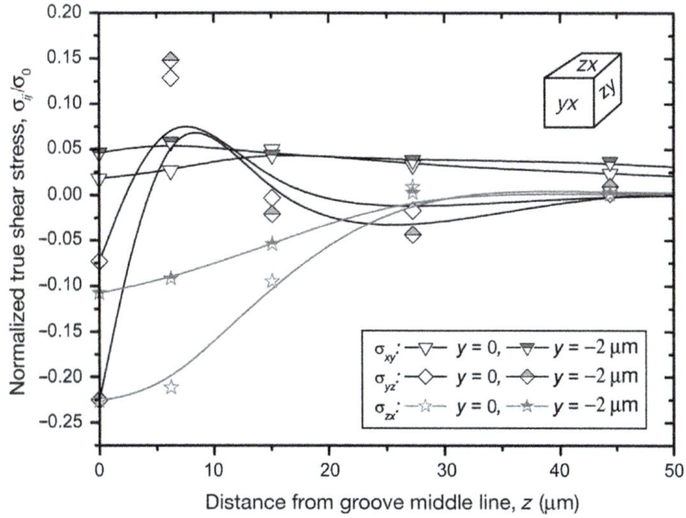

Fig. 3.57. True *shear stress* components normalized with the substrate yield strength on a coated surface ($y = 0$) and at the coating/substrate interface ($y = -2$ μm) in three coordinate planes for a flat steel surface coated with a 2 μm TiN coating. The sliding takes place in the x-direction (Laukkanen *et al.*, 2006).

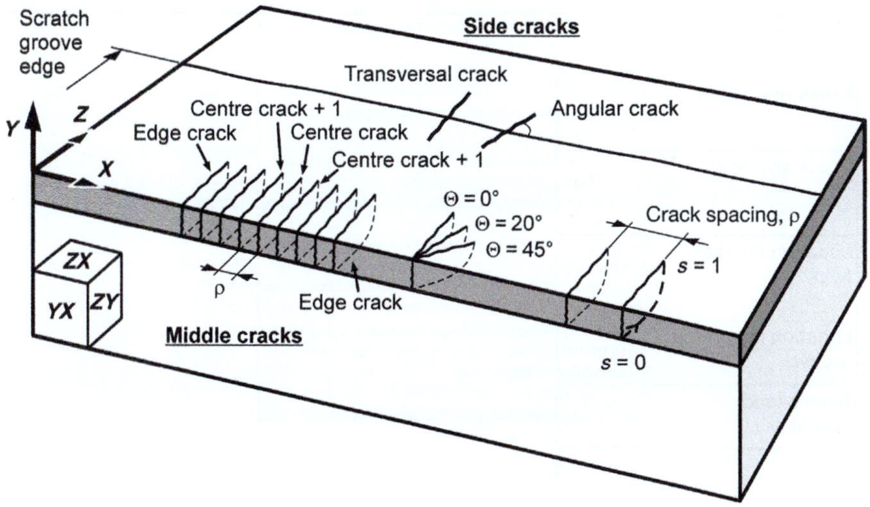

Fig. 3.58. Crack growth parameters.

- crack density: cracks with a larger crack spacing in the regular crack field, here 30 μm, are more prone to grow compared to smaller spacing, 5 or 10 μm, or single cracks,
- location in crack field: the centre crack in a crack field is more prone to grow compared to edge crack or other cracks in the field,
- location in crack groove: transverse cracks in the middle of the groove are more prone to grow compared to both transverse and angular side cracks at the groove edge, and
- tensile load biaxiality: has a marginal influence on crack growth.

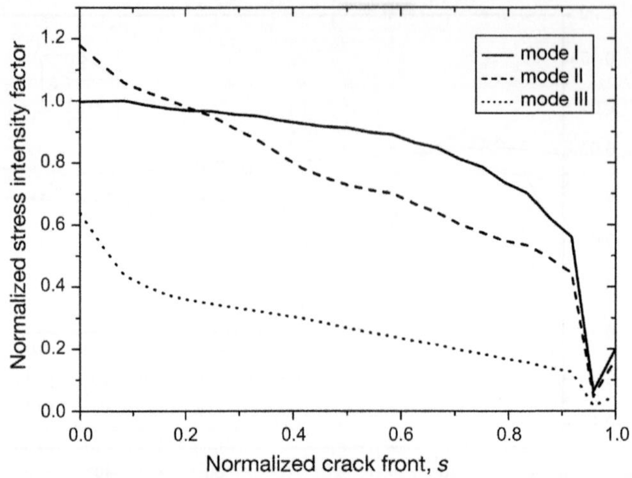

Fig. 3.59. Crack growth stress intensity factor normalized with K value at the deepest point of single crack case under uniaxial tension for the three fracture modes along the normalized crack front defined in Fig. 3.58 for a flat steel surface coated with a 2 μm TiN coating (Laukkanen *et al.*, 2006).

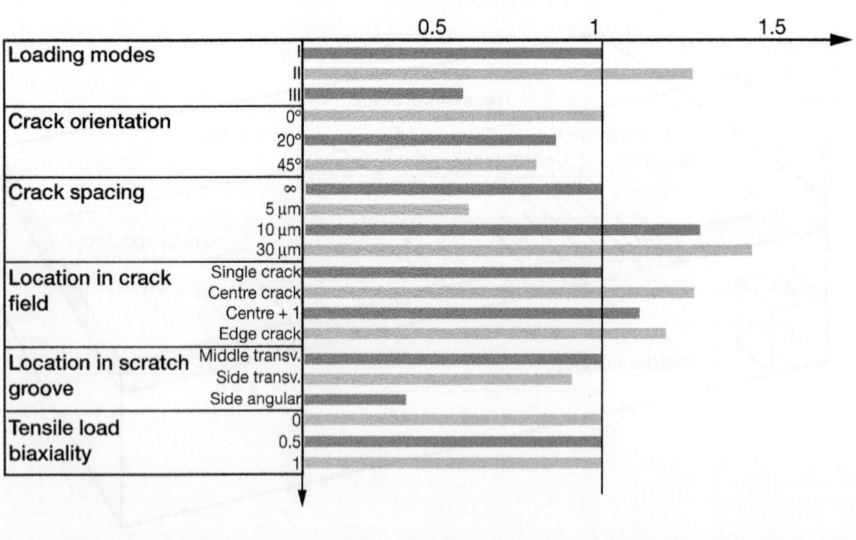

Fig. 3.60. Summary of the effect of different crack parameters on normalized stress intensity factor for a flat steel surface coated with a 2 μm TiN coating and loaded at 11 N with a 200 μm radius rigid sliding sphere (Laukkanen *et al.*, 2006).

In ceramic coatings the thickness and the grain size are important parameters for crack propagation. In indentation tests of 1 to 10 μm thick Al_2O_3 coatings on cemented carbide substrates Yuan and Hayashi (1999) found that the critical load for generation of radial cracks decreased with increasing coating thickness and that the grain size of the coatings decreased with the thickness. A coarser grained microstructure generally displays a lower cohesive strength as compared with a fine grained microstructure. Thus the fracture strength of the coating increases with decreasing grain size and decreasing thickness.

Nanocrystalline-amorphous TiC-amorphous carbon (a-C) composite coatings have been produced by Voevodin and Zabinsky (1998a, b) with a combination of high hardness and toughness. In their coating design, nanocrystals of 10 to 50 nm size were encapsulated in an amorphous phase, separating the nanocrystals at a distance of 5 nm from each other. This permitted the generation of dislocations and grain boundary microcracks and nanocracks, which terminated in the surrounding amorphous matrix. This surface design achieved a self-adjustment in composite deformation from elastic to plastic at loads exceeding the elastic limit, as opposed to deformation from elastic to brittle fracture.

For amorphous diamond-like carbon (DLC) coatings in the thickness range of 0.7 to 2 μm it has been shown that the hardness and apparent fracture toughness of the coating/substrate combination, measured by Vickers indentation, depend on the thickness of the coating (Kodali *et al.*, 1997). The hardness and the apparent fracture toughness of the coating, both of which influence the wear, increased with increasing coating thickness. The hardness and the thickness influenced the initiation of cracks, whereas the residual stress in the film influenced the propagation of cracks.

In a study of very thin DLC coatings in the thickness range of 3.5 to 20 nm, Sundarajan and Bhushan (1999) conclude that the formation of cracks depends on the hardness and fracture toughness of the coating. They suggest that the observed non-uniform failure depends on variations in the coating properties at different locations in the material volume. Surface cracks are developed in weaker regions with lower fracture toughness. As cracks propagate they are forced to expand within the weak region, as the neighbouring strong regions inhibit extensive lateral crack growth. Because of this, cracks propagate down to the interface, where, aided by the interfacial stresses, they get diverted along the interface just enough to cause local delamination of the coating. Simultaneously, the weakened regions experience excessive ploughing. Thus weaker regions fail while stronger regions remain wear resistant. The propagation of cracks along the coating/substrate interface is suppressed due to the strong adhesion of the coatings, otherwise coating delamination would take place.

3.3.4 Toughness and fracture toughness in coated surfaces

There is no widely established way of expressing the resistance to fracture of a coated surface. However, the study of toughness and fracture toughness for thin coatings and coated surfaces is receiving wider support and this is certainly recommended, since they are well-defined and widely used parameters for the crack growth resistance of bulk materials. In the literature there are several other expressions used for the property of a coated surface to resist fracture, such as cracking resistance, higher/lower critical load, etc. The drawbacks of these are that they are related to some certain experimental device and lack generality.

Not to confuse the classical concept of fracture toughness it should be remembered that *fracture toughness* describes the strain energy release rate in the growth of a pre-existing crack, whereas *toughness* describes the total energy input required to cause the material to fail, i.e. both to create the crack and to enable the crack to grow.

When using the expression of fracture toughness of a coated surface it is important to note the distinction between the fracture toughness of the coating and that of the coated surface. Most methods to estimate the fracture toughness of a coated surface relate the K_C value to the coating–substrate system and not purely to the coating. This may not be a drawback since normally it is the fracture toughness of the coated surface, the system that is really of interest and should be optimized for a certain purpose.

The most common ways to estimate the fracture toughness of a coated surface are first to empirically deform the surface by indentation, bending, buckling, scratching or tensile testing and then calculate the K_C value analytically or based on finite element analysis (Zhang *et al.*, 2005a). Some values of the fracture toughness of coated surfaces are presented in Table 3.7. Most authors have calculated the fracture toughness related to cracks going vertically through the coating, K_C, but interestingly Diao *et al.* (1994b) and Blug *et al.* (2001) have also estimated the interface fracture toughness, $K_{C,\ int}$, which is the adhesion-related resistance to crack propagation along the coating/substrate interface.

The four-point bending test has been used to rank the cracking resistance of different PVD coatings. Wiklund *et al.* (1997a) found a significantly higher cracking resistance for their PVD deposited 2.5 μm thick CrN coatings on a hardened HSS substrate compared to 4.3 μm thick TiN coatings on the same substrate. Zoestbergen and De Hosson (2002) found a higher cracking resistance for PVD-deposited TiN coatings compared to TiN/TiAlN multilayer coatings and the lowest cracking resistance for Ti(C,N) coatings all deposited on AISI tool steel substrates and with coating thicknesses of 2.5 to 4.4 μm. Very interestingly they also report that they found no measurable influence of surface roughness on crack density in the roughness range of $R_a = 0.01$ to 1.0 μm for the TiN coated specimen.

3.3.5 Crack patterns

Continuous loading of a surface coated with a thin brittle coating results in a first crack, then crack growth, new cracks, growth of the new cracks and crack merging according to a pattern that is specific for the loading conditions and the surfaces in interaction. Different typical crack patterns are often produced by the scratch test where a spherical stylus is sliding with increasing load over a coated surface. Figure 3.61 shows how first angular cracks are formed at the groove edge behind the sliding sphere (a), some time later at higher loading they are accompanied by parallel cracks along the groove edge (b) and the angular cracks are growing and merging to transverse semicircular cracks (c). A new stage is reached at higher loading when the first coating plates are spalled off due to adhesion failure (e) and finally large areas of the coating are spalled off resulting in coating breakthrough (f). The stress conditions resulting in this process have been analysed by Holmberg *et al.* (2003) and are described in section 3.2.8.

The fracture behaviour is different depending on whether the coating is hard or soft, ductile or brittle or elastic and whether the substrate is hard or soft in relation to the coating. Five such cases where the coating and substrate deformations are shown and the cracking is indicated both for well-adhered and poorly adhered coatings are shown in Fig. 3.62. A soft or elastic substrate results in larger deformation of the coating. If the coating is brittle it cannot deform accordingly and will crack. The coating thickness also influences the toughness of the coating.

The loading of a coated surface results in deformation and failure both in front of the sliding contact and behind it. Spalling and buckling typically take place in front of the slider as a result of the compressive stress field preceding the moving slider and the cracks formed are concentric with the front part of the sphere. Spalling is a result of total detachment of the coating from the substrate and flaking off. Conformal and tensile cracking takes place when the coating remains fully adherent. The conformal cracking takes place in front of the slider. Tensile cracking takes place behind the slider and the cracks are formed concentric with the trailing part of the sphere, as shown in Fig. 3.63. Some of the cracking failures taking place in front of the slider may go through a change in shape as they are ridden over and possibly embedded by the sliding sphere and their appearance in the groove bottom may be quite different from the original failure shape (Burnett and Rickerby, 1987b).

The crack pattern in a scratch test has been studied and classified by Bull (1991, 1997). He shows five different crack patterns for brittle failure modes: tensile cracking, Hertz cracking, recovery spalling, compressive spalling and gross spalling. In ductile failure he identifies four modes: tensile cracking, conformal cracking, buckling and spalling. Somewhat surprisingly it has been experimentally

Table 3.7. Some values for fracture toughness of coated surfaces from literature.

Coating/substrate	Deposition process	Coating thickness (μm)	Empiric assessment	Fracture toughness, K_C ($MPa \cdot m^{-2}$)	Reference
TiN/cemented carbide	CVD	1.5–11.5	Indentation	2.7 ($K_{IC,int} = 0.08$)	Diao *et al.* (1994b)
TiN/steel S6-5-2	PACVD	4.8	Bending + pc	8.7	Jaeger *et al.* (2000)
TiN/high-speed steel	PVD	2	Scratch test	4–7	Holmberg *et al.* (2003) and Laukkanen *et al.* (2006)
TiCN/steel S6-5-2	PACVD	4.5	Bending + pc	7.9	Jaeger *et al.* (2000)
TiAlN/steel S6-5-2	PACVD	7.9	Bending + pc	3.8	Jaeger *et al.* (2000)
TiAlSiN/WC–Co	PVD	2	Indentation	1.55–2.1	Nakonechna *et al.* (2004)
Al_2O_3/cemented carbide	CVD	1.7–12.5	Indentation	2.2 ($K_{IC,int} = 0.05$)	Diao *et al.* (1994b)
Al_2O_3/mild steel	Plasma spray	200–300	Indentation	4.5	Xie and Hawthorne (1999)
TiB_2/cemented carbide	Magnetron sputtering PVD	60	Vickers indentation	4.1	Berger (2002)
DLC/silicon	Sputtering	1.36–1.92	Indentation	1.14–1.57	Kodali *et al.* (1997)
DLC (a-C)/silicon	Cathodic arc PVD	0.4	Nano-indentation	10.9	Li *et al.* (1997); Li and Bhushan (1998)
DLC (a-C:H)/silicon	Ion beam PVD	0.4	Nano-indentation	4.9–5.4	Li *et al.* (1997); Li and Bhushan (1998)
DLC/silicon	PIIP PVD	1.27–1.92	Micro-indentation	10.1	Nastasi *et al.* (1999)
DLC/steel 100Cr6	RF-PECVD	2	Indentation	1–2 ($\Gamma_{int} = 0.55$)	Blug *et al.* (2001)
W/steel	Sputtering	2–16.4	Tensile test	1–2.5	Harry *et al.* (1998)
W(C)/steel	Sputtering	1.8–16	Tensile test	0.2–1	Harry *et al.* (1998)
SiC/silicon	APCVD	3	Indentation	0.78	Li *et al.* (1999)
NiP/aluminium		9	Indentation + pc	15	Tsui *et al.* (2001)
Organic/inorganic film/float glass	Sol-gel	3–20	Indentation	0.07–0.26	Malzbender *et al.* (2000a)

$K_{C,int}$ = interfacial fracture toughness; Γ_{int} = interface toughness; pc = pre-cracked specimen

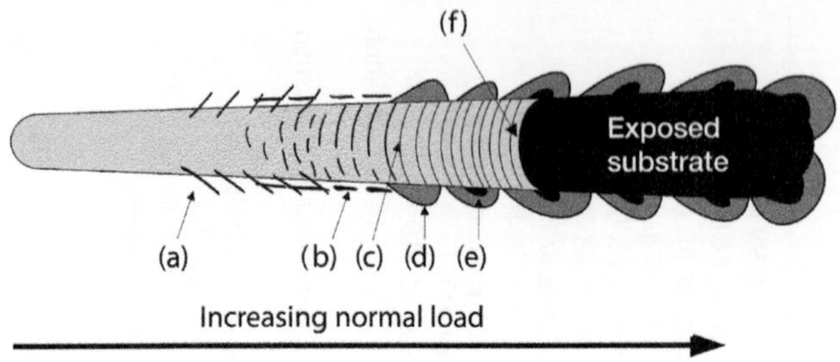

Fig. 3.61. The surface cracks generated in a scratch test track can be classified as (a) angular cracks, (b) parallel cracks, (c) transverse semicircular cracks, (d) coating chipping, (e) coating spalling and (f) coating breakthrough (modified after Larsson *et al.*, 1996).

shown that the surface roughness of the coated surface does not have much influence on the cracking behaviour in scratch testing (Bromark *et al.*, 1992). The cracking of ceramic coatings taking place in a scratch test and classification of the related crack patterns are described in detail in some standards (European Standard, 2000; ASTM, 2005).

There are a number of investigations published where the crack pattern has been experimentally demonstrated and studied both by indentation and scratch testing. They follow the general outline given above and some of them, related to different coating materials, are listed below:

- TiN coating crack pattern described by von Stebut *et al.* (1989), Olsson (1989), Hedenqvist (1991), Bull (1991), and Larsson (1996),
- TiNAlN coating crack pattern described by Lau *et al.* (2007),
- CrN coating crack pattern described by Wiklund *et al.* (1999a, b, c),
- CrO_3 coating crack pattern described by Wang *et al.* (1998),
- Al_2O_3 coating crack pattern described by Kato *et al.* (1991), Wang *et al.* (1998), Yuan and Hayashi (1999) and Xie and Hawthorne (2000),
- SiC coating crack pattern described by Sundararajan and Bhushan (1998),
- DLC coating crack pattern described by Ronkainen *et al.* (1997) and Podgornik (2000),
- W coating crack pattern described by Ligot *et al.* (2000) and Lau *et al.* (2007),
- Al coating crack pattern described by Lau *et al.* (2007),
- nanocrystalline ZnO coating crack pattern described by Prasad and Zabinski (1997),
- TiN duplex coating crack pattern described by Batista *et al.* (2002),
- TiN/TiC multilayer coating crack pattern described by Sue and Kao (1998),
- TiN/CrN multilayer coating crack pattern described by Nordin *et al.* (1999), and
- sol-gel coating crack pattern described by Malzbender *et al.* (2002).

3.3.6 Coating to substrate adhesion

The strength of the adhesive bonds between the coating and the substrate are crucial for the functionality of a coated surface. When they break, some part of the coating will be detached and the substrate material is exposed to the countersurface in the tribological contact. This is called coating delamination, spalling or flaking. The adhesive failure of a coated surface is basically that a crack is formed in the interface between the coating and the substrate and it grows and merges with vertical through-coating cracks and a coating material flake is liberated and removed.

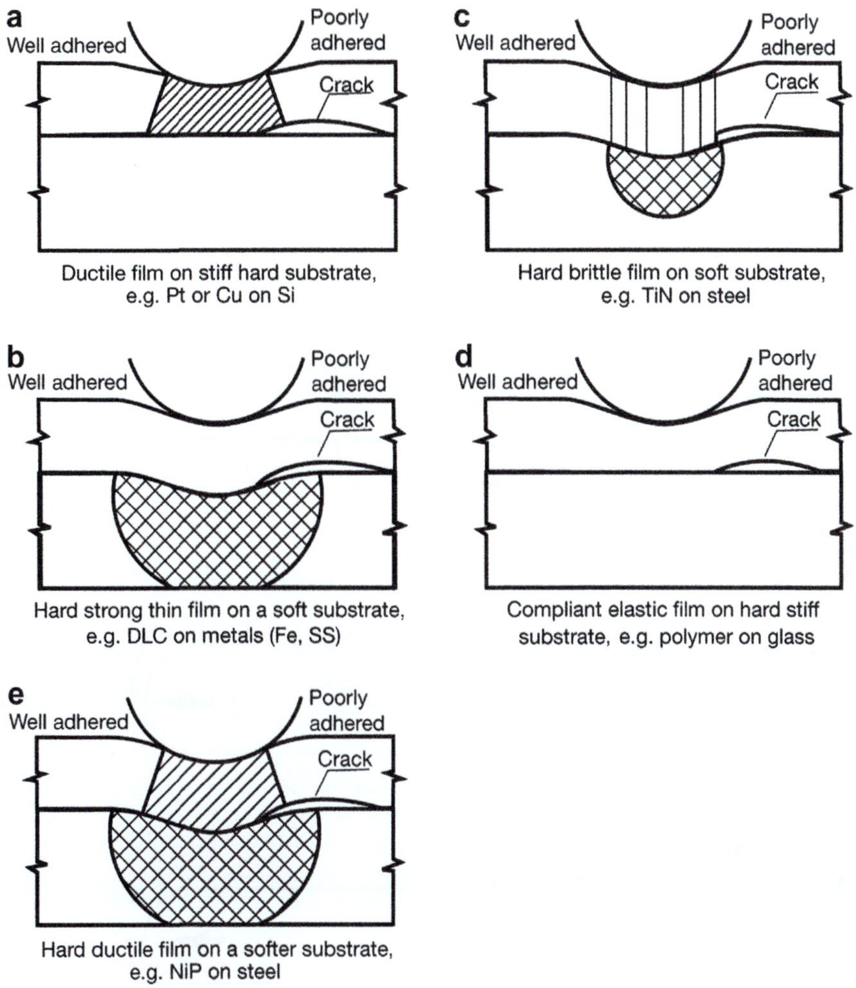

Fig. 3.62. Schematic cross-sections of various coating–substrate materials combinations and the influence of interfacial adhesion on cracking in indentation (after Swain and Mencik, 1994).

Adhesion is defined as the molecular attraction between bodies in contact. According to the fracture mechanics approach the *adhesion energy* or *interfacial fracture energy* of a coating–substrate system is the energy needed to propagate a crack along the interface between the coating and the substrate by a unit area, denoted Γ_{int}. The growth of a crack along an interface can occur under various loading modes, in particular the opening mode I, $K_{I,int}$, and the shearing mode II, $K_{II,int}$. The *adhesion strength* is the stress that the interface can withstand before fracturing. The resistance to crack growth until fracture is called the *interfacial fracture toughness*, $K_{C,\ int}$ (see Table 3.7 in section 3.3.4).

The adhesion strength between the coating and the substrate is governed by various routes for crack propagation depending on the elasticity, hardness and ductility of the coating and the substrate and the coating thickness, and typically results in a spalling or a buckling failure (Swain and Mencik, 1994; Bull, 1997; Diao and Kandori, 2006). In practical systems it is not always clear if a failure at the

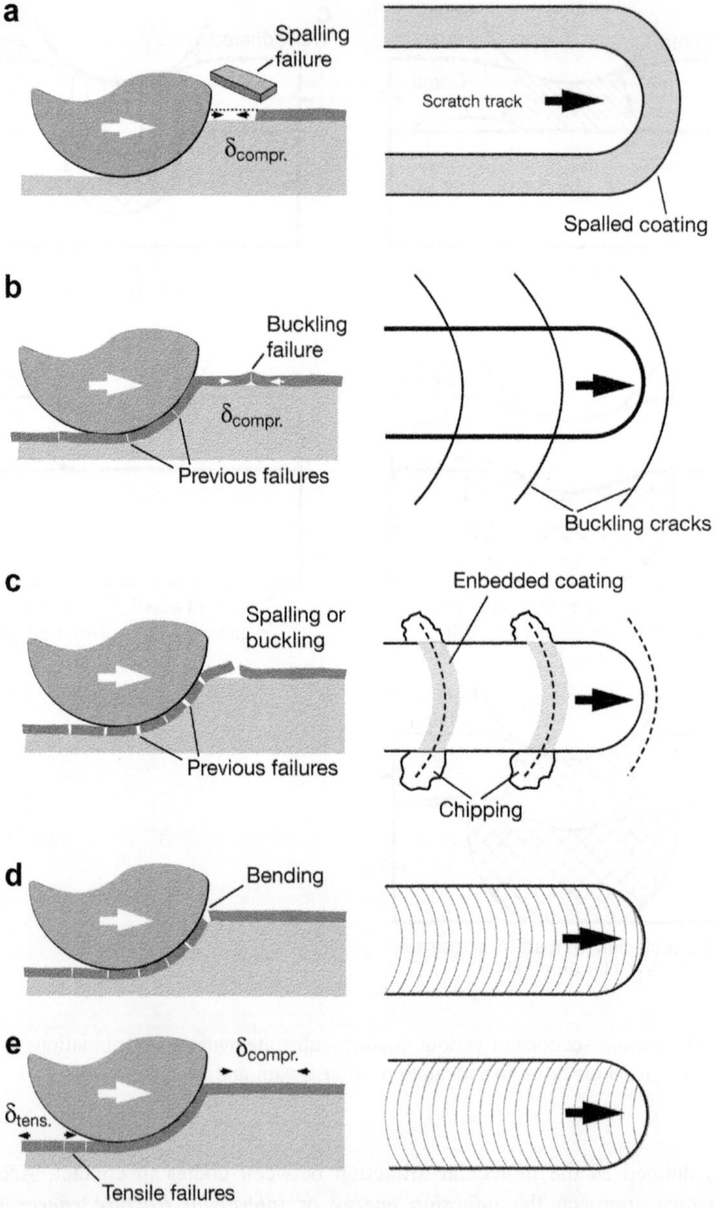

Fig. 3.63. Schematic representation of some of the more common coating failure modes in the scratch test in profile and plan views: (a) spalling failure, (b) buckling failure, (c) chipping failure, (d) conformal cracking and (e) tensile cracking (after Burnett and Rickerby, 1987b).

coating/substrate interface is a truly adhesive failure or a cohesive failure in the coating or in the substrate along and very close to the interface. So in reality cohesive failures at the subinterface between the bulk material and the real interface to the coating are often referred to as adhesive failures (Malzbender *et al.*, 2002). Even if the theoretical definition of adhesion is clear, the assessment of adhesion is very difficult. This is reflected by the fact that there are more than 350 different tests used to characterize adhesion (Mittal, 1995, 2001). Assessment of coating adhesion is discussed in more detail in section 5.3.3.

3.3.7 Debris generation and particle agglomeration

The four wear mechanisms, adhesive, abrasive, fatigue and chemical wear, shown in Fig. 3.11, describe four ways of material liberation from the surface and wear debris generation. The contact conditions influencing the mechanisms vary with regards to adhesion between the two surfaces, hardness of the surfaces, topography, loading, surface fatigue strength, environment chemistry and type of motion. Wear in practical conditions can be classified as:

- severe wear, resulting in wear debris of size dimensions 20–200 μm,
- mild wear, resulting in wear debris of size dimensions 0.01–1 μm, and
- nanowear, resulting in wear debris of molecular and atomic size.

A variety of conditions and mechanisms are involved at the initial stage of the wear process. In sliding contacts with metallic surfaces the plastic strain close to the surface and mechanical mixing at asperity level are crucial (Rigney *et al.*, 1984; Kato and Adachi, 2001a; Cartier, 2003; Scherge *et al.*, 2003a) while the hardness and surface toughness are crucial for ceramic surfaces (Terheci, 1998; Kato and Adachi, 2001a). Normally there is an increase in the coefficient of friction due to wear particles appearing in the contact causing abrasive scratching and particle interlocking (Hwang *et al.*, 1999) but the wear process may also result in surface smoothening and the build-up of lubricious surface layers that decrease friction (Singer, 1998). Particle size and sharpness are major factors influencing friction and wear in contacts with abrasive particles (Stachowiak *et al.*, 2005).

In a coated surface wear debris generation may take place on three different levels: within the coating, as delamination of the coating and as detachment of both coating and substrate material, as shown in Fig. 3.64.

In the case of a diamond sphere sliding on an Al_2O_3 coated flat WC alloy substrate, Kato *et al.* (1991) showed that the wear first takes place as microcutting by asperities but after repeated sliding cycles it changes to coating delamination. The cracking starts with parallel cracks on both sides of the contact groove that grow downwards into the substrate, as shown in Fig. 3.65. The process of crack growth and coating delamination is shown in Fig. 3.66. The width of the delaminated flake is as large as the contact groove. The subsurface lateral cracks propagate parallel to the surface, and then the coating is partially lifted by introducing branched cracks under the combined effect of frictional stress and residual stress. Pre-induced radial cracks on both sides of the groove and tensile cracks at its centre, which are normal to the sliding direction, encourage the lifting and detachment of the delaminated film from the surface.

Cyclic accumulation of the plastic deformation, called ratchetting, in the subsurface within the coating has been suggested as one mechanism to cause coating wear (Yan *et al.*, 2000). Based on ratchetting failure criterion they suggest a wear equation and find that the wear rate increases with contact pressure and depends on coating thickness and the roughness of the counterpart surface. A reduction in friction coefficient and an increase in the coating strain hardening behaviour can considerably improve the wear resistance. Detailed observations of the wear debris generation process has been reported for DLC coatings by Singer *et al.* (2003), Park *et al.* (2004) and Achanta *et al.* (2005); for TiN coatings by Achanta *et al.* (2005); for BC coatings by Singer *et al.* (2003); for CN coatings by Wang and Kato (2003a, b, c) and Wang *et al.* (2003); and for Pb–Mo–S coatings by Singer *et al.* (2003).

Once wear debris has been detached from the surface it will move in the contact. It may be embedded in a softer surface, embedded in the surface topography, involved in the contact shear and deformation process, interlocked between asperities, cause scratching of a softer surface, attach to

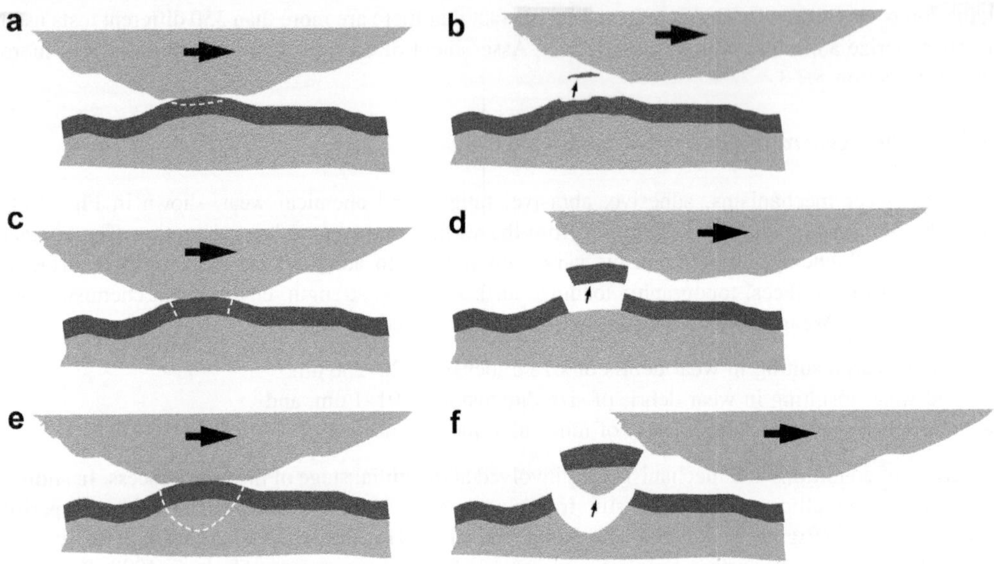

Fig. 3.64. Three levels of wear debris generation in a coated surface: (a, b) flaking of coating material from the top surface of the coating, (c, d) delamination or removal of coating pieces so that the substrate is exposed and (e, f) detachment of debris containing both coating and substrate material.

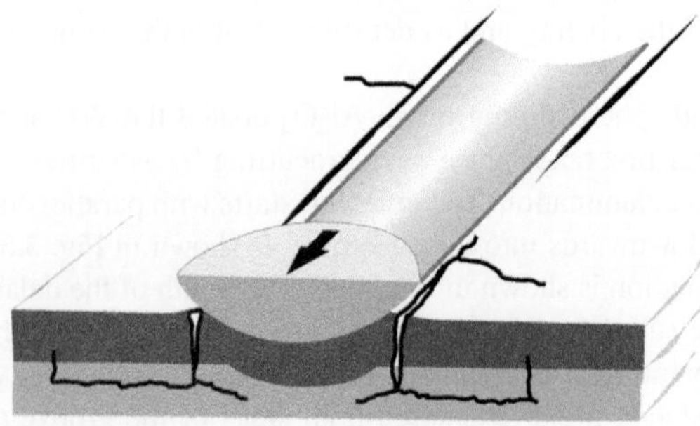

Fig. 3.65. Schematic presentation of early cracks in sliding of a spherical diamond repeatedly on an Al_2O_3 coated flat WC alloy surface (after Kato *et al.*, 1991).

either of the surfaces, be involved in a tribochemical process to form a new compound or just be removed from the contact zone. As the debris has just been detached from a solid it contains pure uncontaminated surfaces and may thus be chemically and physically active to form bonds with other surfaces close to it. This is enhanced by the high local temperatures at the asperity level often involved in sliding processes.

In these active contact conditions, often agglomeration of wear debris to larger material pieces is observed (Rigney *et al.*, 1984; Akagaki and Kato, 1987; Berthier, 2005). In sliding contacts with materials of different hardness (lead, zinc, aluminium, copper, nickel, titanium and steel), Hwang *et al.* (1999) observed different wear particle agglomeration behaviour depending on hardness and sliding direction. Smaller wear particles had a stronger tendency to join together and form larger ones.

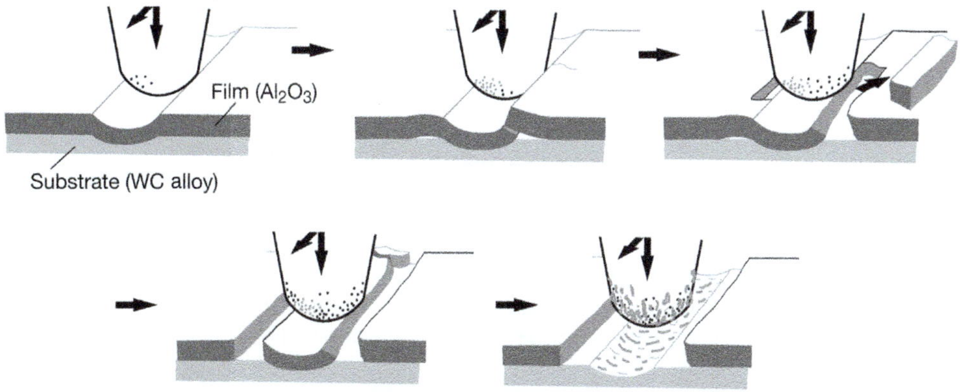

Fig. 3.66. Schematic presentation of crack growth and coating delamination in sliding of a spherical diamond repeatedly on an Al_2O_3 coated flat WC alloy surface (after Kato *et al.*, 1991).

Particles of soft and ductile metals had a stronger tendency to agglomerate than those of hard and brittle metals. There was a difference in agglomeration behaviour depending on whether the sliding was unidirectional or reciprocating and lower friction and wear was measured in the cases of reciprocating sliding. Particle agglomeration was not just limited to one location but occurred simultaneously over a distributed area and the observed particle or flake sizes were in the range of 100–600 μm.

The mechanism of agglomeration is complicated due to adhesion and/or mechanical interlocking. The agglomerated particles can act as larger particles in the contact, detach to either of the surfaces or be rejected from the contact (Yuan *et al.*, 1999). Wear particle agglomeration has been reported in sliding tests with DLC coated surfaces (Park *et al.*, 2004; Wu *et al.*, 2005), TiN coated surfaces (Achanta *et al.*, 2005) and silver coated surfaces (Yang *et al.*, 2001, 2003).

3.3.8 Transfer layers

It is common that wear particles attach to one of the contacting surfaces and continue to be active in the tribological process in the form of a transfer layer. The formation of a transfer layer is the result of a process that starts with surface deformation, then surface cracking, wear debris generation, often also particle agglomeration and finally formation of a transfer layer, as shown in Fig. 3.67. The material transfer layers generated in sliding coated contacts are generally some few micrometres in thickness but may vary in the range of 0.01 to 50 μm (Ko *et al.*, 1992; Blomberg, 1993; Singer *et al.*, 2003; Scharf and Singer, 2003c). Material between the two sliding surfaces in the form of debris or layers is frequently called third bodies. Their rheological, flow and physiochemical behaviour in the tribological contact process has been analysed, described and modelled by Godet (1984), Descartes and Berthier (2001), Berthier (2005) and Fillot *et al.* (2007).

The material transfer mechanism is well known for polymers, e.g. polytetrafluoroethylene (PTFE), sliding on steel (see Fig. 4.3). Surface material from the polymer will wear off and attach by adhesion to the steel counterface to form an extensive PTFE film. This means that after some time of sliding, the tribological pair is actually PTFE sliding against a thin PTFE film on steel, which has very low friction. Similar mechanisms have been observed for sputtered, less than one micrometre thick, PTFE coatings on steel substrates, resulting in film-like PTFE wear debris transfer to the steel counterface; and for sputtered polyimide (PI) coated steel, resulting in fine flake-like wear debris transfer to form a polymer layer on the steel counterface (Yamada *et al.*, 1990). In rolling contact bearings in different

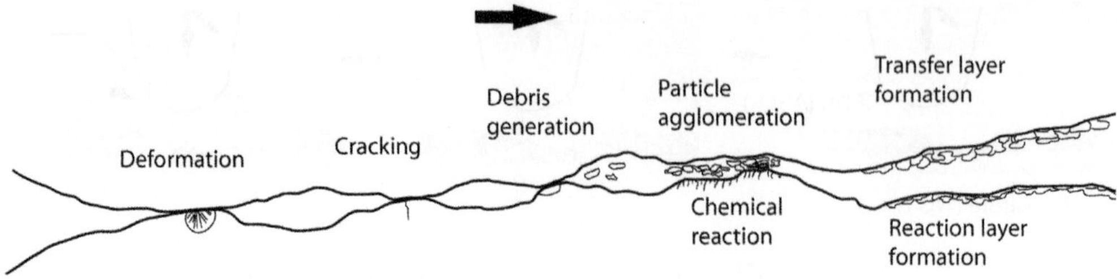

Fig. 3.67. The process of surface layer formation starts with surface deformation and continues with surface cracking, wear debris generation, particle agglomeration, transfer layer formation or reaction layer formation.

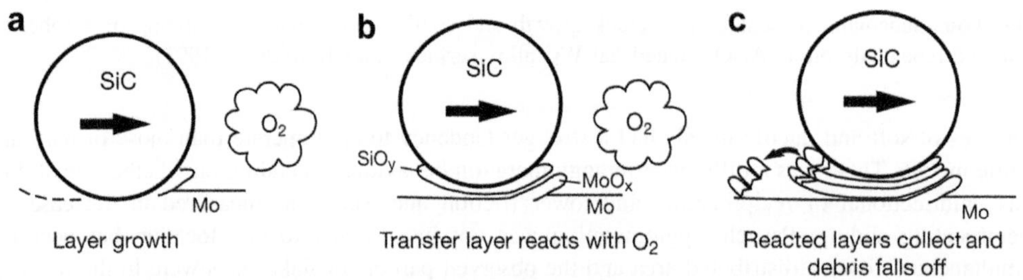

Fig. 3.68. A schematic presentation of detachment and transfer processes that produce transfer layers and debris during sliding (after Singer *et al.*, 1996c and Singer, 1998).

gas environments PTFE films have transferred from the retainer onto CaF_2 layers on the races and balls effectively protecting them from wear and increasing their endurance life (Nosaka *et al.*, 1999).

Based on observations from sliding contacts with MoS_2 and TiN coatings, Singer (1998) presents a three-stage nanoscale model for the formation of reaction layers and for the build-up of transfer layers. First, a thin layer is removed from the coating and transferred to the counterface. Meanwhile both the surface and the transfer layer can react with possible surrounding gases, forming new compounds. The first films to be transferred to the countersurface may be very thin, only of molecular dimensions. As the transfer film thickens, it is extruded from the contact area and may break up to form new wear debris. This process is then repeated and a layer-by-layer build-up takes place. It has been observed for MoS_2 coatings that particle detachment appears to be preceded by deformation, reorientation and sometimes crack propagation within the first 20–50 nm of the coating.

One process of transfer layer formation is illustrated in Fig. 3.68, reported by Singer (1998) and Singer *et al.* (1996a, c). A metallic–ceramic tribocouple of Mo and SiC slides against each other in a reactive gas of O_2 and initially the sliding seems to occur near the oxide–substrate interface (Fig. 3.68a) or at the interface between the transfer film and the track (Fig. 3.68b). Friction is determined by the shear strength of the interface. As debris builds up on the track the sliding occurs between the transfer film and a debris-covered wear track with an interfacial chemistry different from the earlier surface (Fig. 3.68c). A low steady-state friction coefficient value of $\mu = 0.01$ was achieved from the tribochemical reaction products alone. However, without continuous exposure to gas, the friction coefficient immediately rose to $\mu = 1$.

In a series of excellent detailed papers the group led by Singer has described the process of transfer layer formation and its influence on friction and wear for three low-friction coatings: 2 μm thick diamond-like carbon coating, 0.3 μm thick amorphous Pb–Mo–S coating and annealed boron carbide coating (Singer, 1998; Wahl *et al.*, 1999; Dvorak *et al.*, 2002; Scharf and Singer, 2003a, c; Singer *et al.*, 2003). They report a number of very illustrative *in situ* optical microscopy video recordings, Raman, friction and wear measurements in reciprocating sliding contacts between a transparent glass or sapphire sphere and the coated surface. Even if transfer films originate from a parent material, they do not always have the same phase or composition as the parent. With an amorphous Pb–Mo–S coating the transfer film produced was MoS_2, with a DLC coating the transfer film was a graphite-like carbon layer, and with annealed BC the transfer film was either a mix of H_3BH_3 and carbon or, when the H_3BH_3 wore away, carbon alone. An increase in friction with Pb–Mo–S and DLC coatings was observed due to the changes in interfacial shear strength. The transfer films produced with the DLC coatings were measured to be 100 to 200 nm thick. Similar transfer layer observations in sliding contacts with DLC coated surfaces have been reported by Fontaine *et al.* (2004), Konca *et al.* (2005) and Wu *et al.* (2005).

A typical transfer film build-up process is described by Singer *et al.* (2003) with the Pb–Mo–S coatings in the following way: as sliding began, a patchy transfer film attached to the contact area of the sphere; later, thicker, compact pads of debris formed at the leading/trailing edges of the contact; a few debris were also adhered to the coating; the relative motion between the sphere and the wear track on the coating took place by interfacial sliding; as the humidity rose, some of the debris patches in the transfer film began to extrude – moving at speeds of 3% of the sliding speed; loose debris were swept up from the track and collected at the leading/trailing edges of the contact, feeding additional debris into the contact. The tribochemical process converts amorphous Pb–Mo–S into crystalline MoS_2 and the transformation occurs by creating basal-oriented layers at the outermost layers of the wear track.

A typical building-up process of a transferred surface layer in contacts with steel and hard coatings such as TiN, CrN and (TiAl)N has been described by Huang *et al.* (1994). As a result of the ploughing action of hard coating asperities, slider material fragments were first removed and then adhered to some preferential sites on the sliding track of the coating. The preferential sites were the highest asperities that made earliest contact with the countersurface. Repeated sliding resulted in accumulation of fragments, which then united and formed discontinuous layers on the coating surface. After some sliding the highest asperities were covered with transferred material, which was deformed and flattened under continuous sliding, and a transferred layer was built up. Similar processes of transfer layer build-up have been observed and reported for different ceramic and steel contacts, e.g. by Andersson and Holmberg (1994).

Depending on the contact condition the formation of transfer films in the sliding contact of a titanium nitride coating against a steel slider may be very complex. Sue and Troue (1990) have described the process of minute wear fragments adhering to both surfaces, plastic deformation and strain hardening of the transferred layers and patches, cracking and oxidation of the layers, removal of the layers and patches by fracture and finally the formation of very thin films, possibly oxides, on both surfaces. In pin-on-disk experiments with a steel ball sliding on TiN coatings Wilson and Alpas (1998) observed a load effect. At low loads of about 20 N, transfer and build-up of oxidized counterface material on the coating surface and minimal damage to the coating took place; at higher loads up to 100 N increased debris transfer, and polishing and brittle spallation occurred; at even higher loads plastic deformation and ductile ploughing or smearing of the TiN coating prevailed.

3.3.9 Tribochemical reaction layers

In a tribological contact the sliding process brings energy and often high local temperatures into the contacts, and at the same time the wear process results in exposure of pure uncontaminated material surfaces to the environment. This situation is very favourable for chemical reactions to take place on the newly formed or deformed surfaces. The chemical reactions are dependent on what kind of additional material is brought to the contact in gaseous or liquid form (Ruff *et al.*, 1995; Singer *et al.*, 1996c; Ronkainen *et al.*, 1998a; Nakayama and Martin, 2006; Neville *et al.*, 2007; Lindquist, 2008).

In environments containing oxygen, such as air, a thin, typically about 1–10 nm thick, oxide layer is formed very quickly on most metal surfaces. Some oxide layers, like copper oxide, are sheared more easily than a metal, while others, such as aluminium oxide, form a very hard layer. Erdemir *et al.* (1998) studied the shear properties and lubricity of a number of oxides for high-temperature tribosystems. They conclude that a complex mixture of effects, including the kinetics of oxidation and cation diffusion rates, heats of formation, electrostatic electronegativity, surface energy, and other fundamental crystallochemical properties, may influence the adhesion and shear rheology of an oxide film forming on a sliding surface. Mechanistically, the crystallochemical properties relate strongly to the melting point of an oxide and its viscosity, the lowering of the surface energy and melting point of an oxide when a second oxide is present in the system, and the solubility, chemical interactions, and compound forming tendencies between two or more oxides. Tribologically, these phenomena can play a significant role in shear rheology, adhesive interactions, and hence the frictional properties in a sliding contact.

There are a number of reports where different aspects of oxide layer formation on tribological properties have been studied. In contacts with alumina (Al_2O_3) surfaces, Gee and Jennett (1995) found that the tribologically formed hydroxide films are much softer than the remaining alumina and that the films are liable to fracture under concentrated loading conditions. The hydroxide films were composed of small particles, about 20–50 nm in size. Erdemir *et al.* (1997) investigated the formation and self-lubricating mechanism of thin boron oxide (B_2O_3) and boric acid (H_2BO_3) films on the surfaces of boron oxide substrates. They developed a short duration annealing procedure resulting in the formation of a glass-like boron oxide layer for which they measured very low coefficients of friction down to 0.05 when sliding against a steel ball. The tendency of formation of an oxide layer on the surface of different (Ti, Al, Zr, Si)N coatings has been reported by Rebouta *et al.* (1995).

The process of metal cutting includes very high temperatures and is favourable for tribochemical surface reactions. Aizawa *et al.* (2005) demonstrated a self-lubricating process where a tribofilm works as a buffer between work material and tools in dry forming and machining to reduce wear and friction. With the aid of chlorine ion implantation into TiN and TiC coatings, *in situ* formation of lubricious intermediate titanium oxides, TiO and Ti_nO_{2n-1}, resulted in a plastically deforming tribofilm. The self-lubrication process is based on successive, *in situ* formation of lubricious oxide films. In the tribochemical reaction, the initial titanium mono-oxide films gradually transform to magneli-phase oxides by plastic deformation. Since the implanted chlorine atoms diffuse to further depths in the ceramic coatings, the gradual oxidation is sustained in tribochemical reactions in the wear track or on the flank surface.

The formation of environmentally stable and smooth third-body films on the wear surface was believed to be responsible for the improvement in room temperature tribological performance of nanocrystalline zinc oxide (ZnO) films observed by Prasad and Zabinski (1997). The friction coefficient of the ZnO films against a steel ball was between 0.16 and 0.34. Friction was in the beginning sensitive to load and sliding speed, but once a smooth wear scar was formed, it was very stable and the test conditions had no significant effect on it.

Experiments with steel sliding against a chromium nitride (CrN) coating resulted in the formation of a chromium oxide (Cr_2O_3) surface film with very good wear resistance (Lin *et al.*, 1996b). Increases in the applied load or the sliding velocity helped to form a thicker Cr_2O_3 film, thus reducing the wear. However, the growth of a thick TiO_2 surface film in a TiN coated contact under the same conditions did not help to reduce the wear loss.

Additional material, active in the chemical surface reaction, may also be brought to the contact in a liquid form. Li *et al.* (1998) investigated the sliding contact of a plasma-sprayed Cr_3C_2–NiCr surface against a TiO_2 coating under water and ethanol lubricated conditions in a block-on-ring test. They found that water deteriorated the tribological properties of both surfaces by accelerating cracking and fracture. Ethanol reduced the friction coefficient and wear of the Cr_3C_2–NiCr surface, which could be attributed to the formation of a smooth surface film mainly consisting of Cr_2O_3. However, ethanol also increased the wear coefficient of the TiO_2 coating by adsorption-induced cracking as a result of the low fracture toughness of the coating.

3.4 Tribological Mechanisms in Coated Surfaces

Coatings are widely used to control friction and wear in all kinds of contacts. Many excellent results in industrial sliding applications have been achieved. In the most successful solutions the coefficient of friction has been reduced by more than one order of magnitude, e.g. by coating a steel surface, which slides against a steel counterface, with a thin molybdenum disulphide or diamond-like carbon coating; also the wear has been reduced by several orders of magnitude.

The results have mainly been achieved by a trial and error approach concerning both the choice of material and the tribological coating parameters, such as coating thickness, hardness and surface roughness. This is due to the lack of a general theoretical concept that can satisfactorily describe the prevailing friction and wear mechanisms in a tribological contact process with coated surfaces involved. There is, however, a large amount of experimental experience about how coated surfaces behave in tribological contacts and this can be of much use in component and tool design.

3.4.1 Scales in tribology

The tribology of a contact involving surfaces in relative motion can be understood as a process with certain input and output data. Input data that are used as a starting point for the analysis of a tribological contact include the geometry of the contact both at a macro- and microscale; the material properties are based on the chemical composition and structure of the different parts involved and environmental parameters, as shown in Fig. 3.69. Other input data are the energy parameters such as normal load, velocity, tangential force and temperature.

The tribological process takes place as the two surfaces are moving in relation to each other, and both physical and chemical changes occur in accordance with physical and chemical laws with respect to the input data. As a function of time, the tribological process causes changes in both the geometry and the material composition and results in energy-related output effects: friction, wear, velocity, temperature and dynamic behaviour.

The complete tribological process in a contact in which two surfaces are in relative motion is very complex because it involves simultaneously friction, wear and deformation mechanisms at different scale levels and of different types. To achieve a holistic understanding of the complete tribological process taking place and to understand the interactions it is useful to analyse separately four different tribological effects: the macro- and microscale mechanical effects, the chemical effects and the material transfer taking place, as shown in Fig. 3.70. In addition, there has recently been an increasing interest in studying tribological behaviour on a molecular level; i.e. nanophysical effects (Bhushan, 1995, 2001a; Hsu and Ying, 2003).

The macromechanical tribological mechanisms are related to the stress and strain distributions of the whole contact, the total plastic and elastic deformations they result in and the total wear particle formation process and its dynamics. These phenomena are typically of a size level of 1 micrometre or more, up to 1 millimetre.

The micromechanical tribological mechanisms are related to the stress and strain at an asperity level, crack generation and propagation, material liberation and single particle formation. In typical engineering contacts these phenomena are of a size level of about 1 micrometre or less.

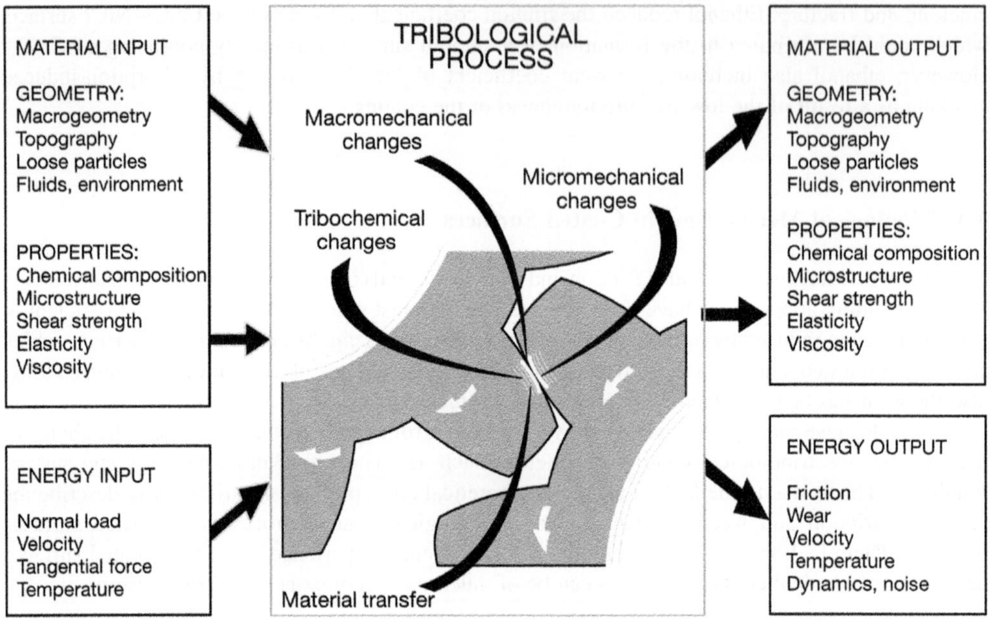

Fig. 3.69. The tribological process in a contact between two surfaces in relative motion includes mechanical and chemical changes that depend on the input data and result in changed contact conditions determined by the output data at a certain time.

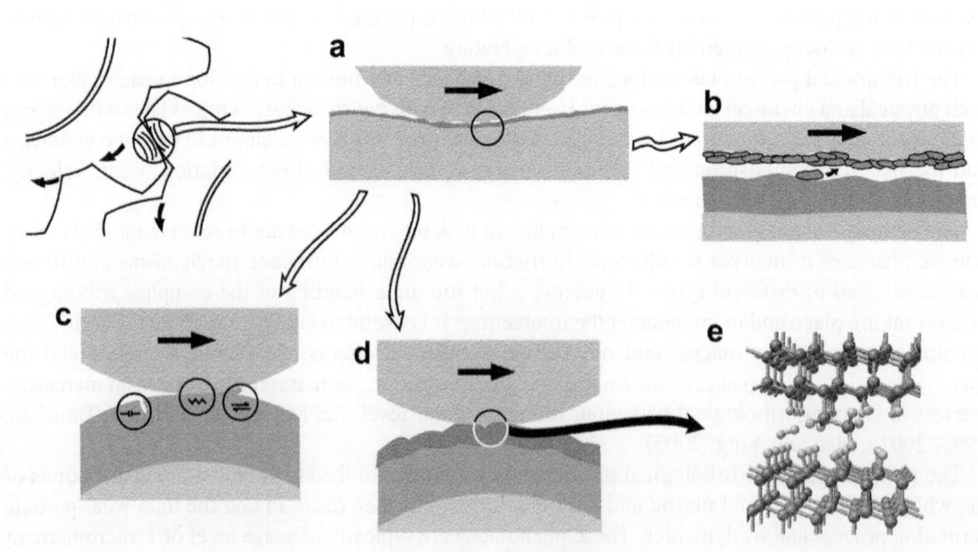

Fig. 3.70. The tribological contact mechanisms: (a) macromechanical, (b) material transfer, (c) micromechanical, (d) tribochemical and (e) nanophysical contact mechanisms (part e with permission from Krim, 1996 and Harrison *et al.*, 1998).

The chemical effects take place on a microscale or less and are referred to as tribochemical mechanisms. Material transfer takes place on both a micro- and macrolevel and both influence the friction and wear behaviour.

The wider range of topics that tribology is involved in has been presented and discussed by Holmberg (2001). Tribology covers a diverse range of scientific problems and technological applications from molecular scale friction to reliability and availability aspects of heavy machinery systems that can be as large as ship machinery, paper machines and automatic production lines in plants. Our improving understanding of tribological phenomena on the nanoscale creates the need to scale up our nanoscale knowledge to conclusions on improved prediction and control of friction and wear that take place on a more macroscopic scale, such as in practically observable everyday life. Here we move over extremely large ranges of size from 10^{-9} m to 10^{3} m and of time from 10^{-15} s to 10^{9} s (Singer and Pollock, 1992). The complexity of the physical, chemical and mechanical phenomena involved in the tribological contact process makes the scaling task extremely challenging as well illustrated by Bhushan (2001a), Israelachvili and Spikes (2001) and Hsu and Ying (2003).

An attempt to illustrate the scaling up of very basic friction and wear phenomena all the way from atomic dimensions to global and universal dimensions is shown in Fig. 3.71.

Nanotribology could also be called molecular tribology, because here the investigations concentrate on phenomena related to the interaction between molecules and atoms, such as the effects of vander Waal's forces and the crystal structures of materials.

Microtribology or asperity tribology was introduced by Bowden and Tabor (1950) in their studies of friction, wear and adhesion that take place at the peaks of the surface topography. Phenomena such as fracture, elastic and plastic deformation, debris formation, surface layer formation and topography effects are of central importance.

Macrotribology or contact tribology was the focus of research at the beginning of the last century. These works are related to contacts between gears, bearing elements and rollers, and phenomena like

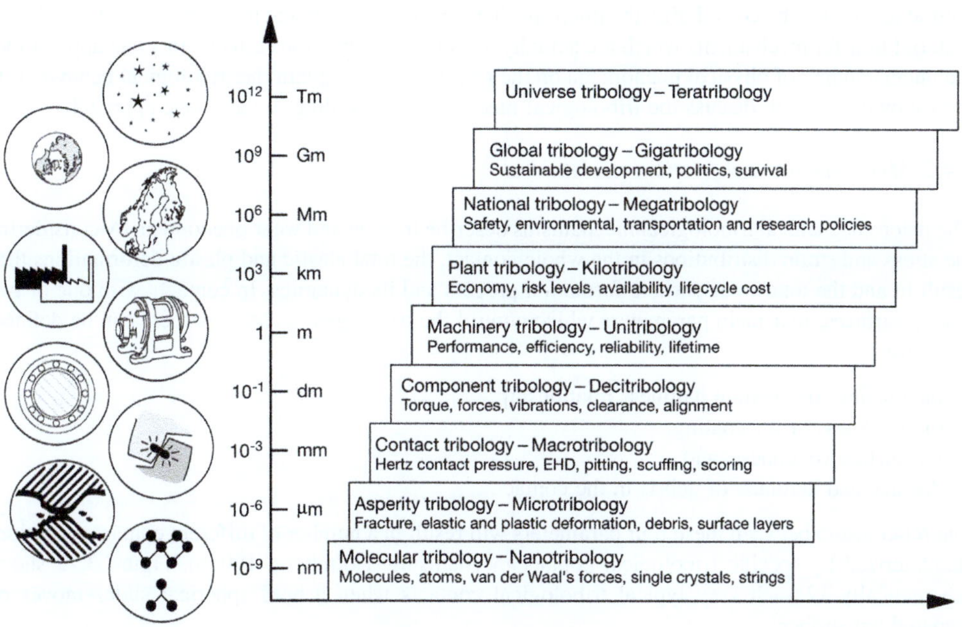

Fig. 3.71. Scales of tribology from nanotribology to teratribology (Holmberg, 2001).

Hertzian contact pressure, elastohydrodynamic lubrication and wear mechanisms clearly observable by the naked eye (such as scuffing, scoring, pitting).

Component tribology or decitribology is related to defining and measuring typical parameters originating from the interaction of components and related to their performance such as torque, forces, vibrations, clearance and alignment.

Machinery tribology or unitribology describes the performance-related phenomena for a system of components assembled in a machine or a piece of equipment. The parameters of interest are performance, efficiency, reliability and lifetime estimation.

Plant tribology or kilotribology deals with a whole system of machinery, structures and equipment – and how parameters such as costs, risk levels, availability and lifecycle costs are used.

National tribology or megatribology extends the effects and consequences on a nationwide perspective to include parameters of relevance such as safety policy, research policy, transportation policy and environmental policy.

Global tribology or gigatribology considers the effects on a worldwide basis as one interacting system; effects dealt with are sustainable development, politics and cultural and human survival.

Universe tribology or teratribology is the largest perspective today that the authors can think of in this scaling-up exercise. But what does this mean? Is it interactions of material in our cosmos? Is it mechanisms for space expansion? Or is it mechanisms for the creation of new life and cultures?

The scaling up from molecular phenomena all the way up to universe tribology is an interesting exercise because it shows that at all levels we have to deal with phenomena that are known to us to some extent. In this sense, then, it shows that talking about 'scaling up' and 'scaling down' is relevant and important.

In the following we will focus on the scales in tribology which are most relevant in the design and use of coated components and tools, which is the range from nanotribology to macrotribology. When the tribological mechanisms in a coated contact are analysed, the macromechanical mechanisms should first be considered in order to achieve an understanding of what kind of contact it is. The effects defined on the macromechanical level dominate the tribological behaviour. Subsequently, it is natural to continue by considering the micromechanical and the chemical mechanisms and, finally, the material transfer mechanisms which are actually a result of the previous effects. In some applications the understanding of physical phenomena on the nanolevel will explain the tribological behaviour. In the following we will discuss the tribological mechanisms according to the above approach.

3.4.2 Macromechanical friction mechanisms

The macromechanical tribological mechanisms describe friction and wear phenomena by considering the stress and strain distributions in the whole contact, the total elastic and plastic deformations they result in and the total wear particle formation process and its dynamics. In contacts with one or two coated surfaces, four main parameters which control the tribological contact process can be defined. They are

1. the coating-to-substrate hardness relationship,
2. the thickness of the coating,
3. the surface roughness, and
4. the size and hardness of debris in the contact.

The relationship between these four parameters will result in a number of different contact conditions characterized by specific tribological contact mechanisms (Holmberg, 1992b). Figure 3.72 shows schematically 12 such very typical tribological contacts when a hard spherical slider moves on a coated flat surface.

It is important to notice that the three geometrically related parameters, the film thickness, the surface roughness and the wear debris, all typically appear in the same dimension range from 0.01 μm

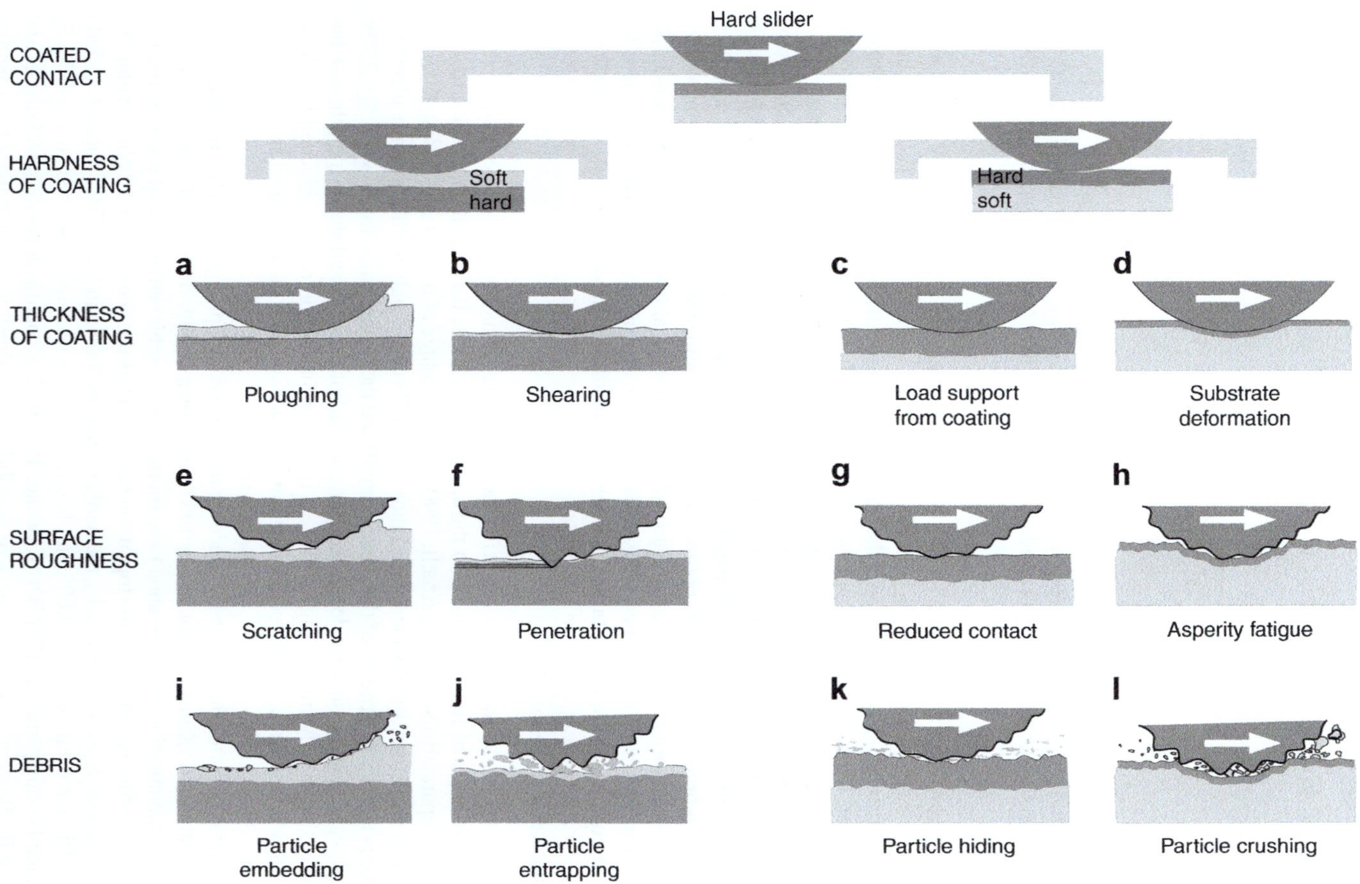

Fig. 3.72. Macromechanical tribological mechanisms in a contact with coated surfaces depend on four main parameters: the hardness relationship, the film thickness, the surface roughness, and the debris in the contact. Characteristic tribological contact phenomena influencing friction are shown schematically in the subfigures a–l.

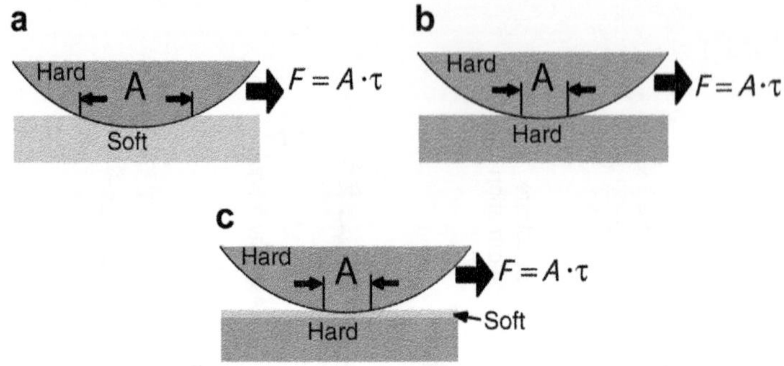

Fig. 3.73. The friction is not greatly dependent on hardness, but low friction can be obtained by depositing a thin soft film on a hard plate in an idealized sliding contact (after Bowden and Tabor, 1950).

to 10 μm. This means that the interrelation between these dimensions for each real case can be considered to have a dominating effect on both friction and wear at the macroscale. The mechanisms involved are very different depending upon if the coating and the substrate are soft or hard (Holmberg and Matthews, 1994; Donnet, 1995; Ramalingam and Zheng, 1995; Bull and Berasetegui, 2006). We will discuss both these separately in the situation with a sphere sliding over a coated surface.

In this text we use the words soft and hard for coatings in a general sense, meaning their deformability and not only their hardness. The loading conditions, both in a vertical and a transverse direction, will influence the stresses and strains in a surface and its friction and wear behaviour. Hardness is one important parameter but not the only one. The influence of elasticity in combination with hardness gives a more comprehensive relationship to tribological behaviour (see section 3.2.8).

The tribological mechanisms can be analysed by a similar approach also in contacts where the slider is soft or coated and the basic contact mechanisms are still the same.

3.4.2.1 Hardness of the coating

One of the most important parameters influencing the tribological behaviour of a coated surface is the hardness of the coating and its relationship to the substrate hardness. It is common to consider hard and soft coatings separately (Suh, 1986; Arnell, 1990). The advantage of using a soft coating to reduce friction has been explained by Bowden and Tabor (1950). When a ball is sliding on a plate, the frictional force is ideally the product of the shear strength and the contact area, as shown in Fig. 3.73a. A harder plate material will result in a decreased contact area but an increased shear strength, and thus the effect on friction is only minor (Fig. 3.73b). A reduction in friction can be achieved by adding a thin soft coating on the plate, which reduces both the contact area and the interfacial shear strength, as shown in Fig. 3.73c.

A hard coating on a softer substrate can reduce wear by preventing ploughing on both a macro- and a microscale (Suh, 1986; Komvopoulos *et al.*, 1987), as shown in Fig. 3.74. Hard coatings are thus particularly useful in abrasive environments. Low friction can be achieved with hard coatings if a low shear strength microfilm is formed on the top of the coating. Thus the shear will take place within the microfilm and the load is well supported by the hard coating, as shown in Fig. 3.74c. This effect, with coefficients of friction even as low as 0.005, has been observed with a thin low shear strength MoS_2 top film on a hard coating, such as B or BN (Kuwano, 1990). New coating deposition techniques and structured modifications offer a large variety of possibilities to adjust coating hardness by structural tailoring, such as by using multilayer, nanocomposite and graded coatings (Zabinski and Voevodin, 2004).

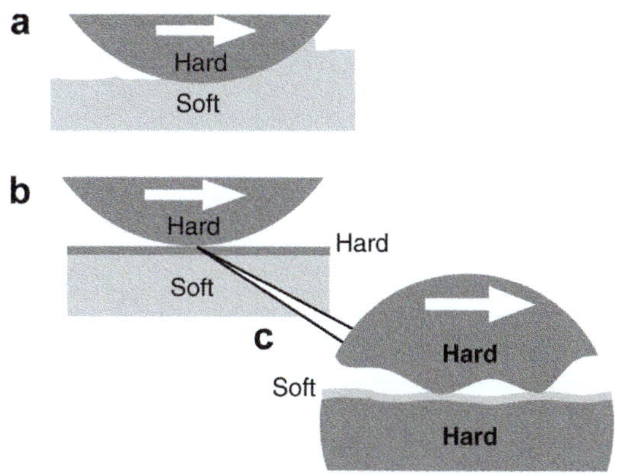

Fig. 3.74. (a) A hard slider moving on a soft counterface results in ploughing. (b) The ploughing can be inhibited by using a hard coating on the soft substrate. (c) A soft microfilm at the top of the hard coating results in decreased friction.

In the following we will discuss in more detail the friction mechanisms present in the 12 contact situations shown in Fig. 3.72. The contacts are characterized by the following typical tribological friction-related contact phenomena: ploughing, shearing, load support from coating, substrate deformation, scratching, penetration, contact area and interlocking, asperity fatigue, particle embedding, particle entrapping, particle hiding and particle crushing.

3.4.2.2 Thickness of the coating

Coating thickness is an important parameter, influencing the stresses in the surface and the wear and friction. The thickness of the coated surface layer is often easy to control during the deposition process but the complexity of the contact mechanisms makes it difficult to find out which coating thickness would be tribologically optimal for a certain application. Below we will discuss the main friction mechanisms related to coating thickness.

Shearing. First, we will consider a thin soft film on a hard substrate with smooth surfaces and no debris present, as shown in Fig. 3.75b. When the film is thin enough, the effect of ploughing of the film is small. The friction is thus determined by the shear strength of the film and the contact area, which is related to the deformation properties of the substrate. For an elastic sphere sliding on a hard plate with a soft coating, the contact area is determined by the Hertzian equation. Roberts (1989, 1990a) has shown that in this situation the coefficient of friction can be expressed by the equation

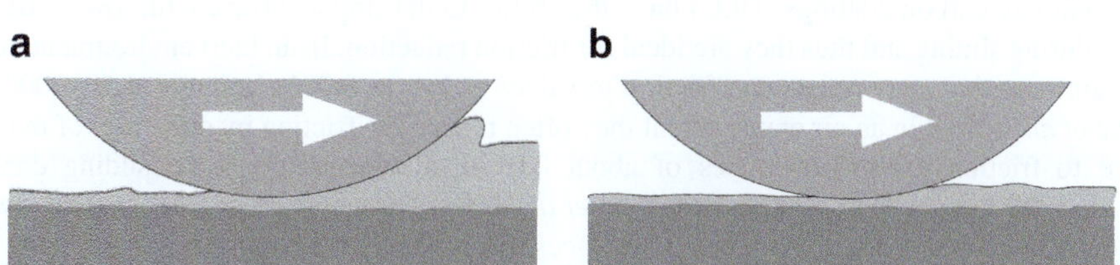

Fig. 3.75. The contact of a hard slider moving on a hard flat surface coated by a soft coating is characterized by (a) ploughing for thick coatings and (b) shearing for thin coatings.

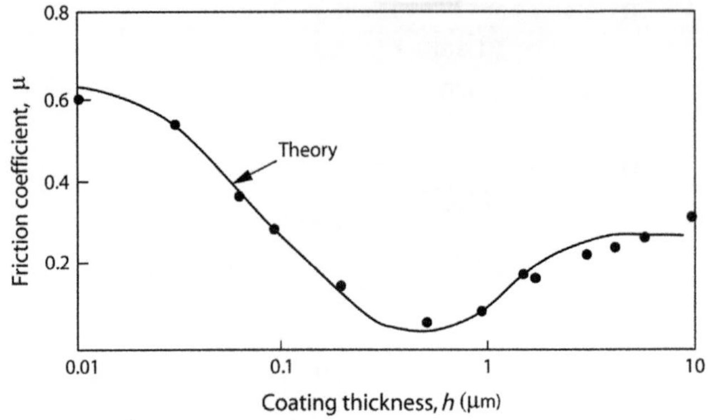

Fig. 3.76. The coefficient of friction as a function of film thickness for lead films deposited on a steel surface sliding against a steel counterface with surface roughnesses of $R_a = 0.75$ μm (data from Sherbiney and Halling, 1977).

$$\mu = \frac{\pi \cdot S}{w^{1/3}} \left(\frac{3\,R}{4\,E'} \right)^{2/3} \tag{3.35}$$

where S is the shear strength of the coating, w the normal load, R the radius of the sphere and E' the reduced modulus of elasticity of the contact materials. The equation shows that there are two main criteria for achieving low friction. The film must possess the property of low shear strength and it must be applied onto surfaces of high hardness or high elastic modulus. In agreement with the equation, Roberts and Price (1989) have reported experimental results that show how the coefficient of friction decreases with an increasing elastic modulus of the contact materials. One typical example of this kind of contact is a steel ball sliding on a smooth steel plate coated with a thin molybdenum disulphide layer, resulting in coefficients of friction as low as 0.02 (Roberts, 1989). A reduction in the coefficient of friction from 0.8 to 0.4 was recorded when a 2 μm thick silver film was applied on a hard ceramic plate sliding against a hard ceramic ball (Erdemir *et al.*, 1991a).

The coefficient of friction is a function of film thickness and it has a minimum value for soft coatings, such as lead, typically at film thicknesses of about 1 micrometre. This has been experimentally verified by Bowden and Tabor (1950) for indium deposited on tool steel and later by Sherbiney and Halling (1977) for both lead and indium films on steel as shown in Fig. 3.76. The increase in friction with decreasing film thickness below optimum is associated with breakthrough of the film by mating asperities. This mechanism will be discussed below in the section entitled 'Penetration'.

Diamond-like carbon coatings (DLC) have the ability to develop a surface with low resistance to shearing during sliding and thus they are ideal for friction reduction. In an inert environment and with smooth surfaces they may reduce the friction to values as low as 0.001 (Erdemir *et al.*, 2000, 2001; Fontaine *et al.*, 2004). In an air environment they often reduce the friction by one order of magnitude or more to friction coefficient values of about 0.05–0.15 depending on the sliding conditions (Miyoshi, 1998; Singer *et al.*, 2003; Watanabe *et al.*, 2004; Holmberg *et al.*, 2007). The low shear conditions can be attained with hard coated surfaces that typically have high friction, such as TiN or TiAlN coated surfaces, by depositing a top layer of DLC on the hard coating (Klafke, 2004c). The mechanisms of low and superlow shear are discussed in more detail in sections 4.3.2 and 4.5.3.

Ploughing. The increase in friction, for soft thick films, with increasing film thickness above optimum in Fig. 3.76 is explained by a decrease in load-carrying capacity of the surface and increased

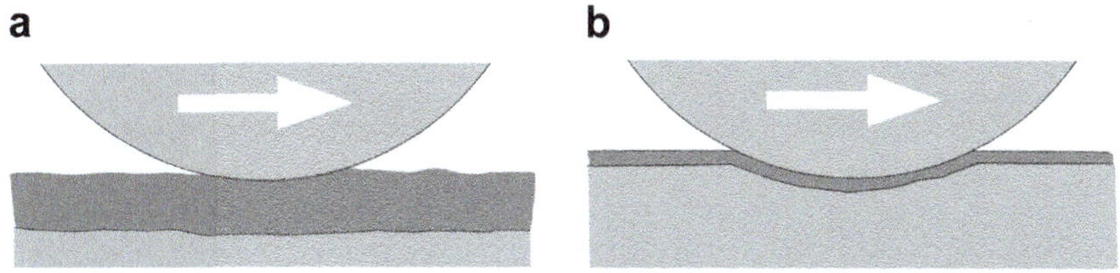

Fig. 3.77. The contact of a hard slider moving on a soft flat surface coated by a hard coating is characterized by (a) load support from the coating for thick coatings and (b) substrate deformation for thin coatings.

friction due to ploughing as shown in Fig. 3.75a. The coefficient of friction is increased due to both the elastic or plastic deformation of the film and to the increased contact area at the interface between the sphere and the coating where the shear takes place. The increase in friction with coating thickness has been experimentally demonstrated for lead films (Tsuya and Takagi, 1964; Sherbiney and Halling, 1977), for gold films (Takagi and Liu, 1967), for silver films (El-Sherbiny and Salem, 1986; Yang *et al.*, 1999), for MoS_2 films (Aubert *et al.*, 1990; Wahl *et al.*, 1999) and for Pb–Mo–S films (Wahl *et al.*, 1999). By assuming that the real area of contact in sliding is virtually the same as that in static loading, Finkin (1969) showed theoretically that the coefficient of friction for thick films is directly proportional to the square root of the film thickness and inversely proportional to the square root of the normal load.

A more general theory for the frictional behaviour of both thin and thick films was developed by Halling and El-Sherbiny (1978) and El-Sherbiny (1984). Even though the theory fits well to their experimental results, as shown in Fig. 3.76 the equation they propose for the coefficient of friction is not very practical because it involves parameters which are not always easily estimated.

For thick, soft coatings the mechanism of ploughing will be very similar to that of soft bulk materials scratched by a hard indenter. Those tribological mechanisms have been described by several authors (e.g. Komvolpoulos, 1991b; Malzbender and de With, 1999; Lafaye *et al.*, 2005; Felder and Bucaille, 2006). In addition to the ploughing mechanism Hokkirigawa and Kato (1988a) also identify a wedge forming and a cutting mechanism depending on the degree of penetration and the shear strength of the contact interface. These will influence the wear particle generation mechanism as shown by Wang and Kato (2003c). Particularly with elastic coatings, like rubber on a hard surface, the hysteresis effects on friction must also be taken into account (Adams *et al.*, 1990).

Substrate deformation. We will now consider the situation with a hard smooth sphere sliding on a soft smooth substrate covered by a hard coating with no debris present as shown in Fig. 3.77. If the coating is very thin, as in Fig. 3.77b, it is unable to support the load. The function of the coating is to separate the substrate from the counterface and to prevent ploughing by increasing the hardness of the top layer of the surface. The latter effect has been considered to be very important (Shepard and Suh, 1982; Suh, 1986; Bull and Rickerby, 1990a; Zabinski and Voevodin, 2004). The prevention of ploughing reduces both the friction and the wear. The higher shear strength introduced at the contact interface by the hard coating has, on the other hand, an increasing effect on friction in sliding if no microfilms are formed as described above (Fig. 3.74). This explains the very high coefficients of friction that often occur in sliding contacts of thin hard coatings. The increase in friction caused by increased shear strength generally seems to be more dominant than the reduction in friction due to decreased ploughing.

When loaded, the coating will deflect in accordance with the deformation of the substrate (Leroy and Villechaise, 1990; Korunsky *et al.*, 1998; Jungk *et al.*, 2008). The substrate will either deform elastically or plastically and, as the coating does not support the load, the deformation can be

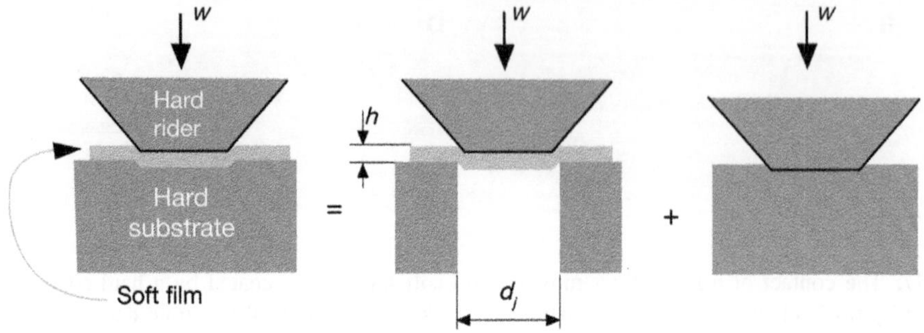

Fig. 3.78. A hypothetical model of a junction where the load, w, is carried partly by the surface layer, which is about to be displaced, and partly by the substrate, which is about to deform plastically (after Rabinowicz, 1967).

calculated as for bulk materials. For softer substrates the deformation will be considerable and will thus add a ploughing or hysteresis effect to friction. The repeated deflection of the coating may cause fracture and fatigue cracks that destroy the coating or the substrate material. The influence of the substrate material's hardness and elastic modulus on friction has been demonstrated for TiN coated surfaces (Komvopoulos *et al.*, 1987; Komvopoulos, 1991b; Podgornik, 2000; Hainsworth and Soh, 2003) and for DLC coated surfaces (Podgornik, 2000; Bandorf *et al.*, 2003; Aoki and Ohtake, 2004; Nakahigashi *et al.*, 2004). The load-carrying capacity was investigated by Podgornik (2000) and Ronkainen (2001).

Load support from coating. When the hard coating is thicker, as shown in Fig. 3.77a, it can, because of its stiffness, carry part of the load, and the deformation of the substrate will be smaller. The frictional situation is more favourable compared with the thin hard coatings because ploughing or hysteresis effects due to substrate deformation will be relatively minor. Ploughing of the coating is prevented by the high hardness, and the contact area between the coating and the sphere, where the shear takes place, is reduced due to decreased surface deformation.

The coefficient of friction has been estimated theoretically by Rabinowicz (1967) as a function of film thickness and junction diameter for soft films when the load is partly carried by a solid film and partly by the substrate as shown in Fig. 3.78.

The influence of coating thickness on the load-carrying capacity of a TiN coated titanium substrate was shown by Komvopoulos *et al.* (1987) in a pin-on-disk test with a 6.35 mm radius spherical pin sliding at low velocity, between 0.0087 and 0.0353 m/s, on a flat surface of the same material. A 0.02 μm thick TiN coating showed high friction up to $\mu = 0.65$ due to the ploughing effect even at low loads of 0.02–0.05 N while a 0.8 μm TiN coating on the same substrate had a coefficient of friction of only $\mu = 0.1$ even at loads of 0.2–2 N. They used boundary-lubricated conditions to minimize the effect of adhesive friction.

3.4.2.3 Roughness of the surfaces

As we already know, ideally smooth surfaces are rare in engineering applications. We will now consider friction effects for coated contacts with rough surfaces.

Scratching. For thick soft coatings the roughness of the substrate can be neglected if it is considerably smaller than the thickness of the film and the film is stiff enough to carry the load, as shown in Fig. 3.79a. However, in situations where the surface roughness is about one order of magnitude smaller than the coating thickness, frictional effects have been observed. Both Roberts and Price (1989) and Aubert *et al.* (1990) have measured a decrease in the coefficient of friction for MoS_2

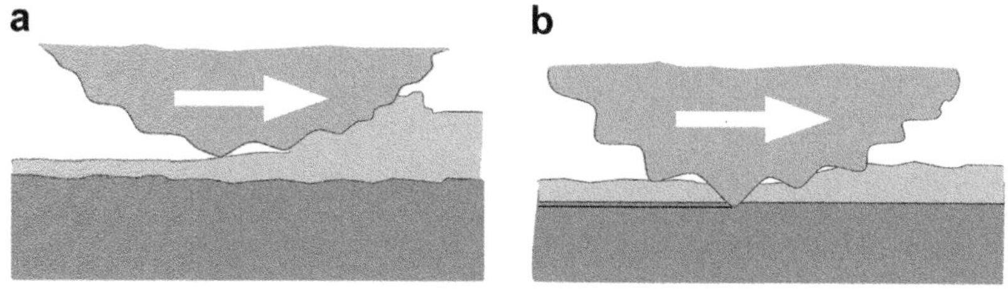

Fig. 3.79. The contact of a hard slider moving on a hard flat surface coated by a soft coating both with rough surfaces is characterized by (a) scratching for thick coatings and (b) penetration for thin coatings.

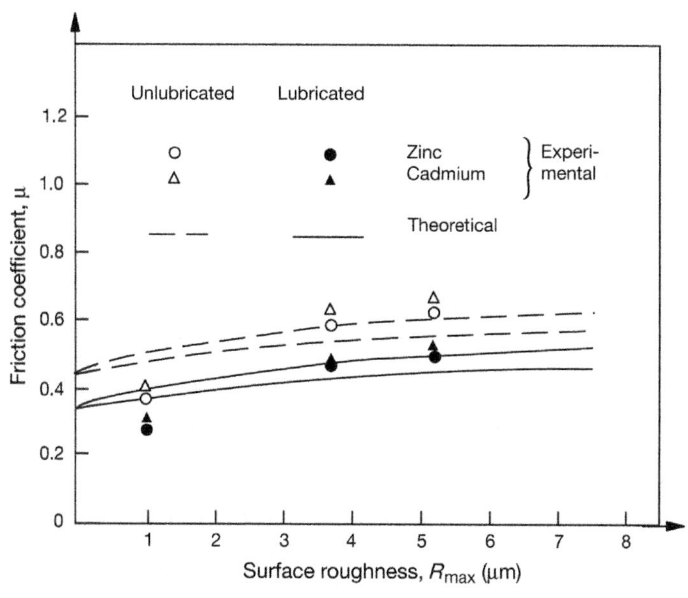

Fig. 3.80. The coefficient of friction as a function of surface roughness when hard asperities are ploughing a soft counterface (data from Hisakado, 1976).

coatings deposited on rough surfaces compared with smooth surfaces, presumably without asperity penetration of the coating. Possible explanations may be an improvement in the load-carrying capacity by the coated substrate asperities or decreased shear strength by increased pressure at the MoS_2 coating between the asperities and the counterface.

The slider with a rough surface will have a similar ploughing effect on friction as described above for smooth surfaces but in addition there will be surface roughness effects to be considered. Because of the softness of the coating there will be no increasing effect on the stress and strain fields at the contact by the surface roughness (Sainsot *et al.*, 1990). However, the asperities on the surface of the sphere will each contribute a microploughing effect on friction. This effect is most probably only minor as the results shown in Fig. 3.80 for bulk materials would indicate (Hisakado, 1976; Hwang and Zum Gahr, 2003).

Penetration. For thin soft films on hard substrates, as in Fig. 3.79b, the influence of surface roughness is considerable. The penetration of asperities through the film will cause increased shear resistance and ploughing of either the substrate or the counterface with an increase in friction as

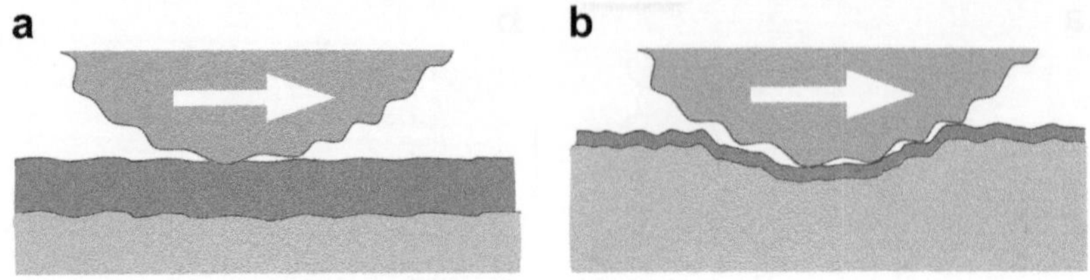

Fig. 3.81. The contact of a hard slider moving on a soft flat surface coated by a hard coating, both with rough surfaces, is characterized by (a) reduced contact area and interlocking for thick coatings and (b) asperity fatigue for thin coatings.

a result. This is illustrated by the left-hand side of the curve in Fig. 3.76. The increase in friction starts when the coating thickness decreases to the same level as the surface roughness (Sherbiney and Halling, 1977).

The effect on friction of asperity penetration through a soft coating has been modelled theoretically by Dayson (1971), Halling and El-Sherbiny (1978) and El-Sherbiny (1984), and the experimental results are in good agreement with the theory as shown in Fig. 3.76. The coefficient of friction is influenced by the hardness ratio of the film and the substrate, the ratio between the real area of contact for the film and the substrate and material properties such as the yield stress to hardness ratio, the yield to maximum shear stress ratio, the shear strength and the deformation mode.

Reduced contact area and asperity interlocking. When a rough surface is covered by a thick hard coating the surface roughness may remain or be altered to some extent depending on the deposition method. When a hard rough slider is moving over such a surface the absence of larger elastic or plastic deformations will result in the contact situation shown in Fig. 3.81a, where the slider is moving on a small number of asperity tops. The area of effective adhesion between the sliding surfaces has been reduced to a few asperity contacts. This situation can only prevail with very hard and tough material combinations because the contact pressure at the asperities will be extremely high (Sainsot *et al.*, 1990; Gong and Komvopoulos, 2003; Lindholm *et al.*, 2006). In sliding with a hard rough surface against a smoother countersurface the initial friction is often low, due to the reduced contact area, and then increases as wear of asperity tops increases the contact area (Wiklund *et al.*, 1999a).

If the shear strength at the small number of asperity contacts is low the coefficient of friction will be extremely low. The suggested formation of a thin slippery microfilm on the top of a diamond coating may result in such contact conditions. Because of the high contact stresses and the brittleness of the coating, a hard substrate is needed to carry the load. This explanation is supported by the very low coefficients of friction of 0.08, measured for a SiC ball sliding on a diamond coating by Jahanmir *et al.* (1989) and 0.05 for a monolithic single crystal diamond stylus sliding on a diamond coating measured by Hayward and Singer (1990).

The friction may be considerably increased by asperity interlocking and breaking mechanisms if both surfaces are rough. This is typically the situation at the initial stage of diamond sliding on a diamond coating before running-in smoothing has taken place (Hayward *et al.*, 1992), as shown in Fig. 4.54.

Asperity fatigue. A hard slider moving on a thin hard coating covering a soft substrate will result in plastic or elastic deformation of the substrate material under the contact, as shown in Fig. 3.81b. In the case of rough surfaces it results in an increased number of asperity contacts and decreased stresses at the asperities because a greater number of asperities are carrying the load.

As the asperities are repeatedly loaded by the sliding counterface, there is a high probability of fatigue failure. Delamination-type surface fatigue wear has been observed by Sue and Troue (1990) for steel sliding against TiN coatings. This fatigue is an important wear mechanism but does not greatly influence friction. Scratching of the hard asperities in the counterface often occurs and this has the effect of increasing friction. Grooves and ploughing in hard coatings have been reported by Sue and Troue (1990) and Miyoshi (1989a, b).

3.4.2.4 Debris at the interface

Loose particles or debris are often present in sliding contacts. They can either originate from the surrounding environment or be generated by different wear mechanisms in the sliding contact. Their influence on friction is in some contact conditions considerable, depending on the particle diameter, coating thickness and surface roughness relationship and the particle, coating and substrate hardness relationship. The role of debris in a tribological contact has been theoretically analysed using the third-body concept, developed by Godet and Berthier (Godet, 1984, 1989; Descartes and Berthier, 2001; Berthier, 2005; Fillot *et al.*, 2007). A scale effect analysis of multiple asperity contacts including the effect of debris was introduced by Nosonovsky and Bhushan (2005).

Particle embedding. Let us first consider the case of hard particles with a diameter considerably smaller than the thickness of a soft coating on a hard substrate as shown in Fig. 3.82a. When the particles are introduced in the contact they will be pressed into the soft coating and be embedded in it without any further contact with the slider as long as the soft coating remains thicker than the particle diameter. The particles will have no great effect on friction, which is mainly controlled by the ploughing mechanisms described above.

Particle entrapping. For thin soft coatings where the dimensions of the particles are of the same magnitude or bigger than the coating thickness and the surface roughness, as shown in Fig. 3.82b, their influence on friction can be considerable. If the particles are harder than the coating but softer than the substrate then they will easily be caught by the roughness of the counterface or partly sink into it and scratch grooves in the soft coating, as in the case of asperity penetration. The friction will increase because of particles ploughing the coating (Hwang *et al.*, 1999). The influence of ploughing wear particles on friction in sliding steel contacts has been shown to be considerable (Kim and Suh, 1991; Komvopoulos, 1991a, b).

An increase in friction may follow if the slider and the substrate are of equal hardness and the loose particles in the contact have a higher hardness (Suh, 1986). Then the particles may partly sink through the coating into the substrate and also into the counterface, and by a kind of anchoring mechanism resist motion. Even if the particles are fairly spherical in shape it is not probable that they will decrease friction by rolling since they will be stuck into the soft coating. Sin *et al.* (1979) have shown that the contribution of ploughing on the friction coefficient is very sensitive to the ratio of the radius of curvature of the particle to the depth of penetration.

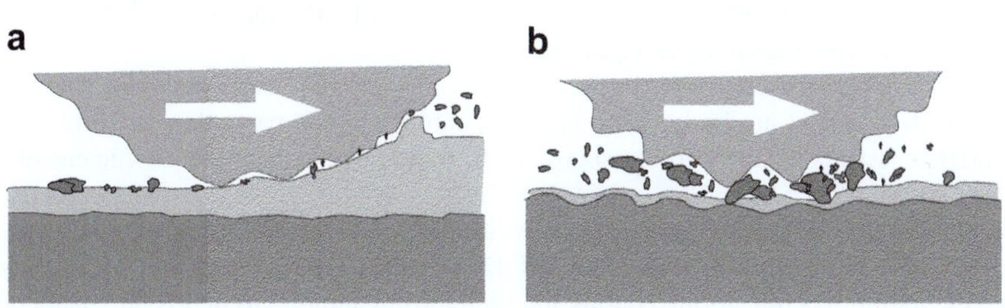

Fig. 3.82. The contact of a hard slider moving on a hard flat surface coated by a soft coating, both having rough surfaces and with hard debris present in the contact, is characterized by (a) particle embedding for thick coatings and (b) particle entrapping for thin coatings.

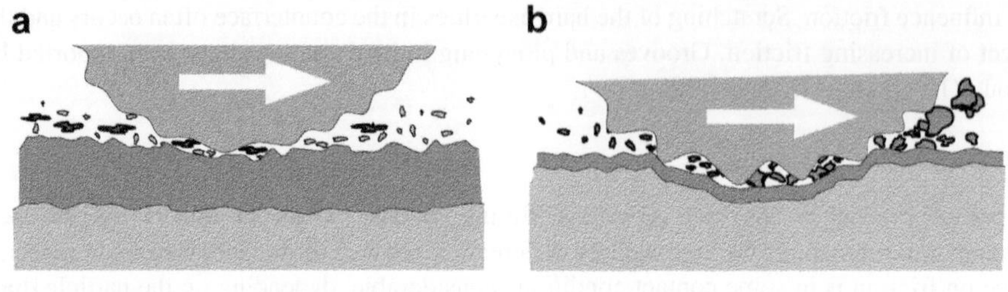

Fig. 3.83. The contact of a hard slider on a soft flat surface coated by a hard coating, both having rough surfaces and with debris present in the contact, is characterized by (a) particle hiding for thick coatings and (b) particle crushing for thin coatings.

If the particles in the contact are soft, then their tribological effect is quite different. Soft particles with low shear strength trapped in the contact can carry part of the load and inhibit direct substrate-to-counterface contact, thus reducing friction (Grill, 1997; Voevodin *et al.*, 1999b). A similar effect has been observed by Yamada *et al.* (1990) who found that polymer particles can act very effectively to reduce friction. In contacts with MoS_2 coatings or when Mo and S are present, MoO_2 wear debris is produced that has a very low shear strength and thus can act as a solid lubricant reducing the friction (Donnet, 1998a; Singer *et al.*, 2003). Increased friction by generated and trapped wear particles has been observed in, e.g., DLC against ceramic Al_2O_3–TiC (Wei and Komvopoulos, 1997) and DLC against steel (Samyn *et al.*, 2006) sliding contacts.

Particle hiding. The introduction of small particles into the sliding contact of hard and rough surfaces, as shown in Fig. 3.83a, does not necessarily make the tribological mechanisms more severe. The particles can be hidden in the valleys formed by the asperities while the sliding takes place at the asperity tops. Thus the particles will have no great effect on either friction or wear. This effect has been demonstrated for wear particles generated in carbon coatings sliding against a ceramic surface (Wei and Komvopoulos, 1997), small diamond onion-like particles in a sliding steel against silicon contact (Hirata *et al.*, 2004) and fullerene particles in steel sliding against steel contact (Rappoport *et al.*, 2005).

It is important to notice that reduced surface roughness of the surfaces may increase both friction and wear if the particles cannot hide in the valleys and instead interact with the surfaces by scratching and interlocking.

Particle crushing. When particles, large in relation to the surface roughness, are introduced between two hard surfaces the result can be particle crushing, scratching or rolling as shown in Fig. 3.83b. If the particles have a lower hardness than the surfaces, then they will be crushed and destroyed under the load of the contact, with smaller debris and some increase in friction as a result. If the particles have a higher hardness than the surfaces, they will be caught by the roughness of the surfaces with ploughing and scratching resulting as described earlier.

The presence of hard particles between hard surfaces may in some cases even reduce the coefficient of friction. If the particles are fairly round in shape, hard enough to carry the load, and at least one of the surfaces is smooth, the particles may act as rollers and reduce the coefficient of friction (Blomberg, 1993).

The debris in the contact may also change in its properties during the sliding action. In the sliding contact of a TiN coating against a corundum ball, De Wit *et al.* (1998) observed that the wear debris produced by fretting was transformed during the sliding from amorphous debris into nanocrystalline rutile-like particles with a drop in the friction as a result. Wear debris formed in a contact with a multilayer (Ti,Al)N/TiN coating on steel substrate sliding reciprocating against a corundum ball was observed to become progressively more and more lubricious at increasing humidity. In sliding a steel ball against two kinds of DLC (a-C:W and a-C:Cr) coated contacts, oxidized coating material was observed to be collected in the surface topography and it increased friction because of the higher shear resistance of the oxidized material (Svahn *et al.*, 2003).

3.4.3 Macromechanical wear mechanisms

The macromechanical wear mechanisms can be analysed by using the same schematic representation for typical tribological contact situations as was used to study the friction mechanisms (Fig. 3.72). The 12 different contact situations can be grouped to represent eight characteristic wear mechanisms (Holmberg *et al.*, 1992b) as shown in Fig. 3.84.

Plastic deformation. When a hard slider moves on a soft thick coating which is stiff enough to carry the load, the result will be a groove in the coating as shown in Fig. 3.84a. The removal of material can be calculated by the geometry of the contact. The situation from the wear point of view is very similar to the contact with a hard slider ploughing in a soft bulk material. That kind of contact has been treated in detail in several textbooks on tribology and, for example, by Hokkirigawa and Kato (1988a) and Kitsunai *et al.* (1990).

Plastic deformation was observed as the main wear mechanism when a slider moved on a plane surface coated with a layer of gold (Tangena, 1987, 1989; Tangena and Wijnhoven, 1988; Tangena *et al.*, 1988). Tangena and his co-workers correlated the calculated von Mises stresses averaged over the contact width and the depth of the coating with the wear volume, using a theoretical model based on low-cycle fatigue. Using finite element calculations they showed that the geometry of the system, the mechanical properties of the substrate and the contact load determine the mechanical deformations of the top layer. The theory was supported by experimental results (Fig. 3.85) which showed that if the average von Mises stress was above the yield stress of the coating, the wear increased enormously, indicating the importance of plastic deformation in the film. This process of low-cycle fatigue wear is related to the mechanisms known as ratchetting and delamination, as described below, which have been previously discussed mostly in the context of uncoated surfaces.

In the sliding of a sphere on a soft coating three phases of plastic deformation have been identified, which represent a process of wear mode transition from ploughing to flowing and finally to delamination wear, as shown in Fig. 3.86. This was shown with a 5 μm radius diamond sphere sliding on a very thin carbon nitride coating by Wang and Kato (2003c). In the flowing wear mode, wear particles were burnished from a plastically deformed CNx coating surface layer by flow wear. In flow wear mode, plastically deformed 'filmy' surface layers extrude from the edges and both ploughing and flowing create fine grooves during the process of horizontal plastic flow. If the filmy layers extrude to cover the fine grooves from the side, filmy wear particles are generated. If the filmy layers curl towards the bottom of the fine grooves, long and slender wear particles are generated. In the delamination wear mode, wear particles were rapidly detached from subsurface layers of CNx by a low-cycle fatigue wear. Surface delamination results from brittle fracture around the maximum shear stress region or it may initially occur through microcracking in the deformed coating subsurface layer, where the maximum shear stress occurs, rather than along the coating/substrate interface. The propagation of microcracks into the deformed coated surface generates wear particles.

If the surface of the slider considered earlier is rough or if small debris is introduced into the contact, the result is that the asperities or the trapped debris cause microploughing and microgrooves in the main groove. This effect can also be calculated (Hisakado, 1976; Sin *et al.*, 1979) but its influence on total wear is normally only minor (Hwang *et al.*, 1999). With a theoretical model El-Sherbiny and Salem (1979) showed that the wear appears to be dependent on the surface texture of the system rather than on material properties during the initial wear when hard conical asperities are ploughing a soft surface coating.

Adhesive and fatigue wear. Adhesive wear is to some extent present in several contact situations. Even if all other wear mechanisms have been eliminated by good tribological design some minor

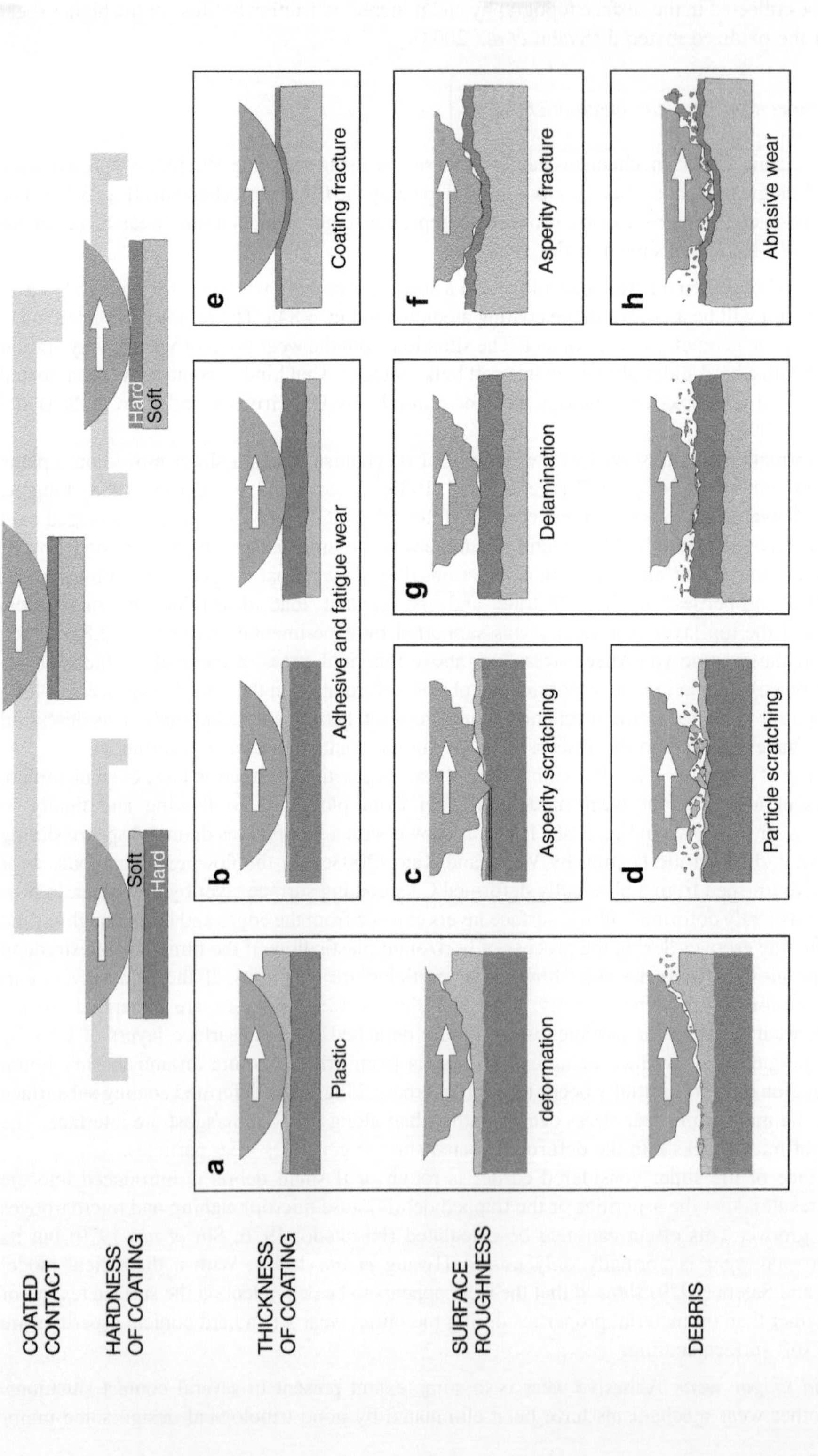

Fig. 3.84. The macromechanical wear mechanisms in a contact with coated surfaces depend on four main parameters: the hardness relationship, the film thickness, the surface roughness and the debris in the contact. Characteristic wear processes are schematically shown in the subfigures a–h.

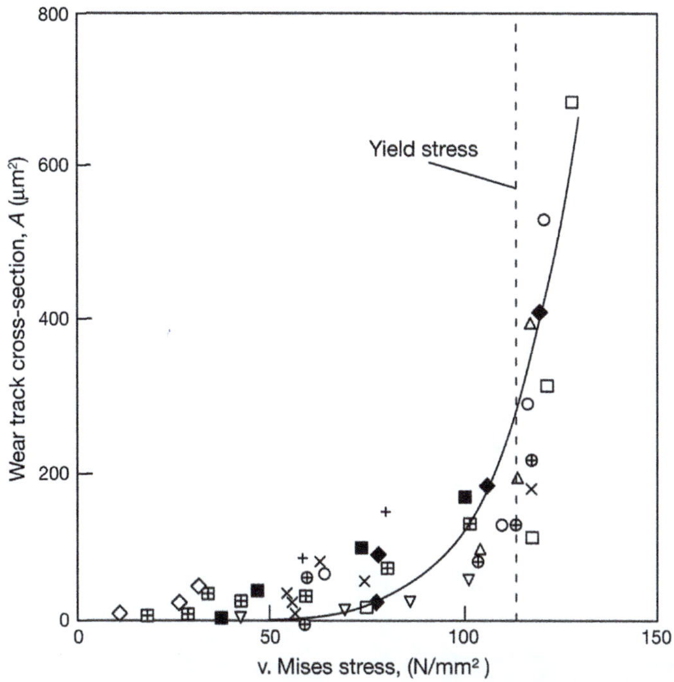

Fig. 3.85. Cross-section of the wear track for both evaporated and electrodeposited gold layers as a function of the von Mises stress averaged over the depth and the contact radius for 56 different contact situations. The solid line is obtained from a wear model (data from Tangena, 1989).

adhesive wear will still exist. Adhesive wear is typical of a hard sphere sliding on a thick hard coating or on a hard substrate covered with a thin soft coating as shown in Fig. 3.84b. With repeated sliding contacts the adhesive wear is often associated with large plastic strain accumulated close to the top surface and low-cycle fatigue (Torrance, 1992; Yan *et al.*, 2000), as shown in Fig. 3.50. The repeated accumulation of plastic strain is also frequently referred to as ratchetting (Kapoor *et al.*, 1996; Kapoor, 1997).

The load-carrying capacity of soft films increases with decreasing film thickness and thinner films have a better resistance to adhesive wear because of the smaller real area of contact (Halling, 1986; El-Sherbiny and Salem, 1986). A thin soft film applied in the sliding contact between two hard surfaces may remarkably improve the wear resistance of the contact. For example, the use of a 2 µm thick silver coating on a hard zirconia ceramic plate sliding against a ceramic ball reduced the wear of the ball by factors of 5 to 3300 and the wear of the coated plate was immeasurable (Erdemir *et al.*, 1991a).

The wear process for thin soft coatings is generally complex. Often, especially for thin polymer coatings, the wear mechanism is a combination of adhesive wear and repeated plastic deformation. This has been observed by Fusaro (1981) for polyimide-bonded graphite fluoride films and by Yamada *et al.* (1990) for polytetrafluoroethylene (PTFE) films. In addition, fatigue wear may be involved depending on the film material. Yamada *et al.* (1990) observed the detachment of thin films from the wear track after a number of passes for PTFE films and the detachment of flake-like wear debris for polyimide films. Very mild adhesive and fatigue wear conditions were observed by Takagi and Liu (1968) in a heavily loaded gold coated cylinder-on-plate test where coatings with

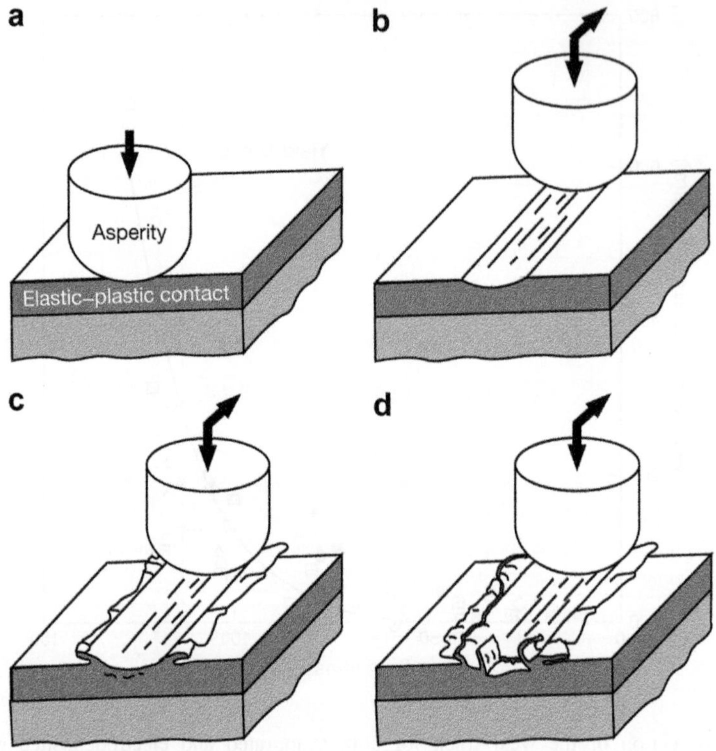

Fig. 3.86. A schematic model for three wear modes in plastic deformation of a soft coating: (a) normal load, (b) ploughing, (c) flowing wear and (d) delamination wear (after Wang and Kato, 2003c).

thicknesses of 5 to 20 μm improved the wear life by up to four orders of magnitude, although the friction coefficient was as high as 0.6 to 0.8.

A soft thin film on a hard substrate can also be used to reduce delamination wear because of its influence on the dislocation density at the outermost surface layer. A thin layer of a soft metal deposited on a hard substrate can retard the wear of the harder substrate by preventing plastic deformation and crack nucleation. If the soft layer is thicker than a critical value, delamination can occur within this layer, forming loose wear particles which cause ploughing and further plastic deformation. This mechanism has been experimentally demonstrated by Jahanmir *et al.* (1976) and Suh (1986) for the soft metals gold, silver, nickel and cadmium. They found that for an optimum film thickness of 1 micrometre the reduction in wear rate can be even more than two orders of magnitude as shown in Fig. 4.21 for a nickel coating on steel. Gold, nickel and cadmium coatings were reported to be effective only in an inert atmosphere, but silver was effective also in air. A soft coating will reduce the surface tensile stresses that contribute to undesirable subsurface cracking and subsequently to severe wear (Spalvins and Sliney, 1994; Matthews *et al.*, 2001).

Diamond-like carbon films are sometimes fairly soft and in a sliding contact adhesive wear is often combined with asperity fracture, chip and flake formation and micropitting. As the wear debris becomes trapped at the contact interface graphitization of amorphous carbon wear products changes the interfacial surface, energy and shear strength and affects the steady-state tribological behaviour (Wei and Komvopoulos, 1997; Jiang and Arnell, 2000).

It seems obvious that very thin films on the top of a surface will affect the work hardening process during sliding at the top layer, but there is little information published on how this influences the tribological mechanisms.

Asperity scratching. In contacts with rough surfaces and thin soft films the mechanism of asperity penetration and ploughing dominates. However, there are interesting examples showing that the surface roughness of the substrate may influence wear even before penetration has taken place. In sliding experiments with a 1 μm thick MoS_2 coating on a steel substrate the lifetime of the coating increased by one order of magnitude and the coefficient of friction decreased by about half when the substrate roughness was increased from $R_a = 0.04$ μm to 0.12 μm (Roberts and Price, 1989) as shown in Fig. 4.26. Possible explanations may be an improvement in the load-carrying capacity by the coated substrate asperities or decreased shear strength caused by increased pressure in the MoS_2 coating between the asperities and the counterface.

Asperities on the surface of a rough slider moving on a hard substrate with a soft coating, as shown in Fig. 3.84c, penetrate the coating and slide directly on the substrate. If the asperities are harder than the substrate, they will also penetrate into the substrate and produce grooves by ploughing both in the coating and in the substrate. A theoretical wear model for this contact situation was developed by El-Sherbiny and Salem (1984) for predicting the wear rate of the coating and the substrate as a function of surface topography, contact geometry, material properties and operating conditions. A fair agreement between theory and experiments with lead, indium and tin films was reported.

When a relatively smooth slider moves on a rough substrate covered with a thin soft coating, the first stage in the wear process is a combination of plastic deformation and adhesive wear, which decreases the coating thickness until there is contact between the sphere and the substrate asperities through the coating. After that, the wear increases considerably because of both adhesive and scratching wear between the substrate asperities and the surface of the slider. This kind of wear situation has been described by Fusaro (1981) for polymer films on steel substrates.

Particle scratching. We will now discuss the situation when hard particles of the same size as the film thickness and the surface roughness are introduced into the contact of a rough sphere sliding on a rough substrate covered with a thin soft coating, as shown in Fig. 3.84d. The particles are caught by the surface roughness and cause microploughing and scratching both in the soft coating and in the hard substrate in a similar way to the effect of hard asperities penetrating the coating described earlier. A difference comes from the fact that the particles are free to rotate in the contact and they are often considerably sharper than the asperity angles in engineering surfaces. The wear rate depends on the particle size. It is small for very small particles, increases rapidly with increasing particle size and then stabilizes when the particles reach a critical size where a transition from a cutting wear mechanism to a sliding mechanism takes place as shown in Fig. 3.87. This mechanism has been described for bulk materials by Sin *et al.* (1979), although it can be assumed that a soft thin coating does not change it very much.

If the particles in the contact are soft, their tribological effect is quite different. Soft particles with low shear stresses trapped in the contact can carry part of the load and inhibit direct substrate-to-counterface contact, thus reducing both wear and friction. This has been observed by Yamada *et al.* (1990) who found that polymer particles can act very effectively to reduce wear and friction.

Coating fracture. The use of thin hard coatings is often a very favourable solution from the wear point of view, but if the substrate is not hard enough to carry the load, plastic or elastic deformation will take place in the substrate under the contact. The increased stresses within the coating or at the interface between the coating and the substrate, as the coating deforms along with the surface of the substrate, may be higher than the tensile or shear strength of the material, resulting in crack nucleation and propagation. Such a contact situation is illustrated in Fig. 3.84e.

The harder the substrate is, the higher the loading which the coating can resist without failure by fracture because of increased resistance to deflection (Hedenqvist *et al.*, 1990a). Increased loads can also be resisted without failure with thicker hard coatings because of their load-carrying capacity

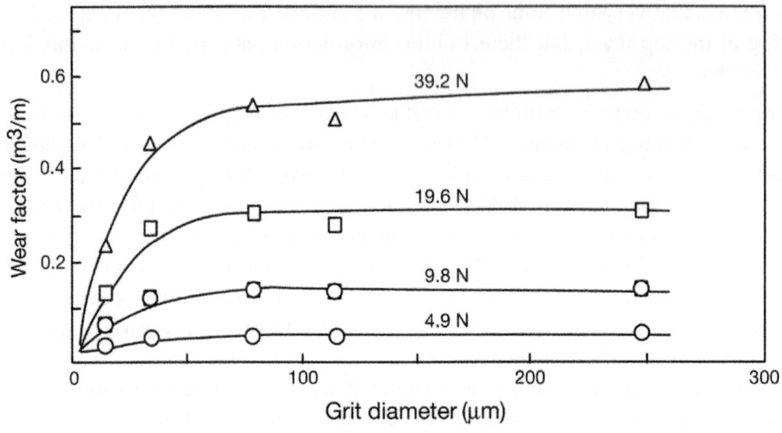

Fig. 3.87. Wear rate for steel as a function of abrasive grit diameter for different normal loads (data from Sin *et al.*, 1979).

which reduces deflection (Roth *et al.*, 1987). With a softer substrate, cracks will occur in the coating both within the contact area and outside at the substrate material pile-up area as shown in Fig. 3.88 (Burnett and Rickerby, 1987b; Miyoshi, 1989a; Page and Knight, 1989; Guu *et al.*, 1996a, b; Lin *et al.*, 1996a, b; Bull, 1997; Korunsky *et al.*, 1998; Yuan and Hayashi, 1999; Ligot *et al.*, 2000; Podgornik and Wänstrand, 2005).

When a highly loaded sphere is sliding on a hard coating the friction originating from both shear and ploughing will result in tensile stresses behind the sphere and compressive stresses and material pile-up in front of the sphere. This can result in several different cracking patterns depending mainly on the geometry and hardness relationships as discussed in detail in section 3.3.5.

Even if the coating does not fracture and fail at the first pass when it is loaded by a sliding counterface, repeated loading may result in fatigue cracks and failure. The lifetime of a thin coating may even be considerably longer than that of a thick coating, for several reasons. First, as shown in Fig. 3.89, under similar deformation conditions, the thicker coating will experience high bend stress levels. Second, since the coatings typically have columnar growth morphologies, any crack normal to the surface will be large in a thick coating, and may exceed the critical crack length, whereas in a thin coating this may not be the case.

There is experimental evidence that in rolling contact fatigue tests hard coatings (TiN) with a film thickness well below 1 micrometre have up to two orders of magnitude longer lifetime than 2 to 3 μm

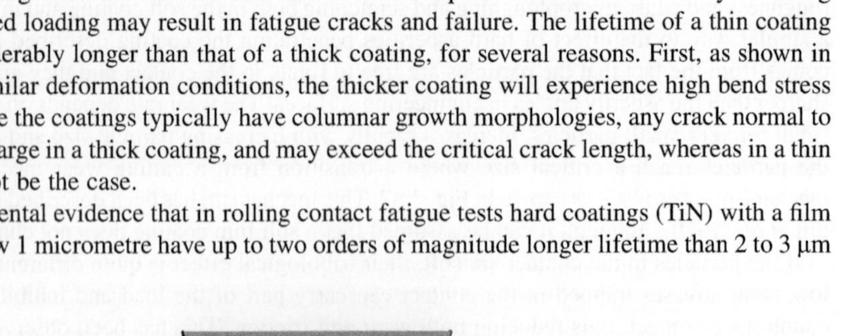

Fig. 3.88. Fracture of a hard brittle coating on a soft substrate takes place in the contact area and at the material pile-up area around the contact. Directions of material flow are indicated by arrows.

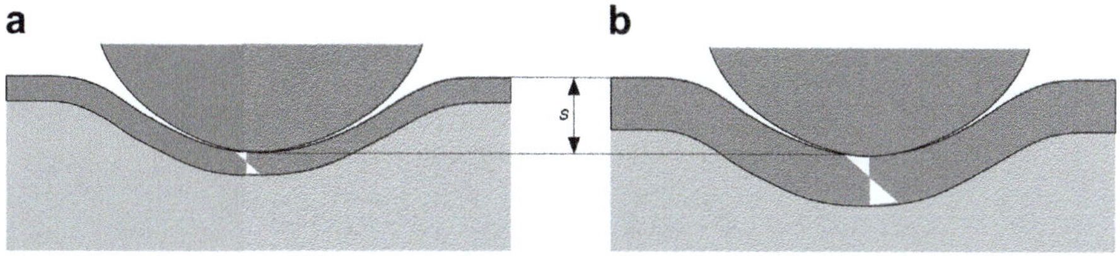

Fig. 3.89. (a) Thin hard coatings on a soft substrate accumulate lower stresses in the coating and at the coating-to-substrate interface compared to (b) thick hard coatings with the same deflections.

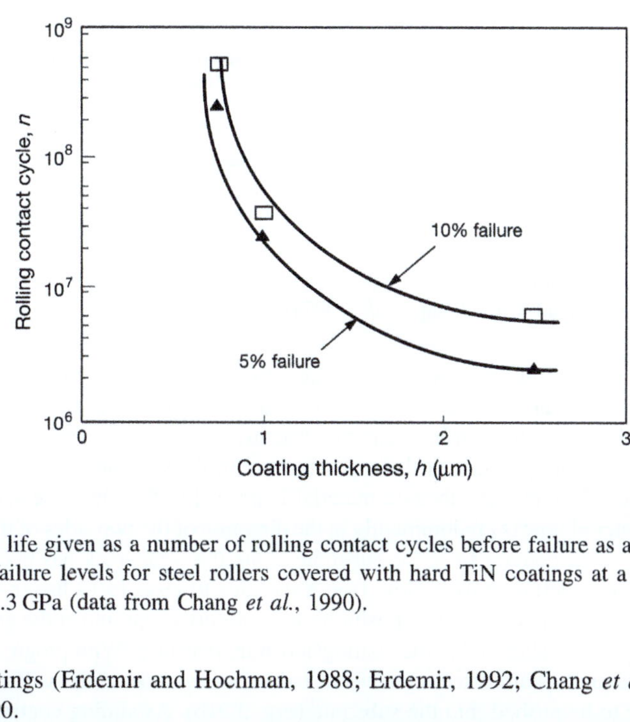

Fig. 3.90. Fatigue life given as a number of rolling contact cycles before failure as a function of coating thickness at two failure levels for steel rollers covered with hard TiN coatings at a maximum Hertzian contact stress of 2.3 GPa (data from Chang *et al.*, 1990).

thick similar coatings (Erdemir and Hochman, 1988; Erdemir, 1992; Chang *et al.*, 1990, 1992) as shown in Fig. 3.90.

It has been shown by Sainsot *et al.* (1990) that high stresses are concentrated at the interface between a hard coating and a soft substrate (Fig. 3.44). The shear stress at the interface is high because the soft substrate allows larger expansions and compressions at the interface plane than the hard coating. The ability to resist the high stresses depends on the strength of adhesion between the coating and the substrate. Single or repeated loading can result in breaking of the adhesive bonds and liberation of parts of the coating in flake-like wear debris (Miyoshi, 1989a; Hedenquist *et al.*, 1990a).

Asperity fracture. In the contact between a hard slider and a hard coating, both with surfaces of high roughness and steep asperity angles as shown in Fig. 4.84f, there is a strong tendency to asperity interlocking with high friction at the initiation of sliding. A considerable initial wear will take place by breaking of the asperity peaks and thus smoothing of the surfaces. The hard and sharp wear debris may considerably affect wear during the continuing sliding. Typical examples of this kind of asperity fracture are the initial wear when sliding a diamond against a diamond coating (Hayward and Singer, 1990; Hayward *et al.*, 1992) shown in Fig. 4.54 and asperity nanofracture observed with DLC coated surfaces (Wei and Komvopoulos, 1997; Borodich *et al.*, 2003).

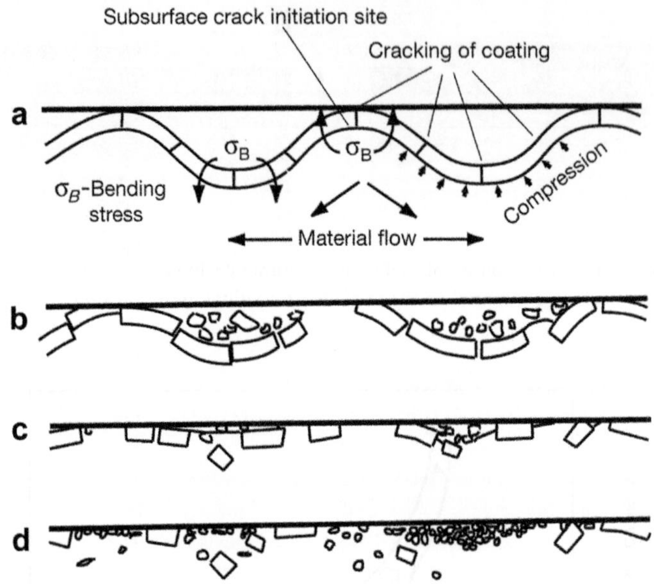

Fig. 3.91. Schematic diagram showing wear processes during sliding wear of hard coatings on a soft rough substrate at a high load (after Jiang *et al.*, 1997a).

The surface asperity fracture process with a rough composite DLC coated soft aluminium alloy substrate sliding against a tungsten carbide ball was analysed by Jiang *et al.* (1997a). Based on their observations they described the surface asperity fracture process as follows and also shown in Fig. 3.91. When the contact pressure at the asperities within the substrate considerably exceeds the yield strength of the substrate, the substrate material is pushed to flow into the surrounding regions (Fig. 3.91a). The material flow is predominantly in the direction of the two sides of the wear track after the first two cycles of the contact since the substrate along the wear track has been work hardened. Locally, the substrate material tends to flow from asperities to grooves to flatten the surface. High bending stresses act on the hard coating at positions at the asperity tops and at the groove bottom, and cracks are initiated at these sites and at the coating/substrate interface. With progress of sliding, wear debris are generated and accumulated somewhere on the wear track which will cause some of the fragmented coating to be pushed into the substrate (Fig. 3.91b). As sliding continues the surface is flattened and more fragments of the coating are either pushed into the substrate or removed from the surface (Fig. 3.91c). Some of the coating fragments may get entrapped within the rubbing interface, grind into finer particles and finally mix up with the substrate material to form a mechanically alloyed hard surface layer (Fig. 3.91d). In parallel with these structural mechanical changes, at the same time a graphitization of the carbon at the top layer may take place.

Delamination. A more common situation is the sliding of a hard rough slider against a hard rough coating with topography of low asperity angles, typically found in engineering surfaces, as represented by Fig. 3.84g. The sliding action takes place at the tops of the contacting asperities. They mainly deform plastically, although the overall contact stress may be less than the yield stress of the materials, because the local stresses at the small asperity areas are much higher.

The high stresses at the asperities will thus generate dislocations, pile-up of dislocations and crack nucleation at only some 10 micrometres beneath the surface as described by Suh (1973, 1986) in the delamination theory of wear and shown in Fig. 3.50. Because of the small plastic deformation of the surface, a large number of cracks must be nucleated before a loose particle can form. The delaminated particles are flake-like and may be some 100 micrometres long. This kind of flake-like wear debris has been observed in the sliding contact between a steel slider and a hard coating (TiN) by Sue and Troue (1990) after about 200 sliding cycles; agglomerated particles being observed after 500 cycles. The generation of flake-like wear debris probably by some kind of delamination mechanism has also been observed for softer materials like polyimide polymers (Yamada *et al.*, 1990) and for DLC coatings (Lackner *et al.*, 2003).

The wear rate in delamination wear can be calculated from an equation developed by Suh (1986) where it is given as a function of hardness, contact length, depth, average crack propagation rate, load, spacing of asperity contacts and crack spacing.

El-Sherbiny and Salem (1981) have worked on theoretically describing on a larger scale the asperity fatigue wear of hard rough coatings. They developed an equation where the wear rate is expressed as a function of normal load, sliding speed, wear track width, standard deviation, mean radius and stiffness of surface asperities and the static yield strength of the multilayered material, but the validity of their equation has not been definitely confirmed experimentally.

Delamination may also refer to the detachment of coating flakes from the substrate in repeated sliding. This process was described by Kato *et al.* (1991) when a 33 μm radius diamond sphere was sliding on a 1 μm thick Al_2O_3 coated WC–alloy surface and is shown in Fig. 3.66. The delamination of the coating starts on both sides of the contact groove. By further repetition of sliding the coating at the groove centre delaminates. Then subsurface lateral cracks propagate parallel to the surface, and the coating is partially lifted. Branched cracks appear from the combined effect of frictional stress and the residual stress. Perpendicular radial cracks on both sides of the groove and tensile-type cracks at its centre help the lifting and detachment of the delaminated coating from the surface.

Wear particles or fragments that are very thin with dimensions less than the magnitude of the surface roughnesses are not expected to influence the wear process, as shown in the lower part of Fig. 3.84g, because they can be hidden in the valleys between the asperities while the sliding action takes place at the asperity tops.

Abrasive wear. Abrasive wear is typified by a hard rough slider sliding on a hard rough coating, with hard particles of sizes larger than the magnitude of the surface roughness present in the contact, as shown by Fig. 3.84h. The trapped particles have a scratching effect on both surfaces, and as they carry part of the load they will cause concentrated pressure peaks on both surfaces as they try to penetrate them. The high pressure peaks may well be the origin of crack nucleation in the coating (De Wit *et al.*, 1998; Tricoteaux *et al.*, 2003). A rolling movement of the particles, which sometimes may even decrease the friction, is often present.

The sliding process will have an influence on the material properties of the counterface material surface. Because of work hardening, phase transitions or third body formation, the microhardness of a steel wear surface can be about three times larger than the initial bulk hardness. In sliding abrasive contacts with uncoated steel surfaces the main part of the wear particles is often smaller than 1 micrometre in size (Kato, 1992). Abrasive wear has been observed in steady-state sliding of a steel slider, e.g. against a hard TiN coating, by Sue and Troue (1990), Malliet *et al.* (1991) and Franklin and Beuger (1992) and against DLC by Jiang and Arnell (2000), Borodich *et al.* (2003), Lackner *et al.* (2003) and Samyn *et al.* (2006).

Rolling contact fatigue. The contact conditions and wear mechanisms discussed above are all related to a sliding contact. The situation in a rolling contact, as shown in Fig. 3.92, has many similarities but there are also clear differences. In a rolling contact the surface sliding component is minor and often limited to microsliding within the contact zone. Thus the adhesive mechanisms are not so dominating and especially in the very common lubricated rolling contacts they are of minor influence. On the other hand, the wear in rolling contacts is typically dominated by the repeated loading and deformation of the contact zone and fatigue wear dominates and limits the lifetime of the contact.

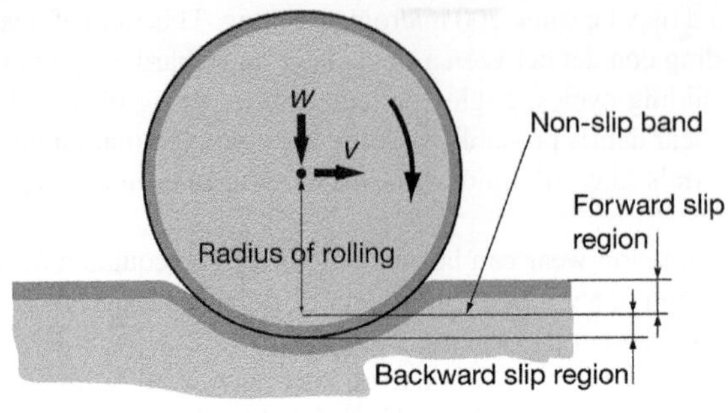

Fig. 3.92. In a coated rolling contact surface deformation and fatigue dominate the wear process and the adhesive part has a minor role.

Several wear mechanisms may be involved in a rolling contact. Typical are macropitting, micropitting, microdelamination, nanodelamination, mild polishing and severe polishing. These mechanisms were classified by Yonekura *et al.* (2005) in a study of rolling contact wear of hardened and tempered steel rollers coated with three kinds of low friction coatings, Cr_2N, CrN + nC and Cr + W + C:H, and rolling against a steel washer in a thrust bearing test.

The classification is shown in Fig. 3.93. Type (a) is delamination of the coating along and/or slightly beneath the coating/substrate interface with no polishing wear of the coating. Type (b) is mild polishing wear of the coating and subsequent delamination along and/or slightly beneath the coating/substrate interface. Type (c) is micro-/nanodelamination of the coating outer layer in combination with polishing wear. Type (d) is macropitting, often referred to as the most typical rolling contact fatigue mechanism, and coating delamination. Type (e) is severe polishing wear and subsequent rolling contact fatigue. Type (f) is micropitting of the coating with pit diameter <10 μm. Micropitting is sometimes observed at an early stage of the rolling and may later reduce or disappear without having any detrimental consequences. Yonekura *et al.* concluded that the most destructive failure modes were the macropitting and the severe polishing with subsequent rolling contact fatigue.

3.4.4 Micromechanical tribological mechanisms

The origin of the friction and wear phenomena that we observe on the macrolevel is found in the mechanisms that take place at the microlevel. The integration of all the micromechanical mechanisms results in the macromechanical mechanisms discussed above.

The micromechanical tribological mechanisms consider the stress and strain at an asperity level, the crack generation and propagation, material liberation and single particle formation, as shown in Fig. 3.94. In typical engineering contacts these phenomena are at a size level of about 1 micrometre or less down to the nanometre level.

Shear and fracture are the two basic mechanisms for the first nucleation of a crack and for its propagation until it results in material liberation and formation of wear scar and a wear particle. These mechanisms were discussed in sections 3.2 and 3.3. In a detailed analysis of the micromechanical mechanisms we need to use the very strict definition of the basic friction and wear mechanisms, as proposed by Holmberg *et al.* (2007). They classify friction in three basic mechanisms, adhesion, ploughing and hysteresis; and wear also in three mechanisms, adhesion + fracture, abrasion + fracture and fatigue + fracture, as shown in Fig. 3.95.

By definition wear always includes material removal. Fracture is here understood widely as a process starting from loss of cohesion between bond structures in the material, continuing as crack propagation and resulting in debris being liberated from the surface. In this classification

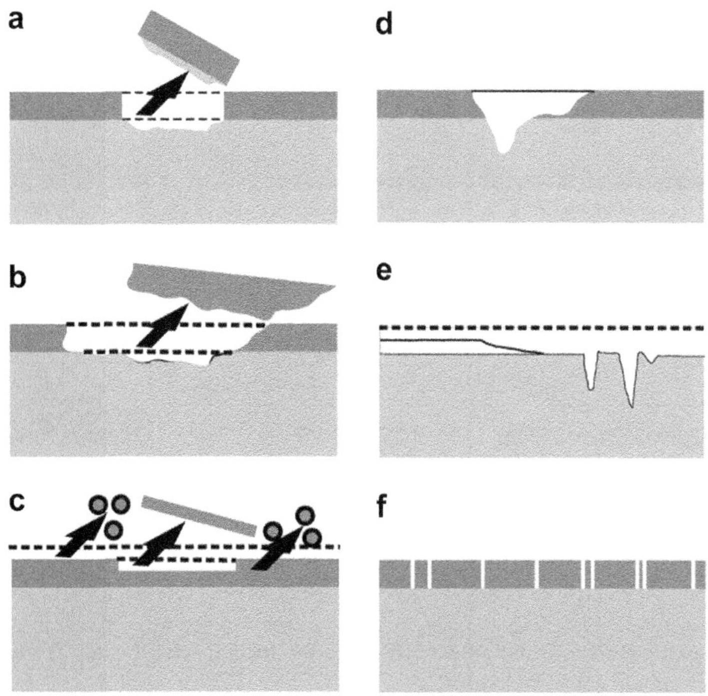

Fig. 3.93. Classification of coated roller wear mechanisms: (a) coating delamination, (b) mild polishing, (c) micro-/nanodelamination, (d) macropitting, (e) severe polishing and (f) micropitting (after Yonekura *et al.*, 2005).

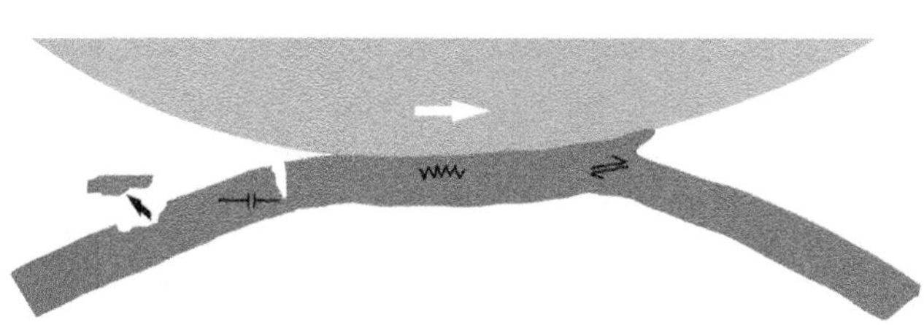

Fig. 3.94. Micromechanical tribological mechanisms consider stress, strain, crack generation and propagation and single particle formation on a microlevel, typically this is what takes place in asperity contacts.

tribochemical wear is not included. The tribochemical reactions are classified as surface modification mechanisms since they represent chemical reactions that cause material modification, either improvement or degradation of mechanical and physical properties, but do not include material removal. In tribochemical wear still one of the three above-mentioned basic wear mechanisms is the one that causes material removal (see Fig. 3.95).

Similarly, fatigue wear is divided into two phases. In the first phase only material modification is taking place, without material removal. During continuous loading of the surface the close-to-surface material properties are slowly changed. The second phase, that is the material removal phase, starts

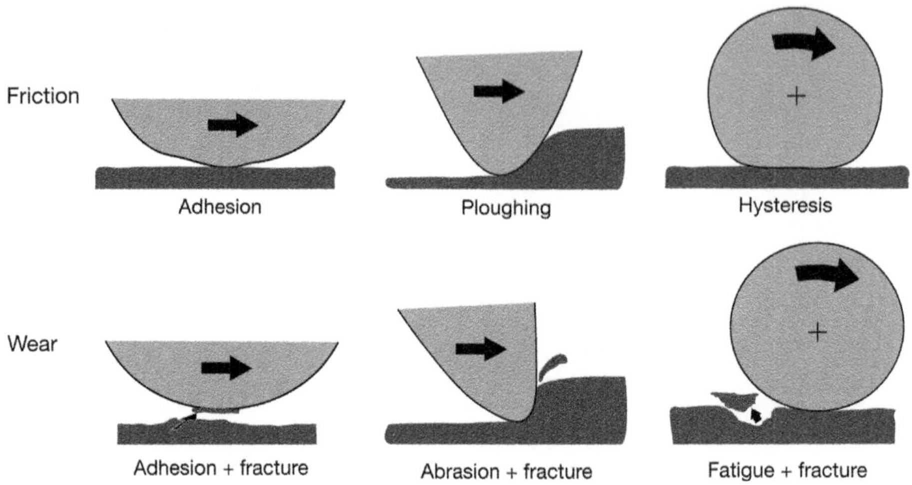

Fig. 3.95. The basic friction and wear mechanisms are related to adhesion, ploughing and hysteresis. In the case of wear these contact mechanisms result in material fracture, detachment and removal.

when the changed material can no longer withstand the loading and a crack is created; it grows, material is liberated and debris is formed.

3.4.4.1 A modelling approach

The above strict classification of basic friction and wear mechanisms is the basis for the modelling approach by Holmberg *et al.* (2007) to understand the micromechanical tribological mechanisms. They analysed different contact situations of coated surfaces and indicated the tribologically dominating parameters that should be the focus in a modelling study.

Friction and wear are governed by the shear taking place at the top surface and in the deformed surface layer, and by the elastic, plastic and fracture behaviour both at the top surface and in the deformed surface layer. A thin coating is typically a part of this deformed surface layer. In addition, surface degradation may take place due to tribochemical and fatigue processes that influence the surface strength and ability to withstand loaded conditions. Thus the crucial material parameters are the elastic modulus, the hardness, shear strength and the fracture toughness on the top surface, in the coating, at the coating/substrate interface and in the substrate under the coating, as shown in Fig. 3.96.

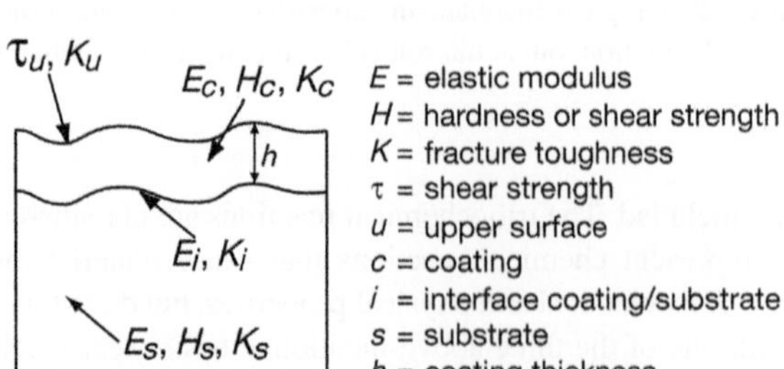

Fig. 3.96. Crucial material parameters and their location influencing on the micromechanical friction and wear behaviour.

In this presentation, hardness, H, is used as symbol representing the resistance to plastic deformation due to its common use, even if referring to the elastic–plastic constitutive response of the material would be the correct expression.

The presentation below is simplified to the conditions of a sphere, e.g. a spherical asperity, sliding over a flat coated surface, ideally smooth surfaces, homogeneous materials and with no contamination or wear debris involved.

Adhesive friction is dominated by the shear taking place in the surface top layer or the shear in between the two interacting surfaces. The coefficient of friction

$$\mu_a = f(\tau_u) \tag{3.36}$$

Ploughing friction is dominated by the elastic and plastic behaviour of the coating and the substrate. The capacity of the substrate coating system to withstand deformation is frequently called load-carrying capacity. The coefficient of friction

$$\mu_p = f(E_c,\ H_c,\ E_s,\ H_s,\ h) \tag{3.37}$$

Hysteresis friction is mainly dominated by the elastic properties of the substrate but also to some extent by the elastic properties of the coating. The coefficient of friction

$$\mu_h = f(E_s,\ E_c,\ h) \tag{3.38}$$

Adhesive wear is dominated by the fracture behaviour in the surface top layer, in the coating, at the coating/substrate interface and in the substrate. The adhesive force, F_a, from the counterface tries to tear off part of the surface material over the contact area A (Fig. 3.97). When surface roughness is included A will decrease and the adhesive wear typically decrease. The wear rate

$$k = V_{adh}/w \cdot s = f(K_u,\ K_c,\ K_i,\ K_s) \tag{3.39}$$

where V_{adh} is the volume of adhesive wear, w is the load and s is the sliding distance.

The breaking of the material may take place at different locations in the surface depending on the material strength at each location as shown in Figs 3.97b–e. Due to the depth of the debris detachment the result is

- top layer fragments dominated by K_u (Fig. 3.97b),
- coating fragments dominated by K_c (Fig. 3.97c),
- coating delamination dominated by K_i (Fig. 3.97d), or
- substrate and coating debris dominated by K_s (Fig. 3.97c).

Abrasive wear is dominated by the geometrical collision of the two moving surfaces resulting in high stresses, material shear and fracture, and debris formation. The collisions may be due to hard asperity or third body ploughing or asperity collisions, as shown in Fig. 3.98. The wear rate

$$k = V_{abr}/w \cdot s = f(K_c,\ K_i,\ K_s,\ H_c,\ H_i,\ H_s,\ h) \tag{3.40}$$

Due to the depth of the debris detachment the result is

- coating fragments dominated by K_c,
- coating delamination dominated by K_i, or
- substrate and coating debris dominated by K_s.

Fatigue wear is a result of material degradation where the strength of the material has decreased to a level so that it can no longer withstand the repeated loading. The result is fracture, cracking and debris formation (Fig. 3.99). It is dominated by the elastic and fracture properties of the coating, the coating/substrate interface and the substrate. The loading may be rolling or sliding. The wear rate

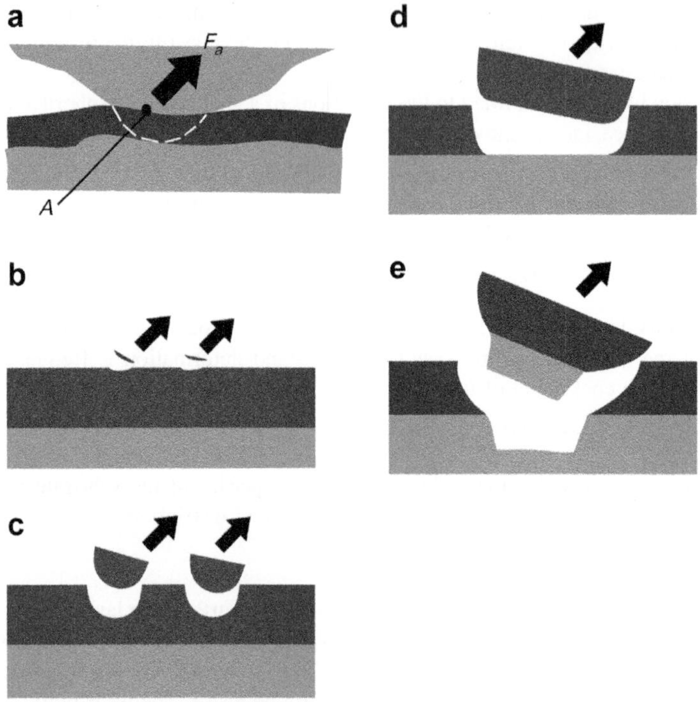

Fig. 3.97. Two surfaces attach to each other by adhesion (a) and the movement of the top surface results in an adhesive force, F_a, that tries to detach material over an area, A, from one of the surfaces. The detachment may take place (b) at the top surface, (c) within the coating, (d) at the coating/substrate interface and (e) in the substrate.

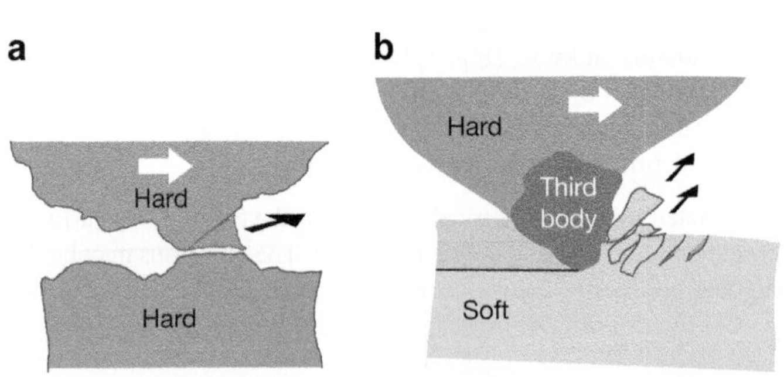

Fig. 3.98. Abrasive wear is characterized by a hard asperity (a) or a third body (b) that deforms the countersurface in a ductile or brittle way resulting in fracture, cracking and debris generation.

$$k = V_{fat}/w \cdot s = f(K_u, K_c, K_i, K_s, E_c, E_i, E_s, h) \tag{3.41}$$

Due to the depth of the debris detachment the result is

- top layer fragments dominated by K_u,
- coating fragments dominated by K_c,

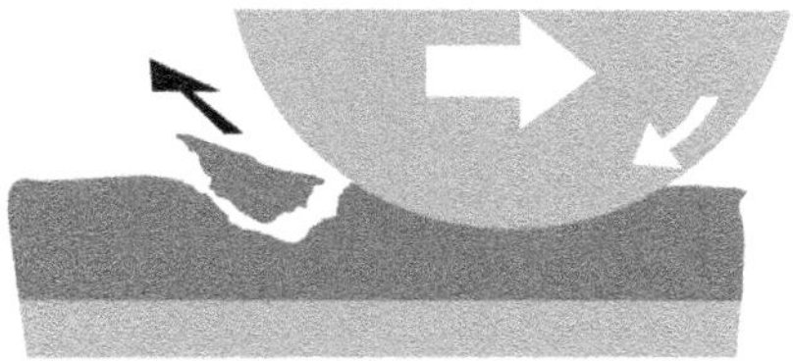

Fig. 3.99. Fatigue wear is characterized by repeated loading of the coated surface resulting in cracking at the surface in the coating, followed by fracture, material detachment and debris generation.

- coating delamination dominated by K_i, or
- substrate and coating debris dominated by K_s.

Surface modification is here considered separately from the pure wear mechanisms since it alone will not result in material liberation. Still it is very important to consider in a wear process since it often precedes the wear event and may be the reason for wear to start. Typical surface modification mechanisms are chemical reactions taking place at the surface and material fatigue from repeated loading.

Chemical reactions and sometimes even physical structural modifications, such as oxidation of metallic coatings, graphitization of DLC and crystallographic reorientation of MoS_2 coatings, take place mainly at the top surface and influence the shear, τ_u, and fracture, K_u, properties and thus the adhesive friction and adhesive wear. Sometimes also the whole coating structure and the coating/substrate interface be affected.

Fatigue is a result of repeated loading on the material that weakens the structure by including cumulative damage to a level that it can no longer take the loading. The maximum stress peaks are normally under the coating in the substrate because the coating layer is so very thin. Thus the substrate elastic properties, E_s, and fracture toughness, K_s, are crucial. If the stress peaks are high and also closer to the surface, the elastic and fracture properties of both the coating and the coating/substrate interface, E_c, K_c, E_i and K_i, may also be important.

This approach to simplify the basic mechanical friction and wear-related phenomena in the tribological contact of a coated surface was used by Holmberg and co-workers (Holmberg *et al.*, 2003, 2006a, b; Laukkanen *et al.*, 2006; Ronkainen *et al.*, 2007) as a basis for their 3D FEM modelling, stress and strain simulations and fracture calculations, also described in section 3.2.8. As the stresses have been determined, their influence on crack growth calculated and the wear debris process defined, the single debris generation process can be integrated into multidebris wear particle generation as discussed in sections 3.3.7 and 3.3.8. This whole process is of course very complex and still today not within the range of being modelled in a holistic way.

3.4.4.2 Velocity accommodation

One other approach to studying the micromechanical wear mechanisms involves using the velocity accommodation mode concept developed by Berthier *et al.* (1989) which is a logical framework for achieving a more complete fundamental understanding of the complex mechanical changes in a tribological contact.

Velocity accommodation is defined as the manner in which the difference in velocity of two surfaces in relative motion is accommodated across the interface (Berthier *et al.*, 1989). This problem formulation is a fundamental question in all tribological studies. The place for velocity accommodation can be either in one of the solids, in the intermediate material or at the interface between them, as shown in Fig. 3.15. The intermediate material may be a liquid, a gas or loose particles. Berthier

a

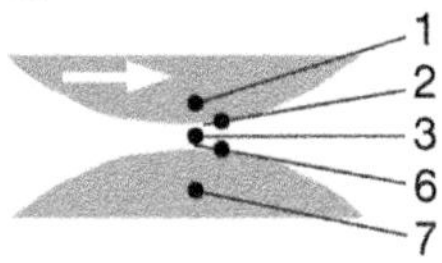

b

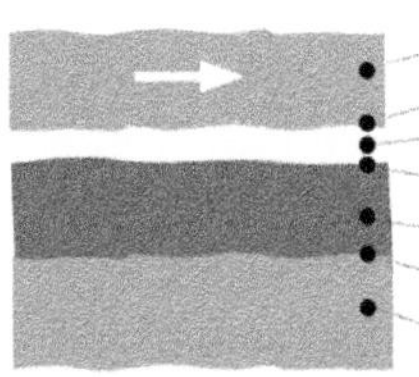

	SHEAR Shear strength	FRACTURE Fracture toughness	ELASTICITY Young's modulus	ROLLING Rolling resistance
1	-	++	++	-
2	+	-	-	-
3	+++	+	+	+++
4	+++	+	+	+
5	+++	+++	+++	-
6	+	++	+	-
7	-	++	+++	-

+++ = very important + = not important
++ = considerable - = generally negligible

Fig. 3.100. The velocity accommodation in (a) an uncoated and (b) a coated sliding contact may take place in (1) the counterface, (2) its interface, (3) the intermediate material, (4) the top surface layer, (5) the coating, (6) its interface to the substrate or (7) in the substrate. The influence of the velocity accommodation mechanisms on friction and wear is indicated in the table.

et al. define four mechanisms describing how the velocity is accommodated; these are shown in Fig. 3.15, and are elasticity, fracture, shear and rolling.

There is a principal advantage with this approach to tribological mechanisms compared with the traditional ones. It is that here the mechanisms causing friction and wear are defined in terms well known from material science and mechanical engineering and the properties can thus be derived to known parameters, such as strength, fracture toughness, Young's modulus of elasticity or rolling resistance. The disadvantage is that when the whole contact system is split up into very small parts, it becomes very difficult to find a way to handle the interactions of the different parts with each other. Another disadvantage is that still today we do not have satisfactory analytical models to describe many of the different mechanisms in velocity accommodation.

In the contact with a coated surface the situation is even more complex because the number of sites for velocity accommodation increases. Figure 3.100 shows that the velocity accommodation can take place also at the interface between the substrate and the coating, within the coating and within a microfilm on top of the coating.

The velocity difference between two surfaces can be accommodated by the elasticity of the surfaces in contact. This is the case when the surfaces adhere strongly to each other and the applied tangential load is smaller than the load needed to cause slip. The junctions do not break but the elasticity allows small movements. This is typically the case in what is called false fretting, when a small amplitude reciprocating movement takes place without sliding.

With increased tangential loads, high tensile stresses at the surface generate cracks by fracture perpendicular to the surface. The cracks more easily allow movements of surface fragments in a tangential direction without shear in the adhered junctions. This may occur in both reciprocating and sliding contacts. The result is a tooth-like surface topography.

Shear of adhered junctions is the most common way for velocity differences between surfaces to be accommodated. This takes place when the shear stress generated by the tangential load exceeds the shear strength at the weakest place of the contact junctions. In brittle solid materials the result is typically a lateral crack at the surface, whereas in fluids, sheets of the material at the molecular level slide over each other just like cards in a pack. With plastic deformation of solids, as in ploughing, shear continuously takes place in the material, producing a new permanent shape for the surface.

With loose particles between the surfaces the velocity difference can be accommodated by the rolling motion of the particles. In certain sliding contact conditions round or cylindrical wear particles are produced and they may act in the contact as rollers if the surface roughness or softness does not inhibit the rolling.

Another approach to investigating the micromechanics of a coated surface is the multiscale approach presented by Yan *et al.* (2000). They used periodic unit cell-type continuum mechanics models to predict localized deformation patterns at the scale of the coating thickness and the rate of materials removal due to repeated sliding contact. The results indicate that the deformation of the coating is controlled by the cyclic accumulation of plastic deformation, or ratchetting, at the coating subsurface. Based on a ratchetting failure criterion, a wear equation is proposed and applied to investigate parametrically the influence of the principal material, loading and surface roughness parameters on the wear rate. The results revealed that the wear rate increases with contact pressure and depends strongly on coating thickness and the toughness of the counterpart surface.

3.4.5 Tribochemical mechanisms

The chemical reactions taking place at the surfaces, during the sliding contact and also during the periods between repeated contacts, change the composition and the mechanical properties of the outermost surface layer. This has a considerable influence on friction and wear because they are to a great extent determined by phenomena like shearing, ploughing, cracking and asperity interlocking, all taking place at the surface top layer. The chemical reactions on the surfaces are strongly influenced by the high local pressures and the flash temperatures, which can often be up to 1000°C and for some hard ceramic surfaces even up to 2000°C, occurring at spots of asperity collisions. The formation of tribochemical surface films in tribological conditions has already been discussed in section 3.3.9.

The tribochemical surface mechanisms influence friction through the increase in contact temperature due to rubbing, increased surface reactivity through exposure of clean surfaces, transformation of surface material, 'triboelectricity' and direct mechanical stimulation of reactions. The influence may be an increase or decrease of friction (Fischer, 1992; Donnet, 1998a; Neville *et al.*, 2007; Wood, 2007).

Mechanical wear mechanisms generally dominate the wear process in severe wear but in conditions of mild wear tribochemical mechanisms become increasingly more important. Thermochemical analysis has been used for describing chemical reactions that accompany low wear and low friction on MoS_2, TiN and TiC coated surfaces (Singer, 1991; Singer *et al.*, 1991).

The tribochemical mechanisms that have been observed in tribological contacts can be divided into four main surface phenomena for which the chemical effects are essential. They are:

1. the formation of thin microfilms on hard coatings,
2. the oxidation of soft coatings,
3. the formation of reactive boundary layers in lubricated contacts, and
4. tribocorrosion of coated surfaces.

The tribological aspects of the mechanisms will be discussed below.

3.4.5.1 Formation of thin microfilms on hard coatings

Earlier we explained theoretically that the deposition of a soft thin coating on a hard substrate will reduce both the contact area and the shear strength at the contact and thus result in very favourable tribological conditions with low friction, as shown in Fig.3.73. This has been confirmed by several experimental investigations into layered materials or soft metals on hard substrates (e.g. Kuwano, 1990; Erdemir *et al.*, 1991a, b).

As a consequence of this, one would expect very unfavourable tribological conditions with high friction when a relatively soft substrate is covered by a hard coating. Surprisingly, however, this is not necessarily the case. Low coefficients of friction down to 0.1 were reported for a hard titanium nitride coating sliding against itself (Mäkelä and Valli, 1985). Also, similarly, low values in dry sliding contacts, down to 0.05, were measured for a DLC coating sliding against a sapphire counterface (Singer *et al.*, 2003). It is obvious that there must be some explanation for this apparent contradiction.

The very low friction behaviour of hard coatings can be explained by the formation of low shear strength microfilms on the hard coating or on the asperity tips of the coating, as shown in Fig. 3.74 and a reduction in the real contact area. Thus, if we are studying the contact on a microscale, we again have a soft coating on a hard substrate, but now it is the coating (e.g. TiN or DLC) which is playing the role of hard substrate and the formed soft microfilm is playing the role of a coating. It is obviously advantageous for the substrate under the hard coating to be as hard as possible to avoid fracture of the brittle coating by deformation, to improve the load support and to reduce the real area of contact.

Experimental results that confirm the above hypothesis can be found in the literature. The formation of a graphite-like microfilm on the hard DLC coating and the countersurface explains the low coefficients of friction reported for ceramic sliders sliding on DLC coatings (Singer *et al.*, 2003), see section 4.5.3.

Very common in practical applications is the formation of oxide microfilms on the top of hard coatings. XRD, XPS and scanning Auger microscopy have indicated the formation of iron oxide (Fe_2O_3) and titanium oxide (TiO_2) in the sliding contact between a TiN coated steel substrate and a steel slider and TiO_2 formation in sliding TiN against TiN contacts (Sue and Troue, 1990; Singer *et al.*, 1991; Franklin and Beuger, 1992). With the presence of ion implanted chlorine atoms inside TiN and TiC coatings, *in situ* formation of lubricious intermediate titanium oxides TiO and Ti_nO_{2n-1} on the worn surfaces was observed in both dry sliding and machining tests (Aizawa *et al.*, 2005). The theoretical background for the formation of oxides on titanium, titanium nitride and titanium carbide presented by Gardos (1989) supports the hypothesis of microfilm formation.

Reduced friction has been observed, which indicates the formation of aluminium hydroxide microfilms, in the sliding contact of sapphire against a diamond coating (Hayward and Singer, 1990) and the formation of chromium oxide (Cr_2O_3) microfilms in the contact of a steel slider on a chromium coating (Gawne and Ma, 1989). Severe wear with spalling and oxidation as the main deterioration mechanisms followed when tribofilms were generated by agglomeration of oxidation products, WO_2 and Cr_2O_3, in a closed ring-on-ring sliding contact with WC/C coated Al_2O_3 sliding on Al_2O_3 (Blomberg *et al.*, 1993).

3.4.5.2 Oxidation of soft coatings

In environments containing oxygen, such as air, a thin (about 1 to 10 nm thick) oxide layer is formed very quickly on most metal surfaces. In particular, freshly cleaned (e.g. lapped or ground) surfaces will acquire a layer of oxide in less than a few seconds. This applies to copper, iron, aluminium, nickel, zinc, chromium and several other metals.

The oxides formed may influence in different ways the tribological properties of the surfaces. Some oxide layers, like copper oxide, are sheared more easily than the metal, while others, like aluminium oxide, form a very hard thin layer. Even if the oxide wear particles are hard, their abrasive effect is not necessarily important because of their small dimensions compared with the roughness of the surfaces or their embedment into soft coatings, as shown in Fig. 3.72i. Sometimes the oxide particles agglomerate to layers strong enough to carry the load.

The oxidation of the top layer of coatings may improve the lubricity of the coating, as has been observed for lead coatings, where oxidation results in a top layer that is a combination of lead and lead oxide (Sherbiney and Halling, 1977) and for molybdenum disulphide coatings, where the oxidation results in layers of MoO_2, MoO_3 and oxides of Mo reacted with the slider material (Fayeulle *et al.*, 1990). Thin surface films of cadmium, silver and nickel have turned out to be ineffective for wear-reducing purposes in air because of oxidation, while the non-oxidizing gold was effective both in air and in inert environments when coated on steel substrates (Jahanmir *et al.*, 1976).

3.4.5.3 Formation of reactive boundary layers in lubricated contacts

Tribochemistry has a central role in boundary lubrication where reactive components in the lubricant react with the rubbing surface material and form lubricious surface layers with good load-carrying capacity. The basic boundary lubrication mechanisms are presented in section 3.1.5 and the tribological effects with coated surfaces in section 3.4.7.

3.4.5.4 Tribocorrosion

In tribocorrosion the coated surface is normally surrounded by a corrosive fluid that causes chemical attack on the surface. The wear and friction-related mechanisms involved have been reviewed by Wood (2007). Three major wear–corrosion mechanisms of coated metallic surfaces were proposed by Dearnley and Aldrich-Smith (2004) to be active in the situation of hard surface coatings like S-phase and CrN on stainless steel substrate. They are schematically shown in Fig. 3.101. Type I relates to mild wear–corrosion interactions due to removal or damage of passive films on coatings and their healing by repassivation. Type II relates to large wear–corrosion interactions due to galvanic attack of

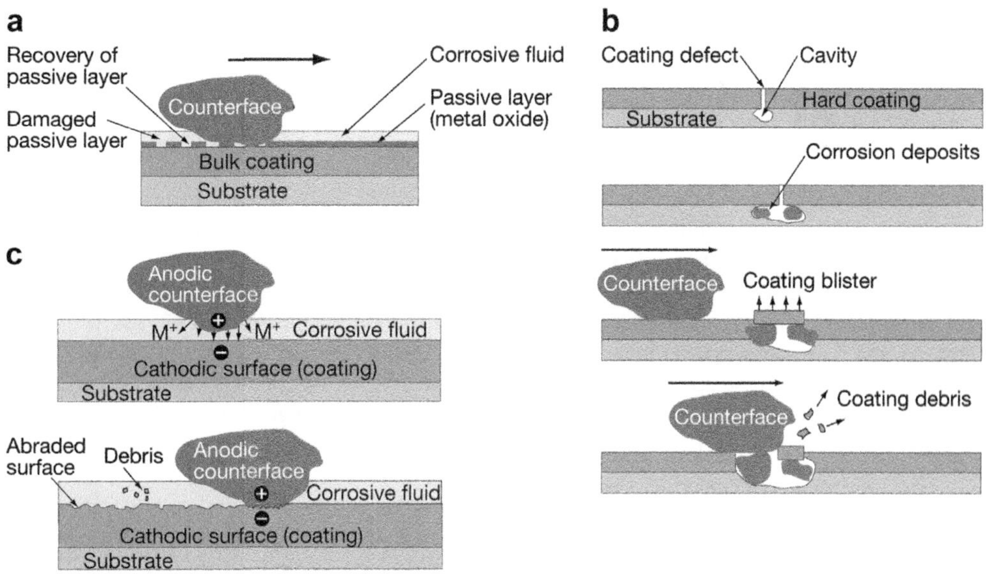

Fig. 3.101. Major types of tribocorrosion are (a) type I: where in hard coatings on metallic substrates the damage of the passive film results in repassivation and healing, (b) type II: where in hard coated metallic alloys pitting or blistering of the coating culminates in mechanical fragmentation and coating removal, and (c) type III: where galvanic roughening of the counterface results in abrasion of the coating (after Dearnley and Aldrich-Smith, 2004 and Wood, 2007).

the substrate leading to blistering and removal of the coating under sliding contact. Type III relates to abrasion of the coating during subsequent sliding due to galvanic attack of the counterface which is subsequently roughened.

3.4.6 Nanophysical tribological mechanisms

Tribology is basically the science of breaking bonds within or between materials being in movement relative to each other. The internal structure of materials comprises atoms organized and linked as molecules, crystals and microstructures. Thus the origin to all friction and wear phenomena is to be found on an atomic or molecular scale. *Nanotribology* is the study of tribological phenomena on length scales less than 100 nm. These phenomena are typically related to adhesion, interfacial friction, molecular scale solid and liquid surface films, molecular scale deformation, bond breaking, material transformation and detachment (Israelachvili, 1992; Dedcov, 2000; Gao *et al.*, 2004; Bhushan, 1995, 1999b, 2005).

Emerging experimental techniques and devices such as the surface force apparatus (SFA), the atomic force microscope (AFM), the friction force microscope (FFM), the scanning tunnelling microscope (STM) and other surface analysis methods have opened the possibility to study friction and wear phenomena on a molecular scale and to measure frictional forces between contacting molecules at the nano-newton level (Israelachveli and Tabor, 1972; Overney and Meyer, 1993; Bhushan, 1999b).

Increased computational power has at the same time made it possible to study adhesion, friction and wear processes by large-scale molecular dynamic simulations (MDS) of sliding surfaces and to investigate the atomic scale contact mechanisms, as shown in Fig. 3.102. The molecular dynamic simulations consist of integration of the equations of motion of a system of particles interacting via prescribed interaction potentials. The interatomic interactions that govern the energies and dynamics of the system are characteristic of the system under investigation. Simulated phenomena include elastic, plastic and yield processes, connective neck formation, wetting, reconstruction, solidification, atomic scale stick–slip and material transfer (Landman *et al.*, 1992; Harrison *et al.*, 1993, 1995; Perry and Harrison, 1996; Zhang and Tanaka, 1997; Gao *et al.*, 2003; Tanaka *et al.*, 2003; Yang and Komvopoulos, 2005; Schall *et al.*, 2007).

3.4.6.1 Nanofriction and nanowear

Only some aspects of the complex nanophysical phenomena related to friction and wear have so far been investigated and the information is fragmented, related to some special cases of contacts and materials. We have some new information about the origin of sliding friction, that it stems from various sources, including sound energy, and that static friction may arise from physisorbed molecules. An interesting observation is that the Amontons' law, stating that the lateral friction force is directly proportional to the applied normal load independently of the apparent contact area, surface roughness and sliding velocity, is amazingly well obeyed for a wide range of materials such as wood, ceramics and metals, and over a wide range of contact scales, all from macro- to nanocontacts (Krim, 2002b).

It has been observed in experimental measurements that friction values are often scale dependent. The values for friction on a nanoscale are much lower than those measured on a micro- and macroscale. The coefficient of friction for a 50 nm radius AFM Si_3N_4 tip sliding over three flat surfaces, graphite, natural diamond and Si(100), was 4 to 16 times lower compared to a 3 mm radius Si_3N_4 ball sliding over the same surfaces (Bhushan, 2005). Bhushan indicates four possible reasons for this: (1) in nanoscale experiments typically the contact stresses are higher, (2) the indentation hardness is higher when measured on a small contact area and low loads, (3) the small apparent area of contact reduces the number of particles trapped at the interface and minimizes third body ploughing and (4) there is a larger range of contacting asperities coming in contact at the macroscale.

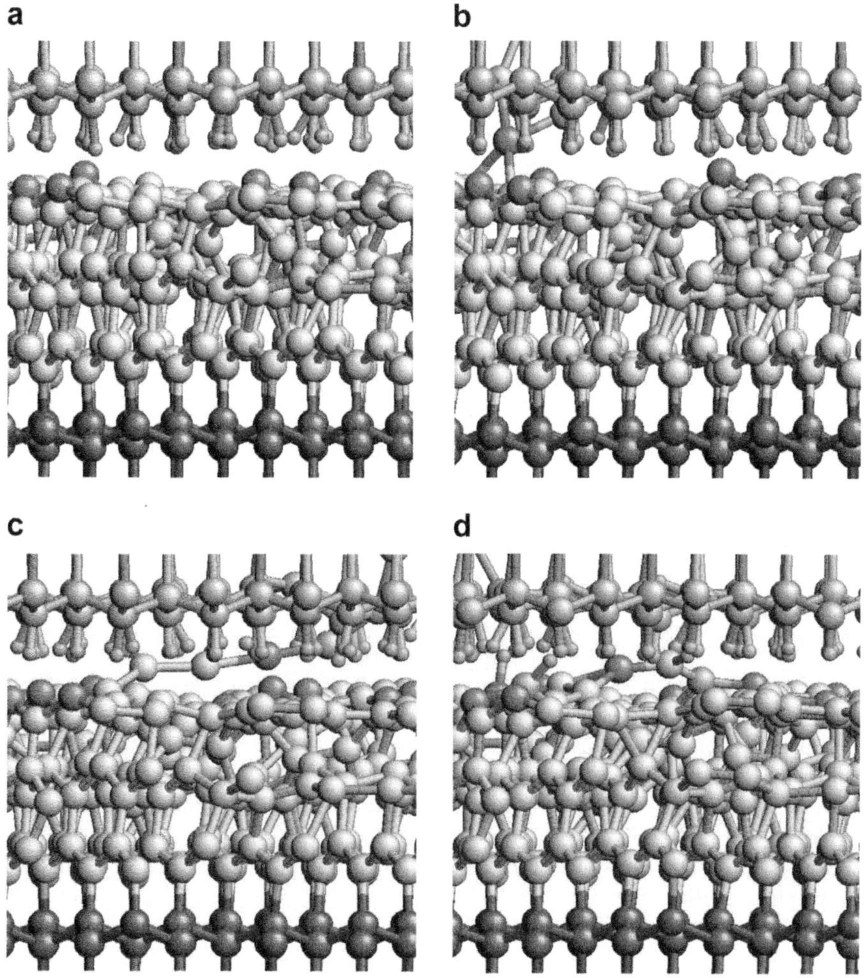

Fig. 3.102. Molecular dynamic simulation showing on an atomic level the interactions in a contact with a diamond surface sliding on a DLC coated diamond substrate. The diamond substrate is in blue, the DLC coating in yellow, the diamond contersurface in grey, hydrogen atoms in green and unsaturated sp-hybridized atoms in the DLC film in red (from Gao *et al.*, 2003, with permission). See Appendix B for coloured picture.

There are very few studies published on wear mechanisms on the nanoscale. For a sliding diamond tip against a copper flat surface system Zhang and Tanaka (1997) used molecular dynamic simulation and found four distinct regimes of deformation on the atomic scale: a defect-free no-wear regime and three wear regimes, the adhering, ploughing and cutting regimes, as shown in Fig. 3.103. The transition between these regimes was governed by some dominating sliding parameters such as indentation depth, sliding speed, asperity geometry and surface lubrication conditions. It was interesting that better lubrication, a smaller tip radius and lower sliding speed enlarge the no-wear regime. The no-wear regime seems to exist for a wide range of contact conditions which indicates that no-wear is possible even under chemically clean sliding conditions. In the cutting regime the frictional behaviour followed a proportional law. In all other regimes the variation of the frictional force was complex and

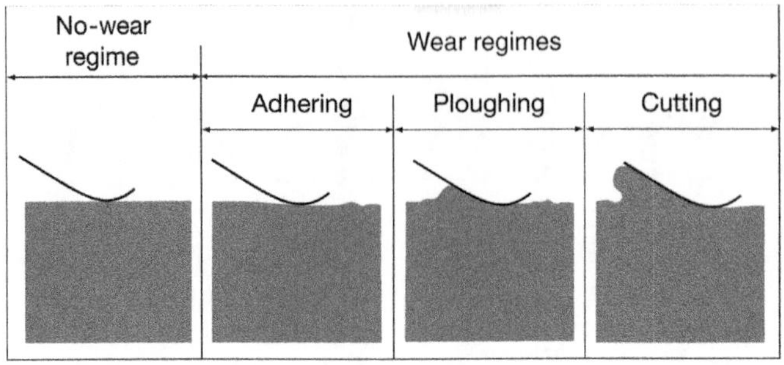

Fig. 3.103. Four atomic scale deformation regimes identified by molecular dynamics simulation of spherical diamond tip sliding on a flat copper surface from right to left (after Zhang and Tanaka, 1997).

not possible to describe by a simple formula. This classification of wear mechanisms appears to be on a general level in harmony with the wear classification used by Kato and co-workers and extended to nanoscale contacts in AFM experiments by Wang and Kato (2003c).

3.4.6.2 Nanolayered coatings

There are very limited data about the tribological behaviour of surface coatings of thicknesses in the nanorange. There is some information from studies carried out by surface force microscopy and environmental scanning electron microscopy with a sliding diamond tip with 20 nm to 20 μm radius sliding against a silicon surface coated with a-C, Cr or TiC coatings in the thickness range of 1 to 500 nm (Lu and Komvopoulos, 2001; Ma *et al.*, 2003; Wang and Kato, 2003a). From the results it appears that in many respects the basic friction and wear mechanisms are similar to those observed on a microlevel.

In the comparison of a 10 nm thick a-C coated Si disk with 100 nm thick Cr and TiC coated Si discs the a-C amorphous carbon coated surface had superior nanowear resistance due to the significantly higher hardness-to-elastic modulus ratio indicating a much higher elastic recovery at unloading (Lu and Komvopoulos, 2001). The deformation of the a-C film was purely elastic while it was elastic–plastic for Cr and TiC. The poorer wear resistance of the TiC films is due to a microscopic fracture process occurring at the grain size. The film nanowear resistance was related to the relative specific energy, i.e. the energy dissipated during the nanowear process in the film divided by that dissipated in the Si bulk material. With a diamond tip radius of 20 μm and 0.4 μm/s sliding speed the coefficient of friction increased at 50 μN load from 0.12 for an uncoated Si surface to 0.15 for Cr and TiC coated surfaces and to 0.2 for a-C coated surface. The increase in friction was less at higher loads up to 400 μN. The tendency was for the coefficient of friction to increase at higher siding speeds, most likely due to strain rate effects. For elastic–plastic sliding, the general trend was for the coefficient of friction to decrease with increasing contact load. The nanowear tests indicate that in addition to nanohardness and H/E ratio, the microstructure, type of atomic bonding and the deposition process controlling film growth and residual stresses in the film exhibit a strong effect on the nanowear resistance for surfaces coated with nanolayered coatings.

In a film thickness range of 1 to 500 nm thick nitrogen-doped amorphous carbon coatings on silicon discs Wang and Kato (2003a) showed that the transition of sliding mode from 'no observable wear particles' to 'wear particle generation' was delayed with increasing film thickness. Two typical kinds of wear particles were observed in the *in situ* examination of the contacting surfaces, feather-like wear particles and plate-like wear particles. The critical number of friction cycles before transition took place increased by 3 to 3.7 times when the coating thickness was increased from 10 to 100 nm.

3.4.6.3 Diamond and diamond-like carbon surfaces

Two common coating materials, diamond-like carbon and diamond surfaces, have been of major interest in nanoscale studies because of their regular structure, inertness and low friction and wear properties. Gao *et al.* (2003) showed by molecular dynamic simulations that the three-dimensional structure of an amorphous DLC film is paramount in determining its mechanical properties (Fig. 3.102). Particular orientations of sp^2 ring-like structures create films with both high sp^2 content and large elastic constants. Films with graphite-like top layers parallel to the substrate have lower elastic constants than films with large amounts of sp^3-hybridized carbon.

Several of the most recent studies of DLC coatings indicate that the general explanation for their extremely low coefficient of friction, down to $\mu = 0.001$ and below (Erdemir *et al.*, 2000; Erdemir, 2002) is related to five effects: (1) the surfaces are extremely smooth so asperity interlocking effects are eliminated, (2) the surfaces and substrates are hard so ploughing effects are eliminated, (3) the dangling bonds of the carbon structure are attached to hydrogen atoms creating inert hydrophobic surfaces that exhibit almost no frictional resistance, (4) strong and stiff carbon–carbon and carbon–hydrogen bonds which do not favour energy dissipation and (5) a positive atomic dipole charge of the hydrogen atoms out from both surfaces gives rise to repulsive forces (Harrison *et al.*, 1995; Su and Lin, 1998; Erdemir, 2002; Dag and Ciraci, 2004; Erdemir and Donnet, 2006). Actually no internal material shear is taking place at all, the hydrogen planes just 'fly' over each other.

The atomic scale mechanisms of friction when two hydrogen-terminated diamond surfaces are in sliding contact has been studied and the dependence of the coefficient of friction on load, crystallographic sliding direction, roughness and methane molecules as third bodies in the contact have been investigated (Harrison *et al.*, 1992, 1993, 1995, 1998; Perry and Harrison, 1996; Schall *et al.*, 2007). Quantum chemical methods were used by Koskilinna (2007) to study the tribology of methylated diamond and cubic and hexagonal boron nitride and tribochemical reactions were found to occur in the contact.

3.4.6.4 MEMS applications

The scaling-up of the nanomechanical explanations of contact mechanisms to practically useful conclusions on a macroscale is a most challenging and complex task and will take many years. Already there exist practical applications on a nanoscale where the increasing knowledge of tribological nanomechansims can be used. This has resulted in the development of microelectromechanical systems (MEMS) such as motors, transducers, gears and bearings of sizes in the micrometre range (Bhushan, 1998b, 2005; Krim, 2002b). For these extremely small components silicon has been used in the early applications for production reasons but studies have shown that tribological improvements can be achieved by using polycrystalline diamond or MoS_2 thin coatings or hydrogenated diamond-like carbon coatings (Donnet, 1995; Beerschwinger *et al.*, 1995; Gardos, 1996).

Today we are only at the very beginning of understanding the nanomechanical tribological contact effects that explain the origin of friction and wear and there is no doubt that in the near future many new theories and explanations for the origin of tribological phenomena will become available.

3.4.7 Lubricated coated contacts

Most of the information available on the tribological behaviour of coatings originates from experiments carried out in a normal laboratory environment and in dry conditions, which means that no lubricant is purposely brought to the contact. However, oil lubrication is common in many practical tribological applications and it offers a convenient way of further decreasing both friction and wear. The general mechanisms of lubrication in contacts in relative motion are described in section 3.1.5.

3.4.7.1 HD and EHD lubrication

In hydrodynamic (HD) lubrication the lubricant film separates the moving surfaces from each other and the effect of coatings on the surfaces is mainly limited to the mixed lubricated start-up and slow-down situations when contact between the surfaces occurs. Typical for elastohydrodynamic (EHD) lubrication is a considerable elastic deformation of at least one of the surfaces and this must be taken into account when using coatings in such contacts. As we have seen earlier in Chapter 3 very thin coatings may well deflect with the surface without fracture and failure. In general, an elastohydrodynamic or mixed lubricated contact will have a more favourable wear behaviour than a similar dry contact (Habig *et al.*, 1980; Roth *et al.*, 1988; Boving *et al.*, 1983; Podgornik *et al.*, 2004). Examples of extended lifetimes of elastohydrodynamically lubricated rolling contact bearings by the use of thin hard coatings is described in Chapter 7.

Soft elastic coatings such as elastomeric or polymeric layers have been suggested to be used in elastohydrodynamically lubricated contacts to improve the lubrication conditions (Cudworth and Higginson, 1976). Gabelli and Jacobson (1990) used a soft rubber layer in the cage contact of a rolling element bearing and observed an improved lubrication performance by an increased lubricant film thickness, which was five to ten times thicker than that in the raceway contact.

3.4.7.2 Lubrication introduced into a coated contact

A very general observation is that when a lubricant is introduced into a coated moving contact in boundary or mixed lubricated conditions both the friction and the wear will decrease (Jamal *et al.*, 1980; Komvopoulos *et al.*, 1987; Bull and Chalker, 1992; Ajayi *et al.*, 1992). This is not surprising because the boundary lubricant film will carry part of the load and the shear strength of a liquid is with few exceptions always lower than the shear strength of junctions between two solids in contact.

3.4.7.3 Coating introduced into a lubricated contact

It is, however, both more interesting and more important to see how the use of a coating affects the friction and wear behaviour in a lubricated contact compared to a similar uncoated lubricated contact. The answer to this question is neither simple nor obvious.

There are several published experimental results that show decreased friction and wear due to the use of coatings in lubricated contacts compared to similar uncoated contacts (e.g. Boving *et al.*, 1983; Liu *et al.*, 1991; Ajayi *et al.*, 1992; Ronkainen *et al.*, 1998a; Jacobson and Hogmark, 2001; Ajayi *et al.*, 2003; Mabuchi *et al.*, 2004; Podgornik *et al.*, 2003, 2004, 2006a; Iliuc, 2006) but the opposite behaviour has also been reported (Jamal *et al.*, 1980; Bull and Chalker, 1992; Podgornik *et al.*, 2003; Kalin and Vizintin, 2006a; Kalin *et al.*, 2004) and in some investigations there was no clear difference observed at all (Komvopoulos *et al.*, 1987; Vercammen *et al.*, 2004).

The complexity of this problem is due to the several tribological contact mechanisms that are simultaneously involved. The tribological behaviour of the contact will depend on which of these mechanisms have a dominating role in the specific contact being studied. In boundary lubricated steel contacts a boundary film will cover the surfaces as described in section 3.1.5. The strength of the chemical or physical bonds between the boundary film and the substrate will have a major influence on the wear and load carrying properties of the contact. This is the reason for using chemically aggressive additives such as sulphur, phosphorus or chlorine in lubricating oil so that they chemically react with the steel surface to form a strong boundary film.

3..4.7.4 Lubricant additive reaction to coated surface

If one of the surfaces is covered by a coating, the strength of the boundary film formed will depend on the strength of the bonds it can form to the coating material. It has been observed that with a sputtered

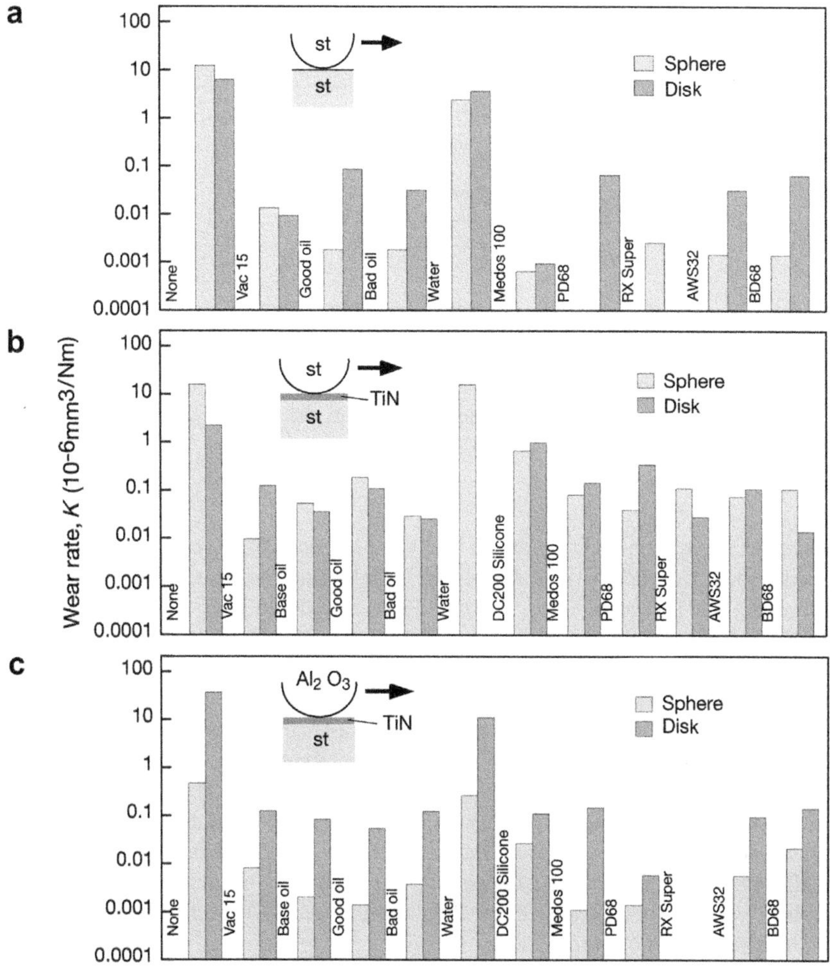

Fig. 3.104. The disk and sphere wear rate for different lubricants in ball-on-disk tests with (a) a steel sphere sliding against uncoated steel, (b) steel sphere against a TiN coated steel disk and (c) an alumina sphere against a TiN coated steel disk with a sliding velocity of 0.1 m/s and load of 2 N (data from Bull and Chalker, 1992).

TiN coating on a steel surface there will be no chemical interaction between the TiN surface and the oil additives (not specified in the article) and thus the formation of a boundary film will be prevented (Ajayi *et al.*, 1992). Figure 3.104 shows that the wear behaviour depends on what kind of lubricant is used (Bull and Chalker, 1992). With most lubricants the use of a TiN coated M2 steel disk has slightly increased total wear compared to uncoated steel discs sliding against a steel ball in ball-on-disk experiments. In contradiction to this, Sander and Petersohn (1992) observed decreased friction and decreased wear rate when using TiN coated steel plates instead of uncoated steel in oil lubricated oscillating ball-on-plate tests.

Several investigations with DLC coated surfaces indicate that the normal extreme pressure and anti-wear additives function in a coated contact, especially if one of the surfaces is steel or the DLC coating is doped by a metal. However, no major difference in the influence of the oil type, mineral,

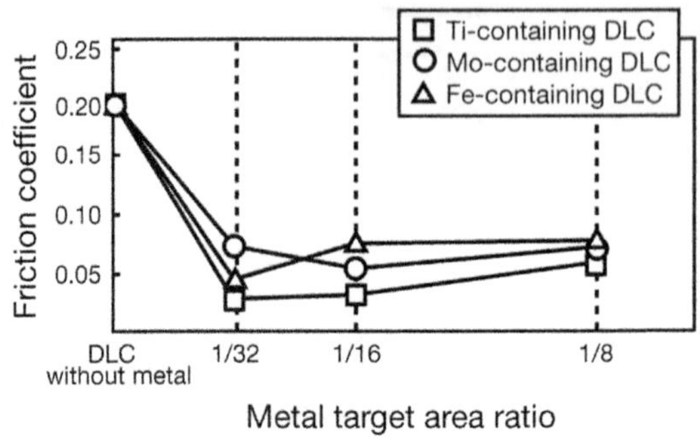

Fig. 3.105. The coefficient of friction for metal doped and pure DLC coatings sliding in lubricant with MoDTC additives. The metal target area is correlating with the quantity of metal added to the DLC coating structure (data from Miyake *et al.*, 2004).

synthetic or vegetable oil, on the tribological behaviour has been observed (Ronkainen *et al.*, 1998a; Jacobson and Hogmark, 2001; Ajayi *et al.*, 2003; Kalin and Vizintin, 2006; Kalin *et al.*, 2004; Vercammen *et al.*, 2004; Barriga *et al.*, 2006; Podgornik *et al.*, 2006a).

The addition of metals into the DLC structure improves the chemical reactivity of the rubbing surface and it can react with the lubricant additives and form lubricious reaction layers. This has been shown by Miyake *et al.* (2004) for Ti, Fe and Mo doped DLC surfaces sliding against a steel ball in zinc dialkyl dithiophosphate (ZDDP) and molybdenum dithiocarbamate (MoDTC) additive containing lubricant environment. By varying the metal target area during the coating deposition process they could control the amount of metal in the coating so that the quantity of metal increased with the increasing area ratio of the target. Figure 3.105 shows that for all three metal doped structures the coefficient of friction decreased from 0.2 to the range of 0.03 to 0.08 with only a very small influence of amount of added metals. Improved wear properties in lubricated conditions of metal doped coatings has also been reported by Ronkainen *et al.* (1998a) for Ti doped DLC and by Podgornik *et al.* (2004, 2006a) for W doped DLC.

3.4.7.5 Boundary lubricated coated contact

Basic mechanisms involved in a boundary lubricated contact with a rigid cylindrical slider moving on a coated elastic or plastic plate have been described in a mathematical friction model by Komvopoulos (1991b). His elastic friction model accounts for the shear properties of the lubricant, the ratio of the layer thickness to the size of the asperity contacts, and the elastic properties of the layer and the substrate. The plastic friction model accounts for interfacial adhesion and ploughing effects and it is shown that the coefficient of friction is a function of the relative size and shape of the ploughing hard wear particles and work-hardened asperities, the interfacial shear stress, the shear strength and strain hardening properties of the substrate and the steady-state surface topography. Predictions of coefficients of friction by the models are in fair agreement with experimental results including 0.01 to 0.13 μm thick oxide coatings on aluminium and chromium and TiN coatings on titanium and steel substrates as shown in Fig. 3.106.

The experimental observations of Komvopoulos (1991a), using oxidized aluminium, oxidized chromium on steel (AISI 1095), TiN coated steel and TiN coated titanium, and coating thicknesses in the range of 0.004 to 0.8 μm, show that the dominating contact parameters are the coating thickness, surface roughness, coating microstructure, the magnitude of the normal and friction traction arising at the asperity junctions, and the mechanical properties of the layer and the substrate. In mild sliding

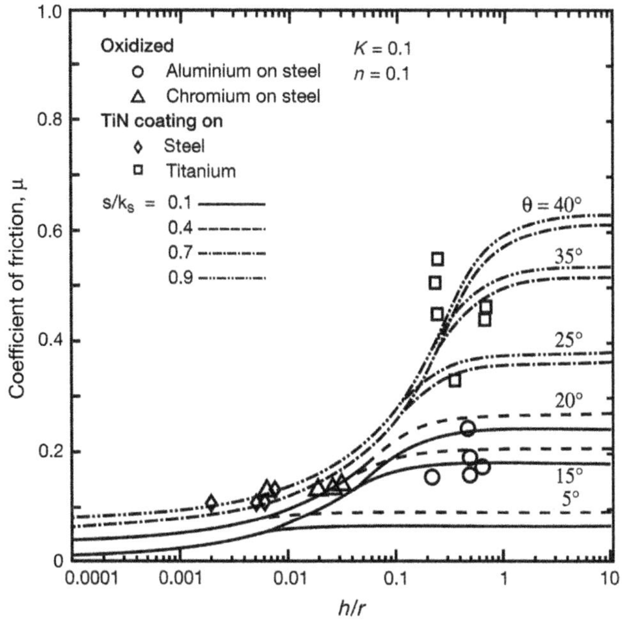

Fig. 3.106. Comparison between theoretical and experimental friction coefficients obtained from plastically deformed boundary lubricated surfaces with oxide and nitride coatings on various substrates and of various thickness ratios sliding in pin-on-disk testing against similarly treated surfaces. Here h is countersurface spherical/conical tip penetration depth, r is tip radius, Θ is sharpness angle, s is the shear stress, k_s is the shear strength, n is the strain hardening coefficient and K is a material strain constant (data from Komvopoulos, 1991b).

conditions without disruption of the coating the friction coefficient was about 0.1 and wear tracks were indistinguishable from the virgin surface. The principal deformation mode at the asperity contacts was elastic and the predominant friction mechanism was the shear of the boundary lubricant. Severe contact conditions with fragmentation and removal of the coating resulted in high friction, extensive surface damage and ploughing due to wear debris formation. The ploughing was due to wear particles trapped between the sliding surfaces, or the work-hardened asperities.

3.4.7.6 One or two surfaces coated

There is very obviously a difference if one or both of the surfaces in a lubricated sliding contact are coated. It would agree with the earlier observations that it is more favourable to have a chemically less-reactive coating on only one of two steel surfaces to allow a strong chemical boundary film to be formed on the surface left uncoated. This assumption is supported by the experimental observations of Boving *et al.* (1983). They tested CVD deposited TiN, TiC and $(Cr,Fe)_7C_3$ coatings with a ball-on-cylinder lubricated test apparatus and found that, in general, combinations of uncoated balls and coated rings perform better than combinations of coated balls and coated rings. By far the best performance in terms of load-carrying capacity was found with a combination of uncoated steel balls with TiN coated rings, as shown in Fig. 3.107.

In contradiction with the above statement are the results of Jamal *et al.* (1980) from lubricated pin-on-disk tests showing low coefficients of friction and low wear with both surfaces coated by TiN or TiC compared with higher values for only one surface coated and the other of uncoated steel. For

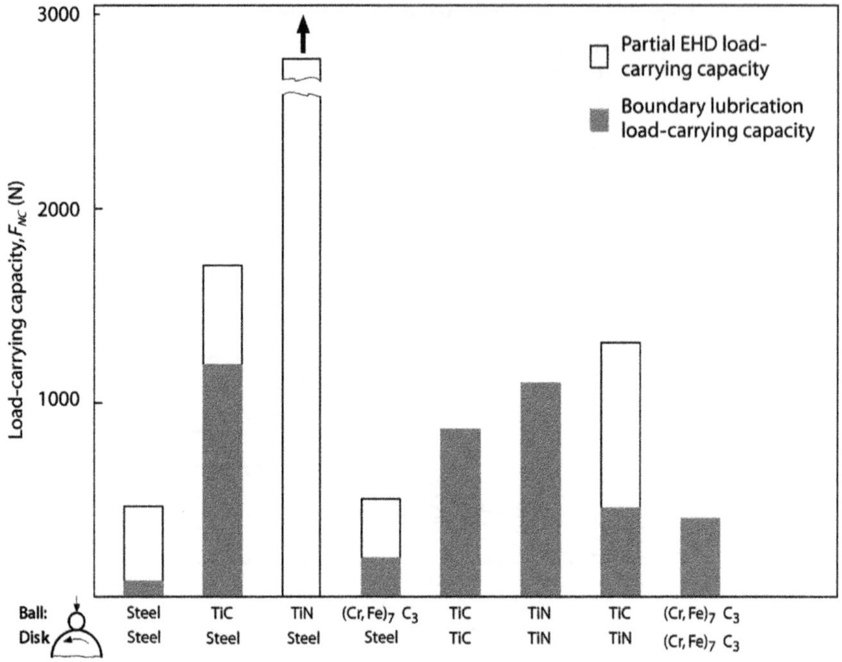

Fig. 3.107. Load-carrying capacity of partial EHD and boundary lubrication films in lubricated IRG ball-on-cylinder tests with an uncoated, TiN, TiC and $(Cr,Fe)_7C_3$ coated steel specimen at a sliding velocity of 0.5 m/s (data collected from Boving *et al.*, 1983).

a-DLC and W-DLC coated steel surfaces Kalin and Vizintin (2006) reported about 50% lower coefficient of friction for the sliding of an a-DLC/a-DLC contact compared to steel/steel, steel/a-DLC, steel/W-DLC and W-DLC and W-DLC contacts while the lowest wear was found for steel/steel, steel/a-DLC and a-DLC/a-DLC contacts. Their results do not really give any clear indication on which of the two contacting surfaces it is more beneficial to coat.

Based on a large series of lubricated cross cylinder sliding tests with 2 μm thick W alloyed laminated WC/C structured DLC coatings on steel substrates with synthetic oil including no extreme pressure or anti-wear additives Jacobson and Hogmark (2001) concluded with the recommendation that it is most advantageous to coat only one of the two surfaces because this gives the best conditions for running-in. They explain the favourable tribological effect of DLC coating in a lubricated contact with the formation of a synergistic tribofilm. Mild wear and transfer of DLC coating material, together with chemical reaction products formed by additives in local areas not covered by coating, work together in reducing friction and wear. This is supported by the results from Podgornik *et al.* (2003) showing the lowest friction and lowest wear in cross cylinder test when only one of two sliding steel surfaces was coated with WC doped DLC coating and using a suitable extreme pressure (EP) or anti-wear (AW) additive, as seen in Fig. 3.108. The friction was higher when using pure poly-alpha-olefine (PAO) synthetic oil and a fully formulated gear box oil (GL-4).

3.4.7.7 Rough surfaces and wear debris

There are, however, other effects that influence the tribological behaviour in coated lubricated contacts. Ajayi *et al.* (1992) found that a major difference between TiN coated and uncoated lubricated

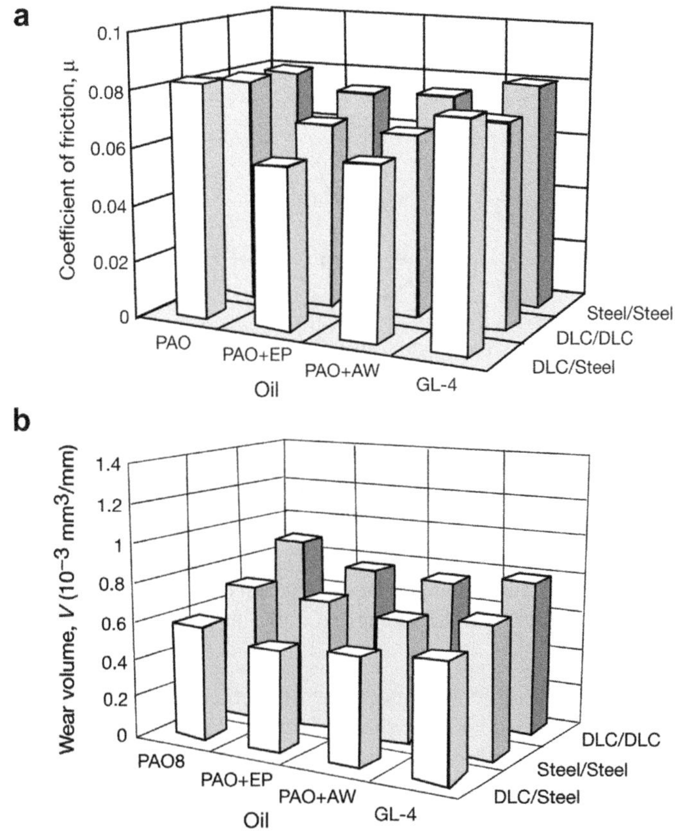

Fig. 3.108. Influence of lubricant and oil additives on (a) coefficient of friction and (b) the wear volume with sliding uncoated, single coated and double coated steel surfaces (data from Podgornik *et al.*, 2003).

tests was the change in surface roughness of the wear track. With uncoated discs there was a considerable roughening of the surface while the TiN coating prevented any change in roughness and there was no wear observed of the disk surface. Wear did occur on the mating pin surface, however. A major difference between the dry and lubricated test was the role of the generated wear debris. The accumulation and reattachment of debris that was prevalent in the dry tests was not observed in the lubricated ones. Surface roughness and cavity-induced microlubrication effects that decrease friction in starved lubricated DLC coated surface sliding against steel were reported by Andersson *et al.* (2008).

3.4.7.8 Soft metallic coatings

The use of thin soft metallic coatings may improve the tribological behaviour in boundary lubricated conditions. Some soft metals, such as gold, silver, lead and indium, possess low shear strength and provide, as thin coatings, a fairly low coefficient of friction in dry sliding conditions, in the range of typically 0.1 to 0.4. Such coatings have resulted in very low wear and a coefficient of friction below 0.04 when sliding at elevated temperatures of about 200°C in the presence of synthetic oil (Erdemir, 2005). With lubricated tests Liu *et al.* (1991) showed that the application of soft metallic brush plated layers of Pb, Sn, or Pb–Sn–Cu on steel substrates can more than double the scuffing load when sliding

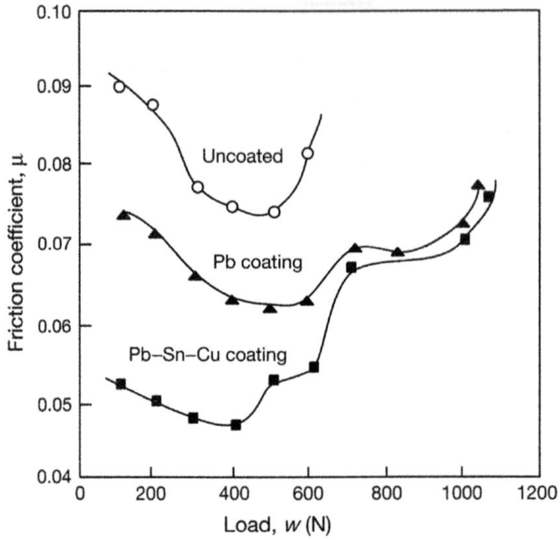

Fig. 3.109. Influence of soft metallic coatings on the coefficient of friction in lubricated ball-on-disk tests. The substrate hardness was 535 H_V and the sliding speed 0.97 m/s (data from Liu *et al.*, 1991).

against a steel counterface and considerably decrease the coefficient of friction, as shown in Figs 4.12 and 3.109. The lubricant was paraffin oil without additives. With higher substrate hardness the coefficient of friction decreased and the scuffing load increased. For Pb coating thicknesses less than 16 μm there was no appreciable effect of the coating thickness on either the coefficient of friction or on the scuffing load. However, for the Pb–Sn–Cu layers thicker than 12 μm they found a significant increase in the scuffing load with increasing coating thicknesses.

3.4.7.9 DLC coated and lubricated sliding contacts

The tribological behaviour of DLC coated surfaces in lubricated conditions are of great interest due to their wide commercial applicability. Still they are not well understood and some of the available information seems even contradictory. This is due to the complexity of the contacting mechanisms. In section 4.5.3 we will show that the dominating friction and wear mechanisms for DLC coated contacts are very different depending in the governing contact scale, see Fig. 4.89. The shear mechanism in the contact may involve asperity interactions, transfer layer generation, surface graphitization and hydrogen attachment to carbon atom dangling bonds. The lubrication adds the possibility of part of the load being carried by and the movement sheared in a pressurized lubricant film or a surface film formed by chemical reaction between surface material and lubricant additives. The outcome of the contact in the form of friction and wear depends entirely on which of these mechanisms in the studied specific conditions can function and dominate over the other.

If a lubricant is introduced in a rough DLC or diamond coated macroscale contact without hydrodynamic effects where the asperity interactions dominate, no larger effect on friction and wear is expected (Fig. 4.89a). On the other hand, in DLC coated nanoscale contacts with superlubricity (Fig. 4.89c) where the shear is taking place between two highly hydrogenated physically smooth surfaces the introduction of a liquid will only disturb the repulsive hydrogen layer interactions and increase friction and wear (Andersson *et al.*, 2003a; Erdemir, 2005).

However, the largest interest with regards to practical applications is related to the mechanisms occurring at the microscale (Fig. 4.89b). Here it is important to study and understand the function and interaction between three partly competing mechanisms: the graphitization of the top layers that offers low shear, the formation of chemical reaction layers offering low shear and high load-carrying capacity, and the build-up of a transfer layer on the counterface, often combined with graphitization, offering a smooth platform for shear. These interactions are not well understood but some detailed studies give indications and are presented below.

In lubricated conditions a boundary reaction film was built up on a steel ball surface sliding against DLC coatings with low friction properties quite similar to dry contacts (Fig. 4.83) with the same surfaces (Ronkainen *et al.*, 1998a; Ronkainen, 2001). In the dry contact the transfer layer formation and graphitization mechanisms had the same low shear effect as the boundary films. With lubrication no transfer layers were observed and it is understood that in these conditions the boundary lubrication mechanisms took the dominating role.

There is also a difference in the behaviour of DLC coatings in lubricated conditions depending on their structure. Figure 4.83 shows that with three kinds of DLC coatings hydrogen-free (a-C) and hydrogenated (a-C:H) carbon coatings had a low coefficient of friction in dry sliding, 0.15 and 0.22, which was further decreased by 10–40% under boundary lubrication. The a-C:H(Ti) coating exhibited good self-lubricated properties with a coefficient of friction of 0.10 in dry sliding and the a-C films had the lowest coefficient of friction in water and oil lubricated conditions, 0.03 and 0.08, respectively. The a-C film showed excellent wear resistance in dry, water- and oil-lubricated conditions. The performance of the a-C:H film could be improved by addition of titanium. In dry sliding conditions, the tribolayer formation on all three DLC films influences the friction and wear performance, but in oil lubricated conditions, boundary layers were formed, which governed the tribological mechanisms in the contact. The introduction of water lubrication resulted in reduced friction for the a-C and a-C:H(Ti) coatings but catastrophic failure for the a-C:H coating.

Some additives like ZDDP and MoDTC react directly with DLC surfaces and seem to be more active with selected hydrogenated DLC surfaces (Barros Bouchet *et al.*, 2005, 2008; Equey *et al.*, 2008). In these conditions the friction and wear performance was improved by coating both surfaces. The composition of the tribofilm appeared to be similar to that obtained on steel surfaces in the same lubrication conditions. It is interesting that the tribochemical reactions can occur without the presence of an iron catalyst element in the tribosystem.

The frictional behaviour of base oil boundary lubricated W-DLC coatings is governed by the transfer of coating material from the coated surface to the steel countersurface or to the revealed steel substrate (Podgornik *et al.*, 2006a). This was observed to reduce the friction with up to 50% as compared with two mating uncoated steel surfaces. In the presence of a sulphur-based EP additive, tungsten in the transferred layer reacts with sulphur in the additive, forming lamellar WS_2 tribofilms on the steel counterface or the revealed steel substrate. These tribofilms have tribological properties similar to MoS_2. Depending on the additive concentration, the tribofilms may increase in thickness and density, thus providing conditions for low friction behaviour of the contact. High additive concentration gives a rapid transfer to the low friction state. On the other hand, if the additive concentration is too high, WS_2 type tribofilms are suppressed by immediate formation of chemical reaction layers typical for steel-to-steel contacts.

The tribological behaviour of DLC coatings in starved, boundary, EHD and HD lubrication regimes has been investigated by Podgornik *et al.* (2004, 2008). They found that although DLC coatings provide reduced friction and better surface protection under boundary lubrication, they do not provide the same conditions for EHD or HD film formation as uncoated surfaces. However, if paired with the steel countersurface faster and smoother transition from boundary to EHD lubrication can be reached, with the coefficient of friction being comparable to steel/steel contact.

Even if there is a large variation in the specific mechanisms controlling the friction and wear in lubricated DLC sliding contacts a general conclusion is that in most conditions DLC coatings can be used in lubricated contacts with conventional lubricant oils, mineral, synthetic and vegetable oils, and conventional lubricant additives. In many cases the difference compared to uncoated is not so large but the DLC coating will always bring a safety layer effect in case of loss of lubrication. In some special cases an optimal design of DLC coating structure and additive may decrease both friction and wear drastically.

As a final point, since most modern lubricants and additives presently in use have been optimized for use with ferrous materials we believe that the increased use of coatings, especially ceramic and carbon based, merits much greater research into their behaviour under lubricated conditions, and the specific development of lubricants with properties suited to these coatings.

CHAPTER 4

Tribological properties of coatings

Contents

4.1 General

We have seen in the previous chapter that considerable theoretical knowledge is available about the mechanical properties of coated surfaces and that there is some agreement at a general level about the basic physical and chemical mechanisms behind the dominant influencing parameters. In general terms, the tribological behaviour of coatings is influenced by:

1. The *contact condition*; which includes parameters such as load, speed, geometry, topography, etc. The contact conditions are very different in, for example, cutting, sliding and erosive particle contacts.
2. The *environmental condition*; which includes parameters such as temperature, lubricant, gases, radiation, contaminants and wear products. The tribological contact behaviour is very different in, for example, sliding in dry humid air, vacuum or in oil.
3. The *contacting materials*; which include parameters such as the physical, chemical and mechanical properties of the coating, the substrate and the counterface. These are hardness, elasticity, toughness, chemical reactivity and thermal conductivity. They are determined by the composition and the microstructure, which include factors such as grain size, density, porosity, etc., which are influenced by deposition process parameters.
4. The *composite coating/substrate system*; which includes parameters such as coating thickness, multilayer structure, gradient properties, doped structures and adhesion between layers.

However, because of the complexity of the tribological contact mechanisms, we are still not in a situation where the coefficient of friction or the wear rate can be accurately predicted for particular contact conditions based solely on theoretical analyses.

An indication of how a coated surface will behave can be found in the numerous published articles describing experimental measurements of the friction and wear properties of coatings. However, this is not very easy, because the measurements are performed with many different types of test equipment, different contact geometries, different test parameters and in different environments, due to a lack of standardization of tribological test methods. This makes it very difficult to find comparable results specific to the set of contact conditions that determine a certain application (Singer, 1989; Bhushan and Gupta, 1991; Sudarshan, 1992; Davis *et al.*, 2001).

In this chapter we will review a large number of experimental tribological tests of coated surfaces, compare the results when possible and discuss the indicated tribological mechanisms. Even though we are aware that in most applications it is the wear resistance that is the more interesting and crucial property, we will frequently use the coefficient of friction when comparing coatings. The reason for this is that it is often the only comparable test data available, because wear data are missing or are given in one of several different non-comparable ways. Where available, speed, load and contact geometry are given to provide the reader with an impression of the severity of the contact condition corresponding to the friction and wear data. The review is structured according to the coating material, starting from polymer, soft metal and lamellar coatings and continuing with hard ceramic, carbon, carbon-based and combined coatings. Tribological test data from the reviewed experiments are presented in tables in the Appendix.

4.2 Soft Coatings

As explained in the previous chapter, a soft thin coating on a harder substrate offers the possibility to reduce sliding friction. The wear and lifetime are particularly critical for thin soft coatings. If the contact is not designed correctly the soft coating might be rapidly detached from the surface. However, for many soft coating materials a range of parameter values can be found where low or very low wear and long lifetimes can be achieved. The parameters to be optimized are typically film thickness, surface roughness, sliding speed, load, environment and counterface material as well as deposition method.

Soft coatings can be classified in three different categories: polymers, soft metals and lamellar solids.

4.2.1 Polymer coatings

Polymers are important tribological materials for many reasons; they have low friction, moderate wear resistance, good corrosion resistance, self-lubricating properties, low noise emission and are cheap to produce. Polymers can be applied as coatings on harder substrates by several different techniques such as adhesive bonding, vacuum deposition, electrochemical methods and pressure casting (Bely *et al.*, 1982). The thickness of the polymer coating produced can vary from a few micrometres to several millimetres. In the following we will limit our attention to the most common thin polymer coatings, which are polytetrafluoroethylene (PTFE), polyimides and elastomers.

4.2.1.1 Tribological properties of polymers

The coefficient of friction for polymers sliding against themselves, against metals or ceramics is typically in the range of 0.15 to 0.6 (Bhushan, 1999b) which is lower than the typical value of 0.5 to 0.7 for unlubricated steel against steel contacts (see Fig. 5.19). The wear resistance of polymers in low or medium loaded contact conditions is usually good. High contact pressures, however, may cause larger elastic and plastic deformations and temperature increases leading to severe wear.

The basic mechanisms that contribute to friction were discussed in Chapter 3. In sliding contacts with polymers, both ploughing of the surfaces by hard particles or counterface asperities and the adhesion between the surfaces in contact contribute to the total friction force. Their relative contribution depends mainly on the hardness and the topographies of the two contacting surfaces.

The main wear mechanisms for polymers in sliding contacts can be classified as cohesive (or deformation) wear and interfacial wear (Briscoe and Tabor, 1980; Lee *et al.*, 2001; Briscoe and Sinha, 2002). The distinction between these arises from the extent of deformation in the polymer caused by the moving counterface. For interfacial wear the frictional energy is dissipated mainly by adhesive interactions at the surface, while for cohesive wear the energy is dissipated by both adhesive interactions and subsurface material deformation interactions related to ploughing and fatigue.

In *cohesive wear*, three basic types of deformation properties and failure modes may occur:

- elastic or viscoelastic with tearing failure,
- plastic or viscous flow with ductile failure, and
- brittle failure accompanied by fracture.

These failure modes are typical for elastomers, ductile polymers and glassy polymers, respectively. The wear processes are generally influenced by the temperature and the rate of deformation. In addition, several of them are sensitive to the chemical nature of the environment.

In *interfacial wear* two basic wear mechanisms are involved:

- adhesive or transfer wear, and
- chemical wear.

These wear processes involve relatively thin layers of the polymer surface and result in higher strains, strain rates and temperatures compared to deformation wear. The transfer of wear particles to the counterface and formation of transfer layers often have a considerable effect on the tribological properties. Important features in this mechanism are:

- the nature of the initial adhesion between the polymer and the counterface,
- the mode of failure at the junction and the criteria for film transfer,
- the structure and the thickness of the transferred layer,
- the bonding between the film and the transferred layer, and
- the mechanism of film removal and the displacement of the debris from the contact region.

Chemical wear is important for thermosetting resins, some thermoplastics and elastomers that suffer appreciable chemical degradation without becoming viscous. Ductile polymers, such as PTFE, HDPE (high-density polyethylene), nylon and polypropylene and brittle polymers such as polystyrene and PMMA (polymethylmethacrylate), melt and soften without significant chemical decomposition (Briscoe and Tabor, 1980).

In many applications, polymers are intended to be used as a replacement for lubricated or unlubricated metals. Detailed information on their relevant physical and mechanical properties is available in appropriate handbooks, e.g. Brydson (1990) and Cartier (2003).

It is illustrative to compare on a general level the physical and mechanical properties of polymers to those of metals. The following comparison is according to Evans and Lancaster (1979), who cite a number of particular advantages when using polymer surfaces in tribological applications:

1. The physical and mechanical properties of polymers can be varied over a wide range by suitable choice of polymer type, fillers and reinforcement.
2. Some polymers, notably thermoplastics, are cheap and easy to fabricate into complex shapes.
3. Many polymers, particularly fluorocarbons, are highly resistant to chemical attack by aggressive media, such as acids and alkalis.
4. Coefficients of friction during unlubricated sliding against either themselves or metals are relatively low and typically within the range 0.1 to 0.4.
5. Wear rates during sliding against smooth metal surfaces are relatively low, and polymers do not normally exhibit scuffing or seizure.
6. Periodic maintenance, e.g. lubrication with fluids, can often be dispensed with. This is particularly important in applications in which access for routine maintenance is not possible.
7. When fluid lubricants are present, polymers undergo elastohydrodynamic lubrication more readily than metals.

The following features are typical limitations for using polymers instead of metals in tribological applications:

1. Polymers are viscoelastic and more susceptible to creep than metals.
2. Ultimate strength and elastic moduli are appreciably lower, typically by a factor of ten. The latter limits stiffness and dimensional retention under load.
3. Thermal expansion coefficients are relatively high, typically ten times greater than that of steel, and this also may introduce problems of dimensional stability.
4. Some polymers, e.g. nylons, readily absorb fluids, including water from the environment, which also may affect dimensional stability.
5. Thermal conductivities are very low, of the order of one-hundredth of that of steel, and the dissipation of frictional heat is therefore poor.
6. Limiting temperatures associated with softening, melting, oxidation and thermal degradation are all relatively low, typically less than 300°C.

Since polymers, in comparison to metals and ceramics, lack rigidity and strength, they are widely used as composites containing different fillers and reinforcements. These may be particles, multidirectional short fibres, unidirectional long fibres or fillers in textile patterns. The tribological characteristics of polymer composites depend on both the material properties of the polymer and the fillers and on the structural system in which these materials have a function. Polytetrafluoroethylene, carbon, graphite and glass are the most commonly used fillers. They are used to reduce the adhesion to the counterpart or to enhance the hardness, stiffness and compressive strength (Friedrich *et al.*, 2002; Bahadur, 2004).

4.2.1.2 Tribology of polymer coatings

There are many parameters which can be expected to influence the tribological properties of polymer coatings. However, this subject has received little attention to date. For example, in a book on friction and wear in polymer-based materials, Bely *et al.* (1982) devote one short section to polymer coatings. They point out that there are two parameters in particular that influence the frictional properties. These are the structural condition of the polymer and the coating thickness. The structural condition is determined by the method used to manufacture the coating. This has an influence on properties such as homogeneity and cohesion, which affect the tribological behaviour.

The elastic–viscoplastic respond of a polymer coating, polymethylmethacrylate (PMMA), to a sliding spherical scratch test indenter has been modelled by the finite element method and simulated by Jayachandran *et al.* (1995). Their simulations illustrate the typical polymer coating behaviour under loading. The substrate/coating interface at the spherical contact is subject to an intensive plastic strain rate, indicating the probable site of shear-induced coating delamination during loading. During unloading, residual tensile normal stresses are generated indicating a possibility of shear-assisted tensile delamination of the coating. The presence of friction at the indenter interface reduces the material pile-up at the indenter periphery. The tensile hoop stresses generated near the surface due to the indenter friction may cause radial surface cracking.

The thickness of the coating is obviously an important parameter. Here, again, there is a difference in the tribological mechanism if we consider very thin coatings with a thickness of a few micrometres or thicker coatings of the order of several hundred micrometres. In both cases Bely *et al.* (1982) refer to reports showing that there is an optimum thickness range that gives minimum coefficients of friction. For different compounds of polytetrafluoroethylene the optimum coating thicknesses have been found to be within the range from 2 to 25 μm. With polyamide resins and polycaproamide, optimum film thicknesses tend to be much higher, about 250 to 500 μm and, in light load conditions, even as high as 1000 μm. It is obvious that parameters determining the contact conditions, like load, speed, geometry and surface topography, also have a significant influence on the range of film thicknesses where minimum friction is obtained. Bely *et al.* conclude that the optimum coating thickness achieves a balance between strain and thermal processes in the layer, in the base, and in the counterbody. The adhesion between a polymer coating and a metal surface is determined by a wide range of interfacial interactions and its strength can be assessed by surface analysis techniques (Van Ooij, 1985).

4.2.1.3 Polytetrafluoroethylene coatings

In general, a fluoropolymer has many outstanding properties including low surface tension, extreme chemical inertness, and high oxidative and thermal resistance. Polytetrafluoroethylene (PTFE) has been widely used as a tribological material because of its low coefficient of friction. The lowest values, of the order of 0.05 or less are, however, only obtained at high loads, low speeds and moderate temperatures (Evans and Lancaster, 1979). Examples of these variations are shown in Fig. 4.1. An advantage is that the lubricating properties of PTFE are similar in both air and vacuum (Sugimoto and Miyake, 1988). A limitation for the wider application of pure PTFE is its relatively poor wear resistance.

The PTFE molecule exhibits extremely high cohesion but the intermolecular strength is not very high (Suh, 1986). For this reason PTFE easily yields in shear. Its load-carrying capacity is low so that it should not, according to Roberts (1990b), be used in coating applications where contact stresses exceed 1200 MPa. Its creep resistance and yield strength can be increased substantially by reinforcing it, for example with glass and graphite fibres.

PTFE as a bulk material is most widely used in a composite form to achieve improved wear resistance. The addition of almost any inorganic, and some organic, fillers to PTFE can reduce its wear rate by a factor of 100 or more. Some examples of the reducing effect of fillers on the wear rates of PTFE are given in Fig. 4.2. Graphite, carbon, glass fibre and bronze are commercially the most widely used PTFE fillers. The wear-reducing effect of fillers is associated with interrupting subsurface deformation and crack propagation that would otherwise produce large wear sheets (Blanchet and Kennedy, 1992). The fillers can improve the adhesion of the transfer film and produce smaller wear particles that are prone to stay longer in the contacting region (Wang and Yan, 2006). A reduction in wear was measured by Osuch-Slomka and Gradkowski (2004) when different amounts of graphite and bronze fillers were introduced in PTFE plates sliding against a steel ring. Molybdenum disulphide fillers increased the wear by 2 to 4 times. A combination of all the three fillers turned out to decrease the wear close to zero level. However, the coefficient of friction was slightly increased in all cases.

The low friction properties, especially at low sliding speeds and moderate temperatures, of PTFE and PTFE composites can be explained by a transfer film mechanism (Briscoe *et al.*, 1974; Suh, 1986). When the polymer is deformed by the sliding action, the molecules near the surface orient along the sliding direction due to the large shear strain gradient near the surface. Because of low cohesion but relatively high adhesion, for example to metals or glass, worn material from the polymer is transferred to the counterface and forms a thin, of the order of 3 to 10 nm, highly oriented PTFE film on the counterface as shown in Fig. 4.3. Thus, after a while, the shear that influences adhesive friction takes place between a thin highly oriented PTFE film on the countersurface and a highly oriented top layer of the PTFE specimen.

If the sliding direction reverses, the coefficient of friction increases because of the orientation in the contact. The wear sheets of PTFE are very thin films, only 5 to 20 nm thick, and are highly oriented in the sliding direction. If the counterface is roughened then, for a surface roughness above 0.1 μm, high friction is observed and torn-out fragments are formed (Briscoe *et al.*, 1974). Higher temperatures and loads increase the deposition rate of the transferred layer on the counterface (Yang *et al.*, 1991b).

The tribological properties of PTFE have been studied very widely and there exist many commercial applications, but surprisingly little information is available on the friction and wear properties of PTFE coatings. Thin PTFE layers, with thicknesses of around 1 μm, have been produced on different substrates by PVD techniques such as sputtering (Biederman, 1981; Nishimura *et al.*, 1989;

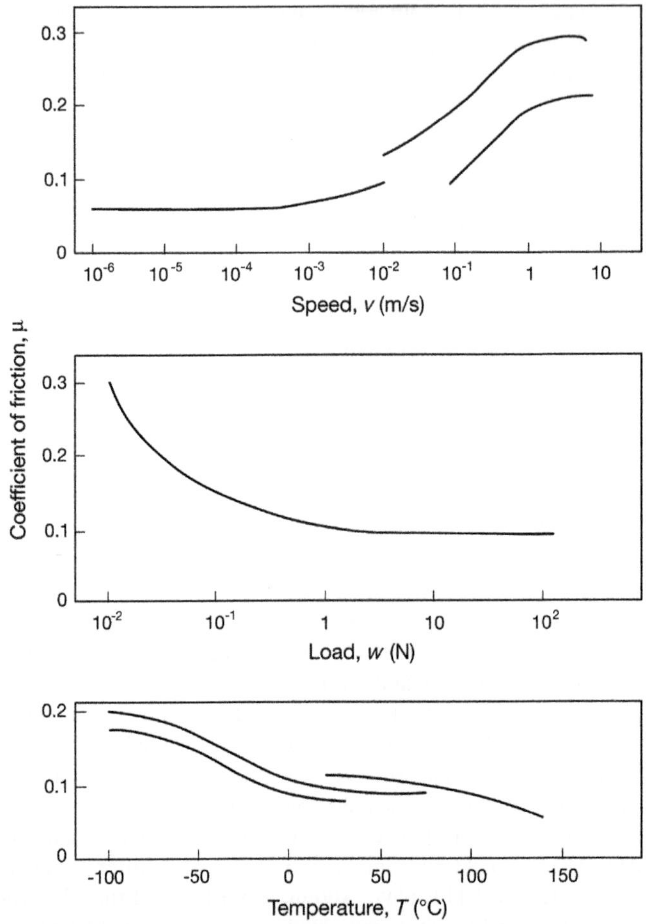

Fig. 4.1. Effects of sliding conditions on the coefficient of friction of PTFE. The curves correspond to different experimental conditions (data from Evans and Lancaster, 1979).

Yamada *et al.*, 1990). In contrast to what could be expected, reports show that the frictional properties of sputtered PTFE films, as well as their thermal stability, are not as good as those of the bulk material. However, the wear resistance of PTFE films can be better than that of the bulk PTFE material.

The friction and wear properties of sputtered, about 1 μm thick, PTFE coatings on steel substrates have been compared with those of the original bulk PTFE material in different environments and at different temperatures by Nishimura *et al.* (1989). They found that the coefficient of friction of sputtered PTFE coatings is almost twice as high as that of a PTFE plate sliding against steel in dry air, nitrogen and vacuum environments, as shown in Fig. 4.4. At the initial stage, the coefficient of friction was between 0.3 and 0.35 up to about 20 to 50 cycles. It then rapidly increased to above 0.4 and again fell to the initial value. However, the wear rate of the PTFE plate was about ten times that of the sputtered PTFE film, without any marked effect due to the presence of gas. The sputtered films were in the form of particles or plates rather than the soft, long films observed with bulk PTFE. The wear particles from the sputtered films were small and brittle.

The above observations indicate that the difference in the tribological behaviour is due to structural changes in the sputtered PTFE film during the coating process. The sputtered PTFE films seem to have

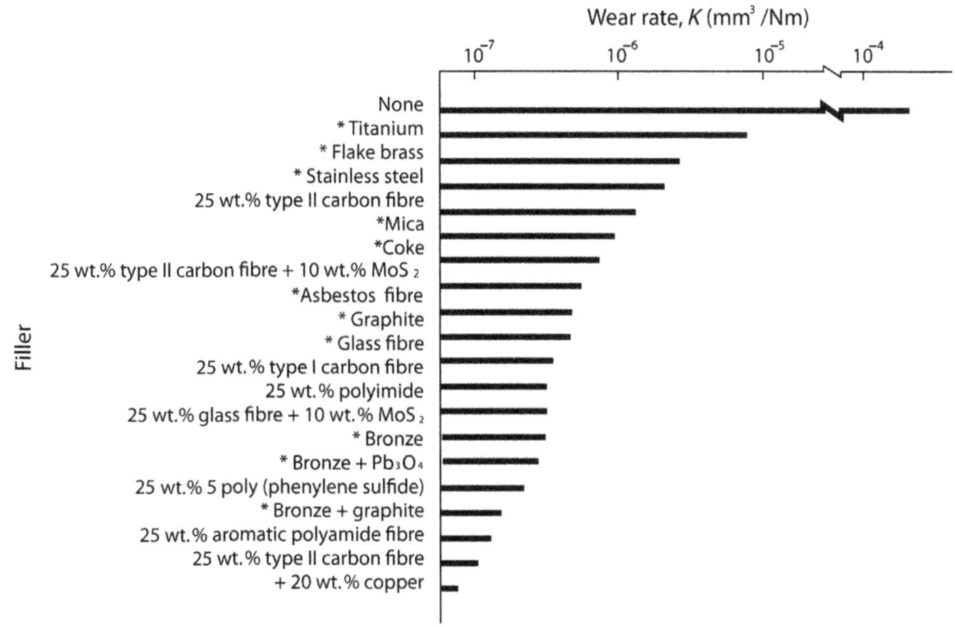

Fig. 4.2. Steady-state wear rates of PTFE composites sliding against smooth steel. The commercially available ones are marked with an asterisk (data from Evans and Lancaster, 1979).

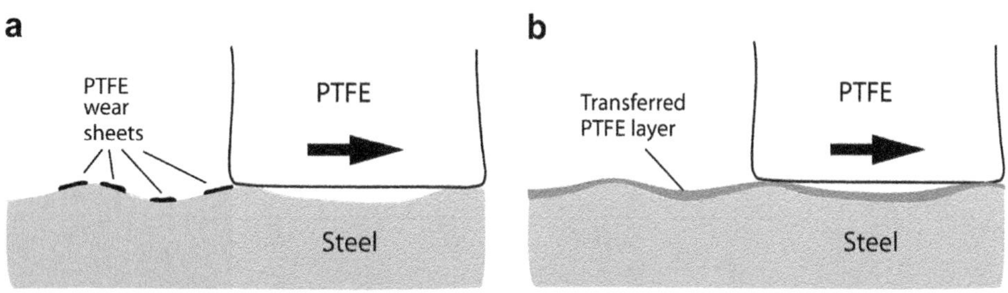

Fig. 4.3. In a sliding PTFE on steel contact (a) thin PTFE wear sheets attach to the steel surface and (b) form, after a number of cycles, a complete PTFE transferred layer.

an amorphous structure; they are harder, more brittle and more chemically active than the original PTFE. The thermal stability of the sputtered PTFE was also considerably inferior to bulk PTFE. The film began to evaporate gradually from about 60°C and its weight loss increased rapidly when the temperature was over 240°C. The bulk PTFE, on the other hand, did not show signs of evaporation until the temperature exceeded 500°C. Nishimura *et al.* (1989) suggest that this indicates that the sputtered PTFE consists of particles having a lower molecular weight than that of the bulk PTFE.

In very slow speed sliding experiments with a steel ball sliding at a speed of 0.0004 m/s on a PTFE coating rf sputtered on to a smooth glass substrate, Yamada *et al.* (1990) found no influence of load within the range 0.5 to 10 N and no influence of coating thickness on the coefficient of friction within the range 0.05 to 1.0 μm. The coefficient of friction was initially 0.2 and then increased after some five

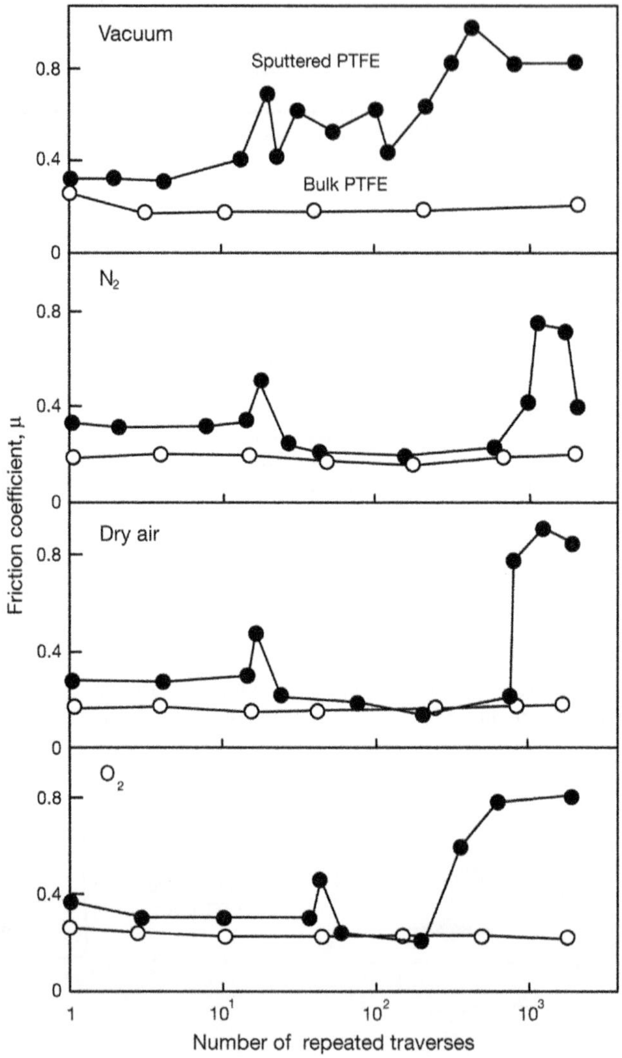

Fig. 4.4. The coefficient of friction for an rf sputtered PTFE coating and a PTFE plate sliding on a steel substrate in pin-on-disc tests with a speed of 0.01 m/s and a load of 5 N in four different environments (data from Nishimura *et al.*, 1989).

passes to 0.3 with a sapphire ball and slightly higher with a steel ball. This coefficient of friction is somewhat higher than that of the bulk polymer target. Yamada *et al.* believe that this is because the surfaces of the deposited films exhibit higher surface energy than those of the target polymer.

The lifetime of sputtered PTFE and polyimide (PI) coatings on a glass substrate sliding against a sapphire pin was observed in pin-on-disc experiments with a speed of 0.1 m/s and a load of 15 N by Yamada *et al.* (1990). The PTFE film failed after 200 passes while PI failed after 4000 passes. It can be concluded that the wear life of sputtered PTFE films seems to be poor, only 1/20 of that of PI films. The PTFE film was plastically deformed and wore away from the track with long film-like wear debris. The lifetime of the sputtered PTFE film could be increased by 3 to 6 times by heat treatment

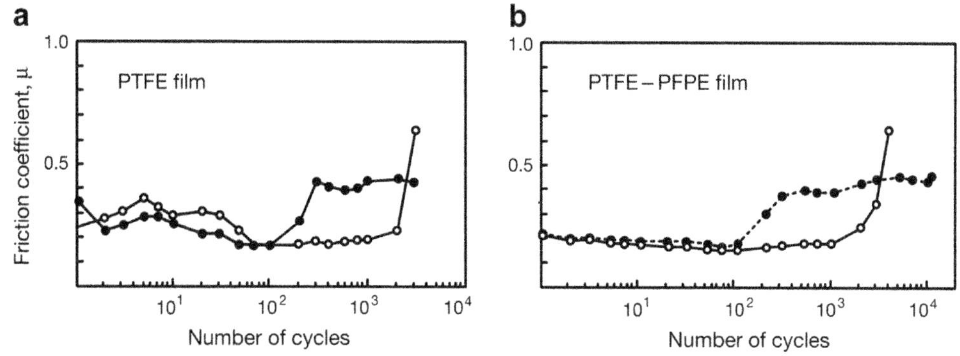

Fig. 4.5. Variation in the coefficient of friction of 3 μm thick sputtered films, deposited on a stainless steel substrate, with number of sliding cycles under circumferential oscillatory sliding movement for (a) PTFE and (b) PTFE–PFPE films with a speed of 0.0015 m/s and a load of 5 N. ○ in air and ● in vacuum (data from Sugimoto and Miyake, 1988).

at 200°C and 300°C, respectively. It was observed that the amount of oxygen in the films increased after heat treatment and this may increase the adhesion strength to the substrate and lead to a longer film life.

Efforts to improve the tribological and the dielectric properties of PTFE by developing a new kind of fluorine-containing thin film produced by rf sputtering using PTFE-perfluoropolyether (PFPE) grease as target material have been reported by Sugimoto and Miyake (1988). They measured a slightly reduced and more stable coefficient of friction down to 0.1 to 0.2 for a PTFE–PFPE film compared to 0.15 to 0.3 for the PTFE film in similar sliding conditions. However, as shown in Fig. 4.5 there was almost no improvement in the wear life as measured by the number of cycles before the coefficient of friction exceeded 0.3.

The addition of hard particles in the PTFE coating may have a beneficial influence on the wear and friction properties mainly due to their hardening effect on the coating. This has been shown by Lee and Lim (2004) who added 1–4 weight% of nanocrystalline diamond particles composed of 4–6 nm size crystalline diamond grains that had been produced by a detonation method. The thickness of the sprayed composite PTFE coating on an aluminium substrate was 30 μm. The best tribological performance was measured for the 2 weight% nanodiamond composite coatings in a reciprocating ball-on-plate wear tester with 0.8 N load and 0.01 m/s sliding speed against a 5.6 mm diameter ball-bearing steel ball. In comparison with pure PTFE coatings the coefficient of friction decreased from 0.16 to 0.08. Both the coefficient of friction and wear were reduced by about 50% as the ambient temperature was increased from room temperature to 150°C.

4.2.1.4 Polyimide coatings

Polyimides (PI) are a group of polymers that are used in tribological applications such as bearings and seals. They have coefficients of friction generally varying within the range of 0.1 to 0.6 depending on composition and additives (Suh, 1986; Fusaro, 1987a). They are interesting for use especially in space applications because they have considerably lower coefficients of friction in vacuum, often less than 0.1, and for some compositions, like 80% PMDA/20% BTDA polyimide, even as low as 0.02 in a vacuum of 0.13 Pa (Fusaro, 1987a).

The coefficient of friction for PI materials is temperature dependent and has a reversible transition region in dry air at 40 ± 10°C. Figure 4.6 shows how the coefficient of friction of bonded polyimide

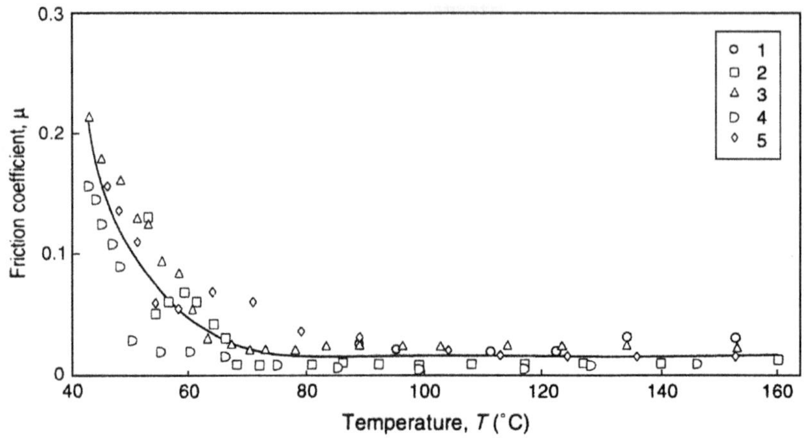

Fig. 4.6. Variation of the coefficient of friction with temperature for bonded polyimide coatings on a steel substrate sliding against a steel pin in dry argon with a speed of 1.6 m/s and a load of 5 N (data from Fusaro, 1977).

coatings sliding against steel increases from a minimum value of 0.03 at temperatures above 70°C to typical room temperature values of 0.2 to 0.3 (Fusaro, 1977). No major influence of sliding speed in the range of 0.027 to 1.33 m/s was observed.

The region of high friction corresponds to high wear and low friction to low wear in dry argon, dry air and moist air environments. However, the transition region has been found to occur at higher temperatures, somewhere between 100°C and 200°C, in moist air (Fusaro, 1978). The wear rate at temperatures above the transition can be up to 600 times less, measured in dry argon conditions, than at temperatures below the transition. In dry air the reduction has been observed to be only about 40 times. The environmental influence on the coefficient of friction at temperatures below and above the transition region is shown in Fig. 4.7.

The changes in the tribological behaviour of polyimide coatings caused by temperature and atmosphere is explained in terms of a mechanism where the hydrogen in the water vapour bonds to the polyimide chains at lower temperatures and constrains their freedom to unfold into an extended molecular chain type of structure. Thus brittle fracture and higher friction occur. By increasing the temperature to a higher level, the water molecules either debond or their influence becomes negligible, and friction and wear drop considerably. When this happens, very thin plastically flowing surface layers have been observed on the film wear tracks and on the pin sliding surfaces.

In a detailed study of the tribological contact mechanisms of bonded polyimide films Fusaro (1981) shows that two different regimes can be identified. In the first, the film supports the load and the lubricating mechanism consists of the shear of a thin surface layer of the film between the slider and the bulk of the film. The second regime occurs after the bonded film has worn to the substrate, and consists of the shear of very thin lubricant films between the rider and the flat plateaux generated on the metallic substrate asperities.

He also found that the wear life is directly proportional to film thickness and that the wear mechanism of the film is strongly dependent on contact stress. The rate and depth of crack propagation in the bulk was a function of contact stress. At light loads the surface stress influenced the spalling of a thin, textured surface layer which was less than 2 μm thick. Failure was characterized by the depletion of the polyimide-bonded graphite fluoride from the valleys between the metallic asperities and by the production of very fine powdery metallic debris. No galling wear of the surfaces was observed.

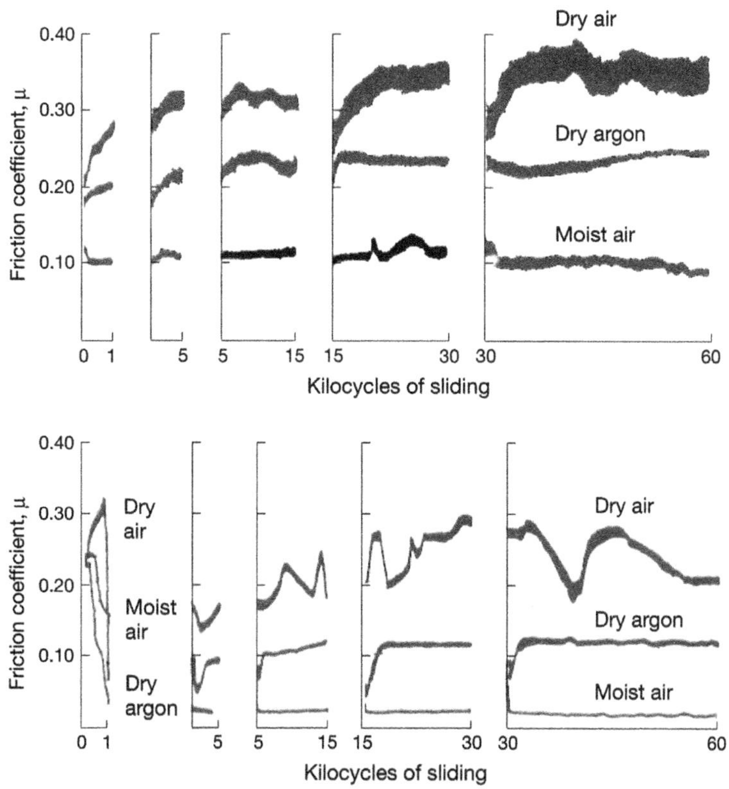

Fig. 4.7. The atmospheric influence on the coefficient of friction at temperatures (a) 25°C and (b) 100°C. The transition region is shown in dry air, dry argon and moist air conditions with a sliding speed of 2.6 m/s and a load of 10 N (after Fusaro, 1978).

A comparison of the tribological properties of three different polyimide coatings, deposited by spraying and heating on steel substrates, with seven polyimide bulk samples was carried out in experiments with a pin-on-disc tribometer with a rotational frequency of 100 rpm and a load of 10 N by Fusaro (1987a). For a polyimide coated surface produced from diamine-hexafluoropane and dianhydride pyromellitic acid (PMDA) he found that when it was applied onto the disc using an artist's airbrush and heated by a certain procedure to produce a 25 μm thick layer on a stainless steel substrate, it had a coefficient of friction of 0.45 which was a little less than the value 0.5 measured for the same material as a solid body. The wear rate of the coated film was slightly higher than for the PMDA solid body. The two other polyimide coatings (PI-4701 and PI-4701 + 50% graphite fluoride) had a coefficient of friction of 0.15 which is lower than the lowest value, 0.2, measured for any of the polyimide solid bodies. However, the wear rates of these coatings were around $10 \cdot 10^{-6}$ mm^3/Nm, which is higher than the average level for the solid bodies. Very thin, flowing transfer films were produced on the 4.76 mm radius hemispherical steel pins. The conclusion for the polyimide coatings studied is that they had very low friction but that the wear rates were slightly higher than for the bulk materials. In vacuum the wear rate of polyimides when sliding against steel is low, typically in the range of 0.01 to $10 \cdot 10^{-6}$ mm^3/Nm.

The tribological properties of sputtered PI and PTFE films on glass substrates sliding with a slow speed of 0.0004 m/s against a steel ball have been compared by Yamada *et al.* (1990). The PI film had a coefficient of friction of 0.2, which was smaller than that for a PTFE film, which was 0.3, and it had a considerably more stable friction curve than the PTFE film. The coefficient of friction was independent of load in the range 0.5 to 10 N and independent of film thickness in the range 0.05 to 1 μm. However, the friction of the PI film was somewhat higher than that for the target material as a solid body.

In the study of contact wear, Yamada *et al.* (1990) observed that the PI film was more brittle and had higher mechanical strength than a PTFE film. The wear lifetime of the PI film was about 4000 passes, which is about 20 times more than for the PTFE film. The longer life of the PI film, according to Yamada *et al.*, could be because very thin lubricant films of PI left on the track may exhibit intrinsically stronger adhesion to the substrate than PTFE. The replenishing effect of the PI film may also be responsible for the long life of the PI film, because a larger amount of wear debris of the PI film was supplied to the track as a result of the film wear at both sides of the track. In contrast to the PTFE film, no increase in wear lifetime for the PI film was observed after heat treatment at 200°C and 300°C.

4.2.1.5 Elastomeric coatings

The characteristic feature of elastomers is their ability to recover from large deformations. The elastic modulus of rubber is extremely small. The elastic deformation of rubber can be up to 1000% and is only 1% for many polymer materials and less than 1% for ordinary metals. The coefficient of friction of rubber is generally high when sliding in dry conditions on other solid materials. Hysteresis phenomena may play an important role in their frictional behaviour. Stick–slip occurs often but there is no single value for the static coefficient of friction as it depends upon the initial dwell time and the rate of starting velocity (Persson *et al.*, 2003). Elastomers have a good erosive and abrasive wear resistance because of their ability to deform elastically (Moore, 1975; SiWei, 2000).

The geometry of a contact on a macroscale also affects the hysteresis friction. When a ball slides on a plane surface covered by an elastomeric coating, the coating continuously deforms elastically and recovers after the contact adding a hysteresis component to the friction. A raised lip is formed in front of the ball. In the opposite situation, when the ball is covered by the coating, the hysteresis friction is minor because there is no continuous elastic deformation and recovery as shown in Fig. 4.8.

The behaviour of elastomeric coatings of different hardnesses and with thicknesses of 4 μm and 500 μm on glass substrates sliding against a glass hemispherical indenter or ceramic cones with different angles at a low speed of only 0.0002 m/s and loads in the range of 0.01 to 0.5 N was studied by Adams *et al.* (1990). They measured a coefficient of friction of 0.31 to 0.45 with a 4.93 mm radius indenter sliding on the thin elastomer film at low loads of 0.01 N and no detectable sliding damage was produced. In thick film experiments with a low load, the ceramic cone gave a coefficient of friction of 0.5 to 1 when sliding on a hard coating and 1.5 to 4 when sliding on a soft coating, without any observation of stick–slip. At a higher normal load of 1 N there was a gross stick–slip response. The coefficient of friction was about 0.55 for the harder coating and 0.9 for the soft coating. Adams

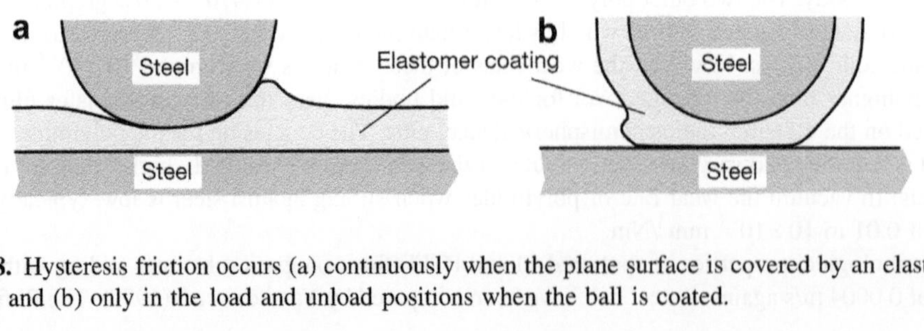

Fig. 4.8. Hysteresis friction occurs (a) continuously when the plane surface is covered by an elastomeric coating and (b) only in the load and unload positions when the ball is coated.

et al. conclude that for low normal loads the non-adhesive component of the coefficient of friction varies with the tangent of the cone angle. The results discriminate between soft coatings with high fracture toughness and hard coatings with low fracture toughness. For higher normal loads the amplitude of the stick phases in the observed stick–slip behaviour decreases with increasing hardness of the coating due to the reduction in fracture toughness. The nature of the damage was analogous to that observed for bulk elastomer specimens.

For coating thicknesses less than 0.2 μm the friction is considered to be independent of the thickness (Adams *et al.*, 1990). This can be explained by the effect of material build-up around the sphere, outside the Hertzian contact area, which has to be under sufficient pressure before it makes a significant contribution to the frictional traction.

In reciprocating friction experiments with cylinders coated with a 300 μm thick silicone rubber layer sliding on a glass plate with mineral oil lubrication at speeds of 0.01 to 0.15 m/s and loads of 42 to 77 N, Gladstone (1988) measured very low peak values for the coefficient of friction in the range of 0.01 to 0.03. The favourable effect of an elastomeric layer in creating elastohydrodynamic lubrication conditions has been shown by Hooke (1986). He developed a method to determine the clearance in lubricated elastomeric layer contacts.

4.2.2 Soft metal coatings

Soft metals like lead, silver, gold, copper, nickel and indium can provide the low shear conditions often needed for good frictional properties of coated surfaces; but this is not enough. Several other parameters also need to be optimized, such as adhesion to the substrate, film thickness, surface roughness, load and sliding speed.

Bowden and Tabor (1950) studied the behaviour of several soft metal films and found that low values for the coefficient of friction, even less than 0.1, can be obtained with coatings in the thickness range of 0.1 to 10 μm, sliding against a metal counterface. However, the wear and lifetime of such films was often a problem. The vacuum deposition methods and especially ion plating developed in the 1960s and 1970s provided a new method to deposit homogeneous metal films with improved adhesion to the substrate.

It was earlier assumed that the metal layer softens during sliding and acts as a lubricant. This explanation may well be correct at high sliding speeds and normal loads when high localized temperatures are produced by the flash heating effects. However, Jahanmir *et al.* (1976) showed that it is not the absolute softness of the coating which is important. It is the ratio of shear stresses between the substrate and the coating which is the determining factor. To support this they demonstrated that even a reasonably high melting point metal such as nickel can be effective in wear reduction under low-speed sliding, and that the thickness of the coating has a pronounced influence.

The film thickness, the surface roughness and the environmental influence on the oxidation of the coating are important parameters in applications with thin metal films. We have earlier mentioned (Fig. 3.76) that the coefficient of friction reaches a minimum value for lead films deposited on steel sliding against a steel counterface at a thickness of about 0.5 μm while friction is higher for both thinner and thicker films. Very thin films suffer from asperity breakthrough while thicker films result in a larger contact area and thus higher friction.

It is interesting that thinner films can produce lower wear rates than thick films (Sherbiney and Halling, 1977). With thick films the wear mechanism is essentially microcutting of the coating by the mating surface asperities. For thin films the wear is based on a fatigue mechanism, with the mating surface asperities breaking through the film and contacting the substrate material.

It appears that there exists an optimum film thickness for soft metal coatings. Jahanmir *et al.* (1976) showed that this is in general less than 1 μm for steel coated with metals like Cd, Ag, Au and Ni. They give a theoretical explanation based on the delamination wear theory. A thin layer of a soft metal coated on a hard substrate can retard the wear of the harder substrate by preventing plastic deformation and crack nucleation in the hard substrate. If the soft layer is thicker than a critical value, delamination can occur within this layer, forming loose wear particles. These wear particles can in turn cause ploughing and further plastic deformation. According to the delamination theory the coated material must be softer than the substrate to minimize the wear rate.

The surface roughness of the substrate is especially important when it is of the same order of magnitude as the thickness of the soft metal coating, as shown in Fig. 4.9. The importance is minor with very thick coatings (Fig. 4.9a) and similar to its influence for uncoated surfaces for very thin coatings (Fig. 4.9c). Very rough counterface surfaces will increase the microcutting wear action for thick coatings (Fig. 4.9a) and with thinner coatings they will more easily interact with the substrate through the coating (Fig. 4.9b and c). In sliding contacts with soft coatings smoother surfaces last longer (Jahanmir *et al.*, 1976).

In rolling contacts, with one of the components covered by a soft metal film, the rolling resistance is associated with elastic hysteresis losses arising from the stress cycling as the rolling proceeds. The rolling resistance is related to the curvature radius of the components and the Hertz contact area. When the value of the Hertz contact area divided by the coating thickness decreases below a certain level plastic behaviour can be assumed and when it approaches zero full plastic deformation of the film ensues (Halling *et al.*, 1985; Halling, 1986).

Rolling contact fatigue fracture in a coated surface may be initiated at the surface and the coating can arrest or at least retard the initiation of fatigue microcracks. Up to tenfold improved lifetimes have been measured when comparing both soft (Cu) and hard (TiN) coated components with uncoated components in rolling contacts (Hochman *et al.*, 1985).

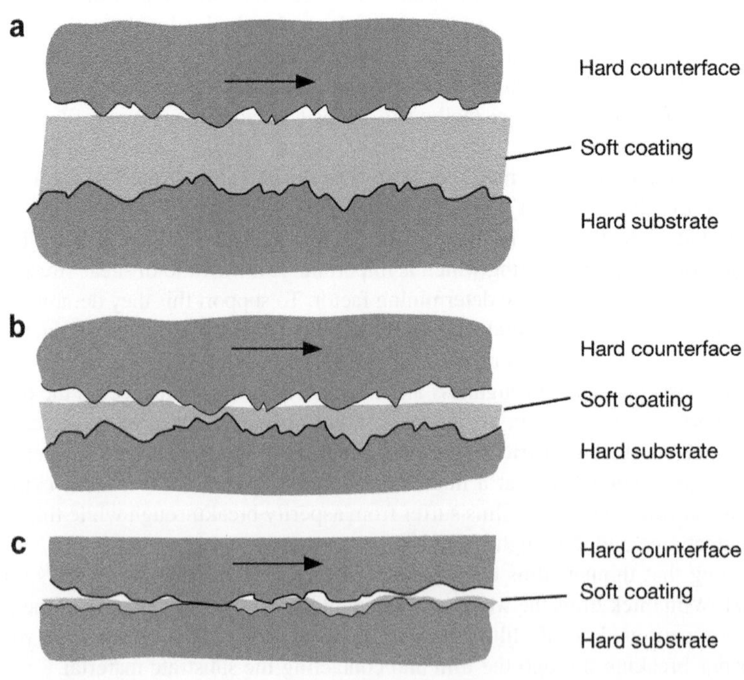

Fig. 4.9. The influence of the substrate and the counterface surface roughness are different depending on whether the thickness of the soft metal coating is (a) thicker, (b) of the same level or (c) thinner compared to the surface roughness.

4.2.2.1 Lead coatings

The coefficient of friction of a soft metal like lead (Pb) or copper (Cu) sliding on itself is extremely high, in the range of 1 to 2, and large-scale seizure and tearing normally occur. For a curved steel slider sliding against these materials the coefficient of friction is typically a little lower, about 0.9, but still very high. Bowden and Tabor (1950) showed that the low shear properties and low melting point of lead, 327°C, can be utilized tribologically by applying a very thin lead layer on a harder substrate. They measured a decreased coefficient of friction to values less than 0.3 when a steel ball was sliding against a 30 μm thick lead surface film that covered a copper substrate. This film thickness was an optimum value, with friction increasing for both thicker and thinner lead films. The optimum film thickness is greater than that observed when the films are deposited on steel, which indicates that it is influenced by the hardness of the metal substrate. It is also greater than that observed for indium films on steel. With softer metal substrates, such as copper, thicker films are generally needed. However, Tsuya and Takagi (1964) found that minimum values for the coefficient of friction for lead coatings on copper substrates corresponded to coating thicknesses below 10 μm when sliding against electro-polished copper.

Sherbiney and Halling (1977) found minimum values for the coefficient of friction down to 0.1 and less for an optimum film thicknesses of 0.2 to 1 μm when a lead coated steel ball was slid on a steel surface, as shown in Fig. 3.76. The thinner the soft films, the more there will be steel to steel interactions between the substrate and the slider through the film and this explains the influence of surface roughness on the coefficient of friction, shown in Fig. 4.10 (Halling and Sherbiney, 1978).

In the load range of 5 to 30 N Sherbiney and Halling found a considerable decrease in the coefficient of friction from 0.28 to 0.12 with increasing load. Only a relatively slight decrease in the coefficient of friction from 0.12 to 0.09 was observed with increasing speed from 0.1 to 1.2 m/s. The lead films were applied by ion plating which provides good adhesion between the film and the substrate and extended film lifetimes. The lifetime was seen to be dependent on the velocity with longer lifetimes for higher sliding speeds. This is believed to be due to the formation of lead oxide, which is a very effective lubricant. At higher velocities the increased temperature results in a larger concentration of PbO and improved lubrication in the contact.

The wear mechanism with thicker films was observed to be mainly microcutting of the films by the mating surface asperities, but with thin films it was believed to be principally a fatigue mechanism due to the mating surface asperities breaking through the film and contacting the substrate material.

The behaviour of lead coatings in vacuum has been of interest because there was a need to find a solid lubricant with low friction and wear to avoid liquid lubricants in space. Lead coatings were the

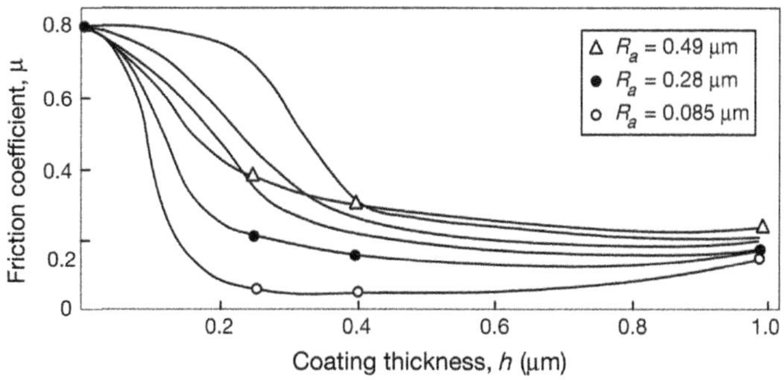

Fig. 4.10. The effect of surface roughness on the coefficient of friction for ion-plated lead coatings sliding against a steel ball at a speed of 0.16 m/s and a load of 15 N (data from Sherbiney and Halling, 1977).

first to be widely used for this purpose. In vacuum Arnell and Soliman (1978) found the same optimum film thickness range of about 0.2 to 1 μm for minimum values of the coefficient of friction, for a surface roughness less than $R_a = 0.5$ μm. With rougher surfaces the coefficient of friction increased with increasing film thickness, reaching 0.8 at film thicknesses of 10 μm, as shown in Fig. 4.11. The increase in the coefficient of friction is remarkable for all values of surface roughness with films thicker than 1 μm. For increasing surface roughness, with R_a values from 0.05 to 0.46 μm, Gerkema (1985) reported an increase in the coefficient of friction but at the same time a considerably increased lifetime for lead coatings.

One explanation for the discrepancy in the influence of surface roughness and film thickness on the coefficient of friction measured in air (Fig. 4.10) and in vacuum (Fig. 4.11) may be due to the different geometry; the stationary pin was the coated component in the former experiment. However, the general conclusion still seems to be that the coefficient of friction increases with very thin films, with thick films and with increasing surface roughness.

It is interesting that Gerkema (1985) reports for a vacuum environment, similar to Sherbiney and Halling (1977) in air, the lowest coefficient of friction for the highest load and observes at the same time a decrease in film lifetime. The same trend – very low coefficients of friction down to 0.06 but very short lifetime – is observed by Gerkema at increased temperatures up to 300°C. The very low friction cannot be explained by formation of oxide layers because the experiments were in vacuum. A more probable explanation could be that increased load, and temperature, causes melting of lead at the contact resulting in lower shear strength but also in increased wear. In conflict with this is the increase in coefficient of friction with increased speed reported by Arnell and Soliman (1978) also in vacuum. An increase of the coefficient of friction at very low speeds of less than 0.001 m/s has been observed by Roberts (1990b).

The low melting point of lead places an upper limit for its temperature range of operation. However, there appears to be no lower limit. Lead lubricated bearings operate equally well in vacuum at cryogenic temperatures, down to 20 K, as well as at room temperature (Roberts, 1990b).

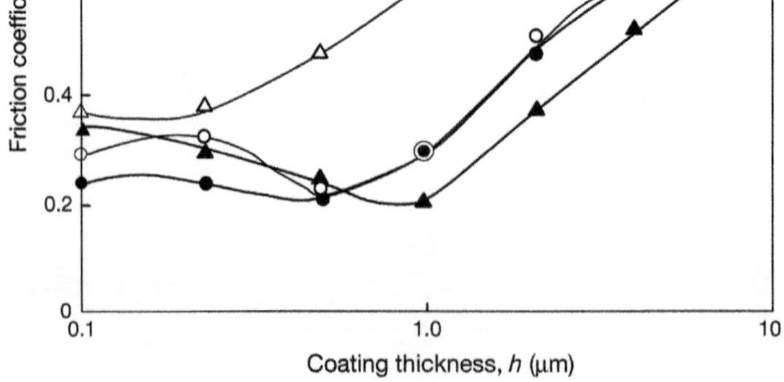

Fig. 4.11. Effects of substrate roughness on the coefficient of friction at different film thicknesses of ion-plated lead films sliding against a steel pin in vacuum with a speed of 0.01 m/s and a load of 1 N (data from Arnell and Soliman, 1978).

Ion plating can provide a well-adhered lead film with a low coefficient of friction but the lifetime of such films under pure sliding conditions is still poor. Gerkema (1985) has studied the possibility of achieving improvements by using metal interlayers, such as Mo, Ta, W, Ag and Cu, and by using films with metal additives such as Mo, Ag, Cu and Pt. Most of the metallic interlayers produced little or no improvement. However, a thin copper interlayer at the lead/steel interface or a small addition of copper or platinum to the lead coating increased the lifetime about fivefold.

Ion-plated thin lead coatings with thicknesses of about 0.2 to 0.5 μm have been shown to provide satisfactory lubrication for ball bearings in slow- and medium-speed mechanisms in space environments. A major drawback in long lifetime and high precision applications is debris generation leading to torque noise. The lead film can produce unwanted ball-speed variations, and therefore ball-to-ball friction, resulting in relatively high torque noise levels (Thomas *et al.*, 1980).

Lead coatings can also be used to improve the load-carrying capacity of lubricated sliding steel contacts. Liu *et al.* (1991) found that the scuffing load of a steel ball sliding under boundary lubrication on a steel disc with a 3 to 16 μm thick brush-plated layer of lead on top was more than double that with an unplated steel surface. They observed that increased substrate hardness results in a decrease in the coefficient of friction and higher values for the scuffing load, as shown in Fig. 4.12. The measured coefficient of friction was in the range of 0.05 to 0.1 for sliding speeds of 1 to 4 m/s and loads of 10 to 1000 N. When the thickness of the lead-plated layer is smaller than 16 μm, the coating thickness has almost no effect on the coefficient of friction and the scuffing load. The scuffing load decreases with increasing sliding speed. The lubricant used was a paraffin oil base stock without additives.

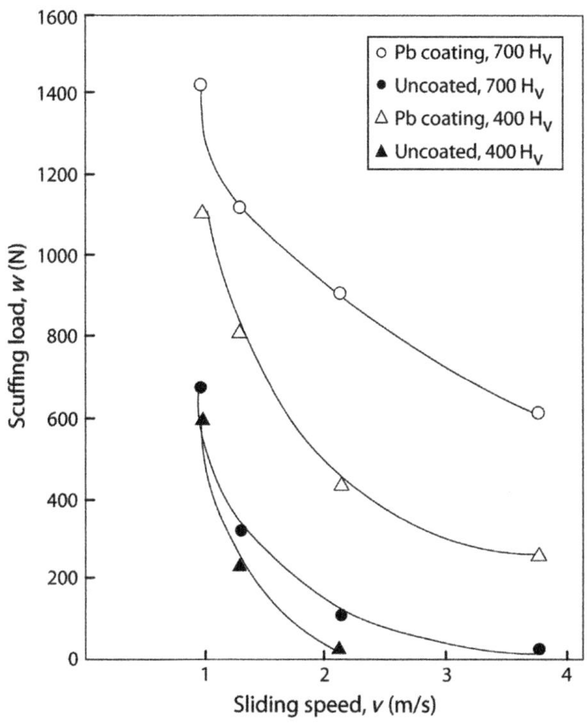

Fig. 4.12. The effect of substrate hardness and sliding speed on the scuffing load when a steel ball is sliding on a lead brush-plated and -unplated steel surface under boundary lubrication (data from Liu *et al.*, 1991).

4.2.2.2 Silver coatings

Thin silver (Ag) coatings can be used to improve the tribological properties of sliding contacts. Silver is relatively soft and can thus easily shear to reduce friction. It is relatively chemically inert, it can generally resist oxidation when used in air and it has a very high thermal conductivity.

If, however, silver coatings do become oxidized, e.g. due to temperature, this can result in increased wear and film removal. The hard oxide particles will cause abrasive wear (Jahanmir *et al.*, 1976). The initial coefficient of friction for a silver coating sliding on a steel surface can be as low as 0.15, but after a period of sliding, when silver oxides have been formed and the wear increases, it rises to high values of about 0.5 as shown in Fig. 4.13. It has been shown that the sliding distance determining the endurance life of silver films in air is constant and independent of sliding speed in the range of 1 to 5 m/s (Sherbiney and Halling, 1977), but that the lifetime becomes shorter at higher speeds.

The good lubrication properties are attributed partly to the low shear strength in the direction of sliding and also to a material transfer mechanism. El-Sherbiny and Salem (1986) have described the different stages in the wear process when an ion-plated silver coated steel ball is sliding on a steel disc. Initially, the surface asperities produce a microcutting form of wear on the coated ball and the wear rate is high. Subsequently, most of the film is removed resulting in non-uniform smearing of silver wear flakes on parts of the track. As sliding continues, the silver wear flakes are smeared out on a much larger area of the track and a material transfer mechanism takes place resulting in a steady mild wear rate. Finally, the bulk of the silver is removed and the asperities of the disc are in contact with the ball substrate. Fine abrasive wear particles of steel and iron oxide are thereby produced. These particles are responsible for polishing the surfaces of both the scar and the track. The polished surface of the scar is generally the original film/substrate interface. The very low wear rates observed during this regime suggest that contact fatigue could be the dominating wear mechanism.

Both the contact pressure and the sliding speed influence the wear of silver coatings and their failure mechanisms. Yang *et al.* (2003) carried out ball-on-disc tests with a 2 μm thick two layer gradient silver coating deposited on steel sliding against a steel ball and produced wear maps with pressure and velocity as dominating parameters, as shown in Fig. 4.14. In the wear map they identify three different wear regimes: (I) elastic–plastic deformation with mild wear but no coating failure, (II) formation of transfer layer with moderate wear and (III) severe wear with no protecting transfer layer.

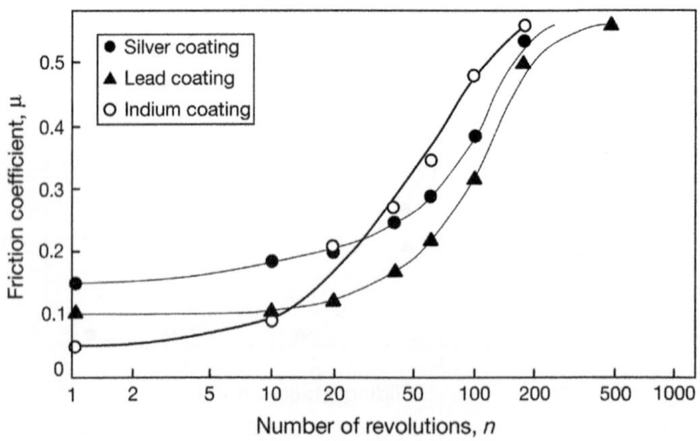

Fig. 4.13. The coefficient of friction during the lifetime of three ion-plated soft metal films sliding against a steel ball (data from Sherbiney and Halling, 1977).

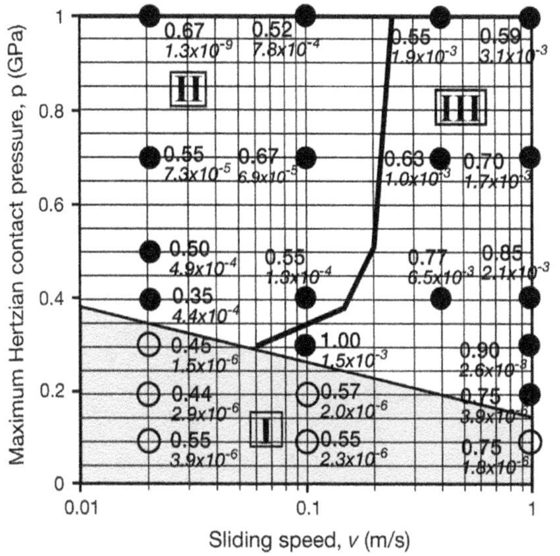

Fig. 4.14. Wear map for 2 μm thick two layer gradient silver coating on steel sliding against a steel ball with the three wear regimes indicated: (I) elastic–plastic deformation without coating failure, (II) transfer layer formation with moderate wear and (III) severe wear with no protecting transfer layer. Coefficients of friction and wear rates are given numerically (data from Yang *et al.*, 2003).

The contact pressures varied in the range of 0.1 to 1 GPa and the sliding speed was 0.02 to 1 m/s. The gradient film consisted of an 8 nm thin IBAD silver bond layer on which a 2 μm relatively soft silver film was thermally evaporated. This functionally gradient film improved the life due to better bond strength. In the mild wear regime the contact surfaces were covered by transfer layers of agglomerated wear particles. The transfer layer acted as a protective layer and resulted in low friction after the initial transfer of coating material.

In a comparison of silver coatings deposited by three different techniques, ion plated, vacuum deposited and gas deposited, El-Sherbiny and Salem (1986) found the ion-plated ones most suitable for tribological purposes. The nucleation of the ion-deposited films is very distinct. The size of the nuclei is small, only 10 nm, and their density is the highest of the three deposits. A continuous film is usually obtained at a thickness of 13 nm with the final grain size being 16% larger than the initial nuclei size. The ion-plated films exhibited good adhesion to the substrate in scratch tests and were harder than those deposited with the two other techniques, as shown in Fig. 4.15. They also showed the largest endurance life and exhibited a minimum coefficient of friction of about 0.15 at film thicknesses in the range of 1 to 8 μm.

Humidity may affect the wear and deformation of silver films. A considerable influence of humidity was measured by Yang *et al.* (2001) in terms of rolling resistance of 7.1 mm diameter steel balls rolling with the rotational surface speed of 0.031 m/s and at a load of 147 N on a 1.4 μm thick silver coated steel surface, as shown in Fig. 4.16. The rolling resistance, expressed as the ratio of measured tangential force divided by the normal force, increased with a factor of six from low values of 0.005 at 30% RH to 0.03 at 100% RH. The initial rolling resistance is for three different silver coatings, one hydrophilic, one unmodified and one hydrophobic, were all in the range of 0.01 to 0.025. However, the steady-state rolling resistance increased from lower values of 0.005 to 0.03 to higher vales of 0.01 to 0.04 and 0.02 to 0.06 for unmodified, hydrophilic and hydrophobic, respectively. Surface analysis indicated that this effect was due to the fact that adsorbed water vapours affected the wear debris

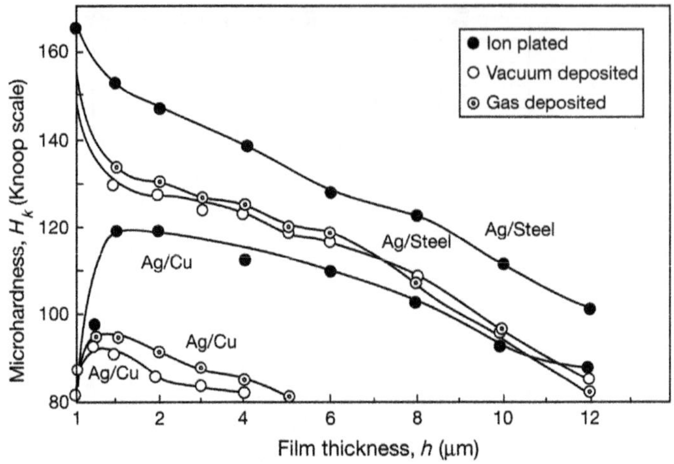

Fig. 4.15. Microhardness of silver coatings as a function of film thickness deposited by three different techniques on steel and copper (data from El-Sherbiny and Salem, 1986).

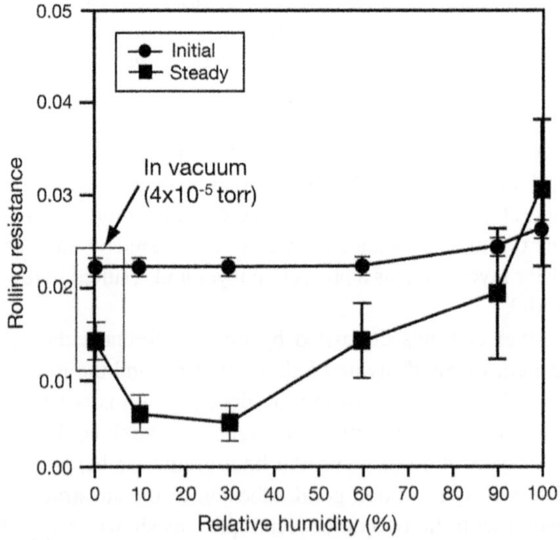

Fig. 4.16. Initial and steady-state rolling resistance of steel balls rolling on a 1.4 μm thick silver coating on a steel substrate in vacuum and at different ambient humidity (data from Yang *et al.*, 2001).

agglomeration and the tendency of contact patch formation. Note that the hydrophobility was controlled by depositing a very thin plasma polymer film on the silver surface.

In a non-oxidative environment like argon, Jahanmir *et al.* (1976) have shown that a thin film of silver on a steel surface can reduce the sliding wear rate by three orders of magnitude at a velocity of 0.03 m/s and a load of 22.5 N. They found that the optimum film thickness for reduced wear is about 0.1 μm. In vacuum Arnell and Soliman (1978) found that the optimum film thickness value for the minimum coefficient of friction of about 0.1 is a little higher, at about 1 μm. For higher and lower film

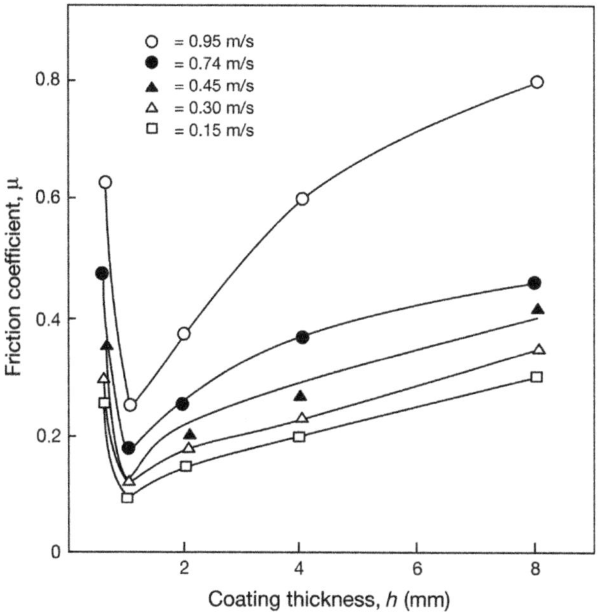

Fig. 4.17. Variation of the coefficient of friction with film thickness and sliding speed for a steel ball sliding on a silver coated steel disc in vacuum with a load of 1 N (data from Arnell and Soliman, 1978).

thicknesses the coefficient of friction increased considerably as shown in Fig. 4.17 but no influence of film thickness on film lifetime was observed. Both the coefficient of friction and the number of cycles to film failure increased with increasing sliding speed.

Silver coatings have also been successful in improving the tribological properties of ceramic sliding components in air. A 1 to 2 μm thick silver layer deposited by vacuum techniques on various ceramic substrates such as Al_2O_3, Si_3N_4, magnesia-partially stabilized zirconia and zirconia has considerably reduced friction and wear. The coefficient of friction dropped by about 50% from high values, typical for ceramic contacts, in the range of 0.7 to 1 down to about 0.3 to 0.4. At the same time the wear rate dropped by 1 to 3 orders of magnitude and in the case of a silver coating on zirconia down to a practically non-measurable level (Erdemir *et al.*, 1990, 1991a, b; Ajayi *et al.*, 1991). The adhesion of silver films to Al_2O_3 ceramic substrates can be improved by applying a 25 nm thick titanium interlayer (DellaCorte *et al.*, 1992).

In pin-on-disc experiments Erdemir *et al.* (1990) studied the contact between a ceramic Al_2O_3 pin sliding on Al_2O_3 disks which had silver coatings applied by PVD or ion-assisted deposition (IAD) techniques. A 2 μm thick silver IAD coating reduced the pin wear rate by an order of magnitude and the coefficient of friction from 0.95 to 0.4, as shown in Fig. 4.18. Stronger coating to substrate adhesion as well as smoother surfaces were observed to result in improved wear life with alumina balls sliding against silver coatings deposited on Al_2O_3 (Erck *et al.*, 1990).

Silver coatings have proved to be successful in reducing the friction of PVD titanium nitride coated steel surfaces sliding against a steel ball at an oscillating sliding speed of 0.002 m/s and a load of 5 N (Sander and Petersohn, 1992). In dry air of RH $< 1\%$ the coefficient of friction dropped from 0.78 to 0.17 and in humid air of RH $= 20$–22% from 0.38 to 0.19. The capability to reduce friction was superior for silver coatings compared to other soft metallic coatings such as indium, copper and lead.

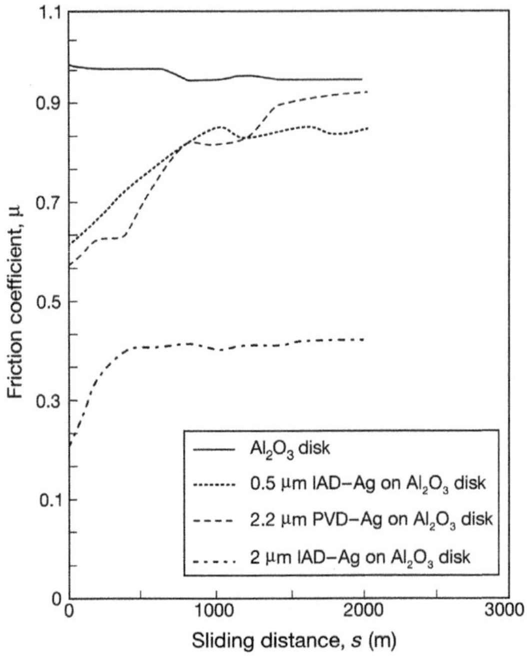

Fig. 4.18. Variation of the coefficient of friction with silver coated Al_2O_3 discs sliding against an Al_2O_3 pin and the ceramic sliding against itself at a speed of 1 m/s and a load of 5 N (data from Erdemir *et al.*, 1990).

4.2.2.3 Gold coatings

Gold (Au) is a noble metal with excellent corrosion resistance and good electrical and thermal conductivities. For this reason electroplated gold coatings are widely used by the electrical industry for electrical connectors and similar components. In these kinds of applications the electrical conductivity and the film lifetime are primary requirements and frictional properties are less important.

It is difficult to make general conclusions on the tribological behaviour of gold coatings because of the very different contact conditions used in the reported experimental work; this seems to have a considerable influence on the results. A thin gold coating plated on steel cylinders with an optimum film thickness of 0.1 μm can reduce the wear rate in sliding against each other by three orders of magnitude compared to uncoated cylinders, according to Jahanmir *et al.* (1976). In their slow cylinder-on-cylinder experiments, with a sliding speed of only 0.03 m/s and a load of 22.5 N, the wear was considerably reduced both in an air and in an argon environment but the coefficient of friction was very high, being up to values from 0.85 to 0.9.

On the other hand Takagi and Liu (1967) measured in a block-on-ring test against steel a low value for the coefficient of friction, 0.1, for 0.1 μm thick electroplated gold coatings on steel at low speeds of only 0.00017 m/s and loads of 2 to 15 N while the best improvement in the wear life, by four orders of magnitude compared to uncoated steel contacts, was observed for thicker coatings in the range of 5 to 20 μm at higher sliding speeds, 0.0132 m/s, and loads, 680 N (Takagi and Liu, 1968). Further improvement of the wear resistance of electroplated gold coatings can be achieved if they contain a small amount of about 0.1% nickel which increases hardness (Solomon and Antler, 1970).

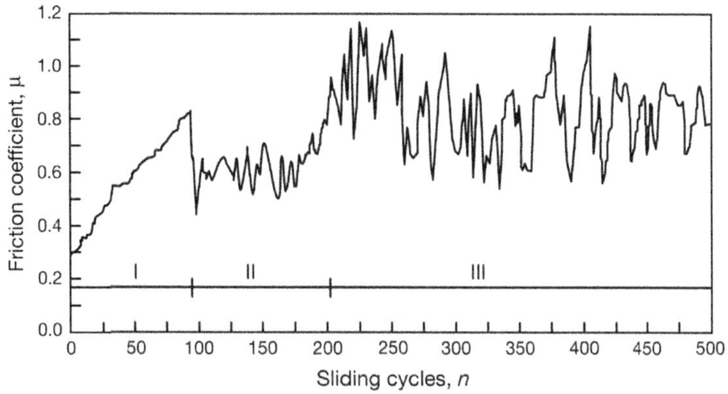

Fig. 4.19. The coefficient of friction as a function of sliding cycles for an electroplated gold coating in reciprocating sliding on itself with a speed of 0.004 m/s and a load of 1 N (after Tian *et al.*, 1991b).

The tribological mechanisms of a gold coating sliding against itself have been studied in detail by Tian *et al.* (1991b). For the purpose of improving electrical connectors they measured friction and wear in a slowly reciprocating contact with a 1.5 μm thick electroplated gold layer on a 2.5 μm nickel interlayer on a brass substrate. The coefficient of friction had an initial value of 0.3 which then increased through three stages reaching a maximum value of 1.2 as shown in Fig. 4.19. The initial linear increase in the coefficient of friction may be due to an increase in the contact area by wear and thus an increased area for adhesion. Lubrication with polyphenyl ether decreased the coefficient of friction down to 0.1.

Tian *et al.* (1991b) present the following description of how the contact mechanism changes during the history of the contact. They distinguish between three different stages as illustrated in Fig. 4.20. In stage 1, a prow forms in front of the slider due to strong adhesion between the gold surfaces (Fig. 4.20a). The wear of the gold film is marginal at this stage. During sliding in stage 2, cracks are initiated and propagate at the nickel/brass interface owing to the large interfacial shear stress. They continue to propagate along the interface and finally emerge at the surface, resulting in thin plate-like wear particles (Fig. 4.20b). These are broken into pieces and adhere to the film surface. By this time the electroplated system has technically failed because the exposure of the substrate will soon lead to severe corrosion, resulting in an unacceptable increase in contact resistance. As the sliding enters into stage 3, wear particles are continually delaminated from the nickel/brass interface of the ball specimen (Fig. 4.20c). The damage zone is enlarged and steady wear of the brass substrate takes place.

Another detailed analysis of the same contacting system but in dry and lubricated fretting conditions with different frequencies and amplitudes was reported by Tian *et al.* (1991a). The coefficient of friction was independent of the fretting amplitude.

Co-sputtering of molybdenum disulphide with gold can improve the tribological properties of gold coatings considerably. Lince (2004) tested co-sputtered Au/MoS_2 coatings with MoS_2 content from 0 to 100% deposited on steel plates in a pin-on-disc tribometer at 5 N load and 0.08 m/s sliding speed sliding against a steel ball and pin in nitrogen gas. The film with 95 at.% Au and pure Au failed after less than 10 m of sliding. However, a 15 at.% of MoS_2 content resulted in good endurance life at the lower contact pressure of 0.1 MPa and the friction dropped to 0.03 for both low and high (730 MPa) contact pressures. Good endurance life was reached at high contact pressure at a MoS_2 content of 41 at.% and then the coefficient of friction was as low as 0.013. At low contact stress the lubrication was provided by a thin, about 1 nm thick, film of relatively pure MoS_2 regardless of the contact stress.

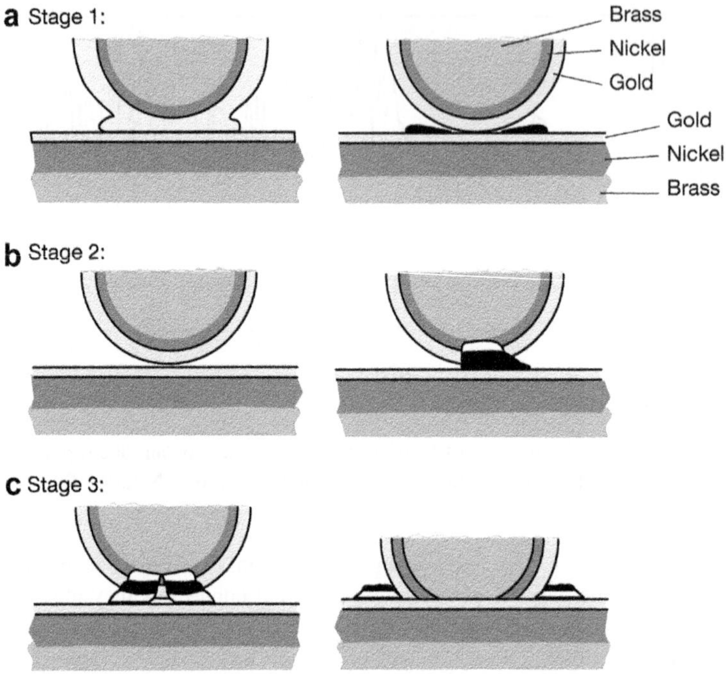

Fig. 4.20. The contact mechanism of an electroplated gold coated ball with a nickel interlayer and a brass substrate in reciprocating sliding on a plate with the same surface coating, showing changes during the history of the contact (after Tian *et al.*, 1991b).

4.2.2.4 Indium coatings

Bowden and Tabor (1950) used indium (In) when they studied the tribological mechanisms of thin and soft metal coatings in their pioneering work. Indium is a soft metal that oxidizes relatively slowly. They found that minimum values for the coefficient of friction down to 0.08 could be obtained by using optimum film thicknesses of 0.1 to 1 μm. This was later confirmed both in air by Sherbiney and Halling (1977) and in vacuum by Arnell and Soliman (1978). The influence of film thickness on film lifetime was only minor in vacuum.

The initial coefficient of friction for a steel ball sliding on an indium coated steel disc was found to be only 0.05 by Sherbiney and Halling (1977) as shown in Fig. 4.13. However, after some 20 revolutions the indium changed from being shiny metallic to a greyish colour and the coefficient of friction then increased to more than 0.5, followed by film breakdown.

Sherbiney and Halling also found that the coefficient of friction decreases with increasing load when an indium coated steel ball slides on a steel surface, but variations in speed within the range 0.1 to 1.2 m/s have almost no effect at all. Increased speed resulted, however, in a shorter film lifetime and a shorter effective sliding distance before failure. In vacuum, Arnell and Soliman observed that both the coefficient of friction and the wear increase with increased sliding speed.

4.2.2.5 Nickel coatings

Nickel (Ni) coatings deposited by electroplating and with a thickness of 40 μm have good corrosion resistance but poor wear resistance, e.g. compared to conventional chromium-plated coatings. The coefficient of friction is high, about 0.9 when sliding against plain carbon steel and 1.2 when sliding against stainless steel in a Falex-type test arrangement (Gawne and Ma, 1989).

However, Jahanmir *et al.* (1976) have shown that a thin nickel coating with an optimum film thickness and a suitable coating/substrate hardness combination exhibits a considerably improved wear resistance and a decreased coefficient of friction in a non-oxidizing argon environment. In their tests the wear rate was minimized when the plated material was softer than the substrate.

This was clearly demonstrated in experiments with a 1 μm thick nickel coated steel cylinder sliding against a stationary steel cylinder. For substrate hardness in the range of 0.84 to 3.7 GPa there was an immediate film failure at the start of sliding. However, with a steel substrate (AISI4140) of 4.6 GPa hardness, the wear rate was three orders of magnitude lower than for unplated materials. A very pronounced influence of coating thickness was observed. The wear rate was very low for nickel coatings with an optimum film thickness of 1 μm but increased rapidly for thicker coatings as shown in Fig. 4.21. At the optimum film thickness of 1 μm also the coefficient of friction had decreased from 0.63 to 0.45, probably because of less ploughing when the layer is thin. The increase in wear rate with thicker coatings is caused by dislocation accumulation leading to strain hardening and delamination within the plate. When the same experiments were performed in air the nickel coatings were not successful, as they oxidized and were removed early in the test.

The mechanical and tribological properties of nickel coatings can be improved by including other materials in the coating during the deposition process. In electroless nickel coatings in the thickness range of 10 to 20 μm the mechanical properties, that is the hardness and load-carrying capacity, have been improved by adding 5 to 20% of SiC particles while an increase in the coefficient of friction was observed at the same time (Grosjean *et al.*, 2001). The addition of PTFE particles in the nickel coating

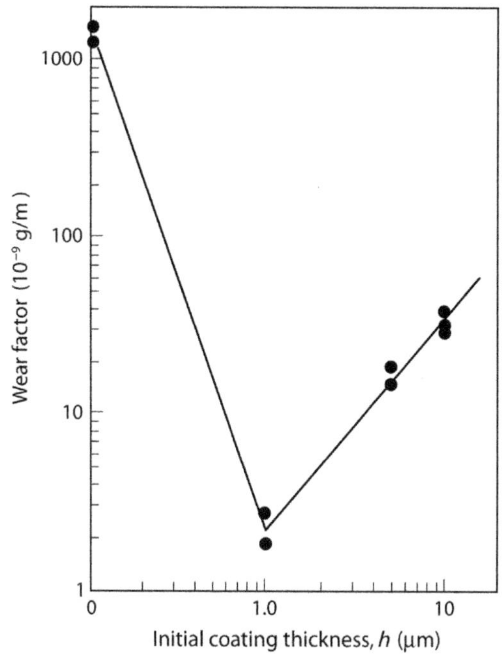

Fig. 4.21. The effect of the initial nickel coating thickness on wear when sliding against steel in cylinder-on-cylinder experiments in argon at a speed of 0.03 m/s and a load of 22.5 N (data from Jahanmir *et al.*, 1976).

has, on the other hand, been used to decrease the friction (Huang *et al.*, 2003; Rossi *et al.*, 2003). The properties and possibilities of such composite coatings are discussed more in detail in section 4.6.2.

4.2.2.6 Chromium coatings

Electroplated chromium (Cr) coatings with thicknesses of about 50 μm are widely used because of their good wear resistance and good oil retention in oil lubricated applications. However, depending on factors such as the counterface and pressure they may exhibit high friction in dry sliding conditions. The conventional way to produce Cr coatings is in a sulphate-ion-catalysed chromic acid bath. Gawne and Ma (1989) compared this type of conventional chromium coating with electroless nickel, heat-treated electroless nickel, electroplated nickel, heat-treated chromium and crack-free chromium coatings and found the conventional coatings superior in terms of wear resistance even though they do not have the highest hardness. All coatings had a thickness of 40 μm and were coated on plain carbon steel substrates. The conventional chromium coating had a coefficient of friction of 0.88 when sliding against plain carbon steel and 0.68 when sliding against stainless steel. The low wear rates of the chromium coatings are believed to be due to the thin passivating, self-healing oxide layer, Cr_2O_3, on their surfaces, which effectively protects the underlying metal. The Cr_2O_3 film acts as a barrier layer, preventing contact between the two metals and suppressing adhesive transfer by presenting an incompatible surface to the steel surface.

4.2.2.7 Copper coatings

Copper (Cu) is a soft metal which in sliding contacts typically has a high coefficient of friction. This correlates directly to the increased real contact area, which is sheared in sliding (Takagi and Tsuya, 1961). However, a thin copper film on a harder substrate can have the ability to reduce friction and increase wear life.

Sander and Petersohn (1992) showed that a thin copper film on a hard titanium nitride coated steel substrate, sliding against a steel ball at an oscillating speed of 0.002 m/s and a load of 5 N, reduces the coefficient of friction in dry air of relative humidity less than 1% from 0.78 to 0.27 and in humid air of relative humidity of 20 to 22% from 0.38 to 0.27. In oil lubricated rolling contact fatigue testing Hochman *et al.* (1985) found that an ion-plated copper film with a thickness of 0.5 μm deposited on stainless steel (440C) increased the endurance lifetime by one order of magnitude at maximum Hertzian stress levels of about 4 and 5.5 GPa. It is, however, somewhat surprising that the same increase in lifetime was not achieved in the same tests when another stainless steel (AMS 5749) was used as a substrate.

4.2.2.8 Cadmium coatings

The behaviour of thin cadmium (Cd) surface coatings is very similar to that which Jahanmir *et al.* (1976) found for the nickel and silver coatings referred to above. The plated material must be softer than the substrate to minimize the wear rate. At an optimum film thickness within the range 0.1 to 1 μm the coefficient of friction has minimum values of 0.25 in argon and 0.4 in air. The wear rate has a very pronounced minimum within the optimum film thickness range in argon. In air the differences are not so big but the optimum film thickness is still within the same range (Suh, 1986). Quite extensive delamination wear occurred in air and this is believed to be caused by oxidation. The oxidation can be minimized by lubrication. Jahanmir *et al.* (1976) studied the influence of substrate surface roughness with a 0.05 μm thick cadmium coating on a fine ground and a metallographically polished AISI 1018 steel substrate and found a very moderate effect on film lifetime. Polishing only increased the life of the coatings from 25 to 31 m of sliding. They suggest that moderate changes of surface roughness may only influence plating/substrate combinations which have low adhesive strength. Cadmium coatings are now being used much less in practical applications because of worries about pollution (Matthews *et al.*, 1992a).

4.3 Lamellar Coatings

A material with a lamellar or layered structure offers interesting surface properties from a tribological point of view. A sandwich-type structure can normally sustain high normal loads and may offer at the same time low shear strength between the planes and thus result in a lubricious low friction surface. The most commonly used lamellar materials in tribology are molybdenum disulphide (MoS_2) and graphite. Other lamellar materials that might be used for tribological coating purposes are graphite fluoride, HBN, H_3BO_3, WS_2, GaSe, GaS, SnSe and $NbSe_2$, but their tribological properties have not been extensively investigated (Roberts, 1990b; Greshham, 1997; Erdemir, 2001a; Teer, 2001). The graphitic coatings will be discussed later in section 4.5 in connection with carbon and carbon-based coatings.

4.3.1 Properties of molybdenum disulphide

Molybdenum disulphide (MoS_2) has a layered molecular structure with individual sheets of molybdenum and sulphur atoms. The low shear strength is explained by its anisotropy. The material is comprised of hexagonally packed planes consisting of a layer of molybdenum bounded on each side by a layer of sulphur as shown in Fig. 4.22. All effective strong bonding is within the resulting sandwich planes. The bond strength between the neighbouring sandwiches is due to van der Waals

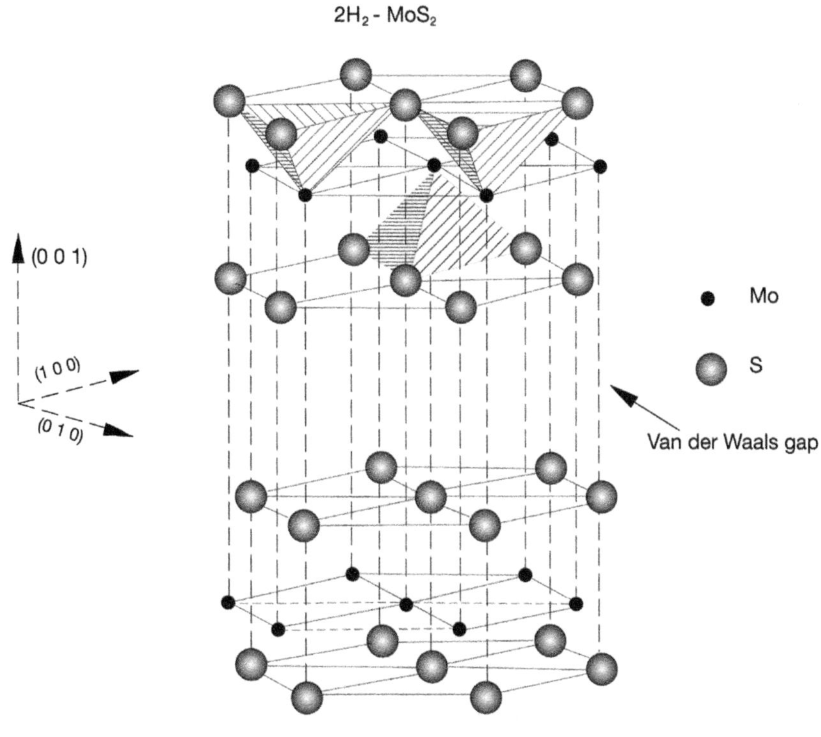

Fig. 4.22. The layered hexagonal crystal structure of molybdenum disulphide.

forces which are very weak. Thus MoS_2 has a strong structure in two dimensions but is weak in the third.

The friction of an MoS_2 surface film depends on the integrity of the film, contact pressure, humidity, film thickness, temperature and presence of contaminants. For a pure, smooth, dense, properly oriented MoS_2 film at high contact pressure in a clean, dry environment coefficients of friction as low as 0.02 have been reported in unidirectional sliding. With impurities, non-aligned orientation, humidity and low pressure, the coefficient of friction may be as high as 0.3 (Lansdown, 1999). For single crystals of MoS_2 the coefficient of friction was about 0.05 to 0.1 in vacuum and slightly higher, between 0.1 to 0.2, in air and in water when measured at slow sliding and 0.5 N load (Uemura *et al.*, 1989). In very controlled ultra-high vacuum conditions at a pressure of $50 \cdot 10^{-9}$ Pa superlubricity properties and a coefficient of friction down to 0.001 and below have been recorded for magnetron sputtered MoS_2 coatings sliding at 0.4 GPa contact pressure against a steel ball (Martin *et al.*, 1993).

Molybdenum disulphide is an attractive lubricant because the favourable tribological properties are retained at high temperatures in air up to about 400°C and in high vacuum at temperatures up to 800°C. It can be applied to a surface by conventional methods such as in powder or grease, or spray form by plasma spraying or by sputtering or other plasma-based PVD methods. Due to its excellent tribological properties in vacuum, it has been used extensively in space applications, both as solid lubricant and as a composite bulk material, such as in bearings, seals, expansion joints, slideways, screw threads, piston rings, valves and splines. The detailed deformation mechanisms of MoS_2 and its behaviour in tribological contacts have been described in several extensive review articles (Salomon *et al.*, 1964; Winer, 1967; Holinski and Gänsheimer, 1972; Farr, 1975; Lansdown, 1999).

4.3.2 Burnished and bonded molybdenum disulphide coatings

In the simplest case, burnished MoS_2 films are produced by applying a smooth sliding pressure to MoS_2 powder against the surface which is to be coated. By this method the MoS_2 crystallites first fill the low spots in the surface texture resulting in surface smoothing. It has been shown by X-ray diffraction of MoS_2 films on copper that the surface consisted of an apparent single crystal layer 2 to 5 μm thick with the basal planes oriented parallel to the sliding surface, but this highly oriented layer was on top of a randomly oriented layer, as shown in Fig. 4.23 (Brudnyi and Karmadonov, 1975; Kanakia and Peterson, 1986). In the single crystal layer there are normally wedge-shaped cavities, as shown in Fig. 4.23.

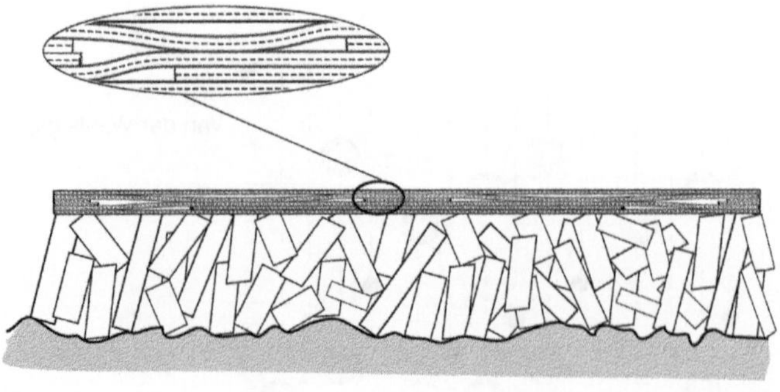

Fig. 4.23. Structure of a burnished molybdenum disulphide coating. The structure consists of a randomly oriented layer attached to the substrate and on top of it a (001) basal plane-oriented layer. This compressed MoS_2 layer includes wedge-shaped cavities.

The burnished film is attached to the substrate mainly by three mechanisms: (1) infilling of low spots on the surface, (2) embedment of crystallites in the surface and (3) chemical bonding between MoS_2 and the surface material. By experience there seems to be little or no advantage in producing burnished films thicker than 20 μm. Other methods used to apply the film to the surface are by flotation, by dipping in dispersion or spraying, or the use of bonding agents (Lansdown, 1999).

The tribological behaviour of pre-rubbed MoS_2 films sliding against a 4.76 mm radius steel rider in pin-on-disc experiments with a velocity of 2.6 m/s and a load of 10 N has been studied by Fusaro (1978). MoS_2 powder with an average particle size of 10 μm was mechanically rubbed over a steel disc surface producing a coating of 1 to 2 μm in thickness. The coefficient of friction was very low, only 0.01 to 0.02, in dry air and dry argon environments and showed good wear resistance with an endurance life up to about 100 kilocycles in dry air and 2000 kilocycles in dry argon.

In a moist air environment, the coefficient of friction increased to 0.08 and the endurance life was only about 10 kilocycles. The results indicate that water and oxygen are detrimental to MoS_2 lubrication. A thin MoS_2 layer was transferred to the steel rider surface but the amount of transferred material did not affect the coefficient of friction. In moist air and dry air, the MoS_2 film on the sliding surfaces was transformed from a bright, metallic-coloured, coalesced film to a dark, black, less continuous film. The rate of transformation depended on sliding time and whether or not the air atmosphere contained water. As the rate of transformation increased the wear life decreased and the rider wear rate increased. Little or no transformation of MoS_2 was observed in dry argon. This supports the conclusion that the predominant chemical reaction taking place is oxidation of MoS_2.

The lifecycle of a sliding MoS_2 film may be divided into the following three periods: running-in, steady state and failure (Bartz and Xu, 1986; Gardos, 1988) as illustrated in Fig. 4.24.

1. In the running-in period under normal load and tangential shear, the film rapidly becomes compacted to a fraction of its original thickness and a transfer film is formed on the counterface. The basal planes of the exposed, hexagonal MoS_2 crystallites become increasingly oriented on the compacted surface of the film, parallel with the direction of sliding. A process of high-temperature oxidation of the MoS_2 film begins.
2. After a few hundred to a few thousand cycles, the reduction in film thickness essentially stops. This is the beginning of the steady-state period which represents the useful lifetime of the film. During this long steady-state running period the coefficient of friction attains its lowest steady-state value and the wear rate is low. The absolute level of friction depends on several factors such as load, velocity and humidity.
3. The failure period starts with the appearance of surface blisters or microfracture of the coating. As the blisters or wear fragments are repeatedly pressed down by the passing load they become progressively more brittle, eventually breaking into fine, powdery debris. Blister formation and delamination is accompanied by a gradual increase in friction. Soon major descaling leads to galling and catastrophic seizure. According to Gardos (1988) the oxidation of the MoS_2 film within the layer and the consequent reduction in molar volume by etch-pitting, combined with gaseous oxidation products being trapped and frictionally heated within the coalesced free-volume regions, is the primary reason for blister formation. Large-scale delamination of blisters is the cause of the characteristic catastrophic failure of MoS_2 films.

The influence of humidity on the frictional behaviour of burnished MoS_2 films has been found to be reversible. Experimental data in the speed range of 0.0005 to 0.03 m/s, load range of 0.5 to 8 N and range of relative humidity from 20 to 90% indicate that absorbed moisture causes the MoS_2 film to mechanically soften and that the film is partly pushed ahead on the sides of moving contact spots (Gao *et al.*, 1993). As a result the film thickness between the contact spots decreases with humidity but increases slightly with sliding speed and contact spot size – that is with load. Accordingly, the shearing part of the frictional resistance is little affected but the frictional contribution due to film ploughing rises with increasing humidity.

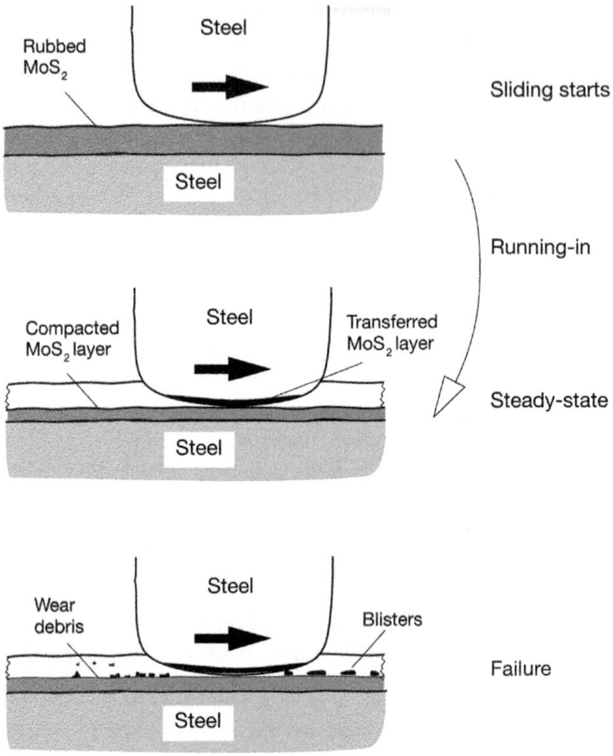

Fig. 4.24. The running-in, steady-state and failure contact mechanisms of a steel pin during the sliding on a steel disc coated with rubbed MoS_2.

Bonded MoS_2 coatings are produced by dissolving a binder in a suitable solvent and dispersing the MoS_2 powder in the solution. The liquid is applied on the surface and the solvent is removed by evaporation. In a bonded coating the MoS_2 crystallites are randomly oriented. The static coefficient of friction is therefore quite high, typically 0.1 to 0.3, and the films can be quite abrasive. The abrasiveness depends on coating hardness, thickness, particle size and loading. As soon as sliding begins there is a tendency for the crystals to become basal-plane oriented at the surface (Lansdown, 1999).

In a study of bonded MoS_2 containing two and three component solid-lubricating films Bartz and Xu (1986) show the dominating influence of blister formation on wear life and that, independent of the original film thickness, the wear life limit is reached at film thickness values of about 2 to 4 μm. Wear rates of about $100 \cdot 10^{-6}$ mm^3/Nm were obtained and characteristic types of friction were described based on pv values – that is pressure multiplied by velocity. Low pv values corresponded to low and stable friction while high pv values corresponded to high friction and very short wear life (Bartz *et al.*, 1984, 1986).

4.3.3 PVD deposited molybdenum disulphide coatings

PVD-based sputtering and ion plating are convenient, although a more expensive, deposition methods for MoS_2 and they provide coatings with excellent tribological properties for use in air, in vacuum and at high temperatures. The closed low-pressure deposition methods, described in Chapter 2, offer good controllability and high purity. Good adhesion, good oxidative stability, low wear and extremely low

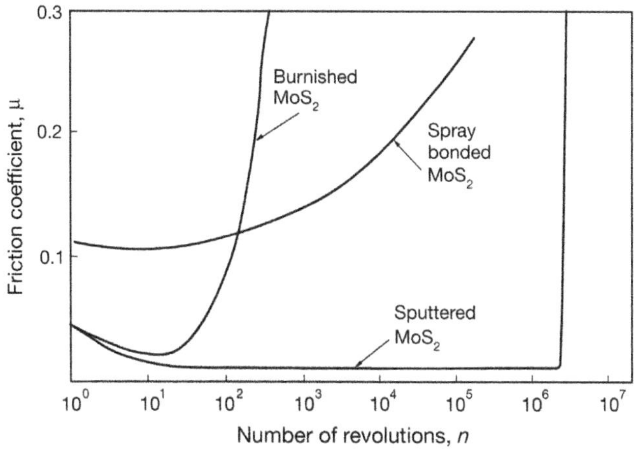

Fig. 4.25. Coefficients of friction of MoS_2 coatings in vacuum environment deposited by various techniques with a sliding speed of 1.2 m/s and a load of 50 N (after Roberts and Price, 1989).

friction have been achieved with sputtered MoS_2 films (Spalvins, 1972; Fleischauer and Bauer, 1988; Roberts, 1990a, b; Lansdown, 1999; Wahl *et al.*, 2000; Voumard *et al.*, 2001). A sputtered MoS_2 coating has a matt black sooty surface appearance and only a very thin film of about 0.2 μm is required for effective lubrication.

The influence of MoS_2 film deposition techniques on performance is well illustrated by Fig. 4.25, based on the experimental results in vacuum by Roberts and Price (1989). The superiority of sputtered films is clear and this is the reason why such films are of great interest to the space industry for bearing and other sliding applications.

Spalvins (1971) reports that there is only a very small compositional change in direct current (dc) and radio frequency (rf) sputtered MoS_2 films compared to the bulk material. The sputtered films have a submicroscopic particle size, possibly of an amorphous nature, and the films are of high density, close to the density of the bulk material, and free from pinholes. The high kinetic energy of the sputtered species, the submicroscopic particle size and the sputter-etched substrate surface are all partly responsible for the strong adhesion and cohesion of the sputtered films. Sputter-deposited MoS_2 films have been found to have similar low initial friction but superior wear life compared to high-energy ion beam-assisted (IBAD) deposited films, which have been found to be partially crystallized with no preferential orientation of the layered structure with respect to the surface (Grosseau-Poussard *et al.*, 1993).

A typical sputtered MoS_2 film has a structure with three distinct layers. First, attached to the substrate, there is a ridge formation zone, in the middle an equiaxed transition zone and on top a columnar fibre zone. When sliding the coating breaks up at the base of the columnar zone and the low friction is provided by the remaining layers. The broken material is reoriented in the sliding direction. It is partly reattached to the surface and partly transferred and attached to the countersurface (Spalvins, 1982; Voumard *et al.*, 2001). The film thickness, substrate material, surface roughness and the environment in which it operates, especially humidity, have a considerable influence on the tribological properties of the coating.

4.3.3.1 MoS_2 films in vacuum

Sputtered molybdenum disulphide coatings have excellent tribological properties in vacuum. In sliding pin-on-flat experiments with various contact geometries and speeds and loads within the ranges of 0.004 to 0.25 m/s and 0.6 to 50 N, respectively, very low coefficients of friction of 0.01 to

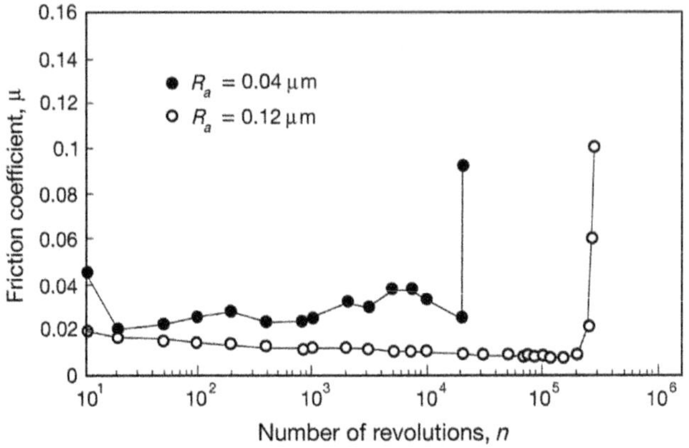

Fig. 4.26. Effect of surface roughness on the coefficient of friction and lifetime of sputtered MoS_2 coatings with a sliding speed of 1.2 m/s and a load of 50 N (data from Roberts and Price, 1989).

0.04, wear rates of only $0.001 \cdot 10^{-6}\ mm^3/Nm$ and excellent film endurances have been measured by Spalvins (1972), Dimigen *et al.* (1985), Roberts (1989), Didziulis *et al.* (1990) and Miyoshi *et al.* (1993). Even lower friction, coefficients of friction down to 0.001 and below, was measured by Martin *et al.* (1993) in very controlled ultra-high vacuum conditions. This is remarkable since this value is the lowest ever measured between two solids in dry sliding. This condition is called superlubricity and is discussed in more detail later in this chapter.

In early experiments Spalvins (1972) observed that for a higher load of 10 N, the film endurance life was short (38,000 revolutions) and the coefficient of friction was as low as 0.02, while at a lower load of 2.5 N the endurance was almost one order of magnitude higher (250,000 revolutions) but the coefficient of friction at the same time was 0.04. A decreasing coefficient of friction with increasing load has also been reported by Roberts. At very low speeds, less than 0.001 m/s, the coefficient of friction decreases with increasing sliding speed (Roberts, 1990b) while it is unaffected at higher speeds (Spalvins, 1972).

The influence of surface roughness has been studied by Roberts and Price (1989) and Roberts *et al.* (1992) and they observed that with a higher roughness of $R_a = 0.12$ μm the coefficient of friction decreased from 0.02 to 0.01 and the film endurance was very good (250,000 revolutions). For a smoother coated surface of $R_a = 0.04$ μm the coefficient of friction was considerably higher, in the range 0.02 to 0.04 and the film endurance only 20,000 revolutions, as shown in Fig. 4.26. This is supported by the observations of Spalvins (1972) that the surfaces he used always had to be roughened to obtain a longer wear life. The wear life is believed to be extended due to an increased surface concentration of the lubricating surface film with the increased volume of surface depressions to contain MoS_2 as long as metal to metal contacts of asperities are still avoided. Roberts (1990b) suggests that for a 1 μm thick film there exists an optimum level of substrate surface roughness at which friction is minimized and film endurance maximized.

4.3.3.2 Superlubricity

It is remarkable that in ultra-high vacuum conditions with a minimum pressure of 10 to $50 \cdot 10^{-9}$ Pa extremely low friction, named superlubricity, with coefficients of friction down to 0.001 and below has been measured by Martin and his colleagues (Martin *et al.*, 1993, 1994; Donnet *et al.*, 1993; Le Mogne *et al.*, 1994; Erdemir and Martin, 2007; Martin, 2007). Since this is the lowest friction measured between two sliding surfaces in dry contact it is of course of interest to study this mechanism in more detail. The experimental conditions used were a pin-on-flat test rig with a 4 mm radius hemispherical steel pin sliding at a normal load of 1.2 N, corresponding to 0.4 GPa contact pressure, at a very slow sliding speed of 0.5 mm/s. The coated surface was a 0.12 μm thick rf magnetron sputtered pure MoS_2 film on cleaned bearing steel. The crystal structure was hexagonal and the S/Mo ratio was 2.04, thus corresponding to a nearly stochiometric structure, with most of the crystallites having their basal planes oriented perpendicular to the surface before sliding.

A relatively high and constant hardness value of 7.8 GPa and high reduced elastic modulus of 165 to 170 GPa was measured for the superlubricious MoS_2 films. Thus the film is very dense and compact with hardness and elastic modulus at least two times higher than the normal coatings containing some porosity. Auger electron spectroscopy analysis did not show any trace of contamination elements such as carbon or oxygen. Very low shear strength values down to 1 MPa were measured for these films (Martin *et al.*, 1994). Initially, the coefficient of friction was 0.01 but drastically decreased to about 0.001 after a few cycles as shown in Fig. 4.27.

The explanation for the extremely low friction seems to be the following:

1. The surfaces were extremely smooth to avoid friction due to asperity deformation.
2. The coating was very hard and dense to avoid ploughing friction.
3. The shear probably almost completely took place in planes of extremely weak bonding by van der Waals intermolecular forces between the MoS_2 planes.
4. The surfaces were atomically clean to avoid structural disturbances in the shear planes.
5. The sliding plane structure was incommensurate atomic lattices. This means that the friction is minimized when a misfit angle shifts in two superimposed sliding crystal planes with respect to each other as visualized in Fig. 4.28.

4.3.3.3 Transfer layers and interfacial shear in MoS_2 films

Surface reaction, wear products and interfacial films play an important role in the wear and friction of MoS_2 coated sliding surfaces. The contact mechanisms in the running-in and steady-state period of

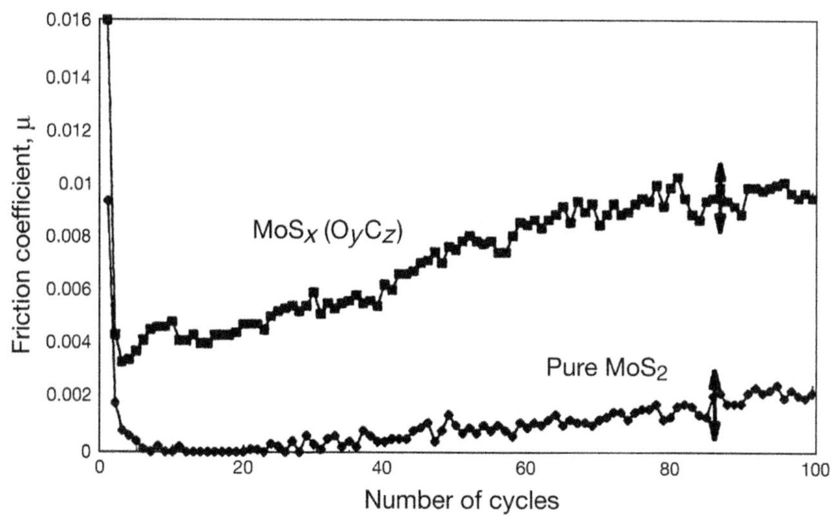

Fig. 4.27. Coefficient of friction versus the number of reciprocating sliding cycles of pure and contaminated MoS_2 coatings. The coefficient of friction for pure MoS_2 coatings is 0.001 and even below (after Donnet, 1998a).

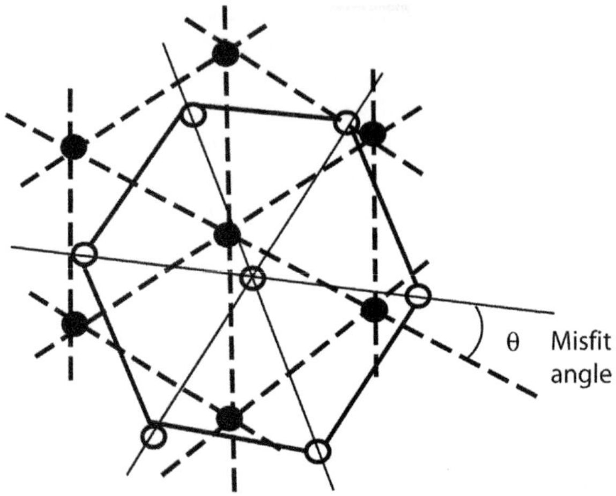

Fig. 4.28. Visualization of two superimposed hexagonal lattices, representing adjacent sulphur planes in MoS_2, with a misfit angle of θ. The frictional force is known to be minimum at $\theta = 30°$ (after Donnet, 1998a).

the lifecycle of the film have a similarity with those for rubbed MoS_2 films shown in Fig. 4.24. In the early stage of an MoS_2 coated surface sliding against a countersurface a thin transferred layer is built up on the counterface. Singer *et al.* (2003) have shown with *in situ* observations how MoS_2 coating material is transferred to and from third body reservoirs on the surface to protect wear spots in the sliding contact. In some cases it has been observed that the wear particles are ejected beyond the wear track instead of remaining inside the contact zone. Why some coatings form reservoirs and give long life, while others eject the wear debris and fail early, is not understood.

The sliding process induces a preferred orientation in the contacting layers, so that basally oriented MoS_2 transfer layers and basally oriented MoS_2 wear tracks are generated during the steady-state sliding (Fayeulle *et al.*, 1990). These orientations are conducive to low shear strength and hence low friction. In favourable contact conditions the wear of the MoS_2 coating may take place even as a layer-by-layer process at the atomic level (Wahl *et al.*, 1999).

The generation of wear products, the formation of transfer films, their development to form material reservoirs and the kinematics of the material transfer process during sliding have been demonstrated by Singer *et al.* (2003) in pin-on-disc measurements with a glass 6.35 mm radius hemisphere sliding against a 0.3 μm thick Pb–Mo–S coating on a steel substrate with 1 mm/s sliding speed and 24 N load. They first observed how patchy transfer films attached to the contact area of the glass hemisphere and later thicker compact pads of debris were formed at the leading and trailing edges. The observation confirmed that the relative motion between the transfer film on the hemisphere and the wear track on the coating took place by interfacial sliding. Analysis showed that the top layers on both surfaces, that is the plane where the shear took place, were layers of MoS_2.

4.3.3.4 MoS_2 films in dry and humid air

It is well known that humidity has an adverse effect on the tribological properties of MoS_2. It has been recommended that the humidity level should be below 5%, or failing this below 20%, to achieve a good tribological performance with traditional MoS_2 coatings in an air environment (Roberts, 1986, 1987). For sputtered MoS_2 films Nabot *et al.* (1990) measured an increase in the coefficient of friction from 0.05 to 0.27 when the relative humidity (RH) was increased from 5 to 60%, as shown in

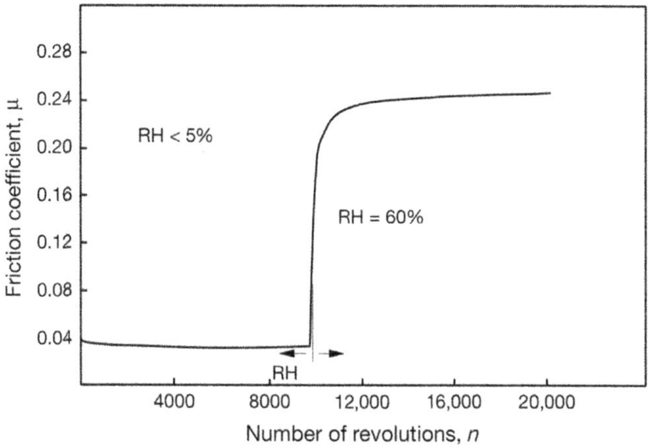

Fig. 4.29. The influence of relative humidity (RH) on the coefficient of friction for a sputtered MoS_2 coated surface sliding against a steel ball with a speed of 0.01 m/s and a load of 5 N (data from Nabot *et al.*, 1990).

Fig. 4.29. A significant increase in wear was observed at the same time. This is in good agreement with the remarkable reduction in MoS_2 film lifetime from 100,000 to 300,000 revolutions in a dry air environment to 700 to 900 revolutions in humid air of RH = 52% at a speed of 0.01 m/s and a load of 5 N, reported by Niederhäuser *et al.* (1983). However, Niederhäuser also showed that the negative effect of humidity can be overcome by using a suitable interlayer such as rhodium.

The humidity effect has further been confirmed in the sliding conditions referred to above by Singer *et al.* (2003). In dry air (RH < 1%) the coefficient of friction was about 0.05 but it increased quickly to 0.17 as humid air (RH = 50%) was introduced. The coefficient of friction soon returned to its original level when changing back to dry air. A similar behaviour has been recorded in fretting tests by Zhang *et al.* (2003a). They measured a coefficient of friction increasing from 0.06 to 0.14 as relative humidity increased from 10 to 90% for MoS_2 coatings with basal orientation and an increase from 0.25 to 0.42 for coatings with random orientation at a normal load of 1 N, oscillation of 10 Hz and with 5 mm radius corundum balls as the counterface. The wear rates were in the range of 5 to $20 \cdot 10^{-6}$ mm^3/Nm for basal oriented and 10 to $80 \cdot 10^{-6}$ mm^3/Nm for randomly oriented coatings.

The effect of humidity on the friction and wear of sputtered MoS_2 coatings has been explained as being due to the formation of oxides in the contact interlayers, as earlier described for bonded MoS_2 films. Fayeulle *et al.* (1990) have discussed in detail the formation process of different Mo and Fe oxides that can be assumed to occur in these kinds of contacts. Experimentally, they observed that the wear rate of MoS_2 films is considerably lower in argon than in air, which can be explained by the fact that molybdenum oxides are formed more slowly in argon. The interfacial films formed in air were much thicker and usually contained more oxygen than sulphur. The mechanism for the change in friction has been suggested to be the effect of the transfer film on the real area of contact (Wheeler, 1993). Thick, smooth films produce large areas of contact having higher friction coefficients than the smaller contact areas produced by thin films. On the other hand, Singer *et al.* (2003) have suggested that the increase of friction in humid air is more likely to be due to an increase in the interfacial shear strength.

Sputtered MoS_2 coatings generally contain oxygen as a surface species and as a bulk contaminant, up to 20 at.%. Surface analytical studies and friction and wear tests with MoS_2 coated steels have shown that surface oxygen participates in both the adhesion and wear of transfer layers. In addition to

MoS_2, MoO_3 and several ternary oxides of the counterface material, such as iron from steel and cobalt from WC:Co, have been found and described by Singer (1991). He also showed that the quaternary phase diagram for an Mo–S–Fe–O system gives a good indication of the phases detected.

When an MoS_2 layer is not subjected to sliding for a certain period the coefficient of friction may be up to four times higher when restarting sliding (Roberts, 1990b). This is thought to result from the adsorption of water molecules within the MoS_2 monolayer. With continued sliding these are removed, probably by desorption induced by frictional heating, and the coefficient of friction decreases.

In dry air ball-on-plate experiments at a speed of 0.02 m/s, in a load range of 1 to 30 N and at room temperature, Singer *et al.* (1990) observed that the coefficient of friction decreased with increasing load and was proportional to $(\text{load})^n$ with the exponent $n = -0.32$.

4.3.3.5 Thickness of MoS_2 coatings

The reported MoS_2 sputtered coatings with favourable tribological properties have, in general, been very thin, with coating thicknesses within the range of 0.1 to 2 μm. Some decrease in the thickness of the film can be assumed to take place during the process of sliding. Spalvins (1982) found that the optimal lubricating film is usually about 0.2 μm thick.

Muller *et al.* (1988) studied thick, compact MoS_2 coatings produced by dc as well as rf magnetron sputtering. Coatings with a thickness up to 30 μm were obtained without a change in morphology. The adhesion was good for coatings on steel, silicon and glass substrates. An improved lifetime for the coating was measured for increasing coating thicknesses up to 3 μm as shown in Fig. 4.30. The tests were accelerated by using a humid environment with RH = 98%. For sputtered MoS_2 films with thicknesses ranging from 0.7 to 28 μm, Aubert *et al.* (1990) measured minimum values for the coefficient of friction at film thicknesses of about 3 μm for smooth surfaces and 1 μm for rough surfaces in pin-on-disc experiments in air at a sliding velocity of 0.01 m/s and a load of 5 N.

Thicker coatings provide an interesting possibility for use on rougher surfaces. The investigation by Muller *et al.* (1988) gives clear evidence that thick MoS_x coatings adhere well in pin-on-disc tests on turned surfaces.

4.3.3.6 Counterface, substrate materials and interlayers

The prevailing environment must be considered in association with the influence of counterface, substrate materials and interlayers on the tribological properties of MoS_2. In humid air the process of

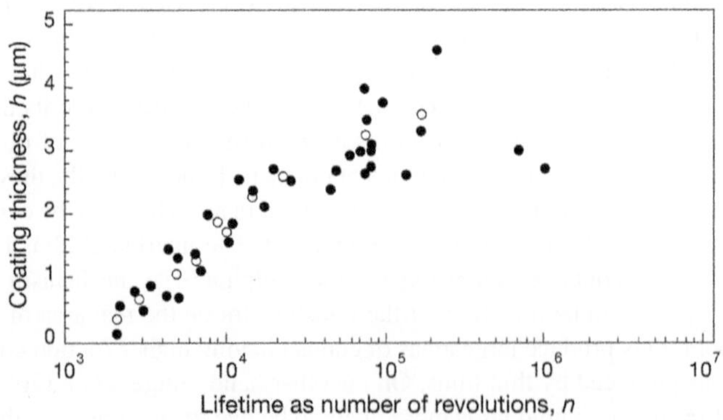

Fig. 4.30. Lifetime in number of revolutions until the friction coefficient exceeds 0.3 for sputtered compact MoS_x coatings of different thicknesses on polished steel plates sliding against a steel pin in pin-on-disc tests with a velocity of 0.1 m/s and a load of 5 N (data from Muller *et al.*, 1988).

oxidation in the contact interlayer and the possible influence of the counterface and substrate materials on this are important.

In pin-on-disc tests with low load but a very humid environment of RH = 98%, Muller *et al.* (1989) measured lifetimes of sputtered MoS_2 on steel, fibreglass and Ti–Al alloy disks sliding against a steel pin within the range 10,000 to 30,000 revolutions. However, it is remarkable that the lifetime was one order of magnitude longer, 100,000 to 325,000 revolutions, when Al_2O_3 ceramic was used as a substrate for the MoS_2 film. This may be due to a substrate hardness effect resulting in increased contact pressure but also to better bonding because the porosity in Al_2O_3 may provide improved nuclei for adhesion.

The influence of substrate material on the tribological properties of MoS_2 coatings has been extensively studied in a vacuum environment because of the potential use in space applications. MoS_2 coatings with thicknesses in the range of 0.2 to 1 μm were sputtered onto Ni, Au, Ag, Co, Mo, W, Ti, Al, a number of steels (e.g. AISI 440 C, AISI 52100), glass and plastics by Spalvins (1972) without showing any stress-induced peeling within the film. He concludes that stress-induced peeling only becomes important for films thicker than 1.5 μm.

In a tribological pin-on-ring comparison of 1 to 5 μm thick MoS_2 coatings deposited by four different sputtering techniques and with 14 different counterface materials, Ruff and Peterson (1993) found the lowest friction and wear rate for the MoS_2 film sliding against gold, bronze and molybdenum.

Spalvins (1973) studied the adhesion of MoS_2 films deposited on soft substrates and found that sputtered MoS_2 had poor adhesion to silver, bronze and copper surfaces that had been mechanically polished, sputter etched or sulphurized. Strong adhesion was found on preoxidized copper and bronze surfaces. Sputtered MoS_2 showed no flaking at all on gold surfaces that had been polished, sputter etched or sulphurized.

Both the lifetime and the coefficient of friction are strongly dependent on the substrate material in vacuum environments. This dependence is well illustrated by Fig. 4.31, which shows how the endurance life increases and the friction decreases when using a titanium alloy (IMI 318), bearing steel (52100, EN31), stainless steel (440C) and silicon nitride as substrate materials in the order mentioned (Roberts and Price, 1989; Roberts *et al.*, 1992). The difference in lifetime is remarkable. The substrate which performed best, silicon nitride, exhibits an endurance which is a factor of 200

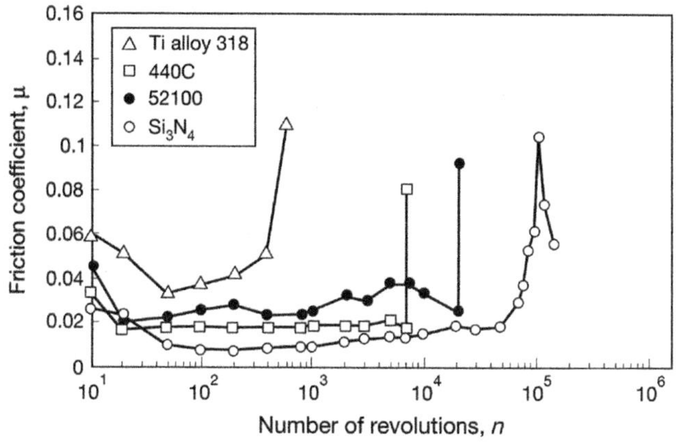

Fig. 4.31. The coefficient of friction and the lifetime for four different substrate materials sputter coated with MoS_2, sliding against a steel pin in vacuum with a speed of 1.2 m/s and a load of 50 N (data from Roberts and Price, 1989).

higher than the poorest substrate, the titanium alloy. Roberts and Price conclude that the endurance of the films is strongly related to adhesion and they find evidence for this by the presence of interfacial chemical bonds on the steel and ceramic substrates and their absence on titanium alloy. However, this does not explain the large difference between the film endurances observed between the ceramic and the steel substrates. On the other hand, it is interesting to note that the film endurance is almost directly related to the hardness of the materials. It seems probable that the substrate hardness might be part of the explanation. In contradiction to this are the vacuum pin-on-disc results by Didziulis *et al.* (1990), showing that sputtered MoS_2 films deposited on silicon carbide have significantly shorter wear lives than films deposited on steel.

Roberts and Price (1989) have observed that for metallic substrate materials, film failure was signalled by an abrupt increase in the coefficient of friction, but for the ceramic substrate, film failure occurred gradually, as evidenced by a slower increase in friction lasting some 50,000 revolutions. The coefficient of friction decreases with increasing elastic modulus of the substrate materials and this behaviour is ascribed to changes in the true contact area.

The tribological properties of sputtered MoS_2 coatings can be improved by the use of thin, hard interlayers that improve the adhesion between the MoS_2 coating and the substrate and increase the load-carrying capacity and thus decrease the contact area. Good results have been achieved with interlayers of Ti, TiB_2, BN and even with rhodium and ruthenium (Niederhäuser *et al.*, 1983; Roberts, 1990b; Teer, 2001; Steinmann *et al.*, 2004). In a comparison of the tribological performance of MoS_2 films deposited by fast atom beam sputtering on a steel SUS 440 C plate with and without a hard B or BN interlayer, Kuwano (1990) observed that the coefficient of friction decreased by a factor of five to a value as low as 0.005 and the endurance lifetime was prolonged by a factor of ten, up to 10,000 to 20,000 sliding cycles, when the hard interlayers were used.

4.3.3.7 Coating composition

The PVD technique offers a possibility to deposit coatings with a large variety of different compositions and structures. For MoS_2 coatings the tribological performance will depend considerably on the sulphur/molybdenum composition in the coating and the use of compound materials in the coating.

The influence of the sulphur/molybdenum ratio in sputtered MoS_x films on their tribological properties when sliding against steel has been studied by Aubert *et al.* (1990) and Nabot *et al.* (1990). A minimum value, 0.025, for the coefficient of friction was found at S/Mo ratios of approximately 1.5 for crystallized (001)-oriented films with a hardness (10 g load) of 5000 MPa assessed via a Knoop device, as shown in Fig. 4.32. These properties correspond to the lowest shear stress. For both higher and lower S/Mo ratios the coefficient of friction rapidly rises to a value of about 0.06. The best wear resistance in collar-shaft experiments was observed for films with an S/Mo ratio of 1.8 exhibiting a hardness of 2000 MPa.

Adding small amounts of some doping material, mainly metals, to the MoS_2 coating can have a considerable positive effect on the friction and wear properties of the coated surface. The effect is complex and not well understood, but it seems to be related to improvement of the hardness and strengthening of the coating structure. The most successful dopants used to build up a compound MoS_2 coating structure are Ti and Pb. In addition other compound materials such as MoS_2 with Ni, Cr, Au, C, TiB_2 and WSe_2 have been applied (Stupp, 1981; Simmonds *et al.*, 1998, 2000; Wahl *et al.*, 1999; Savan *et al.*, 2000; Teer, 2001; Voumard *et al.*, 2001; Singer *et al.*, 2003; Lince, 2005; Steinmann *et al.*, 2004; Kao, 2005; Amaro *et al.*, 2005).

Improved lifetime and a much more uniform sliding friction in tests in air were measured by Stupp (1981) when adding 5 to 7% (assumed vol.%) of Ni or Ti to the MoS_2 coating. Optimal tribological performance both in sliding and drilling tests was observed by Kao (2005) with the addition of 5 to 8 at.% Cr to the MoS_2 coating. The measured average coefficient of friction was between 0.04 and 0.06.

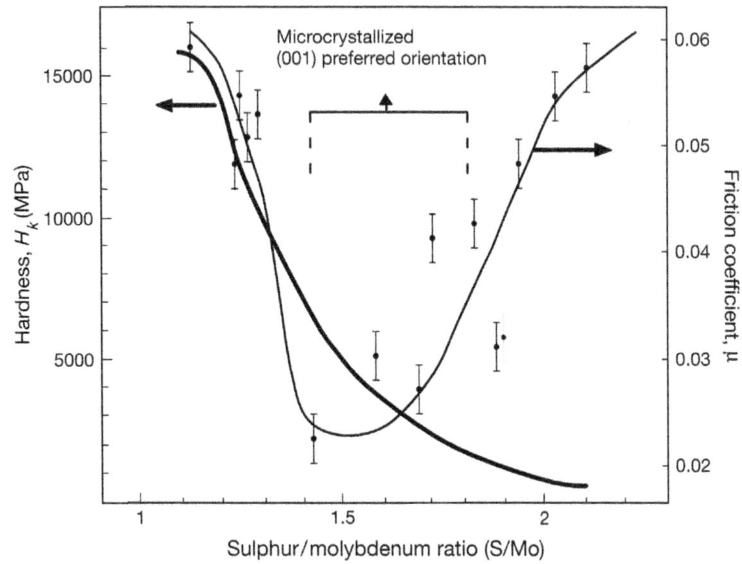

Fig. 4.32. The coefficient of friction and the Knoop hardness for MoS_x coatings with different S/Mo ratios on a steel substrate sliding in ball-on-disc experiments against a steel ball at a speed of 0.01 m/s, a load of 5 N and a relative humidity RH < 5% at 23°C (after Nabot *et al.*, 1990).

In a comparison of adding 0 to 20% of Au, Ti, Cr and WSe_2 separately as dopants in the MoS_2 coating Simmons *et al.* (1998, 2000) found that the coating with about 7 to 11 at.% WSe_2 performed best. In a pin-on-disc test in 50% RH, at a load of 5 N and a velocity of 0.01 m/s, the WSe_2 doped MoS_2 coating had a coefficient of friction of 0.08 to 0.11 compared with 0.16 for pure MoS_2 coating in the same conditions when sliding against a 3 mm radius steel ball. The lifetime increased at the same time with a factor of 5 to 8.

In an investigation of TiN and MoS_2 composite coatings Gilmore *et al.* (1998) and Goller *et al.* (1999) found that compact and well-adhering films of low friction coefficient, between 0.15 and 0.2 against steel and alumina, and high hardness, up to 30 GPa, could be obtained. The wear resistance was reported to be better than that of TiN. These coatings had the TiN cubic structure with a slightly enlarged lattice constant, and the composition was TiN + 8 mol% MoS_2. Various tribological tests were applied by these workers. A typical test configuration was a pin-on-disc with a 3 mm radius 100Cr6 steel or alumina ball sliding on a 0.5 to 3.0 μm thick coating at a load of 1 or 5 N and a sliding speed of 0.05 or 0.1 m/s in ambient air of RH = 40 or 50%.

Steinmann *et al.* (2004) investigated a compound MoS_2-C–TiB_2 coating deposited on a steel substrate. A TiB_2 interface layer was used to improve the adhesion to the substrate and it changed over a graded intermediate layer slowly to the MoS_2:C:TiB_2 top layer structure. In ball-on-disc tests with a 2 mm radius steel ball sliding with 1.86 GPa Hertzian contact pressure and a speed of 0.1 m/s over the coated surface, very interestingly the coefficient of friction was about 0.12 and did not depend on the ambient humidity in the range of RH = 10 to 90%. However, the ball wear rate was slightly decreasing with increasing humidity, from 0.00012 to 0.00006 · 10^{-6} mm^3/Nm while the coating wear rate increased remarkably from 0.1 to 1.6 · 10^{-6} mm^3/Nm with increasing humidity in the same range. These results indicate that the carbon content does not dominate the tribological behaviour which is predominately related to the MoS_2 content. MoS_2 shows good tribological performance in dry atmosphere, but this decreases considerably at high humidity. The remarkably low ball wear rate indicates that there was a strong and continuous transfer layer built up on the steel surface protecting it from wear.

The addition of lead to an MoS_2 coating has been shown to give very good tribological properties. Singer *et al.* coated both steel and Si substrates with amorphous Pb–Mo–S coatings containing 4 to 26 at.% Pb in the thickness range of 0.2 to 0.5 μm (Wahl *et al.*, 1999; Singer *et al.*, 2003). Excellent wear resistance and very low friction with a coefficient of friction down to 0.05 in dry air at $RH < 1\%$ and 0.17 at $RH = 50\%$ was measured when sliding against a 6.3 mm radius glass ball at a load of 24 N and a speed of 1 mm/s in dry air. Surface analysis gave direct evidence that in these low-speed, high-stress sliding conditions a tribochemical process can convert the amorphous Pb–Mo–S coating material into crystalline MoS_2 which will form the active shear layer. The transformation occurs by creating basal-oriented layers at the outermost layers of the track and shows that wear takes place layer by layer at the atomic level. However, it is not clear why the Pb–Mo–S coating has a better resistance to wear than MoS_2 coatings. A microstructure without MoS_2 planes in the bulk might be one explanation, especially if this microstructure inhibits fracture within the coating. However, Wahl *et al.* consider that the high endurance of the Pb–Mo–S coating is mainly due to the slow wear process with MoS_2 layers easily shearing and transferring from and to the ball surface. The role of the Pb in the coating is still not understood.

The addition of Ti both as an interface layer and as a dopant to MoS_2 coatings has given excellent tribological behaviour. Atmospheric testing of 1 μm thick MoS_2–Ti coatings with 15% Ti has shown that the friction is load dependent but is typically 0.02 at high contact pressure of 3.5 GPa while the maximum value was 0.08 and the wear rate was only $0.01 \cdot 10^{-6}\ mm^3/Nm$ (Teer, 2001; Amaro *et al.*, 2005; Michalczewski *et al.*, 2007; Wang *et al.*, 2007). Teer *et al.* explain the excellent adhesion, wear and friction properties by pointing out that previous deposition methods have resulted in structures consisting of a MoS_2 matrix with small islands of metallic inclusions or multilayers, as this new MoS_2–Ti structure contains no metallic inclusions and the titanium is thought to be in solid solution within the MoS_2. Evidence indicates that the Ti atoms are situated between the neighbouring S planes as presented in Fig. 4.33. It is probable that the distortion caused by the Ti atoms is responsible for the

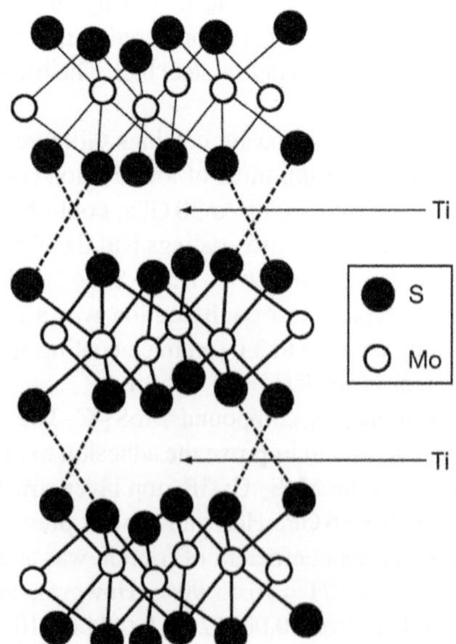

Fig. 4.33. The structure of the MoS_2–Ti coating showing the probable positioning of the titanium atoms in the MoS_2 structure (after Teer, 2001).

increase in hardness. Also the presence of the Ti atoms within the MoS_2 structure may be responsible for reducing the sensitivity of the coating to the water vapour. Still the low friction values indicate that pure MoS_2 crystallites are active at the top surface where shear takes place. When the amount of Ti in the film is increased the hardness also increases to a maximum of about 2000 H_v at around 18% Ti content. Above this level, the solubility level of the Ti is exceeded and multilayers are formed resulting in a decrease in the tribological performance.

The tribological properties in moist air have been improved by co-deposition of PTFE and MoS_2. Niederhäuser *et al.* (1983) have measured in pin-on-disc tests a reduction of the coefficient of friction from about 0.25 to around 0.1 by the addition of PTFE to MoS_2. The lifetime of the film was at the same time prolonged by an order of magnitude. Further improvements in film endurance were achieved by using chemically activated interlayers of rhodium or ruthenium in combination with the MoS_2–PTFE coating.

4.3.3.8 Selection of MoS_2 films

A systematic approach to the design and selection of MoS_2 coatings for different purposes was developed by Fleischauer *et al.* (1990) and is outlined below. Fleischauer and his co-workers consider three primary material properties – composition or phase purity, adhesion and morphology and porosity – to be the primary drivers of the three most important performance parameters, friction, endurance and debris generation.

They claimed that the choice of the proper MoS_2 film for a specific use depends on the contact geometry, stress levels and a trade-off between operational life and environmental factors both in storage and in use. The ideal combination of good adhesion, chemical purity and dense small-grain size in MoS_2 films provides the best lubrication and endurance in most relatively high-contact stress conditions with moderate operating lifetimes (10^8 to 10^{10} revolutions) such as ball bearings. This combination of optimal film properties can be obtained by high-rate, ambient temperature (less than 70°C) sputtering. Optimal properties can also be obtained by post-deposition ion processing with high-energy ions or by low-energy ion processing during deposition.

Other applications such as telescopic joints, release mechanisms, actuators and solar array drive mechanisms may not require the optimal endurance but may need protection from environmental effects. Ultra-low friction can be obtained with resistance toward oxidation, by using high-temperature growth conditions that induce larger crystallite and grain formation. Such films generate relatively large wear particles during operation that can cause noise in torque sensitive applications and are thus recommended only if torque stability is not critical.

4.4 Hard Coatings

A thin hard coating on a softer, tough, substrate has proved to be a tribologically very beneficial material combination. Many new coatings have been investigated and developed since the 1970s and also taken successfully into commercial use.

A hard layer on a softer substrate will give improved protection against scratching from a hard counterface or debris. Hard coatings have thus been especially useful in applications involving abrasive or erosive wear. An early successful application was ceramic coatings, especially aluminium oxide, titanium nitride and titanium carbide, on cutting tools where they give good protection against a combination of diffusion and abrasive wear at high temperatures. This has often resulted in extensions of tool lifetimes by ten times and more. However, these coatings normally exhibit high friction and are not suitable for low-friction sliding applications.

The second successful new coating taken widely into commercial use were the hard carbon-based ones, especially the group of coatings named diamond-like carbon (DLC). These coatings form a hard film on the surface reducing scratching and offer a good load-carrying capacity, and in addition they have the ability to form low shear strength reaction and transfer layers on the top surface and the countersurface, resulting in weak shear planes and a remarkably low friction. The coefficient of friction is about one order of magnitude lower than for most other coated contacts and the wear resistance is excellent. Coatings based on diamond-like carbon are widely used commercially on magnetic disks in the computer industry, in precision devices, on mechanical components in machines and engines and on special tools for moulding and machining applications.

Other hard coatings developed and used in various applications are nitrides such as (Ti,Al)N, BN, CrN and ZrN; carbides such as TiC, CrC and BC; oxides such as Al_2O_3; and borides such as TiB and FeB; and their combinations in several different forms. Much research and development work is focused on producing hybrid coatings, microstructured and nanocomposite coatings as well as gradient and multilayer structures to achieve even better tribological properties.

One limitation on the use of hard ceramic coatings on relatively soft substrates is the high stresses that are often induced in the coating and at the interface to the substrate when the surface is loaded and the substrate deflects. In addition, internal stresses are generally created when the deposition temperature is higher than the operational temperature. However, these stresses are often compressive in the plane of the surface and can be beneficial from a wear and fatigue failure point of view. The nature and tribological behaviour of thin hard coatings have been discussed by many authors (Hintermann, 1984; Halling, 1986; Singer, 1989; Bhushan and Gupta, 1991; Knotek *et al.*, 1991; Franklin and Beuger, 1992; Sander and Petersohn, 1992; Matthews and Eskildsen, 1993; Erdemir, 2001b; Holmberg and Matthews, 2001, 2005; Zabinsky and Voevodin, 2004; Erdemir and Donnet, 2005).

4.4.1 Titanium nitride coatings

Titanium nitride (TiN) was first coated commercially on tools by the CVD method. In the 1980s it triggered a commercial breakthrough of the plasma-assisted PVD coating process. It is mainly used on high-speed steel tools for metal cutting but has also found other tribological applications, such as in forming tools, bearings, seals and as an erosion protection layer. One important attraction with titanium nitride is its golden colour which has also encouraged its use in decorative applications.

The TiN coating is very hard, with hardness values of 17 to 30 GPa, measured for coatings in the thickness range of 1 to 12 μm on steel or ceramic substrates. The elastic modulus of the TiN coatings depends on the deposition method and may range from 160 to 600 GPa. There is little information available on the fracture toughness of thin coatings but some measurements indicate that the fracture toughness of TiN coating is about 1 to 7 $MPa \cdot m^{0.5}$ and the interfacial fracture toughness to a steel substrate about 0.1 $MPa \cdot m^{0.5}$. The Poisson's ratio is typically 0.22, but values in the range of 0.2 to 0.232 appear in the literature. Compressive residual stresses in TiN coatings are often very high, in the range of 1 to 6 GPa (Diao *et al.*, 1994b; Polonsky *et al.*, 2002; Zoestbergen and de Hosson, 2002; Bull, 1997; Holmberg *et al.*, 2003; Laukkanen *et al.*, 2006).

Because of its excellent wear-resistant properties, titanium nitride has attracted considerable research and it is, in tribological terms, the first largely explored hard, thin coating. A considerable number of reports on the tribological performance of TiN coated surfaces have been published, beginning in the late 1970s, and detailed and systematic analyses of the tribology of TiN coatings started to appear in the 1980s. Because of this, TiN coatings will be treated here in more detail. The suitability of titanium nitride as a tribological coating is explained by its high hardness, good adhesion to steel substrates and good chemical stability.

4.4.1.1 Adhesion to the substrate

Poor effective adhesion between the coating and the substrate may be due to a low degree of chemical bonding, poor interfacial mechanical contact, low interfacial fracture toughness or high internal stresses. These properties depend on the coating process and thus coating technology development has aimed at optimizing the deposition process to achieve the most favourable combination of the above-mentioned properties – resulting in highly effective adhesion. The mechanisms operating are not always well understood, but the optimization has nevertheless been quite successful.

It has been shown that high levels of ionization efficiency during the plasma-assisted PVD process produce films with improved structural, compositional and hardness properties (Matthews, 1980; Matthews and Sundquist, 1983; Matthews, 1985; Matthews *et al.*, 1993). A titanium interlayer, normally with a thickness of about 100 nm, between the TiN coating and the steel substrate can improve the adhesion and result in better tribological properties (Valli *et al.*, 1985a; Cheng *et al.*, 1989; Rickerby *et al.*, 1990). The scratch test method has been widely used to evaluate the adhesion between the coating and the substrate and the mechanisms involved have been discussed by Bull *et al.* (1988a) and Hedenquist *et al.* (1990a). Recent extensive modelling and stress simulation of TiN coated surfaces has been carried out, as reported in section 3.2.8. Bending tests have shown that TiN coatings have a high crack resistance and this can be improved by increasing the interface roughness (Zoestbergen and De Hosson, 2002).

Remarkable improvements in the wear behaviour have been achieved by depositing a 1 μm thick pure titanium layer between a cast iron substrate and a titanium nitride coating to reduce the disturbing influence of graphite on the growth of the hard coating (Matthes *et al.*, 1990b). Improvements in cutting wear of a TiN coating were observed when using an intermediate TiN layer of nitrogen concentration of about 35 at.% (Harju *et al.*, 1990).

In a recent study Gerth and Wiklund (2008) compared the practical adhesion strength of 3 μm thick TiN coatings deposited on high-speed steel surfaces with seven different interlayers, namely W, Mo, Nb, Cr, Ti, Ag and Al, in the thickness range of 100 to 150 nm. The adhesion testing was carried out by scratch and Rockwell adhesion tests. They found that Mo and Nb interlayers gave the best adhesion to the substrate closely followed by the more conventionally used Ti and Cr. The elastic modulus of the interlayers did not affect the adhesion as much as their hardness. Interestingly, they question the necessity to use interlayers between TiN coatings and HSS substrates as the reference samples, without interlayers, showed adhesion properties comparable with the highest ranked interlayer containing samples.

4.4.1.2 Sliding against steel

There is a variation in the contact mechanism and the wear process in sliding contacts between titanium nitride and steel depending on contact parameters such as geometry, speed, load and roughness. However, it is possible to identify four regimes in the wear process that are typical for a wide range of contact conditions. The mechanism of wear is different in each regime. These four regimes, illustrated in Fig. 4.34, are: (1) initial wear, with increasing friction and wear, (2) layer formation, with high friction and wear, (3) steady-state wear, with low friction and wear and (4) layer destruction, with high friction and wear. The initial wear and layer formation regimes represent what is frequently called running-in.

Initial wear mechanisms are present in the first few contact events between two surfaces that have not previously been in contact. Very small wear fragments are produced in the collisions between steel and titanium nitride asperities. The wear rate is higher on the steel surface because it is softer than TiN and thus the iron content in the wear products dominates. The wear fragments adhere mainly to the steel surface forming patches and layers. The transferred material is active in the contact, carries part of the load and will thus undergo both plastic deformation and strain hardening. The surfaces become rougher, the coefficient of friction is initially low, about 0.1–0.2, but increases continuously to higher values.

In the regime of layer formation a more stable surface layer with agglomerated material is built up mainly on the steel surface. Frictional heating in the contact results in oxidation of the layer and surfaces. Cracking occurs in the layers, and layer segments may move between the surfaces and either

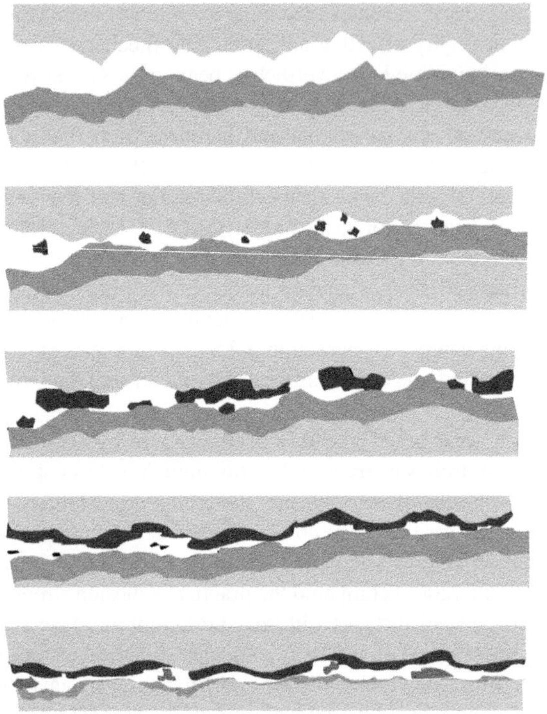

Fig. 4.34. The four different wear regimes in a typical wear process in sliding contact between TiN coated steel sliding against steel.

be reattached to the surface or transported away from the contact. Larger material segments may roll or be dragged between the surfaces and cause abrasive wear. At this stage the coefficient of friction is at a high level with many fluctuations and high wear.

Steady-state wear represents the third regime where a stable layer or thin film has been produced on the steel surface and some wear products may also be present on the TiN surface as well. The transferred material consists predominantly of Fe_2O_3 with small amounts of TiO_2. Both the wear and the coefficient of friction have decreased to lower but stable values. The explanation for the lower friction is probably that the thin oxidized surface layer inhibits direct adhesion between steel and TiN and the ploughing effects are minor because of the absence of large wear particles and the smoothened surfaces. Dominating wear mechanisms are typically oxidative and delamination wear but the wear rate is low.

In tribologically well-designed contacts, the steady-state wear regime lasts very long, which gives a long lifetime for the component. However, even though the wear is low it gradually makes the TiN coating thinner until finally it becomes so thin that it can no longer protect the underlying substrate from wear.

In the layer destruction regime, fracture in the thin TiN layer occurs producing larger wear particles that cause third body abrasive wear, and chipping is common. The thin protective oxide layers are destroyed and direct contacts between the steel surface and the steel substrate under the TiN layer occur resulting in increased friction and wear. There are, however, reports (Sundquist *et al.*, 1983) indicating that even after the TiN layer is worn away the friction and wear remain lower than in similar steel-on-steel contacts. This is probably due to the friction-reducing and adhesion-preventing actions of the wear products from the TiN layer still left in the contact.

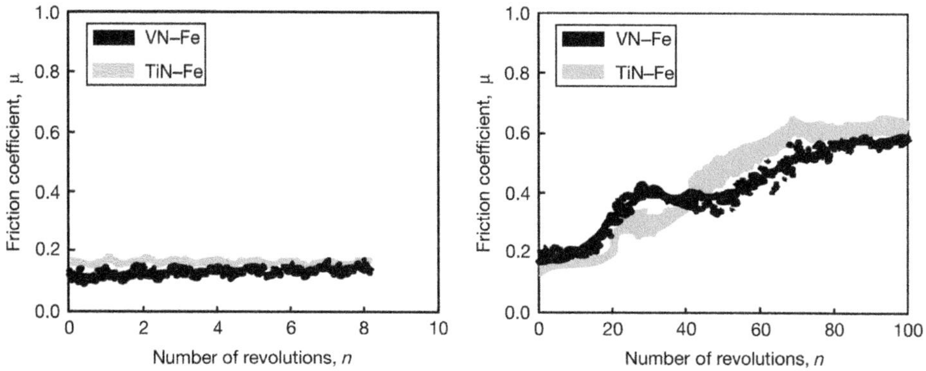

Fig. 4.35. Coefficient of friction for a steel ball sliding against a TiN coated plate at 30 N load and 0.01 m/s speed (a) during the first eight revolutions and (b) up to 100 revolutions (after Wiklund *et al.*, 2006).

The titanium nitride wear process described above is similar to that described by Sue and Troue (1990) and by Singer (1998) and is in general agreement with the observations of Shimazaki and Winer (1987), Bull *et al.* (1990) and Singer *et al.* (1991). The composition of the produced transfer film depends on both counter material and contact conditions. For TiN sliding against sapphire the expected rutile TiO_2 is produced while with TiN sliding against steel it has been observed that a ternary compound, ilmenite $(Fe_xTi_{1-x})_2O_3$ where $0 < x < 1$, is produced (Singer, 1998).

The steady-state coefficient of friction in TiN and steel sliding contacts is often reported to be fairly high, about the same level as in steel-to-steel contacts which is about 0.6, or even higher (Shimazaki and Winer, 1987; Franklin and Beuger, 1992; Wiklund *et al.*, 1999a, 2006). Wiklund *et al.* (2006) showed in pin-on-disk tests with a 6 mm diameter steel ball loaded by 30 N and sliding with 0.01 m/s speed against a TiN plate how the coefficient of friction is initially low, only 0.15 to 0.2, during the first 20 revolutions but then starts to increase and reaches a value of 0.6 after 70 revolutions, as shown in Fig. 4.35.

On the other hand, there are reports, such as from Peebles and Pope (1989) and Habig *et al.* (1980) that give low coefficients of friction, down to 0.17, for very similar contact conditions at high speeds. Figure 4.36 represents a collection of published data from well-documented experiments which show the trend that high speed can give low friction, but there is scatter, caused by differences in load, contact geometry, materials and environment. For example, even at speeds above 0.2 m/s extremely high-friction coefficients up to 1.15 have been reported (Suri *et al.*, 1979). On the other hand, low-friction behaviour with coefficients of friction even down to 0.03 has been measured by Kelly *et al.* (2006, 2007) with TiN coating sliding against phosphated steel plates in thrust washer tests.

The surface roughness has a remarkable influence on the coefficient of friction. Singer *et al.* (1991) measured, for a steel slider moving on a magnetron sputtering TiN coated tool steel surface with a roughness of $R_a = 0.06$ to 0.1 μm, friction coefficients in the range of 0.5 to 0.7 in contact conditions characterized by wear and transfer of the steel. After the TiN coating was polished to $R_a = 0.004$ μm, transfer was reduced and the initial friction coefficient range was as low as 0.15 to 0.2, as shown in Fig. 4.37. This is in agreement with results reported by Iliuc (2006).

It is also well documented that the friction and wear properties depend on which of the surfaces is coated; the one in continuous contact, like the pin in a pin-on-disk device, or the rotating or reciprocating one, where each place on the surface is only in contact once in every cycle (Mäkelä and Valli, 1985; Bull *et al.*, 1990a, b). This is illustrated by the friction curves in Fig. 4.38.

The film thickness does not seem to have any influence on the coefficient of friction under heavy load but in light load conditions Shimazaki and Winer (1987) observed that the friction increases with

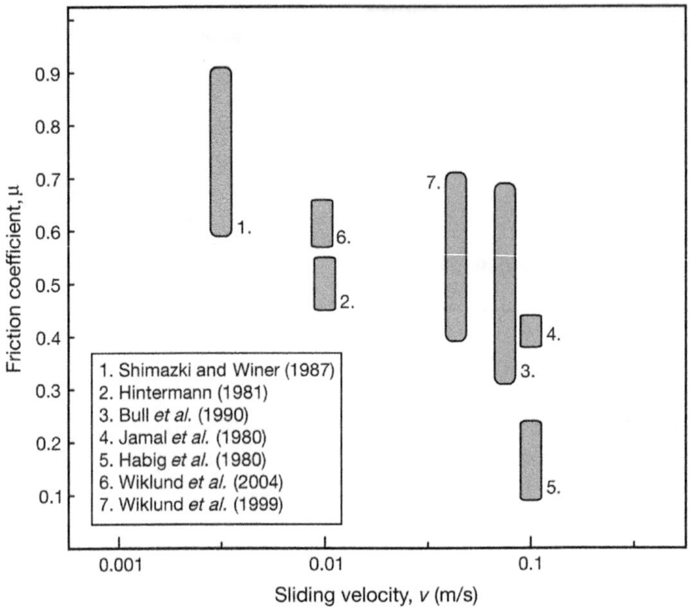

Fig. 4.36. Coefficients of friction measured at different sliding speeds for a steel slider on a TiN coated substrate collected from the literature.

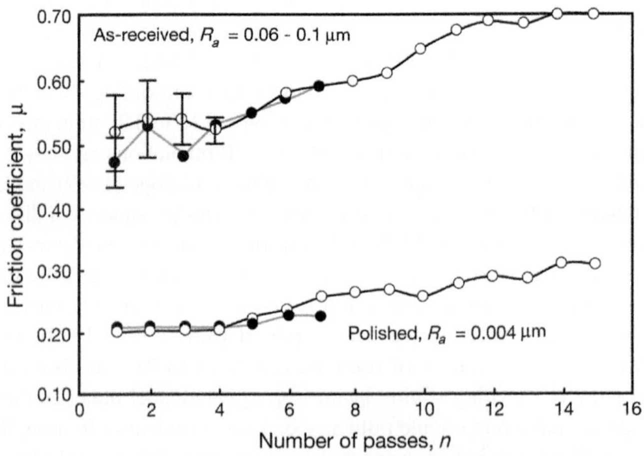

Fig. 4.37. Friction coefficient as a function of pass number for a hardened AISI 52100 12.7 mm diameter steel ball sliding on TiN coated M2 tool steel with as-received and polished surfaces at a speed of 0.1 m/s, a load of 10 N and with a relative humidity of 30%. Vertical lines indicate stick slip (data from Singer *et al.*, 1991).

the absolute value of film thickness. According to Peebles and Pope (1989) neither the coefficient of friction nor the wear should be dependent on the contact stress, in the range of 60 to 1060 MPa.

The microstructure of the titanium nitride coating and thus the coating method has an influence on the tribological properties of the coating. This has been well demonstrated by Bull *et al.* (1990). They show that the adhesive wear of the TiN coating is determined by two factors, the openness of the

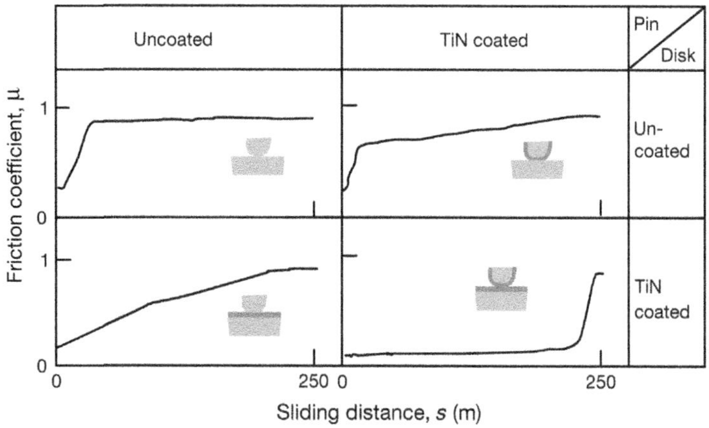

Fig. 4.38. The coefficient of friction depends on which one of the surfaces is coated, the pin or the disc. Results from pin-on-disk experiments with a steel pin and a steel disk uncoated and TiN coated sliding at a speed of 0.2 m/s and a load of 5 N (after Musil and Jirout, 2007).

microstructure and the dissipation of heat at the sliding contact. The mechanisms of adhesive wear are based on the local transfer of iron, which subsequently becomes trapped in the open regions of the coating microstructure, increasing the coating-counterface adhesion at these positions. Consequently, the best sliding wear properties result from hard, dense coatings. However, oxidative wear of these coatings may be relatively more important.

The wear performance of a titanium nitride coating can be improved by the deposition of, e.g., a titanium interlayer to improve the adhesion to substrate as discussed above. This has been observed by Peebles and Pope (1989) to result in negligible wear of the coating without failure at low contact stresses and to cause no change in the coefficient of friction. However, the titanium interlayer must not be too thick.

4.4.1.3 Sliding against non-ferrous materials

A tribologically excellent counterface for titanium nitride coatings is one coated with either titanium nitride or titanium carbide. The coefficient of friction for TiN–TiN and TiN–TiC contacts is often reported to be less than 0.2 and in some conditions even as low as 0.05 and the wear on both surfaces is very low (Jamal *et al.*, 1980; Hintermann, 1981; Mäkelä and Valli, 1985; Matthews *et al.*, 1987; Habig, 1990; Ghosh and Kohler, 1992; Stott *et al.*, 1992; Franklin and Beuger, 1992; Wiklund *et al.*, 1999a). This can to some degree be expected theoretically according to the explanation by Gardos (1989) that a low shear strength oxide layer (TiO_{2-x}) is formed on the hard coating, which now plays the role of a substrate, resulting in very favourable low-friction conditions. The hard coating supports the load and the shear takes place in the oxide layer or at the TiO_{2-x}–TiN interface as suggested by Singer *et al.* (1991). Some of the low-friction values reported may be due to their initial friction conditions. Higher values for the steady-state coefficient of friction ranging from 0.6 to 0.9 were measured by Wiklund *et al.* (1999a).

The wear rates of TiN–TiN sliding contacts are approximately the same in air and in vacuum, although the wear mechanisms operating are different, and the friction coefficient is higher, about 0.5, in vacuum (Habig, 1990). At a higher temperature of 300°C in a carbon dioxide environment the coefficient of friction has been observed to increase to a value of about 0.4 and failure to occur more rapidly (Stott *et al.*, 1992).

Jamal *et al.* (1980) have found that other titanium-based variations of the counterface coating can produce favourable tribological conditions, such as TiC/Ti, TiC_2N/Ti and Ti + TiC/Al. Tribologically less favourable counterfaces are titanium, aluminium (Jamal *et al.*, 1980), copper (Matthews *et al.*, 1987), alumina (Huq and Celis, 1997) and chromium and stainless steel (Wiklund *et al.*, 2006). A nickel ball as counterface, on the other hand, resulted in very stable sliding with a coefficient of friction of 0.35 in pin-on-disk tests by Wiklund *et al.* (2006).

Very low-friction coefficients were reported from pin-on-disk tests with single-crystal sapphire balls of various diameters sliding against polished TiN coatings at speeds of 0.02 to 0.08 m/s and loads of 2 to 10 N. The initial coefficient of friction was as low as 0.05 and then slowly increased to 0.1 to 0.15 after about 1500 revolutions (Singer *et al.*, 1991). The very smooth sliding surfaces used seem to be one explanation for the low friction. The higher friction coefficient corresponded to more debris in the wear track.

Titanium nitride in sliding contact with graphite gives reasonably low wear but when the coating gets thin spalling occurs, which results in a very significant increase in wear rate, owing to the presence of hard, abrasive third bodies between the contacts as shown by Kennedy *et al.* (1990) in seal contact experiments.

The ambient humidity can considerably influence the tribological properties in sliding TiN contacts. This is well illustrated by measurements carried out in reciprocating fretting contacts, with a 10 mm diameter corundum ball of surface roughness 0.2 μm sliding at 10 Hz, 1 N load and 100 μm sliding distance reported by Zhang *et al.* (2003a). They record a coefficient of friction of 0.15 at 90% relative humidity, 0.2 at 50% RH and as high as 0.8 at 10% RH. However, the wear rate was the same, $20 \cdot 10^{-6} \cdot mm^3/Nm$, for all three humilities. The friction results are very similar to the results from fretting tests of an Al_2O_3 ball sliding against a TiN plate in the same humidity range measured by Klafke (2004c). However, Klafke observed a slight decrease in wear with increasing humidity.

4.4.1.4 Abrasive wear resistance

The wear resistance against abrasion of steel components can be considerably improved by applying a 2 to 6 μm thick TiN coating on the surface (Sirvio *et al.*, 1982; Rickerby and Burnett, 1987; Matthes *et al.*, 1990a; Ghosh and Kohler, 1992; Bromark, 1996; Batista *et al.*, 2002). The high hardness is indeed one parameter which improves the abrasive wear resistance. The initial surface roughness influences the wear resistance so that smoother surfaces result in less wear (Sirvio *et al.*, 1982). Improved toughness and abrasion resistance can be produced by thicker films with a multilayer structure with titanium and titanium nitride layers, and with a duplex structure where the TiN film is deposited on plasma-nitrided surfaces (Leyland, 1991; Leyland and Matthews, 1993; Batista *et al.*, 2002).

The load-carrying capacity of the coating/substrate system is important, as has been pointed out by Rickerby and Burnett (1987). Under mild abrasive wear they found that thin coatings eventually fail by localized detachment of the film at the intersections of the plastic grooves formed by the abrasive particles. Thicker coatings, on the other hand, have the ability to support the contact stresses elastically and degrade by a microchipping or polishing mechanism with little or no deformation of the underlying substrate, as shown in Fig. 4.39. However, in thicker titanium nitride coatings differences in wear rate have been observed related to microstructural effects. These can be attributed to a decrease in coating density and increasing grain size with increasing thickness (Burnett and Rickerby, 1987a, b). Under severe abrasive wear, thick coatings fail mainly by a cohesive fracture mechanism.

The abrasive wear resistance of titanium nitride coatings can be further improved by introducing titanium either as interlayers or dispersed within the bulk of the coating and this can also promote good adhesion (Matthews, 1980). Some techniques, especially arc evaporation, produce discrete droplets of titanium in the film. At low stresses the presence of metallic titanium may have a beneficial

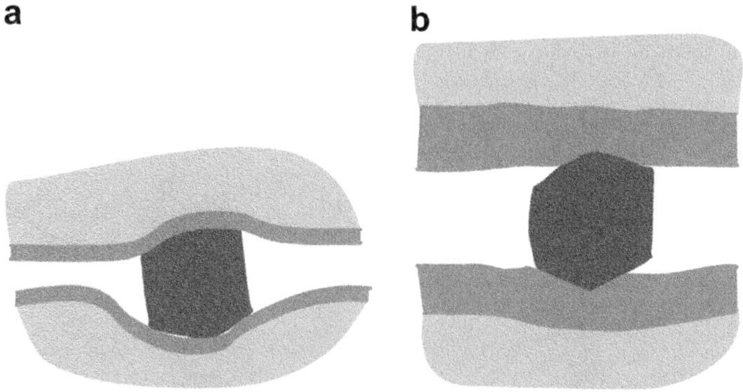

Fig. 4.39. For hard coatings on softer substrates the substrate (a) deforms elastically with thin coatings but (b) is undeformed with thick coatings, because the coating can carry the load.

effect by reducing the internal stresses and improving the toughness. At higher contact stresses the penetration of the metallic titanium droplets in the coating by the abrasive particles may result in increased tearing and abrasion between the abrasive and the coating and in increased wear.

4.4.1.5 Erosive wear resistance

The erosive wear resistance of steel coated by a layer of 1 to 10 μm titanium nitride is considerably better than for uncoated steel. It has been shown that the wear resistance depends on the hardness of the coating/substrate composite system, the shape of the eroding particles and the angle of bombardment (Burnett and Rickerby, 1988; Hedenquist and Olsson, 1990; Rogers *et al.*, 1992; Iwai *et al.*, 2001). Higher composite hardness, i.e. the hardness of the coating/substrate combination, in general gives reduced erosive wear, as shown in Fig. 4.40.

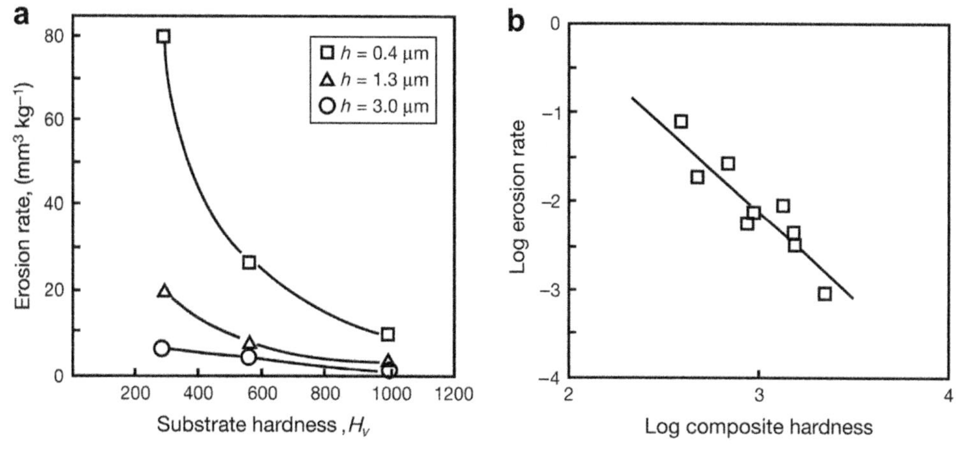

Fig. 4.40. Erosion tests show that the erosion rate decreases (a) with increasing coating thickness, h, and substrate hardness, H_V, and (b) with increasing surface composite hardness (after Hedenquist and Olsson, 1990).

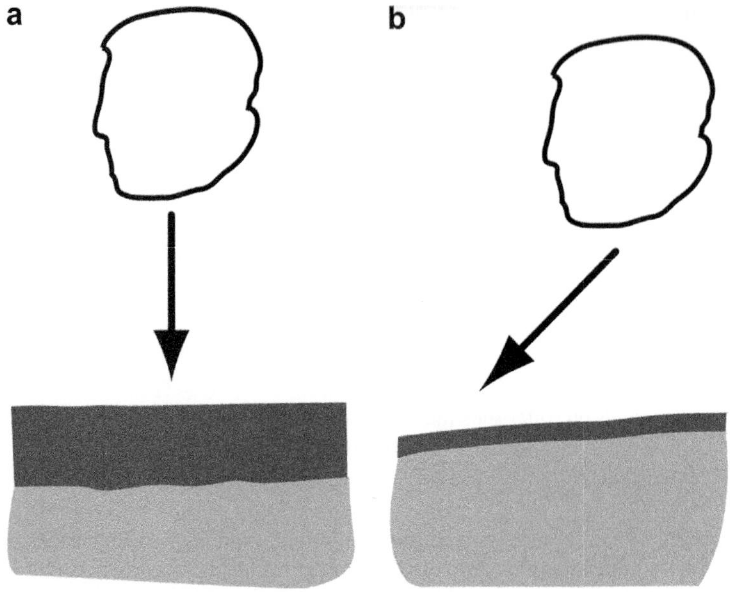

Fig. 4.41. Better erosion resistance is achieved (a) with thick TiN coatings at perpendicular sharp particle bombardment and (b) with thin TiN coatings at a low angle (45°) round particle bombardment.

Thicker coatings can be expected to provide greater erosion resistance than thinner ones when subject to normal impact by sharp particles. On the other hand, thinner coatings have shown to provide better erosion resistance when subject to impact from round particles arriving at a 45° angle to the surface, as shown in Fig. 4.41 (Burnett and Rickerby, 1988). This effect is attributed to the formation of Hertzian cone fractures and elastic–plastic indentation fractures once a critical coating thickness has been exceeded.

Three types of coating erosion mechanisms which may typically occur can be defined: erosive fatigue wear, cutting erosion and spalling. Erosive fatigue wear occurs very commonly while cutting erosion predominantly occurs on surfaces with low composite hardness, and spalling occurs with high composite hardness (Jönsson *et al.*, 1986; Hedenquist and Olsson, 1990).

At elevated temperatures Rogers *et al.* (1992) found for 2 μm thick TiN coated steel surfaces that the volume wear rate increased with temperatures between 200°C and 400°C, primarily due to a decrease in coating hardness with increasing temperature. Above 400°C, the coatings provided minimal protection against erosion and corrosion, due to an increase in their oxidation rate. However, TiN coatings still provided some protection against oxidation of the steel substrate in unworn regions at 500°C.

4.4.1.6 Rolling contacts

As discussed in section 3.4.3 the rolling contact fatigue life can be increased by the use of thin coatings in contacts similar to those in gears, rolling bearings and cam-and-tappet contacts (Hochman *et al.*, 1985). In heavily loaded mixed lubricated contacts, with a maximum Hertzian pressure of 2.3 GPa, and a surface speed of 3.1 m/s, Chang *et al.* (1990) found that the maximum lifetime can be achieved with 0.25 μm thick TiN coatings. The relatively thick coatings, up to 5 μm, produced large initial spalling. This is thought to be a function of the coating thickness and the compressive stress. Podgornik (2001) compared the load limits of DLC, TiN and TiAlN coated surfaces in reciprocating sliding and rolling tests with a speed of 0.1 m/s and load of 10 N and found that TiN could take 3 to 5 times higher loads compared to TiAlN in similar conditions but the DLC coated surface could take even a 30% higher load than TiN without failure.

In a comparison of a steel surface with TiN, CrN and WC/C coated surfaces Remigiusz *et al.* (2004) received poor lifetime results in lubricated rolling tests for both TiN and CrN coatings compared with the other two. Thus the experience documented in the literature is not entirely clear. There seems to be evidence that the rolling contact fatigue life can be increased by using TiN coated steel surfaces. On the other hand, the stress and strain conditions are so complex and it is so difficult to optimize the coating dimensions and mechanical properties, so poor results can easily be encountered.

4.4.1.7 Cutting contacts

Titanium nitride coatings deposited on high-speed steel (HSS) by physical vapour deposition have successfully been used in metal cutting applications for a number of years. It has been documented in several experimental reports (Sundquist *et al.*, 1983; Matthews, 1989; Posti and Nieminen, 1989a, b; Harju *et al.*, 1990; Spur *et al.*, 1990; Kowstubhan and Philip, 1991; Bromark, 1996; Grzesik, 2000) that the tool life in cutting can be increased by even orders of magnitude, by TiN coated HSS inserts. This is due to a unique combination of properties of the TiN coated tool, such as sufficient adhesion to the substrate, high hot hardness, high wear resistance and an ability to improve the contact conditions at the cutting edge (Hedenqvist *et al.*, 1990b). Equations for the prediction of tool life of TiN coated tools have been developed by Kowstubhan and Philip (1991).

The increased wear resistance of the tool is mainly achieved by the high hardness of the TiN coating, 2000 to 2500 H_V, which gives a good resistance to abrasive wear, and the high chemical stability, resulting in a high resistance to solution wear. The oxidation of the TiN surface is slow at temperatures below 600°C. The good adhesion to the substrate can inhibit interfacial cracking even when the substrate is plastically deformed due to surface stresses. An increased production rate can be achieved in cutting with TiN coated tools because of increased cutting speeds and feeds. Decreased cutting temperatures have been reported, which probably is due more to improved contact conditions and reduced friction than to the TiN coating acting as a thermal barrier (Hedenqvist *et al.*, 1990b). It has been shown that even when the coating is locally worn through, the remaining coating still protects the substrate and is capable of carrying the full load of the chip without collapsing. This applies even on reground tools, where some benefit still derives from the remaining side-face coating.

The optimum thickness of the coating depends on the cutting methods, as shown by Posti and Nieminen (1989a). In continuous cutting methods, like turning, a thicker coating gives a longer tool life. Good results were achieved with a coating thickness of 6 μm. In discontinuous cutting, like planing, on the other hand, an optimum coating thickness of 2 to 3 μm was observed as shown in Fig. 4.42. Both thicker and thinner coatings reduced the lifetime of the tool. Thin coatings are worn out faster by mechanical wear and with thicker coatings local detachment occurs due to the reduced flexibility of the hard coating on the softer substrate.

In dry machining conditions the flank wear can be reduced with the aid of chlorine ion implantation in titanium-based ceramic coatings (Aizawa *et al.*, 2005). Titanium mono-oxide is formed *in situ* to act as a plastically deforming tribofilm. The tribofilm works as a lubricious buffer between work materials and tools to reduce the wear and friction. Dry drilling tests have shown that the combination of MoS_x with TiN in the deposition process resulting in a composite TiN–MoS_x coating, can increase the average drill lifetime by more than one-third (Cosemans *et al.*, 2003). The MoS_x component reduces the friction.

Similar remarkable improvements in lifetime have not been achieved in wood cutting as with TiN coated tools in metal cutting. However, Usenius *et al.* (2004c) noticed some improvements, e.g. a 20% reduced rake face wear in cutting of spruce with high-speed steel TiN coated tools compared with uncoated tools and 50% reduced rake face wear when sawing hardboard.

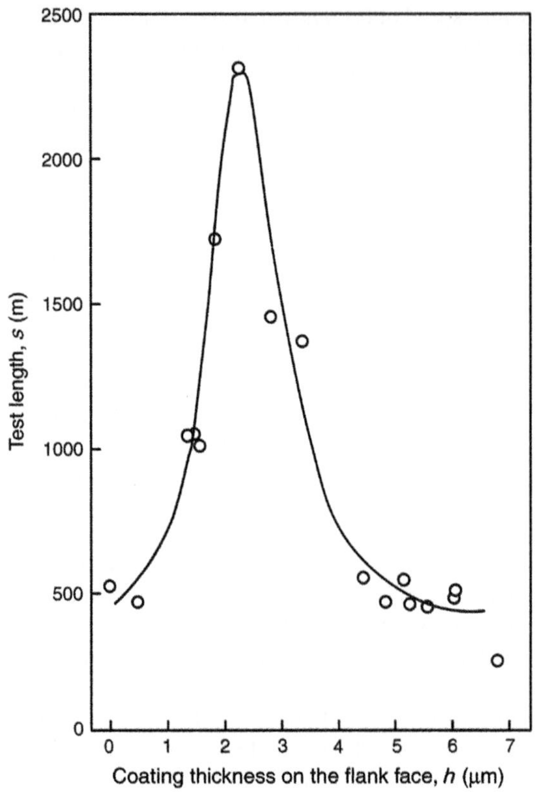

Fig. 4.42. TiN coated ASP 30 tool lifetimes in planing tests as a function of coating thickness (data from Posti and Nieminen, 1989a).

4.4.2 Other nitride coatings

Even though titanium nitride coatings have a very strong market position there have been a variety of other nitride coatings developed for tribological applications. In addition to titanium aluminium nitride (Ti,Al)N, and boron nitride (BN), other nitrides have been studied, such as chromium nitride (CrN), carbon nitride (CN), vanadium nitride (VN), hafnium nitride (HfN), zirconium nitride (ZrN), as well as multicomponent nitrides like Ti(B,N), Ti(B,C,N), Ti(C,N), (Ti,Nb)N, (Ti,Zr)N, (Ti,V)N, (B,C)N and (Zr,Hf)N. However, for many of these there is only a very limited amount of information available about their tribological properties and the tribological mechanisms are seldom explained. The main uses of these other nitride coatings have been in applications such as cutting tools, turbine blades, forming tools and pump components.

Recently, the cubic form of boron nitride (c-BN) has attracted a great deal of interest because of its extreme hardness, chemical inertness, high thermal conductivity and transparency. The different PVD and CVD techniques available for the deposition of c-BN have been reviewed by Karim *et al.* (1991), Veprek (1999) and Erdemir and Donnet (2001).

4.4.2.1 Nitride coatings in sliding contacts

The nitrides are generally hard coatings with a typical microhardness in the range of 2 to 4 GPa. In sliding contacts with steel their coefficients of friction are often fairly high, in the range of 0.6 to 0.9.

Here boron nitride forms an exception for which coefficients of friction down to 0.05 have been measured for the hexagonal phase.

In sliding pin-on-disk tests with a steel pin sliding on a *titanium aluminium nitride* (Ti,Al)N coating on a steel substrate with a speed of 0.1 m/s and a load of 10 N, Ronkainen *et al.* (1990a) measured coefficients of friction as high as 0.7 to 0.9, with pin wear rates of 3 to 5 $\cdot$ 10^{-6} mm^3/Nm and disk wear rates of 20 to 60 $\cdot$ 10^{-6} mm^3/Nm. Wear debris and transferred layers of reaction products were observed. In comparison with the wear of TiN, Podgornik (2001) measured in a pin-on-disk test with 0.1 m/s speed and 60 N load and sliding against a steel ball that (Ti,Al)N showed about 10% lower wear rate than TiN. In rolling tests the same (Ti,Al)N had a considerably lower load-carrying capacity, the load limit was only 10 to 30% of that of TiN depending on the substrate hardness.

Roos *et al.* (1990) found in a comparative test that a ball coated with (Ti,Al)N was worn twice as much as a ball coated with TiN in a pin-on-disk test at speeds of 0.4 m/s and loads of 4 N. Coll *et al.* (1992) studied the influence of the titanium–aluminium ratio in (Ti,Al)N coatings and found in pin-on-disk tests, with a speed of 0.5 m/s, a load of 2 N and a steel counterface, a coefficient of friction of 0.74 and a wear rate of 0.092 $\cdot$ 10^{-6} mm^3/Nm for a Ti:Al ratio of 53:47 wt%. The coefficient of friction decreased to 0.54 and the wear rate to 0.056 $\cdot$ 10^{-6} mm^3/Nm with a lower aluminium content of Ti:Al = 83:17 wt%. The coefficient of friction decreased with increasing speed and was as low as 0.37 at 5 m/s. Increasing wear with higher aluminium content in the coating has also been found in pin-on-disk tests with an aluminium oxide (corundum) ball as counterface (Vancoille *et al.*, 1993a). In fretting tests Klafke (2004c) measured coefficients of friction of about 0.7 in dry air of 3% RH and normal 47% RH air when sliding against an Al_2O_3 ball while the increase of the humidity to 100% dropped the coefficient of friction to 0.5.

Boron nitride (BN) coatings on different substrates and in different environments have been studied by Miyoshi (1989a). He describes the ion beam-deposited coatings as mainly amorphous containing a small amount of hexagonal BN. The boron nitride films deformed elastically and plastically and surface films on the coatings affected their tribological behaviour. In very slow single pass contacts with a speed of 0.0002 m/s he measured very low coefficients of friction from 0.05 to 0.2 for a diamond stylus sliding on a boron nitride coating and almost equally low-friction coefficients from 0.08 to 0.12 for a steel pin sliding on the BN coating. Similar results for diamond sliding on BN have been reported by Elena *et al.* (1988). The presence of the BN film decreases adhesion and plastic deformation in the contact and accordingly the friction. Similarly, low values for the friction in vacuum have been measured by Miyoshi (1989a). The sliding combination of a 80 mN lightly loaded textured BN slider sliding with 0.5 m/s in 50% RH against a DLC coated silicon disk resulted in a higher coefficient of friction of 0.41 and a slider wear rate of 0.01 $\cdot$ 10^{-6} mm^3/Nm (Zhou *et al.*, 2000).

Boron nitride with a *cubic crystal structure* (c-BN) is an extremely hard material. It is the hardest known material after diamond and thus a potential candidate for wear-resistant coating applications. Unlike diamond, it does not dissolve in or interact with Fe-based alloys, and thus can be used to machine ferrous materials. It has a high melting point, excellent thermal and chemical stability and thus excellent wear and oxidation resistant properties at high temperatures. Tribological studies have shown that the performance of c-BN is highly dependent on the environment and the counterface material. The wider commercial use of c-BN is limited by its film thickness, high internal stresses, poor adhesion to the substrate and poor reproducibility of structure and properties (Erdemir and Donnet, 2001; Koskilinna, 2007).

Cubic boron nitride coatings have, in general, poor adhesion to various substrates but this can be considerably improved by applying an interlayer such as silicon nitride (Inagawa *et al.*, 1989). They found that in a sliding ring on disk contact loaded at 10 N the coefficient of friction was 0.35 but at a low pressure of 1.7 $\cdot$ 10^{13} Pa and a high temperature of 400°C it was extremely low, only 0.02. The c-BN film had a thickness of 1.5 μm and microhardness of 5.5 to 6.3 GPa H_V. Good results have also been achieved with a titanium interlayer.

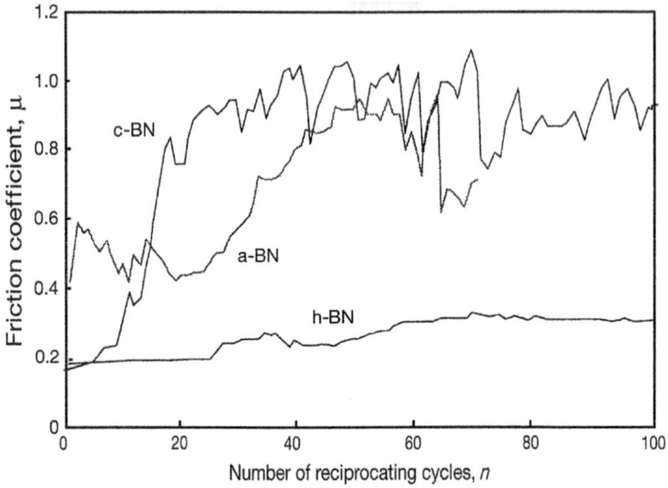

Fig. 4.43. The coefficient of friction as a function of number of reciprocating cycles for cubic (c-BN), amorphous (a-BN) and hexagonal (h-BN) boron nitride coatings on silicon substrates sliding against a 6.35 diameter steel ball in reciprocating pin-on-flat tests with a speed of 0.0017 m/s, load of 5 N, stroke length of 0.01 m and relative humidity of 50 to 60% (after Miyake *et al.*, 1992).

Watanabe *et al.* (1991) and Miyake *et al.* (1992) have compared the different forms of BN films deposited on silicon wafers by using a reciprocating friction and wear tester and found that the tribological properties of cubic boron nitride are superior to *amorphous boron nitride (a-BN)* and *hexagonal boron nitride (h-BN)*. The steady-state coefficient of friction for a-BN and h-BN was in the range of 0.8 to 1.0 while it was about 0.3 for c-BN as shown in Fig. 4.43. In a continuously sliding ball-on-disk test with 1 N load, 0.01 m/s speed and 40–50% RH Watanabe *et al.* (2004) measured a rapidly increasing coefficient of friction up to 0.6 and 0.7 after only some 10 to 50 revolutions with c-BN sliding against itself and against a-BN. With c-BN and a-BN sliding against a DLC coating the coefficient of friction reached a steady-state value of about 0.14 and no damage was observed.

Theoretical calculations have predicted that a new form of *carbon nitride* (C_3N_4), the cubic beta phase, would have very low compressibility and extremely high hardness exceeding that of natural diamond. This has attracted many researchers to develop new methods to synthesize crystalline C_3N_4 but this has not yet been fully realized. However, many groups have been able to produce amorphous C_3N_4 coatings with very good tribological performance even if they are softer than diamond. Due to the difference in the deposition methods the amorphous C–N coatings exhibit large variations in mechanical and tribological properties (Erdemir and Donnet, 2001). High hardness of amorphous C–N coatings in the range of 15 to 50 GPa has been reported.

Based on results from a ball-on-disk test with Si_3N_4, balls sliding against 100 nm thick CN_x coatings deposited on silicon wafers sliding in the load range of 80 to 750 mN and speed range of 0.004 to 0.4 m/s in different environments Kato *et al.* (2003) could conclude the following. The coefficient of friction was high, 0.2 to 0.4, in air and O_2, but low, 0.01 to 0.1 in N_2, in CO_2 and in vacuum. The lowest coefficient of friction, less than 0.01, was measured in N_2. The wear rates of the CN_x coatings were in the range of 0.001 to $10 \cdot 10^{-6}$ mm^3/Nm depending on the environment. The major running-in wear mechanism of CN_x in abrasive sliding against hard asperities was low-cycle fatigue which generated thin flaky wear particles of nanometre size (Wang and Kato, 2003b, c). It is interesting that good wear resistance properties have been achieved with extremely thin, only 1 nm thick, CN coatings on magnetic disks sliding against a friction force microscope (FFM) diamond tip with a sliding length of 5 μm and up to 2000 sliding cycles (Wang *et al.*, 2005).

Chromium–nitrogen (Cr–N) compound coatings have been developed by the PVD method for the purpose of achieving superior properties to electrodeposited chrome. The tribological properties of various Cr–N coatings were studied by Bull and Rickerby (1990a) with a pin-on-disk tribotester at a speed of 0.1 m/s and a load of 2.75 N. They found that the coefficient of friction for a steel ball sliding on a Cr–N coating on a steel substrate was 0.6 to 0.8 and the wear was not particularly low, being in the range 1 to 100 $\cdot$ 10^{-6} mm^3/Nm for the coated disk and 0.1 to 10 $\cdot$ 10^{-6} mm^3/Nm for the ball. They noticed that as the hardness of the coating increases, plastic deformation in the substrate is reduced and this leads to a decrease in the disk wear rate until it is comparable with that of a similar titanium nitride coated disc. In this situation a visible flat is formed on the ball and the wear of the ball material dominates. A considerable transfer of material from the disk to the ball was observed.

At a higher load of 10 N, same speed but with a ceramic Al_2O_3 ball as the counterface Tuszynski *et al.* (2003) measured a similar high coefficient of friction of 0.8 but the wear rate of the coated disk was only 0.05 $\cdot$ 10^{-6} mm^3/Nm. The wear resistance was two orders of magnitude better than that of TiN coatings in the same conditions. The CrN coatings improved the abrasive wear resistance considerably, as found also by Leyland *et al.* (1993). In lubricated sliding the CrN coating has been found to considerably improve the scuffing wear resistance compared with that of steel surfaces (Remigiusz *et al.*, 2004).

Vanadium nitride (VN) is a ceramic compound closely related to TiN. Quantum calculations have shown that it should have very different adhesive properties against pure iron compared to TiN (Wiklund, *et al.*, 2006). Wiklund *et al.* compared the friction properties of VN and TiN in a ball-on-disk test with 3 mm radius balls sliding at 30 N load and 0.01 m/s speed against different ball materials. When sliding against steel and chromium balls both coatings showed the same friction behaviour, with a coefficient of friction starting initially at 0.15 but soon increasing to the level of 0.6 (see Fig. 4.35). With pure nickel balls the coefficient of friction for VN was 0.3 while it was slightly higher, 0.35, for the TiN coating. With stainless steel balls both coatings very quickly reached a coefficient of friction of 0.6 but there were indications that VN had better galling protection properties.

Zirconium nitride (ZrN) coatings have been of interest since some reports show that they have better mechanical, wear and corrosion properties than TiN coatings. Thus they have been increasingly used in industry for specific applications such as coatings for tools for machining non-ferrous materials (aluminium and nickel alloys) and thermal barrier coatings in microelectronics, as well as for decorative coatings (Atar *et al.*, 2004).

The tribological properties of *titanium boron nitride Ti(B,N)* coatings produced by ionization-assisted electron beam PVD and by rf magnetron sputtering have been studied by Ronkainen *et al.* (1990a). The coefficient of friction against M50 steel was in the range of 0.6 to 0.9, the pin wear 0.8 $\cdot$ 10^{-6} mm^3/Nm and the disk wear 15 $\cdot$ 10^{-6} mm^3/Nm measured with a pin-on-disk test at a speed of 0.1 m/s and a load of 10 N. The sliding wear resistance is of the same order as for titanium nitride and slightly better than for titanium aluminium nitride coatings in the same contact conditions. White powder-like wear debris and transferred layers of reaction products on the surfaces were observed. The Ti(B,N) coatings had a thickness 4 to 8 μm and a microhardness of 2 to 3.5 GPa. The hardness, elastic modulus and also the internal residual stresses increase with the relative boron content in the coating (Kullmer *et al.*, 2003).

The tribological properties of *titanium carbonitride Ti(C,N)* are better or similar to titanium nitride depending on the contact conditions. The residual compressive stresses in Ti(C,N) coatings can be very high, values up to about 7 GPa have been measured for 4 μm thick coatings (Zoestbergen and De Hosson, 2002). In abrasive wear tests as well as in cross-cylinder tests Ti(C,N) coatings have shown a better wear resistance compared to TiN coatings but the result was the other way round in reciprocating and ball-on-disk wear tests (Ghosh and Kohler, 1992; König and Kammermeier, 1992; Tuszynski *et al.*, 2003). In a comparison of tribological properties of TiN, (Ti,Al)N, (Ti,Nb)N, and

Ti(C,N) deposited on high-speed steel and sliding against aluminium oxide (corundum) Vancoille *et al.* (1993b) found that titanium carbonitride coatings were the most promising at a speed of 0.1 m/s and a load of 5 N. The steady-state coefficient of friction was in the range of 0.1 to 0.4 while for TiN it was 0.8 to 1.2, for (Ti,Nb)N it was 0.8 to 1.4 and for (Ti,Al)N it was 1.0 to 1.4. The wear was also low for Ti(C,N). The tribological behaviour of Ti(C,N) coatings in fretting tests has been very similar to that of TiN (Klafke, 2004c; Liskiewicz *et al.*, 2004). In a comparison of six different commercial Ti(C,N) coatings Bull *et al.* (2003a, b) found that both the friction and wear properties varied a lot and were critically dependent on the coating structure produced by the different deposition processes. Coating adhesion was improved by the use of a thin TiN interlayer.

4.4.2.2 Nitride coatings in metal cutting contacts

Major improvements in the lifetime of cutting tools have been achieved by coating them with a titanium aluminium nitride layer. The fundamental advantage of (Ti,Al)N is its characteristic of forming a dense, highly adhesive, protective Al_2O_3 film on its surface when heated, preventing diffusion of the coating material into the workpiece. A second advantage for machining applications is its low thermal conductivity. Considerably more heat is dissipated via chip removal. This enables correspondingly higher cutting speeds to be selected, since the thermal loading on the substrate is lower (Leyendecker *et al.*, 1989).

Finding the optimal coating thickness is important for cutting tool performance. An increase in the coating thickness usually deteriorates the coating mechanical strength. The typical cone-shaped column growth leads to increased grain size at the top surface for thicker coatings. Due to the coarser microstructure at the top surface, it can be expected that a diminishing of the film mechanical properties occurs with thicker coatings. On the other hand, thicker coatings lead to higher effective cutting edge radii, thus generating lower stresses on the cutting edge. The substrate is better protected against abrasive wear and thermal loads occurring during the cutting process with thicker coatings. Based on FEM calculations and with milling tests Bouzakis *et al.* (2003) investigated $(Ti_{46},Al_{54})N$ coated cemented carbide inserts in the thickness range of 2 to 10 μm and demonstrated the enhanced cutting performance of thick coatings especially over 6 μm, in comparison to thinner ones, as shown in Fig. 4.44.

It is relevant to compare the properties of (Ti,Al)N coated tools with TiN coated tools because the latter represents the commonly used commercially available state-of-the-art level. In turning tests on steel with (Ti,Al)N coated high-speed steel inserts the tool lifetime has been increased by two to three times compared with TiN (Molarius *et al.*, 1987; Leyendecker *et al.*, 1989; Ronkainen *et al.*, 1991b; Byeli *et al.*, 1992). However, the lifetime was only half that of TiN when cutting aluminium alloy as shown in Fig. 4.45 (Ronkainen *et al.*, 1991b). A superior performance of (Ti,Al)N coated drills was also reported by Leyendecker *et al.* (1989) who drilled 12 times as many holes in grey cast iron as was possible with a TiN coating before tool failure. When drilling in an aluminium alloy the tool life was increased by three times (Leyendecker *et al.*, 1989) and when drilling in steel by two times (Ronkainen *et al.*, 1991b) compared to TiN coated drills. Improvements in lifetimes of two to five times for (Ti,Al)N coated hobs in gear production tests compared to uncoated hobs has been demonstrated by Zlatanovic and Munz (1990).

The improvements in tool lifetimes by several other varieties of nitrides have also been investigated: zirconium nitride (ZrN) by Molarius *et al.* (1987), hafnium nitride (HfN) by Harju *et al.* (1990), titanium boron nitride Ti(B,N) by Ronkainen *et al.* (1991a), titanium vanadium nitride (Ti,V)N by Knotek *et al.* (1990a, 1992b), titanium zirconium nitride (Ti,Zr)N by König and Fritsch (1992), titanium niobium nitride (Ti,Nb)N by Roos *et al.* (1990) and chromium nitride in wood cutting by Beer *et al.* (2003) and Noveau *et al.* (2005). The general result is that the lifetimes of tools with these coatings have often been about the same as or slightly better than TiN coated tools. However, in drilling tests in steel with a (Ti,Nb)N coated tool a lifetime of three times that of TiN coatings has been reported.

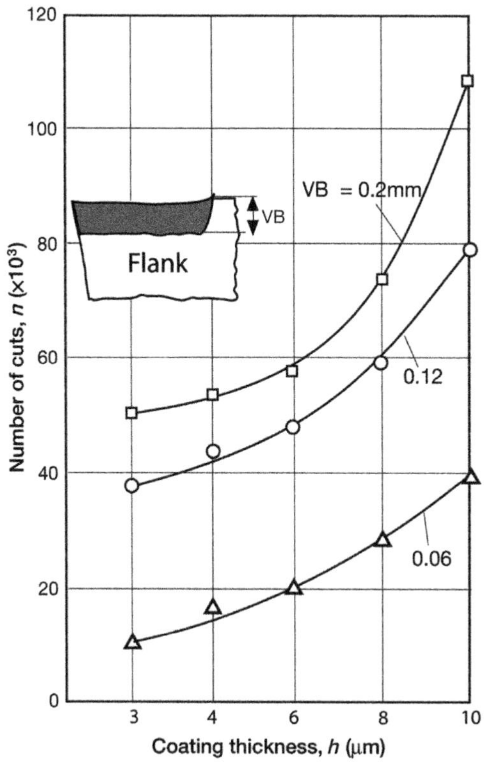

Fig. 4.44. Number of cuts up to small (0.06 mm), medium (0.12 mm) and large (0.2 mm) flank wear depth values (VB) vs the coating thickness (data from Bouzakis *et al.*, 2003).

4.4.2.3 Nitride coatings in erosive and abrasive contacts

Titanium aluminium nitride coatings have an overall good tribological performance in erosive, abrasive and sliding contacts and this performance is normally better than that of other nitrides. Bromark *et al.* (1994) compared the performance of CrN, TiN, (Ti,Al)N and Ti(C,N) coated surfaces and found that in order to minimize erosive wear of coated parts, a coating–substrate composite with high substrate and coating hardness and sufficiently high coating thickness should be chosen. The influence of coating hardness increases with substrate hardness. High coating fracture toughness improves the erosive wear resistance. The difference in abrasive wear resistance of the investigated coatings was small.

4.4.2.4 Nitride coatings in metal forming

For very soft metals with a high degree of plastic deformation during forming, such as for aluminium and aluminium alloys, nitrided forming tool steels and nitride-type coatings, such as TiN and VN, show a good resistance to galling and a relatively stable friction (Podgornik *et al.*, 2006b). Titanium carbonitride Ti(C,N) coatings have excellent wear resistance when applied on cold forming ejector dies for pressing the outer cases of high-speed shuttles used in industrial sewing machines. The lifetime was increased by 40 times using TiN coatings but with Ti(C,N) coatings even more, 80 times according to Shizhi *et al.* (1990). In sheet metal forming the lifetime of grey cast iron drawing edges

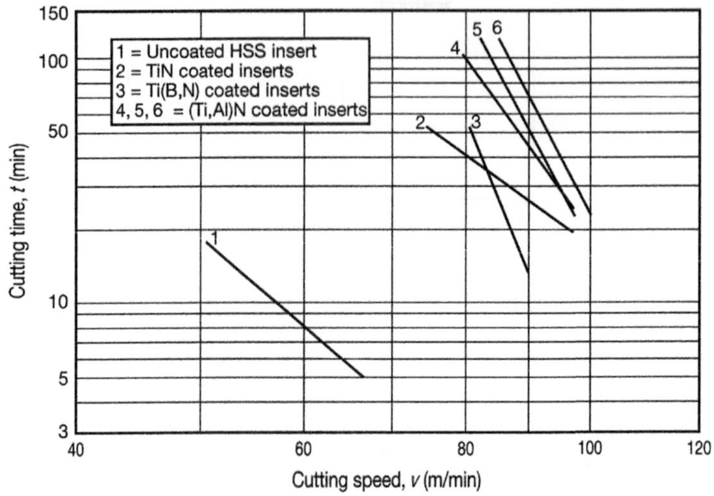

Fig. 4.45. Tool lifetimes for (Ti,Al)N and TiN coated tools vs cutting speed, in turning tests (data from Ronkainen *et al.*, 1991b).

has been increased two times by TiN coatings and by almost as much with Ti(B,N) and TiB_2 coatings (Matthes *et al.*, 1990b).

4.4.2.5 Rolling contact fatigue performance

As mentioned for TiN coatings the influence of nitride coatings on the surface fatigue life of PVD coated surfaces has not been extensively investigated and those results that are available do not give unambiguous indications. Yonekura *et al.* (2005) studied the fatigue life and mechanisms of three Cr-based PVD coated and one uncoated 100Cr6 surfaces in a thrust roller bearing-type fatigue tester. CrN performed best of the coated surfaces but the authors still conclude that it is unclear if the ultimate fatigue life of the coated rollers was significantly different to those displayed by uncoated rollers. The CrN coating failed by simple exfoliation or deep pitting-type rolling contact fatigue. However, TiAlN did not perform as well as TiN or DLC in reciprocating sliding and rolling tests, as discussed earlier (Podgornik, 2001).

The resistance to fatigue loading for (Ti,Al)N and Cr–N coatings has been improved by five to 10 times when deposited on a duplex-type surface with a plasma nitrided layer between the steel substrate and the PVD coating compared to non-duplex coated surfaces (Batista *et al.*, 2003). They found a good correlation between impact and scratch test results which indicates that adhesion is one of the key parameters that influence the fatigue resistance of PVD coatings.

4.4.2.6 Improvements by using intermediate layers

A thin intermediate surface layer between the substrate and the surface coating can improve the tribological properties for several reasons. Adhesion between the coating and the substrate can be improved by a suitable intermediate layer. Effects of material mismatch such as differences in thermal expansion can be minimized by using an elastic intermediate layer which allows some deformation and decreases the internal stresses. The intermediate layer can improve the growth process of the coating by modifying the substrate–coating chemistry.

The wear properties of the extremely hard cubic boron nitride coating on a tool material like WC–Co alloy or high-speed steel can be improved by introducing a titanium intermediate layer (Murakawa and Watanabe, 1990). Improvements have also been achieved by applying a silicon nitride intermediate layer between the c-BN and different substrates (Inagawa *et al.*, 1989).

4.4.3 Carbide coatings

Metal carbides are generally hard and have a high melting point and good tribological properties. The coefficient of friction for bulk carbides sliding on themselves is low, typically about 0.2 (Mordike, 1960). This is about one-third of the coefficient of friction value for steel sliding against steel.

Many different kinds of carbides have been used as thin tribological coatings. They have in general shown low friction, often in the range of 0.15 to 0.4 and good wear resistance. After titanium nitride, titanium carbide (TiC) is one of the most widely investigated thin ceramic coatings. Good tribological properties can be expected because of its high hardness, about 3.5 GPa H_V, which is higher than the typical value for titanium nitride of 2 GPa. Other extremely hard carbides are silicon carbide (SiC), $H_v \approx 4$ GPa, and boron carbide (B_4C), $H_v \approx 5$ GPa. Interesting tribological coating materials are also tungsten carbide (WC), chromium carbide (CrC), vanadium carbide (VC) and tantalum carbide (TaC).

4.4.3.1 Sliding contact with TiC against steel

There are considerable advantages in using conventional steel as a counterface material. In a wide range of contact conditions the friction is low, in the range of 0.1 to 0.4, and low wear has been observed for both PVD and CVD produced TiC coatings sliding against steel (Habig *et al.*, 1980; Hintermann, 1981; Boving *et al.*, 1987; Matthews *et al.*, 1987; Franklin and Beuger, 1992). However, Suri *et al.* (1979) and Jamal *et al.* (1980) have reported experimental results, both performed in the same laboratory using activated reactive evaporation TiC coatings, for the coefficient of friction that are at a higher level of 0.5 and 0.7. The reason for this discrepancy is not clear but it indicates that low friction contact conditions are not necessarily always present with this material combination.

Boving *et al.* (1987) measured coefficients of friction around 0.3 in dry air and vacuum for a CVD TiC coating sliding against steel, while it decreased to 0.2 when the experiments where carried out in 50% humidity. This indicates that an environmental effect is present; a layer of TiO_2 is built up on the top of the coating in humid air and this decreases friction. In lifetime tests Maillat and Hintermann (1985) compared the fretting wear resistance of several soft and hard coatings and found that TiC in this respect was one of the best wear-protecting coatings. The build-up of a tribolayer that works as a lubricious buffer between work materials and TiC coated tools in cutting contacts has an important friction and wear-reducing effect (Hedenqvist and Olsson, 1991; Aizawa *et al.*, 2005).

4.4.3.2 Sliding contact with TiC against other materials

Excellent tribological properties have been reported for two hard coating combinations, TiC–TiC and TiC–TiN, sliding against each other. The coefficient of friction is low, 0.15 to 0.3, and the wear very low (Jamal *et al.*, 1980; Hintermann, 1981; Matthews *et al.*, 1987; Habig, 1990; Franklin and Beuger, 1992). The low friction might be explained by the formation of a low shear layer, for example an oxide, on top of the hard coatings. Low coefficients of friction have also been observed with TiC sliding on other hard materials like aluminium oxide and silicon carbide coatings (Hintermann, 1981) and diamond (Knight *et al.*, 1990). Figure 4.46 shows that TiC and TiN coatings sliding on each other in different combinations in general give very low friction and wear. It is interesting that the friction and wear values in several cases are not decreased, as would be expected, but increased when a kerosene oil lubricant is introduced.

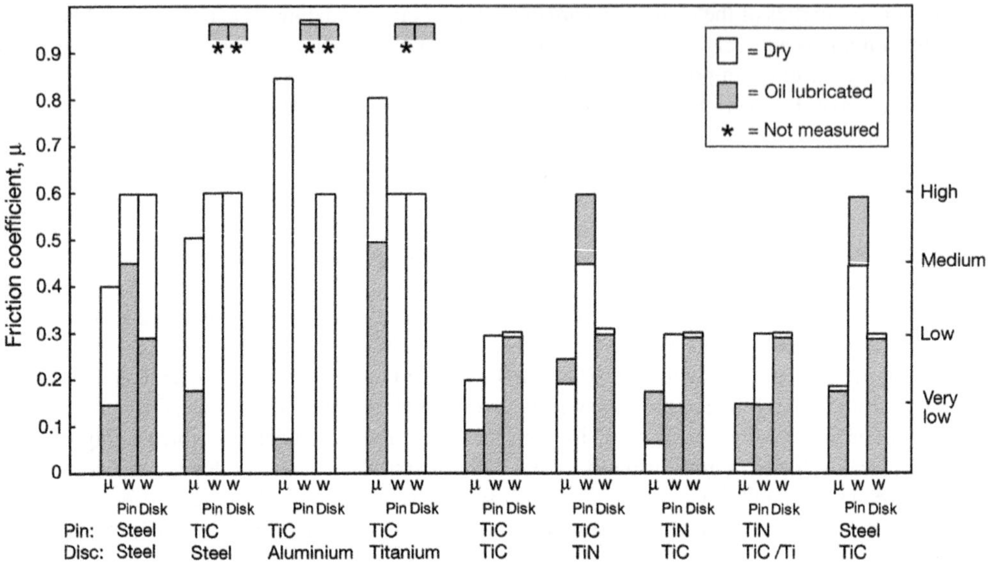

Fig. 4.46. The coefficient of friction and the wear of the pin and the disk in dry and oil lubricated sliding contacts with reactively evaporated TiC coatings sliding against different surface materials at a speed of 0.1 m/s and a load of 40 N (data from Jamal *et al.*, 1980).

4.4.3.3 Sliding contact with chromium carbide coatings

According to dry sliding experiments by Habig *et al.* (1980) the coefficient of friction of a chromium carbide coating sliding against steel is about twice as high, 0.3 to 0.5, and the wear about one order of magnitude higher than for titanium nitride or titanium carbide in the same sliding conditions. This is probably attributable to a higher adhesion between the chromium carbide and the steel. Even higher coefficients of friction in the range of 0.4 to 0.6 have been measured by Lindgren and Johnson (1987) for chromium carbide sliding against steel in different helium environments at high temperature.

4.4.3.4 Sliding contact with boron carbide coatings

Boron carbide is an extremely hard coating. Experiments performed at very low sliding speeds by Rey *et al.* (1988) with a chemically vapour-deposited trigonal boron carbide ($B_{13}C_2$) coating sliding against itself, against various ceramics and against steel showed that it has very good friction and wear properties especially in abrasive conditions. The lowest friction coefficient, 0.15, was obtained for $B_{13}C_2$ sliding against itself and the wear was very low. With $B_{13}C_2$ sliding against aluminium oxide, zirconia, silicon nitride, silicon carbide and steel the coefficient of friction was in the range 0.2 to 0.4 but considerably more wear in the form of deep scratches occurred on the ceramic and steel surfaces compared to sliding conditions with $B_{13}C_2$ as a counterface. Only the hardest ceramics (SiC, Al_2O_3) managed to produce some wear on the $B_{13}C_2$ surface. However, in general the nature of the counterface material does not seem to have much effect on the friction coefficient or on the wear phenomena.

Similarly, low friction coefficients, in the range of 0.08 to 0.2, were measured by Dvorak *et al.* (2002) for annealed boron carbide coatings in reciprocating sliding against a transparent sapphire hemisphere loaded at 6.4 N at a sliding speed of 0.001 m/s. Raman spectroscopy of the formed surface layer revealed a mixture of boric acid and carbon initially. As friction increased from a lower initial value, the boric acid intensity decreased to zero while the carbon intensity increased. Boron carbide B_4C is also well known for its high hardness and wear resistance. If one of the surfaces in a sliding contact is B_4C, it may polish the countersurface and act as a running-in surface. This process has been studied and described by Harris and colleagues (Harris *et al.*, 2002; Siniawski *et al.*, 2003; Borodich *et al.*, 2004).

The remarkably good tribological behaviour of boron carbide coatings is probably related to the mechanical properties of the material itself and to its structure due to the processing technology. The absence of any bonding material, of impurities at grain boundaries, and of porosities can also have a beneficial effect on the behaviour of boron carbide. The value of the coefficient of friction obtained is in fact lower than the value obtained for the sintered material.

4.4.3.5 Sliding contact with vanadium carbide coatings

In sliding pin-on-disk tests Wang (1997) found that depositing different steel substrates with 6 to 14 μm thick VC coatings in dry sliding with loads in the range of 30 to 90 N and a speed of 0.5 m/s the VC coating improved the wear resistance by nine to 30 times compared to uncoated surfaces. With a VC coating no running-in wear was observed.

4.4.3.6 Abrasive and erosive wear resistance of carbides

In abrasive and erosive wear the hardness of the surface is an important parameter. Thus the hard carbide coatings can be expected to have good resistance to these wear mechanisms (Borodich *et al.*, 2003; Bose and Wood, 2005). In an abrasive wear test the wear resistance of chemically vapour-deposited TiC and W_2C coatings was compared to that of much thicker electroplated and sprayed coatings and anodized and bath nitrided surfaces (Archer and Yee, 1978). The best wear resistance was measured for a hard 12 μm thick TiC coating and a 30 μm thick W_2C coating.

In another comparison test performed by Suri *et al.* (1979) considerably higher abrasive wear resistance was observed for PVD TiC coatings in comparison to boron ion implanted and untreated stainless steel surfaces. On the other hand, both boron and nitrogen ion implanted, as well as untreated, surfaces had a better erosive wear resistance compared to the hard TiC coating. These observations indicate a ductile erosive behaviour for untreated and ion implanted materials but brittle behaviour for the overlay coated substrates.

Both hardness and thickness have been found to influence the erosion resistance of 2 to 11 μm thick chemically vapour-deposited TiC, Al_2O_3 and $B_{13}C_2$ coatings in erosion experiments performed by Olsson *et al.* (1989). The hardest coating, $B_{13}C_2$, had a very high erosion resistance and the second hardest coating, TiC, had a significantly better erosion resistance than Al_2O_3. For TiC and Al_2O_3 in particular the erosion rate increased approximately linearly with increasing coating thickness.

For TiC, Olsson *et al.* found that erosion predominantly occurs by an intercrystalline fracture mechanism. The erosion proceeded through both the outer and inner TiC zones and into the cemented carbide substrate without any evidence of interfacial spalling. These observations indicate that the erosion of TiC in principle is by an elastic–plastic mechanism, but that the formation of intercrystalline cracks modifies the crack pattern. The authors suggest that in order to obtain high erosion resistance for TiC coatings the porosity, grain size and coating thickness should be kept as low as possible. The dominant material removal mechanism for the $B_{13}C_2$ coating seemed to be micro-chipping of small, about 10 μm in diameter, fragments due to indentation of sharp corners of the impinging alumina particles. This was followed by extensive spalling.

4.4.3.7 Rolling contact wear of carbides

Boving and Hintermann (1987) have shown that for thin CVD TiC coatings, with a thickness of 4 μm, on steel substrates the TiC coating follows the substrate deformation in a rolling point contact, provided that the stress applied on the coated substrate remains below the elastic limit of the substrate. One conclusion from this work is that TiC coatings can exhibit excellent adherence to the substrate, as well as good rolling contact fatigue resistance.

Dill *et al.* (1984) showed that in addition to CVD TiC coatings, reactively sputtered TiN coatings also have an excellent adhesion and spalling resistance in rolling contacts. However, they obtained the best rolling contact fatigue results for a CVD produced CrC–TiC coating where the substrate was first layered with 2 μm CrC and then with 3 μm TiC. This coating showed an excellent resistance to rolling contact fatigue in rolling point contacts with Hertzian contact pressures up to more than 4 GPa, which is remarkable.

It is interesting to note, on the contrary, that when Archer and Yee (1978) performed rolling contact fatigue life tests with 30 μm thick W_2C coatings on high-speed tool steel the coating flaked rather rapidly. This might be due to the fact that the higher thickness of the hard coating creates higher stresses in the coating and makes it more brittle and less deformable with the substrate. Also in the rolling contact fatigue tests carried out by Remigiusz *et al.* (2004) there was no clear improvement in fatigue life for WC–C coated bearing steel cones rolling against steel balls compared to uncoated.

4.4.4 Oxide coatings

Oxide coatings can be produced by several coating techniques to improve friction and wear properties. One of the industrial applications of such thin hard coatings is the use of aluminium oxide (Al_2O_3) layers produced by chemical vapour deposition on cemented carbide inserts used as metal cutting tools. Coatings a few micrometres thick can resist the high forces and temperatures at the tip and improve the tool lifetime by a factor of 10 (Lux *et al.*, 1986; Larsson, 2000).

4.4.4.1 Aluminium oxide coatings

Most reported tribological studies on alumina coatings are on those deposited by CVD at high temperature. Typically, these are crystalline of the κ or α phase (or a mixture). Alumina films formed by other techniques are usually amorphous. In sliding contacts aluminium oxide coatings normally have good wear resistance and medium friction. Typical values for the mechanical properties of CVD Al_2O_3 coatings are hardness of about 20 GPa, elastic modulus of 250 GPa and a fracture toughness of about 2 $MPa \cdot m^{0.5}$ (Diao *et al.*, 1994b).

In pin-on-disk experiments Hintermann (1981) measured a coefficient of friction of 0.45 for a CVD Al_2O_3 coated pin sliding against steel at a speed of 0.01 m/s and a load of 20 N. With a titanium carbide coated disk as counterface the coefficient of friction was only 0.19 and the wear was very low. The disk wear rate was $0.5 \cdot 10^{-6}$ mm^3/Nm and the pin wear rate as low as $0.005 \cdot 10^{-6}$ mm^3/Nm. Roth *et al.* (1987) have studied the structure, internal stresses and mechanical properties of sputtered Al_2O_3 coatings and found that the adhesion increases with increased coating thickness and reaches a high level for coatings thicker than 3 μm as shown in Fig. 4.47. In ring-on-plate experiments they measured good wear resistance for 4 μm thick Al_2O_3 coatings sliding against steel.

For CVD Al_2O_3 coatings with a TiC interlayer on cemented carbide substrates exposed to bombarding erosive particles, Olsson *et al.* (1989) found that an increase in coating thickness decreases the wear resistance. They found that chipping by transcrystalline fracture is the dominant erosion mechanism and subsurface cracks were observed in cross-sections. For thicker Al_2O_3 coatings, interfacial spalling at the Al_2O_3/TiC interface occurred at an early stage.

The basic mechanical aspects of very thin oxidized surfaces on aluminium and chromium sliding against a similar countersurface in lubricated pin-on-disk tests with a load range of 0.02 to 2 N and at speeds of 0.006 to 0.035 m/s were investigated by Komvopoulos (1991a, b). He found that under these sliding conditions the coating deforms elastically and the friction is mainly due to shear of the boundary lubricant film. The experimental results verified that the elastic and plastic deformation at

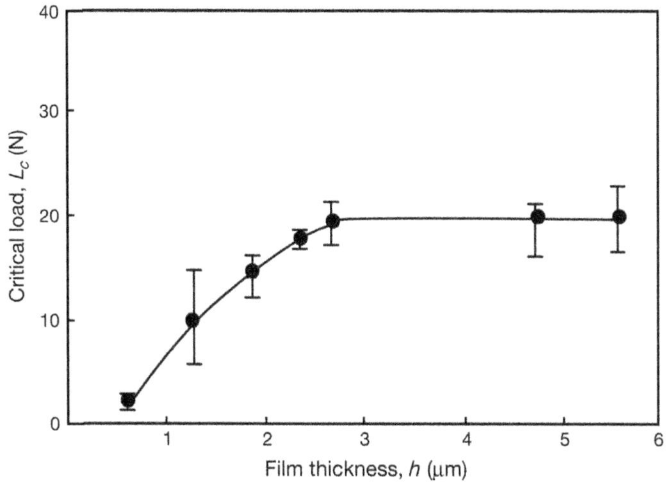

Fig. 4.47. Variation of the adhesion related critical load L_C for sputtered Al_2O_3 coatings vs coating thickness (data from Roth *et al.*, 1987).

the sliding interface strongly depends on coating thickness, surface roughness, microstructure of the coating and the mechanical properties of the coating in relation to those of the substrate. Surface profilometry and optical and scanning electron microscopy showed that when the coating was not removed the surface damage was virtually zero.

4.4.4.2 Titanium, cadmium and chromium oxide coatings

Other oxide coatings, such as the oxides of titanium, cadmium and chromium, have not been extensively investigated for tribological purposes.

For a titanium oxide PVD coated pin sliding on a steel disk at a speed of 0.2 m/s and a load of 5 N Matthews *et al.* (1987) measured a fairly low coefficient of friction of only 0.3, and a wear rate of $0.8 \cdot 10^{-6}$ mm^3/Nm, of the same order of magnitude as that for TiN or TiC coatings in the same conditions. With copper as the counterface the coefficient of friction increased to above 0.6.

Sputtered cadmium oxide disks sliding against a steel ball in a pin-on-disk device with a speed of 0.26 m/s and a load of 10 N showed a very low coefficient of friction in the range of 0.17 to 0.23, in experiments performed by Fusaro (1987b). The friction was of the same level as for graphite surfaces in the same conditions. However, even though the wear rate in the contact was reasonably low the endurance life was short – only about 15 kilocycles. The endurance life was defined as the sliding time in kilocycles to reach a friction coefficient of 0.25.

Chromium oxide layers deposited by CVD on layers of silicon oxide and aluminium oxide (Cr_2O_3/ SiO_2/Al_2O_3) on steel forming a 30 μm thick uniform and continuous coating showed very good scratch adhesion properties, almost at the same level as TiN, and good abrasive wear resistance in experiments performed by Bhansali and Kattamis (1990). In the scratch test the coefficient of friction gradually increased with increasing normal load to about 0.3, which is approximately equal to the value measured for TiN. Three different wear modes were identified for 400 μm thick plasma sprayed Cr_2O_3 coatings on stainless steel substrates when a spherical diamond tip was repeatedly sliding over the surface at a load of 0.07 to 1.5 N and a speed of 0.0004 m/s with increasing load: (1) crack and powder formation correlating to low wear rate; (2) flake formation correlating to high wear rate; and (3) ploughing and powder formation correlating again to a low wear rate (Kitsunai *et al.*, 1991).

4.4.5 Boride coatings

Boride coatings are generally very hard and can be used mainly in tribological applications where good abrasive wear resistance is needed. The coefficient of friction in contacts with boride coatings is generally fairly high. Tribologically interesting boride coatings are titanium diboride and iron boride. Iron boride coatings may be useful as brake materials because of their peculiar tribological property of giving very low wear and high friction at the same time.

4.4.5.1 Titanium diboride coatings

Titanium diboride (TiB_2) is a ceramic material with a hexagonal structure which has various attractive properties, such as high hardness, excellent corrosion resistance, good thermal oxidation resistance and good wear resistance. Very low wear has been obtained in sliding contacts with brittle and hard titanium diboride coatings. However, the adhesion to the substrate has often been a problem and the friction is high when sliding against steel, often in the range of 0.4 to 0.9 (Kullmer *et al.*, 2003; Podgornik *et al.*, 2006b). For thick PVD TiB_2 coatings Berger (2002) measured the following mechanical properties: hardness 50 GPa, elastic modulus 600 GPa and fracture toughness 4 $MPa \cdot m^{0.5}$.

In dry sliding contacts at a variety of temperatures up to 540°C a very low incremental wear rate was measured for 30 to 40 µm thick CVD titanium diboride coatings in sliding contact against a chromia–silica–alumina coating in experiments performed by Levy and Jee (1988). However, the coefficient of friction was fairly high, about 0.6 to 0.8, at room temperature and increased with temperature to slightly over 1 at 540°C. It was a little lower than for the other plasma sprayed thick coatings tested, Cr_3C_2–Mo, WC–Mo, TiC–Mo and Al_2O_3–TiO_2. The high coefficient of friction made the TiB_2 coating, as well as the other coatings tested, unsuitable for sliding component applications like bearings or piston ring vs cylinder wall contacts. A compound layer was formed on both the TiB_2 and the counterface and probably played a major role in the tribological behaviour of the contact.

In pin-on-disk tests with a TiB_2 coated surface sliding against itself at a speed of 0.1 m/s and a load of 2.5 N Habig (1990) found a high and constant wear rate both in air and in vacuum but a higher initial wear in vacuum. The wear volume was about six times that of TiN against TiN in similar contact conditions and the coefficient of friction was in the range of 0.4 to 0.6 in air and 0.75 in vacuum. The relative humidity of the air environment was 50%.

In scratch adhesion tests, lower critical load values have been obtained for TiB_2 coatings compared to TiN coatings (Bhansali and Kattamis, 1990; Matthes *et al.*, 1990b; Podgornik *et al.*, 2006b) which suggests that while the adhesion of the coating is reasonable, it is not of the level of TiN. The stresses and strains in the scratch test contact between the diamond stylus and a TiB_2 coated high-speed steel surface were FE modelled and simulated by Panich and Sun (2006). They also found that sputter cleaning of the substrate helps to improve the TiB_2 coating hardness and adhesion strength.

In metal forming wear tests with PVD deposited TiB_2 coatings on grey cast iron Matthes *et al.* (1990b) measured good lifetime results at the same level as TiN coatings. The wear performance of the TiB_2 coated tools was limited by cracking under high loads on the drawing edge. In abrasive wear tests Bhansali and Kattamis (1990) found that the wear resistance of TiB_2 coatings was not as good as that of TiN coatings.

4.4.5.2 Iron boride coatings

Extremely good sliding wear resistance has been obtained for CVD iron boride coatings on steel sliding against itself. Habig *et al.* (1980) have compared the tribological properties of CVD coated FeB, TiN, TiC and CrC sliding against themselves. Iron boride showed the very lowest wear rate; it was one order of magnitude less than for TiN and TiC and two orders of magnitude less than for CrC. However, the coefficient of friction was not so very low. For FeB it was 0.38 to 0.63, which is considerably higher than the very low values measured in the same conditions for TiN, 0.19, and TiC, 0.13. The coefficient of friction is also high, 0.76, when a CVD FeB coated steel specimen is slid with steel as the counterface, according to Hintermann (1981). The coefficient of friction was three times greater than for TiC and half as high as for TiN sliding against steel.

4.5 Carbon and Carbon-based Coatings

4.5.1 Diamond as a coating material

Diamond and diamond-like carbon-based coatings have attracted a rapidly increasing number of researchers since the middle of the 1980s and these layers are today the most intensively researched tribological coatings. Work on the deposition of diamond films was already under way in the 1950s but the research boom was triggered in the early 1980s when deposition rates reached the order of 1 μm/h. The commercial interest in diamond and diamond-like coatings is considerable, not only because of their great potential in tribological applications but also because of their potential as new and more efficient semiconductor and optical thin film materials. Diamond-like carbon coatings are extensively used in the computer industry. Practically all magnetic hard disks produced are covered with a multilayer film including a wear protective carbon-based coating. Other successful tribological applications are diamond-like carbon coatings on tools for cutting of non-ferrous materials and on various sliding components.

Diamond is attractive as a tribological material due to its unique material properties: high mechanical strength, high hardness, chemical inertness, excellent thermal conductivity, extremely low thermal expansion, low friction and good wear resistance.

Polycrystalline diamond and amorphous diamond-like coatings are two specific forms of carbon surface layers. Carbon occurs in them as diamond or graphite allotrope or as amorphous carbon. Diamond is a metastable phase of carbon at room temperature and pressure. Diamond has a cubic crystal structure with a fourfold sp^3 covalent bond structure (Fig. 4.48a). Graphite has a hexagonal lattice structure with sp^2-bonded two-dimensional planes interconnected by weak forces (Fig. 4.48b). A third synthetically produced structure of pure carbon are the fullerenes that may have a ball or a tube form (Fig. 4.48c). Amorphous carbon is a transitional form of carbon which has no long range order and the local structure varies depending on the growth procedure (Fig. 4.48d).

The stable phase of carbon at 300 K and 10^5 Pa is graphite. The diamond phase is stable at 300 K for pressures greater than $1.6 \cdot 10^9$ Pa, but the conversion of graphite to a diamond structure demands pressures and temperatures of much higher magnitude. The structure and properties of diamond coatings have been described by Derjaguin and Fedoseev (1975), Bachmann and Messier (1989),

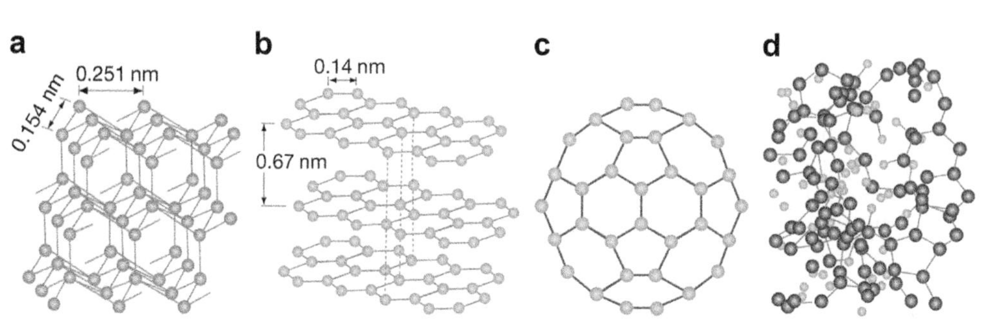

Fig. 4.48. Carbon structures in (a) diamond, (b) graphite, (c) fullerene ball and (d) amorphous diamond-like carbon.

Spear (1989), Angus (1992), Gardos (1994), Erdemir and Donnet (2001) and Miyoshi (2001). Detailed tribological studies on diamond and diamond-like carbon coatings carried out in numerous research and industrial laboratories in the 1990s has confirmed that they are inherently self-lubricating and exhibit an excellent resistance to adhesive, abrasive and erosive wear. In addition, they are resistant to corrosion and oxidation.

It is possible to deposit diamond and diamond-like coatings at low pressure by using a large variety of deposition techniques mainly based on chemical and physical vapour deposition. The most common are hot filament CVD and microwave plasma CVD for depositing diamond coatings, and radio frequency, laser or arc-assisted PVD or CVD for diamond-like coatings. Details on the different deposition techniques are found in Spitsyn *et al.* (1988), Lux and Haubner (1989), Rossi (1992) and Andersson (2004). Perhaps the most surprising deposition technique for diamond coatings is the combustion flame technique. Based on the combustion of an acetylene oxygen mixture from an oxy-acetylene torch at atmospheric pressure this simple procedure results in nucleation of diamond crystals at high deposition rates of 1 μm/min (Ravi *et al.*, 1990; Murakawa *et al.*, 1990; Alahelisten, 1994; Hollman, 1997a).

In a tribological sense diamond coatings behave very much like natural diamond but also amorphous diamond-like coatings have very similar properties. Only in certain respects, like the influence of humidity on friction, is the behaviour of diamond-like coatings more like that of graphite (Miyoshi, 1990; Erdemir *et al.*, 1991c; Feng and Field, 1991; Miyoshi *et al.*, 1992; Donnet and Erdemir, 2008). There are several different techniques for identification and characterization of diamond and diamond-like surfaces (Tsai and Bogy, 1987; Williams and Glass, 1989; Robertson, 1992a, b; Miyoshi, 2001; Ferrari, 2008). Raman spectroscopy is particularly suitable for identification of diamond structures and is commonly used (Knight and White, 1989; Huong, 1991; Blanpain *et al.*, 1992; Ferrari, 2008).

4.5.1.1 Bulk diamond

Diamond has many unique properties. The one which makes it particularly interesting from a tribological point of view is that it is the hardest known material. In addition, it is chemically inert and non-toxic, can withstand extreme temperatures, conducts heat faster than any other solid substance, is an excellent electrical insulator and has low thermal expansion. The topography of a diamond surface may vary from extremely smooth to rough with pyramidal asperities, depending on its crystalline orientation and how it has been polished. Typical values for the mechanical properties of diamond are: hardness 60 to 100 GPa H_v, elastic modulus 910 to 1250 GPa and Poisson's ratio 0.1 to 0.29 (Miyoshi, 2001).

4.5.1.2 Diamond sliding against diamond

The very favourable tribological properties of diamond were recognized and studied by Bowden and Tabor (1950). They reported coefficients of friction as low as 0.05 and extremely small surface damage. They suggested that friction arises from the low adhesion of the two diamond surfaces which are each covered with a contaminant layer across which van der Waals forces act (Bowden and Tabor, 1964). However, this theory cannot alone explain the observed frictional anisotropy (Brookes and Brookes, 1991; Gardos, 1994) and the fact that friction is reported to be unaffected by the addition of a lubricating oil on the surfaces (Samuels and Wilks, 1988; Feng and Field, 1991; Grillo and Field, 2003).

The tetrahedral bonding structure of a diamond results in a situation at the surface where there will be one vacant electron coming out from the surface that has no carbon atom to attach to, as shown in Fig. 4.49a. This is called a dangling bond orbital (Tabor, 1979). If one hydrogen atom with its own electron is put on each such carbon atom, it will bond with the dangling bond orbital to form a

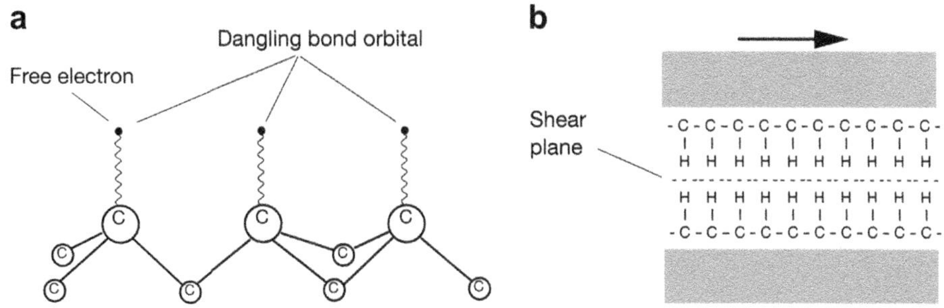

Fig. 4.49. (a) Carbon atoms at the surface can make three strong bonds leaving one electron in the dangling bond orbital pointing out from the surface. (b) Hydrogen atoms attach to a polished diamond surface which becomes hydrophobic and exhibits low friction.

two-electron covalent bond. When two such smooth surfaces with an outer layer of single hydrogen atoms slide over each other, as shown in Fig. 4.49b, shear will take place between the hydrogen atom layers. There is no chemical bonding between the surfaces, only very weak van der Waals forces, and the resistance to shear is very low (Pepper, 1982; Goddard, 1985; Koskilinna, 2007).

This mechanism would explain the extremely low friction of polished diamond surfaces and the observation that the addition of oil has no effect because there are no vacancies for additional hydrogen atoms. At high temperatures, above 850°C, the surface hydrogen leaves the surface and the dangling bonds are vacant again, resulting in high adhesion and high-friction coefficients up to a level of 0.7 (Gardos and Soriano, 1990). The surface-bonded hydrogen atoms are subject to a repulsive force which is affected by the dielectric permittivity of the environment. This repulsive force was calculated by Huu *et al.* (1995) to be 80 times stronger in vacuum than in humid air. Thus the surface hydrogen atoms become weakly bonded in vacuum and can desorb during frictional contact or heating to 300 to 400°C. A profound effect of adsorbed species in the sliding contact on friction has been shown by molecular dynamic simulations by Harrison *et al.* (1993) and Perry and Harrison (1996).

In very controlled lightly loaded sliding conditions in a liquid environment the friction of natural diamond sliding against natural diamond can reach values of the order of 0.001, which is also called superlubricity. In sliding experiments with a load of 0.5 N, sliding speed of 0.00005 m/s and a Hertzian contact pressure of 11 GPa between the two diamond surfaces, one spherical with a diameter of 50 μm and one flat, Grillo and Field (2003) measured coefficients of friction down to 0.003 in the presence of water with different pH values. They propose that the sp^2 hybridized material produced during sliding acts as a lubricant, due to the weak resistance to sliding of the graphitic basal planes. The transformation of diamond to sp^2 hybridized carbon during sliding of diamond on diamond at high pressure is consistent with theoretical work based on pseudopotential total energy calculations (Jarvis *et al.*, 1998).

The adhesion of two diamond surfaces is remarkably low even if the surfaces are thoroughly cleaned. This is partly due to a lack of strong interfacial bonds. One main influencing factor has been explained to be the presence of small surface asperities which push the surfaces apart when the joining load is removed. Even extremely small surface asperities are sufficient to reduce the adhesion to negligible values because of the high stiffness and high elastic modulus of diamond. If one of the adhesion pairs is ductile, much stronger adhesion is observed because of a larger contact area, as shown in Fig. 4.50 (Tabor, 1979).

With rough surfaces other contact mechanisms affecting friction must also be considered. One such mechanism arising from a roughness effect has been suggested by Casey and Wilks (1973). When two diamond surfaces slide over each other, work is required to move the surfaces against the normal load,

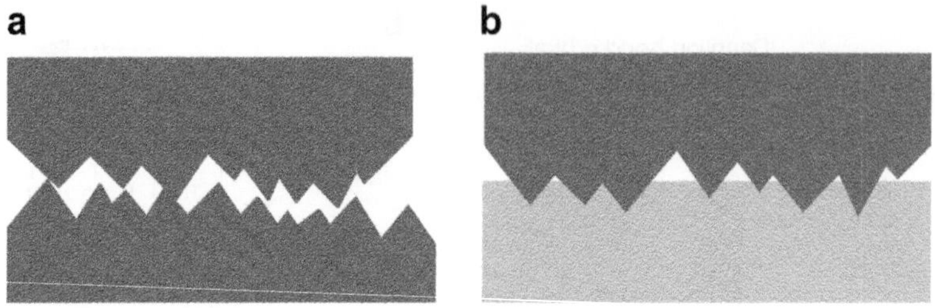

Fig. 4.50. The contact geometry for (a) a rough diamond surface on another diamond surface and (b) a rough diamond surface on steel.

as they are forced apart when asperities ride over each other. As the asperities are rough and jagged, this transference of contact may take place abruptly and irreversibly, with some of the work of separation degenerating into heat. This friction mechanism is frequently called the ratchet mechanism. The main problem with this theory is that the energy used in ascending the asperities to create the measured friction values would be larger than is reasonably expectable.

The effects of surface roughness, fracture, the ratchet mechanism and elastic losses have been analysed by Hayward and Field (1987), Samuels and Wilks (1988) and Hayward (1991). There are indications that microscopic plastic deformation occurs at diamond contacts but it is not clear how this effect should be taken into account in the friction models (Feng and Field, 1992). It has been speculated that there may be some graphitization at the contact tips.

When a rough diamond surface is sliding over another, small asperity fragments are chipped away and the resulting surface probably consists of a series of asperities whose faces are easy-cleavage planes. This is in agreement with the observation that the wear rate does not increase with temperature, and that the wear rate is very dependent on the direction of sliding (Hayward, 1991). The wear rate of diamond sliding on diamond is extremely low but after a longer period of sliding small quantities of debris are collected at the sides or the end of the wear track. The debris, with observed sizes of up to 20 μm in diameter, has been analysed by various methods which indicate that the debris material consists mainly of hydrocarbons, with a little graphite but no diamond (Hayward, 1991; Feng and Field, 1992). It is also possible that a portion of the carbon in the debris does not originate from the diamond, but from the polymerization of adsorbed hydrocarbons.

It is possible that the debris formed has a considerable role in the mechanism producing low friction with sliding diamond surfaces and that it is a consequence of some slippery debris layer or microfilm on the top of the surface. The erosion resistance of diamond has been studied by Hayward and Field (1990) and they found that it is very high, even when compared with the erosion resistance of ceramics.

Experimental tribological studies of diamond sliding on diamond at fairly moderate speeds result in the following conclusions:

1. The coefficient of friction is in the range 0.05 to 0.15 over a wide range of experimental conditions. It rarely exceeds 0.2 in air and room temperature.
2. The friction depends on crystal face and orientation. A variation in the coefficient of friction from 0.05 to 0.10 through a 360° rotation has been reported by Casey and Wilks (1973).
3. At low loads the influence of sliding speed and load is very small. At very low loads and sliding speeds in a water environment a coefficient of friction as low as 0.003 has been measured (Grillo and Field, 2003). At high loads appreciable surface damage and higher friction has been observed.
4. There is almost no effect on friction when a lubricant such as a mineral oil is introduced but there are conflicting results reported for the influence of moisture and water on friction (Feng and Field, 1991; Miyoshi *et al.*, 1992).

5. The influence of surface roughness is crucial.
6. Diamond contacts are affected by extreme environmental conditions. At temperatures above 850°C the coefficient of friction increases to a level of 0.7 and in ultra-high vacuum it is approximately 1.0 (Gardos and Soriano, 1990).

It is important to note that many of these observations were a result of testing with single-crystal diamond samples. Polycrystalline diamond films, however, often have a variety of crystallographic orientations present in the contact interface and the surface film is microscopically rough, which can be expected to change the above observations.

4.5.1.3 Diamond sliding against metals

When a diamond surface slides against a metal the surface topography of the diamond has a significant influence on friction and wear. This indicates that ploughing or abrasive effects caused by diamond surface asperities or loose wear particles have an important role in the contact mechanism in addition to adhesive effects.

A diamond surface does not undergo significant wear in sliding contact with metals except for small polished areas and fine scratches. In contacts with smooth diamond surfaces the metal wear is generally low and the coefficient of friction can be less than 0.1.

Severe wear occurs if the surface roughness of the diamond exceeds about $R_a = 0.3$ µm and high values of the coefficient of friction up to 0.7 occur. Mehan and Hayden (1981–82) claim in a detailed experimental study of diamond and metal contacts that oxidational wear with the production of hard Fe_2O_3 abrasive particles can have a dominating effect on the wear process. When they inhibited the formation of surface oxides by performing the experiments in a nitrogen atmosphere very low wear and friction was observed. The introduction of lubricating oil into the contact decreased the friction both for smooth and rough diamond surfaces.

The study by Mehan and Hayden (1981–82) shows that diamond has better tribological properties than other hard materials such as alumina, cemented tungsten carbide, silicon nitride and hard metals. Yust *et al.* (1988) have reported an extremely low coefficient of friction of only 0.02, with diamond sliding on ion-implanted titanium diboride (TiB_2).

4.5.2 Diamond coatings

Chemical vapour deposition offers an elegant way to produce thin diamond layers on surfaces. It includes the decomposition of hydrocarbon, such as methane, with some form of activation technique, in the following way:

$$CH_4 \rightarrow \text{Activation} \rightarrow C\ (\text{diamond}) + 2\ H_2$$

The activation techniques may be high temperature and/or plasma, both of which require energy input. The carbon species in the hydrocarbon molecule must be activated, since at low pressure graphite is thermodynamically stable. Without activation only graphic material would be formed.

Depending on the predominant electron state of carbon atoms, the degree of crystallinity and other factors, three kinds of carbon films are identified: diamond, diamond-like and graphite-like carbon films. Diamond coatings have a polycrystalline structure which predominantly consists of the crystalline tetrahedral sp^3 bondings and only a small number of sp^2 and sp bonds at the boundaries of diamond crystallites. As the amount of sp^2 and sp bonds increases, the structure changes to become more amorphous (Spitsyn *et al.*, 1988; Hirvonen, 1991; Gardos, 1994; Erdemir, 2002). Crystalline diamond coatings have an extreme hardness of about 100 GPa and a Young's modulus over 1000 GPa.

A detailed review of the structures and mechanical properties of diamond and diamond coatings has been written by Miyoshi (2001). Diamond films can be grown on some ceramic and metallic substrates such as Si and W and their carbides by CVD methods. Typical substrate temperatures for diamond film deposition are 600 to 950°C. For deposition of diamond below 600°C the growth rate and diamond quality is substantially reduced. Because of the high thermal conductivity and electrical insulation, diamond coatings are used as a heat sink in X-ray windows, circuit packing and high-power electronic devices (Erdemir and Donnet, 2005). Carbon coatings, including basic knowledge, film classification and their properties, are described in the standard VDI-Richtlinien (2005).

4.5.2.1 Stresses in the coating

The tribological behaviour of a diamond coating is influenced considerably by the residual stresses in the coating. They consist of two main components, the thermal stresses and the intrinsic growth stresses. The thermal stresses can be calculated from the differences in thermal expansion between the substrate and the coating. The thermal expansion coefficient of diamond is in the range of 1.5 to $4.8 \cdot 10^{-6}\,K^{-1}$, which is low compared to many ceramics and very low compared to most metals. Thus compressive thermal stresses will be induced when, for example, aluminium oxide, silicon carbide or zirconium oxide are used as substrates and the stresses will increase in magnitude as the substrate temperature increases during deposition. The expansion coefficient of WC is of the same level as diamond, which makes it a suitable substrate material because the diamond film can be deposited in a virtually stress-free condition (Anthony *et al.*, 1991; Lindbauer *et al.*, 1992; Alahelisten, 1994; Hollman, 1997).

For most metals and metal alloys used as substrates the thermal expansion coefficient is high. This can result in coating spallation because of the large residual compressive stresses if the deposition temperature is high. Conversely, it has been observed that the intrinsic stresses for many diamond films are tensile, which will counteract the thermal stresses (Guo and Alam, 1991; Kamiya *et al.*, 1999). The tensile stresses measured by Kamiya *et al.* (1999) for diamond coatings on Si substrate decreased slowly from 0.2 GPa at the top surface 4 μm from the coating/substrate interface to 0.1 GPa at 0.02 μm from the interface where they drastically changed to highly compressive stresses of more than −0.6 GPa even closer to the interface. The origin of the tensile intrinsic stresses is often explained by the attractive interatomic forces acting across grain boundaries or micropore gaps. This will induce an elastic strain in the anchored grains and is referred to as grain boundary relaxation.

Based on strain measurements by Raman spectroscopy and FEM calculations Gunnars and Alahelisten (1996) estimated the stresses in diamond coatings on flat geometries and over substrate edges. Far from the edge they found compressive stresses of −2.3 GPa in the coating and low tensile stresses of 0.1 GPa in the cemented carbide substrate. The tensile stresses in a 5 μm thick diamond coating deposited over a substrate edge with 5 μm radius had tensile normal stresses of 0.80 GPa, shear stresses of 0.46 GPa and tangential compressive stresses of −0.56 GPa. With a larger edge radius the normal tensile and shear stresses decreased while the tangential compressive stresses increased to values up to 2 GPa.

4.5.2.2 Environmental effects

At high temperatures, around 600°C, natural diamond crystals begin to oxidize. The temperature at which the oxidation begins, as well as the oxidation rate, depends on factors such as the morphology of the film and the nature and the amount of defects and impurities in the carbon. Usually, diamond coatings have a significant amount of crystal defects such as twins and dislocations. Microwave and hot filament CVD diamond coatings often contain hydrogen and other impurities affecting the oxidation process. According to measurements by Tankala *et al.* (1990) oxidation of hot filament CVD diamond films in air is noticeable at about 250°C and it becomes pronounced at 800°C. The oxidation of microwave CVD diamond films starts at a lower temperature. A hot filament CVD film was somewhat more oxidation resistant than a microwave CVD film. The oxidation behaviour of diamond films has been observed to be kinetically similar to that of natural diamond. One conclusion is thus that diamond films are not suitable for extended exposure to an air atmosphere at temperatures above 600°C. However, in an inert environment they are stable at much higher temperatures and remain optically transmissive (Johnson *et al.*, 1990).

The chemistry of the test environment may play a significant role in the friction and wear performance of diamond coatings. The very low friction and wear performance with diamond films are achieved only in air and with relatively dry inert gases and at low ambient temperatures. In vacuum and at high temperatures the friction and wear are higher. The difference in film structure due to the different deposition techniques results in different environmental influences on the tribological behaviour.

In tribological tests carried out in humid air, dry nitrogen, water and ultra-high vacuum, Miyoshi *et al.* (1992) and Miyoshi (1998) found that, regardless of the environment, nitrogen ion-implanted CVD diamond films had low steady-state coefficients of friction, less than 0.1, and low wear, less than $10^{-6}\,m^3/Nm$ when sliding against a CVD diamond pin at a 0.49 N load, 2 GPa Hertzian contact pressure and 0.031 to 0.107 m/s speed. These films can be used as wear-resistant, self-lubricating coatings regardless of the environment. On the other hand, as-deposited, fine grain CVD diamond films, polished coarse grain CVD diamond films and polished and then fluorinated, coarse grain CVD diamond films performed similarly tribologically well in humid air, dry nitrogen and water but had a high coefficient of friction and a high wear in ultra-high vacuum. Also interesting was that the polished, coarse grain CVD diamond film had an extremely low wear rate, less than $0.0001 \cdot 10^{-6}\,m^3/Nm$, in water. This correlates with the experiences reported by Erdemir and Donnet (2005) from tests carried out in water and oil environments. The friction and wear performance was reported to be exceptional with smooth diamond films under oil lubricated sliding. After some initial running-in wear there was essentially no wear at all.

The variation of surface conditions and friction with the environment and the material in sliding with very smooth surfaces of diamond against diamond coatings, where the ploughing component can be neglected, has been illustrated by Miyoshi (1998). With adsorbed contaminants, such as water, carbon oxides or hydrocarbons, the hard diamond film provided a small contact area A (Fig. 4.51a), minimized penetration of the diamond pin and either reduced the shear strength S or maintained the low surface energy γ of the contact area. When the contaminant surface layers were removed by repeatedly sliding the pin in the same track in ultra-high vacuum the result was stronger interfacial adhesion between the diamond pin and the diamond film and accordingly an increase in the coefficient of friction, as shown in Fig. 4.51b. Removing the contaminant layer from the contact area enhanced adhesion and increased either the shear strength or the surface energy of the contact area.

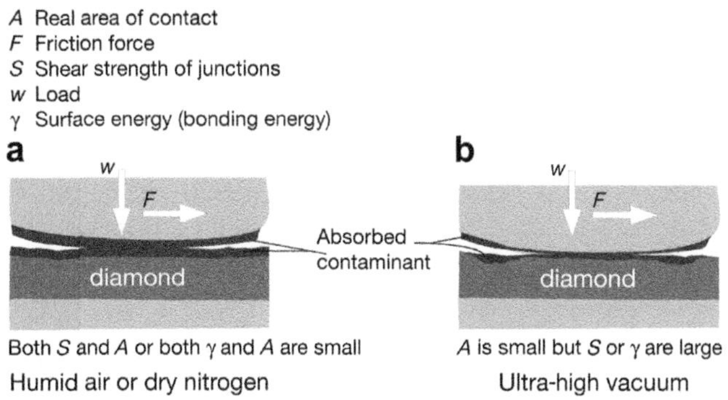

Fig. 4.51. Contact mechanisms influencing friction in (a) humid air or dry nitrogen and (b) ultra-high vacuum, when diamond is sliding on diamond coatings (after Miyoshi, 1998).

4.5.2.3 *Surface topography*

As most diamond films are both rough, because of their facetted structure, and very hard, it is clear that surface roughness and topography have a crucial effect on the tribological behaviour in sliding contacts. This has been experimentally confirmed for diamond coatings in contact with single-crystal diamond, ceramics and steel. The surface topography of a thin diamond film produced by CVD methods can vary very much. Typical topographies are pyramidal, microcrystal and plate-like (Blau *et al.*, 1990) as shown in Fig 4.52a, b and c.

Rougher surfaces and pyramidal topographies produce higher friction and more wear because of abrasive cutting and ploughing effects. On smoother microcrystalline surfaces, on the other hand, effects of material transfer and wear debris in the contact will be more pronounced. Smooth plate-like films are tribologically favourable because of fewer abrasive effects. Hard wear particles can be embedded in the spaces between the plates and thus their abrasive actions are reduced (Blau *et al.*, 1990; Gardos, 1998). In contacts between diamond coatings and softer materials, rougher surfaces will cause more ploughing and cutting effects.

Another mechanism to consider is the running-in wear. In the initial stages of contact there will be quite considerable numbers of asperity collisions because of the roughness of the diamond surface, producing wear debris. After a running-in period the surfaces become smoother and the possible transfer layers are stabilized, generally resulting in lower wear and friction. The development of the deposition techniques opens new possibilities to control the surface topography of diamond films. Both extremely smooth diamond surfaces as well as textured surfaces with, e.g., well-controlled protruding pyramidal abrasive tips have been designed and manufactured for special purposes (Hollman, 1997; Gåhlin *et al.*, 1999; Pettersson and Jacobson, 2006).

4.5.2.4 *Surface graphitization*

One proposed explanation for the extremely good tribological properties of diamond films is the formation of a thin graphite film in the actual contact areas (Jahanmir *et al.*, 1989; Holmberg and Matthews, 1994; Gardos, 1994; Grillo and Field, 2003; Dienwiebel and Frenken, 2007; Miura and Sasaki, 2007), as indicated in Fig. 4.52. As diamond is a thermodynamically unstable phase of carbon, at elevated temperatures of 600 to 1500°C, depending on the environmental conditions, it can transform to graphite. These high temperatures may occur as very short flash temperatures in the asperity collisions. On the other hand, there have been extensive surface analyses carried out by different techniques on worn diamond coated surfaces, but very little clear evidence of a graphite layer on top of the coating has been found. It seems likely that there must be some low shear strength microlayer on the top of the diamond coating to explain the extremely low frictional behaviour but it may also be some hydrocarbon polymer-like slippery film as suggested by Gardos (1994).

The explanation that the extremely low friction and wear properties of diamond coatings are caused by the transformation of diamond to graphitic carbon and this sp^2 hybridized material produced during sliding acts as a lubricant, due to its weak resistance to sliding of the graphitic basal planes, is supported by Grillo and Field (2003). They carried out a theoretical analysis based on pseudopotential total energy calculations and proposed a model where the dependence on crystal direction and orientation, in terms of strains induced at the interface during sliding, is included.

4.5.2.5 *Abrasive wear resistance*

It is well known that harder materials generally have better abrasive wear resistance so it is no surprise that coatings of diamond exhibit excellent resistance to abrasive wear (Alahelisten, 1994; Hollman, 1997; Andersson, 2004; Bose and Wood, 2005). CVD diamond coatings have been found to have better abrasive wear resistance than diamond-like amorphous carbon films (a-C:H) and metal containing carbon films (metal-C:H) by Klages (1990).

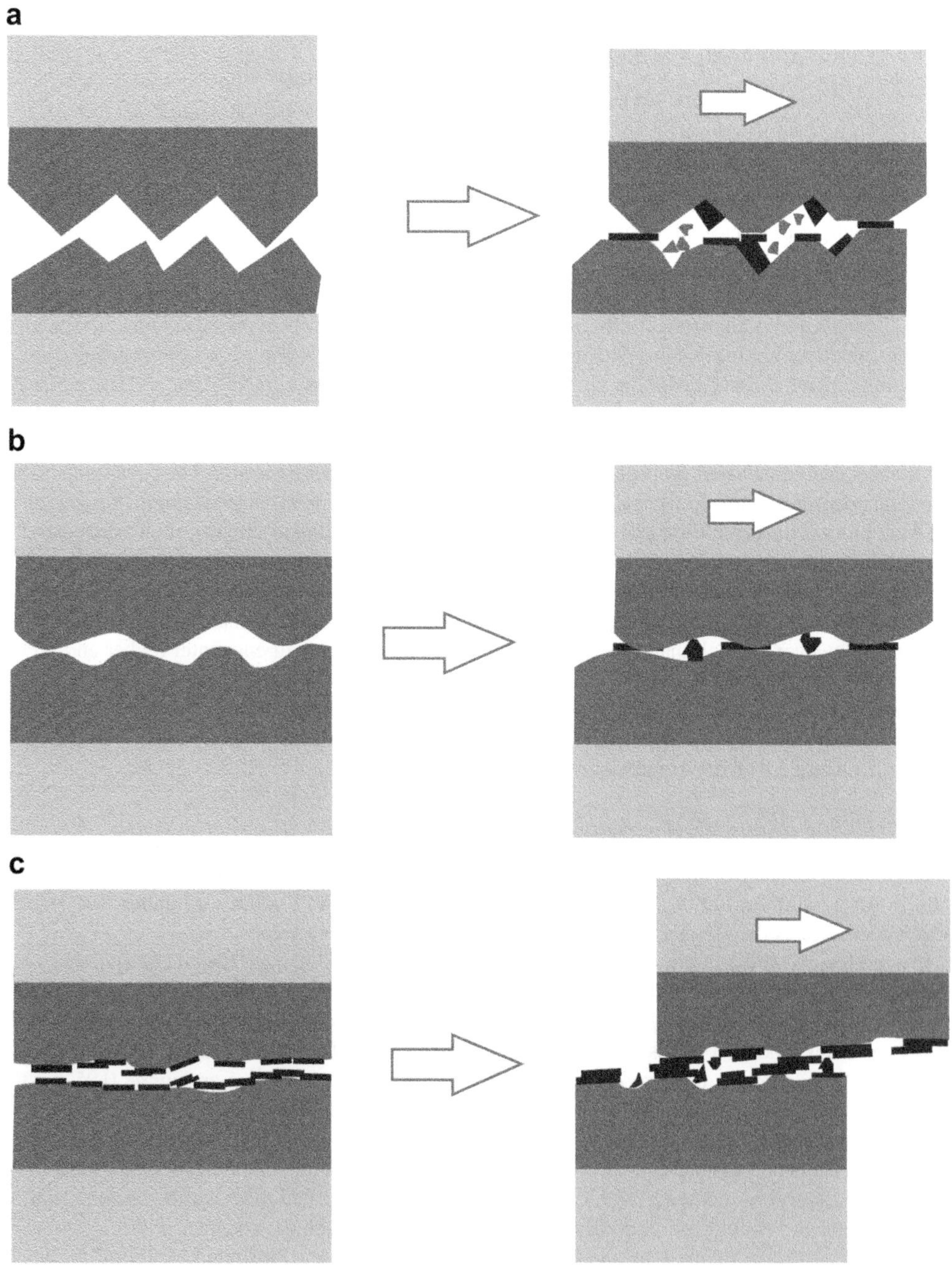

Fig. 4.52. The contact mechanisms for three typical diamond coating topographies: (a) pyramidal, (b) microcrystal and (c) plate-like, when sliding on themselves.

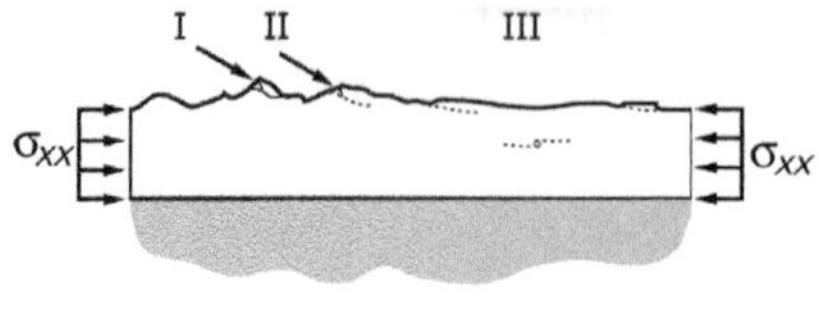

Fig. 4.53. Model for the wear process of brittle diamond coatings subjected to high compressive biaxial stresses resulting in surface smoothing (after Gunnars and Alahelisten, 1996).

When a rough diamond coated surface is sliding against another hard surface, smoothing can take place. Gunnars and Alahelisten (1996) observed this process with CVD diamond coating on cemented carbide substrates sliding against SiC abrasive paper. The wear process started by fragmentation of protruding diamond asperities. This proceeded until the surface was relatively flat and thereafter the wear rate diminished to a very low level. By using diamond grains as abrasives the same process occurred, although the final coating surface was smoother.

Gunnars and Alahelisten propose the following model for the wear process. The stress state in the diamond coating is assumed to be dominated by high biaxial compressive stress. In the beginning of the wear process the protruding grains will take up the externally applied stresses, shown as region I in Fig. 4.53. The result is that high local stress concentrations and fragmentation occur. The residual stresses are low in the protruding grains since the geometry permits relaxation. A crack nucleated in a protruding grain will propagate preferentially along easy cleavage planes, which are (111) for diamond. However, a crack that propagates deeper into the coating will encounter higher and higher compressive residual stresses which will deflect the crack into a path, irrespective of cleavage planes, region II in Fig. 4.53. The large residual stresses will prevent cracks initiated at the surface from propagating towards the interface and thus promote crack paths parallel to the surface. The result is a smooth surface free from asperities, region III in Fig. 4.53.

4.5.2.6 Sliding against single-crystal diamond

The coefficient of friction for diamond sliding on diamond coatings was measured to be in the range of 0.03 to 0.5 by Hayward *et al.* (1992). They found a crucial effect of surface roughness on friction, which decreases as the coating and slider are smoothed by abrasive wear.

In experiments with 1 μm thick CVD diamond films coated on single-crystal silicon substrates and sliding against diamond, the initial coefficients of friction, 0.1 to 0.5, were generally higher than those measured when natural diamond slides on natural diamond, 0.05 to 0.15 (Hayward and Singer, 1990). The initial wear produced debris that caused smooth patches on the coating and resulted in smoother surfaces. The coefficient of friction decreased at the same time to values as low as 0.05 for the smoothest surface after about 100 cycles, as shown in Fig. 4.54a. The diamond coatings were sufficiently hard to wear the single-crystal diamond sliders.

The low friction coefficients achieved after smoothing of the coatings are similar to those of polished single-crystal diamond. When the surfaces were first polished there was no region of initial high wear and the coefficient of friction was low from the beginning, as shown in Fig. 4.54b. The influence of surface roughness on the coefficient of friction is well illustrated by Fig. 4.55 where experimental data from friction measurements with six different coatings are shown.

The adhesion component of friction in a sliding diamond on diamond coating contact was studied by Feng and Field (1991). They added water to the contact during the experiment and found that the coefficient of friction decreased from a level of 0.05 to less than 0.02. This shows that friction has, in addition to the ploughing component demonstrated by the roughness effect, an adhesion component also. This changes when water between the surfaces reduces the area for adhesive bonding. No hydrodynamic effect would be expected because of the very low sliding speed of 0.0002 m/s. These

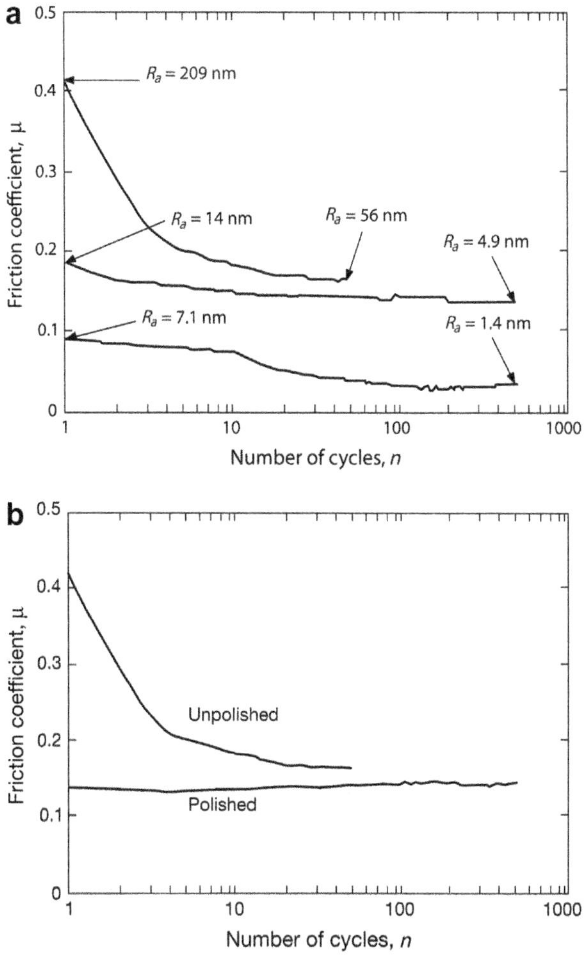

Fig. 4.54. The variation of the coefficient of friction for (a) three diamond coatings with different surface roughness and (b) an unpolished and a polished diamond coating, sliding against diamond (after Hayward *et al.*, 1992).

results correlate well with the friction measurements of Miyoshi (1998) carried out in air, nitrogen and water, discussed earlier.

4.5.2.7 Sliding against diamond coating

When two smooth diamond coated surfaces slide against each other the friction coefficient is normally very low, in the range 0.05 to 0.15, and the wear rates are equally low, about $0.1 \cdot 10^{-6}$ mm^3/Nm in vacuum and a little higher, $1 \cdot 10^{-6}$ mm^3/Nm, in air (Gardos and Soriano, 1990; Mohrbacher *et al.*, 1993a, b; Gardos, 1994; Krauss *et al.*, 2001). Even lower coefficients of friction down to 0.03 and a wear rate of only $0.01 \cdot 10^{-6}$ mm^3/Nm have been measured by Abreu *et al.* (2005) for CVD diamond coated silicon nitride substrates in self-mated reciprocating sliding in air at loads of 10 to 105 N corresponding to a Hertzian pressure of 5 to 11 GPa. A self-polishing mechanism resulted in a surface roughness of only approximately 8 nm after 16 hours of sliding and this may explain the very

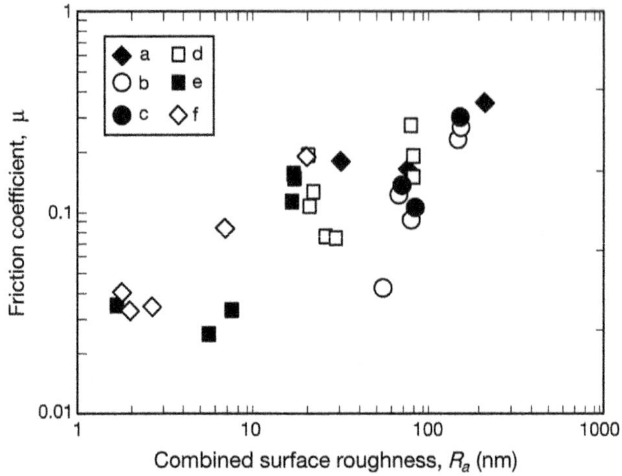

Fig. 4.55. Coefficients of friction as a function of combined surface roughness for a series of experiments with diamond sliding on six different diamond coatings in a reciprocating test with a speed of 0.0003 m/s and a load of 2 N (data from Hayward *et al.*, 1992).

low friction and wear. However, no signs of surface graphitic phase transformation were observed by micro-Raman surface characterization.

Friction and wear tests of 1.5 μm thick dc PACVD diamond coatings on α-SiC substrates sliding against each other, both in vacuum and in air at temperatures ranging from room temperature up to 850°C and back to room temperature again, were carried out by Gardos and Soriano (1990). They confirmed Goddard's (1985) hypothesis that the adhesion and friction of the sliding couples is mainly controlled by the surface chemistry of the exposed crystallites. Dangling bonds are formed when the contact is heated, due to desorption of hydrogen which results in high friction up to a level of 0.8. When the contact is cooled in a relatively hydrogen-rich atmosphere, hydrogen atoms are again attached to the dangling bonds which decrease the coefficient of friction back down to about 0.1.

At high temperatures the coefficient of friction is substantially lower in air than in vacuum. This reduction is most probably caused by the generation of oxidation products combined with the temperature and shear-induced phase transformation of diamond to graphite.

4.5.2.8 Sliding against ceramics

There is a considerable similarity in the friction and wear behaviour when different ceramics slide against diamond coatings. This is remarkable because the tribological behaviour of ceramics differs considerably from one to another when they slide against steel or against themselves. When a ceramic slides against a diamond coating the initial coefficient of friction is normally high, in the range of 0.3 to 0.7, depending on the surface roughness. During sliding, smoothing and polishing take place and microfilms and transfer layers are built up. The coefficient of friction drops to values of 0.02 to 0.15 and wear is extremely low. This kind of behaviour, which has been reported for sapphire, corundum Al_2O_3, Si_3N_4, SiC and α-SiC, is illustrated by Fig. 4.56 and will be briefly discussed below.

For diamond coatings sliding on sapphire the initial friction and wear is high and decreases after running-in. Hayward and Singer (1990) measured an initial coefficient of friction of 0.6 which decreased to 0.1 after 500 cycles. No wear of the diamond coating was observed. There was considerable wear of the sapphire which was influenced by the surface roughness of the diamond coating so that smooth surfaces gave low wear. Blau *et al.* (1990) have shown how the initial friction is dependent on

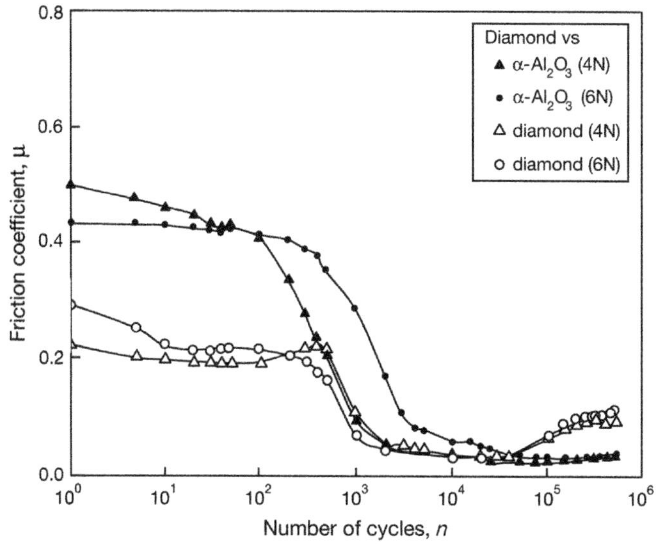

Fig. 4.56. The coefficient of friction as a function of number of cycles in reciprocating sliding for a 1 μm thick CVD diamond coating sliding against corundum and against CVD diamond at a frequency of 10 Hz and a stroke length of 0.1 mm (data from Morbacher *et al.*, 1993a).

the surface topography. The initial coefficient of friction was approximately 0.4 for pyramidal diamond surfaces but only 0.25 for plate-like surfaces. Blau used CVD diamond films coated on thin silicon wafers and sapphire sliders.

In a sliding contact between silicon carbide and silicon carbide coated with CVD diamond films with thicknesses in the range from 2.6 to 4.3 μm, Jahanmir *et al.* (1989) measured a low coefficient of friction of 0.08 and low wear rates. A large quantity of wear particles less than 1 μm in diameter was observed. The wear scar on the SiC surface contained tribochemical reaction products of silicon oxide. The coefficient of friction was reduced by almost an order of magnitude compared to the same SiC–SiC contact without the diamond film, and the wear was reduced by four orders of magnitude.

For CVD diamond coated α-SiC samples sliding against α-SiC Gardos and co-worker (Gardos and Ravi, 1989; Gardos, 1994, 1998) report equally low friction and low wear especially with a pure diamond coating. For a more sp^2 contaminated diamond coating higher wear of the coating and also of the counterface was observed, due to increased self-generation of fine wear debris trapped between the frictional surfaces. They conclude that tensile cracking of the diamond film under high friction conditions and high loads appears to be the initiating step in the wear mechanism of the diamond film, under tangential shear conditions. The higher wear of the more contaminated diamond coating may be attributed to its composite-like structure, in that the grain and columnar growth boundaries within the film are the most contaminated and also the weakest regions.

In reciprocating sliding experiments with microwave CVD diamond coatings sliding against an Si_3N_4 slider Miyoshi *et al.* (1992) observed high initial friction of 0.6 followed by ploughing and cutting actions of the higher facet tips of the diamond grains and a drop in the coefficient of friction to a level of 0.1 in humid air and to 0.02 in dry nitrogen after a total of 30,000 passes. The wear scars produced were very smooth and had an extremely fine polished appearance. The wear rate was $0.05 \cdot 10^{-6}$ mm^3/Nm in humid air and $0.15 \cdot 10^{-6}$ mm^3/Nm in dry nitrogen. Moisture increased the coefficient of friction but decreased wear. The increase in friction was attributed to the silicon oxide film produced on the surfaces of the Si_3N_4 pins in humid air.

Extremely good wear resistance has been observed for a corundum ball in reciprocating sliding against a diamond coating deposited on cemented carbide by the hot filament technique (Blanpain *et al.*, 1992, 1993). Even after five million cycles no measurable wear of the diamond coating was observed and the friction coefficient had dropped to a level below 0.05. Zeiler *et al.* (2000) also measured good tribological performance of a diamond coating sliding against alumina in a reciprocating contact but the friction was higher, about 0.13 and the wear also relatively high.

4.5.2.9 Sliding against metals

As for bulk diamond, abrasive cutting and ploughing effects are more dominant when a hard diamond coating slides on softer materials like metals. At the initial stage of the contact Blau *et al.* (1990) have shown tribological topography effects. They observed microcutting of a 52100 steel surface by sharp diamond points and found that steel debris tended to accumulate in pockets in the surface on the pyramidal and plate-like films. On the microcrystalline film, evidence for transfer of steel to the diamond surface was obtained. The initial coefficient of friction for the pyramidal topography was approximately 0.5, for the microcrystalline topography 0.3 and for the plate-like topography 0.2.

Diamond coatings deposited by the hot flame method may have very different surface topographies that influence friction in sliding. Hollman *et al.* (1994) tested 5 μm thick diamond coatings with a surface roughness of $R_a = 0.53$ μm in sliding against a flat tool steel surface at a speed of 0.062 m/s and a load increasing to 1000 N and measured coefficients of friction in the rage of 0.06 to 0.12. When Hollman (1997) tested close to atomically smooth hot flame-deposited diamond coatings of a surface roughness of only $R_a = 0.006$ to 0.030 μm with a ball-on-disc device against different materials at a load of 2 N and a speed of 0.1 m/s, the friction was very low, in the range of 0.055 to 0.095 for cemented carbide, ball bearing steel and austenitic steel countersurfaces while it was considerably higher for Ti and Al counter surfaces, as shown in Fig. 4.57. The diamond coated ball radius was 6 mm and the countersurfaces were flat. The initial friction was typically about 0.3 but dropped to the low value after some 5 to 10 m of sliding. Water slightly reduced the friction but the influence of both water and oil was only small. A very low wear of the diamond surfaces was detected, about 100 times lower than for steel against steel in lubricated sliding, and the countersurfaces displayed practically no wear at all.

4.5.2.10 Cutting tool contacts

A variety of diamond coated tools are today offered by commercial coaters at a reasonable cost. The advantage of diamond coatings in machining are the extreme hardness, good fatigue strength, high

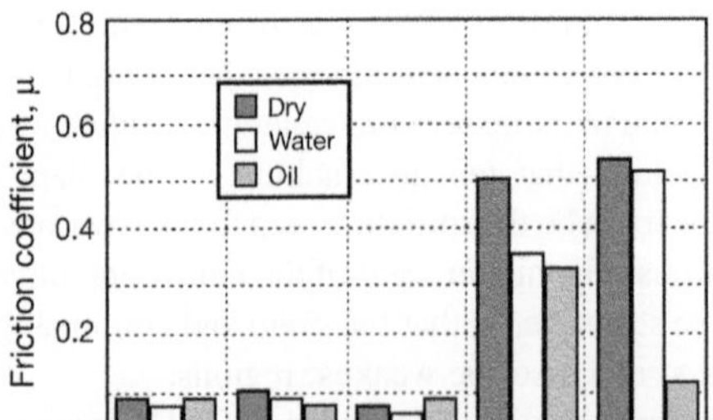

Fig. 4.57. Friction coefficients for very smooth hot flame-deposited diamond surfaces with a roughness of only $R_a = 0.03$ μm sliding at 2 N load and 0.01 m/s speed against different counter material in various environments. The materials are cemented carbide (CC), ball bearing steel (BBS), stainless steel (SS), titanium (Ti) and aluminium (Al) (data from Hollman, 1997).

chemical inertness, high thermal conductivity, low friction, good resistance to abrasion, erosion and corrosion. CVD-deposited diamond coatings of high quality are successfully deposited on cemented carbide tool inserts. Even if other substrate materials have been tried there has been no large-scale commercial utilization. Problems in the commercial manufacturing of diamond coatings are related to high deposition temperatures, poor adhesion, oxidation at high temperatures and high fabrication costs (Erdemir and Donnet, 2005).

Diamond coatings are mainly used as cutting tools for non-ferrous alloys, such as aluminium, copper, magnesium and lead alloys; metal-matrix composites; non-metals, such as plastics, epoxy resins, fibre-reinforced polymers and other composites; semi-sintered ceramics, cold-pressed cemented carbides, graphite, silicon, hard rubbers and wood. Diamond coatings are more problematic to use for cutting ferrous metals because of the high chemical solubility of carbon in ferrous materials at elevated temperatures, making diamond unsuitable for machining steels, cast iron and superalloys. The very high temperatures, around 1000°C, generated at the tool/chip interface, result in extremely rapid wear of the diamond tool by chemical dissolution (Söderberg *et al.*, 1991; Lux and Haubner, 1992).

Another limitation for using diamond coatings more extensively on cutting tools has been the difficulty in achieving good adhesion between the coating and the substrate. Methods to remove the cobalt from the surface of cemented carbide substrates (Anon., 1990) or to remove carbon (Saijo *et al.*, 1990) have been suggested to solve this problem.

Söderberg *et al.* (1991) report that they have largely overcome the adhesion problem when machining Al–Si alloys by using silicon nitride or SiAlON substrates. In their experiments CVD diamond films exhibited superior flank wear resistance as compared to conventional PCD tools. The end of tool life was reached when the coating was worn through due to accelerated wear of the underlying substrates. Hence, the performance of CVD diamond coated tools is determined by the flank wear resistance and the thickness of the coating.

A considerable decrease in flank wear and an increase in lifetime have been achieved when using CVD diamond coated cemented carbide, tungsten carbide and SiAlON tool inserts in cutting Al–Si and Al alloys (Shibuki *et al.*, 1988; McCune *et al.*, 1989; Saijo *et al.*, 1990; Takatsu *et al.*, 1990; Kikuchi *et al.*, 1991; Yazu and Nakai, 1991; Alahelisten, 1994; Schmitt and Paulmier, 2004). The film thicknesses were in the range 5 to 10 μm. Similar wear performance has also been observed when cutting hard carbon with CVD diamond coated tungsten carbide inserts (Saijo *et al.*, 1990).

4.5.2.11 Erosive wear resistance

The erosion resistance of diamond coated components is excellent when optimal surface parameters are achieved. The thickness of the coating is one crucial parameter. A problem with a diamond coated component is that when its erosion resistance is overcome the coating may crack and quickly flake off, resulting in catastrophic failure.

Solid particle erosion testing of 10 μm thick diamond coatings deposited by microwave plasma CVD on silicon, tungsten and Ti–6Al–4V substrates has shown excellent erosion resistance both with 30° and 90° angles of particle attack and velocities of 4 and 15.5 m/s (Kral *et al.*, 1993), as shown in Fig. 4.58. The diamond coating increased the erosion resistance considerably for all coating–substrate combinations and the best erosion resistance was measured with silicon and tungsten substrates. Diamond films on silicon substrates eroded by a gradual chipping or cleavage of individual grains until the coating was penetrated at a single location. Once a coating had been penetrated sudden catastrophic failure occurred.

The coating thickness was the crucial factor when Wheeler and Wood (1999) tested the erosion resistance of 10 to 47 μm thick CVD-deposited diamond coatings deposited on W and SiC substrates in water–sand slurry and high-velocity air–sand rigs with sand sizes of 135 to 235 μm diameter, in the velocity range of 16 to 268 m/s and at 90° impact angle. Coating thicknesses of less than 20 μm were

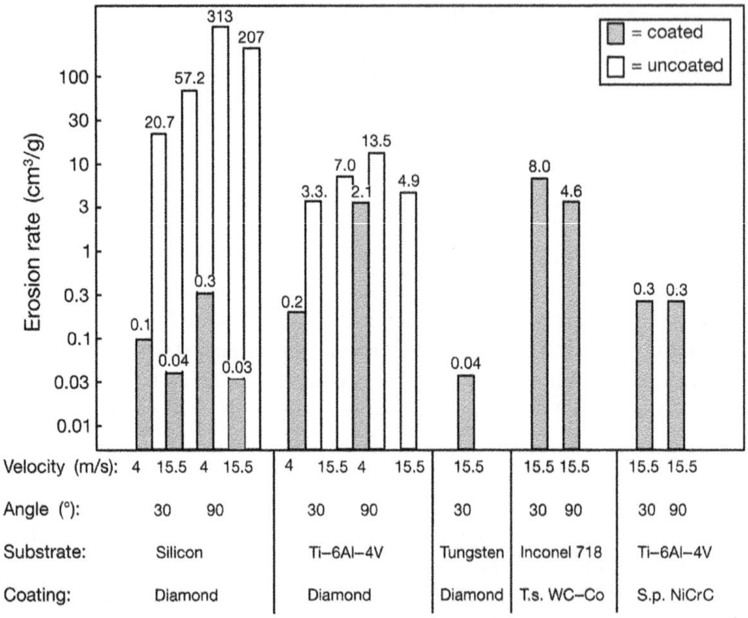

Fig. 4.58. Steady-state erosion rates for diamond coated and uncoated silicon, Ti-6Al-4V and tungsten, and as reference coatings thermally sprayed (ts) WC–Co on Inconel 718 and sputtered (sp) NiCrC on Ti-6Al-4V, for different angles of particle attack and velocities (data from Kral *et al.*, 1993).

insufficient to withstand erosion and failed catastrophically. However, in high-velocity erosion the erosion rate of 39 to 47 μm thick diamond coatings on tungsten was six times lower than for cemented tungsten carbide surfaces.

Good erosive wear protection and a clear correlation with coating thickness was measured for diamond coatings deposited on cemented carbide substrates by the hot flame technique in erosion tests with SiC erosive particles of 121 to 423 μm diameter sizes, at velocities of 11 to 87 m/s and with a 90° impact angle (Alahelisten *et al.*, 1993; Hollman *et al.*, 1994). A coating thickness of 6 μm was needed for a wear protection that outperformed the erosion resistance of the reference materials that were cemented carbide, TiN/TiC and Al_2O_3/TiC coated cemented carbide samples. The product of the erodent mass and the velocity, called the impulse, needed for coating failure increased almost linearly with coating thickness, as shown in Fig. 4.59.

4.5.2.12 Scales of diamond coating tribological mechanisms

From the above it can be seen that diamond or diamond coatings sliding against themselves or against various materials may have extremely low friction down to the range of superlubricity, but the friction may also be very high in other conditions. The wear may be zero or it may also be very high. In some conditions there are environmental effects and in others they are negligible. In some conditions the surface roughness is crucial and in others it can be neglected. How can all this be understood?

We believe that the key to an overall understanding of the tribological behaviour of diamond and diamond coatings lies in the scaling effects. In the literature some experiments and theories relate to contact conditions representative for macroscale contacts, others for microscale contacts and some others for nanoscale contacts. This is illustrated in Fig. 4.60.

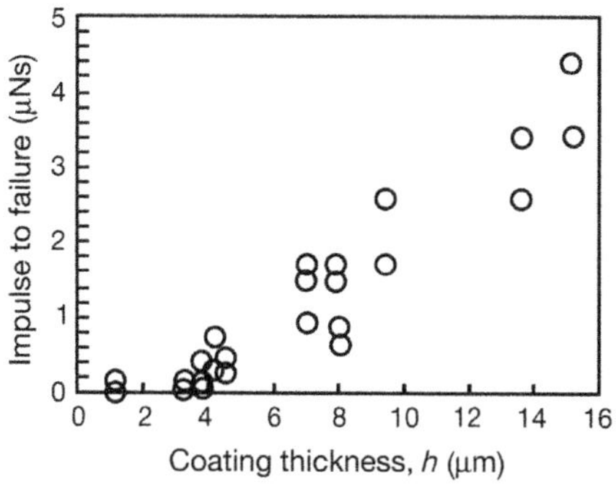

Fig. 4.59. Impulse to failure vs coating thickness of hot flame-deposited diamond coatings on cemented carbide substrates in erosion tests with 90° impact angle. The impulse to failure is the product of the erodent mass and the impact velocity (data from Hollman *et al.*, 1994).

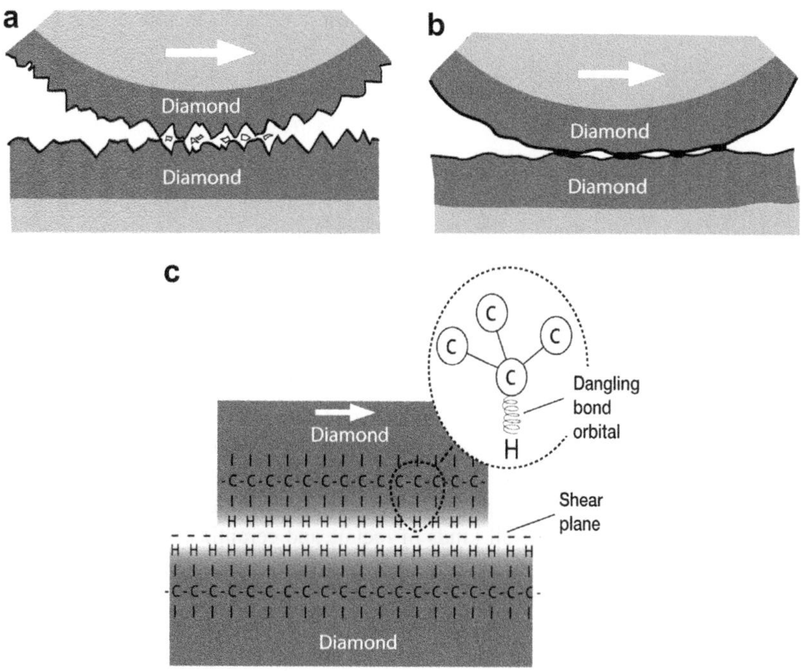

Fig. 4.60. Friction and wear mechanisms of a diamond coated surface sliding against another diamond coated surface at (a) macroscale, (b) microscale and (c) nanoscale.

At the *macroscale* and especially with rougher diamond surfaces it is obvious that the surface roughness effects, including asperity interlocking, asperity breaking, asperity ploughing and similar effects from hard wear debris between the surfaces, are dominating both the friction and wear, as shown in Fig. 4.60a. The surface roughness is typically in the range of $R_a = 0.1$ to 1 μm, the coefficient of friction of the order of 0.1 to 0.7 and the wear in the range of 0.1 to $100 \cdot 10^{-6}$ mm^3/Nm. The environmental conditions are typically air. With smoother surfaces, high temperatures above

600°C or high vacuum may also result in similarly high friction and wear. Then it is not due to surface topography effects but to increased adhesion between the two surfaces because the low shear sliding mechanism has been disturbed and is inhibited from functioning.

At the *microscale*, with surface topographies typically in the range of $R_a = 10$ to 100 nm, the dominating tribological contact mechanism is graphitization. The shear takes place within the basal planes of the sp^2 hybridized graphitic carbon that has been formed by transformation from sp^3 at the high flash temperatures and high pressures between colliding surface asperities in sliding, as shown in Fig. 4.60b. The friction is very low, of the order of 0.01 to 0.1, and the wear is in the range of 0.0001 to $0.1 \cdot 10^{-6}\,mm^3/Nm$. The environmental conditions are typically air but similar results are obtained also in water or oil environments.

At the *nanoscale* with atomically smooth surfaces of the range of $R_a = 1$ to 30 nm the conditions of superlubricity with coefficients of friction down to 0.003 can be achieved. The contact mechanism which explains this is that the shear now takes place between two perfectly flat surfaces of single hydrogen atoms attached to the dangling bond orbitals from the crystalline diamond carbon structure, as shown in Fig. 4.60c. There are thus no chemical bonds in the shear plane but only weak van der Waals forces to be broken. This mechanism is inhibited by high temperatures above 600°C and high vacuum. The presence of some gas or liquid may have disturbing effects, while water or other oxygen containing liquid environments with optimized pH values may result in superlubricity.

4.5.3 Diamond-like carbon coatings

Diamond-like carbon (DLC) is the name commonly used for hard carbon coatings which have similar mechanical, optical, electrical and chemical properties to natural diamond, but which do not have a dominant crystalline lattice structure. They are amorphous and consist of a mixture of sp^3 and sp^2 carbon structures with sp^2-bonded graphite-like clusters embedded in an amorphous sp^3-bonded carbon matrix, as shown in Fig. 4.48d. Generally, they contain significant quantities of hydrogen originating from the use of hydrogen for controlling the initial nucleation and subsequent growth in the deposition process.

Diamond-like carbon coatings were the most widely studied tribological coatings in the 1990s. Their attraction is that they have excellent tribological properties, very similar to the diamond coatings discussed above, they are chemically inert and have excellent biocompatibility. However, because of their amorphous structure they are less brittle than diamond coatings and due to their lower deposition temperature, down to room and even subzero temperatures, they can be deposited on a large variety of materials, including all kinds of metals and ceramics, with good adhesion.

Several CVD and PVD deposition techniques such as plasma-enhanced CVD, sputtering, ion plating, laser ablation, etc., have been successfully used. The substrate temperature during deposition is mostly below 200°C. The adhesion to the substrate has been improved by depositing an interlayer some nanometres thick of Si or Ti between the DLC coating and the substrate. DLC coatings are typically used in the thickness range of 0.01 to 2 μm but they have been prepared with thicknesses down to only a few nanometres and as nanosmooth coatings with intrinsic roughness less than 0.1 nm (Grill, 1997; Bhushan, 1999a, 2008).

The largest commercial application is probably their use as a wear protective layer in the multilayer system deposited on practically all computer hard disks. This means that for this purpose only there are one million DLC coatings deposited every day! There is a large variety of other commercial applications where the excellent low friction and good wear-resistant properties of DLC are beneficial such as: sliding bearings, valve rockers, gears, injection pumps, tappets, other automotive sliding components, cutting tools, mechanical seals, textile industry parts, optical windows, flat panel displays, photomultiplier and microwave power tubes, acoustic wave filters, sensors, heat spreaders, razor blades, surgical tools and biological implants (Hauert, 2004; Erdemir and Donnet, 2005).

The mechanical properties of hydrogen-containing amorphous DLC are hardness in the range of 5 to 60 GPa, Young's modulus of 60 to 400 GPa, fracture toughness of 1 to 12 MPa · $m^{0.5}$ and Poisson's ratio of 0.2 to 0.4. The hardest amorphous DLC coating is a tetrahedrally bonded carbon (ta-C) with hardness in the range of 40 to 90 GPa and Young's modulus of 200 to 900 GPa. Internal stresses are normally compressive and in the range of 0.2 to 3.6 GPa. The coefficient of friction for DLC is generally in the range of 0.02 to 0.4 and the wear rate 0.0001 to 1 · 10^{-6} mm^3/N · m in normal air, while the coefficient of friction in a vacuum or inert environment is in the range of 0.001 to 0.02 and the wear rate may be as low as 0.00001 · 10^{-6} mm^3/N · m (Robertson, 1992b; Holmberg and Matthews, 1994; Tsui *et al.*, 1995; Grill, 1997; Kodali *et al.*, 1997; Gangopadhyay, 1998; Bhushan, 1999a; Nastasi *et al.*, 1999; Schaefer *et al.*, 2000; Ronkainen, 2001; Erdemir, 2002, 2004a, b; Fujisawa *et al.*, 2003; Singer *et al.*, 2003; Gupta *et al.*, 2004; Klaffke *et al.*, 2004a; Hauert, 2004; Erdemir and Donnet, 2005; Klaffke *et al.*, 2005a; Rha *et al.*, 2005; VDI-Richtlinien, 2005; Neuville and Matthews, 2007; Donnet and Erdemir, 2008; Fontaine *et al.*, 2008; Lemoine *et al.*, 2008).

4.5.3.1 Microstructure and hardness

The CVD and PVD deposition techniques offer good possibilities to tailor DLC coatings with various structures corresponding to the required properties. The main categories of DLC coatings and their abbreviations are (VDI-Richtlinien, 2005):

- a-C: hydrogen-free amorphous carbon coating,
- ta-C: tetrahedral hydrogen-free amorphous carbon coating,
- a-C:Me: metal-containing hydrogen-free amorphous carbon coating (Me = W, Ti...),
- a-C:H: hydrogenated amorphous carbon coating,
- ta-C:H: tetrahedral hydrogenated amorphous carbon coating,
- a-C:H:Me: metal-containing hydrogenated amorphous carbon coating (Me = W, Ti...), and
- a-C:H:X: modified hydrogenated amorphous carbon coating (X = Si, O, N, F, B...).

Since small quantities of hydrogen, e.g. from residual gases, can still be incorporated in the films even when there is no use of hydrogen gas in the deposition process, a hydrogen limit of about 3 at.% is quoted (VDI-Richtlinien, 2005) as the transition between hydrogen-free and hydrogenated carbon coatings.

The most important structural parameters are the sp^3/sp^2 ratio and the hydrogen content in the coating. The balance between them and the regions of the various DLC coatings is illustrated in the ternary phase diagram in Fig. 4.61a. At higher temperatures a recrystallization of the carbon structure to graphite takes place and this transition depends on the sp^3/sp^2 ratio, as shown in Fig. 4.61b.

In hydrogenated amorphous diamond-like carbon coatings (a-C:H) the proportion of sp^3 bonds in the structure is typically in the range of 30 to 60% for hard coatings and in the range of 50 to 80% for softer coatings. For the hard coatings the hydrogen content is in the range of 10 to 40% and for the softer coatings 40 to 65%. Here hard coatings refer to hardness values of 10 to 40 GPa and soft coatings to hardness below 5 GPa (Franks *et al.*, 1990; Ham and Lou, 1990; Hirvonen, 1991; Robertson, 1992a, b; Raveh *et al.*, 1993; Neuville and Matthews, 2007). The influence of total hydrogen content on surface microhardness of a-C:H coatings deposited in various plasma conditions is shown in Fig. 4.62.

Non-hydrogenated amorphous diamond-like carbon coatings (a-C) typically have about 85 to 95% sp^3 bonds in their carbon structure and the hydrogen content is often less than 1%. There are large variations reported in their hardness, from 30 to 100 GPa, which is a range that covers coatings even as hard as diamond (Hirvonen, 1991; Robertson, 1992a, b; Ronkainen *et al.*, 1993). The very hard a-C coatings are normally referred to as ta-C but they have also been referred to as amorphous diamond, because of their extreme hardness, and because their local structure is tetrahedral, as for diamond, even if their structure is disordered on a large scale (Green *et al.*, 1989).

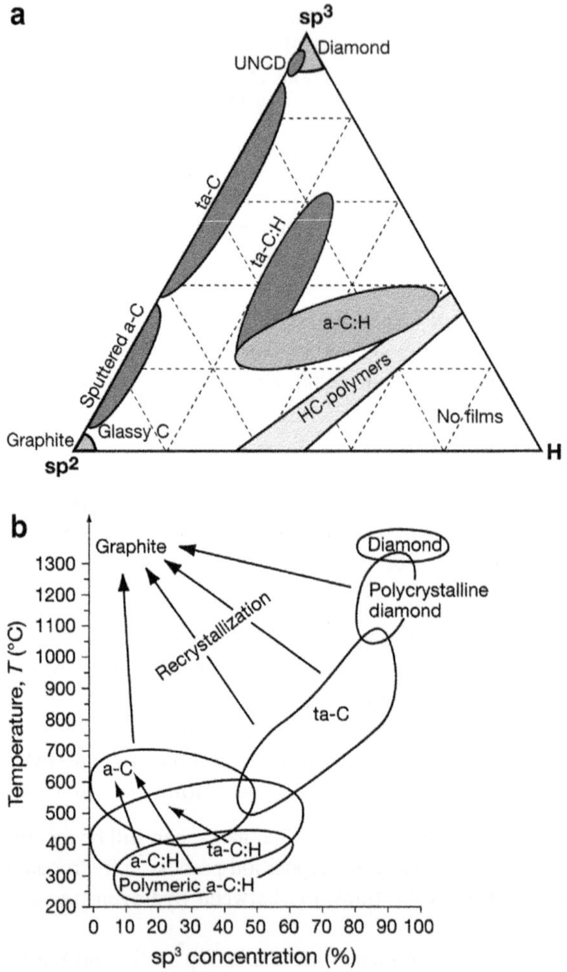

Fig. 4.61. (a) Ternary phase diagram for DLC coatings showing the balance of the sp^3/sp^2 ratio and the hydrogen content in the coating (after Ferrari and Robertson, 2000; Erdemir and Donnet, 2005). UNCD refers to ultra-nanocrystalline diamond and HC to hydrocarbon. (b) Simplified overview of thermal stability limits of different categories of hard carbon materials. The arrows indicate the effect of heating on changes to the phases (after Neuville and Matthews, 2007).

Molecular dynamic simulations have shown that the three-dimensional structure, not just the sp^3/sp^2 ratio, is important in determining the mechanical properties of the coatings (Gao *et al.*, 2003). Particular orientations of sp^2 ring-like structures create coatings with both high sp^2 content and large elastic constants. Coatings with graphite-like top layers parallel to the substrates have lower elastic constants than coatings with large amounts of sp^3-hybridized carbon. The layered structure of the hydrogen-free coatings results in a mechanical behaviour that influences the friction-to-load relationship. The atomic-scale structure of the coating at the interface has turned out to be of critical importance in determining the load at which tribochemical reactions and wear between the coating and the counterface are induced. Unsaturated sp-hybridized atoms on the surface of a coating serve as initiation points for such reactions, as shown in Fig. 4.63.

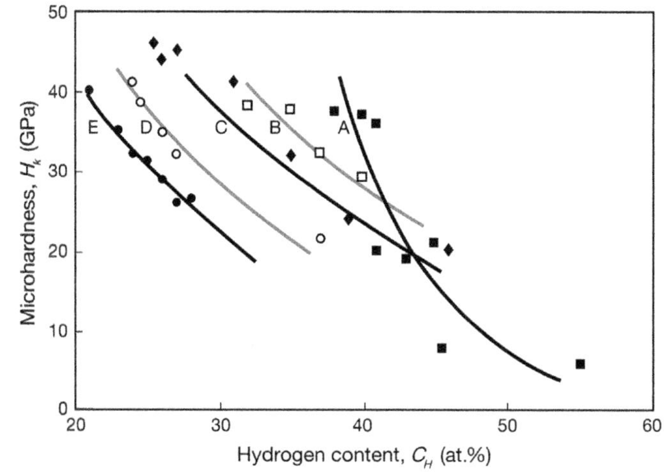

Fig. 4.62. Surface microhardness (H_k) as a function of total hydrogen content (C_H) in a-C:H diamond-like coatings deposited in various plasma conditions: (A) $CH_4 + H_2$ in rf or mwrf, (B) rf CH_4, (C) mwrf CH_4, (D) rf CH_4 + Ar and (E) mwrf CH_4 + Ar (data from Raveh *et al.*, 1993).

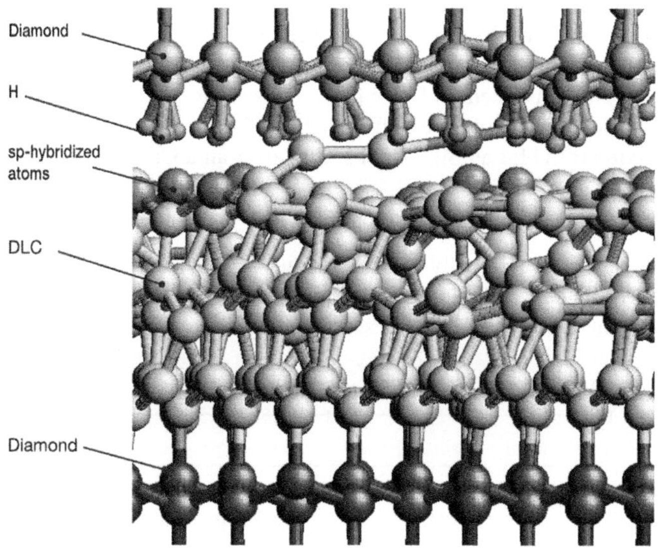

Fig. 4.63. A hydrogen-terminated (H) diamond counterface in sliding contact with an amorphous carbon layer (DLC) attached to a diamond substrate. Sliding-induced chemical reactions have occurred between sp-hybridized film atoms and the counterface (from Gao *et al.*, 2003 with permission). See Appendix B for coloured picture.

4.5.3.2 Deposition techniques

The most important feature, in addition to the excellent tribological properties, that makes this group of coatings so interesting, is that they can be produced at low temperatures, even below room temperature. This is a considerable achievement compared to the diamond coatings normally produced at temperatures in the range of 400 to 900°C, PVD TiN coatings in the range of 400 to 600°C and CVD coatings in the range of 900 to 1100°C. The low coating temperature offers the possibility to coat substrates that cannot withstand high temperatures, such as polymers and heat-treated metals. On the other hand, diamond-like carbon coatings are not suitable for high-temperature applications because they are thermally unstable at temperatures above about 450°C. This makes them less suitable for use as coatings for cutting tools where the contact temperature between the chip and the insert can be up to 1000°C, although as discussed in Chapter 7 there are some promising results reported in cutting tests using DLC coated tools.

4.5.3.3 Temperature dependence

The diamond-like carbon structure is a metastable form and can only exist up to a certain temperature level (Fig. 4.61b). Thermal graphitization of the graphite film takes place, starting from temperatures in the range of 300 to 600°C, and may be accompanied by the effusion of hydrocarbons and hydrogen. As hydrogenated DLC coatings are heated to temperatures above 300°C dehydrogenation transforms sp^3 bonded carbon atoms into sp^2 bonded atoms, resulting in a decrease in hardness and elastic modulus (Gangopadhyay, 1998). Hydrogen-free DLC coatings, on the other hand, are more temperature stable. Koskinen (1988) reported no change in hardness for hydrogen-free amorphous carbon coatings when heated up to 600°C. The influence of temperature on the mechanical properties of the coatings will also influence their friction and wear behaviour. Erdemir and Donnet (2005) report how the wear rate is increased by four orders of magnitude as the temperature increased from 29 to 250°C, as shown in Fig. 4.64.

With increasing plasma deposition energy, the mass of the released molecules decreases and the thermal stability increases considerably. In a softer coating both the hydrogen and the hyrocarbons diffuse at a low temperature, while only hydrogen diffuses at a higher temperature for hard coatings (Franks *et al.*, 1990). The consequence of such thermal instability was observed by Miyoshi (1991) as a sharp increase in the coefficient of friction from 0.2 to 0.8 when the ambient temperature in vacuum increased from 500 to 600°C for an Si_3N_4 ball sliding over an a-C:H coating.

4.5.3.4 Intrinsic stresses

Diamond-like carbon coatings normally exhibit significant compressive in-plane internal stresses; in the range of 0.2 to 3 GPa depending on the film thickness, but even higher values up to 5 GPa have been reported. They may be thermal stresses originating from the coating–substrate mismatch in

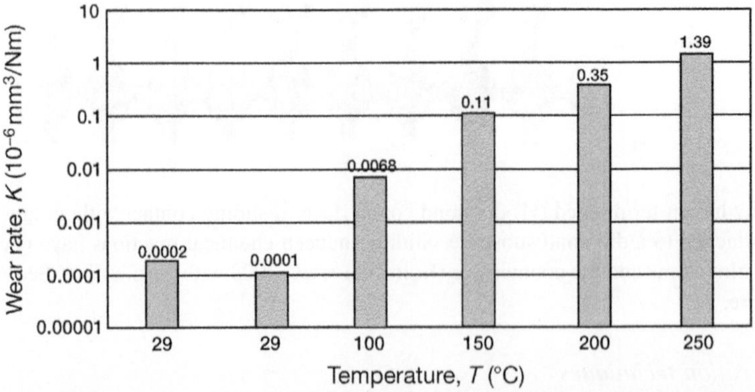

Fig. 4.64. Effect of temperature on wear performance of hydrogenated diamond-like carbon coatings (data from Erdemir and Donnet, 2005).

thermal expansion, and intrinsic stresses originating from changes in the film density as the film growth proceeds. In addition Raveh *et al.* (1993) have demonstrated the ability of DLC to accommodate trapped gases as a principal cause for internal stresses, when Ar or other gases, such as unbonded hydrogen, are incorporated into voids or microstructural boundaries in the coatings.

These effects improve the coating hardness but also cause the DLC film to become more brittle. The intrinsic stresses increase with coating thickness and coatings thicker than 2 μm tend to be so stressed that they easily delaminate due to the high stresses (Ham and Lou, 1990; Gangopadhyay, 1998). Modification of the film composition by adding Si or Ti into the coatings reduces the internal stresses, due to fewer interconnections in the random carbonaceous network, by a factor of two to four and this allows the deposition of thicker films. The internal stresses can also be relieved by annealing in an oxygen-free environment. Grill *et al.* (1990) showed that annealing in vacuum for 4 hours at 390°C reduced the internal stress from 1.34 GPa for the DLC film to 0.30 GPa without any significant change in film structure. At higher annealing temperatures the films may start to graphitize and lose their diamond-like properties.

Another approach to reduce the high intrinsic stresses in DLC coatings is to deposit them as multinanolayers, with consecutive soft and hard DLC nanolayers. In general, a multilayer coating will exhibit lower intrinsic stresses, higher hardness and elastic modulus than the average values of their two parent films. However, the influence was not as big as expected when Gupta and Meletis (2004) deposited six different 2 μm thick multinanolayers with consecutive soft and hard DLC nanolayers each of 50 to 200 nm thickness. The internal stresses in the soft and hard parent a-C:H films were 0.7 and 1.2 GPa, respectively, but the stresses in the six multilayer films varied from 0.8 to 1.5 GPa and were on average slightly below 0.95 GPa, which would be the average of the internal stresses in both constituent films. The influence on the internal stresses was neither as large nor as clear as one would expect. There are large variations in the internal stresses in DLC multilayers reported, varying from 0.8 to 6 GPa. The internal stresses in DLC films have been reviewed by Pauleau (2008).

4.5.3.5 Adhesion to the substrate and interface layers

The effective adhesion determines the strength of the interface between the coating and the substrate and has a dominating influence on the wear properties and especially on the maximum load the surface can sustain before delamination and failure. The effective adhesion depends on the strength of the adhesive bonds between the coating and the substrate. These bonds may be pre-stretched and thus weakened by intrinsic stresses in the coating system. The strength of the bonds between the coating and the substrate is influenced by the substrate material and by the deposition method. Diamond-like carbon coatings have a widely varying adhesion to different substrates. Good adhesion of DLC coatings to carbide and silicide substrates has been found.

A wide variety of substrate materials were coated by rf plasma CVD by Ham and Lou (1990). They observed that, of all materials coated, silicon was most effective as substrate material. It was virtually impossible to remove the DLC from Si. The adhesion to most other substrates was good. However, no adhering films could be deposited on Au, Cu, W, or on stainless steel containing Cu. Other substrate materials tested were Al, Ni, Mo, GaAs, Ti, TiN, steel, stainless steel, Pyrex glass, quartz, ceramics and Lexan.

The adhesion of 0.2 μm thick DLC coatings deposited by a pulsed arc discharge method on 19 different substrate materials such as metals, ceramics and polymers was studied by a pull test, Vickers indentation and a scratch test by Koskinen *et al.* (1992). In general, materials with high affinity to carbon had the best adhesion as shown in Fig. 4.65.

There are different ways of improving the adhesion of a coating to its substrate of which the use of interface layers is the most common. In general, coatings deposited by ion beam-assisted deposition or by cathodic arc discharge techniques do not require any interlayer for good film adhesion to aluminium alloys or steel because the high energy of the impacting ions provides good interfacial

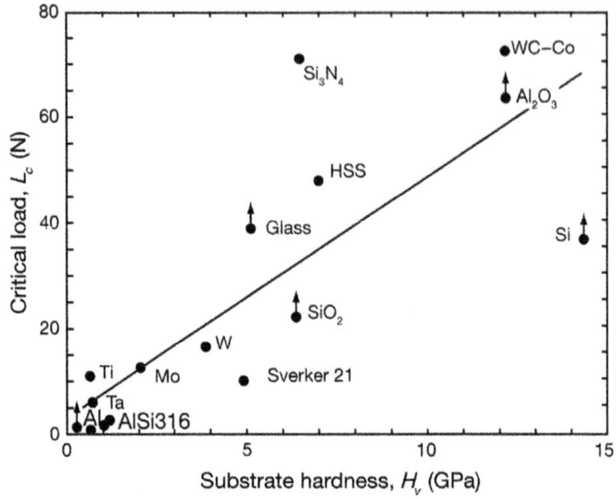

Fig. 4.65. The adhesion of a-C DLC coating to different substrate materials given as scratch test critical load related to substrate hardness (data from Holmberg *et al.*, 1994). The values marked with arrows indicate values of adhesion when failure of substrate material occurred.

mixing. However, coatings deposited by plasma-assisted CVD techniques often require an interlayer for good film adhesion (Gangopadhyay, 1998). Si, Ti, TiN, TiC, Ni and Cr have been successful interlayer materials for improving the adhesion between the DLC coating and the substrate (Ronkainen *et al.*, 1997; Vercammen *et al.*, 2004; Maurin-Perrier *et al.*, 2005).

Figure 4.66 shows how the adhesion of rf plasma-deposited DLC on cemented carbide can be considerably improved by using a TiC intermediate layer and low sputter-cleaning power (Vihersalo *et al.*, 1993; Ronkainen *et al.*, 1993, 1997). TiN has also been found to have an improving effect on the scratch adhesion of a-C:H layers to steel substrates (Holiday *et al.*, 1992), though this result was not always borne out in real sliding tests. This is because the scratch test does not accurately simulate a normal sliding contact and emphasizes the load support under the coating (Matthews and Eskildsen, 1993; Ollivier and Matthews, 1993).

Hirvonen *et al.* (1992) have shown that the adhesion of pulse arc discharged a-C DLCs can be improved when deposited on carbon steel substrates by combining ion bombardment with a silicon carbide intermediate layer. This bombardment resulted in a mixed layer with a thickness of 1.2 to 1.7 nm. The improvement in the adhesion was interpreted as originating from a chemical interaction between silicon and iron. Ion bombardment without an interlayer did not improve the adhesion.

An interlayer has two functions, one is to increase the bonding strength between the coating and the substrate and the other is to reduce the interfacial intrinsic stresses. The fulfilment of the first function does not require a thick interface layer; a very thin one of some nanometres thickness may be enough. However, an interlayer can better accommodate the stresses at the coating/substrate interface. A thicker coating may also affect the load-carrying capacity of a coated surface and prevent plastic deformation of the substrate. This is well illustrated by the experiments of Gangopadhyay *et al.* (1997) who deposited a series of successively thicker silicon interlayers between an, about 1.0 μm thick, amorphous hydrogenated carbon coating and an aluminium–silicon alloy substrate. Figure 4.67 shows the measured coating durability, defined as time to coating failure, when sliding against steel balls, as a function of coating thickness. The thickness of the interface layer is the given coating thickness minus 1 μm. A clear increase in coating durability was achieved with a thicker interlayer. The friction coefficient decreased from 0.36 to 0.12 with increasing thickness.

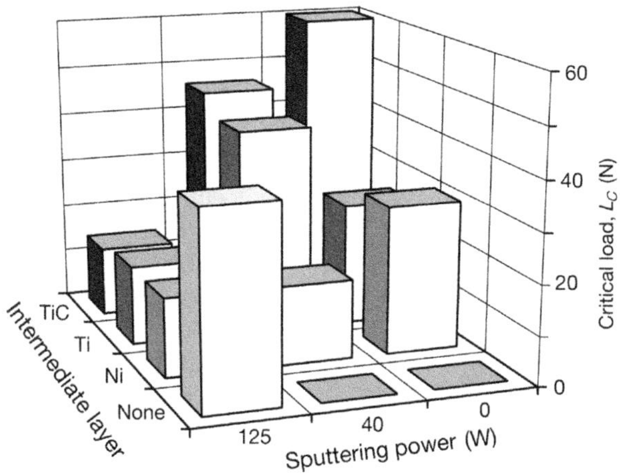

Fig. 4.66. Adhesion measured as critical load for rf plasma diamond-like carbon coatings on cemented carbide substrates with different intermediate layers and for different sputtering powers (after Vihersalo *et al.*, 1993).

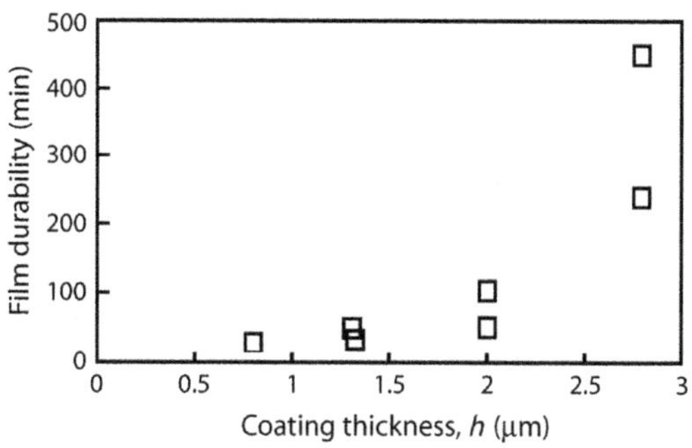

Fig. 4.67. Durability of a DLC coated aluminum–silicon alloy surface with silicon interlayers of successively increasing thicknesses while the thickness of the DLC coating is all the time constant, about 1 μm (data from Gangopadhyay *et al.*, 1997).

4.5.3.6 Biocompatibility

Diamond-like carbon coatings are considered to be suitable for use in the biomedical field, particularly in orthopaedic implants. They can be made to adhere strongly to a range of materials used for bio-engineering and surgical purposes, including metals such as stainless steel, titanium, cobalt–chrome and various plastics including polyethylene and polyurethane. The diamond-like coating can be used as a chemically protective, impermeable layer on such materials in a biological environment. Thomson *et al.* (1991) have carried out *in vitro* tests with mouse macrophage cells and fibroblast cells. No inflammatory response or loss of cell integrity by contact with the carbon coating occurred. Morphological examination confirmed the biochemical results of no cellular damage. Other studies on the biocompatibility of DLC have been reviewed by Cui and Li (2000). Very promising results have been reported by Ajayi *et al.* (2005) on the tribological behaviour of amorphous carbon coated stainless steel sliding against uncoated stainless steel to replace the commonly used stainless steel against UHMWPE in artificial hip joint implants. The wear of the DLC coating was four orders of magnitude lower than for the UHMWPE surface and no wear of the uncoated stainless steel counter surface was observed.

4.5.3.7 Surface graphitization

As for diamond, it has been speculated that the very low coefficient of friction of diamond-like carbon coatings in sliding contacts is due to graphitization and the formation of a thin graphitic layer on top of the hard coating (Jahanmir *et al.*, 1989; Erdemir *et al.*, 1991c; Klaffke *et al.*, 1992; Holiday *et al.*, 1992; Holmberg and Matthews, 1994). Some uncertainty on this mechanism has been due to the problem of managing to detect clear traces of graphite on the sliding surfaces, although many have postulated this phenomenon.

Erdemir *et al.* (1991c) slid a sapphire pin in argon over an old wear track on a DLC coating that was formed in humid air. The initial friction coefficient was about 0.09 and decreased after about 100 passes to a value of 0.03. When they slid the same pin over a new wear track the friction coefficient level of 0.03 was reached after only ten passes. The authors explained this behaviour with the hypothesis that a thin graphitic film is formed on the contact interface of the DLC coating during sliding in humid air. A decrease in friction with increasing humidity is typical for graphite.

Wei *et al.* (1993) found a very clear indication of the formation of graphite in the wear track on a diamond-like a-C:H coated disc after a 100 million cycle rolling contact fatigue test. The growth in the Raman peak, which corresponds to a disordered mode in the small graphite crystallite, compared to that measured outside the wear track implies that the effect of fatigue wear of the coating is to break the interlayer bonds which characterize hard-carbon films, thereby allowing the top layer of the coating material to return to graphite which is the more stable carbon form.

A number of further studies have pointed to the appearance of graphite in the wear tracks. They have also shown another important contact mechanism that influences friction and wear – the formation of a carbon-rich transfer layer on the countersurface (Hirvonen *et al.*, 1990a, b; Ronkainen *et al.*, 1993; Grill, 1997; Wei and Komvopoulos, 1997; Voevodin and Zabinsky, 2000; Ronkainen, 2001; Scharf and Singer, 2003a, b, c, 2008; Singer *et al.*, 2003; Zhang *et al.*, 2003b; Field *et al.*, 2004; Erdemir and Donnet, 2005; Kano *et al.*, 2005; Konca *et al.*, 2005; Neuville and Matthews, 2007). This transfer layer covers more or less completely the countersurface giving it a smooth load-carrying platform ideal for low shear in combination with the graphitized top layer of the DLC coating, as shown in Fig. 4.68. Transfer layers consisting of carbon, oxygen, chromium, aluminium and iron have been observed. The result is that the friction occurs between two surfaces with amorphous carbon top layers, which have a structural chemistry quite different from the original DLC coating, but similar to that of crystalline graphite.

The process when the transfer layer is built up on the countersurface has very convincingly been illustrated by *in situ* experiments with a transparent sapphire hemisphere of 6.35 mm diameter sliding

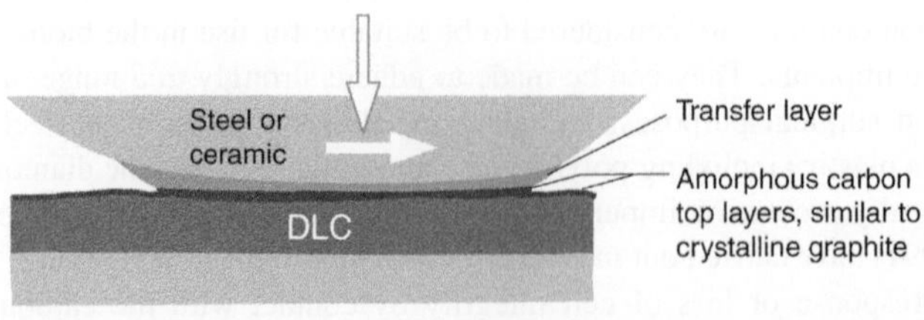

Fig. 4.68. In a steel or ceramic surface sliding against a diamond-like carbon surface a transfer layer is formed on the steel/ceramic surface and the shear is taking place between two amorphous carbon top layers which are similar to crystalline graphite.

against an amorphous DLC film by Singer and co-workers (Singer *et al.*, 2003). In sliding tests in dry air of RH = 4% at a speed of 0.001 m/s, load of 6 N and a mean Hertz stress of 0.7 GPa the coefficient of friction dropped rapidly from an initially high value of 0.25 down to 0.05 and then remained constant. A transfer film of 100 to 200 nm thickness was built up on the sapphire surface. As the film became thicker and patchy its Raman spectra showed peaks characteristic of a graphite-like transfer film. Similar tests were performed at higher stresses, up to 1.1 GPa and in moist air. The friction coefficient in humid air was about 20% higher than in dry air. The transfer films were thicker and the coefficients of friction were more stable in moist air.

The graphitization process in hydrogenated a-C:H and in hydrogen-free ta-C carbon coatings was studied by Ronkainen and co-workers (Ronkainen *et al.*, 1996, 1998, 1999b, c, 2001; Ronkainen, 2001). They found clear evidence of graphite formation using micro-Raman in the a-C:H coating wear track while the amount of graphite detected in the wear track of the ta-C coating was smaller after tribotesting in the same conditions. They also calculated the flash temperatures for both contacts with smooth surfaces. In low-load and low-speed conditions of 5 N and 0.1 m/s, resulting in a high coefficient of friction up to 0.4 and high wear rate up to $0.2 \cdot 10^{-6}\,mm^3/Nm$, the calculated flash temperatures were generally less than 100°C for the a-C:H coating sliding against alumina or steel balls. On the other hand, in high-load and high-speed conditions of 40 N and 3 m/s, resulting in a low coefficient of friction of 0.1 and lower wear rate of $0.2 \cdot 10^{-6}\,mm^3/Nm$, the calculated flash temperatures were about 350°C. The calculations showed much higher flash temperatures for the ta-C coating, up to 750°C.

As most contacts have some level of surface roughness with higher pressures at the asperities temperatures as high as 350°C can normally easily be reached at asperity tops. At this temperature, out-diffusion of hydrogen occurs in the a-C:H film. After hydrogen out-diffusion, the sp^3 structure may be transformed to an sp^2 structure and forms graphite. Thus the graphite would be formed on the top of the asperities to provide low friction. Graphite requires hydrogen to provide low-friction properties and this may be provided from the environmental humidity, water or from the out-diffused hydrogen that is provided in larger amounts from a-C:H coatings but only very sparsely from ta-C coatings. Important conclusions are that the graphitization process is enhanced by sliding at higher loads and higher speeds and ta-C coatings have an increased need for receiving hydrogen from the environment to provide low friction and low wear.

In a relatively slow graphitization process of smooth and dense DLC coatings the graphitization takes place mainly on the top layer of the coating. However, if the graphitization process is very rapid, as it may be at high temperatures, then the carbon material will not only be transformed into amorphous graphitic carbon at the surface, but also in the bulk of the coating and may also affect the coating-to-substrate interface underneath the coating. This may reduce the mechanical strength of the bulk and the coating materials and result in increased friction and wear (Neuville and Matthews, 2007).

The very good tribological behaviour of hard carbon-based coatings, with a dense and amorphous structure dominated by almost entirely sp^2 bondings, was considered to be due to the formation of a graphite-like lubricious film between the coating and the counterpart in ball-on-disc testing (Field *et al.*, 2004). A tungsten carbide 5 mm diameter ball was slid against the coating at loads of 40 and 80 N. The coefficient of friction was initially about 0.15 but decreased to a steady-state value of 0.05 to 0.07 and the wear rate was 0.021 to $0.043 \cdot 10^{-6}\,mm^3/Nm$. The authors suggest that also reorientation and recrystallization of the coating top layer takes place so that the basal planes become parallel to the sample surface. Higher normal loads would result in more reorientation and lower friction, as was observed.

4.5.3.8 Effect of hydrogen

The hydrogen content in a diamond-like carbon coating has a most significant influence on its hardness, modulus of elasticity, density and the sp^3/sp^2 ratio (Dunlop *et al.*, 1991). The presence of

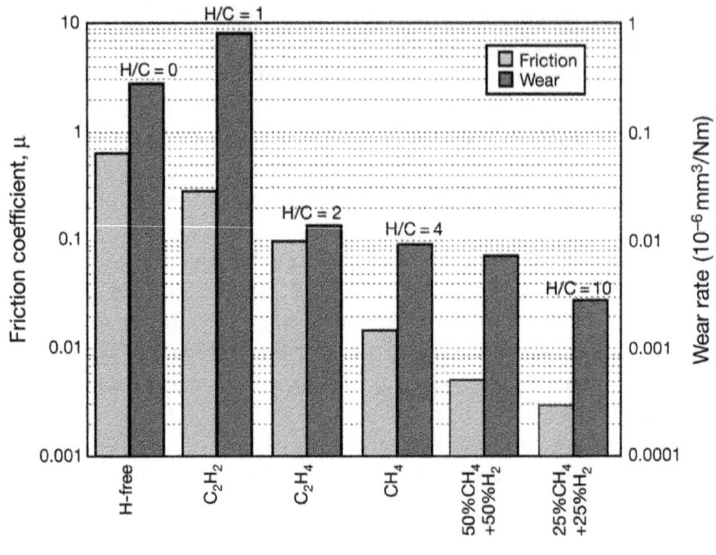

Fig. 4.69. Coefficients of friction and wear rates in sliding in dry nitrogen for DLC coatings derived from different carbon sources with various hydrogen-to-carbon (H/C) ratios (after Erdemir and Donnet, 2005).

hydrogen is also important for the development of favourable friction conditions in diamond and DLC coatings according to the dangling bond orbital and graphitization mechanisms described earlier. The decrease in microhardness for a variety of DLC coatings with increasing hydrogen content in the coating is shown in Fig. 4.62 (Raveh *et al.*, 1993; Gupta *et al.*, 2004). The influence of hydrogen content in the coating as well as the presence of hydrogen in the surrounding environment has been shown by several studies (Ronkainen *et al.*, 1993; Andersson *et al.*, 2003b; Fontaine *et al.*, 2004; Erdemir and Donnet, 2005; Li *et al.*, 2007; Borodich *et al.*, 2008; Ronkainen and Holmberg, 2008). The coefficients of friction and wear rates for DLC coatings derived from different carbon sources with various hydrogen-to-carbon (H/C) ratios are shown in Fig. 4.69. For higher hydrogen content in the coating, the surface is covered by hydrogen atoms, leading to weak van der Waals bonds between the sliding surfaces. For lower hydrogen content, on the other hand, there are not enough hydrogen atoms to shield the strong interactions between the free orbitals of carbon bonds.

In pin-on-disc tests Ronkainen *et al.* (1993) have shown that, in a load range of 5 to 40 N and a sliding velocity range of 0.1 to 3 m/s, the coefficient of friction for a steel pin sliding against an rf plasma a-C:H coating is in the range 0.10 to 0.42 and against a pulse arc discharge a-C coating is in the range 0.14 to 0.23. Figure 4.70 shows the results. Corresponding friction coefficient ranges for an Al_2O_3 counterface were 0.02 to 0.13 for a-C:H and 0.10 to 0.14 for a-C. The wear rate of the coated disc was smaller for the a-C coating but the wear rate of the counterface was smaller for the a-C:H coating. The a-C:H coating had a hydrogen content of 26 at.% and the a-C coating less than 1 at.%.

4.5.3.9 Topography and its influence on friction and wear

The amorphous structure of the diamond-like carbon coatings does not produce the sharp peak topography that the facetted structure of some diamond coatings may result in. The DLC coatings are generally smooth and the roughness of the coated surface is significantly influenced by the roughness of the substrate. Depending on the deposition technique and the coating thickness the surface roughness after deposition may be unchanged, or increased or decreased, by some 20 to 50%.

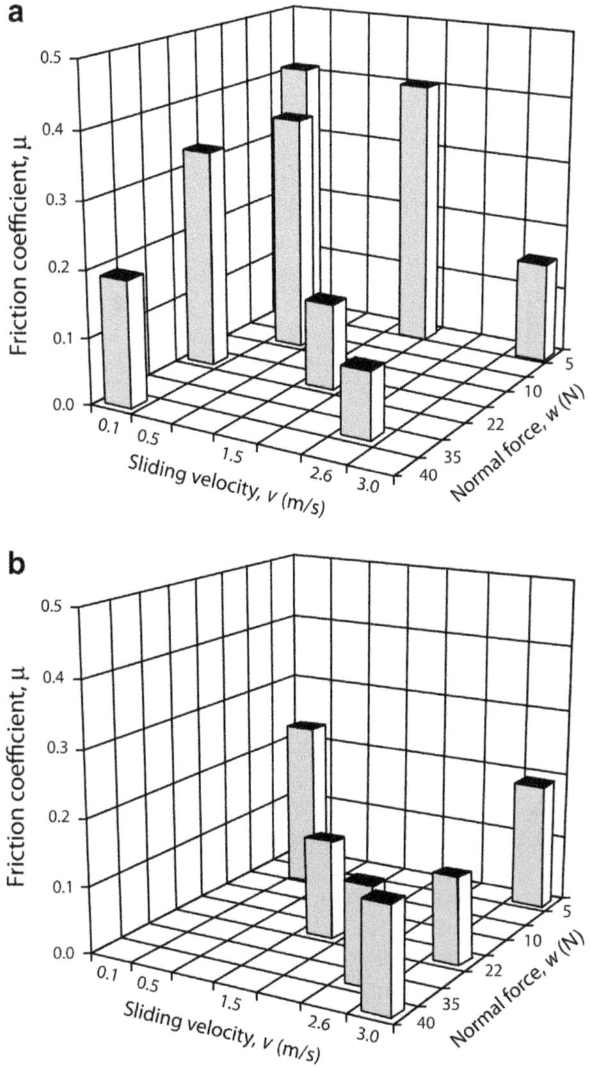

Fig. 4.70. Coefficient of friction for a steel pin sliding over (a) a hydrogenated a-C:H DLC coating and (b) a hydrogen-free a-C DLC at different loads and speeds. The coating thickness was 0.5 μm and the substrate material was stainless steel AISI 440B (after Ronkainen *et al.*, 1993).

The surface roughness and topography is still a very important factor influencing its tribological performance due to the hardness of the DLC coated surface. A rough surface may scratch the countersurface resulting in high friction and wear whereas a very smooth surface contact may increase the adhesion between the surfaces and have a similar effect. Thus the challenge is to find the optimal surface roughness for the contacting surfaces to achieve optimal friction and wear performance.

The vast majority of the investigations on carbon coatings have been made on coatings deposited on very smooth polished substrates. In large-scale production, however, the smoothing can be both time consuming and costly. On a macroscale both the friction and wear normally increase with surface roughness. In ball-on-disc tests with 1.3 μm thick PACVD coated diamond-like carbon coatings on

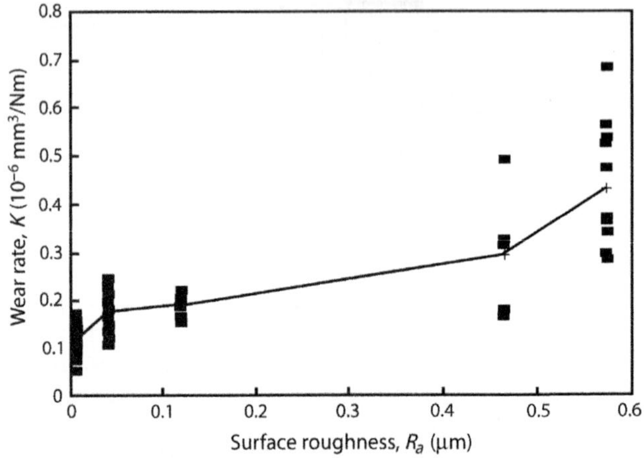

Fig. 4.71. The effect of surface roughness of a DLC coated surface on the wear rate of diamond-like carbon coatings (data from Jiang and Arnell, 2000).

tool steel surfaces in the surface roughness range of $R_a = 0.04$ to 0.7 μm sliding against tungsten carbide balls of 6.3 mm diameter at a speed of 0.05 m/s and at loads of 20 to 40 N in dry air of RH = 7% Jiang and Arnell (2000) measured a slight increase in the wear rate from 0.1 to $0.5 \cdot 10^{-6}\,mm^3/Nm$, as shown in Fig. 4.71. The frictional behaviour was not apparently affected. After deposition the surface roughness of the coating was approximately half of the original substrate surface roughness, which was in the range of $R_a = 0.08$ to 1.15 μm. A transition was observed at the roughness of 0.93 μm, then the wear increased rapidly and the wear mechanisms changed from adhesion to chip and flake formation and fragmentation of the coating possibly including plastic yielding.

A larger effect of surface roughness on wear and especially on friction was measured by Svahn *et al.* (2003) for magnetron sputtered doped 1.2 μm thick a-C:W and 3.5 μm thick a-C:Cr diamond-like carbon coatings deposited on steel with surfaces in the roughness range of $R_a = 0.007$ to 0.39 μm. The surface roughness after deposition was unchanged for the a-C:W coating and increased by 20% for the a-C:Cr coating. In ball-on-disc test sliding with a speed of 0.07 m/s, load of 4 N and at RH = 30 to 40% against steel balls of 3 and 9 mm diameter there was a clear surface roughness-related increase in the coefficient of friction from 0.06 to 0.21 for the small ball with high contact pressure and from 0.11 to 0.31 for the larger ball with lower pressure with the a-C:W coating, as shown in Fig. 4.72a. The wear rate increased from 0.2 to $0.5 \cdot 10^{-6}\,mm^3/Nm$ with roughness, as shown in Fig. 4.72b. The roughness influence of the a-C:Cr coating on friction was slightly smaller, an increase from 0.14 to 0.32 was observed while the increase in wear rate was larger, it increased from 0.1 to $1 \cdot 10^{-6}\,mm^3/Nm$ in the same conditions. The observed wear mechanisms were mainly gentle polishing of surface peaks where larger grooves collected wear debris. Some fatigue pits were found in the a-C:Cr coating deposited on rougher surfaces. The increase in friction with surface roughness is in this case not considered to be due to surface peaks ploughing in the steel countermaterial, since ball wear was very small compared to coating wear. The reason for the higher wear for rougher surfaces may be that the scratches are filled with oxidized coating material and the shearing of this third body material causes higher friction.

The effect of surface roughness on friction and wear may be the opposite. Several investigations with microscale smooth contacts have shown that a textured surface with some roughness can decrease friction when DLC is sliding against alumina, steel, silicon and sapphire surfaces compared

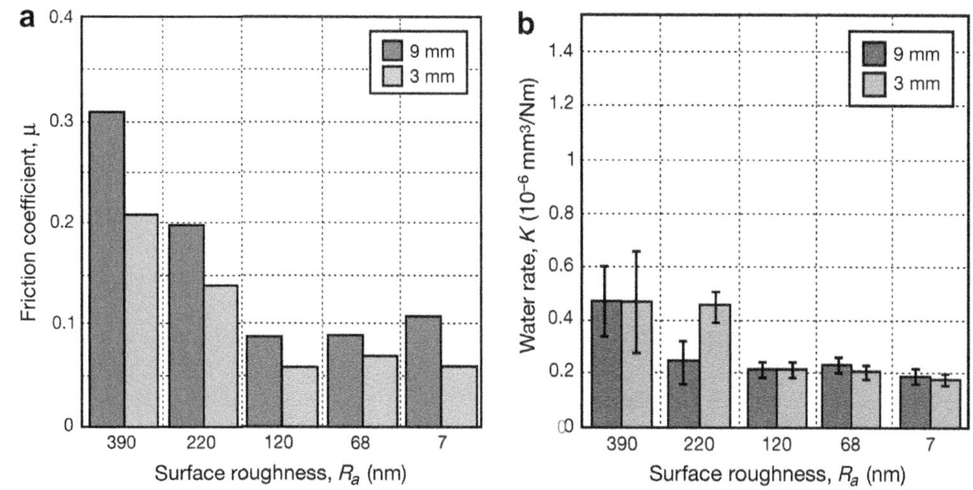

Fig. 4.72. Influence of surface roughness of an a-C:W coated steel surface sliding against steel balls with 9 and 3 mm diameter at 0.07 m/s speed and 4 N load (a) on the coefficient of friction and (b) on the wear rate (data from Svahn *et al.*, 2003).

to a very smooth surface (Zhou *et al.*, 2000; Ahmed *et al.*, 2003; Bandorf *et al.*, 2003), but also the opposite effect has been reported (Erdemir, 2004b). This shows that to understand the influence of surface roughness on friction and wear a thorough investigation of the prevailing contact mechanism is needed.

4.5.3.10 Humidity dependence

In non-oxidizing controlled environments like vacuum, argon or dry nitrogen the tribological behaviour of diamond-like carbon coatings is often similar to that of bulk diamond. However, water vapour influences the DLC coating and in humid air the coating behaves more like bulk graphite (Miyoshi, 1990; Erdemir *et al.*, 1991c; Ronkainen and Holmberg, 2008). It is well documented that graphite requires moisture or other vapours to achieve low friction and low wear. The adsorbed vapour covers the edge sites, thereby lowering the surface energy and reducing the adhesion to other surfaces (Singer, 1989).

Covalent bonds or σ-bonds cause the strongest interactions across sliding diamond and DLC coatings. In ultra-high vacuum when the σ-bonds of surface carbon atoms are exposed and free to form strong covalent bonding across the surfaces, the coefficient of friction may be even above 1.0. If moisture or gaseous species such as oxygen, hydrogen, fluorine, etc., are introduced the coefficient of friction of the diamond or hydrogen-free DLC surface drops to as low as 0.05, presumably due to the passivation of the σ-bonds and elimination of the strong covalent interaction across the sliding interfaces (Erdemir *et al.*, 2000).

However, it can be confusing to try to find answers from the literature on how the humidity influences friction and wear for DLC coatings. There are several investigations showing an increase in friction and wear with increasing humidity (Gangopadhyay, 1998; Andersson *et al.*, 2003a; Zhang *et al.*, 2003b; Park *et al.*, 2004; Tanaka *et al.*, 2004a, b; Erdemir and Donnet, 2005; Klafke *et al.*, 2005a), but there are also several showing a decrease with humidity (Andersson *et al.*, 2003a; Tanaka, 2004a, b; Klafke *et al.*, 2005a) and then again others showing really no influence at all (Miyoshi, 1998; Klafke *et al.*, 2005a). These investigations are all performed in very competent laboratories so there is really no doubt about the reliability of the measurements. We believe that the explanation is more due to the fact that there are several mechanisms related to sliding DLC surfaces that have a humidity influence, and these effects are sometimes contradictory.

The explanation relating to a specific humidity behaviour lies with the dominating contact mechanisms in that specific experimental setup. The following mechanisms of humidity influence on DLC coatings have been suggested:

- the presence of moisture or adsorbed gases is required for low friction of graphitized top layers on coatings,
- the addition of small amounts of elemental species, such as N, F, Si, Ti, etc., into the DLC coatings may reduce their tribological sensitivity to humidity,
- high relative humidity can interfere negatively with the transfer film formation ability on steel surfaces,
- hydrogen is required for superlubricity based on the dangling bond orbital mechanism for atomically smooth surfaces; adding water destroys the superlubricity conditions,
- addition of moisture passivates strong covalent bonds between surfaces in ultra-high vacuum and the friction decreases,
- at high humidity the saturation of microcontacts with water molecules can take place and during sliding the water molecules exert a force across the opposing surfaces resulting in higher friction, and
- in DLC coatings sliding against steel in increased humidity results in increased oxidation of the steel surface and incorporation of Fe in the wear debris and increased friction.

Some general conclusions from the published work can be summarized as follows:

- highly hydrogenated DLC coatings have a low coefficient of friction in dry air and inert gases, even down to 0.003 when sliding against themselves; in humid air their coefficient of friction may increase to above 0.2 while sliding against themselves, metals or ceramics,
- hydrogen-free DLC coatings have low coefficient of friction in humid environments, down to 0.05; in dry air or vacuum it may be very high, in the range of 0.3 to 0.6,
- adding water normally decreases the coefficient of friction of hydrogen-free DLC coatings, to values below 0.1; adding water to hydrogenated DLC coatings may in some cases increase the friction but in others the coefficient of friction may drop to very low values around 0.01, and
- the coefficient of friction remains very low in contacts with DLC sliding on diamond over a wide relative humidity range.

The wear rate correlates generally with the behaviour of friction for DLC coatings. A collection of published values for coefficients of friction and wear rates for diamond, various diamond-like carbon, graphite and polymeric coatings is presented in Table 4.1.

The very low coefficient of friction, often below 0.05, when a steel slider slides over a hydrogenated DLC surface in dry air gradually increases with increasing humidity and reaches values in the range of 0.15 to 0.3 at relative humidities close to 100% (Enke *et al.*, 1980; Memming *et al.*, 1986; Franks *et al.*, 1990). Ronkainen and Holmberg (2008) have reviewed the environmental and thermal effects on the tribological performance of DLC coatings and they show the different influences of humidity on different categories of DLC coatings and the mechanisms controlling their tribological performance as presented in Fig. 4.73. The different mechanisms influencing on the frictional behaviour of hydrogenated and hydrogen-free DLC coatings in different humidity environments have been discussed by Meunier *et al.* (2005), Li *et al.* (2007) and Liu *et al.* (2007).

4.5.3.11 Sliding against metals

A transfer layer is often formed on a steel surface when it slides over a diamond-like carbon coating and the process of layer formation has a crucial influence on the friction and wear properties.

Table 4.1. Friction coefficients and wear rates collected from the literature for diamond, diamond-like carbon, graphite and polymer coatings.

	Diamond coatings	Hydrogen-free DLC	Hydrogenated DLC	Modified/doped DLC	Graphite coatings	References used
Structure	CVD diamond	a-C ta-C	a-C:H ta-C:H	a-C:Me a-C:H:Me a-C:H:X Me = W,Ti.... *x* = Si,O,N,F,B...		Andersson *et al.* (2003 a, b); Donnet (1998); Erdemir (2004a, b); Field *et al.* (2004); Fontaine *et al.* (2004); Gangopadhyay (1998); Grill (1997); Grillo and Field (2003); Harris (1997); Hirvonen *et al.* (1990a); Klafke *et al.* (2005a); Konca *et al.* (2003); Koskinen *et al.* (1992); Miyoshi (1998); Miyoshi *et al.* (1992); Park *et al.* (2004); Ronkainen *et al.* (1993, 1996); Ronkainen and Holmberg (2008); Ronkainen (2001); Svahn *et al.* (2003); Tanaka *et al.* (2004a, b); Voevodin and Zabinski (2000); Zhang *et al.* (2003b)
Atomic structure	sp^3	sp^2 and sp^3	sp^2 and sp^3	sp^2 and sp^3	sp^2	
Hydrogen content	–	>1%	10–50%			
μ in vacuum	0.02–1	0.3–0.8	0.007–0.05	0.03	–	
μ in dry N_2	0.03	0.6–0.7	0.001–0.15	0.007	–	
μ in dry air 5–15%	0.08–0.1	0.6	0.025–0.22	0.03	0.05–0.07	
μ in humid air 15–95%	0.05–0.15	0.05–0.23	0.02–0.5	0.03–0.4	0.05	
μ in water	0.002–0.08	0.07–0.1	0.01–0.7	0.06	–	
μ in oil	–	0.03	0.1	0.1		
K in vacuum	1–1000	60–400	0.0001	–	–	
K in dry N_2	0.1–0.2	0.1–0.7	0.00001–0.1	–	–	
K in dry air 5–15%	1–5	0.3	0.01–0.4	–	0.02–0.04	
K in humid air 15–95%	0.04–0.06	0.0001–400	0.01–1	0.1–1	–	
K in water	0.0001–1	–	0.002–0.2	0.15	0.01–0.05	
K in oil	–	–	–	0.1	–	

μ = coefficient of friction; *K* = wear rate (10^{-6} mm^3/Nm)

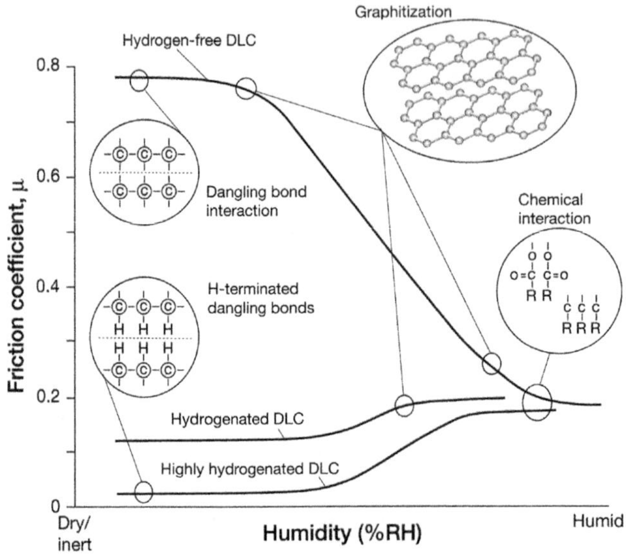

Fig. 4.73. Schematic presentation of friction performance of DLC coatings as a function of humidity and the related mechanisms controlling their tribological performance (from Ronkainen and Holmberg, 2008).

Ronkainen *et al.* (1993) have shown that a fairly strong and stable layer is formed on the steel pin as it is slid over an rf plasma a-C:H diamond-like carbon coating deposited on a stainless steel substrate. In SIMS analysis the layer showed chromium depletion, iron enrichment, oxygen and hydrogen but less carbon than the pin. It seems that the layer was mainly a mixture of iron and chromium oxides. The formation of the transfer film on the pin surface correlated with low-friction values and low wear rates of the pin in the load range of 5 to 40 N and sliding velocity range of 0.1 to 3 m/s. The pin wear rate was lowest at high load and high speed, as shown in Fig. 4.74a and the thickness of the tribofilm increased with load and speed. The coefficient of friction was in the range of 0.10 to 0.42 as shown in Fig. 4.70a. The 0.5 μm thick coating contained 26% hydrogen and was smooth with a surface roughness of $R_a = 0.03$ μm.

Ronkainen *et al.* (1993) further investigated, in the same experimental conditions, a pulsed arc discharge-deposited hydrogen-free a-C diamond-like carbon coating and observed differences in the layer formation. Again a transferred layer was formed on the pin surface but this was less stable and thinner and not strong enough to give the same wear protection for the pin. The coefficient of friction was reduced to about the same level as for the a-C:H coating but had a narrower variation range for different loads and speeds, as shown in Fig. 4.70b. The areas with thicker layer formation consisted mainly of iron and chromium oxides. At low loads the transfer film contained some hydrogen but at high loads no hydrogen was detected. The reason for the disappearance of hydrogen may be due to evaporation of water at the higher temperatures corresponding to the higher loads. The coefficient of friction was in the range of 0.14 to 0.23. The 0.5 μm thick coating contained less than 1% hydrogen and had a surface roughness of $R_a = 0.03$ μm. Similar transfer layers on steel sliders, but also on the a-C:H coated disc, were observed by Holiday *et al.* (1992) and on steel sliders when sliding against an a-C coated disc by Hirvonen *et al.* (1990a, b) and Ronkainen *et al.* (1992a, b).

It can be seen that there is a general similarity between the tribological behaviour of hydrogenated and hydrogen-free DLC coatings and the friction and wear values are of about the same level. It is also important to note that the tribological properties of the coatings show only a fairly small variation with

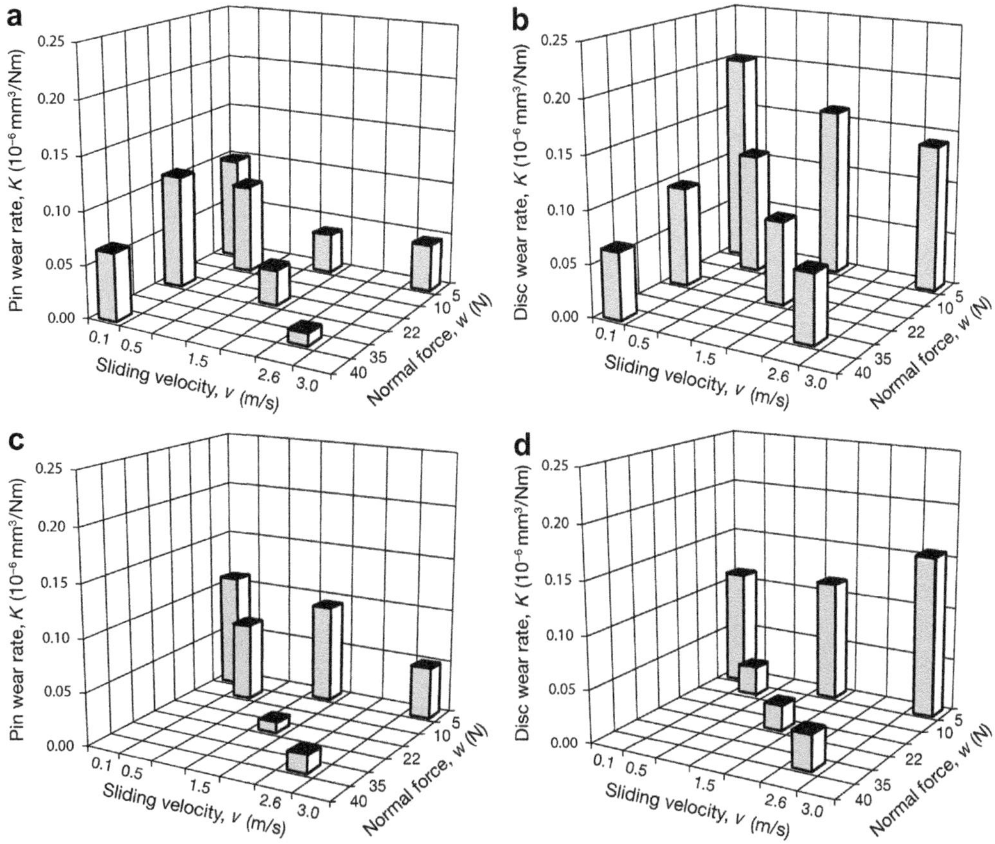

Fig. 4.74. Wear rates at different loads and speeds from sliding contacts with (a) a steel pin sliding on (b) a hydrogenated a-C:H diamond-like coating, and (c) a steel pin sliding on (d) a hydrogen-free a-C diamond-like coating. Coating thicknesses were 0.5 µm and substrate material was stainless steel AISI 440B (after Ronkainen *et al.*, 1993).

experimental parameters and the DLC coating can thus be considered as robust from a design point of view.

A considerable number of later studies on friction and wear in DLC coated surfaces sliding against steel provide much additional data that in general is in agreement with the above observations (e.g. Erdemir *et al.*, 1995; Gangopadhyay, 1998; Singer, 1998; Podgornik, 2001; Ronkainen, 2001; Ahmed *et al.*, 2003; Fujisawa *et al.*, 2003; Klafke *et al.*, 2004a; Park *et al.*, 2004; Podgornik *et al.*, 2006b; Cho *et al.*, 2005; Klafke, 2005; Maurin-Perrier *et al.*, 2005; Fontaine *et al.*, 2008). Friction and wear data from ball-on-disc experiments with a variation of DLC coatings sliding against different metal surfaces is found in Klafke *et al.* (2004b, 2005a) and Blanpain *et al.* (1993).

4.5.3.12 Sliding against ceramics

For hydrogen-free DLC films (a-C) deposited on silicon nitride (Si_3N_4) produced by PVD ion-beam deposition, Hirvonen *et al.* (1990a) observed a very low coefficient of friction of 0.06 in a sliding contact against silicon nitride, as shown in Fig. 4.75. This is considerably lower than for the same coating sliding in the same conditions against metals, when it was 0.14. However, the contact

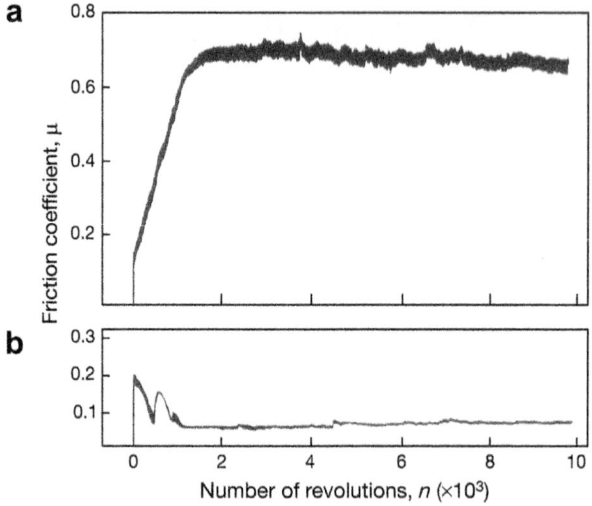

Fig. 4.75. The coefficient of friction of Si_3N_4 sliding on (a) WC–Co cemented carbide and on (b) a hydrogen-free diamond-like coating on Si_3N_4 at a speed of 0.13 m/s and a load of 1.55 N (after Hirvonen *et al.*, 1990a).

mechanism with ceramics is different from the steel contact because now no carbon transfer layer on the ceramic surface could be observed. Thus it was not clear what tribological mechanism caused the very low friction in the contact. The DLC coating was very thin, only 0.5 μm, but it managed to reduce the coefficient of friction between Si_3N_4 and Si_3N_4 – which normally is about 0.67 – down to 0.06. The wear was at the same time greatly reduced. Koskinen *et al.* (1992) have reported, for the same kind of coatings tested in similar conditions, friction coefficients in the range 0.045 to 0.085 and wear rates in the range 0.05 to $0.17 \cdot 10^{-6}$ mm^3/Nm.

In slow sliding contacts with plasma-deposited diamond-like hydrogenated carbon films (a-C:H) on Si_3N_4 sliding against Si_3N_4 Miyoshi (1989b, 1990) recorded higher coefficients of friction in the range of 0.2 to 0.3 in air but extremely low values, even down to 0.01, in a nitrogen environment. Miyoshi suggests that the explanation for the extremely low friction values may be due to the generation of a substance, probably a hydrocarbon-rich layer, on the surfaces. The low friction values were recorded after about 1000 passes which indicates that it takes some time to build up this layer. A very similar behaviour with low friction values could also be observed if the coated sample was replaced by bulk diamond. In comparison with other results it must be remembered that Miyoshi performed his experiments at extremely low speeds, only about 0.0001 m/s.

Transfer layers on both the diamond-like carbon coating and an Si_3N_4 counterface have been observed by Kim *et al.* (1991). They found that the surfaces were oxidized by a tribochemical reaction to form carbon–oxygen bonds replacing the carbon–hydrogen bonds that had been mechanically broken by shearing. In general, water vapour increases friction and wear of diamond-like films when sliding against ceramics, as shown in Fig. 4.76 (Kim *et al.*, 1991; Miyoshi *et al.*, 1992).

An increase in the surface temperature tends to cause chemical changes which influence the friction and wear behaviour. This was shown by Miyoshi (1989b) who measured coefficients of friction in a very wide range from 0.1 to 0.8, with hydrogenated DLC coatings (a-C:H) sliding on Si_3N_4 in vacuum at temperatures of 100 to 800°C. He also found that when his DLC coating was thermally annealed at 700°C in vacuum, the annealed surface of the coating behaved like graphite. The coating then needed, just like bulk graphite, a certain amount of water to show low friction. Other work has

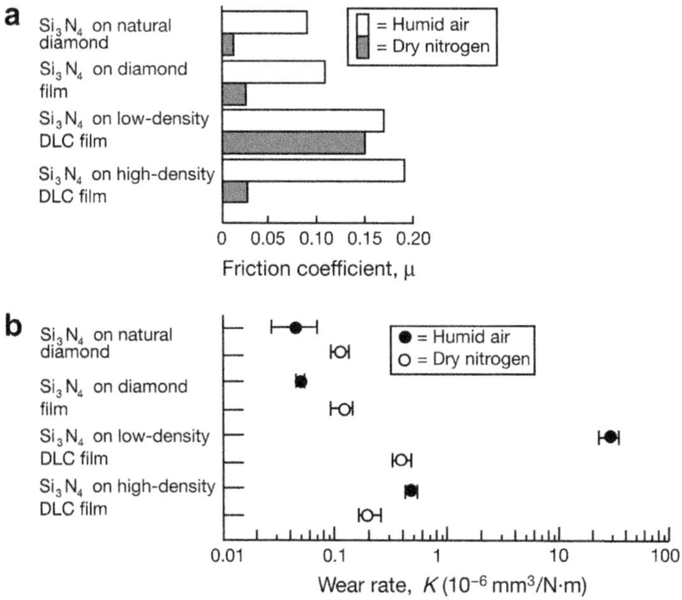

Fig. 4.76. (a) Coefficients of friction and (b) wear rates for silicon nitride pins sliding on natural diamond, a diamond coating, and low- and high-density diamond-like carbon coatings in humid air and in dry nitrogen at a reciprocating sliding speed of 0.0014 m/s and a load of 1 N (data from Miyoshi *et al.*, 1992).

shown, however, that rapid thermal annealing has only a minimal effect on the coefficients of friction both in humid air and in dry nitrogen but the wear rates are strongly dependent on the annealing temperature and the moisture in the environment (Wu *et al.*, 1992).

The formation of a strong and low shear-strength tribofilm has been observed to have a considerable effect when Al_2O_3 pins slide over a diamond-like carbon coated surface. This was investigated by Ronkainen *et al.* (1993) with the same coatings and experimental setup as referred to earlier with metal counterfaces. For hydrogenated a-C:H DLC coatings they found that a transparent layer was formed on the pin wear surface and the thickness of the film seemed to increase with higher loads and sliding velocities. Simultaneously, the coefficient of friction decreased drastically from 0.13, corresponding to 5 N and 0.1 m/s, to 0.02, corresponding to 22 N and 1.5 m/s, when the normal force and the sliding velocity were increased. On the other hand, both the pin and the disc wear increased at higher normal forces. Alumina was transferred to the surface of the disc wear track.

The hydrogen-free a-C DLC coating behaved in a very similar way. The coefficient of friction was slightly higher, in the range of 0.10 to 0.14, and extremely low values were not reached. The wear of the pin was slightly higher but the disc was worn less. The tribofilm formed on the pin was weaker and could not protect it from wear, and again alumina was transferred to the disc wear track.

A considerable number of studies on friction and wear in DLC coated surfaces sliding against ceramic surfaces provide much additional data that in general is in agreement with the above observations (e.g. Gangopadhyay, 1998; Miyoshi, 1998; Singer, 1998; Lu and Komvopoulos, 2001; Ronkainen, 2001; Ahmed *et al.*, 2003; Gupta and Meletis, 2004; Field *et al.*, 2004; Klafke *et al.*, 2004a, c; Tanaka *et al.*, 2004a, b; Watanabe *et al.*, 2004; Achanta *et al.*, 2005; Klafke, 2005; Maurin-Perrier *et al.*, 2005). Friction and wear data from ball-on-disc experiments with a variation of DLC coatings sliding against different ceramic surfaces is found in Blanpain *et al.* (1993) and Klafke *et al.* (2004b, 2005a).

4.5.3.13 Sliding against polymers

Polytetrafluorethylene (PTFE) is a polymer with superior tribological properties. When sliding on itself the coefficient of friction is very low, down to 0.08, and the wear can be reduced by the use of suitable fillers. When PTFE is sliding on steel a PTFE transfer layer is built up on the steel surface and after some time the contact mechanism is mainly PTFE sliding on a PTFE transfer layer on steel, as shown in Fig. 4.77a. The coefficient of friction is higher than in the PTFE–PTFE contact because of the presence of occasional PTFE–steel contacts in places where the layer is not complete as shown in Fig. 4.78, based on experimental results from Yang *et al.* (1991a).

The same authors showed, however, that when a PTFE pin is slowly sliding on a diamond-like carbon coating (a-C) very little PTFE is transferred onto the DLC coating and no transfer film is formed on it, as shown in Fig. 4.77b, because of low adhesion between PTFE and the diamond-like film. This low adhesion results at the same time in a remarkably low coefficient of friction of only 0.035. A good wear resistance for diamond-like coatings when sliding against polymers with reinforced fillers has been observed (Yang and Hirvonen, 1993).

Sliding of a PACVD hydrogenated (a-C:H) DLC coated flat steel surface in reciprocating sliding against a 5 mm diameter polyamide cylinder at a load of 50 N, a Hertzian surface pressure of 35 GPa and in a speed range of 0.3 to 1.2 m/s in humid air, RH = 60%, the friction was very high, with

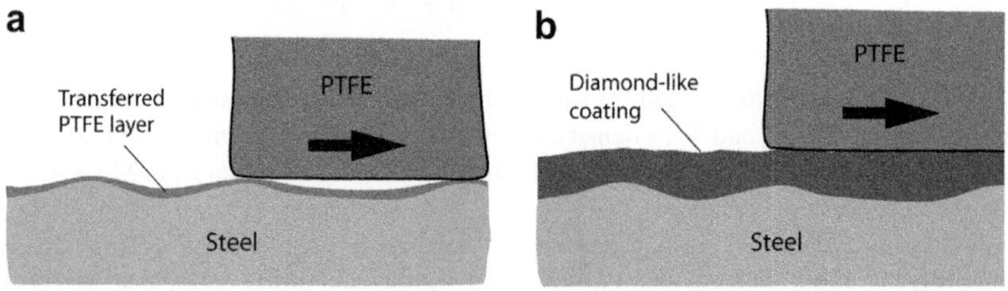

Fig. 4.77. The contact mechanism when a PTFE pin is sliding on (a) a steel surface and (b) a diamond-like carbon coating.

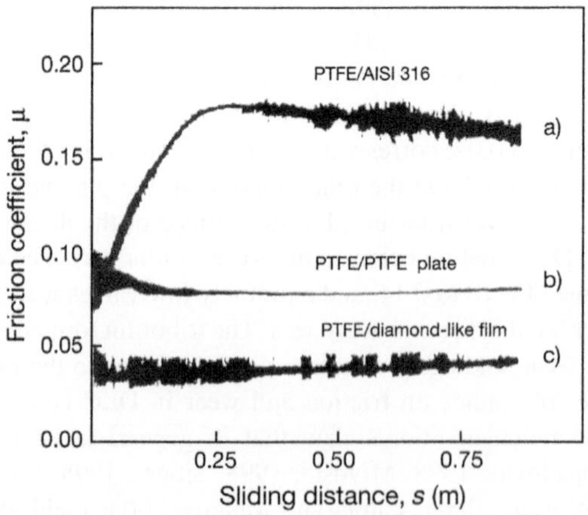

Fig. 4.78. The coefficient of friction of PTFE sliding on (a) a steel AISI 316 plate, (b) a PTFE plate and (c) an a-C diamond-like carbon coated plate at a speed of 0.006 m/s and a load of 3.05 N (after Yang *et al.*, 1991a).

coefficients of friction in the range of 0.55 to 0.8 (Samyn *et al.*, 2006). This was clearly higher than for polyamide sliding in the same conditions against carbon steel, high alloy steel and a nanocomposite DLC, for which the coefficient of friction was in the range of 0.29 to 0.51. The wear rate for the a-C:H DLC was in the range of 10 to 36 $\cdot$ 10^{-6} mm^3/Nm, which was similar to the other surface materials. The explanation for the high friction against the DLC surface is suggested to be thermal degradation due to high bulk and flash temperatures in the contact, above the polyamide degradation temperature which is 180°C.

4.5.3.14 Doped diamond-like and carbon alloy coatings

The structure and the properties of the diamond-like carbon coating can be further improved by the introduction of alloying elements or dopants into the coating during the deposition process. This has been especially attractive since it has turned out that two of the main problems in connection with the practical use of DLC coatings, the moisture sensitivity and the high intrinsic compressive stresses, can be modified by the use of dopants.

The following effects of using dopants in DLC structures have been found:

- reduction in intrinsic compressive stresses due to fewer interconnections in the random carbonaceous network,
- normally the hardness and the elastic modulus decrease but also the opposite has been observed,
- improved adhesion strength to the substrate,
- changes in surface energy due to changed density of the network structure,
- reduction in moisture sensitivity,
- normally the coefficient of friction is reduced, sometimes even very drastically down to low values like 0.03; this has been attributed to improving the surface graphitization process, formation of lubricating oxides (SiO_2) and influences on the transfer layer build-up process,
- often the wear rate increases slightly but reduced wear can also be achieved,
- improved rolling contact fatigue life,
- improved or unchanged boundary lubrication performance in lubricated contacts, and
- improved biocompatibility.

A very wide spectrum of different dopants has been tried. The most common are light elements like silicon, fluorine, nitrogen, boron, oxygen and then a wide variety of metals including Ti, Ta, Cr, W, Fe, Cu, Nb, Zr, Mo, Co, Ni, Ru, Al, Au and Ag. Reviews of their influence on the tribological properties has been published by Donnet (1998b), Gangopadhyay (1998) and Sánchez-López and Fernández (2008). The incorporation of metals should be limited to 30 to 40 at.%. Higher concentrations produce an increment in hard carbide and metallic phases that cause increased abrasive and adhesive wear together with higher friction coefficients.

The variation in reported doped DLC coating structures is huge and there is also a large diversity in their effect of the mechanical properties and tribological performance. Thus is it not possible to present a definitive description of the effects of all dopants on the tribology related properties. Below is a collection of typical observations from the literature (Harris *et al.*, 1997, 1999; Harris and Weiner, 1998; Donnet, 1998; Gangopadhyay, 1998; Olofsson *et al.*, 2000; Strondl *et al.*, 2001; Nilsson *et al.*, 2003; Miyake *et al.*, 2004; Vercammen *et al.*, 2004; Barriga *et al.*, 2006; Zhang *et al.*, 2006; Michalczewski and Piekoszewski, 2007; Renondeau *et al.*, 2007; Vevrekova and Hainsworth, 2008; Sánchez-López and Fernández, 2008), but note that in many cases opposite observations have been reported:

- *Silicon dopants* in DLC coatings have been very successful since they have been found to decrease the coefficient of friction considerably in ambient humid air, even down to levels of 0.03 (Oguri and Arai, 1990, 1991). The wear resistance has also been good, but these favourable tribological properties have been lost at higher contact pressures above 1 GPa.

- *Flourine dopants* in DLC coatings result in lower degrees of surface energy and lower stress levels, improved adhesion to ceramic substrates and low friction and better wear performance.
- *Nitrogen dopants* in DLC coatings result in an increase in surface energy, reduced stresses, preservation of hardness, small influence on wear resistance and friction.
- *Metal dopants* in DLC coatings result in reduced intrinsic stresses, reduced sensitivity to humidity, improved adhesion to steel substrate, reduced friction down to coefficients of friction of 0.04, various influences on sliding wear, improved rolling contact fatigue life and improved boundary lubricating performance. Low friction behaviour has especially been measured for Ta dopants in dry sliding.

Some of the early investigations on improving DLC coatings by dopants are described below. Matthews *et al.* (1989) observed some improvements in the wear resistance when adding boron nitride or tungsten to a diamond-like a-C:H carbon coating. Coatings with the added elements showed, however, a slightly higher coefficient of friction in the range of 0.1 to 0.15 as minimum values compared to 0.08 for pure a-C:H carbon films.

Oguri and Arai (1990 and 1991) found that considerable improvements can be achieved by alloying a-C:H carbon coatings with silicon. They formed amorphous carbon–silicon hydrogenated coatings by a dc glow discharge CVD method from reactant gases of CH_4, $SiCl_4$, H_2 and Ar. With a hydrogen content of 30 to 50 at.% they recorded coefficients of friction as low as 0.04 with a ball-on-disc test at speeds of 0.2 to 3 m/s and loads up to 33 N in a dry air environment. As a comparison, the coefficient of friction with the same experimental parameters was 0.12 for amorphous carbon coatings, 0.18 for titanium carbide and 0.8 for an uncoated steel M2 substrate when sliding against a steel surface, as shown in Fig. 4.79. Both the amorphous carbon–silicon and carbon coatings had an excellent wear resistance with wear rates down to 10^{-7} mm^3/Nm, as shown in Fig. 4.80. No major influence of load on friction could be observed in the load range of 10 to 50 N. A layer of silicon dioxide (SiO_2) wear debris was formed on the steel surface and this possibly plays a role in producing the excellent tribological contact conditions.

Meneve *et al.* (1993a, b) studied systematically rf PACVD-deposited amorphous hydrogenated silicon carbide (a-$Si_{1-x}C_x$:H) films ($x = 0.65$ to 1) in dry and humid N_2 conditions. They found that

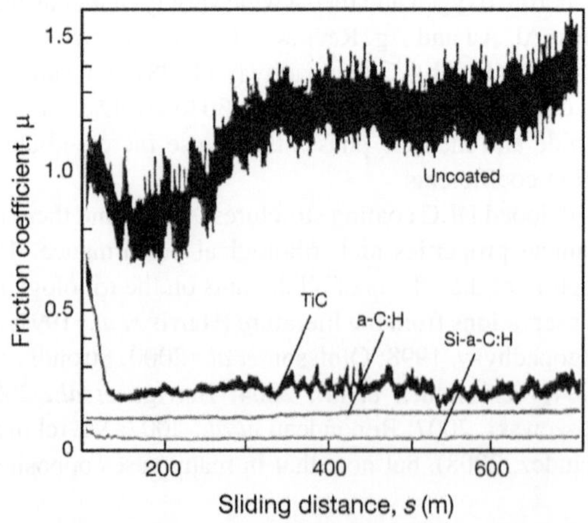

Fig. 4.79. Coefficient of friction as a function of sliding distance for a steel ball sliding against steel disks coated by PACVD Si-a-C:H, ion-plated a-C:H, PACVD TiC, and against an uncoated steel disk at a sliding speed of 0.2 m/s and a load of 6.4 N (after Oguri and Arai, 1990).

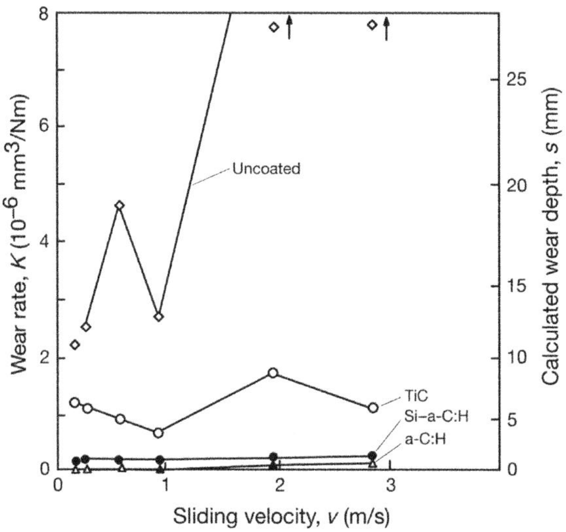

Fig. 4.80. Wear rates as a function of sliding velocity for steel discs coated by PACVD Si–a-C:H, ion-plated a-C:H, PACVD TiC, and an uncoated steel disk (data from Oguri and Arai, 1990).

adding silicon to a diamond-like a-C:H coating reduces the hardness, elastic modulus and internal stresses by values of 15 to 30%. In pin-on-disk tests with a low load of 1 N and in dry N_2 environment, the 2 μm thick silicon-containing coating showed low wear of both the coating and the counterface and a steady-state friction coefficient below 0.1. But at higher loads in dry N_2 a pure a-C:H coating performed better than the a-$Si_{1-x}C_x$:H films in all respects.

Excellent tribological performance has been reported for metal-containing a-C:H coatings (Me-C:H) deposited by rf or dc magnetron sputtering (Dimigen and Klages, 1991). The metal dopants they used were tantalum, tungsten, titanium and niobium. The Me-C:H coatings always exhibited compressive intrinsic stresses increasing with the metal content and attaining values of about 0.3 and 1 GPa for Nb–C:H and Ta–C:H, respectively, with metal contents of 20 to 40%. The friction when sliding against an uncoated WC ball showed a pronounced dependence on relative humidity of the ambient gas, normal load and metal content. Under optimal conditions very low coefficients of friction of 0.04 for Ta–C:H and 0.02 for W–C:H coatings were observed.

The wear of the coated disc was two orders of magnitude lower than for TiN and TiAlN PVD-deposited coatings and the wear of the counterpart was drastically reduced. The favourable tribological properties are attributed to a combination of ceramic-like properties, such as high hardness, and polymer-like properties, such as high elasticity and low surface energy. With regard to elasticity and surface properties the Me-C:H coatings with low metal contents bear more resemblance to polymers than to metals or ceramics, as shown in Fig. 4.81.

Improved tribological performance has also been achieved by doping zirconia, tantalum, tungsten, titanium and silicon into hard amorphous carbon coatings. The low friction coefficients and wear rates of such coatings deposited on steel and sliding against a steel ball are shown in Fig. 4.82. It was concluded in this case that the carbide material has no great influence on friction and wear because the upper layer of about 1 to 2 μm consisted of pure carbon.

The positive effect of tantalum and tungsten as metal dopants in hard carbon coatings on tribological performance is confirmed in both oscillating and pin-on-disk tests against silicon carbide by Klafke *et al.* (1992). They observed that the coating performed best in conditions of high humidity.

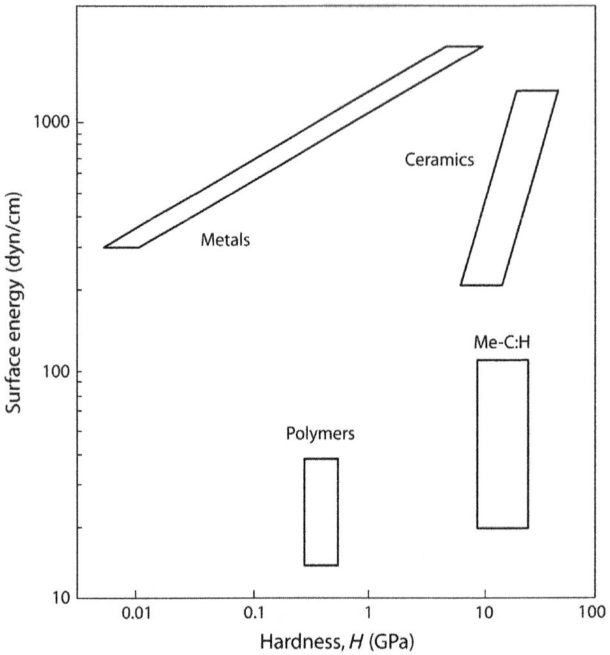

Fig. 4.81. Plot of surface energy vs hardness for metals, ceramics, polymers and the metal-containing hydrogenated carbon coatings (Me-C:H) (after Dimigen and Klages, 1991).

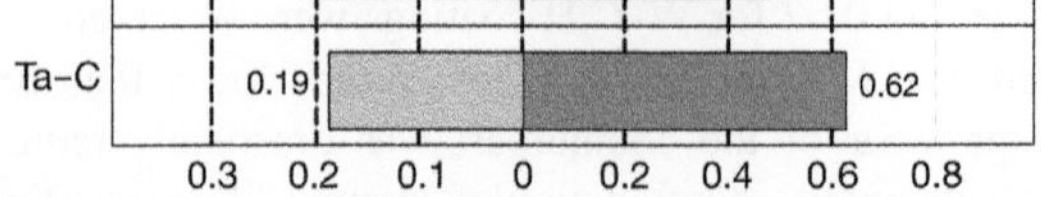

Fig. 4.82. Coefficients of friction and wear rates for doped amorphous carbon coatings deposited on bearing steel in reciprocating sliding against a steel ball (100Cr6) with a load of 5 N in air with a relative humidity of 30% (data from Sander and Petersohn, 1992).

Transfer of coating material took place to the sliding steel ball. They state that at a temperature of 250°C graphitization of the layer became the predominant wear mechanism. The friction coefficient then increased to values of 0.5 to 0.9. Improved wear resistance was reported by Freller *et al.* (1991) when using zirconium-doped carbon coatings (Zr–C) deposited on steel with aluminium and tungsten intermediate layers, compared to uncoated and TiN coated.

The early studies mentioned above laid the foundations for the present state-of-the-art for metal-doped DLC coatings, which now tend to emphasize W, Cr and Ti additions but which are continuously developing to meet the needs of specific contacts and applications.

4.5.3.15 Sliding in oil lubricated conditions

The very good tribological properties of diamond-like carbon coatings in dry sliding conditions can also be beneficial in lubricated conditions. There has been interest in investigating their interaction with lubrication oils to improve the wear resistance especially in starved lubrication conditions. Then they would be used as a kind of safety layer to prevent drastic failures as oil feed is inhibited in, e.g., automotive and machinery applications (Doll, 2003). When a thick elastohydrodynamic lubrication film separates the two sliding surfaces from each other there is really no need for a protective surface coating. In start and stop situations, on the other hand, and in slow sliding, the solid surfaces are in direct contact and only protected by a boundary lubrication film, the strength of which is critical to the tribological performance of the contact. Here a thin coating can be useful for improved surface failure protection.

In dry sliding the formation of graphitized top layers on the coated surfaces and the formation of a transfer layer on the countersurface are of crucial importance for good tribological performance of diamond-like carbon films, as discussed earlier. These mechanisms are both influenced by the presence of oil or chemically active lubricant additives.

The mechanisms of interaction of diamond-like carbon coatings and normal mineral oil with and without extreme pressure (EP) additives were shown by Ronkainen *et al.* (1998). In ball-on-disc tests with a 10 mm diameter steel ball sliding against hydrogenated (a-C:H), hydrogen-free (a-C) and Ti-doped (a-C:H:Ti) diamond-like carbon coatings at a speed of 0.004 m/s, load of 10 N and a Hertzian pressure of about 1 GPa transfer layer formation was observed and low coefficients of friction of 0.21, 0.15 and 0.10 were respectively measured. Low wear of the a-C:H and a-C:H:Ti coatings was observed but the wear of the steel counterface was higher than for the a-C coating, as shown in Fig. 4.83a. The introduction of a mineral oil reduced the friction by 10% for the a-C and a-C:H coatings while the friction slightly increased for the a-C:H:Ti coating, as shown in Fig. 4.83b. The introduction of the mineral oil with EP additives decreased the friction by 10 to 40% for the first two coatings and again slightly increased the friction for the Ti-doped one.

The oil prevented the formation of a transfer layer on the steel balls both for the a-C and the a-C:H coated contacts while it was observed to be formed for the a-C:H:Ti coated contact. For the a-C and a-C:H coated contacts the low friction was now due to normal boundary lubrication mechanisms where a reaction layer is formed on the steel surface. This layer provides lower friction and better wear protection when EP additives are present. The EP additives contained Zn, P and Ca, the ball surface roughness was 0.02 μm and the disk roughness was 0.09 μm.

The effect of different lubricant additives in DLC lubricated contacts varies. The very common anti-wear zinc dialkyldithiophosphate (ZDDP) additive was observed to have no significant influence on the wear behaviour of either hydrogenated (a-C:H), hydrogen-free (a-C) or Ti-doped (Ti-C:H) DLC coated contacts in reciprocating cylinder-on-flat testing in self-mated contacts with a load of 350 N, Hertz pressure of 0.6 GPa and speed of 0.2 m/s (Barros Bouchet *et al.*, 2005). Hydrogenated DLC surfaces were better lubricated in the presence of the friction modifier additive molybdenum dialkyldithitiocarbomate (MoDTC) through the formation of MoS_2 solid lubricant compared to hydrogen-free DLC. Both additives were observed to react directly on the amorphous carbon surfaces and seemed to be more active with the hydrogenated DLC surface. The composition of the tribofilms appeared to be similar to those obtained on steel surfaces in the same lubrication conditions. Interesting is that in this case the tribochemical reactions can occur without the presence of an iron catalyst element in the tribo-system.

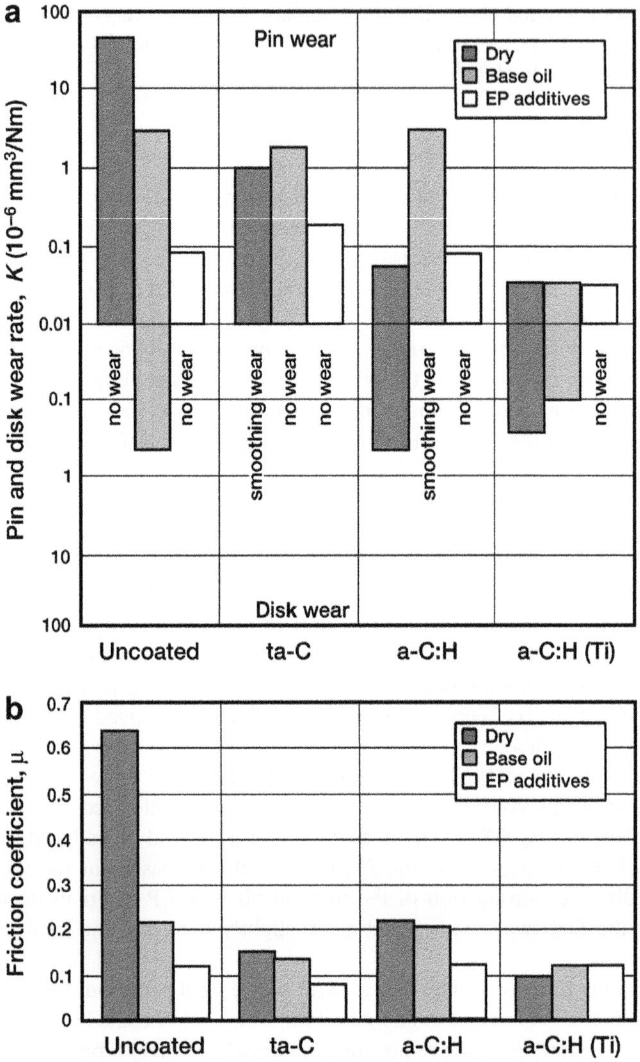

Fig. 4.83. Wear rates (a) and coefficients of friction (b) for uncoated, a-C, a-C:H and a-C:H:Ti coated diamond-like carbon surfaces sliding against steel balls at 10 N load and 0.004 m/s speed in dry, mineral base oil and EP additive containing base oil conditions (data from Ronkainen *et al.*, 1998).

The boundary lubrication properties of diamond-like carbon coatings can be further improved by the use of metal dopants in the coating. This has been reported for W, Ti, Mo and Fe dopants (Miyake *et al.*, 2004; Podgornik and Vižintin, 2005; Podgornik *et al.*, 2003, 2006a, 2008; Kalin and Vižintin, 2006a; Podgornik, 2008).

The influence of the oil lubricant polyalphaolefin (PAO) including a commercial sulphur-based EP additive and a ZDDP-based anti-wear (AW) additive in boundary-lubricated and starved conditions with three W alloyed DLC coated surfaces sliding in a cross cylinder test with a 0.1 m/s speed over the load range of 150 to 1500 N was studied by Jacobson and Hogmark (2001). The endurance life was considerably improved by the DLC coatings compared to uncoated surfaces, as shown in Fig. 4.84.

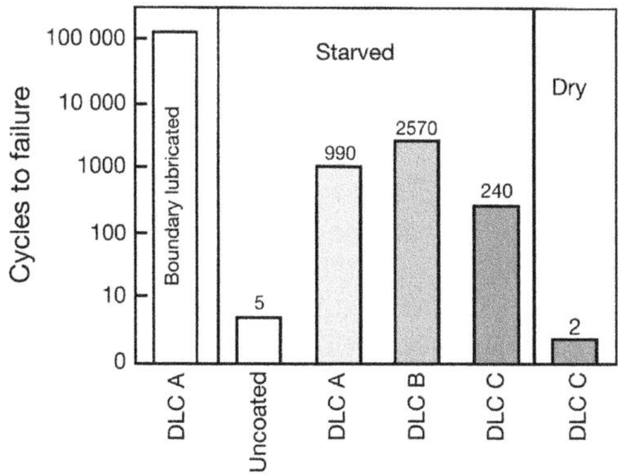

Fig. 4.84. Endurance life given as cycles to failure for three W alloyed DLC coatings sliding against steel in boundary-lubricated, starved and dry conditions with PAO lubricant (data from Jacobson and Hogmark, 2001).

The additives used decreased the coefficients of friction and the lowest values were recorded for steel against DLC surfaces. This was slightly lower than for DLC self-mated surfaces as shown in Fig. 4.85. Based on the experimental investigation the authors give the following design recommendations for lubricated DLC surface applications:

- it seems to be advantageous to deposit the DLC coating only on one of the contacting surfaces, because this gives the best prerequisites for running-in,
- in sliding against an uncoated surface, the substrate under the DLC coating should be harder than the counter material,
- an uncoated material must not contain any hard constituents that can cause wear of the coating by abrasion,
- the substrate roughness should not exceed 0.1 μm; however, if the substrate is too smooth, the beneficial action of the lubricant may be reduced, but

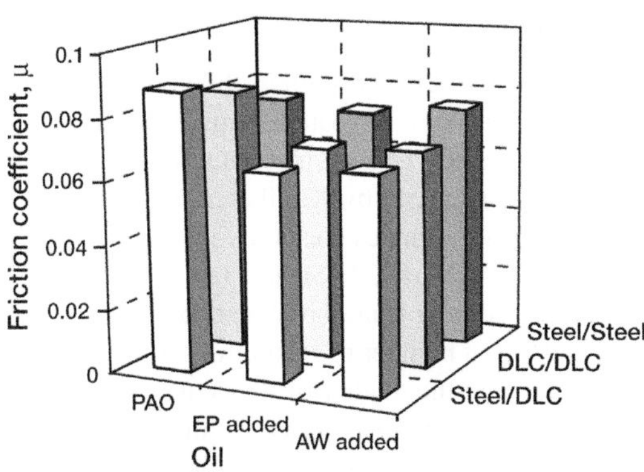

Fig. 4.85. Steady-state friction coefficients in sliding contacts with three W alloyed DLC coatings and steel surface pair combinations and different lubricant and additive conditions measured after 8000 cycles at 700 N load (data from Jacobson and Hogmark, 2001).

- for a contact with both surfaces coated by DLC the recommendation is less straightforward: plastic deformation of the substrate and associated coating failure may occur if the substrates are too soft; soft substrates accordingly require deformable coatings; but on the other hand a too hard substrate might reduce the ability for running-in.

The lubrication, friction and wear mechanisms related to lubricated DLC contacts with various counterface materials and various lubricants and lubricant additives has been reported by Podgornik *et al.* (2003, 2008), Podgornik and Vižintin, (2005), Kalin and Vižintin (2006a, b), Kano (2006), Equey *et al.* (2007, 2008), Neville *et al.* (2007), Andersson *et al.* (2008), Barros'Bouchet and Martin (2008), Manier *et al.* (2008) and Podgornik (2008). A general conclusion is that DLC coatings can be used in most lubricated contacts, they provide an additional protective layer; depositing only onto one of the mating surfaces shortens the running-in of the surfaces and may be beneficial for boundary lubrication layer formation. When alloyed with suitable dopants they can interact beneficially with lubricant additives and provide excellent tribological conditions. It is obvious that the tribological performance deeply depends on the nature of the DLC material, the additive composition and the contact conditions.

4.5.3.16 Abrasive wear resistance

The diamond-like carbon coatings are most known for their extraordinarily low sliding friction properties and for this reason most of the investigations are also focused on behaviour in sliding conditions. However, their abrasive wear resistance seems to be similar to that of other hard coatings based on the few investigations carried out.

With the ball crater test method using hard TiO_2 particles to cause the abrasive wear Van Acker and Vercammen (2004) compared the wear resistance of three PACVD DLC coatings and six other hard coatings. The DLC coatings were about 2 μm thick, the surface roughness was 0.008 μm, the hardness in the range of 11 to 24 GPa and the Young's modulus in the range of 107 to 162 GPa. The hard DLC coating had the best wear resistance of all the coatings while the soft DLC coating had the poorest wear resistance, as shown in Fig. 4.86. In the comparison of all the tested coatings there was no hardness vs wear resistance correlation. However, for the three DLC coatings the abrasive wear resistance correlated directly with the hardness of the coatings. Good abrasive wear resistance of metal-doped DLC coatings has been reported by Harris *et al.* (1997).

4.5.3.17 Rolling contact fatigue wear resistance

In contrast to sliding tests, diamond-like carbon coatings did not perform well in early rolling contact fatigue testing and often failed earlier than uncoated samples or showed no improvement (Rosado *et al.*, 1997). However, recent slip-rolling tests in lubricated conditions with Hertzian surface pressures in the range of 1 to 1.5 GPa carried out by Klaffke *et al.* (2005b) and Löhr *et al.* (2006) show promising results. They tested 40 roller samples coated by seven different DLC coatings including a-C:H, a-C:H:W, a-C:H:Ta/Ti + W, a-C:H, a-C:H:Si and ta-C. The variation between the different coatings, but also between samples coming from the same deposition batch, varied significantly, even more than three orders of magnitude in number of revolutions to failure, e.g. from 300 to one million revolutions. This shows first of all that the quality of the coated surfaces is not at all on the level of what is required for components used in highly loaded rolling fatigue contacts.

In sliding the contact process is less severe and more forgiving but in rolling contact fatigue even one small defect close to the surface may immediately have a fatal consequence. However, the promising part of the tests was that there were two DLC coated rollers, the hard a-C:H and ta-C coated, that showed very good rolling fatigue life and rolled without failure up to 10^7 revolutions, which is a good result. When in the future similar coated components of good homogeneity and

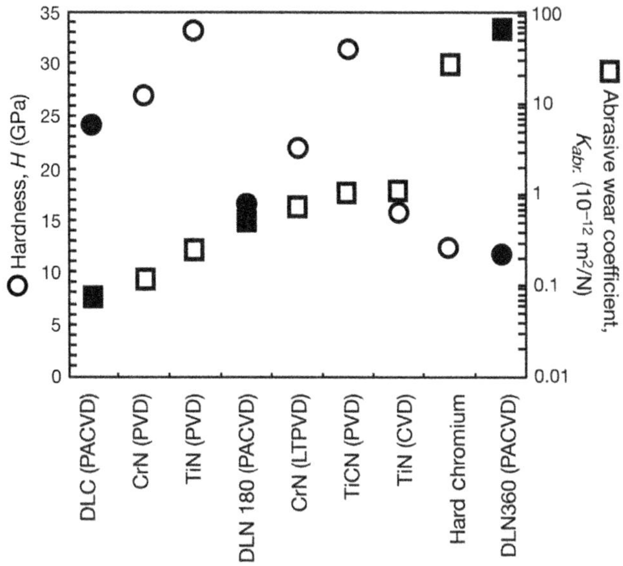

Fig. 4.86. Coating hardness and coating abrasive wear resistance for thin hard coatings tested with the ball crater method with TiO_2 abrasives. Filled balls and squares indicate the DLC coatings (data from Acker and Vercammen, 2004).

quality can be deposited then DLC coatings will also have potential in rolling contact fatigue applications. This is supported by investigations carried out by Manier *et al.* (2008).

4.5.3.18 Low-friction mechanisms of dlc coatings

In very different contact conditions low coefficients of friction, typically in the range of 0.05 to 0.2, and low wear, in the range of 0.01 to $0.5 \cdot 10^{-6}$ mm^3/Nm, have been measured with DLC coatings of various structures and produced by various deposition methods on metal and ceramic components. The common feature of all these is that at least one of the surfaces is amorphous carbon consisting of a mixed structure of crystalline sp^3 bonds and graphitic sp^2 bonds. The reason for the low friction is that this structure has the ability to create easy shear planes between two moving surfaces with very low interfacial strength. The mechanism to produce low friction shear planes is different for DLC than for soft metals, such as In, Pb, Ag, etc., which have a kind of multiple slip system or shear paths resulting in gross plastic shear or flow; and it is also different from lamellar solids, such as MoS_2, graphite and boric acid, which have extremely low shear strengths between their atomic layers (Erdemir, 2004b).

Extensive research has resulted in a widely accepted view that the low friction for DLC surfaces is due to shear primarily taking place between the sliding contact interfaces and that the shear planes are only weakly bonded due to graphitization in microcontacts such as on asperity tops. This mechanism is very relevant when both surfaces are covered with a DLC coating and no debris accumulation or transfer film build-up occurs at the interface. In situations when only one of the surfaces is covered by a DLC coating, the uncoated surface will often, due to the rubbing process, be covered by a carbon-rich mixed oxidized and metal-containing transfer layer that forms a rigid and smooth platform for the partly graphitized top interface layer that is active in the low shear process.

This process has been observed to occur in DLC sliding contacts with many metals and ceramic materials and it explains why the friction and wear in DLC contacts is not so dependent on the

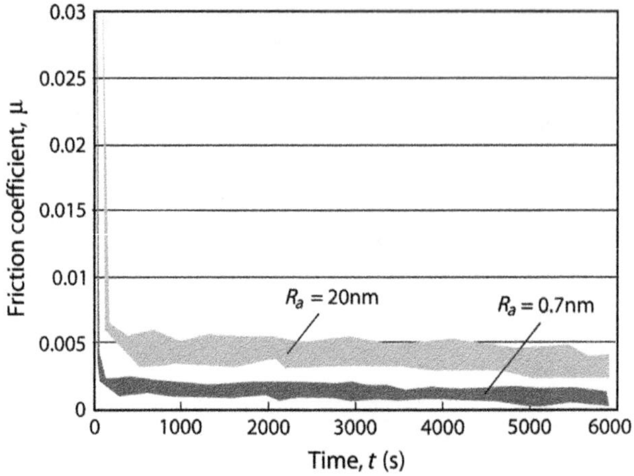

Fig. 4.87. Superlubricity coefficients of friction of highly hydrogenated DLC coatings deposited on a smooth steel surface ($R_a = 20$ nm) and on a very smooth sapphire surface ($R_a = 0.7$ nm) sliding self-mated in dry nitrogen with a load of 10 N and a speed of 0.03 m/s (data from Erdemir, 2004b).

counterface material as is normally the case in tribological contacts. Here the counterface material has the role of only being a platform for the formed transfer layer. The main parameters that influence this contact mechanism and thus the friction and wear performance are the hydrogen content of the coating, the surface roughness of the surfaces, mechanical strength (hardness and elasticity) of the surfaces, normal load and speed of the contacting surfaces and the environment, as discussed above.

4.5.3.19 Superlubricity of DLC coatings

In dry nitrogen extremely low friction of specially prepared DLC surfaces has been measured, with a coefficient of friction in the range of 0.001 to 0.005 and even below 0.001, which represents superlubricity and is shown in Fig. 4.87 (Erdemir *et al.*, 2000, 2001). The wear rate was also low, in the range of 0.00001 to $0.001 \cdot 10^{-6}\ mm^3/Nm$. These DLC hydrogen-rich coatings, called nearly frictionless carbon (NFC), were developed and tested by Erdemir and his co-workers and their superlow friction mechanism has been extensively investigated and described (Erdemir *et al.*, 2000, 2001, 2008; Dickrell *et al.*, 2004; Erdemir, 2004a, b; Erdemir and Eryilmaz, 2007; Fontaine and Donnet, 2007).

To reach the conditions of superlubricity, first macro- and microscale friction increasing effects must be eliminated. Ploughing friction can be avoided by using hard substrates, and asperity collisions can be avoided by using atomically smooth surfaces, such as Si wafers, sapphire or cleaved mica. The loads and speeds are low and the inert environment excludes humidity and contamination effects. Now when disturbing friction increasing effects are eliminated the next challenge is to create the conditions for extremely weakly bonded shear planes. The mechanism for this in the NFC coatings is believed to be the dangling bond orbital mechanisms presented earlier in section 4.5.2 and shown in Fig. 4.49.

The shear is taking place between two atomically smooth planes of hydrogen atoms bonded perpendicularly from the shear plane to the free dangling bond orbitals of carbon in the DLC coating. The coatings are highly hydrogenated to provide enough hydrogen for all free dangling bonds to avoid the possibility that some strong covalent bonds are formed between the surfaces. Some unbonded free hydrogen can serve as a reservoir and replenish or replace those hydrogen atoms that may have been lost due to thermal heating and mechanical grinding during sliding. A molecular dynamic simulation of an NFC coating structure including bonded, atomic and molecular hydrogen within the coating is

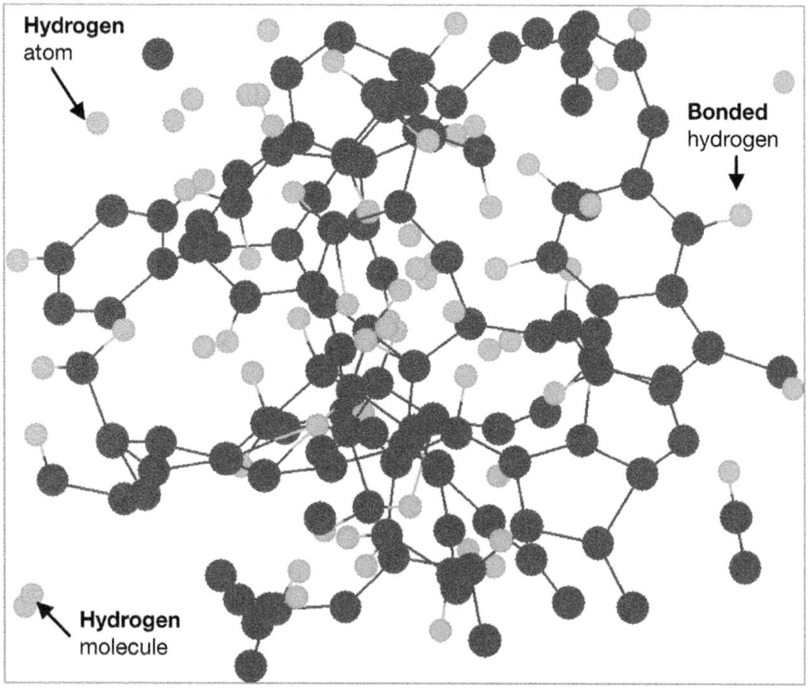

Fig. 4.88. Tight-binding molecular dynamic simulation of highly hydrogenated DLC coatings, showing bonded, atomic and molecular hydrogen within the structure (from Erdemir, 2004b, with permission).

shown in Fig. 4.88. The effect of the hydrogen-to-carbon ratio on friction and wear is shown in Fig. 4.69 and the paramount role of hydrogen in superlow friction has been confirmed by Fontaine *et al.* (2004).

Erdemir (2004b) proposes as the genesis of superlow friction a mechanistic model of superlubricity of DLC coatings. In addition to the above, he presents that in the highly hydrogenated gas discharge plasmas some carbon atoms can be di-hydrogenated during deposition and the existence of di-hydrogenated carbon atoms on the surface will further increase the hydrogen density of the surfaces and thus provide better shielding or passivation of the surfaces. One more reason for these DLC surfaces to provide extremely low friction is that, when the free electrons of hydrogen atoms pair with the dangling σ-bonds of surface carbon atoms, the electrical charge density is permanently shifted to the other side of the nucleus of the hydrogen atom and away from the surface. This allows the positively charged hydrogen proton to be closer to the surface than the electron. A dipole configuration is created at the sliding interface and it gives rise to repulsion rather than to attraction between the hydrogen-terminated sliding surfaces of the DLC coatings. The two atomically smooth surfaces are in a way flying over each other on repulsive forces without any material-internal shear (Su and Lin, 1998; Dag and Ciraci, 2004; Erdemir and Donnet, 2006).

4.5.3.20 Scales of DLC coating tribological mechanisms

Diamond-like carbon surfaces have been found to have excellent tribological properties in many different sliding situations and they are today already in wide commercial use. Their connection to natural diamond surfaces and also the possibility to achieve the lowest coefficient of friction measured

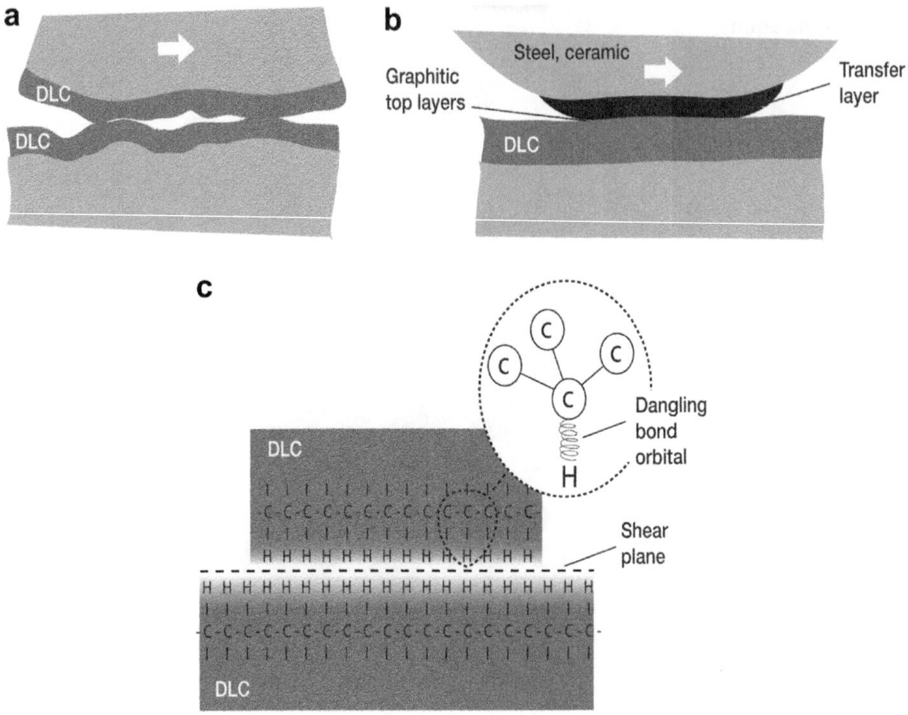

Fig. 4.89. Friction and wear mechanisms of a diamond-like carbon (DLC) coated surface at (a) macroscale, (b) microscale and (c) nanoscale.

for any material has attracted many academic researchers and resulted in extensive and detailed studies of their contacting mechanisms from physical, chemical, material, mechanical and engineering points of view during the last two decades.

Similarly as for diamond coatings discussed earlier (see Fig. 4.60), for diamond-like carbon coatings it is also important to have in mind that the different tribological mechanisms are often typical for a certain length scale level. The dangling bond mechanism that may result in superlubricity appears on the nanolevel, the graphitization and transfer film formation-related mechanism is typical on the microlevel but may also appear at macrolevel if asperity or debris ploughing effects are not involved. This is illustrated in Fig. 4.89. The description of the mechanisms occurring on different scale levels below summarizes observations reported in the literature referred to earlier in this section.

At the *macroscale* and especially with rougher DLC surfaces the contact mechanism is graphitization of the top surfaces with shear within sp^2 graphitic basal planes resulting in low shear resistance, as shown in Fig. 4.89a. In some cases the surface roughness may inhibit graphitization, resulting in high friction and wear. The surface roughness is typically in the range of $R_a = 0.1$ to 1 μm, the coefficient of friction of the order of 0.1 to 0.6 and the wear in the range of 0.0001 to $10 \cdot 10^{-6}$ mm^3/Nm. The environmental conditions may be air, water or oil.

At the *microscale*, with surface topographies typically in the range of $R_a = 10$ to 100 nm, the contact mechanism is first smoothing of the countersurface by building up a transfer layer containing typically Al, C, Cr and Fe. Graphitization occurs on both the DLC top surface and the countersurface transfer layer. The shear takes place within the basal planes of the sp^2 hybridized graphitic carbon that has been formed by transformation from sp^3 at the high flash temperatures and high pressures between

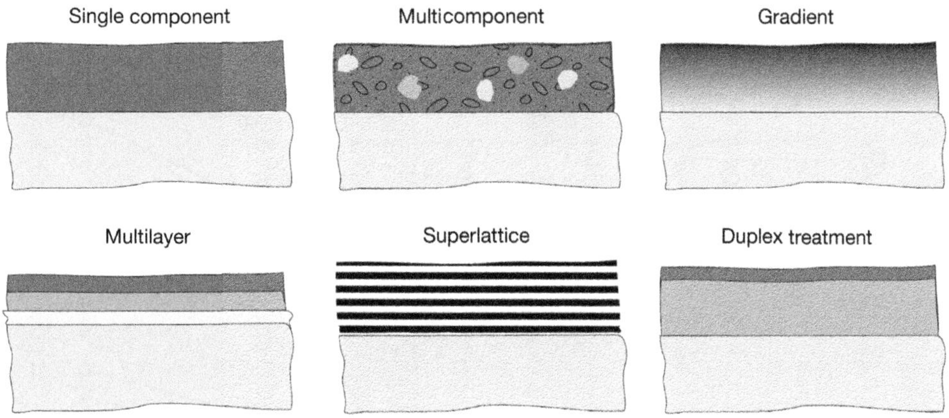

Fig. 4.90. Possible complex structures of tribological coatings.

colliding surface asperities in sliding, as shown in Fig. 4.89b. The friction is low, of the order of 0.05 to 0.3, and the wear is in the range of 0.0001 to $1 \cdot 10^{-6}$ mm^3/Nm. The environmental conditions are typically air.

At the *nanoscale* with atomically smooth surfaces in the range of $R_a = 1$ to 30 nm the conditions of superlubricity with coefficients of friction down to 0.001 and wear rates down to $0.00001 \cdot 10^{-6}$ mm^3/Nm can be achieved. The contact mechanism is shear between two flat, highly hydrogenated layers of single hydrogen atoms at dangling bonds. Positive atomic dipole charge of the hydrogen atoms out from the surface at both surfaces gives rise to repulsive forces, as shown in Fig. 4.89c. The environmental conditions are vacuum, dry nitrogen, argon or helium.

4.6 Combined Coatings

In the previous sections we have mostly discussed the tribological properties of surfaces with single layered coatings of various materials. For many years this was the approach used to find coatings-related solutions to specific application problems. In addition to the use of single component materials we have seen that improved tribological properties can be achieved by adding into the coating elements of other materials to produce a doped or composite coating. The new deposition techniques provide us with further possibilities to produce coated surfaces with different kinds of complex structures and architectures, such as multicomponent coatings, gradient coatings, nanocomposite coatings, multilayer coatings, superlattice coatings, duplex coatings and adaptive coatings, as shown in Fig. 4.90 (Subramanian and Strafford, 1993; Hogmark *et al.*, 2000).

4.6.1 Multicomponent coatings

A wide range of possibilities to further improve the tribological properties of surface coatings is offered by new multicomponent coatings. These have a mixed composition of several materials. By changing the relative concentration of non-metallic elements in a coating composition, the valence electron concentration will change and thus produce a change in mechanical and physical properties. Another possibility is to substitute the metal lattice of the compound phase with another compatible metal; e.g. in TiN coatings the lattice position of Ti can be partly substituted by Al, V, Zr or Hf. The PVD deposition techniques are especially suitable for producing such multicomponent coatings with

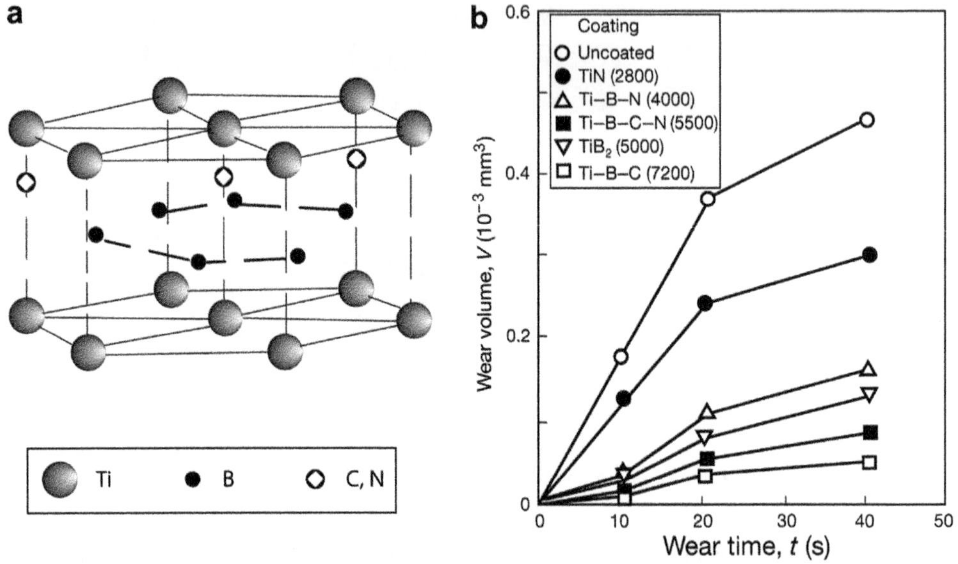

Fig. 4.91. (a) Possible incorporation sites of the metalloid atoms carbon and nitrogen in the hexagonal TiB_2 cell. (b) Wear volume as a function of time for multicomponent and reference coatings from abrasive wear tests. The microhardness (H_v 0.05) of the surface is given in the figure. The abrasive medium was an Al_2O_3 suspension and counterpart 100Cr6 (data from Knotek *et al.*, 1990b).

a large variety of different compositions (Freller and Haessler, 1988; Holleck, 1991; Knotek *et al.*, 1992c; Subramanian and Strafford, 1993; Hogmark *et al.*, 2000).

Superhard multicomponent coatings with improved abrasive wear resistance were developed by Knotek *et al.* (1990b). By magnetron sputter ion plating they incorporated carbon and nitrogen atoms in the titanium–boron structure, as shown in Fig. 4.91a, and produced Ti–B–N, Ti–B–C and Ti–B–C–N coatings. The thickness of the coatings was in the range of 6 to 25 μm on cemented carbide, high-speed steel and austenitic stainless steel substrates. Considerably improved wear resistance was observed for Ti–B–C coatings in abrasive wear tests, as shown in Fig. 4.91b. The volumetric wear rate of the different coatings was in direct relation to the microhardness of the coatings.

A multicomponent composition of Ti–Al–V–N–C was optimized for wear protection of carbide indexable inserts used for turning (Knotek *et al.*, 1988). The choice of deposition target compositions and configurations of the complex coatings are shown in Fig. 4.92 and the results of tool life determined by crater depth in turning tests are shown in Fig. 4.93. It can be seen that a specific composition of (Ti,Al,V)C,N resulted in significantly improved tool life. The optimized coating possesses good hardness, good adhesion and adequate thermal stability, even after lengthy high temperature annealing. The results are supported by good wear behaviour in tool life turning tests in terms of both crater and flank wear.

It is now realized that the early promise of multicomponent coatings of the kinds referred to above was usually due to the influence of small grain sizes on mechanical, physical and chemical properties. This realization has resulted in an increase in research into nanocomposite multicomponent coatings.

4.6.1.1 Gradient coatings

Gradient coatings are a type of multicomponent coating in which the compositions or structural morphologies continuously change when moving from the substrate to the surface. This gives the

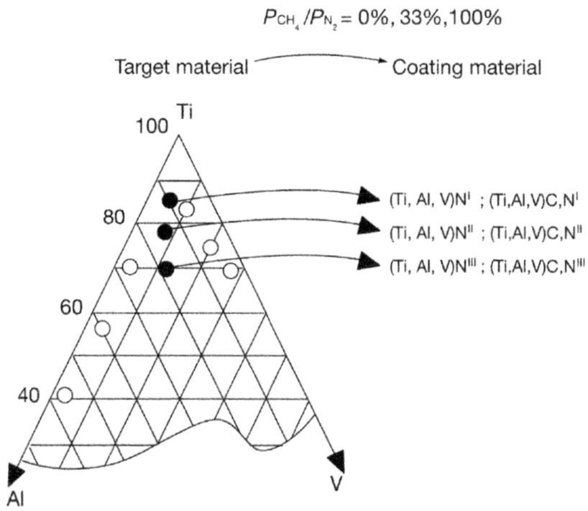

Fig. 4.92. Deposition target composition and configurations in the complex multicomponent (Ti,Al,V)N and (Ti,Al,V)C,N coatings (after Knotek *et al.*, 1988).

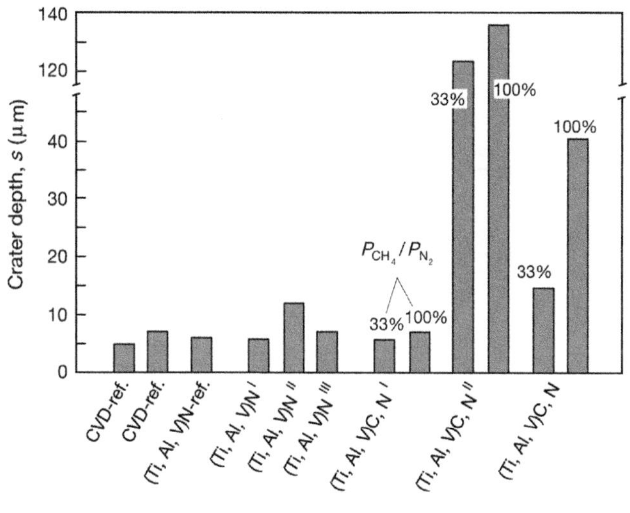

Fig. 4.93. Tool life determined by crater wear depth of multicomponent (Ti,Al,V)N and (Ti,Al,V)C,N coatings on carbide indexable inserts in turning tests (data from Knotek *et al.*, 1988).

possibility to vary the properties of the coating at different depths from the surface. WC–TiC–TiN gradient coatings, where the composition changes from WC over TiC to TiN, were deposited by magnetron sputtering on hard metal substrates and tested in turning conditions by Fella *et al.* (1988). Figure 4.94 shows the PVD phase diagram and the change in composition of the gradient coatings in the WC–TiC–TiN system. For comparison, similar coatings with a reversed order of composition were produced. Results from turning tests of steel (AISI 1042) in Fig. 4.95 show that flank and crater wear are reduced to a level almost as low as for TiN or TiC coated tools.

Fischer and Oettel (1997) have discussed how microstructural gradients can be achieved in TiN coatings by time-dependent variations in the bias voltage during deposition, and this can beneficially

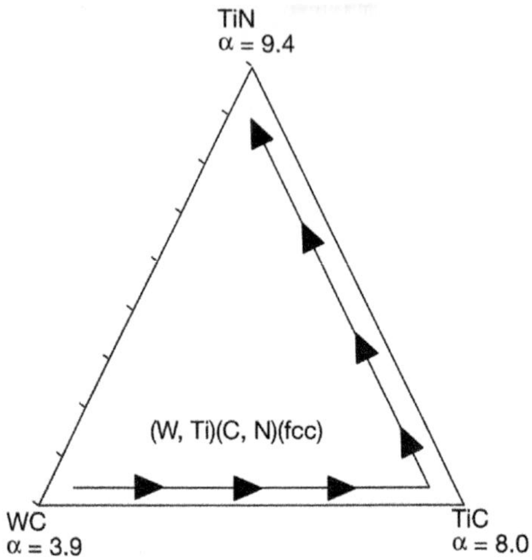

Fig. 4.94. PVD phase diagram at 500 K showing the change in composition of the multicomponent gradient WC–TiC–TiN coating when moving from substrate to surface. α is the coefficient of thermal expansion in $10^{-6}\,K^{-1}$ (after Fella *et al.*, 1988).

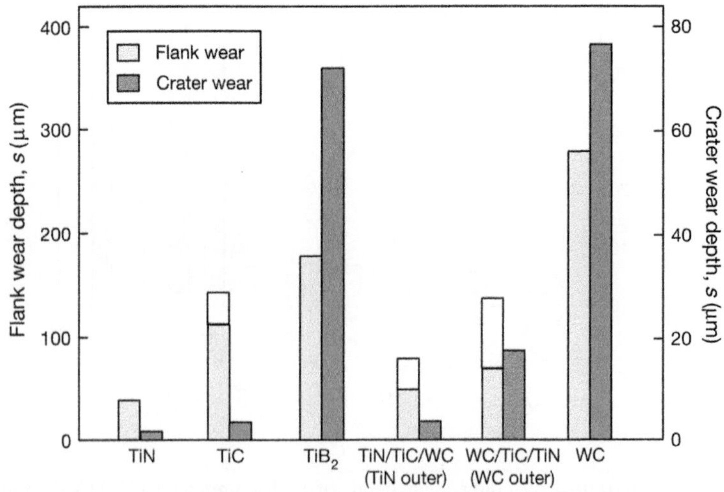

Fig. 4.95. Flank and crater wear after turning of steel with multicomponent gradient coated tools and reference coated tools (data from Fella *et al.*, 1988).

modify the stress state and the adhesion. It is, however, more common to utilize compositional gradients to achieve such benefits. For example, commercial TiCN coatings are provided with an initially deposited layer of TiN to promote adhesion to the substrate and then the nitrogen content of the process gas is gradually increased to make TiCN with a graded composition running from pure TiN to TiCN over a few 100 nm of coating growth. This graded layer may or may not be overcoated with a layer of TiCN with fixed composition. The hardness, adhesion and the tribological behaviour of such coatings have been studied (Bull *et al.*, 2003a, b), and their findings suggest that by using predictive hardness modelling, coating architectures can be designed to give improved hardness responses.

Uglov *et al.* (2005, 2008) have deposited gradient Ti–Cr–N coatings and have shown that these can have reduced internal stress compared with coatings of constant through-thickness composition. They found that the mechanical properties, hardness and elastic modulus of the gradient coatings are the superposition of the solid solutions Ti_xCr_{1-x} N of different concentrations ($0.6 < x < 0.84$ and $0.25 < x < 0.67$). Many researchers have found that ternary coatings are harder than either of the parent binary nitrides. Boxman *et al.* (2000) reported this for (TiZr)N, (Ti, Nb)N and (Zr, Nb)N, which explains why compositional variations can be used affectively to control not only chemical and structural characteristics but also mechanical properties. Indeed, gradient coatings have proved to be particularly effective in enhancing thermal barrier coating (TBC) performance (Movchan and Marinski, 1998; Movchan, 2002). In that case a gradient composition through an NiAl bond coat at the coating/substrate interface, through to a yttria-stabilized zirconia (YSZ) at the outer surface provides the desired adhesion and the capability to accommodate strain caused by differential thermal expansion between the coating and substrate, as well as the required thermal, tribological and hot-corrosion barrier properties. As well as YSZ, other functionally gradient TBC coatings are being investigated, including NiCrAl + $MgZrO_3$ (Tekmen *et al.*, 2003). Malzbender (2004) has modelled the effects of mechanical and thermal stresses on TBCs, to aid in predicting mechanical behaviour and lifetimes of coatings.

Many examples of compositionally gradient coatings can be found in the coatings literature. Most of these relate to coatings for cutting tools, with particular emphasis on the Ti–Al–N system and its derivatives (Qiao *et al.*, 2000; Manaila *et al.*, 2002; Zhou *et al.*, 2004). Other gradient systems reported include several which include carbon, such as Ti–TiN/CN_x (Donnet *et al.*, 1999; Stüber *et al.*, 1999; Fernández-Ramos *et al.*, 2004).

4.6.2 Nanocomposite coatings

Several of the gradient coatings mentioned above have nanocomposite structures. Nanocomposite coatings are coatings with structural features, like the grain size, with dimensions below 100 nm. Such coatings can be achieved by controlling deposition parameters when mutually insoluble phases are simultaneously deposited or by subsequent heat treatment. The yield strength, hardness and toughness of polycrystalline materials all generally improve with decreasing grain size. A similar phenomenon seems to be valid for thin coatings down to nanometre size grains. In addition to mechanical properties, nanocrystalline materials can exhibit higher thermal expansion, lower thermal conductivity, and unique optical, magnetic and electronic properties. Nanocomposite hard coatings may be formed by a nitride-forming transition metal and a metal component or another compound which is not dissolved in the nitride. In a second class of nanocomposite coatings the nitride crystallites are embedded in an amorphous phase of another compound (Hogmark *et al.*, 2000; Zhang *et al.*, 2005b; Galvan, 2007; Jehn, 2006).

Because of its excellent properties, nc-TiN/a-Si_3N_4 is a widely studied hard coating system. It is composed of the nanocrystalline nitride and amorphous silicon nitride, where the nc-crystallites are embedded in the amorphous Si_3N_4 phase. Additionally, a third phase is formed, a- and nc-$TiSi_2$, which forms smaller crystallites in the amorphous boundary phase. Examples of other nanocomposite hard coating systems are: Ti–Ag–N, Ti–Cu–N, Ti–Ni–N, Zr–Ni–N, Cr–Cu–N, Mo–Cu–N, Mo_2N/Ag, nc-ZrN/a–Si_3N_4, Ti–Al–V–Si–N, nc-$(Al_{1-x}Ti_x)$N/a-Si_3N_4 and h-BN/c-BN (Jehn, 2006).

The detailed friction and wear mechanism of nanocomposite coatings in tribological contacts is quite complex and merits extensive investigation. There are interesting tribological studies of nanocomposite coatings that demonstrate the potential of these films, such as the investigation of WC/DLC/WS_2 nanocomposite coatings (Voevodin *et al.*, 1999a), nanocrystalline yttria-stabilized zirconia (YSZ) imbedded in an amorphous YSZ/metal matrix (Muratore *et al.*, 2006), TiC/a-C:H

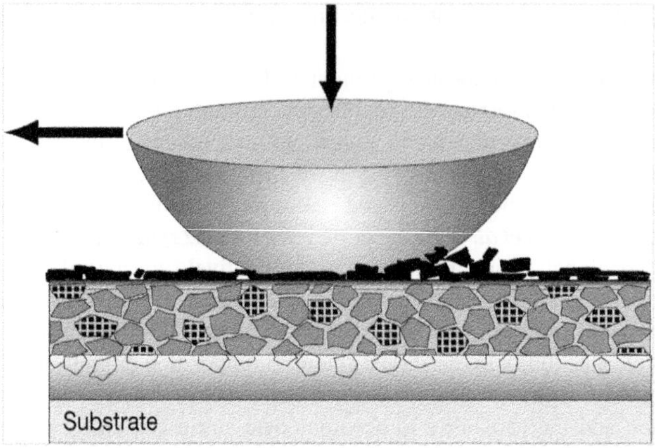

Fig. 4.96. Schematic of a tough nanocomposite coating, featuring a nanocrystalline/amorphous design in the main coating layer for cohesive toughness and a functionally gradient interface layer for adhesive toughness. DLC is used as an amorphous phase in this design together with nanocrystalline carbide phases (after Voevodin, 2008).

nanocomposite coatings (Pei *et al.*, 2006), TiC/DLC nanocomposite coatings (Galvan, 2007), nanostructured tungsten carbide particles/polymer composite coatings (Wang and Lim, 2007), tough nanocrystalline carbide/amorphous DLC coatings (Voevodin, 2008) and Cu-doped TiN, CrN and MoN nanocomposite coatings (Öztűrk *et al.*, 2008).

A combination of a nanocrystalline/amorphous DLC design with a functionally graded interface can provide high cohesive toughness and high interface adhesive toughness, as shown in Fig. 4.96.

Interesting new tribological features were reported by Pei *et al.* (2006) in their study of TiC/a-C:H nanocomposite coatings. They report self-lubricating effects resulting in ultra-low friction and superior wear resistance, transfer film formation on the counterpart and a breakdown of the Coulomb friction law in the nanocomposite coating contact mechanism. In addition, they recorded that the coefficient of friction decreased with decreasing relative humidity. The tests were carried out with a pin-on-disk device, the counterpart was a 100Cr6 steel ball with 6 mm diameter, normal load was 5 N and sliding speed in the range of 0.1 to 0.5 m/s in room temperature and this resulted in friction coefficients of 0.013 to 0.047 and a wear rate of $0.01 \cdot 10^{-6}$ mm/Nm.

Nanocomposite coatings are mentioned frequently elsewhere in this book, such as the metallic and ceramic materials-related aspects of nanocomposites discussed in section 2.6.2, and the various carbon-based coatings containing nanocrystals discussed in section 4.5. A book edited by Cavaleiro and De Hosson (2006) provides a particularly useful insight into the current research state-of-the-art of this topic, and the paper by Veprek and Veprek-Heijman (2008) provides a review of industrial applications of superhard nanocomposite coatings.

4.6.3 Multilayer coatings

Many of the gradient layer coatings mentioned above, and even some coatings classified as nanocomposites, are in fact multilayers. This may arise due to the nature of the deposition process, involving, for example, planetary rotation of components past a series of vapour sources. One of the earliest exponents of the benefits of multilayering of PVD coatings for tribological applications was Holleck (Holleck, 1986, 1991; Holleck *et al.*, 1990; Holleck and Schier, 1995) who particularly emphasized the use of a large number of repeated layers. There are three main reasons why it may be advantageous to use layered coatings.

1. Interface layers. These are increasingly used to improve the adhesion of a coating to the substrate and to ensure a smooth transition from coating properties to substrate properties at the coating/substrate boundary. Additional effects may be to inhibit the substrate material from influencing the coating deposition process, to add a substrate wear-protecting layer or to decrease the probability of pinholes in the coating system.
2. Large number of repeated layers. By depositing several thin layers with various mechanical properties on each other the stress concentration in the surface region and the conditions for crack propagation can be improved.
3. Diverse property layers. The properties of the surface can be improved by depositing layers of coatings that separately have different kinds of effects on the surface, such as corrosion protection, wear protection, thermal isolation, electrical conductivity, diffusion barrier and adhesion to the substrate.

The plasma-assisted PVD process is most suitable for the production of both multilayer and multi-component coatings because of its high flexibility in terms of materials which can be deposited. With this process surface layers can be formed from any desired metal or metal-based component, and the coating can be deposited on virtually any substrate material (Matthews, 1988a; Holleck, 1991; Knotek *et al.*, 1992c; Subramanian and Strafford, 1993; Hogmark *et al.*, 2000; Jehn, 2000, 2006).

4.6.3.1 Interface layers

In tribological applications of surface coatings, very commonly the adhesion of the coating to the substrate is one of the crucial problems to overcome before a satisfactory surface behaviour can be achieved. One technique that may be helpful, and which is increasingly being used, is to deposit a very thin interface layer on the substrate before the coating is deposited. By a correct material choice such an interface can replace weak coating-to-substrate bonds with a strong coating-to-interface layer and interface layer-to-substrate bond. Another effect of the interface layer can be to reduce the internal stresses in the interface region. By the use of at least two layers on each other the probability of pinholes or defects going from the surface to the substrate is decreased as shown in Fig. 7.41 (Knotek *et al.*, 1992e).

Thin pure titanium interface layers have successfully been used by Valli *et al.* (1985a); Cheng *et al.* (1989); Peebles and Pope (1989); Matthes *et al.* (1990b) and Kim *et al.* (2003) to improve the wear resistance of TiN coatings on steel substrates in different wear situations. The characteristics of different reported coatings and interface layers are given in Table 4.2. Cheng *et al.* (1989) show that the interface layer increases the attachment of the TiN coating to the substrate via a better interface contact and stronger chemical bonds. Matthes *et al.* (1990b), on the other hand, show that the wear resistance improvement in their experiments originates from improved coating quality due to a reduced influence from graphite in the grey cast iron substrate on the growth of the hard coating.

A thin chromium layer has been used by Farges *et al.* (1989) to improve the adhesion of a layered tungsten–carbon/tungsten carbide coating to a steel substrate. Chromium also proved to be a suitable interface layer for improved erosion resistance when depositing hard coatings such as ZrB_2 on polyimide substrates (Chambers *et al.*, 1990). The use of titanium interface layers under TiC/Ni coatings on polyimide was not, however, successful.

The effect of ambient, thermal and mechanical stresses on the adhesion of a titanium layer to a polyimide substrate in different multilayer systems has been analysed by Seshan *et al.* (1989). With pull tests they showed that the titanium–polyimide bonding degrades by a thermochemical mechanism when the composite is thermally cycled. Through finite element techniques they demonstrate

Table 4.2. Examples of interface layers for coating-to-substrate adhesion improvement.

No.	Substrate	Interface layer		Coating			Testing device	Reference
		Material	Thickness nm	Material	Thickness μm	Deposition method		
1	Steel (AISIM50)	Ti	150	TiN	2	Ion-plating	Scratch tester	Cheng *et al.* (1989)
2	Steel (304SS)	Ti	16	TiN	0.1	Reactive evaporation	Pin-on-plate wear tester	Peebles and Pope (1989)
3	Grey cast iron	Ti	1000	TiN	5–7	Rf sputtering	Sheet metal forming tester	Matthes *et al.* (1990b)
4	Tool steel	Cr	4	W–C	16	Reactive sputtering	Pin-on-disk tester	Farges *et al.* (1989)
5	Polyimide	Cr	12	ZrB_2	20–30	Sputtering	Erosion tester	Chambers *et al.* (1990)
6	Cemented carbide	TiC, TiN or Ti (C, N)	20–40	Al_2O_3	1–3	CVD and PVD	Scratch tester	Lhermitte-Sebire *et al.* (1986)
7	Silicon, iron, cemented carbide	Si_3N_4		C–BN	1.5	Reactive evaporation	Ring-on-disk tester	Inagawa *et al.* (1989)
8	Steel	Cu	50–100	Pb	100	Rf sputtering	Ball-on-disk tester	Gerkema (1985)
9	Cemented carbide	TiC	5000	Al_2O_3	1–2	CVD	Scratch tester	Hintermann and Laeng (1982)
10	Ceramic (ZrO_2)	SiC	10–20	DLC	2	Ion beam deposition	Ball-on-disk	Erdemir *et al.* (1991c)
11	Steel	Zr:C, W, Ta, CrN, Al	100–4400	DLC		PVD-PCVD	Scratch tester	Freller *et al.* (1991)
12	Steel	TiC	1000	DLC	2	Rf CVD	Scratch tester	Yoshino *et al.* (1991)
13	Ceramic (MgO–PSZ)	Ti	25	Ag	1.5	IBAD	Pin-on-disk	Erdemir *et al.* (1991b)
14	Ceramic (A_2O_3)	Ti	25	Ag	1.5	Sputtering	Pin-on-disk	DellaCorte *et al.* (1992)
15	Steel (440C)	Au–Pb	1.5–4.5	MoS_2	10	Rf magnetron sputtering	Two-disk rolling test	Hopple *et al.* (1993)
16	WC–Co	TiC	1.5	Al_2O_3	10	CVD	Scratch tester	Echigoya *et al.* (1994)
17	Steel Si_3N_4	DLC	20	Diamond	30–80	Microwave PACVD	Topple tester	Itoh *et al.* (1999)
18	AISI H13	Ti	0.24	TiN	0.24–0.75	Arc ion-plating	Impact	Kim *et al.* (2003)
19	WC–Co	Cu and Cu/Ti	Diffused	Diamond	10	Microwave PACVD	Scratch tester	Huang *et al.* (2007)

that when loads and displacements are imposed at the coating-to-substrate interface, large localized stresses are developed.

Possibilities of improving the adhesion of alumina coatings on cemented carbide substrates by using 20 to 40 μm thick PVD or CVD intermediate TiN, TiC or Ti(C,N) layers were studied by Lhermitte-Sebire *et al.* (1986). They found that when alumina is deposited at low temperatures, e.g. by PVD methods such as ion beam sputtering, the adhesion remains poor and failure of the bond occurs, owing to insufficient chemical interdiffusion. In contrast, when interdiffusion takes place, i.e. for alumina deposited by CVD at high temperatures, the strength of the adhesive bond is very high. Oxidation of the TiC surface prior to the alumina deposition resulted in thin layers of oxide being responsible for a significant lowering of adhesion of the alumina films. Echigoya *et al.* (1994) quantified a considerable improvement in the bond strength of alumina films to WC–Co substrates by using a TiC interlayer under non-oxidizing conditions.

The adhesion of hard cubic boron nitride (c-BN) coatings to different substrates, such as silicon, iron or cemented carbide, can be improved by using a silicon nitride interlayer (Inagawa *et al.*, 1989). Other interlayers such as carbon, germanium, iron, titanium, chromium and aluminium have not proved to be successful. The attachment of diamond-like coatings to ceramic and steel substrates has been improved by SiC, TiC and Zr:C interface layers (Erdemir *et al.*, 1991c; Yoshino *et al.*, 1991; Freller *et al.*, 1991).

Many writers have discussed the use of interlayers to improve the bond strength of diamond films, using DLC (Itoh *et al.*, 1999) and Cu or Cu/Ti (Huang *et al.*, 2007).

A copper film was found to be beneficial as an interlayer for a soft lead coating on a steel substrate, giving an improvement in sliding wear resistance compared to silver, molybdenum, tantalum and tungsten interlayers (Gerkema, 1985). This was believed to be due to an additional layer protecting direct substrate/counterface contacts when the lead coating had been removed by wear. The adhesion of a soft, low-friction silver layer to ceramic substrates has been improved by depositing a very thin, only 25 nm thick, interface layer of titanium (Erdemir *et al.*, 1991b; DellaCorte *et al.*, 1992).

Ollivier and Matthews (1993) and Matthews and Eskildsen (1993) discuss, with particular reference to DLC coatings, how a hard interlayer, such as TiN or Ti(C, N), can apparently give an improvement in adhesion in the scratch test. This is due to the improved load support provided. In pin-on-disk tests these interlayers may lead to adhesive bond failure between the coating and the interlayer, due to the higher contact stresses induced by the smaller contact area. In that case a softer, more conformal, interface material such as titanium or chromium may be better.

Ni *et al.* (2004b) discuss how the appropriate selection of an interlayer can influence the mechanical response of the surface in terms of the ratio of the hardness to the elastic modulus. They found that a superelastic NiTi interlayer enhanced the pin-on-disk dry sliding friction and wear performance of a CrN coating on an aluminium alloy substrate, against a cemented carbide counterface.

4.6.3.2 Large number of repeated layers

The mechanical properties of a surface can be manipulated by coating it with a multilayered structure which has a great number of thin layers of two or more different materials deposited one on another, as shown in Fig. 4.97. When the layers have different elastic and plastic properties the composite characteristics of the whole coating can be controlled in a unique way to give behaviour which is unachievable in normal metals, alloys or ceramics (Koehler, 1970; Springer and Hosford, 1982; Springer *et al.*, 1974; Bunshah *et al.*, 1980; Subramanian and Strafford, 1993; Wiklund, 1999; Nordin, 2000; Hogmark *et al.*, 2000; Carvalho, 2001).

Coatings comprising many thin layers can often possess greater strength, elasticity and also plasticity than thick coatings comprising one material (Holleck, 1991; Knotek *et al.*, 1992c). The formation and propagation of cracks in the coating under critical tribomechanical stresses can thus be inhibited by using multilayer coatings.

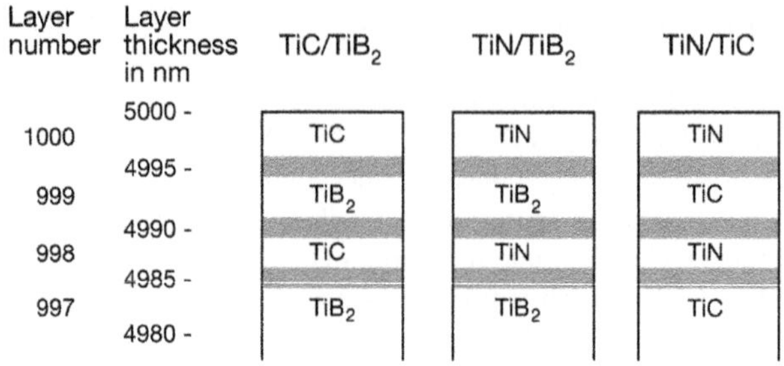

Fig. 4.97. Schematic coating microstructure with intermixed zones of various thicknesses for the outer part of TiC/TiB_2, TiN/TiB_2 and TiN/TiC multilayers consisting of 1000 single layers (after Holleck *et al.*, 1990).

The term 'superlattice' is sometimes used to describe multilayered coatings of small layer thickness. Strictly speaking, the expression should only be used when each layer is a single crystal and there is a crystallographic relationship between each alternate layer. However, the term is sometimes also used loosely to indicate structures consisting of polycrystalline layers. In order to qualify for the title superlattice each layer should be of the order of 5 to 25 nm thick. This type of multilayered coating structure can improve the hardness as well as the toughness, compared to single layers of the same materials. Examples of quoted superlattice coatings are the combination of TiN/VN (Helmersson *et al.*, 1987) and TiN/NbN (Shinn *et al.*, 1992; Hogmark *et al.*, 2000). As pointed out by Sproul and co-workers (Sproul, 1994; Yashar and Sproul, 1999), many reported superlattice coatings could more correctly be termed polycrystalline superlattices – although nanometer-scale multilayers would be a scientifically correct term; these include nitride systems such as CrN/NbN (Purandare *et al.*, 2006) and TiAlN/VN (Zhou *et al.*, 2004). The cited papers illustrate the use of such coatings in diverse tribo-contact conditions, including erosion–corrosion.

The usefulness of interposing repeated ceramic and metallic layers to reduce interface shear forces has been demonstrated with a mathematical model for stress analysis and failure possibility assessment of multilayer coatings developed by Lyubimov *et al.* (1992a, b). The model takes into account thermal, intrinsic and operational stresses and enables the determination of optimal thicknesses and deposition conditions especially for ion-assisted PVD coatings. The failure assessment is carried out according to a mechanism including failure of a single layer under stresses existing in it, interface crack propagation and finally peeling above an area with imperfect adhesion.

Djabella and Arnell (1993a) showed by using a finite element analysis that double-layer coatings generally have less severe stress distributions than monolayers having the same thickness as the double layer and the same modulus of elasticity as the outer layer of the double-layer coating. The profound effect of the tangential load on the pressure profiles both for soft and hard coatings has been shown by a numerical solution for dry sliding line contacts of multilayered elastic bodies by Elsharkawy and Hamrock (1993).

In a study of Mo/Ni, Pt/Ni and Ti/Ni multilayer coatings with bilayer thicknesses of 1 to 20 nm and total coating thicknesses of 200 to 300 nm Clemens and Eesley (1989) show that a measured hysteresis effect, also called elastic softening, can result from the interfacial strain. Thus, in spite of the considerable differences in the interfacial structure of the investigated metal pairs, it is the interfacial strain that is the cause of apparent changes in the bulk elastic response. The plastic work done in the soft substrate under a multilayer coating has been analysed by Engel *et al.* (1993), and they found, not surprisingly, that it was proportional to wear and inversely proportional to friction.

There is much experimental evidence confirming the advantages of multilayer coatings compared to single layers. In ball-on-disk tests with a load of 5 N and sliding velocities of 0.05 to 1 m/s, 2 μm thick TiC/Ti(C,N)/TiN multilayer coatings deposited by PVD on high-speed steel (AISI M2) and sliding against corundum balls had three to four times lower wear rates than a PVD TiN single layer coating in the same conditions (Vancoille *et al.*, 1993b). The multilayer coating consisted of nine separate sublayers.

Improved wear resistance by using multilayer coatings compared with single layers has been reported by Renji *et al.* (1991) for 5.5 μm thick $TiC/TiC_xN_{y...}/TiC/TiC_xN_y/TiN$ coatings on a steel substrate, by Lyubimov *et al.* (1992a), for 9.5 μm thick Ti/TiN/Ti/TiN/Ti coatings on cast iron substrates and by Hilton *et al.* (1992), for $Ni/MoS_2/Ni...Ni/MoS_2$ or $Au–Pd/MoS_2/Au–Pd...Au–Pd/MoS_2$ coatings on steel or silicon substrates. Leyland and Matthews (1993) have shown that by multilayering Ti and TiN much greater thicknesses can be achieved than possible with TiN alone, and the abrasive wear resistance is considerably improved. The ideal was a 2:1 ratio in TiN to Ti thickness in each layer, typically 2 μm and 1 μm, respectively. Bemporad *et al.* (2006) demonstrated that Ti/TiN multilayers could be produced at greater thicknesses than TiN layers alone, and demonstrated improved adhesion, toughness and wear resistance.

Multilayers deposited by magnetron sputtering and composed of 10, 100, 250, 500 or 1000 individual layers with an overall coating thickness of 5 μm for TiC/TiB_2, TiN/TiB_2 and TiN/TiC layer systems were studied by Holleck *et al.* (1990). They found that a coating composed of about 100 to 200 single layers has optimum characteristics with regard to crack propagation resistance and wear resistance, especially under interrupted cutting conditions. The microhardness of the coatings changed only slightly with the number of layers.

There are indications that the cracks are deflected at the interface zones and in this way they do not destroy the coating. Figure 2.22 shows a possible mechanism of crack deflection and stress relaxation in the multilayer coating system. However, interfaces can increase crack propagation resistance and provide opportunities for energy dissipation. It is probable that the periodically changing stress and strain fields occurring in the multilayers play an important role in strengthening the layered surface. Holleck *et al.* (1990) conclude that in view of the marked changes in microstructure of coatings prepared, in one case with ion bombardment and in the other case without, it seems unlikely that the deviations in chemical composition and the degree of crystallization or orientation play predominant roles in determining the differences in properties and wear behaviour.

Improved tribological properties have been achieved by ion or laser beam mixed multilayered Fe–Ti–C structures on stainless steel (Hirvonen *et al.*, 1989b) and by magnetron sputter-deposited multilayered tungsten–carbon/tungsten carbide coatings with a chromium interface layer on a tool steel substrate (Farges *et al.*, 1989). Significantly prolonged tool lives have been achieved by the use of Ti–Zr–N, Ti–C–N and Ti/N–C–N multilayer coatings on carbide inserts, especially in interrupted cut machining (Knotek *et al.*, 1992d).

Kuruppu *et al.* (1998) describe a variety of Cr/CrN multilayer coatings on different substrates, including some temperature-sensitive ones such as bearing steel and aluminium alloy, and proved that they give better wear resistance than monolithic coatings. Martinez *et al.* (2003) also demonstrated a superior performance for CrN/Cr multilayers compared to CrN, in both abrasion and sliding tests. Berger *et al.* (1999) explain how Cr/CrN multilayers can provide enhanced abrasion resistance, by combining the respective mechanical properties of each material. Rutherford *et al.* (1996) examined the abrasion resistance of TiN/NbN multilayers. In their case the performance was poorer than the monolayer coating, possibly due to the lack of a tough metallic layer.

While the majority of repeated-layer multilayer coatings include a metallic or hard ceramic constituent, there is also scope to use this approach for solid lubricant lamellar materials like MoS_2. Watanabe *et al.* (2004) achieved significantly improved tribological performance using WS_2/MoS_2 multilayers than obtained with just MoS_2 or WS_2 alone.

The systems mentioned above are examples of reported results using the multilayering technique to improve the tribological properties of coatings. Many possibilities exist to enhance friction, wear and corrosion behaviour using this approach. There already is a widespread use of multilayered DLC and metal carbide films in many automotive applications.

4.6.3.3 Diverse coating layer properties

It is not always possible to find one coating which, to a satisfactory level, has all the required surface properties for a certain application. A possible solution can then be to produce a layered coating with the desired properties in such a way that each layer contributes with one or a set of specific properties. This technique is in common use in the semiconductor and electronics industries. Properties that can be separately influenced on the surface are, for example, wear resistance, corrosion resistance, chemical diffusion resistance, electrical conductivity, shear strength, thermal conductivity and adhesion to the substrate, as shown in Table 4.3.

Diverse multilayer coatings are used in various practical applications. Kleer and Doell (1997) describe the use of an $AlN/Al_2O_3/TiAlN$ system to enhance the performance of glass moulding tools.

A typical layered coating structure in the electronics industry is a silicon or ceramic substrate covered by a polyimide layer on which is deposited first an adhesion layer, then the functional layer and on the top a passivation layer, as shown in Fig. 4.98 (Seshan *et al.*, 1989).

Another illustrative example of coatings with diverse layered surface properties are the layered coatings used in magnetic recording devices as described in section 7.9. In, for example, the rigid thin film recording disk schematically shown in Fig. 7.32c the aluminium–magnesium substrate is sometimes covered by a 10 to 20 μm thick Ni–P electroless plated layer to improve the surface

Table 4.3. Properties of layers in multilayer coatings.

Required property	Type of coating	Coating material examples	Typical coating thicknesses, μm
Wear resistance	Hard coatings	TiN, TiC, Al_2O_3, diamond	0.5–5
Low friction (shear)	Soft or layered coatings	DLC, MoS_2, PTFE, Pb, diamond	0.5–3
Lubricity	Liquid coatings	Perfluoropolyether	0.001–0.05
Corrosion resistance	Dense, inert coatings	Au, Zn, Cd, Al_2O_3	3–20
Adhesion to substrate	Soluble or bonding coatings	Ti, Cr, Si, Si_2N_3	0.01–1
Diffusion barrier	Dense, inert coatings	Ni, NiAl, Al_2O_3	1–30
Thermal barrier	Insulating coatings	ZrO_2, PSZ	50–250
Magnetic	Metallic coatings	γ-Fe_2O_3, Co–Ni, Co–P	0.02–5
Electrical conductivity	Noble or pure metal coatings	Au, Cu	1–10

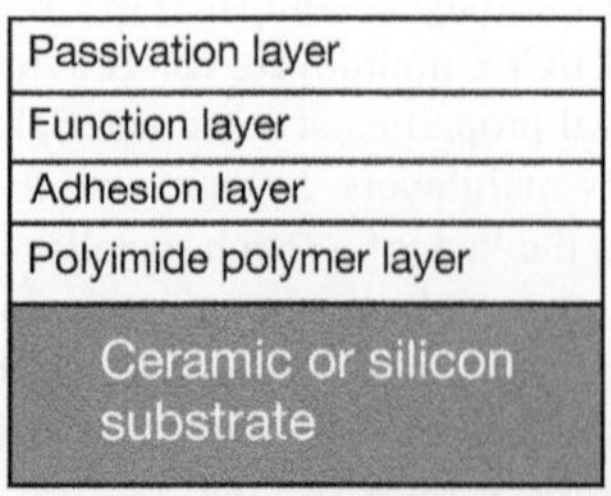

Fig. 4.98. Typical multilayer structure used in the electronics industry.

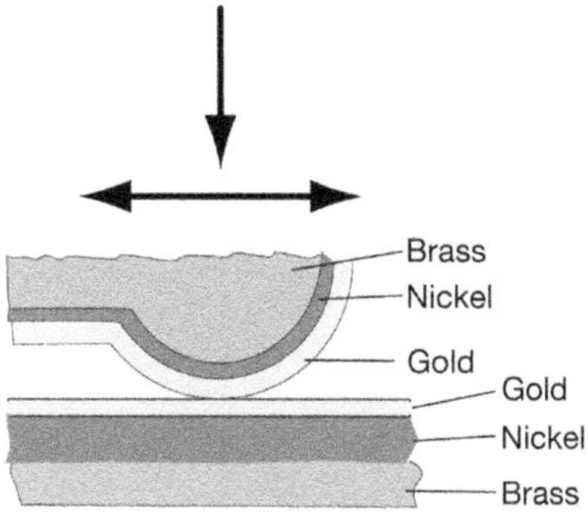

Fig. 4.99. An electroplated nickel layer is used as a diffusion barrier between the gold layer with good electrical and thermal conductivities and the brass substrate in electrical connectors (after Tian *et al.*, 1991b).

hardness and smoothness. Then comes a 15 to 75 nm metallic or oxide magnetic layer on which is deposited a 3 to 10 nm protective layer of diamond-like carbon. Finally, on the top is a 0.5 to 2 nm thin film of perfluoropolyether to provide the low shear lubrication properties (Bhushan, 1996, 2001a; Gellman and Spenser, 2005).

Multilayered coating structures have been used to form a diffusion barrier between the functional coating and the substrate. In electrical connectors a 2 to 3 μm thick nickel layer has been shown to be a good diffusion barrier between the 1 to 2 μm thick gold coating and the substrate of copper, brass or bronze, as shown in Fig. 4.99. The coefficient of friction is originally as low as 0.2 in such contacts but increases with time to high values in the range of 0.4 to 0.8, while typical failure mechanisms are characterized first by mild wear, then prow formation, later by coating crack initiation and propagation and finally by the formation of plate-like delaminated wear particles adhering to the surface and resulting in severe wear (Tian *et al.*, 1991b).

The diverse property layered coating structure can incorporate a friction-controlling material. This is well illustrated by the experiments of Hayward and Singer (1990), where they deposited two 1 μm thick microwave CVD diamond coatings, one rough and one smooth, by ion beam-assisted deposition (IBAD) with 0.5 μm thick MoS_2 layers. The upper MoS_2 layer has a friction-decreasing function because of its low shear strength due to the lattice structure, while the hard diamond layer has a load supporting function due to its extreme hardness. Repeated sliding friction and wear tests, with a sliding velocity of 0.3 mm/s and a load of 5 N, showed a decrease in the coefficient of friction by one order of magnitude, from 0.6 to 0.05–0.06 and a decrease in wear volume from values of about 10^5 to 10^6 μm^3 to none detectable when comparing the results of a sapphire ball sliding against uncoated diamond coating and a MoS_2 coated diamond coating. As discussed in section 4.3, there are other examples in which lamellar solid lubricant coatings are combined with harder materials. In the case of metal cutting tools, the addition of a lubricious outer layer can benefit cutting performance, especially in dry machining.

4.6.4 Duplex treatments

Duplex treatment means that surface treatment is combined with coating deposition to produce a surface with optimal tribological properties. The surface treatment is applied prior to the coating deposition to enhance the surface performance, usually by enhancing the load support or load-bearing capacity provided to the coating to prevent it from cracking or debonding under load due to substrate

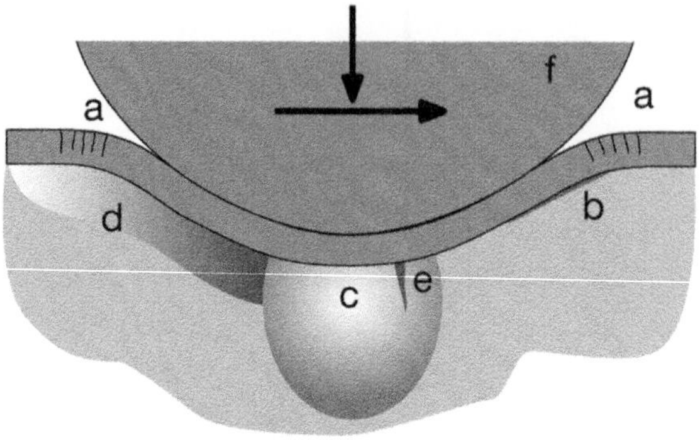

Fig. 4.100. The surface load support is influenced by: (a) coating fracture, (b) coating debonding, (c) substrate elastic deformation, (d) substrate plastic deformation, (e) substrate fracture and (f) ploughing friction.

surface deformation (see Fig. 4.90). The objective of using a duplex combination of a surface treatment and a coating is thus to provide a synergistic effect such that the resultant performance is better than either of the individual entities could provide alone.

Plasma nitriding was combined with an ion-plated hard TiN coating and the improved load support could be demonstrated as increased higher critical load in scratch testing by Leyland *et al.* (1991). Good load support for the coating is crucial for achieving long lifetime in many applications. Different kinds of surface responses can influence the limit of load support in a coated surface, as shown in Fig. 4.100. It seems that a static loading test for the limit of load-bearing capacity would not be adequate, as it does not provide evidence for an increase in ploughing friction. It is therefore likely that a controlled scratch test is the most effective way to estimate the limit of load support. That was the approach adopted by Matthews and co-workers in evaluating the improved performance achieved by using a plasma-nitriding process on steel and titanium alloys prior to PVD coating (Leyland *et al.*, 1991; Matthews *et al.*, 1995; Matthews and Leyland, 1995).

Our recent work has indicated that for the evaluation of the limit of elastic load-bearing capacity a larger sliding impressor may be needed than the standard Rockwell C indentor used in the scratch test. With a larger radius counterface the contact area, and therefore the contact pressure, can be more accurately assessed. Also, whereas in the scratch adhesion test the aim is usually to penetrate the surface, this may not be desirable in a test to evaluate the load-bearing ability. Furthermore, a larger radius counterface more accurately simulates the real contact geometry in most practical applications. Leyland *et al.* (1991) have demonstrated with a standard Rockwell C indenter in scratch tests that the load-carrying capacity of a hardened high-speed steel (grade ASP23) is improved by more than 25% by applying a plasma-nitriding treatment prior to TiN coating, as shown in Fig. 4.101.

Other early exponents of the duplex approach were the groups of Bell and Rie, with the latter emphasizing plasma nitriding as a precursor to a vapour deposition process. Bell and co-workers have published widely on the evaluation and modelling of duplex treated surfaces (Salehi *et al.*, 1991; Bell *et al.*, 1998a, b; Dong and Bell, 1999; Mao *et al.*, 2007; Zhang *et al.*, 2006a). They presented diagrams to illustrate schematically the concepts behind, and benefits of, duplex surface engineered systems. A prior diffusion treatment combined with a coating can provide a composite effect to improve the tribological response of a surface as illustrated by Fig. 4.102, based on their ideas.

Bell *et al.* (1998a) showed how the maximum Hertzian contact pressure limit in a twin-disk test can be extended by the use of a duplex treatment approach, as shown in Fig. 4.103. In their early papers they held one disk stationary to give a sliding speed of 0.3 m/sec, and they ran tests in oil under increasing increments of load until accelerated wear took place, as evidenced by an increase in

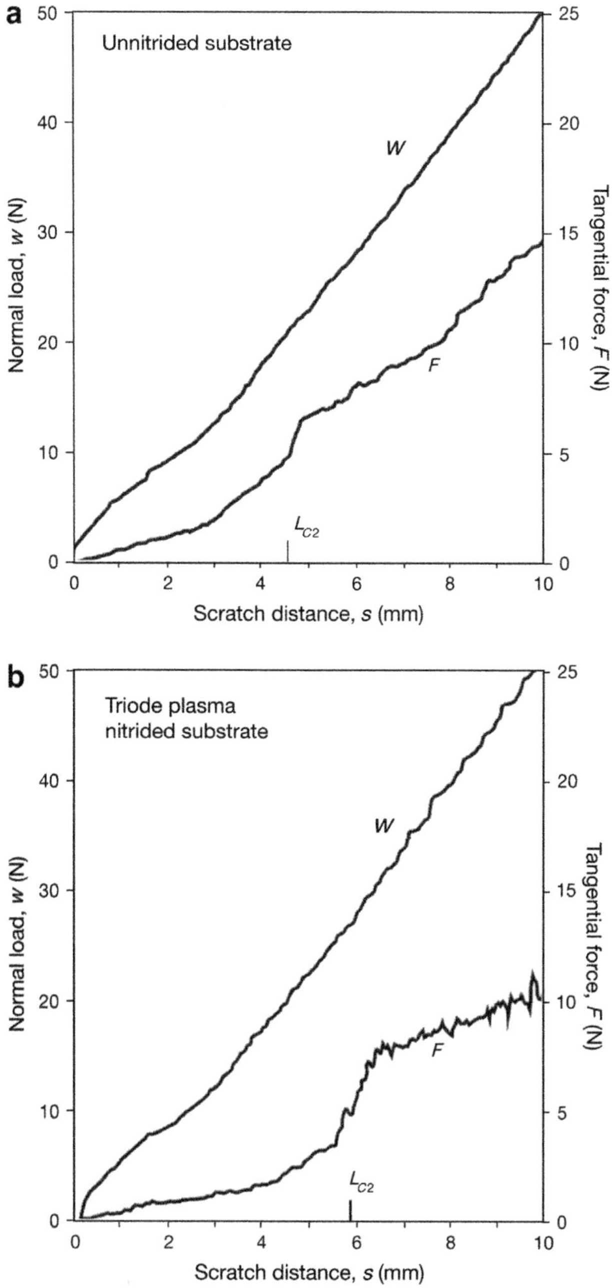

Fig. 4.101. Normal load (w), tangential force (F) and upper critical load (L_{C2}) as a function of sliding distance in scratch test for (a) non-nitrated substrate and (b) triode plasma-nitrided substrate (data from Leyland *et al.*, 1991).

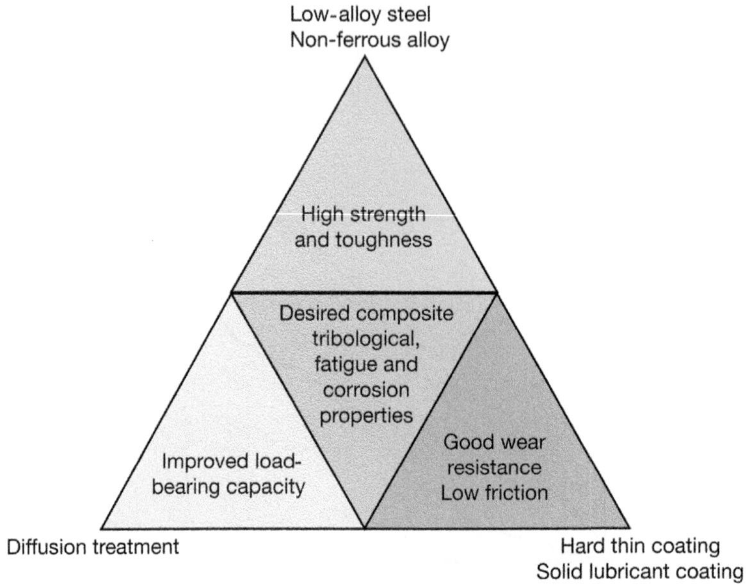

Fig. 4.102. The duplex treatment concept and its benefit (modified from Bell *et al.*, 1998a).

frictional force, increased noise or surface colour change (Salehi *et al.*, 1991). In that paper the failure load was used as a ranking parameter for different duplex and non-duplex treatments on Ti6Al4V titanium alloy.

A ranking of the tribological surface performance of duplex treated and other state-of-art coated and uncoated surfaces was carried out based on wear volume measurements, with the worst performing surface exhibiting severe subsurface plastic deformation in a ball-on-wheel test with an alumina ball, by Bell *et al.* (1998a). The duplex treated surface outperformed the untreated steel surface by two orders of magnitude, the PVD TiN coated by tenfold and the plasma nitrided surface by fourfold less wear. They also utilized a friction monitored scratch tester and an indentation tester to evaluate the load-bearing capacities of different treatments and coatings on Ti6Al4V alloy. In the case of the static indentation tests, ring cracks around the indentation imprints were used to signify the limiting critical load which the surface could support.

Scratch tests were reported by Bell *et al.* on an oxygen diffusion treated Ti6Al4V alloy with a DLC coating, and also on a DLC coated surface without the prior treatment. In the latter case, an increase in friction was observed at a load of 8 N and further increases in load led to the collapse of the DLC coating at 45 N. For the duplex treated specimen, tensile cracking occurred at a load of 30 N, but there was neither spalling nor chipping. At 40 N lateral flaking failure occurred, followed by forward lateral flaking at 45 N. Bell *et al.* (1998a) evaluated the dynamic load-bearing capacity under conditions which were similar to those found in a bearing, and these tests, evaluated in terms of the appearance of the wear track, demonstrate that the load-carrying capacities of duplex-treated materials show an improvement by a factor of 2.6, compared to factors of improvements of 4 indicated by indentation or 5.6 by scratch tests.

Bell *et al.* recognized the need for knowledge of stress distributions when designing with duplex surfaces and they discussed the benefits of finite element modelling, which Bell and Sun (1990) applied to the earlier twin-disk tests. Their model was used to predict the subsurface crack failure of spur gears (Mao *et al.*, 1996). In later publications they reported the development of testing and

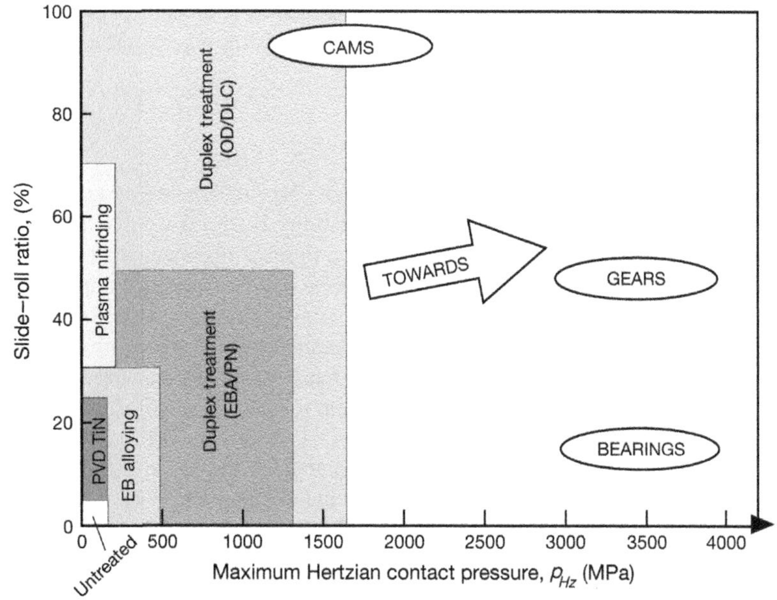

Fig. 4.103. Duplex treatment extends the contact pressure limit for a coated surface in rolling contacts like gears, bearings and cam mechanisms (modified from Bell *et al.*, 1998a). PN = plasma nitriding, EB = electron beam, OD = oxygen diffusion and EBA = electron beam alloying.

modelling methods to evaluate the benefits of duplex treatments, especially on titanium alloy (Zhang *et al.*, 2006; Kwietniewski *et al.*, 2001; Mao *et al.*, 2007).

A wider range of materials, including austenitic steel, Ti alloys and Co-based alloys, were investigated in a study of the benefit of duplex treatment by Rie and co-workers (Rie and Broszeit, 1995; Menthe and Rie, 1999; Rie, 1999). They used plasma nitriding or nitrocarburizing as a precursor to subsequent coating by PVD or PACVD techniques. They reported success by plasma nitriding or plasma nitrocarburizing different electroplated chromium coatings on steel. Those duplex treatments are now finding successful application on press tools (Klimek *et al.*, 2003). There was practical success for plasma-nitrided steel followed by PACVD (Ti(BN), TiB_2) coatings on die casting and forging tools (Klimek *et al.*, 2003, 2006). Pfohl and Rie (2001) even found friction and wear benefits for TiBN and TiB_2 coatings on plasma-nitrided satellite in dry sliding against 100Cr6, in a pin-on-disk test.

Batista *et al.* (2003) examined the impact resistance of duplex and non-duplex treated hardened AISI HI3 steel coated with (Ti,Al)N and Cr–N. The duplex (i.e. pre-plasma-nitrided) samples exhibited at least a doubling in the scratch test upper critical load, and the number of impacts before adhesive failure occurred was increased by a factor of about 50 for the (Ti,Al)N coatings and around 5 for the Cr–N coatings.

The possibility to plasma nitride and coat in one process was demonstrated by Michler *et al.* (1999), who deposited an a-C:H:Si coating onto a hardened X115CrVMo121 cold work tool steel, which was pre-plasma-nitrided in the same chamber under similar process conditions. This provided enhanced film adhesion and wear performance, under optimized conditions.

A novel duplex treatment combining the boost diffusion oxidation (BDO) treatment with carbon-based coatings (BDO/Cr–DLC) has been shown to significantly improve the tribological properties of the Ti6Al4V titanium alloy and increase the load-bearing capacity on the basis of scratch and pin-on-disk testing (Zhang *et al.*, 2006a).

There is no doubt that the idea to combine plasma treatment and coating in one process is now gathering pace in industrial usage, especially for titanium and titanium alloy materials (Kaester *et al.*, 2001a, b).

4.6.5 Adaptive coatings

The general concept of smart or intelligent coatings which adapt to their operating environment is one which has great appeal in tribology. In fact, such adaptability is a property that has existed in many coatings for decades. One example is coatings designed to resist high-temperature corrosion in gas turbine engines. There the coating forms a protective oxide scale of varying composition, depending on the environment, on its surface during operation. A similar situation exists for some of the most recently developed cutting tool coatings, based on, for example, TiAlN, which produces a stable oxide during high-temperature cutting operations. This effect has been further enhanced in TiAlN coatings that contain additions of yttrium and/or chromium to help to extend oxide film stability over a wider range of temperatures (Munz *et al.*, 1997; Pfluger *et al.*, 1999; Savan *et al.*, 2000). In some ways, diamond and DLC films can also be regarded as adaptive, in the sense that the formation of a temperature-induced low shear strength layer in the contact is thought to be the source of the low friction demonstrated by many such coatings.

Another issue in the context of adaptive coatings is the recently renewed interest in the use of Ni–Ti, and other, e.g. copper-based, shape memory alloys as wear-resistant coatings. They are proposed as being suitable for MEMS devices and micro-actuators (Kneissl *et al.*, 2003; Bhattacharya and James, 1999; Fu *et al.*, 2003; Ni *et al.*, 2004c), due to their perceived desirable mechanical, tribological and corrosion-resistant properties (Sheldon *et al.*, 1988; Bouslykhane *et al.*, 1991; Lin *et al.*, 1997; Ni *et al.*, 2004a, b; Zhang *et al.*, 2005a). Such materials are known as superelastic materials. The term superelastic is perhaps somewhat incorrect since the elastic strain behaviour is essentially non-linear, hysteretic and strain-path/temperature dependent. A more correct term might be pseudoelastic.

These materials can exhibit desirable wear-resistant characteristics and are claimed to have very low sliding friction coefficients (Ni *et al.*, 2004c). These properties are usually represented as being due to a reduced ploughing term during sliding contacts, that is, the unrecoverable plastic deformation, which also absorbs frictional energy during sliding. It is nonetheless clear that a family of martensitic materials can be produced which exhibit thermally recoverable, if not necessarily fully temperature-reversible, high-strain characteristics, which can provide enhanced tribological properties in certain types of contacts and environments.

Some of the most promising of the these alloys are produced as coatings (Kneissl *et al.*, 2003; Bhattacharya and James, 1999; Fu *et al.*, 2003; Ni *et al.*, 2004c) or surface-modified bulk materials (Sheldon *et al.*, 1988; Bouslykhane *et al.*, 1991; Lin *et al.*, 1997). They comprise Ni–Ti-based compounds with additions of other elements – either substitutional (Kneissl *et al.*, 2003; Bhattacharya and James, 1999; Zhang *et al.*, 2005c), or interstitial (Bouslykhane *et al.*, 1991; Lin *et al.*, 1997; Zhang *et al.*, 2005c). They utilize the recoverability of the shape memory alloy, and also provide enhanced hardness; with the latter, in some cases, being provided by grain-size related Hall–Petch hardening. It does appear that, to gain the most effective control of friction and wear behaviour, it can still be beneficial to deposit a coating with the desired tribochemical and hardness characteristics on top of the Ni–Ti-based alloy (Ni *et al.*, 2004a, b).

Zabinski and his co-workers have developed solid lubricant coatings which are capable of adapting to temperatures from ambient up to 800°C (Zabinski *et al.*, 1995a, b). At such high temperatures, conventional oils and greases cannot be used, and graphite and metal dichalcogenides such as MoS_2 and WS_2 have historically been employed as alternatives. Their lubricious nature is generally attributed to weak interplanar bonding and low shear strength. Favourable conditions may, however, not prevail at higher temperatures, where such materials oxidize. They reported that this oxidation may become a problem at temperatures as low as 350°C, and have carried out research into the deposition of coatings with additives such as graphite fluoride (CF_x) in order to extend the permissible operating temperature up to 450°C. They utilized a pulsed laser deposition method, showing that WS_2/CF nanocomposite films were less sensitive to moisture than WS_2 alone, achieving friction coefficients of 0.01 to 0.04 against a stainless steel counterface in dry air. Although not fully adaptive, in the sense of being reversible, such coatings nevertheless point the way forward toward coating materials that can be effective over a wider operating range.

To provide lubricity in the higher temperature range (500–800°C), other materials must be considered. Oxides such as ZnO, along with fluorides such as CaF_2 and BaF_2, are lubricious in this regime, due to their low shear strength and high ductility. The oxides and, to a lesser extent, the fluorides are also chemically stable in air at high temperatures. At ambient temperatures, however, these materials are brittle and subject to cracking and high wear rates. Thus an answer to producing a lubricant coating that can operate over a broad temperature range is to combine low- and high-temperature lubricant materials uniformly dispersed throughout the coating in either a nanolayered or nanocomposite structure.

Zabinski *et al.* (1995b) examined both layered and composite structures of CaF_2 and WS_2 grown by pulsed laser deposition. Subsequent work on steel and TiN coated steel substrates illustrated that these films were lubricious up to 500°C. The CaF_2 and WS_2 materials interacted to form $CaSO_4$, among other compounds. The coatings, in effect, adapted to their environment by forming a low-friction outer layer. The adaptive or chameleon coating concept was applied to a range of DLC-based coatings, and it was demonstrated that nanocomposite coatings can be produced which are adaptive to load, environment and temperature (Veovodin and Zabinski, 1998a; Voevodin *et al.*, 2002).

The concept of combining coating materials to extend the operating range has been applied using several other compounds; for example MoS_2 with PbO and WS_2 with ZnO, to provide lubrication over a wide temperature range (Donley and Zabinski, 1994; Zabinski *et al.*, 1996, 1992; Walck *et al.*, 1994, 1997; Zabinski and Donley, 1994). In these cases, the metal dichalcogenides reacted with the oxides to form $PbMoO_2$ and $ZnWO_4$, which were found to be lubricious at high temperatures. The problem with these coatings was that they were adaptive only on heating; that is, the low-temperature lubricant materials reacted to form a high-temperature lubricant – but the reaction was irreversible. After returning to room temperature, the lubricious properties had disappeared.

A promising way toward reactions is to use diffusion barrier coatings with layered structures, and thereby to develop fully chameleon-like coatings, which will function reversibly through multiple temperature cycles. The ideal structure for such adaptive coatings may well be a nanodispersion of non-percolated, or perhaps partially percolated, active constituents. One key issue is controlling the extent of interpenetration of the constituent phase networks and thus the rates of supply and reaction of the species. This is best achieved by coatings exhibiting nanocomposite structures. Such structures have two significant benefits for adaptive coatings. They allow control of the ingress of environmental species (e.g. humidity, oxygen, etc.) and also provide a degree of control of the availability of active constituents. In other words, the release rate can be adapted *in situ*, to suit the operating life requirements.

CHAPTER 5

Coating characterization and evaluation

Contents

5.1 The Requirements

In previous chapters we discussed several of the methods available to produce coatings and described how they behave tribologically. In this chapter we turn our attention to the techniques used to ascertain what has been produced, its physical properties and how it performs. The goals of surface characterization and evaluation can be addressed by a wide range of techniques. The relative importance given to different properties depends on a number of factors, such as a coating's level of development, e.g. the need for further optimization, and the intended application. A coating in full production for a number of years may be subjected to less rigorous testing than one which is newly introduced. Indeed, the emphasis moves towards non-destructive methods when the coating is applied to a product.

The relative importance of each surface coating property to be characterized and evaluated is thus usually different depending on whether the coating to be studied is in production or is still under research and development. The emphasis in the former case is on quality control and repeatability, whereas the latter is aimed towards the evaluation of specific properties and the way in which they are influenced by process variables.

Thus, for example, adhesion and thickness become key properties once a coating has been optimized and is in production, whereas the friction and wear performance and composition are usually of most interest when a tribological coating is under development. Ultimately, there will be a need for a standard specification for any coating, since only with a defined standard can the design engineer, or other person who specifies a coating, actually communicate what is required.

Intrinsic coating properties need to be measured due to the move towards computer databases on coatings which aid both in their selection and in the modelling and prediction of coating behaviour. These factors all increase the need for effective, precise and repeatable property measurement (Godet *et al.*, 1992; Strafford and Subramanian, 1997; Jennet and Gee, 2006).

5.2 Coating Characteristics

5.2.1 The surface

The surfaces of coatings, like the surfaces of bulk materials, have particular attributes which may differ from those of the material within the coating. For example, a chromium coating will generally have an oxide film at the surface. Similarly, as with a bulk material, a coating which has been subjected to a final mechanical treatment, such as polishing, may exhibit the effects of near-surface deformation. It is important to be aware of these localized variations in the nature of coatings, especially since this is where the immediate contact with the counterface occurs. Effects such as wettability, adsorption and chemisorption are controlled by the immediate surface, or more correctly the interface between the coating and its counterface – be that solid, liquid or gas (Iliuc, 1980; Abel and Ferrante, 2001).

Equally important are the geometrical properties of surfaces and especially their topography. With coatings, the subject takes on a new significance, since the deposition method may sometimes significantly influence the surface topography. Some coating methods, such as electroplating, can smooth the surface. Other methods such as some thermal spraying techniques can lead to a roughening, while others may impart a specific and distinct type of topography, such as the faceted type exhibited by CVD diamond films.

Given this range of topographies it is particularly important to utilize appropriate surface roughness parameters when characterizing coated surfaces. Indeed, for certain types of coatings, such as those used in optical devices, it has been necessary to develop new techniques for the investigation of surface topography, even on an atomic scale, and these are referred to later.

5.2.2 Thickness

This, apparently simple, property has a criticality which many practitioners have failed to recognize, especially in tribological applications. There is a complex interplay between effects such as the intrinsic stress within a coating, the hardness of the coating and the substrate and the adhesion to the substrate which must be optimized, essentially by the appropriate choice of thickness (see Chapter 3). It can also be represented by concepts such as 'mass thickness', which seeks to recognize that films have varying densities and porosities (Pulker, 1984). Thus simple step-height measurements of thickness could in many applications be meaningless without a quantification of density or porosity.

When dealing with surface treatments, such as nitriding or ion implantation, the definition of thickness or treatment depth is even less clear. Usually, in the former case the depth is taken to be that at which the layer hardness falls to a level one-third above the core hardness of the bulk material.

5.2.3 Adhesion

Adhesion is the ability of a coating to remain attached to the substrate under the required operating conditions and this is clearly vital. Many use the words 'adhesion' or 'bond strength' to describe this ability, but the subject is much more complex than the description of a single strength parameter might suggest. The adhesion between the coating and the substrate should not be mixed up with the adhesion between the coating top surface and the sliding counterface, which is related to adhesive friction and adhesive wear.

Failure of coating adhesion to the substrate results in interfacial debonding which for brittle materials is often called cracking. The crack nucleation and propagation to larger surface failure is due to local load levels and directions, resulting stress conditions, strain and deformation. Sometimes a failure that is referred to as adhesive may in fact be a cohesive failure when the cracking occurs in the coating or in the substrate. Indeed, the complexity is such that it is unlikely that any one measure or test can satisfactorily differentiate between the ability of different coatings to stay bonded to different substrates under different application conditions. We are even further away from a test which will assess this non-destructively.

As an example of the above statement, consider a highly columnar and relatively porous PVD coating, compared to a more dense and cohesive one. In a test designed to assess adhesion by stretching or bending the substrate to put the coating in tension, the columnar coating is more likely to remain adherent, since it will simply crack along column boundaries. The dense cohesive film will, however, experience greater interfacial stress levels between coating and substrate, and is more likely to debond, even if the absolute bond strength between coating and substrate is greater. Conversely, in

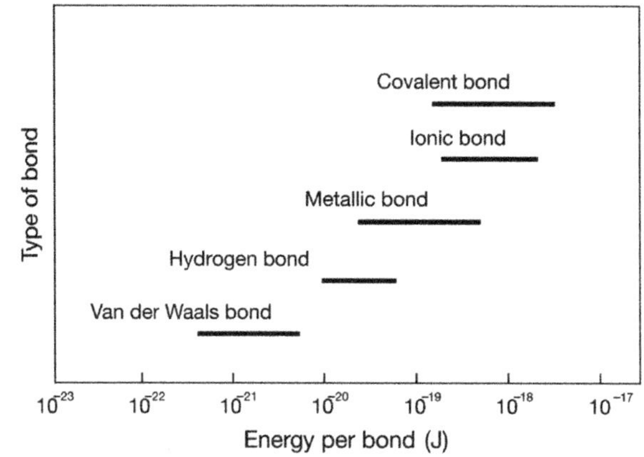

Fig. 5.1. Relative bond strengths for a range of possible interfacial binding forces (after Bull, 1992).

a sliding contact the latter coating is likely to remain in place, whereas the former will be more friable and will probably be rapidly removed from the substrate.

From a scientific point of view it is preferable to think of adhesion as a fundamental property which can be quantified according to known data about atomic binding forces, as shown in Fig. 5.1. Breaking the bonds between the coating and the substrate will produce a lateral crack. The growth of this crack is equivalent to crack growth in fracture mechanics and thus the term interfacial fracture toughness is increasingly used to describe the strength of coating adhesion, see section 5.3.3.

The definition of adhesion used by the American Society for Testing Materials (ASTM, 1970) is 'the state in which two surfaces are held together by interfacial forces which may consist of valence forces or interlocking forces or both'. It was the recognition of the inadequacies of a theoretical or idealized view of adhesion that led Mittal (1976) to define the term 'practical adhesion', as opposed to 'basic adhesion', to take account of extraneous factors such as the internal stress within the film or the presence of defects. However, as the practical adhesion will be application or test method dependent, it has been suggested that an alternative term might be 'effective adhesion' or even 'surface system adhesion', to take account of the fact that different surface loading conditions can result in different mechanisms for coating debonding from the substrate. Thus, in establishing any measure of relative ability for a coating to remain attached to a substrate it is important to define precisely the contact conditions. The basic mechanisms of coating adhesion to substrate have been the subject of several in-depth studies such as Thouless (1988), Bull (1992), Bhushan (1999b), Maugis (2001) and Malzbender *et al.* (2002).

5.2.4 Morphology

The various coating methods produce structures or morphologies which are specific to themselves. The vapour deposition methods, due to their widely variable process conditions, which include the possible use of plasma or ion assistance, were from an early date studied in terms of the influence of deposition conditions on structure (Movchan and Demchishin, 1969; Anderson, 1971; Thornton, 1977). The model put forward by Thornton is now widely referred to, in terms of the structure zones which he classified as shown in Fig. 2.7. With the homologous temperature – the substrate temperature divided by the melting point of the coating in absolute units – as one axis, and pressure as the other, Thornton described the following zones for sputter-deposited films.

Zone 1 consists of tapered crystallites and is not fully dense, having longitudinal porosity between the grains. Zone T is a transition zone consisting of fibrous grains, still with weak boundaries. Zone 2 is a fully dense columnar structure and Zone 3 is an equiaxed grain morphology. The temperature ratios cited by Thornton are not appropriate for plasma- or ion-assisted deposition methods, which can often produce extremely dense films at very low homologous temperatures, e.g. <0.2 for ceramics such as titanium nitride. For reasons of its apparent non-applicability in many cases, the Thornton model has been criticized in the literature. However, as a means of describing structures it remains a useful tool which is still widely used to categorize the structure of vapour-deposited coatings.

The structures exhibited by the coating processes carried out using molten or semi-molten state precursors differ considerably from those in the Thornton diagram. This is described by Rickerby and Matthews (1991a). The molten or semi-molten state processes typically produce coatings with splat-like features whose boundaries tend to be parallel rather than perpendicular to the substrate surface. Coating properties such as hardness, elastic modulus and residual stress are all influenced by the structure, as indeed is the tribological response.

The influence of the zone structure on sliding wear behaviour for a coating, such as TiN on steel, has been postulated (Bull *et al.*, 1988b; Rickerby and Matthews, 1991b; Kelly *et al.*, 2006). This demonstrates that composition is not the only important parameter in determining the effectiveness of coatings, an observation also made by Matthews and Sundquist (1983), who reported several titanium nitride-based coatings which showed quite different tribological properties, influenced not only by the phase composition but also by the morphology and even the crystallographic preferred orientation.

5.2.5 Composition

There are many methods available which provide chemical composition and other information about coatings. As pointed out by Ferrante (1982) the range of techniques has led to a plethora of acronyms, some of these will be elaborated in section 5.3.4. However, simple chemical composition is not a sufficient indication of a coating's properties to enable its tribological behaviour to be predicted with a satisfactory degree of confidence. This is especially true in the case of ceramic coatings.

As we have seen, coatings can be composed of grains, i.e. they can be polycrystalline, either random or textured and have a preferred orientation, they may even be epitaxial, or in some cases they may be amorphous (Bhushan, 1999b).

The appropriate analysis technique for a particular coating will thus depend on its morphological nature, as well as aspects such as its thickness, surface finish and actual composition. In tribology, increasing attention is being focused on tribochemical reactions, and the choice of surface analytical technique will also be determined by its ability to detect and evaluate the reaction products and transfer layers which often control contact behaviour.

5.2.6 Wettability

In addition to the chemical reactivity of a surface, the surface tension, or surface energy may be of importance. This is frequently measured as the contact angle between the air and the liquid of a droplet on a surface, which defines the wettability of the surface. This has been recognized as sometimes being an important factor influencing its tribological characteristics. It influences the behaviour of any liquid lubricants present. It can affect thermal transfer, such as in heat exchanger systems, and increasingly it is being studied in terms of the ability of surfaces to 'self-clean' (Fürstner *et al.*, 2005).

The physical background to wettability and its measurement has been explained by Iliuc (1980), Bhushan (1999b) and Vizintin (2004). Although usually simply determined by the contact angle of a droplet of liquid in contact with a solid surface, in practice the wetting angle is a thermodynamic variable which Iliuc states can be used to determine a so-called 'wetting energy' which is given by the adhesion energy of the solid/liquid interface minus the liquid surface tension. The wettability is influenced by the surface state, e.g. the roughness, its degree of cleanliness and the degree of oxidation, and it can be assumed that both friction and wear can be affected by changes in the wettability or hydrophobicity of a surface.

Especially, since the surface energy of a material depends on the nature of the medium on either side of the material boundary, it is clear that the wettability of a surface is subject to changes in the operating environment.

When considering the subject of wettability in tribology, there is a tendency to think of liquid/solid contacts at the non-elevated temperatures experienced by lubricants (Podgornik *et al.*, 2007; Borruto *et al.*, 1998). In fact wettability can also be important at very high temperatures, for example when dealing with coatings which must resist attack from molten metals, as in die-casting operations (Salas *et al.*, 2003). In that case, there is a need to consider the wettability performance as well as the wear resistance, since the degree of wetting influences the degree and extent of attack by the molten metal on the die surface.

However, Podgornik *et al.* (2007) found in their investigation of DLC and W–DLC coated steel surfaces, sliding against a similar surface on steel crossed cylinders, that there was no direct correlation between surface energy, wettability of the surface and tribological performance of the oil lubricated surface.

A correlation between ultra-low friction, $\mu < 0.05$, low wear and high contact angle of carbon coatings has been reported by Rha *et al.* (2005). Changes in wetting angle had a remarkable influence on silver wear debris agglomeration and the tendency of contact patch formation in ball-bearing testing of silver coated upper and lower bearing races (Yang *et al.*, 2001).

In the field of biological applications for tribological coatings, the degree of wetting, for example by the lubricant used to mimic the *in vivo* conditions, has been shown to influence the friction and wear performance (Kwok *et al.*, 2005; Gispert *et al.*, 2006). This indicates that the selection of a coating with the appropriate wetting characteristics is important to the life and functionality of prosthetic devices. Indeed, the coating selection issues in such cases are particularly stringent due to the need to utilize biocompatible coatings (Sunny *et al.*, 2005) and avoid toxic materials which may be harmful, as discussed in section 7.11.

5.2.7 Residual stress

Tribological coatings often contain residual stresses, which can be advantageous or disadvantageous, depending on the level of stress and the application. For example, a high compressive stress in-plane with the coating can lead to coating detachment, particularly in PVD coatings if the stress increases with thickness, and this can limit the thickness of coating which can be used. Conversely, an in-plane compressive stress can ensure that any incipient cracks normal to the surface do not open when the component is subjected to stress, particularly cyclic stresses, and this can extend the fatigue life. The role of residual stresses in coated surfaces on their tribological performance is discussed in detail in section 3.2.10.

5.3 Property Characterization and Evaluation

To achieve an optimal design of a coated surface we need to know its key characteristics and how it will behave in different loading conditions. Theoretical models are a good start but they need to be confirmed empirically. In production, adequate test methods are needed for quality assurance of the coated surface before it is taken into use. Various aspects of surface property evaluation have been discussed by Bunshah (1982), Hintermann (1983b), Matthews and Holmberg (1992) and Jennet and Gee (2006) and problems with quality assurance assessment by Bull *et al.* (1991) and Weiss (1991).

5.3.1 Roughness

Surface roughness is typically characterized using a stylus profilometer. This method is widely used despite its limitations, brought about chiefly by the need to have a relatively blunt stylus, as discussed by Hutchings (1992). The main roughness parameters are defined below and their usefulness in a tribological sense has been discussed in Chapter 3.

The main parameters used are the peak to valley height, R_t, and the centreline average value, R_a, which is the arithmetic mean value of the vertical deviation of the profile from the centreline. Also used is the root mean square value, R_s, which is the square root of the arithmetic mean of the squares of the deviations from the centreline. Although most engineering surfaces tend to give a Gaussian distribution of texture heights, other distributions are possible. Typical fine ground surfaces used for PVD coatings tend to be Gaussian but can be negatively skewed due to gouging features caused by the action of welded grits (Stout *et al.*, 1990). The coating will tend to fill such troughs, thereby improving the finish, but the exact effect of the coating depends on its own topography.

For very soft films the profilometer stylus may actually scratch the surface. In order to avoid such stylus-related problems, non-contacting optical methods of profile assessment have been developed. One such is described by Bhushan and Gupta (1991). It consists of a microscope, an interferometer, a detector array and support electronics. The surface profile measurement is based on two-beam optical interference and requires a reference surface which is mounted on a piezoelectric transducer permitting the use of phase shifting techniques to measure the phase of the interference pattern, which is proportional to the height at that location. Another means of profile assessment based on optical principles is confocal scanning microscopy (Gee and McCormick, 1992), which uses a tandem scanning microscope and the principle of convocal imaging to provide a 3D topographic image. Both of these optical methods thus provide a highly sensitive means of characterizing surface profiles.

Even more sensitive is the Scanning Tunnelling Microscope (STM) (Kumar Wickramasinghe, 1989). This senses atom-scale topography by monitoring the flow of electrons which 'tunnel' across the gap between a probe and the surface. Piezoelectric ceramics can be used to manoeuvre the probe in three dimensions, thereby allowing an image of the surface to be built up on an atom by atom basis. Another sensitive device is the Atomic Force Microscope (AFM), which, unlike the STM, can probe insulating surfaces. This also involves scanning the surface with a probe, but in this case a repulsive force which varies with relief deflects the tip, the movement of which is monitored by a laser beam which is in turn monitored by a feedback control to provide near atomic-scale images. Various other scanning probe microscopes are under development and considerably improve our understanding of surfaces and their behaviour (Bhushan, 1999b, 2001e).

5.3.2 Thickness

The thickness of a thin layer on a surface can be measured by a large variety of different methods as shown in Fig. 5.2. The bottom three of these give a composite reading which relates to both thickness and density.

Even if some of the thickness measurement methods can achieve great accuracies, they all have both strengths and weaknesses. X-ray fluorescence requires careful calibration, but is particularly useful in a coating facility where nominally identical coatings are being produced. Multilayered coatings can present problems due to differences in defining the substrate surface. Direct observation of sections by optical or scanning electron microscopy can avoid this, but this suffers from the drawback that the component must be cut to produce the section. One of the most accurate methods is to use a step-height measurement facility on a stylus profilometer – which of course necessitates leaving or producing an uncoated region, over which the step can be measured.

Optical techniques are also available, but they are usually for transparent films. An interference technique for thickness measurement of opaque coatings is sometimes used. This also requires a step

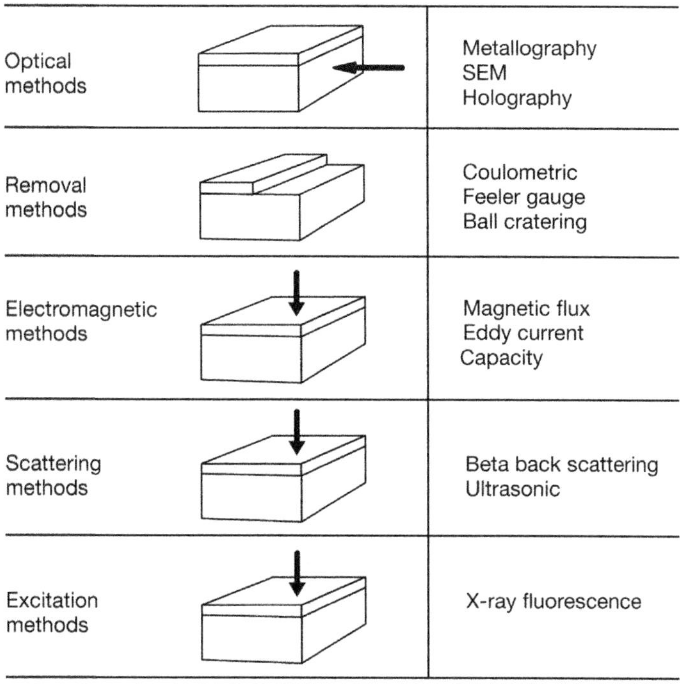

Fig. 5.2. Different methods of measuring the coating thickness (after Weiss, 1991).

to be left by masking during coating. There are various other methods of thickness assessment available commercially, based on electrical, magnetic or electromagnetic principles. These are generally suitable for thicker films, as are techniques based on gravimetric measurements. Further information on coating thickness measurements is given in Bhushan and Gupta (1991).

A comparison of four common coating thickness measurement methods shows that the ball crater method (see later), the step profilometer and the scanning electron microscope (SEM) all give thickness results with satisfactory accuracy, while weighing usually underestimates the thickness – as shown in Fig. 5.3.

The subject of laboratory coating thickness measurement repeatability and reproducibility is further demonstrated by a set of trials which was carried out at three different laboratories: the University of Hull (UH), the Technical Research Centre of Finland (VTT) and the Technical University of Darmstadt (THD) (Ronkainen *et al.*, 1990a, b). Three coating types, A, B and C, were produced at these laboratories on ASP23 tool steel hardened to 64 Rc and polished to 0.4 μm Ra. Coatings A and B were titanium boron nitride compound films produced by dc plasma-assisted electron beam reactive evaporation and rf magnetron sputtering, respectively. Coating C was a titanium aluminium nitride film produced by dc plasma-assisted electron beam and resistive evaporation.

For thickness measurements the ball crater method was chosen as a simple, cheap and easy to use technique. The principle of the ball crater measurements are shown in Fig. 5.4. The ball diameters used were 25.4 mm at UH and THD, and 30.0 mm at VTT. A kerosene–diamond suspension was used as the abrasive medium. The following expression was used to assess the thickness:

$$\text{Thickness} = \frac{x \cdot y}{\text{ball diameter}} \tag{5.1}$$

Measurements of x and y were made at UH using a Shinko profile projector, at VTT using an Olympus BH-2 microscope and at THD using the measurement facility on a Leitz Miniload microhardness tester.

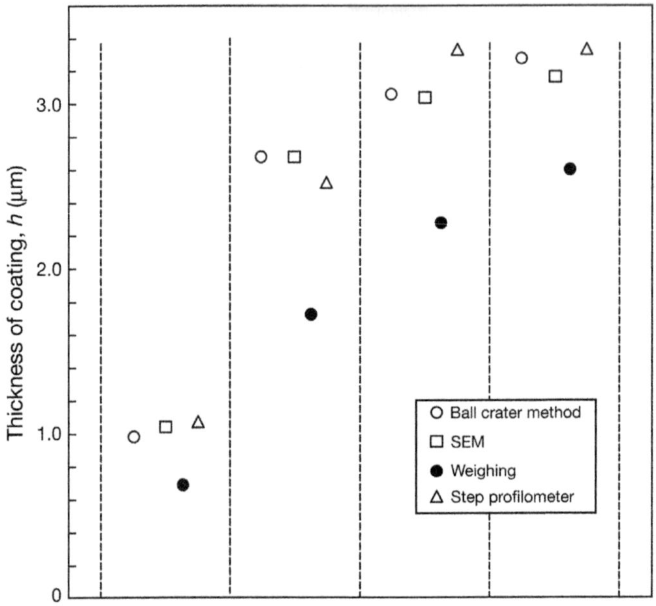

Fig. 5.3. Comparison of thickness measurement results of four different coatings performed by the ball crater method, SEM, weighing and step profilometer (data from Valli *et al.*, 1985b).

Each laboratory produced one crater, from which four determinations of thickness were made, to obtain a mean and standard deviation value. All craters were located within a 4 mm by 5 mm area on each sample. Further details of the ball crater technique are given in Valli *et al.* (1985b). Figure 5.5 shows the mean thickness values for each coating found from ball cratering. The error bars represent standard deviations. Agreement between the mean values was considered good, being typically within 4%.

5.3.3 Mechanical evaluation

5.3.3.1 Hardness

The material parameter to which we give the title hardness is not an intrinsic material property. However, this parameter reflects well the resistance to plastic deformation when a well-defined and sharp tip, the indenter, is forced into the surface. In real engineering materials the hardness is influenced by grain boundary effects, flaws and a complex interaction of response behaviours which result in the value measured varying around a mean. The nature of the basic test, involving the measurement of the dimensions of an indentation left after the removal of an indenter, also introduces some subjectivity to the reading. Nevertheless, hardness is one of the simplest properties to measure and it is widely used as a ranking parameter when comparing coatings.

The hardness test is a useful tool for process control, being especially useful for comparing similar films, without having to carry out special specimen preparation. It can also, in suitably developed form, provide information on the elastic properties (Bull and Rickerby, 1991; Jennet and Gee, 2006).

The most common hardness measures used for coatings are the Vickers or the Knoop micro-hardness, typically using loads of between 10 and 500 g. These differ in the shape of the diamond used for indentation, as shown in Fig. 5.6.

Many authors use different conventions when measuring or reporting hardness levels. In this book, when quoting hardness levels, we are mainly using the unit GPa or we adopt the units used by the originating author.

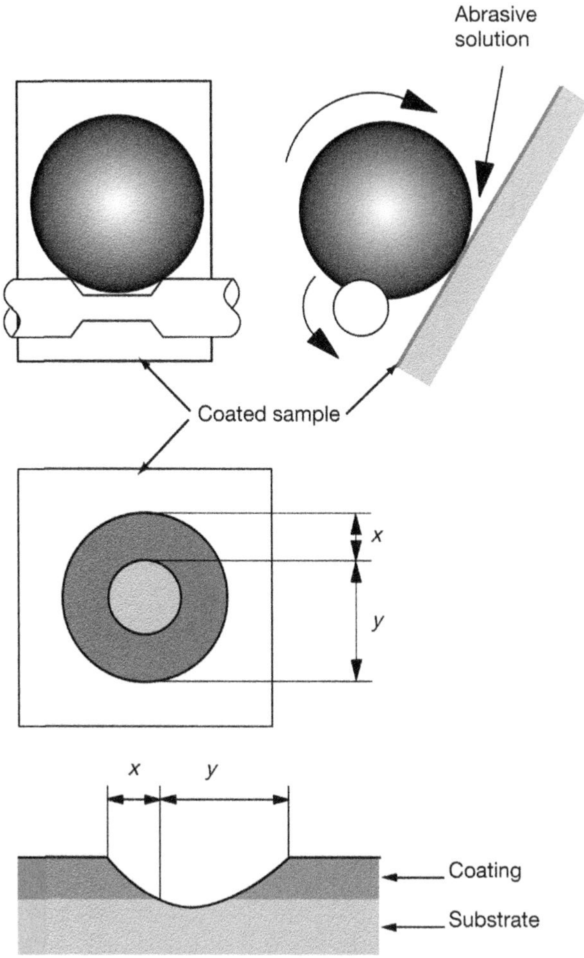

Fig. 5.4. Schematic views of the ball crater apparatus and the crater profile.

As an illustration of the variability which can occur in hardness test readings we shall refer to the round-robin experiment mentioned earlier in section 5.3.2. Hardness tests at UH and THD were carried out using a Leitz Miniload microhardness tester, and at VTT using a PMT3 machine. A Vickers microindenter was used in each case. Loads of 15, 25, 50 and 100 g were used, with an indentation dwell time of 10 seconds. Each laboratory made five indentations at each load on each of the three different coatings, within a specified area of 20 mm × 5 mm. The mean and standard deviation of the hardness number was determined from each group of five indentations by one person at each laboratory.

The results are shown in Fig. 5.7. The error bars represent standard deviations, and as can be seen there are considerable variations in the absolute values in some cases and even in the mean values measured at different institutions, under nominally identical conditions. These test results endorse the view expressed earlier, that hardness readings should be treated with caution, and exact test conditions – especially the load – should be specified. In this particular case, the substrate had an influence on the measured values, which may have increased the variability.

A general rule of thumb often applied is that, when measuring coating hardness directly on the surface, the indentation depth should be less than one-tenth of the coating thickness, to ensure that the

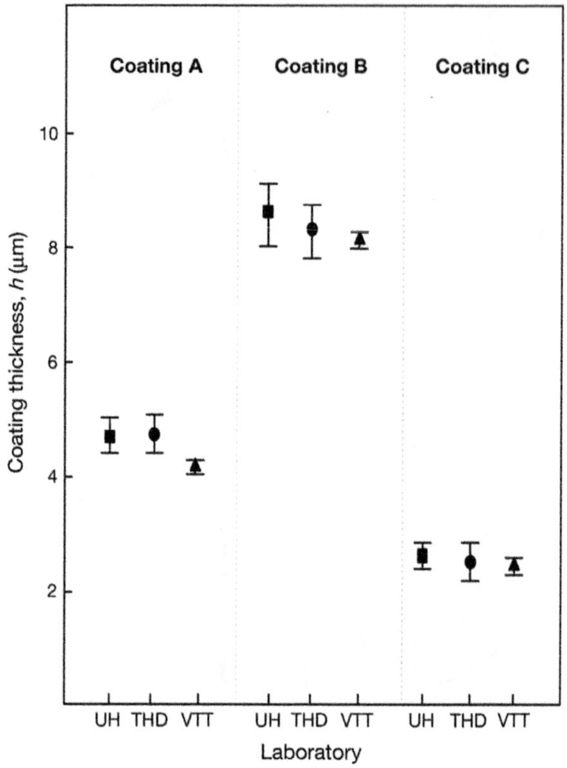

Fig. 5.5. Coating thickness measurements by the ball cratering method in three different laboratories (data from Ronkainen *et al.*, 1990b).

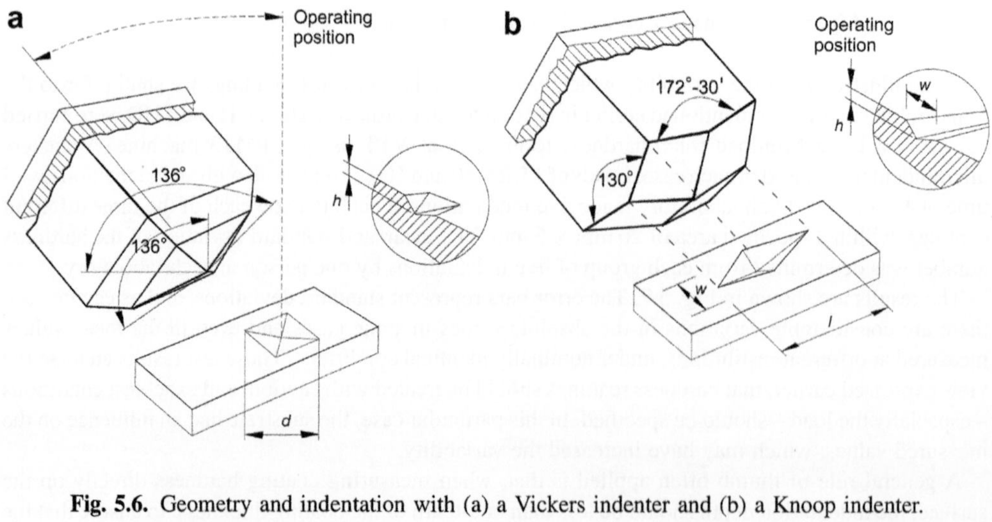

Fig. 5.6. Geometry and indentation with (a) a Vickers indenter and (b) a Knoop indenter.

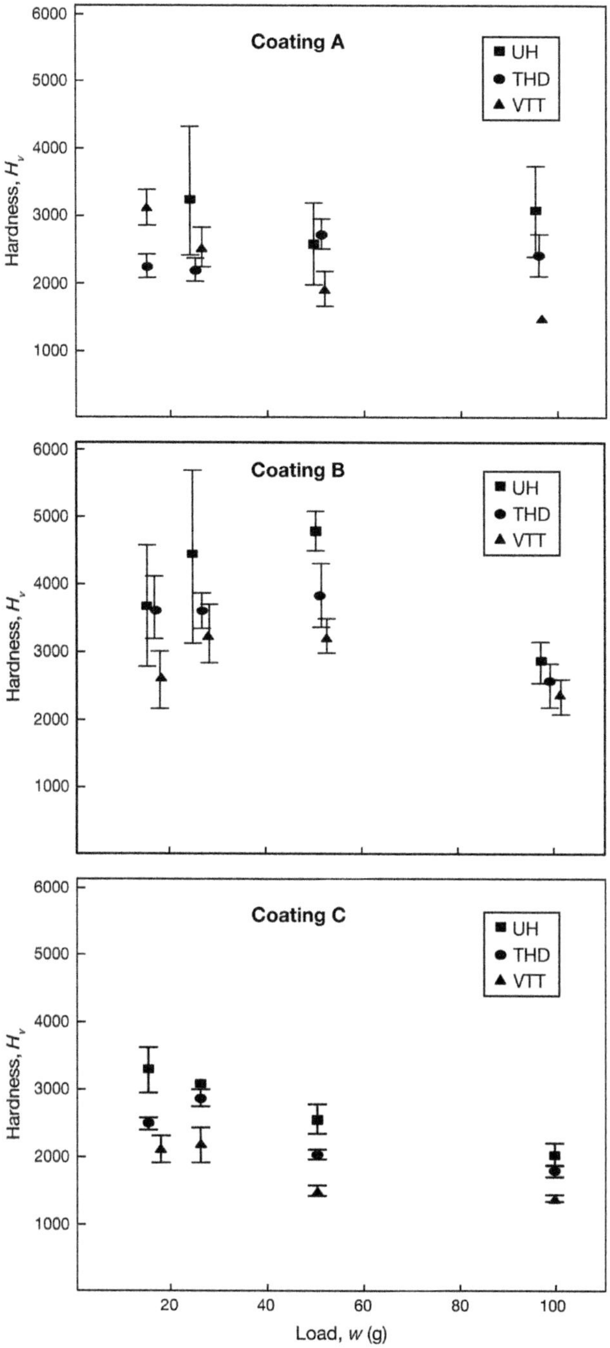

Fig. 5.7. Vickers hardness measurements carried out in three different laboratories for two titanium boron nitride films (A and B) and one titanium aluminium nitride film (C) deposited on tool steel by PVD methods (data from Ronkainen *et al.*, 1990b).

substrate effect is minimized. Since the Knoop indenter penetration is shallower than the Vickers for a given load, this means that this test should have a wider applicability, especially for thin coatings.

A number of authors (Sargent and Page, 1978; Jönsson and Hogmark, 1984; Vingsbo *et al.*, 1986; Burnett and Rickerby, 1987b) have noted an indentation size effect (ISE) on the hardness readings measured, which tends to show a higher hardness with decreasing load. One reason for this can be elastic recovery of the indentation, another is the decreasing effect of grain boundaries and fracture.

For the Vickers pyramid indenter the hardness value is defined as the load divided by the surface area of the impression, whereas for Knoop it is the load divided by the projected area of the impression. Ignoring elastic recovery can lead to large errors in the hardness value measured at low loads (Bull and Rickerby, 1991). A general expression for hardness is

$$H = \frac{c \cdot w}{d^2} \tag{5.2}$$

where w is the load applied, d is the indentation diagonal length measured and c is a constant. When w is in kg and d in mm then $c = 1.854$ for the Vickers test and 14.23 for Knoop.

An expression quoted by Bull and Rickerby to quantify the influence of indentation size is

$$H = k \cdot d^{m^{-2}} \tag{5.3}$$

In this case k is a constant and m is known as the indentation size effect (ISE) index.

For a more comprehensive model of coating hardness, which attempts to eliminate the influence of the substrate, it is useful to adopt a law of mixtures approach. Jönsson and Hogmark (1984) used an area law of mixtures to model the composite hardness H_{comp} measured in the Vickers test thus

$$H_{comp} = \frac{A_c}{A} \cdot H_c + \frac{A_s}{A} \cdot H_s \tag{5.4}$$

where A_c and A_s are the load-support areas of the coating and substrate and A is the projected area of the indentation.

Sargent (1979) suggested a volume law of mixtures, which was extended over the years by Burnett and Page (1984), Burnett and Rickerby (1987b), Bull and Rickerby (1990b), Korsunsky *et al.* (1998) and Bull *et al.* (2003a, 2004). In this model the deformation beneath the Vickers indenter is modelled as a hemisphere and the volumes of the portions of the total deforming volume that lie in the coating, V_c, and the substrate, V_s, are taken into account to give

$$H_{comp} = \frac{V_c}{V} \cdot H_c + \frac{V_s}{V} \cdot \chi \cdot H_s \tag{5.5}$$

where H_c and H_s are the hardness of the coating and substrate. χ is an empirical interface parameter representing the modification to the deforming volume in the softer substrate, due to the constraint from the harder coating.

The total deforming volume is

$$V = V_c + \chi^3 \cdot V_5 \tag{5.6}$$

In practice of course it would be preferable to have a means of assessing the coating hardness without the need for such models. One way to do this is to ensure that the load applied is so low that absolutely no substrate effect is possible. Also an ideal test would allow the influence of elastic recovery to be assessed.

These goals are sought in the nanoindentation test, which senses penetration depth during testing, rather than relying on the measurement of a residual impression left after the load has been removed (Pethica *et al.*, 1983; Ross *et al.*, 1987; Pollock, 1992; Tayebi *et al.*, 2004a, b). Typically, nanoindentation machines operate in the mN load range and can continuously monitor load and

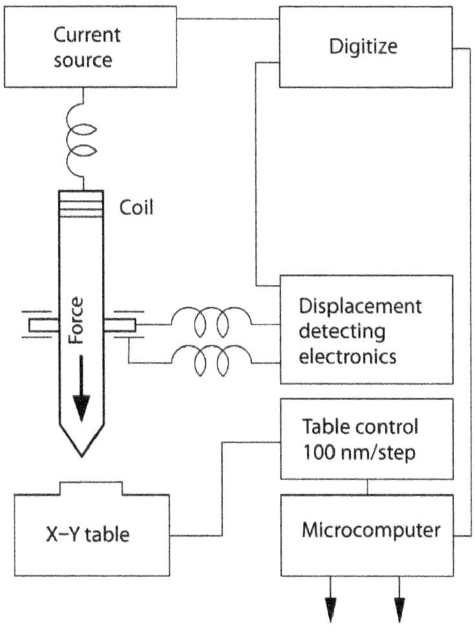

Fig. 5.8. Schematic diagram of a depth-sensing nanoindentation hardness tester.

displacement during loading and unloading. Depth resolution must be subnanometer in order to allow the load vs depth curve to be plotted. A typical system layout is shown in Fig. 5.8. Usually a Berkovich-type diamond indenter is used, which has triangular pyramid geometry, permitting accurate tip profiling. With knowledge of the indenter geometry it is possible to obtain a value for hardness from the depth of plastic deformation taking away the elastic contribution from the total depth. By using a sinusoidally varying displacement superimposed on that caused by loading, the hardness and modulus of the coating can be evaluated as a function of depth (O'Hern *et al.*, 1989). The development of techniques such as this is important in allowing the accurate assessment of mechanical properties of coatings – an essential prerequisite for modelling studies of the tribological response behaviour of coated surfaces.

A hardness analysis based on finite element simulation results and contact constitutive models for homogeneous and layered elastic–plastic surfaces has been presented by Ye and Komvopoulos (2003b). The analysis provides criteria for obtaining the real material hardness from indentation experiments performed with spherical indenters. A relation between hardness, yield strength and elastic modulus derived from finite element simulations of a homogeneous half-space indented by a rigid sphere is used in conjunction with a contact constitutive model for coatings to determine the minimum interface distance needed to produce sufficient plasticity in order to ensure accurate measurement of the material hardness.

Advanced nanoindentation techniques offer the possibility to investigate even extremely thin films attached to the surface down to a film thickness of 1 to 3 nm (Wang *et al.*, 2005). The interpretation of the indentation results are supported by theoretical analyses such as analytical modelling (Bull *et al.*, 2004), finite element modelling (Blug *et al.*, 2001) and even atomic-scale FEM modelling (Jeng and Tan, 2004). The testing of very thin surface coatings by nanoindentation is a very challenging task and needs extremely careful procedures. Commonly occurring sources of error in using this technique have been reviewed by Fischer-Cripps (2005).

5.3.3.2 Elastic modulus

The elastic modulus or Young's modulus, E, is normally calculated based on an analysis of the unloading curve of the nanoindentation test. It is assumed even for materials which exhibit plastic deformation during loading that the initial unloading is elastic. Thus the initial slope of the unloading curve is directly related to the elastic modulus. The following formula developed by Oliver and Pharr (1992, 2004) can be used for the reduced elastic modulus,

$$E_r = \frac{\sqrt{\pi} \cdot S}{2 \cdot \sqrt{A}} \tag{5.7}$$

where S is the slope of the unloading curve at the point of maximum load and A is the projected area of contact between the indenter and the material at that point, as illustrated in Figs 5.9 and 5.10. The slope of the unloading curve S is estimated as shown in Fig. 5.10.

The projected contact area is obtained from the contact depth, h_c, which can be expressed as

$$h_c = h_{max} - \varepsilon \cdot w_{max}/S \tag{5.8}$$

where h_{max} and w_{max} are as shown in Fig. 5.10 and ε is a constant that depends on the geometry of the indenter. $\varepsilon = 0.72$ for a conical, 0.75 for a spherical and 1 for a flat indenter. The contact area is generally a function of the contact depth

$$A = A(h_c) \tag{5.9}$$

where for a perfectly sharp Berkovich or Vickers indenter $A(h_c) = 24.5 \cdot h_c^2$. The elastic modulus, E, can be determined from the above equations and using the available information of the Poisson's ratio of the materials from the equation

$$\frac{1}{E_r} = \frac{1 - \nu_i^2}{E_i} + \frac{1 - \nu^2}{E} \tag{5.10}$$

where E_i and ν_i are the elastic modulus and Poisson's ratio of the indenter material.

Some limitations of this method have been pointed out by Malzbender *et al.* (2002). When a material exhibits pile-up or sinking-in, the estimated contact depth, and hence the projected area, is either underestimated or overestimated. This is typical for metals that show strain hardening. Also, the equations are not applicable for materials that creep substantially.

For a coated surface the effective elastic modulus is a combination of that of the coating and the substrate. As for hardness, the indentation depth should be less than one-tenth of the coating thickness to obtain E values for the coating that are without significant influence of the substrate. The most

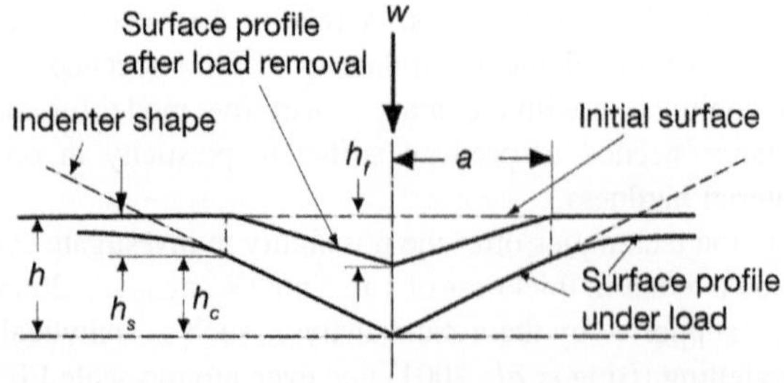

Fig. 5.9. Schematic representation of a section through an indentation using a conical indenter. h_c = contact depth, h_s = sink-in depth and h_f = final depth.

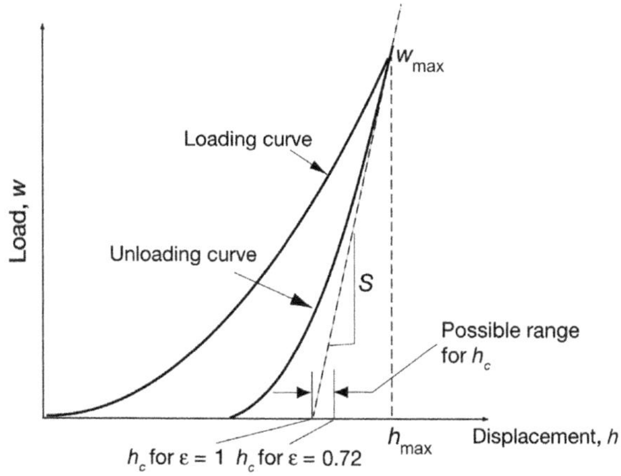

Fig. 5.10. A typical load–displacement indentation curve.

accurate measurements are done with small indentation depths and a sharp indenter. Models used for measuring elastic modulus of a coated surface have been described by Malzbender *et al.* (2002).

Cantilever and beam methods can be used in the determination of elastic modulus of a coating (Taube, 1992). Typically, free-standing cantilevered microbeams of the substrate with the film on it are prepared. The load/deflection data from measurements in a nanometre-indention system can be used to determine both the in-plane elastic modulus and the yield strength of the coating material. The in-plane elastic modulus can differ from that measured by indentation. Also, the yield stress of the coating can be measured directly by this technique. Recently, laser acoustic methods have been used increasingly for measurement of the elastic modulus of coated surfaces (Schneider and Schultrich, 1998; Schneider *et al.*, 2002).

Although much of the literature on hardness and elasticity evaluation relates to hard ceramic coatings there is also interest in softer films, especially solid lubricants, such as MoS_2 (Gardos, 1990; Muller *et al.*, 1989). Gardos in particular has investigated a number of layered hexagonal lubricants for their mechanical properties and has demonstrated their anisotropic behaviour, which of course is critical to their tribological properties.

5.3.3.3 Fracture toughness

At present there is neither a standard test procedure nor a standard methodology for assessment of fracture toughness of thin coatings. In the literature there have been several different attempts to develop such measurement techniques but so far they all have limitations in some respect. Some of the methods can also be used to estimate the residual stress in the coating. The indentation and scratch testing have commonly been used for qualitative assessment. Quantitative measurements have been done by bending, buckling, indentation, scratching and tensile testing. The methods are either stress based or energy based and today the latter approach seems to be more advantageous. Recent critical reviews of these methods have been published by Malzbender *et al.* (2002), Zhang *et al.* (2005a) and Cheng and Bull (2007).

In an indentation test of a brittle surface a pattern of cracks is generated on the surface. Radial cracks are more likely to occur at low loads when the tip is sharp, while picture-frame cracks are formed at high loads or when a blunt tip is used (Palmqvist, 1957, 1963; Cheng and Bull, 2007). The fracture toughness of the coating or the coated surface can be calculated by combining observations of the cracks, load–displacement curves and stress or energy models. It has been shown for bulk materials (Lawn *et al.*, 1980) that there exists a simple relationship between the fracture toughness, K_c, and the crack dimensions in Vickers and Berkovich indentation

$$K_c = \varepsilon \cdot \sqrt{\frac{E}{H}} \cdot \frac{w_p}{\sqrt{a^3}} \tag{5.11}$$

where w_p is peak indentation load, a is the length of radial cracks and ε is an empirical constant which depends on the geometry of the indenter. Lawn and Evans (1977) analysed the indentation of median cracking for bulk materials with a Vickers indenter at elastic–plastic indentation. They developed the following equation for the critical load, w_c, for crack initiation

$$w_c = 21.7 \cdot 10^3 \cdot K_{Ic}^4 / H^3 \tag{5.12}$$

where K_{Ic} is the fracture toughness and H the hardness of the surface material. The factor $21.7 \cdot 10^3$ was theoretically derived by considering a Vickers indenter geometry. The same factor can be used for a Berkovich indenter since the projected contact area is the same at same indentation depth.

Another commonly used relationship (Broek, 1997) for the fracture toughness and the residual stress includes the length of the radial cracks due to indentation by Vickers or Berkovich indenter and can be expressed as

$$K_{Ic} = \chi_r \cdot \frac{w}{\sqrt{a^3}} + Z \cdot \sigma_{res} \cdot \sqrt{a} \tag{5.13}$$

where a is the length of the cracks, Z is a crack shape factor and σ_{res} is the residual stress in the surface. The parameter χ_r is related to the size of the plastic zone and for Vickers and Berkovich indenters is

$$\chi_r = 0.016 . \sqrt{\frac{E}{H}} \tag{5.14}$$

The results are only reliable when the radial cracks are significantly larger than the lateral size of the plastically deformed zone. The parameter Z that depends on the depth of the radial cracks is approximately

$$Z = \frac{1.12 \cdot \sqrt{\pi} \cdot \frac{d}{a}}{\frac{3 \cdot \pi}{8} + \frac{\pi}{8} \cdot \frac{d^2}{a^2}} \tag{5.15}$$

where d is the depth of the crack. It is commonly assumed that the crack has a semicircular shape resulting in $d = a$ and $Z = 1.26$.

The above methods have been developed for bulk materials and can strictly only be applied for coatings if the stress intensity is not influenced by the substrate. Malzbender *et al.* (2002) has shown how the methods can approximately be applied also for coated surfaces. A different approach has been developed by Li *et al.* (1997) and Li and Bhushan (1998). They introduce a method for calculating the fracture toughness for thin coatings by analysing the energy release rate in indentation cracking and estimate the energy release from a step observed during the loading cycle of a load–displacement curve in indentation. They show that this method can be used for coatings even as thin as 400 nm. Further analysis and developments of the energy-based models for fracture toughness calculation of thin coatings has been presented by Cheng and Bull (2007).

In scratch testing a spherical indenter moves with increasing load over the surface and results in a more complicated crack pattern as discussed in section 3.3.5. Malzbender and de With (2000c) tested various theories for prediction of the critical load for formation of cone cracks in scratch testing of hybrid organic–inorganic coatings on glass. They found that a simple brittle fracture model was sufficient to estimate the depth of a flaw that leads to fracture.

Holmberg *et al.* (2003) developed a three-dimensional finite element model for calculating the stresses resulting in surface cracking in the scratch test, described in section 3.2.8. They analysed the contact conditions and concluded that the fracture toughness of the coating can be calculated in scratch testing by the simple equation

$$K_c = \sigma \cdot \sqrt{\frac{\rho}{2}} \tag{5.16}$$

where σ is the tensile stress in the coating calculated by the FEM model and ρ is the crack spacing (see Fig. 3.58). This applies when the crack spacing is of the same order as the crack length which means that the density of the crack field is high enough to dominate over the effect of crack length.

Bend testing has been used for fracture toughness measurement of thick freestanding films in a very similar way as for bulk materials. In some investigations a precrack has been introduced into the coating of a coated surface (Jaeger *et al.*, 2000) and this has enabled calculation of the fracture toughness of the coating in bend testing. However, the introduction of a precrack of a known size in a coating as thin as typical CVD and PVD coatings is not easy. Another approach is to carry out bending without a precrack and measure the 'cracking resistance' that is indirectly used as a measure for fracture toughness. The cracking resistance is defined as the threshold strain over which density of the cracks sharply increases. Zoestberger and De Hosson (2002) used the four point bend test for measurements of crack resistance in TiN, Ti(C,N) and TiN/TiAlN multilayer coated steel surfaces.

Wiklund *et al.* (1997a, b) developed a four point bend test small enough to be placed into an SEM and be observed *in situ*. They used both acoustic emission and direct observations to detect the coating cracking of TiN and CrN coated high-speed steel surfaces and obtained with both methods virtually identical values for cracking resistance. However, the acoustic emission method is much faster and easier to perform. SEM observation, on the other hand, allowed a more straightforward interpretation. The development of acoustic microimaging of cracks makes it possible to detect and image surface cracks with very small dimensions down to crack widths in the range of 10 to 100 nm (Thomas, 2005).

The measurement of the interfacial fracture toughness, that is, the resistance to crack growth of a lateral crack in the interface between the coating and the substrate, has been carried out by a buckling test where a buckling load was applied on an axially compressed plate of changing width (Malzbender and de With, 2002). Tensile testing of coated surfaces has been performed by some investigators reviewed by Zhang *et al.* (2005a).

5.3.3.4 *Residual stress*

The residual stress in a coating can be estimated by measuring the length of radial cracks in indentation as discussed in the section above. The limitations for residual stress measurement are the same as for estimating the fracture toughness.

A common method to separately measure the residual stresses in a coating is the deflection method. The coating is deposited on a thin plate, e.g. a silicon wafer, and the curvature of the plate after deposition is compared with its curvature before deposition as shown in Fig. 5.11. The residual stress, σ_{res}, in the coating can be calculated from the equation

$$\sigma_{res} = -\frac{E_s \cdot h_s^2 \cdot (r_a^{-1} - r_b^{-1})}{6 \cdot h_c \cdot (1 - \nu_c)} \tag{5.17}$$

where E_s and h_s are the elastic modulus and the thickness of the substrate, r_a and r_b are the measured substrate curvatures after and before deposition, h_c the coating thickness and ν_s the Poisson's ratio of the substrate (Stoney, 1909; Sue, 1992; Larsson *et al.*, 1996; Wiklund, 1999).

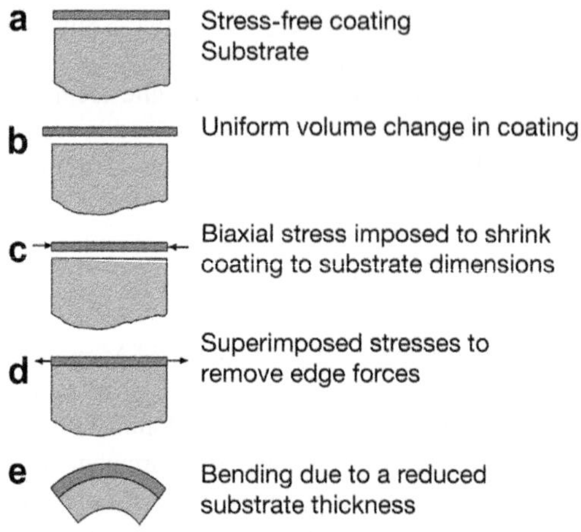

Fig. 5.11. Schematic illustration of deflection technique for determining residual stress in a coating (after Larsson *et al.*, 1996).

Residual stress in the coating can also be measured by the X-ray diffraction $\sin^2 \psi$ method (Chollet and Perry, 1985; Rickerby, 1986; Perry *et al.*, 1989; Borland, 1997). The stress measurement is based on the change in the interplanar spacing (strain). If a biaxial stress state exists within the coating, the strain in a direction inclined to the surface normal of the coating by an angle ψ and the stress acting in the surface plane of the coating in the direction φ are related as derived in theoretical calculations found, e.g., in Noyan and Cohen (1987).

A good agreement between the two methods was reported for TiN coatings by Sue (1992). Larsson *et al.* (1996) compared the deflection and the X-ray technique for measuring residual stresses in a TiN coating on high-speed steel and a diamond coating on cemented carbide and received only about 5 to 10% lower compressive stress values for the X-ray technique. The major advantages of the deflection method are that no information about the coating elastic properties is required for the calculation of the residual stress and it can be applied to most coated substrates, including most technically important materials.

5.3.3.5 Adhesion

This may be the most important coating property, since if the adhesion of the coating to the substrate is inadequate all functionality of the surface may be lost. Many techniques have been suggested for adhesion assessment. Some of these have their origins in technologies in which the bonding forces are comparatively low, such as painting, and are therefore limited in their scope. Others are deficient for other reasons, as outlined below. The theoretical background to coating adhesion and adhesion assessment has been reviewed by Bull (1992), Matthews (1988a), Bhushan (1999b) and Ollendorf *et al.* (2001).

Pull-off tests. Several techniques have been reported which rely on forming a bond to the coating and applying a force to pull it off. A very basic test is the one reported by Kuwahara *et al.* (1978), where the end face of a cylindrical mild steel rod is stuck at right angles to the coating surface with epoxy cement. After curing, a load is applied at the end of the rod, parallel to the coating. The load at which the rod falls down and strips the film gives a measure of the adhesion. Instruments are available commercially employing a similar principle; others apply a pull normal to the coating surface. It is important in the latter case to ensure that the loading is purely normal and that no bending moments

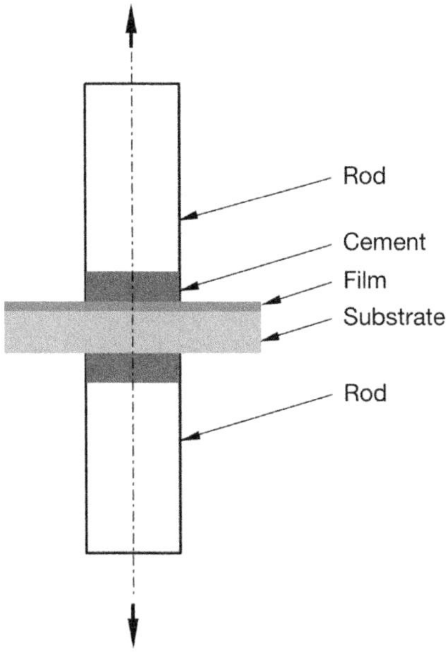

Fig. 5.12. The direct pull-off adhesion test method.

are applied at the interface. El Shabasy (1981) has described the arrangement shown in Fig. 5.12 in which the film and substrate are sandwiched between two rods which are aligned and should therefore not apply any bending or shearing force parallel to the coating surface.

The limitation of all these techniques which use epoxy or other bonding agents is clearly that the maximum adhesion that can be measured must be less than the strength of the bonding material itself. This fact is particularly pertinent in the case of tape tests which were used for many years as a means of adhesion assessment (Campbell, 1970). The technique used was to apply a pressure-sensitive adhesive tape to the surface and then pull it off to reveal the degree of coating detachment.

Indentation tests. Indentation tests are in fairly widespread use in industry, as a quick and convenient means of obtaining a qualitative indication of adhesion. This usually takes the form of a simple Rockwell C indentation at 150 kg and microscopic (100×) observation of the cracking around the resulting indent, which is compared with a chart displaying acceptable and unacceptable results on a 54 Rc hard substrate (VDI Richtlinien, 1991). Jindal *et al.* (1987) described how static indentation using this kind of Brale indenter could be used to obtain a quantitative measure of adhesion. They produced a series of indentations and determined the average extent of lateral cracking as a function of load. They showed that the slope of the indentation load–lateral crack length function provides a better measure of relative adhesion of hard coatings on cemented carbide substrates than the approximate measurement of the crack initiation load. The method suffers from the deficiency that it relies on the cracks propagating to the surface, so that they can be observed. It is quite possible for debonding to arise without such surface cracks being visible.

Scratch test. A technique that has received widespread scientific and industrial acceptance for adhesion assessment is the scratch test, also sometimes known as the stylus method. Typically, an indenter is pulled across the coating surface, under increasing normal load, until coating detachment occurs. The load corresponding to failure gives a guide to the adhesion strength and is often referred to as the critical load.

This method was first used to study the adhesion of metal films on glass by Heavens and Collins (1952). Subsequently, Benjamin and Weaver (1959, 1961, 1963) undertook further studies into metal on glass adhesion. Matthews (1974) used a scratch test to investigate the adhesion of silver coatings on mild steel and copper, deposited by both ion plating and vacuum evaporation. In this work both hemispherical and pyramidal indenters were used.

Currently, the most widely used version of the test involves a conical indenter with a 0.2 mm radius spherical tip (Hammer *et al.*, 1982; European Standard, 2000; Jennet *et al.*, 2003; ASTM, 2005). The definition of failure can be the onset of cracking round the indenter, spalling of the film, or the production of a channel in which all the coating has been removed. The type of failure will depend to some extent on the ductility of the film. As discussed later, this test has received extensive attention, partly because it also is a very accurately controllable device for investigations of basic tribological contact phenomena (Hedenquist *et al.*, 1990a; Bull, 1991; Malzbender *et al.*, 2002; Holmberg *et al.*, 2003; Zhang *et al.*, 2005a; Bull and Berasetegui, 2006; Jennett and Gee, 2006; Lau *et al.*, 2007). Various aspects of scratching both bulk and coated surfaces are presented in Sinha (2006).

In its simplest original form, the scratch test gave only comparative information about adhesion. It is commonly agreed that similar substrate hardnesses and similar coating hardnesses must be used in comparing coating adhesion. Lau *et al.* (2007) investigated four different kinds of coatings, Al, W, TiN and $Ti_{1-x}Al_xN$, with coating/substrate hardness ratios in the range of 0.05 to 41 and found eight different failure mechanisms in scratch testing. Based on these they constructed scratch failure maps with normal load vs hardness ratio as the axes, and different regions of failure mode could be identified. They call the regions zero failure, mixed failure and coating breakthrough, which means total delamination. They found that when the hardness ratio is larger than 23 the hard coatings will delaminate even at small loads. In contrast, when hardness ratio was in the range of 1 to 3, the coatings had superior adhesive strength and damage resistance in the scratch test.

The test is also thickness dependent, as indeed may be the effective adhesion. By this we mean that the adhesion may be apparently satisfactory if the coating is thin, but if the internal stresses in the coating increase with thickness, then the adhesion may not be adequate to withstand these. The internal stresses act as a preload on the interfacial bonds and thus the effective adhesion that can withstand an additional load is apparently lower. This introduces a further complication which often applies to other test methods more than the scratch test, in that the adhesion can apparently be improved by producing an inferior coating structure. One that is porous or voided can accommodate deformations more readily, but is usually less serviceable in mechanical contacts. It may, however, be beneficial in thermally loaded situations where the coating must expand and contract with the substrate.

The results obtained by the scratch test are dependent on the indenter material, its surface finish and geometry. Also, the coating surface condition, its geometry, and even the relative humidity in the air during testing, can influence the results.

In spite of the problems outlined above, several workers have been engaged in improving the accuracy and sensitivity of the scratch test, with the result that it is now much better understood. Valli *et al.* (1985a, 1986) carried out studies into friction monitoring coupled with acoustic emission detection and found that in many cases the former method is more sensitive in detecting failure. In particular, it was shown that on very thin coatings, less than 1.0 μm thick, coating failure could be detected only by a change in the friction force. The equipment layout is shown in Fig. 5.13.

Figure 5.14 shows a plot of the friction coefficient as a function of the normal force. In this plot lower and upper critical loads are indicated, corresponding to different degrees of coating failure. Normally, the lower critical load is taken to indicate cohesive failure within the coating, and the upper critical load signifies adhesive failure between the coating and the substrate.

Various attempts have been made to model the scratch test mathematically, beginning with the work of Benjamin and Weaver (1959) who used an analysis based on fully plastic deformation. Recently, more realistic elastic–plastic characteristics have been used, specifically in relation to the study of the change in stored energy due to buckling or spalling. This allows a Griffiths energy balance

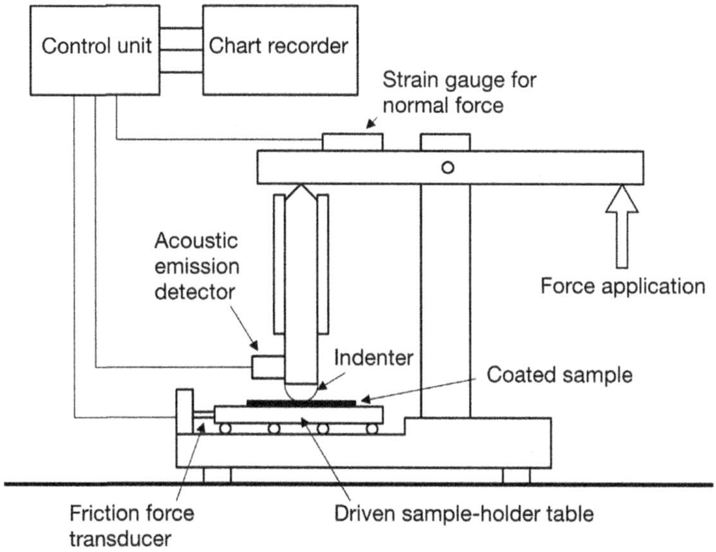

Fig. 5.13. Equipment layout of a scratch tester (after Valli *et al.*, 1985a).

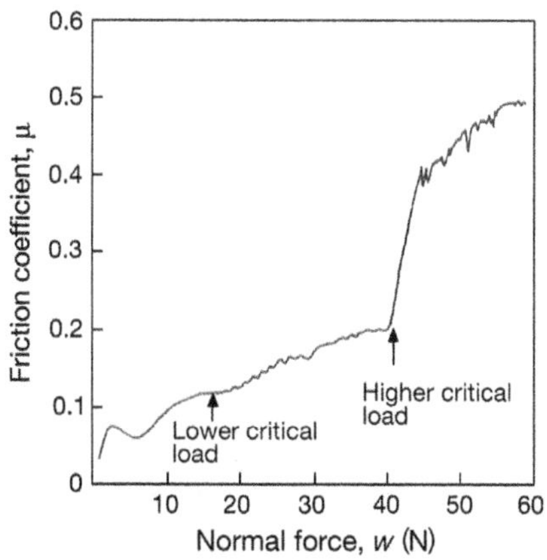

Fig. 5.14. The coefficient of friction as a function of normal force measured as a diamond stylus is sliding on the coated surface with a linearly increasing normal load (after Valli *et al.*, 1985a).

approach to be used, which Laugier (1984) analysed using elasticity equations. Bull *et al.* (1988a) identified the following three contributions to the stresses responsible for coating detachment:

- an elastic–plastic indentation stress,
- an internal stress, and
- a tangential friction stress.

Based on an analysis of these contributions the following equation was developed for the critical load, L_c, to cause detachment

$$L_c = \frac{A}{\nu_c \cdot \mu} \sqrt{\frac{2 \cdot E_c \cdot W_a}{h}} \tag{5.18}$$

where A is the cross-sectional area of the track, ν_c is the Poisson's ratio for the coating, μ is the coefficient of friction, E_c is the elastic modulus of the coating, W_a is the work of adhesion and h is the coating thickness.

This confirms the empirically validated observation that L_c will increase with reducing friction and increasing substrate hardness, as shown in Fig. 5.15. The latter reduces the cross-sectional area. On the other hand, it appears contrary to the trend that the critical load increases with thickness before a fall-off is reached, as shown in Fig. 5.16. This analysis can be criticized as not being universally applicable, due in part to the complex and very different failure mechanisms which can occur in practice.

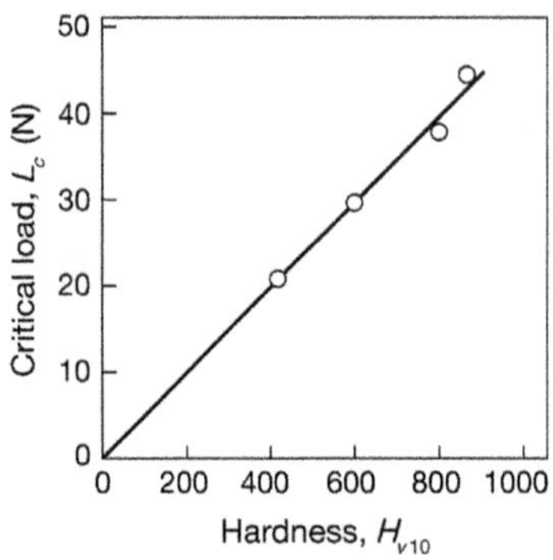

Fig. 5.15. The critical load as a function of substrate hardness from scratch test measurements. A 8.1 μm thick TiC coating was deposited by CVD on steel substrates (data from Valli, 1986).

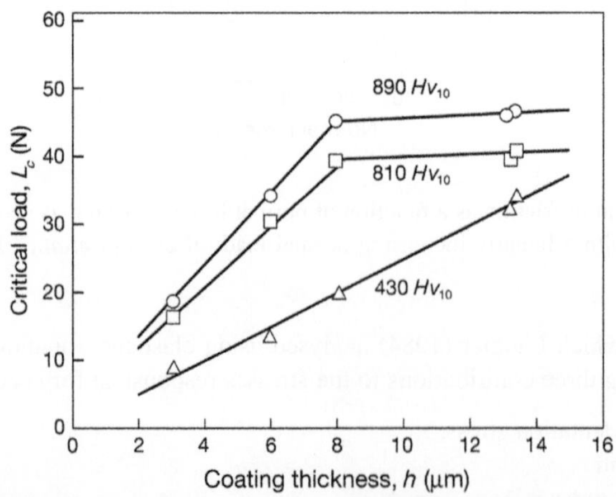

Fig. 5.16. The critical load as a function of coating thickness from scratch test measurements. A TiC coating was deposited on substrates with three different hardnesses (data from Matthews, 1988a).

It can be considered that the scratch test, at least in part, assesses the ability of the coating to deform with the substrate. In other words, a coating which can deflect with the substrate under load will exhibit a higher critical load. Bromark *et al.* (1992) put this in slightly different terms by stating that pure adhesion failure seldom occurs in a scratch test. They claim that the critical load is a measure of the ability of the coating to 'resist deformation'; by this they actually mean the ability of the coating to accommodate substrate deformation without failure, due to its ductility.

A round-robin scratch adhesion test study with six sets of samples being circulated and measured in five different laboratories showed, for all sets but one, very good reproducibility for a given instrument–stylus combination, about ±5%. The statistical variation was greater between different styli and the overall variation was about ±20%. The samples were 1 to 10 μm thick TiN deposited by CVD or PVD on to cemented carbide, TiN deposited by PVD on to high-speed steel and TiC deposited by CVD on to cemented carbide (Perry *et al.*, 1988). The variation appeared to increase for thicker PVD TiN coatings on cemented carbide.

In a critical analysis of failure in scratch and wear testing Rezakhanlou and von Stebut (1990) showed that by increasing the tip radius of the indenter the measured critical load for crack generation becomes increasingly sensitive to coating/substrate interface brittleness and flaws. They propose the test to be used in combination with studies of surface damage and recommend multipass tests as more realistically modelling sliding wear contacts. Bennett *et al.* (1992) have shown that multipass tests can produce debonding after a very small number of passes even at loads less than a third of the critical load. Multipass scratch testing at subcritical loads simulates more accurately real tribological contacts, where it is unusual for one of the counterfaces to remain elastic while the other is fully plastic, as in the scratch test.

The surface deformation and stress and strain generation process taking place in a scratch test has been modelled by a microfinite element method and parameters influencing the coating fracture behaviour were analysed based on stress simulations by Holmberg *et al.* (2003, 2006a, b, 2008) and Laukkanen (2006). This work is described in detail in section 3.2.8.

Further developments of the scratch test method are reported by Xie and Hawthorne (2002) where they study the effect of contact geometry on the failure modes of coated surfaces and conclude that an indenter with large radius and high normal load should be used. This results in high compressive coating stress and low bending-induced stress and thus the results correlate better to the adhesion of the coating to the substrate and suppress cohesive coating failure. One future development of the scratch test might be to use two indenter geometries: one sharper to measure the coating cohesion failure and one with larger radius to measure coating to substrate adhesion.

More detailed examinations of scratch test results can be done by cutting the scratched sample in the vertical sliding direction and polishing it until the middle of the scratch channel is reached. After polishing, the sample can be nital etched and analysed by high-resolution scanning electron microscopy. By knowing loading rate and position of the analysed area relative to the scratch channel end, it is possible to correlate the SEM picture, findings and features to the load (Podgornik and Wänstrand, 2005). Another interesting approach is to produce a crater by a ball cratering technique over the scratched track. This was used by Panjan *et al.* (2003) to study propagation of microcracks in different multilayer coatings.

Laser techniques. A number of adhesion test methods have been developed which utilize laser impingement on to the coating or substrate to detect or produce detachment.

Vossen (1978) describes a technique known as laser spallation in which a high-energy pulsed laser is directed on to the reverse side of a substrate, remote from the coating. A compressive shock wave is generated which propagates through the substrate to the film/substrate interface, causing detachment if sufficiently intense. By gradually increasing the incident power, a threshold level is found which is related to the bond strength.

Aindow *et al.* (1985) have reviewed laser-based non-destructive testing techniques for the ultrasonic characterization of subsurface flaws and Cooper *et al.* (1985) extended this study to cover the detection of laminar defects such as debonded regions. They apply a laser pulse on the coating surface, and detect the resultant acoustic waveform on the reverse side of the substrate. Satisfactory images of adhesion defects as small as 0.2 mm in width have been obtained. The coating thickness was about 0.5 mm, and the substrate was about 3.0 mm thick.

There have been attempts to permit the laser photoacoustic examination of coatings using a transducer on the same side of the coating as the laser beam. This removes substrate thickness limitations and is intended to allow thinner coatings to be studied also. By using an appropriate modulation frequency and a photoacoustic signal detector, films less than 2.0 μm in thickness have been successfully investigated. It could be argued that laser-based techniques of this type offer advantages over other available methods. They are effectively non-destructive on well-adhered coatings. Also, by imposing a thermal gradient at the interface, they more accurately simulate many real operating conditions for coatings such as those in metal cutting.

Body-force methods. The adhesion of thin films can be assessed by imparting acceleration or deceleration forces into the coating, as in ultra-centrifugal and ultrasonic methods (Valli *et al.*, 1986), though these do not appear to have found widespread adoption. Some producers have adopted the non-quantitative method of immersing coated objects in an ultrasonic cleaning bath to detect poor adhesion.

Acoustic imaging. It has been found that if a coating is observed in an acoustic microscope and also in an optical microscope, then the images produced can be compared and regions of poor adhesion can be seen in the acoustic image. The acoustic microscope operates by applying a short radio frequency pulse to a piezoelectric transducer, producing an acoustic plane wave which is focused by a lens and coupled to the reflecting object. The reflected pulse of sound returns through the lens to the transducer, which reconverts it to an electrical signal. By scanning the object mechanically a micrograph of the surface can be obtained on a CRT screen. This method does not provide quantitative information (Bray, 1980; Ollendorf *et al.*, 2001).

5.3.4 Physico-chemical evaluation

The developments in coating processes have been paralleled by improvements in the range and capabilities of techniques for surface analysis. These are important to the tribologist not only since they permit identification and optimization of new coatings but also because they allow an improved understanding of friction, wear and corrosion behaviour. Physico-chemical analysis techniques used in combination with mechanical characterization are the key to future models which correlate production parameters, structure and material properties to achieve optimized performance.

There are many books and articles that focus especially on the subject of surface analysis, such as Ferrante (1982), Chalker (1991), Vickerman (1997), Riviere and Myhra (1998), Briggs and Grant (2003), Flewitt *et al.* (2003), Werner (2003), Adams *et al.* (2005) and Birkholz *et al.* (2006). Here we shall confine ourselves to an overview of the subject.

Ferrante (1982) takes as a starting point the notion that almost all of the available methods utilize the following process cycle:

Excitation (Interaction)
↓
Dispersion
↓
Detection
↓
Spectrogram

The first step involves the production of the particles of the material to be analysed, for example in photon emission spectroscopy a spark may produce photons. Alternatively, the particle analysed could be an atom or cluster of atoms obtained by sputtering. The dispersion stage represents the mechanism by which the required information is filtered out from the available data. In light spectroscopy, this could be represented by a grating or prism; for an ion or electron it could be an electrostatic analyser. Next comes the detection stage, which could be a photographic plate for light or an electron multiplier for ions or electrons. Finally, spectrograms can provide information about the chemistry (qualitative, quantitative and chemical state), morphology and microstructure.

It is common to categorize the various methods in a matrix form similar to that in Table 5.1. This table includes incident species in the left-hand column and emitted species along the top row and extends on an earlier table due to Chalker (1991).

The basic methods of surface analysis used for coatings are SEM, TEM, XRD, AES, XPS (ESCA), RBS and SIMS. The features of methods commonly used in tribology are shown in Tables 5.2 and 5.3.

AES has been used extensively to measure elemental compositions as a function of depth. In that case it is usual to prepare a section through the coating. The ball cratering method described earlier provides a very convenient means of producing a tapered section. An alternative is *in situ* inert ion sputtering, with spectra being recorded after sequential etch cycles.

Both AES and XPS/ESCA are suitable techniques for elemental analysis. Sputter depth profiling and composition depth profiling are possible for both techniques (Quaeyhaegens *et al.*, 1992; Bishop, 1989; Christie, 1989). Chemical state information can be achieved by XPS/ESCA, in some cases also with AES (Bishop, 1989). In terms of the elemental analysis AES provides the complementary information to X-ray microanalysis (EDX).

SIMS examines charged particles sputtered from a surface and is used in tribology extensively to analyse the composition of thin surface transfer and reaction films. SIMS provides good data about very thin surface films but there is also a large averaging effect. The data are based on an area of normally more than 1 mm in diameter of the surface and cannot show the variations in the composition within this area that in tribological films may be significant (Stachowiack *et al.*, 2004).

X-ray diffraction (XRD) is one of the most widely used techniques in surface engineering. It has the potential to provide a very diverse range of information, such as phase composition, grain size, residual strain/stress and texture/grain orientation. Chalker (1991) provides a useful overview on the interpretation of XRD data.

The third generation coatings are mostly multiply layered or multiply treated. Depth resolution is therefore of fundamental importance. The so-called low angle diffraction (LAD) technique can be used to achieve depth-related information simply by altering the angle of incidence, and therefore the penetration depth (Schneider, 1994). This can be done using a regular Bragg–Brentano diffractometer with a suitable hardware configuration on the secondary side: usually a long collimator, for angular resolution, and a monochromator, permitting discrimination in favour of the source $K\alpha$ line. The LAD technique is not only of interest for layered or multiply treated samples, it is also of great potential in detecting compositional and microstructural gradients as a function of depth in a single layer. For example, Schneider (1994) has identified strain and phase gradients in plasma-nitrided stainless steels.

Other useful tools are the scanning and transmission electron microscopes. The former is used routinely for observing fracture and polished sections of coatings, permitting characterization of the morphology. Energy dispersive X-ray analysis (EDX) and wavelength dispersive X-ray analysis (WDX) can be used for elemental composition analysis, with suitable calibration. The transmission electron microscope (TEM) is in increasing use to study coating grain structures, and specifically to provide hitherto unobtainable information about interfacial effects including heteroepitaxial growth through the technique of cross-sectional TEM (Håkansson *et al.*, 1991).

New coatings, such as diamond and diamond-like carbon have led to the further development of methods such as Fourier transform infra-red spectroscopy (FTIR) and Raman spectroscopy to provide information unobtainable by other techniques. FTIR utilizes the mechanism that photons with a range

Table 5.1. Classification of analytical techniques in terms of the incident excitation source and analysed emission.

Emitted incident	Optical (photons)	X-rays	Electrons	Ions
Optical (photons)	IR, VIS, UV, PL, EL		UPS	
X-rays		XRF, XRD	ESCA, XAES, ESD, XPS, EXAFS	ESD
Electrons		EMPA, EDX	LEED, EELS, TEM, SEM, AES, SAM	
Ions		PIXE	IAES	SIMS, RBS, NRA, ERDA, NIMS

AES: Auger electron spectroscopy
EDX: energy dispersive X-ray analysis
EELS: electron energy loss spectroscopy
EL: ellipsometry
EMPA: electron microprobe analysis
ERDA: elastic recoil detection analysis
ESCA: electron spectroscopy for chemical analysis
ESD: electron stimulated desorption
EXAFS: extended X-ray absorption fine structure
IAES: ion-induced Auger electron spectroscopy
IR: infrared spectroscopy
LEED: low-energy electron diffraction
NIMS: neutral ion mass spectroscopy
NRA: nuclear reaction analysis

PIXE: particle-induced X-ray emission
PL: photoluminescence
RBS: Rutherford backscattering spectroscopy
SAM: scanning Auger microscopy
SEM: scanning electron microscopy
SIMS: secondary ion mass spectrometry
TEM: transmission electron microscopy
UPS: ultraviolet photoelectron spectroscopy
UV: ultraviolet spectroscopy
VIS: visible spectroscopy
XAES: X-ray induced AES
XPS: X-ray photoelectron spectroscopy
XRD: X-ray diffraction analysis
XRF: X-ray fluorescence

Table 5.2. Evaluation of surface analytical tools (after Ferrante, 1982).

	Surface sensitive	Commercially available	Analysis of practical systems	Elemental analysis	Compound analysis	Analysis quantitative
1. Elemental and chemical results by electron levels						
AES	Yes	Yes	Yes	Yes	To a degree	To a degree
ESCA-XPS	Yes	Yes	Yes	Yes	Yes	Yes
UPS	Yes	Yes	No	No	No	No
EMPA	No	Yes	Yes	Yes	No	Yes
XRF	No	Yes	Yes	Yes	No	Yes
2. Chemical and elemental analysis by mass						
RBS	No	Yes	Yes	Yes	No	Yes
SIMS	Yes	Yes	Yes	Yes	To a degree	To a degree
3. Elemental analysis by vibrational state						
EELS	Yes	Yes	No	No	No	No
4. Structural analysis, macroscopic features						
SEM	Yes	Yes	Yes	Yes with EDX	To a degree	To a degree
EL	Yes	Yes	Yes	No	No	No
5. Structural analysis microscopic						
EXAFS	Yes	No	Yes	No	No	No
LEED	Yes	Yes	No	No	No	No

Table 5.3. Comparison of features of surface analytical techniques (after Buckley, 1978; Ferrante, 1982; Vickerman, 1997; Watts and Wolstenholme, 2003; Walton and Fairley, 2004; Adams *et al.*, 2005; McArthur, 2006).

	AES	ESCA/XPS	SIMS
Spatial resolution imaging	3 μm Elemental maps	1 μm Quantitative imaging possible	Large area to 100 μm Inorganic material 100 nm, organic material 300 nm, ion microprobes 50 nm
Depth resolution profiling	20 Å Continuous profiles	20 Å Step profiles	1 atomic layer Continuous profile necessary
Chemical information	Possible, not well understood yet	Yes	Yes but interpretation is complex
Quantitative	Semiquantitative in some systems	Yes	Difficult, requires standards to establish sputter yields
Sensitivity	0.1% bulk 0.01 monolayer	0.1% bulk 0.1 monolayer	ppm
Sample consideration	Intense electron beam, modifies chemistry of surface	X-ray beam may have effect on some samples. Polymers and biological samples possible	Destructive technique. Accumulated damage depends on ion source. Polymers and biological specimens possible
Speed and simplicity	Fastest elemental data, easy quantitative and chemical data harder	Rapid data collection, easily interpreted	Rapid data collection, most difficult to interpret

of frequencies can be absorbed by chemical bonds with corresponding frequencies, and this allows the study of, for example, C–H bonds. Raman spectroscopy utilizes the inelastic scattering of photons by molecules to provide chemical structure information (Blanpain *et al.*, 1992).

However, none of the techniques mentioned above is able to solve all problems in characterizing coatings. It is therefore usually necessary to utilize several different techniques in order to fully characterize a surface or coating.

5.3.5 Tribological evaluation

Throughout this book we refer to different tribological tests which are in use in various laboratories. As is typical for tribology, even today there are a very wide variety of such tests with little standardization.

Some typical configurations are shown in Table 5.4 partly adapted and further modified from HMSO (1986). Tribological test methods and equipment are also presented in an ASTM book (Bayer, 1982) devoted to the subject of wear tests for coatings and by Bhushan and Gupta (1991), Czichos and Habig (1992), Bhushan (1999b), Spikes (1999), Szczerek *et al.* (2001), Hogmark *et al.* (2001), Blau (2001) and Hsu *et al.* (2005).

5.3.5.1 Sliding tests

The variety of tribological test methods used only for sliding contacts is illustrated by Bhushan and Gupta (1991) who list details of eight test methods. Of these, the pin-on-disk test is very suitable as a material screening test and for friction and wear mechanism analysis. It is by far the most widely used. Indeed, it was the subject of an international interlaboratory exercise aimed at standardization, under the auspices of VAMAS (Versailles Project on Advanced Materials and Standards). The lessons learned from this exercise have been published by Czichos *et al.* (1987, 1989) and used as the basis for an ASTM wear testing standard (G-99) and a DIN standard (DIN 50324). The other test methods may be very useful in cases where they better simulate the actual contact conditions and the results thus become easier to transfer to a specific application.

A typical pin-on-disk layout complete with humidity and temperature control, which is important in dry sliding experiments, is shown in Fig. 5.17. In the authors' laboratories a spherically ended pin (conveniently an AISI 52100 hardened ball, 10 mm in diameter) contacts a coated disk. The sliding conditions are typically a 10 N normal force and a sliding speed of 0.1 m/s in an environment of 50% RH, according to the VAMAS specifications. A total sliding distance of 100 to 1000 m is typical for tests with coated samples. The wear is assessed by stylus profilometry of the wear track and by direct microscopic observation.

As part of the round-robin study referred to in section 5.3.2, the authors carried out pin-on-disk tests on the three coatings, under the unlubricated conditions referred to, for a sliding distance of 250 m. The pin material in this case was M50 steel hardened to 62 Rc. The relative humidity at laboratories designated UH and VTT was 50% $\pm$ 2%, and at THD it was 68% $\pm$ 4%. The temperature for the VTT and UH tests was 21 $\pm$ 2°C, while at THD it was 23 $\pm$ 1°C. THD also had a vertical disk orientation, whereas the others used a horizontal disk.

The results, given in Fig. 5.18, show that even if the repeatability of the tests may be good within each laboratory, there can be a large variation in the reproducibility from one laboratory to another.

Engineers frequently tend to regard the friction coefficient, like hardness, as a kind of intrinsic material property. This is, however, not the case, since both are the result of quite complex interactive behaviour, and are subject to statistical variability. A measured friction coefficient is the result of both physical and chemical interaction in a discrete system, which is quite difficult to repeat from one laboratory to another because of the large number of influencing parameters. In the case of pin-on-disk testing it has been observed that even the vibrational characteristics of the test machine can have a marked influence on friction and wear behaviour. The level of reproducibility reported here is quite typical and has also been found in tests on bulk materials, as shown in Fig. 5.19.

Table 5.4. Wear and friction test methods for tribological evaluation of coated surfaces. Typical parameters are given but many variations of these are in use.

	Pin-on-flat (ASTM)	Pin-on-disk (VAMAS)	Block-on-ring (ASTM)
Test configuration	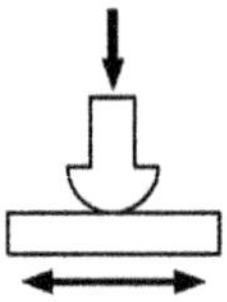	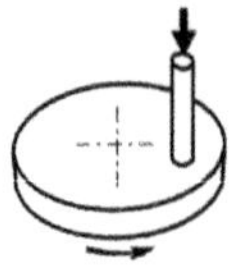	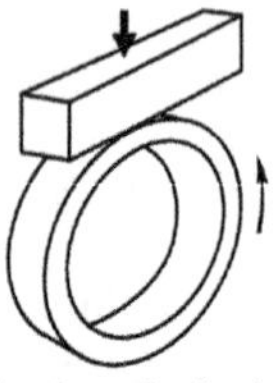
Test purpose	Determination of wear rate and coefficient of friction at low loads and speeds and with small specimens.	Determination of sliding wear rate and coefficient of friction.	Determination of adhesive wear rate of materials.
Specimens	A pin or a ball rubbing against a plate.	A pin or a ball rubbing on a rotating plate.	A steel ring is rotated against a stationary block.
Test conditions (typical)	Reciprocating sliding frequency 5 Hz. Stroke length 10 mm. Load 25 N. Temperature 18–23°C. Humidity 40–60% RH. Specimen can be dry or fully immersed in lubricant.	Sliding speed 0.1 m/s. Load 10 N. Sliding distance 1 km. Temperature 23°C. Humidity 50% RH.	Ring speed (max) 7000 rev/min. Load (max) 6000 N. Duration 5400–24,000 cycles. Specimen can be dry or immersed in oil or other fluid.
Measurements	Weight loss of disk. Height loss of pin. Wear track profilometry. Ball wear area measurement. Friction force.	Height loss of pin. Wear track profilometry. Ball wear area measurement. Friction force.	Test block weight loss. Scar width. Ring weight loss. Friction force. Lubricant film load-carrying capacity.
Wear types/comments	Mild adhesive wear suitable for tribological coating evaluation.	Mild and severe adhesive wear. Suitable for tribological coating evaluation.	Mild and severe adhesive wear. Lubricated wear.

	Block-on-pin (Falex)	Twin disc (Amsler)
Test configuration	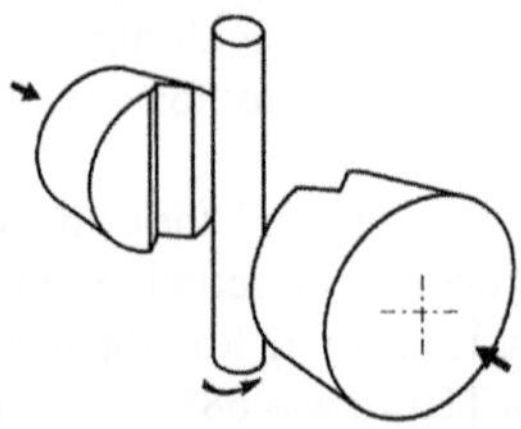	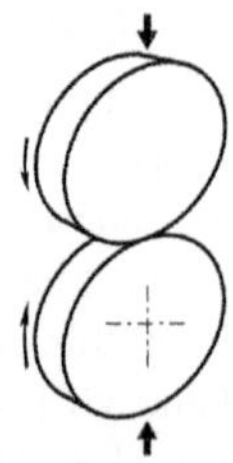
Test purpose	Determination of wear rate, load-carrying capacity and friction coefficient for sliding contacts.	Determination of sliding or rolling wear rate of test materials, treatments or coatings.
Specimens	A 6.35 mm diameter cylindrical journal is rotated within two loaded stationary V-blocks to give a 4-line contact.	Normally, two disks 40 mm diameter and 10 mm thick. Alternatively, the disks may be 30 and 50 mm in diameter.
Test conditions	Journal speed 290 rev/min. Sliding velocity 0.1 m/s. Load (constant or increasing) 89–20,000 N. Hertz stress 242–3450 MN/m^2. Specimens can be immersed in oil or other fluid.	Lower disk speed 0–400 rev/min. Upper disk speed 440 rev/min. Load 200–2000 N. Dry or lubricated.
Measurements	Journal driving torque. Progressing wear depth. Wear life. Seizure failure load. Final specimen weights.	Individual disk weight loss or profile. Total disk weight loss. Wear life.
Wear types/ comments	Mild and severe adhesive wear. Scuffing.	Mild and severe adhesive wear. Rolling contact fatigue (pitting).

Table 5.4. Continued

	Hammer wear	Fretting	Rubber wheel abrasion (ASTM)
Test configuration	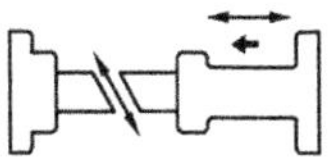	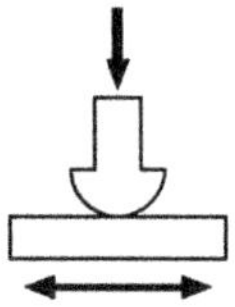	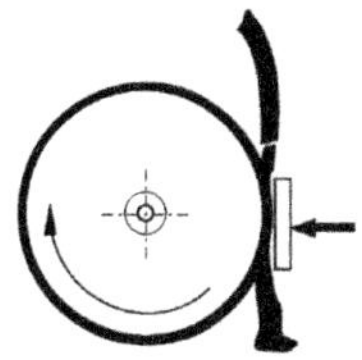
Test purpose	Determination of cohesion and adhesion qualities of coatings and also their wear characteristics under arduous vibrating conditions.	Determination of wear rate and friction in small amplitude oscillating contacts.	Determination of wear resistance of materials to third body-scratching abrasion.
Specimens	Obliquely cut specimens of 23 mm^2 contact area.	Ball or pin against a flat plate.	Flat (coated) pad is loaded against a rotating rubber wheel. The abrasive particles are a rounded quartz grain sand as typified.
Test conditions	Combined impact and sliding. Sliding distance 0.76 mm. Impact frequency 60–70 Hz. Impact load 60–223 N. Temperature 600°C. Duration 6–10 h.	Sliding amplitude 0.02–0.4 mm. Sliding frequency 5–20 Hz. Load 2–50 N. Temperature 22°C. Specimen can be dry or fully immersed in lubricant.	Wheel speed 2000 rev/min. Load 130 N. The abrasive media is dry sand particles or sand particles in wet slurry.
Measurements	Volume loss. Fretting resistance.	Volume loss by wear scar measurements. Friction force.	Weight loss or wear scar depth by profilometry.
Wear types/comments	Adhesive wear. Fatigue wear. Fretting.	Fretting wear.	Dry or wet low stress abrasion.

	Dry abrasion (Taber)	Wet slurry abrasion
Test configuration	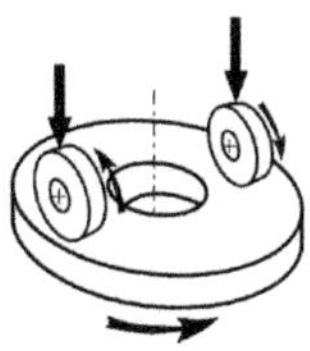	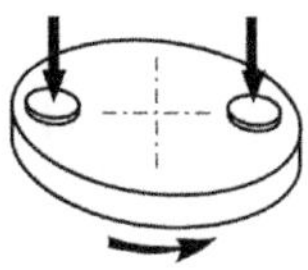
Test purpose	Determination of the abrasive wear rate and the Taber wear index of material surfaces.	Determination of the abrasive wear rate and the abrasion resistance of materials and treatments.
Specimens	Flat face of the disk abraded by two rubber bonded abrasive wheels (CS-10).	Two specimen pads loaded against a rotating circular plate.
Test conditions	Load 9.81 N. Wheels cleaned with abrasive paper every 1000 revolutions. Test is run without lubricant.	Load 112 N. Runs in bath of abrasive slurry.
Measurements	Weight loss. Taber wear index = weight loss/1000 revs (mg).	Volume loss.
Wear types/ comments	Low stress abrasion.	High stress grinding abrasion. Low stress abrasion.

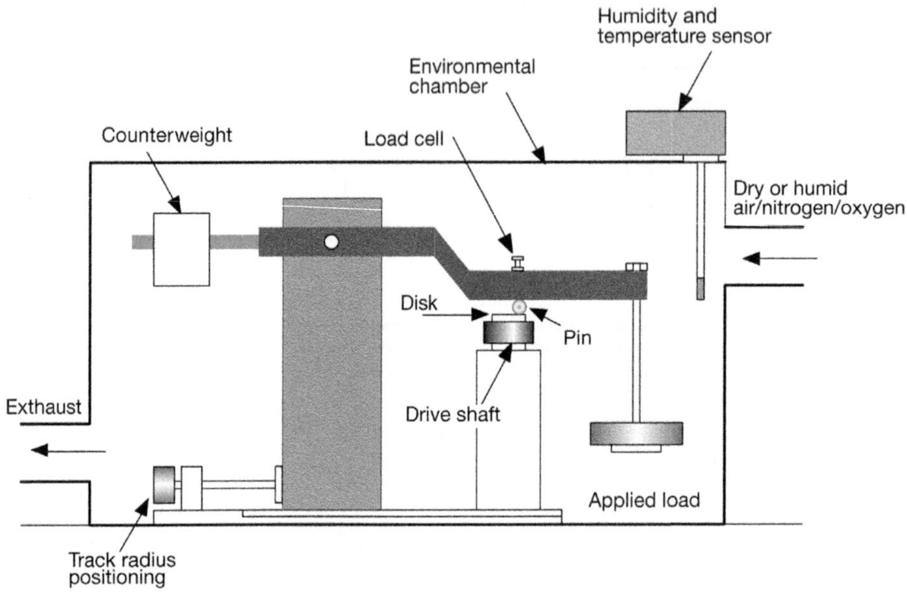

Fig. 5.17. Equipment layout for a pin-on-disk tester with humidity and temperature control.

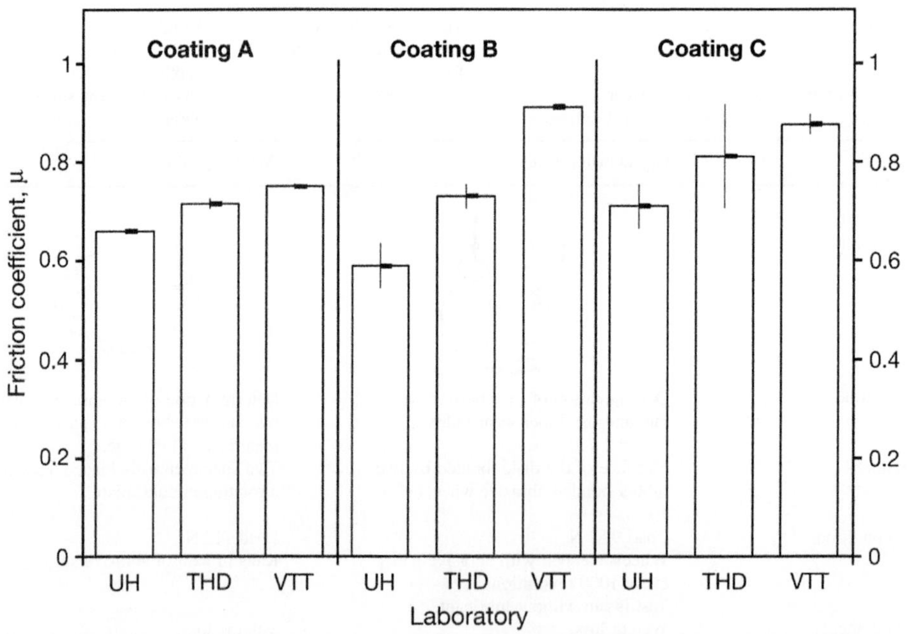

Fig. 5.18. Friction coefficients of a steel pin against coated surfaces measured by pin-on-disk testers in three different laboratories at a sliding speed of 0.1 m/s and a load of 10 N (data from Ronkainen *et al.*, 1990b).

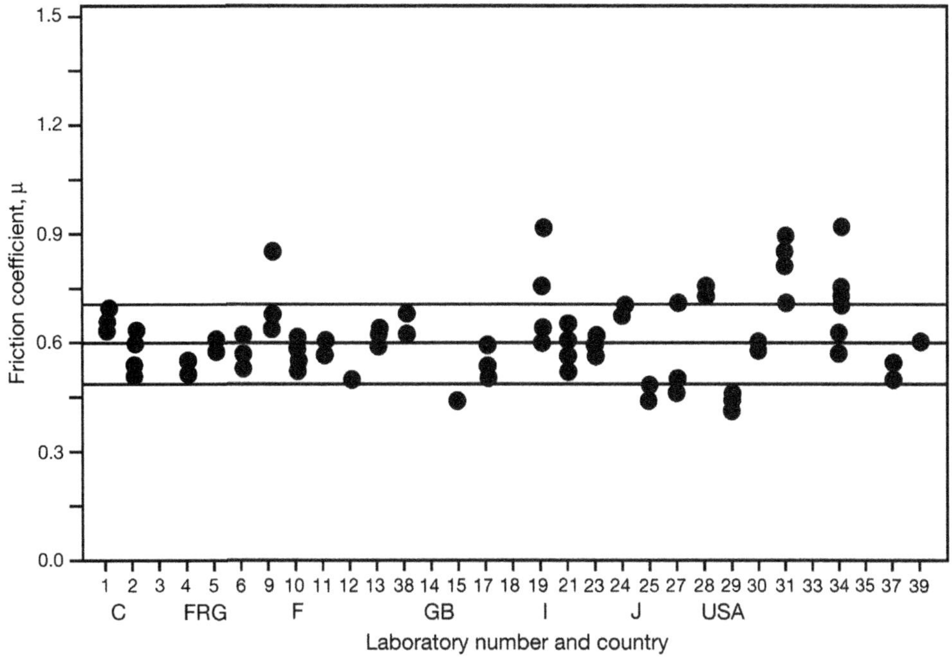

Fig. 5.19. Friction coefficients from pin-on-disk tests with a steel ball sliding on a steel disk in the VAMAS round-robin study carried out in 39 different laboratories. The mean value was 0.6 and the standard deviation 0.109 at the test conditions with a sliding speed of 0.1 m/s and a load of 10 N (data from Czichos *et al.*, 1987).

As with the friction coefficient, wear rates also show considerable variability. It is not unusual that wear rates may even vary over orders of magnitude. This indicates that there is some difference in the contact conditions that change the wear mechanisms from one wear regime to another. The wear rate values and standard deviations in the previously mentioned round-robin study carried out by UH, THD and VTT are summarized in Fig. 5.20, with the wear rates of the discs following the same order as the steady-state friction coefficients. It should be noted that wear rate results are typically displayed with a logarithmic scale. Some differences between THD and the others may be attributable to the different relative humidity and the vertical disk orientation, which would lead to debris being lost from the system more readily. However, observations of the worn surfaces and transfer layers indicated that the wear mechanisms were similar in the case of each coating at each site.

The above interlaboratory study illustrates the variations which can occur in tribological testing, even when identical coatings are tested and similar worn surfaces are observed. Other work has shown that if parameters such as the relative humidity or operating environment change then the variations in friction and wear can be even greater.

Thus there is always a need in pin-on-disk testing, as in all tribological tests, to control and monitor the test conditions as thoroughly as possible, to ensure identical test conditions when comparing different coatings, and to perform repeated testing to ascertain the level of reliability. Total repeatability should not, however, be expected, since friction and wear are due to contact phenomena on surfaces with properties which themselves exhibit some variability, such as crack formation and composition.

Another sliding test that has proved to be very useful as a screening test for coated surfaces is the reciprocating pin-on-flat test, shown in Table 5.4. The advantage is that the samples are of simple

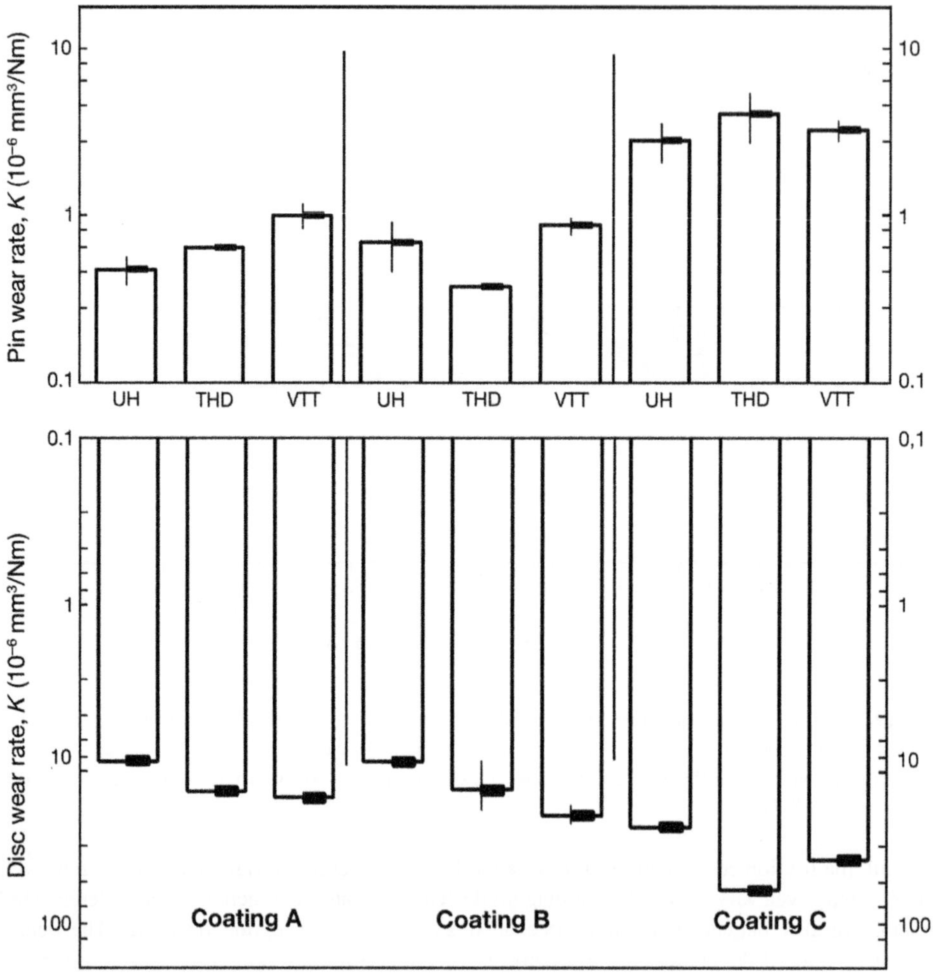

Fig. 5.20. Wear rates of coated surfaces measured by pin-on-disk tests in three different laboratories at a sliding speed of 0.1 m/s and a load of 10 N (data from Ronkainen *et al.*, 1990b).

geometry and the test dynamics is well controlled. If the test is used with small sliding amplitude the contact mechanism will be fretting wear. The fretting test has been developed to be almost a tribological reference test (Klafke, 2004b) and fretting maps are available (Zhou *et al.*, 2006).

A new range of requirements for tribotesting is coming from the development of microdevices related to nanotechnology research and development, such as micro-electro-mechanical-systems (MEMS). In such devices the loads are typically in the milli-newton range, the wear depth in the nanometre range and the dissipated energy in the microjoule range. Reciprocating sliding contact tribotesting devices have been developed to meet these requirements both on a micro- and nanolevel (Achanta *et al.*, 2005).

5.3.5.2 Abrasion testing

The best known tests in this category are the Taber test and the rubber wheel test, shown in Table 5.4 and Fig. 5.21. The former relies on the abrasiveness of the rotating wheels, while the latter utilizes

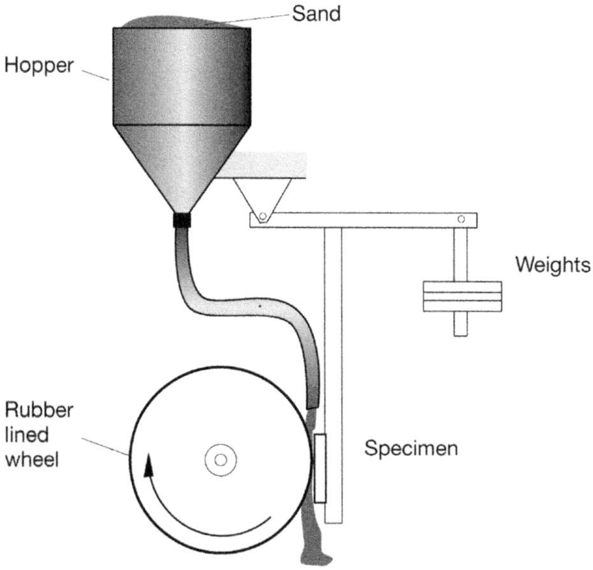

Fig. 5.21. Layout of the dry sand rubber wheel abrasion test apparatus.

a loose third-body particulate abrasive, either dry or in a water slurry. A typical medium is quartz sand, and the specification code for the dry sand test is ASTM 65 while the wet test is known as the SAE wet sand rubber wheel test (Bayer, 1982).

A good surface finish on a coating is important in such contact conditions, where asperity removal can lead to accelerated wear by revealing the substrate. This was demonstrated by Sirviö *et al.* (1982) for titanium nitride coatings deposited on plasma-nitrided steel substrates and tested with a dry sand abrasive wheel. The wet slurry test has also been useful, for example, in evaluating the wear resistance of carburized steel surfaces (Stevenson *et al.*, 1992). The tests are quick, taking only a few minutes to produce measurable wear, and assessment is easily carried out by profilometry of the worn regions or by weight loss.

The ball cratering test first used as a simple way of producing an inclined section and measuring coating thickness, as described in section 5.3.2 and shown in Fig. 5.4, has increasingly been used and further developed as a microabrasion wear test (Rutherford and Hutchings, 1996, 1997; Van Acker and Vercammen, 2004; Bose and Wood, 2005). It is well suited for wear resistance evaluation of coatings and can be performed on very small specimens because only a small free surface is needed for the crater. Still the results from this method have been shown to provide both good reliability and repeatability.

For purely erosive wear a very effective method is to use a particle air-blast test, as described in Bayer (1977). This has the benefit of providing a controlled erodent supply, including speed and direction. The method can be used as a single particle test, whereby the damage due to one impact can be determined (Stephenson *et al.*, 1986). Another erosion test, described by Jonsson *et al.* (1986) and Olsson *et al.* (1989), is based on a centrifuge principle, with particles fed into the centre of a rapidly rotating disc. Jonsson *et al.* also cite this method as providing a useful means of adhesion measurement.

5.3.5.3 Rolling contact fatigue tests

The most common apparatus used to simulate rolling contacts is the twin-disk machine, which can usually be operated at varying slide–roll ratios. It has been used, for example, to evaluate the performance of carburized surfaces for gears (Nieminen, 1990). Other rolling contact tests are used for evaluating lubricants and the high cycle fatigue behaviour of ball bearings. The resistance to cyclic loading can also be determined without rolling by applying a dynamic impact load on a coated sample (Knotek *et al.*, 1992a). This also provides an effective test of coating adhesion.

As mentioned in Chapters 4 and 7, rolling contact conditions are not the traditional domain of coatings. However, there is increasing evidence that very thin PVD coatings can enhance bearing and gear life under rolling contact fatigue conditions even if conflicting results have also been reported (Chang *et al.*, 1990; Amaro *et al.*, 2005; Yonekura *et al.*, 2005).

5.3.5.4 Laboratory simulation tests

There are a number of examples of research which have simulated real operating conditions in a laboratory environment in an effective way.

Matthews (1980) and Matthews and Teer (1980) devised a multistroke plane–strain compression test to simulate a cold forging operation between two coated dies. The test was run for 5000 indentations in strip steel at 50% strain, and provided an excellent assessment both of coating adhesion and of the retention of hardness in the dies during coating. Industry often needs to have this kind of evidence of coating reliability before it will risk putting a new coating into its production facilities.

Another example of a test, designed to simulate a specific industrial contact condition, is bending-under-tension (Sundquist *et al.*, 1983). This replicates a press forming operation and permits the evaluation of coatings for sheet press tools. As shown in Fig. 5.22, the method involves pulling strip metal over a coated die and is a severe test of anti-galling, i.e. severe adhesive wear resistance of the coating. This test was successfully used to show that it is possible to avoid the need for press forming lubricants by using hard PVD coatings.

Another common tribocontact in industry is the metal cutting tool to workpiece interface. This is more difficult to simulate, because of the high contact forces and heat generated. Nevertheless, methods have been developed to evaluate coatings for these conditions (Hedenquist and Olsson, 1991; Hedenquist *et al.*, 1990b; Habig and Meier zu Köcker, 1992).

Ultimately, the only truly satisfactory test is to run the coating under its intended operating conditions, and for metal cutting trials it is common to use the ISO 3685:1983 standard (Tool Life Testing with Single-Point Turning Tools). Ronkainen *et al.* (1991b) report the results of tests using this standard. For milling the standards are ISO 8688-1:1989 and ISO8688-2:1989 (Tool Life Testing in Milling, Part 1: Face Milling and Part 2: End Milling). For interrupted cutting conditions, in which there is a repeated impact action on the coating Knotek *et al.* (1992d) have devised a coating impact test which allows quantification of the number of impacts which a coating can withstand and the definition of a critical fatigue force.

There have been a large number of tribological tests reported of coated real products in laboratory conditions, such as testing of coated gears by Amaro *et al.* (2005).

Very advanced laboratory testing of a high power clutch system was carried out by Watson *et al.* (2005) and their analysis and interpretation of results includes developing a dynamic clutch model,

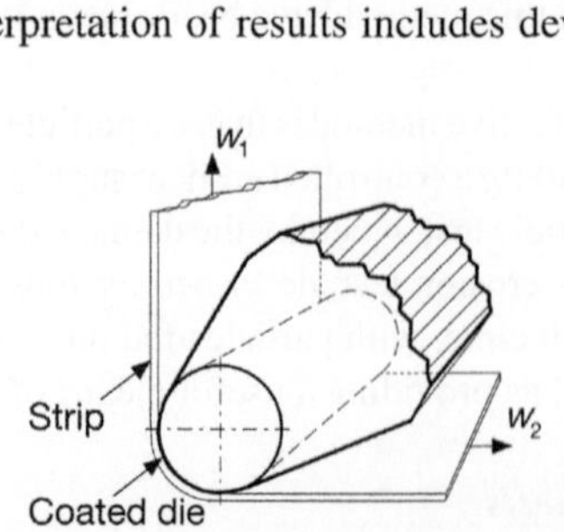

Fig. 5.22. Tribological contact in the bending-under-tension test.

a control system model, virtual temperature and thermal models, a multimechanism wear model, remaining useful lifetime prediction and statistical treatment of the results and uncertainty estimation. They extend Archard's law by using multiple stochastic models to determine a wear coefficient for three relevant wear mechanisms: abrasive, adhesive and oxidative wear. These models consider the physical wear process, including wear particle generation and protective surface layer formation, and use parameters such as surface roughness, particle size and surface temperature. Finally, they demonstrate a model-based clutch fault progression module. This work is an example of very advanced and carefully performed tribotesting and represents the forefront of the state-of-the-art in the field of real component laboratory testing.

5.3.6 Accelerated testing

Much of the information presented so far relates to techniques which are designed to provide performance information in a shorter time than would be required in a real-life situation. However, in such tests there is always a problem in establishing the extent to which they can accurately predict in-service performance (Holmberg, 1991; Franklin, 2008).

Tribological accelerated testing of components can be performed by increasing load, speed, contact pressure, temperature, rate of contamination and operation intervals or by decreasing the lubricant viscosity or additive content. Many of these parameter changes are not easily achieved in real machines. It is often more appropriate to perform the accelerated testing in a laboratory environment on rigs where the real contact condition can be simulated and overloaded with more controlled parameter variation possibilities.

The simplification of a tribological test can be divided into six levels of simulation according to Uetz *et al.* (1979). The complexity of the test decreases when moving from field test level down to rig test, component test, miniature test, contact simulating test and simple sample test, as shown in Fig. 5.23. When moving to more simple tests, the interpretation of the results gets easier, the repeatability and the statistical significance of the results are improved and the cost of each test declines. On the other hand, the tests will show less well the interaction effects between different parameters and the practical utilization of the results is less straightforward.

Based on a classification of the different strategies that can be used in accelerated testing, it can be concluded that simplified tests are especially suitable when the cost of one component, including the surrounding equipment, is high (Onsoyen, 1990). The use of simplified accelerated tests enables a large number to be made, which makes it possible to treat the results statistically. Box *et al.* (1978), Montgomery (1991) and Bergman (1992) have shown that more information can be gained from multiple tests when they are planned by factorial design techniques.

There are many advantages in using simplified accelerated tests, but there is also one considerable risk. If the acceleration introduced by overloading the contact changes the wear or lubrication mechanism to be simulated, then the results are of little significance for the application considered.

Results from simplified and overloaded accelerated tests are reliable only if measures have been taken to ensure that the simulated contact conditions are similar to, or in a controlled relationship with, the contact in the real component.

According to Holmberg (1991) there are three ways to ensure that laboratory tests are performed in conditions relevant to the application:

- by using the wear and lubrication maps that have been devised on the basis of large collections of published data,
- by monitoring as many parameters as possible during the laboratory tests, parameters such as friction, wear, film thickness and temperature, and
- by analysing and comparing the surface damage and the wear debris from both the laboratory tests and from worn component surfaces in real machines.

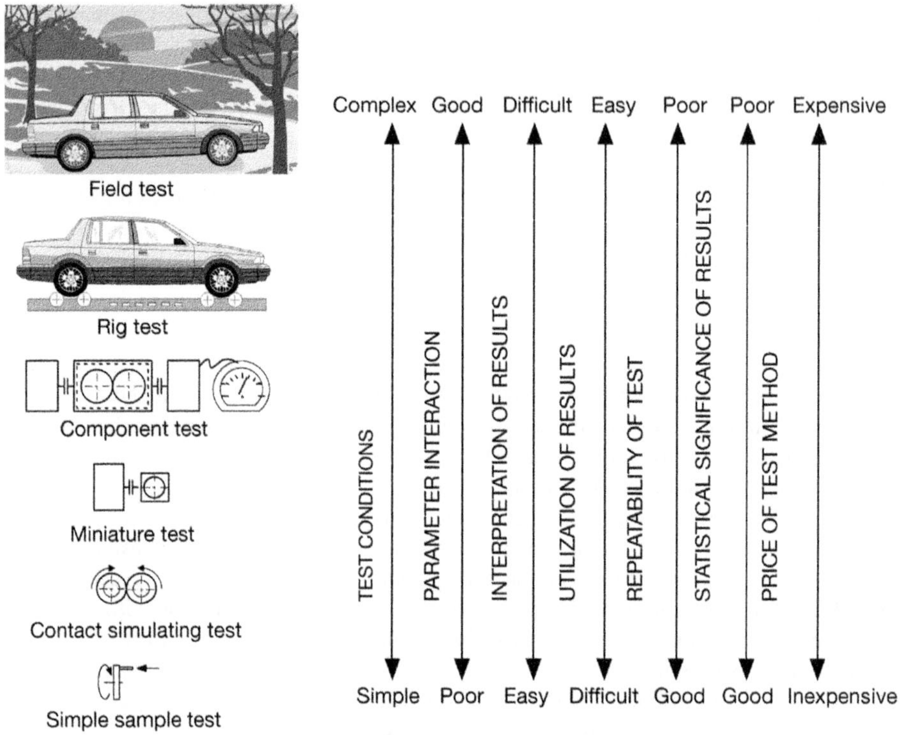

Fig. 5.23. Levels of simulation in tribological testing.

5.3.6.1 Wear and lubrication maps

A wear-mechanism map, or a wear map, is a figure which shows the regimes of different wear mechanisms depending on two parameters along the coordinate axes. This is of course a simplified and approximate way to present wear mechanism regimes, because only two parameters are varying and all the others are kept constant. If, however, the two parameters are those which have a dominating influence on wear, the wear map gives a good indication of how the wear mechanisms change and where the more uncontrolled transition regimes between two mechanisms can be expected to appear.

Lim and Ashby (1987) have shown how wear maps can be developed for steel pairs using a pin-on-disk configuration. The map, which is shown in Fig. 3.16, is produced on the basis of a detailed analysis of a large number of published wear test results and on a theoretical analysis of the wear mechanisms. It shows how the dominating wear mechanism changes with the speed and load given as normalized velocity and normalized pressure, respectively. Levels for the normalized wear rates are given in the map as curves of constant wear rates.

The wear mechanisms used by Lim and Ashby in their map representation are ultra-mild wear, delamination wear, mild and severe oxidational wear, melt wear and seizure. They give a simplified wear rate equation for each mechanism. These wear mechanisms are mainly combinations of some of the basic wear mechanisms described earlier and can be related to these as shown in Table 5.5.

It is important to note that Lim and Ashby have limited their wear map to the rotational sliding pin-on-disk configuration and to steel pairs. For this reason this map is only valid in applications with steel and with similar contact conditions. A method for developing similar wear maps for ceramics has been reported by Hsu *et al.* (1991). Wear maps for alumina ceramics have been published by Wang

Table 5.5. The wear mechanisms used in the Lim and Ashby (1987) wear maps are combinations of basic wear mechanisms.

Wear mechanisms	Adhesion	Abrasion	Fatigue	Chemical wear
Seizure	X	X		
Melt wear	X			
Severe oxidational wear	X			X
Mild oxidational wear	X			X
Delamination wear	X		X	
Ultra-mild wear	X			

et al. (1991). Vingsbo *et al.* (1990) have developed fretting maps to show the wear behaviour in reciprocating moving contacts. They use displacement amplitude and frequency or tangential force amplitude and normal load as coordinate axes in their maps. Further development of fretting wear maps was done by Fouvry and Kapsa (2001) and Zhou *et al.* (2006). Wear maps for uncoated high-speed steel cutting tools with feed rate and cutting speed as coordinate axes have been published by Lim *et al.* (1993a, b).

In a review paper of wear-mechanism maps Lim (1998) covers the development of wear maps including maps for metals, ceramics, metal–matrix composites, polymers, coatings, tools, fretting, erosion and maps for time-dependent wear transitions. He sites wear maps especially for coated surfaces related to wear of abradable coatings on gas-turbine components and flank-wear as well as crater-wear maps for TiN coated high-speed steel inserts. In a discussion on the tribological fundamentals behind wear mapping Williams (2005) concludes that wear maps are useful as they enable implications of wear on changes in design, material and operating parameters to be assessed and allow sensible correlation to be made between laboratory-based experimental investigations and observations in the field.

Similarly, Begelinger and de Gee (1981) developed a kind of lubrication mechanism map to show the changes from one lubrication regime to another. They represented the lubrication regimes as elastohydrodynamic (I), boundary (II) and scuffing (III) in a map with velocity and load as coordinate axes, as shown in Fig. 5.24. Scuffing is not really a lubrication mechanism but can be considered as a contact mechanism for failed lubrication. These lubrication maps are also known as IRG transition diagrams (IRG = International Research Group for Wear of Materials under the sponsorship of OECD).

Methods for construction of both wear and lubrication maps have been described in detail by Hsu and Shen (2005). They show how wear maps can be used as material selection guides and to support wear modelling and wear prediction.

Wear and lubrication maps are very useful when performing an accelerated test. By comparing the actual test parameters to these maps, it is possible to identify what kind of contact mechanisms can be expected and how close to the transition regions the test is, even if the map cannot be considered as exact. The accelerated test results are likely to be relevant to a real contact if they are both within the same wear or lubrication regime.

5.3.6.2 Test parameter monitoring

It is possible to obtain an indication of the dominating lubrication or wear mechanism by monitoring suitable test parameters such as friction, lubricant film thickness, temperature, wear and surface roughness. The advantage of parameter monitoring is that an indication of the contact mechanisms during the whole process of wear is obtained and possible changes in the mechanisms and the time when they take place are identified.

Friction is measured in most accelerated tests. The coefficient of friction in lubricated contacts gives an indication of the lubrication regime, as shown in Fig. 5.24, but this information may not

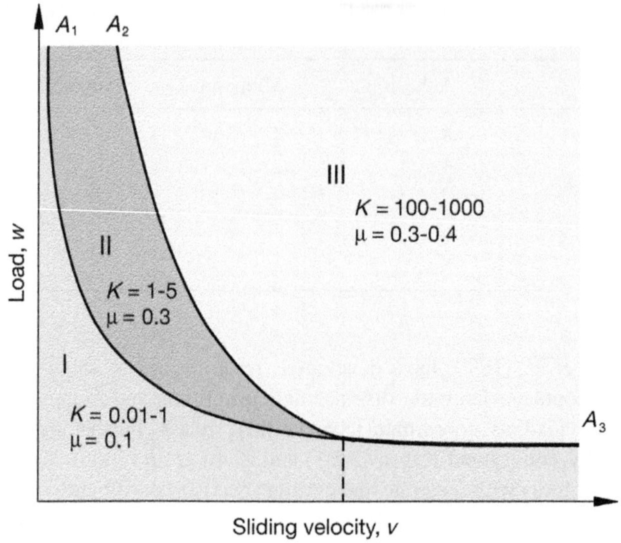

Fig. 5.24. Lubrication map showing (I) elastohydrodynamic, (II) boundary and (III) scuffing lubrication regimes for sliding concentrated steel contacts operating fully submerged in a liquid lubricant. K is the wear rate (10^{-6} mm^3/Nm) and μ the friction coefficient (data from Begelinger and de Gee, 1981).

be very accurate. Lim *et al.* (1989) have also shown that typical values for the coefficient of friction can be plotted on the wear map for dry steel contacts but this estimation may not be very accurate either.

A better way to define the lubrication regime in a contact is to measure the film thickness. This can most easily be achieved by passing an electrical current through the contact and measuring the electrical resistance or the capacitance, to give information about the amount of metallic contact (Dowson and Whomes, 1967–68; Georges *et al.*, 1983; Suzuki and Ludema, 1987). This gives a good indication of the transition region between boundary and full film (HD and EHD) lubrication, but not an absolute value for the film thickness. An absolute value for the film thickness can be determined by measuring the capacitance between the two parts, but this technique and especially the calibration is somewhat more complicated (Schmidt, 1985; Alliston-Greiner *et al.*, 1987). An alternative innovation is to use ultrasound (Dwyer-Joyce *et al.*, 2002, 2004).

Accurate on-line wear measurements are difficult to perform. One approach is to measure the acoustic emission signal from the tribological contact. The source of the signal is a rapid release of stored energy caused by cracking or crack propagation and thus this relates to the origin of wear debris generation (Löhr *et al.*, 2006). This is a very sensitive technique and if the sensor is not mounted properly and close enough to the contact the signal is easily disturbed. Löhr *et al.* report very good wear monitoring results and in their experiments the acoustic emission sensor was mounted 1 mm from the centre of the contact area.

Another approach is the radionuclide technique (RNT) where the samples are made radioactive and then the amount of wear debris by radiation measurements is registered (Ivkovic and Lazic, 1975; Scherge *et al.*, 2003b). The handling of the irradiated samples is not a problem since the radiation levels used with the techniques are very small. This method has an extremely high resolution and accuracy and is thus especially suited to mechanical systems showing very low wear rates; e.g. running-in wear in real combustion engines can be precisely assessed. Scherge *et al.* (2003b) present the modern RNT and conclude that this technique is very useful for continuous measurement and display of wear as a function of time on machine parts in operation as well as simultaneous measurement of wear for different selected machine parts or different locations on the same machine part. The high sensitivity is capable of resolving wear rates as low as 0.01 nm/h or 1 μg/h.

Wear particle analysis of oil samples from the contact by the ferrography method is a well-established technique which gives a good indication of changes in the wear mechanism or the wear rate (Jones *et al.*, 1978; Lansdown, 1985; Corso and Adamo, 1989; Roylance, 2005a, b; Williams, 2005). The shortcoming of this method has been the amount of laboratory work that the analysis of each sample requires but promising attempts to automate the procedure by on-line measurements have been reported (Sato *et al.*, 1987; Chambers *et al.*, 1988; Kuoppala, 1988; Roylance, 2005b). A future challenge lies in establishing through systematic tribological research the true origins of particles and the extent to which they influence the wear process.

Monitoring the temperatures in the contact gives useful information but is not sufficient to draw conclusions about the wear or lubrication mechanisms.

5.3.6.3 Wear track analysis

After interaction between two components, marks are usually left on the sliding surfaces. These marks can convey useful information about the contacting process. Each wear mechanism leaves its own features and the mechanism can often be identified by these. Descriptions of how wear mechanisms can be identified from wear scars have been published by Vingsbo (1979) and Engel and Klingele (1981) and of how they appear on machine components by Jones and Scott (1981) and Booser (1997).

A good method for ensuring the reliability of tribological accelerated tests is to compare the wear scars on the tested specimen with those from real components that have been working in industrial conditions similar to those being simulated. The applicability of the results may be assumed to be good if the wear scars indicate the same contact mechanisms.

5.3.7 Industrial field testing

The ultimate way to ensure that a coating will have the expected beneficial effect is to test it under real operating conditions. In general, this kind of testing is much more expensive than laboratory testing, especially if the same statistical certainty is expected. Thus in-field trials are mainly carried out to confirm earlier rig or component tests performed in the laboratory, as shown in Fig. 5.23. Methods for undertaking and analysing tribological field tests are discussed by Uetz *et al.* (1979) and Czichos and Habig (1992).

It is important to perform a thorough characterization and quality assurance assessment of coatings before they are released for field testing, to ensure that the coatings are what they are expected to be. Surface design to optimize the coated contact should be carried out and possible effects of the coating on the component or tool design, or on operating parameters, should be considered. There are several examples, unfortunately, of field tests being carried out in a hurry without previous coating quality assessment or design considerations and in such cases the probability of achieving good results is very low. This can cause lasting damage to the potential user's perception of coatings.

5.3.8 Standardization

The standardization of test procedures is continuously being pursued on a number of fronts. This should help to solve many problems which exist today for those interested in developing and using coatings. Although enormous amounts of tribological coating evaluation and testing work is carried out, it often contributes very little to our general understanding, due to the lack of standardization. It is seldom possible for one worker to take full advantage of tests carried out and published by others as reference data, because there will usually be some differences in the sliding conditions, such as speed, load, specimen geometry or surface roughness.

Table 5.6. Some test methods which have proved to be useful in evaluating the tribology-related properties of coatings.

Test method	Purpose
1 Profilometry	Surface roughness, topography and wear depth measurements.
2 Ball crater test	Coating thickness measurement.
3 Scratch test	Coating/substrate adhesion, coating cohesion and load-carrying properties evaluation.
4 Microindentation test	Coating hardness and elastic modulus measurements.
5 Pin-on-flat reciprocating test	Friction and wear properties at low load and sliding speed. Suitable for small coated areas and boundary lubricated conditions.
6 Pin-on-disk test (with ball as pin)	Friction and wear at normal loads and speeds.
7 Rubber wheel test	Abrasive wear.
8 Twin-disk test	Rolling contact fatigue.

There are strong scientific and economic arguments for further standardization of the characterization of coatings, measurements of coating properties and tribological test methods and devices.

In particular the microhardness test and the scratch adhesion test are considered by several authors to be important tools for the control and characterization of mechanical properties of surfaces and for quality assurance assessment (Schmutz *et al.*, 1989a, b; and Bull *et al.*, 1991).

Experiences from the authors' laboratories show that the ball crater method for coating thickness measurements, micro-Vickers for coating hardness measurements, scratch testing for effective adhesion evaluation and pin-on-disk or reciprocating sliding testing for wear and friction measurements provide useful information on the mechanical and tribological properties of coatings. They should be taken into consideration as the basis of standardized coating evaluation methods (Matthews *et al.*, 1991a). Table 5.6 summarizes the test methods that have so far emerged being particularly useful for the evaluation of mechanical and tribological properties of thin coatings.

Table 5.7. Examples of standards related to test methods of thin ceramic coated surfaces.

Standard	Title
EN 1071-1	Determination of coating thickness by contact probe profilometer.
EN 1071-2	Determination of coating thickness by the crater grinding method
EN 1071-3	Determination of adhesive and other mechanical failure modes by a scratch test.
EN 1071-4	Determination of chemical composition by electron probe microanalysis.
ENV 1071-5	Determination of porosity.
EN 1071-6	Determination of the abrasion resistance of coatings by a microabrasion wear test.
CEN/TS 1071-7	Determination of hardness and modulus by depth-sensing indentation.
CEN/TS 1071-8	Evaluation of adhesion by Rockwell indentation.
CEN/TS 1071-9	Determination of fracture strain.
ENV 1071-10	Coating thickness by cross-section.
CEN/TS 1071-11	Determination of internal stress by Stoney formula.

The standardization work is continually in progress. Two important bodies in the standardization of test methods related to ceramic thin surface coatings are CEN/TC 184/WG5: Advanced Technical Ceramics – Test methods for ceramic coatings and ISO/TC 206/WG17 and WG38: Fine Ceramics – Advanced Technical Ceramics – Test methods for ceramic coatings. Table 5.7 shows examples of some published standards.

CHAPTER 6

Coating selection

Contents

6.1 Problems of Selection

It has often been stated that to optimize the benefits achievable from coatings it is vital to consider their implementation at the design stage. However, no widely recognized procedure or system exists to assist those who select and specify coatings. This situation arises for many reasons. One is the vast range of coatings and treatments that exist, and the very different approaches that practitioners in different technologies use in order to select a coating. Thus, for example, electroplated coating or weld deposit guides exist, but they usually present information in different ways, making comparison between techniques very difficult. Another reason is that most coatings available have been developed empirically by trial and error and thus structured and generic information about their performance is lacking and data are not available to permit a general reasoned surface design and selection. Furthermore, many tribological coatings have come on to the market in recent years, often developed to fulfil specific applications requirements, but are untried in other situations.

In this chapter we present some approaches to surface coating selection and discuss ideas about the important questions and relevant methods for optimized design of coated surfaces. Before discussing the different approaches to coating selection, it is interesting to review a number of important considerations with regard to design which are sometimes overlooked (Holleck, 1991; Rickerby and Matthews, 1991b).

1. The bulk material must be able to withstand the processing steps required to deposit the coating. One of the major advantages of physical vapour deposition techniques over chemical vapour deposition is the ability to coat hardened steels below their softening or tempering temperatures, thereby eliminating the need for a rehardening stage. In some cases, even polymers can be coated with a hard wear-resisting coating by PVD, though these only function well if contact loads are low.
2. The coating should not impair the properties of the bulk material. It is not useful to apply a coating for wear protection if the corrosion or fatigue properties of the bulk material are adversely affected. In many applications it is not sufficient to consider the coating in isolation. One must take a systems approach to coating design, whether it is applied for wear resistance or some other purpose such as oxidation or corrosion protection.

3. Some changes in material specification may be necessary in order to accommodate the coating. It may not be sufficient to coat an existing component. It may be necessary to redesign in order to realize the full engineering benefits of the coating. This may involve a change of the counterface material, contact geometry or the loading conditions.
4. The deposition process must be capable of coating the component, in terms of both size and shape. Physical vapour deposition is partly a line of sight process, and rotating the component during the deposition may be necessary to achieve a uniform coating. In contrast, chemical vapour deposition and electroless nickel plating are noted for their ability to coat complex-shaped components with a layer of uniform thickness.
5. The coating method must be cost effective, but a comparison between methods should include factors such as reduced downtime, longer service life, more favourable working conditions, higher product quality, field of application and resource conservation, in addition to the coating costs. For example, the improvement of surface finish, which occurs through the use of a coated tool, may give an additional engineering benefit in that a subsequent finishing operation may be omitted.

In the past various writers have suggested checklist approaches to coating selection, but these are only helpful provided the information required by the checklist is readily available. Smart (1978) and James (1976) were probably the first to identify the problems of achieving an appropriate selection and they proposed various ways by which this might be achieved. Subsequent writers have elaborated further on the coating selection issue (Kramer, 1983; Holleck, 1986; Bell *et al.*, 1998b; Holmberg *et al.*, 1998; Davis, 2001; Cartier, 2003; Hogmark *et al.*, 2000, 2001; Jacobson and Hogmark, 2001; Becker, 2004; Erdemir, 2004a; Tucker, 2004; Holmberg and Matthews, 2005; Matthews *et al.*, 2007).

Smart presented a checklist based on two subheadings, process selection and materials selection. Process selection includes:

1. General factors such as process availability and quantity required.
2. Job factors such as size, weight and machinability.
3. Surface preparation factors such as undercutting and tolerances.
4. Finishing factors such as required finish and post-surfacing actions.

Materials selection includes:

1. General factors such as knowledge of similar applications and cost.
2. Operating environment, including wear types and lubrication.
3. Properties required, including wear resistance, hardness and machinability.
4. Substrate factors such as previous surface treatments and size.

The above provides a rather global approach which begs many questions. For example, the properties of the same materials deposited by different techniques can be very different. The lists do not necessarily make the designer's job easier. James (1978) made an effort to define the criteria and their interdependencies, and thereby make selection more straightforward. He did this by dividing the selection method into four distinct headings: requirements, limitations, interactions and economics and design. He described these as follows.

Requirements. Before any selection can begin, it is important that the function of the part and its surface is understood. James suggests that by comparing surface and bulk requirements the designer will be able to determine whether a simple material will be sufficient or if a composite structure such as a coated surface is needed. Some requirements may be changed by design, including, perhaps, surface requirements. By comparing requirements with resources it will be possible to make preliminary suggestions for materials and processes, but various limitations and interactions must be considered in order to select the most suitable ones.

Limitations. Here the design requirements are applied to the preliminary choices, based on five factors: environmental, use, social, supply and process. Use constraints refer to the operating location in which the coating will be used. A food factory or a nuclear power plant, for example, may impose certain specific constraints on what can be used, often based on safety legislation. Similar constraints may be associated with social aspects. Environmental, supply and process limitations are more straightforward to interpret.

Interactions. These are split into three parts:

1. Coating/substrate: the interaction of these two can be important, for example on adhesion, or under corrosive or stressed conditions.
2. Coating/process: some materials deposited by different processes give different properties. Examples are cobalt- or nickel-based alloys, which can differ in performance if applied by spraying or welding.
3. Process/substrate: the effect of the process on the substrate must be considered, e.g. overheating during weld cladding or hydrogen embrittlement during electroplating.

Economics and design. James emphasized that these factors need not necessarily be considered as constraints. A design should not be considered as frozen at too early a stage, as a change may bring benefits in allowing a wider range of surfacing processes. He thus defines the ability to redesign as a resource, although the problem of how the requirements can be matched against limitations and interactions, in a methodical way, still remains.

In essence James established a philosophy for selection, but which left many questions unanswered. This was developed by Smart, emphasizing wear mechanism identification – an approach which has been traditionally used by several writers.

6.2 Traditional Approaches

Traditionally, when considering the tribological requirement of a coating, an approach based on identification of the dominant wear mechanism was used. Smart (1978) used a designation based on three wear types (adhesion, abrasion and fatigue) to illustrate some general rules of thumb on selection. He contended that adhesive wear in the presence of lubrication is best resisted by the use of soft, dissimilar metal coatings with little tendency to mutual solubility. Where lubrication is marginal or absent, he suggested that harder surfaces are required, the precise hardness being dictated by the contact pressure, the hardness of the contacting surface and its roughness. He claimed that abrasive wear is best resisted by a coating with high shear strength – translated into hardness 1.3 times that of any abrasive particles present.

The thickness of the surface deposit should, according to Smart, be determined by the likely depth of abrasive penetration. In practice, this will be related to the degree of load support provided by the substrate. When low angle particle impact erosion is involved, hard materials give good resistance, even if they are brittle. With high angle impact, however, materials that are tough and ductile and can absorb substantial amounts of energy before fracture are preferred.

If cavitation erosion prevails, Smart states that it is desirable to use a material with a high ultimate resilience, defined as half of the ultimate tensile strength squared divided by the modulus of elasticity. For contact fatigue resistance Smart prescribes a coating with a high yield strength and hardness coupled with adequate toughness. Others have specified the *elastic strain to failure* as a determinant parameter in such situations. This has been equated to a low ratio of elastic modulus to yield strength (Halling, 1982; Matthews, 1984; Leyland and Matthews, 2000, 2004).

It is clear that there is never likely to be a universal wear-resistant coating to meet all requirements. Indeed, desirable properties may be counteracting. As Smart pointed out, high hardness tends to be associated with low impact resistance, notch sensitivity and poor resistance to crack propagation.

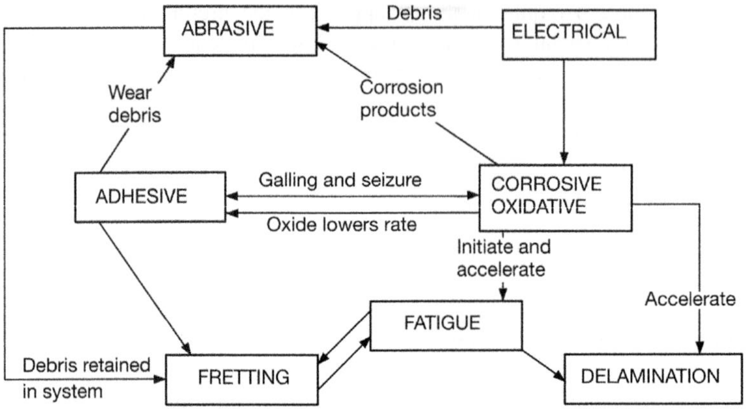

Fig. 6.1. Interacting mechanisms in wear (after Farrow, 1986).

Thus selection is largely a compromise between hardness and toughness. Simplistically, there should be adequate strength to resist the imposed loads, but sufficient toughness to avoid failure due to surface imperfections or incipient fatigue cracks. Smart favours an emphasis on toughness, since a higher than optimal wear rate will be preferable to catastrophic and unpredictable failure.

One drawback with an approach to coating selection which is based on identification of one dominant wear mode is that in almost all cases a number of wear modes operate, as illustrated by the sketch in Fig. 6.1 (Farrow and Gleave, 1984; Farrow, 1986). The message from this is that wear is a process which often involves at least three or four interdependent wear mechanisms. Thus, a complete determination of the wear process by a systems approach (Czichos, 1978), in which all input and output parameters are separately defined, will be a complex task. Similarly, approaches based on the quantification of wear coefficients under laboratory conditions have been shown to be prone to uncertainty. The ingress of small amounts of non-specified contamination can occur in practical situations, and this may change the wear coefficients by orders of magnitude.

There thus exists a need for a more robust methodology for coating selection, one which is built on current knowledge of tribological contact behaviour, bringing in the latest information about the characteristics of coatings produced by different processes, including their mechanical and chemical properties. Nowadays, especially since the introduction of advanced ion- and plasma-based vapour deposition techniques, ceramic, cermet and metallic films can be produced with specifically designed properties.

Kramer (1986) described how ceramic film performance can be predicted for specific wear situations. Holleck (1986) produced another model, suggesting mixed-phased deposits, produced by layering. Many mixed ceramic and metallic phases exhibit improved hardness and corrosion resistance compared to single-phased materials. Multilayered coatings are now widely deposited by vapour deposition methods. They offer a number of advantages. By incorporating layers with different properties, such as low heat conduction, high hot hardness, or low counterface compatibility, the coating can possess a combination of properties not possible in a single layer.

As discussed in Chapter 4, compositional or structural layering can also impart a degree of stress equalization through the film. According to Almond (1984) a layered film can also arrest crack propagation, a problem which can otherwise increase with thickness. Matthews (1980) proposed mixed-phase metal/ceramic coatings produced by layering. The particular laminated structure studied in this case was Ti/TiN, which was shown to have beneficial tribological properties in sliding contacts. This type of composite coating has been tested at increased coating thickness under three-body abrasive wear, and erosive conditions, confirming the beneficial combination of toughness and hardness which multilayered systems can provide (Leyland and Matthews, 1993).

The traditional wear mechanism approach to selection of coatings included for many years a lack of suitable coatings to fulfil the stringent requirements of a rigorous selection procedure. Now the availability of improved coatings and an improvement in the theoretical understanding of tribological contacts means that there is a better basis for finding more elegant solutions to the selection problem and the development of aids for the designer. This is particularly true given the widening availability of computer systems which can provide support for selection by allowing design simulations, database compilation and even logic-based selection.

In the traditional *positive selection* approach a possible solution is tried, and if it gives a satisfactory improvement, it is used. This often results in better solutions being rejected by default. A more advanced approach is *progressive elimination*, in which all solutions are considered, and those that are less feasible are eliminated (James, 1976, 1978). Progressive filtering is carried out until the optimum solution is identified. In the following a methodical approach is described. The discipline inherent in this technique encourages the acquisition of information, rather than the assumption of knowledge, and promotes the investigation of alternative ways of solving minor problems, which might otherwise be obstructing a highly desirable solution.

6.3 A Methodology for Coating Selection

An improved approach to coating selection should allow progressive elimination and lend itself to a computerized implementation. Furthermore, an approach which avoids the identification of one dominant wear mechanism would be desirable. A systematic framework fulfilling these goals is described below. This is intended to provide a basis for minimizing the probability of tribological problems by indicating the material property limitations and characteristics needed (Matthews *et al.*, 1992b, 2007).

The approach is based on matching the requirements of the application – what is needed? – with the combinations of properties that can be offered by different coatings – what is possible? – as shown in Fig. 6.2. The possible solutions are found by progressively eliminating those coatings that do not possess the required combination of properties. In order to achieve a successful marriage between the needs and the possibilities, it is necessary to express the requirements in a way that can be directly compared to the known properties and characteristics of coatings and coating processes. In effect, the component design engineer must translate the requirements into a language that can be understood by the coating process engineer. The different stages of the methodology and how the selection procedure is implemented are shown in Fig. 6.2.

The purpose of stages 1, 2 and 3 is to derive the correct tribological requirements needed for the match by successively refining the design specification. It is convenient to make a distinction between surface functional requirements (stage 4), non-functional requirements (stage 5) and economic and procurement requirements (stage 6). Various aspects that should be taken into consideration concerning the functional tribological and coating requirements (stages 3 and 4) have been critically discussed by Godet *et al.* (1991). Brief descriptions of the stages are given below. Examples of the contents of each stage are provided in Table 6.1.

Stage 1: Application and design study. This stage is similar to the early part of all engineering design procedures and involves selection of the global design requirements leading to a general design specification and a definition of the expected working conditions for the complete design, also referred to as embodiment design (Pahl and Beitz, 1984). The tribologically critical components are identified.

Stage 2: Component specification. The second stage involves a more detailed analysis of the critical components. For each of these, the specific contact and service conditions, such as contact geometries, forces, velocities, environment, etc., and the applicable constraints, such as overall component dimensions and dimensional accuracies, costs, desired lifetime, etc., are specified. It is necessary to distinguish between the part to be considered for coating (designated part 1) and any counterpart

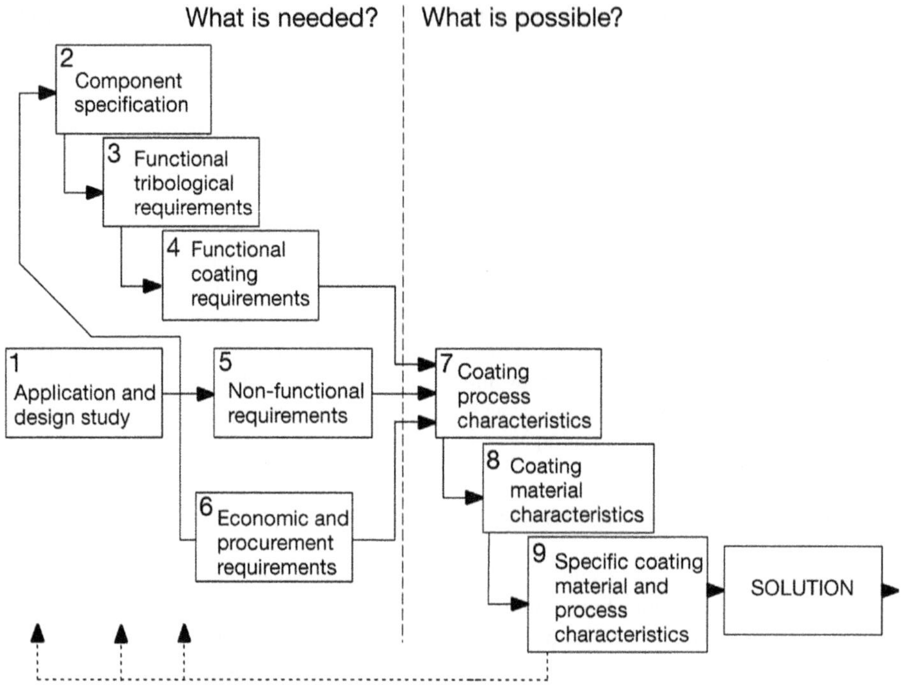

Fig. 6.2. Methodology for tribological coating selection.

(designated part 2) that is involved in the tribological contact. It is assumed that the material of part 2 is pre-specified. The next stage in the methodology is to convert the specifications in stage 2 into requirements of the coatings and coating processes to be considered.

Stage 3: Functional tribological requirements. The functional tribological requirements are those that are essential to the tribological performance of the design and will include noise and vibration limitations as well as friction or traction force requirements and allowable wear rates.

Stage 4: Functional coating requirements. These are the required limits on the appropriate surface properties, needed in order to fulfil the functional tribological requirements under the applicable service conditions. Typical limits are yield strength and elastic modulus, hardness, surface roughness, thickness of the coating, wear factor, etc.

Stage 5: Non-functional requirements. These requirements are not essential to the tribological function of the component but, nonetheless, represent important constraints on the coating materials and processes that can be applied. Examples are the dimensions required to be coated, maximum allowable dimensional changes associated with the coating or coating process, the need to mask off certain parts of the component from coating, desired surface colour, visual perception, preferred coating materials and processes, etc.

Stage 6: Economic and procurement requirements. Aspects concerning the allowable costs of the final component and the ease with which a coating can be obtained, either in-house or externally, are covered by these requirements. Also included are legal, environmental and availability factors which may prohibit or favour certain coatings or coating processes.

Stage 7: Coating process characteristics. Stages 7, 8 and 9 can be thought of as filters, positively eliminating those coating processes or coatings that are not able to meet the requirements in stages

Table 6.1. Examples of the information provided at the different stages in the coating selection methodology.

Stage	Examples (3 and 4 dependent on the applicable tribological situation)
1. Application and design study	General working conditions, appraisal of design requirements, overall machinery design, identification of critical components.
2. Component specification	Geometry of the contact between parts 1 and 2. Material and surface roughness of part 2. Working environment: abrasive particles, liquid, gas, temperature. Forces on the contact. Velocities involved. Nature of the motion involved. Desired lifetime. Maximum allowable tolerance.
3. Functional tribological requirements	Friction or traction force requirements: initial and steady-state. Noise and vibration. Avoidance of localized plastic deformation, fracture, surface fatigue and fretting. Wear rate.
4. Functional coating requirements	Yield strength of substrate in relation to the reduced elastic modulus. Thickness requirement for coating. Wear rate in relation to the counterpart. Required coating and substrate hardness. Surface roughness. Preferred coating types.
5. Non-functional requirements	Dimensions and weights of parts to be considered for coating. Maximum allowable dimensional changes associated with the coating or coating process. Thickness tolerance. Finishing requirements. Requirement for the masking-off of certain parts of the component from coating. Requirement for coating to penetrate holes and recesses. Desired surface colour. Preferred coating materials and processes. Any limitations on the substrate material for the coating.
6. Economic and procurement requirements	Allowable costs. Delivery time. Availability in-house, locally and worldwide. Legal and environmental constraints.
7. Coating process characteristics	Partial-coating ability. Penetration ability. Environmental friendliness. Ranges of possible thickness, precision of thickness, deposition rate, process temperature and relative distortion of workpiece, maximum coatable dimensions, cost per unit surface area, surface roughness change, relative adhesion and permissible substrates.
8. Coating material characteristics	Ranges of colour, elastic modulus, Poisson's ratio, hardness, fracture toughness, chemical reactivity, wear rate and friction coefficient.
9. Specific coating material and process characteristics	Narrower and combined ranges or specific values for the properties given in 7 and 8.

4, 5 and 6. Stages 7 and 8 can be considered as coarse filters because they are concerned with the invariable characteristics of particular coating processes or coating materials. Stage 7 deals with the process characteristics and property ranges that are valid for all coatings applied using a certain process such as CVD, PVD or plasma thermal spraying. In this way, if a requirement such as a maximum allowable coating thickness or maximum as-coated surface roughness does not fall within the range that is practically feasible for a particular process, then all coatings applied using this process can be eliminated. If a particular coating material is required, only those processes capable of depositing this material need be considered.

Stage 8: Coating material characteristics. Similar to stage 7, stage 8 concerns coating material property ranges and characteristics that, for practical purposes, can be considered to be independent of the coating process used. For example, the hardness of titanium nitride typically lies between 17 and 30 GPa, irrespective of whether a CVD or a PVD process is used. The ranges of engineering material properties have been collected in numerous design charts to support material selection in mechanical design by Ashby (1999). The charts cover material characteristics like Young's modulus, strength, hardness, fracture toughness, thermal conductivity, dry wear rate, density, relative cost, energy content, but are available only for bulk materials.

Stage 9: Specific coating material and process characteristics. This stage is in some respects a refinement of stages 7 and 8, dealing with the property ranges and characteristics for specific coating material–process combinations. In this way, a differentiation can be made between the property ranges of similar coatings deposited by different processes or varied conditions.

The most difficult aspects of the methodology involve the specification of the required tribological behaviour of the surface (stage 3) and the conversion of this information into surface property requirements (stage 4). We have earlier seen that tribological quantities such as friction coefficient and wear rate are strongly influenced by the tribological system, depending not only on the natures of the interacting surfaces involved in the tribological contact, but also on system conditions including forces, velocities, temperature, atmosphere, etc.

The system parameters are often variable and are not mutually independent, adding further complexity to the problem of predicting tribological behaviour. It is therefore not surprising that, in spite of intensive research, there are many questions in connection with the relationships between material surface properties, tribological system conditions and tribological behaviour which are still today unclear.

However, it is still possible to define which surface characteristics influence the tribological mechanisms occurring and are therefore important in determining the friction and wear behaviour of coated surfaces. This subject was reviewed in Chapter 3 and on that basis the following parameters are of major importance:

1. Yield strength and plasticity of the coating and the substrate (related to abrasive wear resistance and load-carrying capacity; quantified by hardness).
2. Elasticity of the coating and the substrate (related to stress and strain generation; quantified by Young's modulus and Poisson's ratio).
3. Tensile strength of the coating and the substrate (related to fracture resistance; quantified by fracture toughness).
4. Shear strength of the coating/substrate interface (related to the coating delamination resistance; quantified by the interfacial fracture toughness or adhesive strength).
5. Shear strength of the coating or any microfilms present on the coating surface or generated during service (related to friction).
6. Thickness of the coating (related to adhesive wear allowance and stress generation).
7. Roughness of the surfaces (related to friction and wear).

Ideally the design values for the above properties would be derived from knowledge of the required wear rate and friction coefficient and the known service conditions. However, the general understanding of the tribological phenomena involved, in addition to the theoretical models concerning the contact mechanics of layered surfaces, are not yet at a stage that enables comprehensive relationships to be derived from fundamental principles. Moreover, data on several of the properties listed above, in particular interface shear strengths and properties of microfilms, are not available for the majority of coatings. Unlike bulk materials, for which comprehensive and freely accessible databases exist, a general database for coatings is not available.

An alternative approach to theoretical models is to combine some of the more practically applicable contact mechanics theories with rules of thumb that, although not yet fully explainable on theoretical grounds, are nonetheless well established with tribological design engineers. Such an approach to tribological design has been suggested previously (Thijsse, 1989) and has been further developed (Matthews *et al.*, 1993; Franklin, 2008). The intention of this approach is not to calculate the exact property requirements but rather to provide, with a reasonable safety margin, a reliable indication of the material property limitations and characteristics needed. This enables the design engineer to effectively minimize the probability of tribological problems occurring in service; this should be supported by testing to simulate in-service conditions.

6.4 Selection Rules

The design methodology described above is founded on a number of practical design rules, each of which has one or more materials selection criteria associated with it. Depending on the design situation, one or more, commonly several, of these rules is appropriate. By applying the rules successively, the materials selection requirements are progressively refined. The methodology can accommodate many of the possible general tribological problems that occur in practical design situations. The selection of which rules are appropriate for a particular design situation is based on the following factors:

- the service conditions under which the component is required to perform, including the type of relative motion between the surfaces,
- the limitations of the rules, and
- the constraints imposed by the application of the materials used.

It is very important to note that the methodology is based on the rules which relate to the classification of the contact type or types, not to the identification of a dominant wear mechanism. This approach recognizes that more than one wear mechanism may occur in a given contact.

The following rules have been identified using contact conditions illustrated in Fig. 6.3.

Rule 1 – Contact stresses: Stresses generated by the surface loads should be below the yield strength and the fracture limits of the surface.

The most important criterion that must be satisfied by the contacting materials in many potential wear situations is that plastic deformation or fracture in the contact region must be avoided under both stationary and moving conditions. In the case of coated and surface-treated materials, this also applies to the substrate material. If considerable plastic deformation or fracture does occur, severe wear is likely to take place from the onset of motion (Tangena, 1987). This does not include failure of roughness peaks, which is inevitable in most contact situations.

Coated and surface-treated surfaces require special consideration because the mechanical properties are non-isotropic. The variation in material properties such as yield strength, hardness, elastic modulus, Poisson's ratio and fracture toughness with the depth from the surface must be taken into account.

The rule is based on the theory and analytical solutions originated by Hertz (1882) and developed by Johnson (1985) and Hamilton and Goodman (1966). For coated surfaces, use is made of the analytical solutions provided by Leroy and Villechaise (1990) and finite element analysis by Laukkanen *et al.* (2006). The H/E ratio can also be used to indicate the hardness and elastic modulus requirements of the surface, illustrating the benefits of a suitably low elastic modulus and a suitably high hardness (Leyland and Matthews, 2006). A coating which can deflect with the substrate's deformation under load is vital, and this requires a long elastic strain to failure, i.e. a high H/E ratio of the coating. It is important to note that the minimum allowable strengths for materials calculated in this rule are not intended to be used to provide accurate estimates of the maximum stresses actually occurring. Because the rule is intended for design purposes, the worst case situation is used as the basis for the calculations.

Rule 2 – Sliding: Contact temperatures, surface roughness and wear particle generation should be at a level which avoids effects such as accumulated surface melting, asperity and debris interlocking and ploughing. Then the conditions are characterized by stable sliding, typically with generated surface layers having low shear strength and minimum wear.

The tribological mechanisms involved in sliding are fairly well known and have been described earlier. In the case of conformal contacts, e.g. flat sliding on a flat surface, the contact is characterized

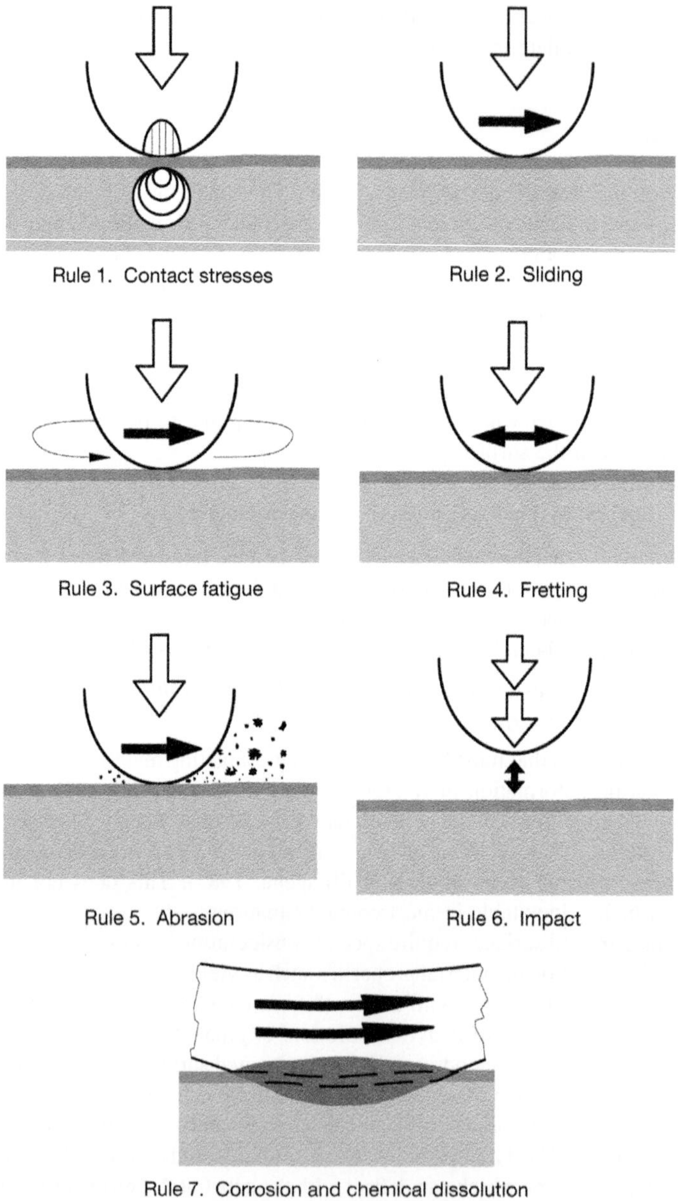

Fig. 6.3. Tribological contact conditions covered by the coating selection methodology rules.

by low pressure, but high contact temperatures may be generated and wear particles may remain within the system. The coating must then minimize wear and create a long lifetime and stable contact condition. This is often related to the formation of suitable transfer layers on the surface that allow the shear to take place with minimum wear.

In some non-conformal sliding contacts, such as a sphere sliding against a flat surface, continuous heat generation occurs only at the location of the sphere contact and not the flat counterface, which may see intermittent heating as the pin repeatedly passes over it. The contact conditions and the stresses and strains generated have been modelled and analysed by Holmberg *et al.* (2003, 2006a, b, 2008b). The aim is usually to reduce both friction and wear, and the need to prevent plastic deformation is paramount. This is the reasoning behind the use of the plasticity index Ψ (Greenwood and Willamson, 1966), which was originally devised to estimate the surface roughness required to ensure that asperity plastic deformation does not take place:

$$\psi = (E'/H)\cdot(R_{asp}/\zeta)^{1/2} \tag{6.1}$$

where E' is the reduced modulus of the contacting surfaces, H is the hardness of the softer material, R_{asp} is the typical radius of an asperity and ζ is the standard deviation of the asperity height distribution. Recent development of coatings such as DLC and MoS_2 make it possible to achieve superlubricity and close to zero wear with friction coefficients down to 0.001 and wear rates of only $0.0001 \cdot 10^{-6}$ mm^3/Nm. However, such tribological performance requires extremely smooth surfaces and a strictly controlled environment. Design criteria for such contacts are basically the same as presented above and the special conditions that must be considered have been presented by Erdemir (2004a) and Erdemir and Martin (2007).

Rule 3 – Surface fatigue: Coating elasticity and thickness should be of a level to allow deflection and avoid high-stress peaks and simultaneously distribute the stresses. The coating structure should inhibit crack growth. The substrate and coating strength should be above the level needed to avoid fatigue fracture. The surface should have optimized roughness to allow polishing.

In contact fatigue the surface is repeatedly loaded, and this can be a sliding or rolling action. This results in deterioration of the surface strength and crack initiation and propagation at or below the surface. The aim of a surface coating is mainly to distribute and decrease the surface stresses and to inhibit crack propagation. Surface fatigue is a complex phenomenon influenced by a large number of factors such as surface roughness and defects, shape, coarseness and distribution of carbides and inclusions. However, both theoretical and empirical relationships between the contact stress, crack propagation, coating thickness, surface roughness, the composition and the hardness of steels, the lubrication regime and the number of cyclic load changes have been presented in the literature (Niemann, 1960; Rowe and Armstrong, 1982; Kral and Komvopoulos, 1996a; Klafke *et al.*, 2005b; Yonekura *et al.*, 2005; Lindholm and Svahn, 2006; Zhao *et al.*, 2006a, b). Practical experience has shown that these relationships can be used successfully as a design tool to minimize the chances of failure occurring due to surface fatigue.

The maximum shear stress in a rolling contact is, in the majority of engineering design situations, located some 10 to 100 micrometres below the surface. In addition, when the roughness of the surfaces is taken into account, it has been shown by Sainsot *et al.* (1990) that for hard thin coatings with a thickness less than 15 μm on soft substrates, the highest stresses tend to be concentrated at the interface between the coating and the substrate. However, the fatigue life of a thin coating may be considerably longer than that of a thick coating for different reasons. Under similar deformation conditions the thicker coating will experience higher bend stress levels. Furthermore, since many coatings typically have columnar growth morphology, any crack normal to the surface will be large in a thick coating, and may exceed the critical crack length, whereas in a thin coating this may not be the case. It has been shown experimentally that in rolling contact fatigue tests hard TiN coatings with a film thickness well below 1 micrometre have up to two orders of magnitude longer lifetime than 2 to 3 μm thick similar coatings (Erdemir and Hochman, 1988).

Thickness effects in rolling contact fatigue for a-C:Cr and DLC coatings has been reported by Lindholm and Svahn (2006) and Löhr *et al.* (2006). Polonsky *et al.* (1997) claim that they have been able to demonstrate that the rolling contact fatigue life increase achieved with less than 1 μm TiN coatings was entirely due to polishing of the steel loading balls by the significantly harder TiN coating. They changed the test procedure and eliminated the polishing effect and observed a somewhat negative effect of the TiN coating on the rolling contact fatigue life. Similar mechanisms have been reported for DLC-based coatings in rolling contacts by Harris and Weiner (1998) and Lindholm and Svahn (2006). Doll (2001) and Doll *et al.* (2004, 2005) give further information on the benefits of metal-doped DCL coatings in rolling bearing applications, emphasizing the benefits of nano-composite structures.

Surface treatments, such as carburizing or shot peening, which can induce compressive in-plane stress in the surface, can provide benefits to fatigue life, by ensuring that in-service surface tensile stresses are reduced and crack opening effects are restricted. Promising results on improvements in rolling contact fatigue life by the use of thin surface coatings have been reported for mechanical components like gears (Amaro *et al.*, 2005), bearings (Sjöström and Wikström, 2001; Doll *et al.*, 2004, 2005; Yonekura *et al.*, 2005; Vanhulsel *et al.*, 2007) and valve lifters (Lindholm and Svahn, 2006).

Rule 4 – Fretting: The coating should enhance the generation of a low-friction surface layer to accommodate the movement or be elastic to accommodate deflection and reduce high stresses. The coating should increase surface inertness in relation to surface destructive chemical reactions. The coating should hinder crack initiation and crack growth.

Fretting is a special case of fatigue wear at the surface, where the amplitude of reciprocating sliding is typically smaller than the contact length. In the contact the reciprocating friction load produces surface stresses that can result in cracks and asperity detachment. The wear products stay within the contact region and influence the contact conditions, e.g. concentrating the surface load due to the released wear particles and increasing the localized surface stresses. On steel surfaces the contact process wears off the oxide layers on the surface, which is then more exposed to chemical reactions. Oxidized wear particles can cause increased abrasion, and the associated volume increase can lead to seizure in closed mechanical systems with a close fit. Energy analysis of both uncoated and coated surfaces in fretting wear has been useful in structuring the involved wear and friction mechanisms (Fouvry *et al.*, 2003; Liskiewicz *et al.*, 2004; Liskiewicz and Fouvry, 2005b; Kapsa *et al.*, 2005).

Fretting wear or fretting corrosion can occur in electrical switch contacts, mechanical detachable joints such as press-fitted, bolted or riveted joints, and in precision-engineering constructions which are digitally controlled and exhibit limit cycling. The results of fretting are loss of fit in mechanical joints, seizure or high friction in moveable joints and high electrical resistance in electrical contacts.

The low-amplitude movements necessary for fretting to occur can originate from undamped machine vibrations, electronic position control or thermal expansion–contraction effects. In order to minimize the chances of problems occurring in a potential fretting wear situation, use can be made of certain coatings and surface treatments which have been proven to provide effective solutions. Examples are phosphating and sulphiding treatments, MoS_2 coatings and ceramic coatings. In the latter case, both surfaces would normally require coating (see section 7.8.2). Even elastic rubber coatings have been applied with good results (Baek and Khonsari, 2005).

It is possible in some cases to build up favourable transfer of reaction layers on the surfaces that decrease the shear strength and friction. Fretting wear is often most efficiently reduced by effective macromechanical design measures, e.g. to control the displacements and stresses induced in the contact. However, it is also possible to modify the frictional conditions to reduce friction, e.g. by means of a solid lubricant coating such as MoS_2 (Zhang *et al.*, 2003a), PTFE (Langlade and Vannes, 2005) and layered or composite coatings like DLC on TiN (Klafke, 2004c), MoS_2 doped TiN coating (Cosemans *et al.*, 2003) and MoS_2 doped in Cu–Ni–In coating (Fridrici *et al.*, 2003). There can be advantages in providing such additions as phases in a nanocomposite coating.

Rule 5 – Abrasion: In abrasion the coating should have high hardness to inhibit groove formation and be thick enough to prevent abradant penetration.

In abrasion, typically, a hard counterbody moves against and ploughs through a softer material and thus the hardness of the softer material is of considerable importance. The aim of the surface material choice is to inhibit plastic deformation, such as grooving, on the surface. Usually thick coatings are used in abrasive contact conditions, and the performance of thin coatings such as PVD TiN is often poor under such conditions. This is partly due to the thickness limitations of hard PVD and CVD coatings, their lack of toughness and the need for effective load support, especially under three-body abrasion with sharp and hard abrasive particles.

The load-support requirement, to prevent adverse macromechanical mechanisms, can be fulfilled by utilizing interlayers such as electroless nickel (Matthews and Leyland, 1995). It has been found that by multilayering Ti and TiN films it is possible to produce a composite coating which is both hard and tough, thus controlling micromechanical mechanisms and performing well in abrasive conditions (Leyland and Matthews, 1994). In comparisons of the abrasive wear resistance of different coatings DLC, CrN, TiN, TiCN and ZrN have been successful (Van Acker and Vercammen, 2004; Tucker, 2004). The abrasive mechanisms related to TiN coatings are presented in section 4.4.1, DLC coatings in section 4.5.3 and diamond coatings in section 4.5.2.

The rule includes the specific case when foreign particles, not originating from wear, are present in the contact. This is based on simple rules of thumb (Eyre, 1992), which are applied by practising tribological design engineers. It provides a minimum surface hardness requirement and, in the case of coatings, a thickness requirement. When this hardness requirement cannot be met by the available coatings, the non-proportionality between hardness and abrasion resistance, which is observed when comparing different material types, is approximated using empirical relationships compiled from published data (Khruschov, 1974).

Rule 6 – Impact: The coating and substrate should be sufficiently elastic to deflect and absorb impact energy and a flexible well-adhered coating/substrate interface layer is required. In erosion with normal impacts the coating should be sufficiently elastic to avoid high-stress peaks. In inclined erosion the coating should be hard enough to avoid grooving.

Impact resistance is required under conditions of repetitive dynamic loading. It may be regarded as another form of contact fatigue, sometimes in combination with static contact loading. In such conditions, a coated surface must possess a high toughness in order to absorb the impacts that result in repeated stress states and energy input to the surface. Ideally, this energy should be absorbed elastically, so that no permanent deformation results. This ability to absorb energy is governed by the elastic resilience of the material, which is defined as $Y^2/2E$, where Y is the yield stress. Since the hardness is typically about $3Y$, we can see that this means that we need to maximize H and minimize E in order to increase the elastic resilience. The maximum value of the contact pressure occurring in a contact as a result of an elastic impact between two massive bodies can be estimated from contact mechanics theory (Johnson, 1985).

We stated above that, ideally, the impact energy should be absorbed elastically and that is generally the case. However, since fracture toughness can be enhanced by having a ductile constituent present, there are arguments for having a nanocomposite coating structure with a small amount of a ductile phase present, if fracture toughness is to be enhanced, i.e. to increase the 'ultimate' resilience of the surface – as defined by the total area under the stress–strain curve.

The impact mechanism is one component in erosion wear. Erosion can take several forms – for example solid particle, liquid droplet and cavitation – and the rules pertinent to its avoidance may combine effects of abrasion, impact, fatigue, corrosion and dissolution and take into account parameters such as particle velocity, size and shape, etc. Hard and thin PVD coatings like TiN, DLC, CrN, TiAlN have been investigated for erosive resistance (Iwai *et al.*, 2001; Batista *et al.*, 2003; Fujisawa *et al.*, 2006) and good results has been achieved especially with TiN based multilayer coatings (Iwai *et al.*, 2001) and duplex coatings with (Ti,Al)N and Cr–N on pre-nitrided steel (Batista *et al.*, 2003).

Rule 7 – Corrosion and chemical dissolution:

Corrosion condition: The coating should be inert, dense and pore free to prevent chemical attack of it and the substrate or it should have suitable sacrificial properties.

Moving mechanical contacts within a corrosive environment is likely to see both accelerated wear and corrosion, as the corrosion products are continually being removed to reveal a clean surface. This was mentioned above for fretting, but can also occur for the other contact types mentioned here. To reduce and control corrosion, coatings need to be dense and pore free, which typically applies for nanocomposites. The fine grain size of nanocomposites leads to a more uniform corrosion, rather than localized pitting, which can be a particular problem with thin ceramic coatings, often producing accelerated substrate corrosion at pin-hole defects. It has also been suggested that the fine grain structure can suppress galvanic corrosion effects which can occur within conventional polycrystalline metallic and composite coatings. Certain coatings, such as zinc, can provide sacrificial protection to steel due to their more negative electrode potentials – which means that they will protect the substrate from corrosion, e.g. in an aqueous environment.

Chemical dissolution condition: The coating should have high thermal and chemical stability, high toughness and high hardness, low adhesion to the counterface material and high adhesion to the substrate and be dense to inhibit dissolution from the substrate.

The rule above applies especially to the case of typical chemical dissolution dominated tribochemical wear when a workpiece material is moving over a cutting tool. The aim of a coating on the cutting tool surface is to resist the high-temperature conditions, to reduce friction and thus heat generation, to inhibit dissolution of the surface material and reduce scratches produced by hard elements in the workpiece material (see section 7.5). Thin hard coatings of nitrides and carbides match these requirements well and are used successfully. In metal-cutting applications the resistance of TiN films to oxidation has been improved by employing mixed-phase nanolayered and nanocomposite ceramics, sometimes with phase additions to help form a stable oxide on the surface (Holmberg and Matthews, 2005). Wear due to the chemical effects of dissolution and diffusion can be accommodated by using rules such as those developed by Kramer and Suh (1980) and applying equations for forces and stresses at the interface listed by (Grzesik, 2000).

Rule 8 – Lubricated condition: The coating surface should function beneficially with the lubricant. It should either react with the lubricant additives to generate a low shear strength boundary lubricant surface film or it should have the ability to generate a low shear strength surface layer despite the presence of the lubricant.

Most of the current research on surface coatings deals with their performance in dry conditions, some even in vacuum or in an inert gas. However, there are numerous applications, e.g. related to engines, transmissions and seals, where the coating is expected to perform in a liquid, often in oil. In the full-film lubrication regime, the oil lubricant carries the load and separates the surfaces from each other and the shear takes place in the lubricant. The role of the coating is then to act as a safety layer that comes into action when the lubricant film fails for one reason or another. Thus, the coated surface is expected to have good tribological performance in boundary-lubricated conditions where surfaces are partly in direct contact with each other and surrounded by oil.

It is important to consider the influence that the lubricant has on the effectiveness of the tribofilm formation process. When the lubricant contains reactive additives and the contact is in the boundary lubrication regime, a reaction film is formed on top of the surface and the shear takes place in this film. The film is typically formed by chemical reaction between oil additive elements such as Zn, S, Cl, P or Mo and the steel surface and is enhanced by high temperature (see section 3.1.5). The basic mechanical aspects related to boundary-lubricated coated surfaces including effects of inherent stress in the coating, its thickness, structural integrity and surface roughness on the tribological performance have been derived by Komvopoluos (1991a, b).

For a coating to be effective in such conditions it is necessary either that it has the ability to react chemically with the additives and form a strong low shear strength reaction film, or that it has the ability to produce a strong and low shear strength surface layer that protects the contact from destruction. Ronkainen *et al.* (1998a) showed how a reaction layer is built up on a steel ball sliding against a steel surface in lubrication oil with additives. Very similar tribological performance was achieved with the ball sliding in dry conditions against a DLC coating and then a transfer layer was formed on the steel ball. When the steel ball was sliding against the DLC layer in oil lubricated conditions no visible transfer layer was formed but the tribological performance was in the same range as in the previous cases. This indicates that this DLC coating did not really improve the friction and wear in an oil lubricated contact, but could act as a safety layer in the case of lubricant loss or failure (Ronkainen, 2001).

Metal doping of DLC coatings has turned out to improve the ability to build up a boundary lubrication film on the surface and result in improved tribological performance. In combination with conventional extreme pressure (EP) additives this has been reported for W-doped DLC (Jacobson and Hogmark, 2001; Podgornik *et al.*, 2006a), Cr-doped DLC (Lindholm and Svahn, 2006) and Si-, Ti-, W-doped DLC (Vercammen *et al.*, 2004). Enhanced tribological performance was also observed for Ti-, Mo- and Fe-doped DLC in oil containing ZDDP and MoDTC additives (Miyake *et al.*, 2004). Similar improvements has been achieved with nanocomposite coatings based on carbon with metal additions, designed to create a transfer layer, and thus the ability to run both dry or with a lubricant was enhanced (Doll *et al.*, 2005).

6.5 Design Guidelines

In various papers the present authors and their collaborators discussed the empirical and theoretical design guidelines that are appropriate to the above contact conditions (Matthews *et al.*, 1992b; Franklin and Dijkman, 1995). A comprehensive résumé of eight design guidelines pertinent to the selection of materials and coatings to resist wear was presented. Rather than providing a guarantee for the prevention of wear problems, the guidelines are intended to assist design engineers in minimizing the likelihood of wear problems occurring when the design goes into service. The guidelines presented in Franklin and Dijkman (1995) cover the following phenomena:

- Contact overloading (CO): intended to avoid plastic deformation in the contact.
- Fretting (FR): designed to assess the propensity for fretting to occur, so that suitable treatments or coatings can be applied.
- Excessive material transfer (MT): based on knowledge of material pairings which should be avoided in order to minimize the chances of adhesive material transfer occurring.
- Two- and three-body abrasion (2B and 3B): based essentially on the well-known rule of thumb that if a surface is to resist abrasion its hardness should be at least 30% harder than any abradant present.
- Negligible wear (NW): based on the work of Bayer *et al.* (1969). This guideline checks whether the maximum shear stress in the contact region exceeds a critical value and also considers the possible presence of a lubricant.
- Surface fatigue (SF): based on empirical data and thus restricted to materials for which data has been assembled (mostly steels).
- Steady-state wear factor (WF): this guideline examines the wear rate data and the scatter available for the selected materials pairings to predict wear life.

In effect, the boundary conditions for applying the above guidelines are determined by the dominating phenomena listed in Table 6.2, and they are typically applied in order to ensure that the required properties are achieved in the surface and near-surface regions (Matthews *et al.*, 2007).

The contact overload guideline states basically that the elastic limit should not be exceeded in the contact. This particular rule, although simple in concept, is complex in application, particularly when applied to coatings. This is because the limit of load-carrying capacity may in fact be determined by

Table 6.2. Guidelines on surface requirements for different contact conditions.

Contact condition	Dominating phenomena	Typical required properties of the surface and near-surface region	Examples of coating solutions (application dependent)
Static vertical loading	Hertzian stress distribution	Adequate yield strength and hardness	Thermal hardening, thermochemical diffusion treatments
Two-body sliding, tangential surface loading, ~ adhesive sliding	Low-friction sliding Mild wear Hertzian stress distribution	Low shear strength at surface top layer Good load support	DLC, MoS_2, diamond, (electroless) NiP-PTFE, (electrolytic) Co–Cr_3C_2 (at $T > 300$°C), thermal spray WC–Co
Three-body sliding, tangential surface loading with local scratching, ~ abrasive sliding	Third particle indentations Two-body ploughing	Good microtoughness and load support High hardness to resist plastic deformation Sufficient coating thickness	Weld surfacing, hardfacing, thermal spray hard coatings, TiN, TiAlN, TiC, Al_2O_3, CrN, CrC, hard chrome, thermochemical diffusion treatments
Impact loading	High-energy impacting stress	Good macrotoughness Good elasticity	Thermochemical diffusion treatments, tough resilient coatings (sandwich, graded, duplex, multilayer, nanocrystalline, multicomponent)
Fatigue loading	Cyclic surface stress waves	Good macrotoughness Good load support	Thermochemical diffusion treatments, tough coatings (sandwich, graded, duplex, multilayer, nanocrystalline, multicomponent)
Vibrating low-amplitude tangential loading, ~ fretting	High-frequency large stress waves Wear debris continuously trapped within contact	Good elasticity Low shear strength surface layer Not producing hard wear debris Surface chemistry: inertness or lubricious reaction product	MoS_2, Cu–Ni–In multilayer, DLC
Chemical dissolution ~ metal cutting	High temperature	Non-soluble Thermally conductive	TiN, TiAlN, TiC, WC, CrAlN, DLC, diamond
Corrosion	Corrosive/aggressive environment	Galvanic protection or inert barrier layer	Zn-based coatings, Ni-, Al- and Cr-based coatings that provide hermetic sealing, CVD TiN, polymer coatings
Lubricated	Coating giving load support for lubricant film and acting as emergency layer	Interaction with lubricant additives Texturing to support lubricant availability	DLC, TiN, TiC, CrN, hard chrome, thermochemical diffusion treatments

several different effects, not just the need to avoid exceeding the elastic limit. In addition, the values of appropriate properties relating to the failure criteria of coatings are often poorly known to the design engineer. Figure 4.100 illustrates that the contact load limit may be defined in several ways, such as coating debonding, coating cracking, substrate cracking, substrate or coating plastic deformation or even elastic deformation which might take a surface, or component, outside its designed operating dimensions. The hardness to elastic modulus of the coating has a key role in determining the ability of the coating and substrate to accommodate elastic strains without failure. Without adequate mechanical properties at the surface the inability to carry the required contact pressure can result in a rise in ploughing, coating fracture, plastic yield or fracture under the coating.

Even if the basic boundary conditions for applicability are met, the extent to which these guidelines may be applicable to a specific contact situation is left to the judgement of the user. It would be highly unlikely that one rule would suffice in any one particular contact condition, and it is most likely three or more would have to be applied in order to gain a reasonable level of confidence that a particular surface material pairing will perform satisfactorily.

It should be noted that the guidelines presented in Franklin and Dijkman (1995) have their origin in tribological theories relating to uncoated surfaces. Only now are robust models being developed which permit a thorough understanding of coated surface response to various contact types. One such model has been developed by Holmberg *et al.* (2006a, b) and Laukkanen *et al.* (2006) for a sliding contact on a coated surface.

6.6 Expert Systems

The availability of computer hardware and software with ever greater capabilities will enable considerable strides to be made in overcoming the coating selection problem. In particular, the development of computer expert systems – also called knowledge-based systems – is expected by some to have an increasing impact. While the early promise of such systems has not yet been fulfilled, they still represent a potentially powerful tool in the engineer's armoury. Expert systems are defined as computer programs which are specifically developed to encode expert knowledge and make it easily available to the user, often in a conversation mode. Unlike conventional hardwired programmes, most expert systems have a separate reasoning module or interpreter, which manipulates knowledge to reach a conclusion. This allows the knowledge base, e.g. containing coating property experience, to be stored separately in the programme.

Because the knowledge is separate from the inference procedure the information can be updated easily without rewriting the whole system. The knowledge used for a consultation and the decision path can be monitored to provide an explanation for any decisions reached. This latter point is especially important if designers are to have confidence in the system.

There are two main kinds of knowledge in an expert system. The first is the hard data, for example material properties. The second knowledge comprises the heuristic rules or specialist expertise used in the methodology of selection. An everyday illustration of this distinction is that of the taxi driver. The map of the city is the basic data, while the heuristics are represented by his knowledge of the relative merits of alternative routes under the likely conditions of traffic at various times.

There are many computer programs which are described as expert systems but which are merely database systems with only the first type of knowledge. Of course, one of the problems encountered in constructing a knowledge base is acquiring or eliciting the expert knowledge, most of which will be diffuse and in the head of the expert. Given that the expertise can be obtained, the problem remains of how to represent it in the computer. Many forms of representation techniques are available to the system builder, but rules are the most common method of representing knowledge.

Despite the large volume of literature published about expert systems, very little has been written about the use of such systems for coating selection. There exist several database systems for materials selection but considerably fewer knowledge-based ones (Tallian, 1987). Database systems generally only perform searches of materials based on data sets entered by the user, returning materials whose specifications match or exceed those entered. Expert or knowledge-based systems are generally characterized by their ability to avoid asking for unnecessary data input and the inclusion of some form of selection procedure rather than a simple search.

Some examples of expert systems in the tribology field are described below.

TRIBSEL. A number of coating selection expert systems were developed in the 1980s and 1990s. Each one was to some extent a development of earlier systems (Matthews and Swift, 1983; Syan *et al.*, 1986, 1987; Matthews *et al.*, 1991b; Robinson *et al.*, 1993). The initial work tended to emphasize the identification of wear mechanisms, and to apply mathematical models. In practice, the number of occasions when a strictly mathematical approach to selection can be utilized was found to be very limited. Even the metal-cutting situation, which is well documented and was thought to be covered by only two dominant wear mechanisms, has been found to be difficult to model (Kramer and Judd, 1985). Thus it was necessary to devise a system that could avoid the need to identify the wear mechanisms.

The system used an approach based on 15 main criteria which were found to be inherent in the reasoning processes carried out by human experts. These criteria are listed in Table 6.3.

Of course, these cannot be fully specified by just 15 questions, and 31 factors were identified which together sufficiently define all of the criteria. These are broken down into five factor groups:

- operating constraints,
- processing constraints,
- geometrical constraints,
- topographical constraints, and
- economic constraints.

Three sample rules are given in Table 6.4.

The approach used has been to consider coating and treatment technologies in generic groups. A consultation will start with the hypothesis that all of these will be suitable. As the user supplies information about a particular application the system uses the rules in its knowledge base to reject any that are unsuitable. Thus the process is one of progressive elimination, which as mentioned earlier is considered desirable

Within the heading of operating constraints come factors such as the maximum or minimum operating temperature, the operating environment, contact type (point, line, etc.) and loading type

Table 6.3. Criteria used in coating selection.

Operating temperature
Operating environment
Counterface material
Counterface hardness
Substrate material
Substrate hardness
Contact pressure
Contact geometry
Relative motion type
Relative speeds
Surface finish
Component size and shape
Coating thickness and uniformity
Quantity of parts
Economics, including versatility

Table 6.4. Sample rules from TRIBSEL coating selection expert system.

Operating environment rule	Material constraint rule	Geometrical constraint rule
if operating temperature is <200°C and not solid erosion and not liquid erosion or cavitation then mechanical surface working is a possible choice	if ferrous material and substrate surface finish <6 microns and counterface hardness <800 VPN then carburizing is a possible choice	if melting temperature >1000°C and re-entrant capability <0.5 and not complex shape category then welded carbide is a possible choice

(rolling, sliding, impact, etc.). The system is based on an enormous amount of case history information, as recalled by several human experts after a lifetime of experience. This information was extracted by asking them when particular coating or treatment types were or were not successful. An advocate of the wear mechanism identification approach might argue that the questioning in this factor group category is merely identifying the likely wear types. At no point, however, is such information requested from the computer, neither does it possess any knowledge of wear mechanisms *per se*.

Certain coatings are eliminated because of their poor suitability for the operating constraints imposed by the processing method. These include factors such as the maximum dimension of the component which can be coated, the processing temperature and its possible influence on the component, and production capability factors such as the throughput required.

Next come geometrical constraints, such as the coating uniformity and re-entrant penetration capability of the process. These are followed by questions relating to the topographical constraints which cover aspects such as the surface finish requirements.

When particular coatings or treatments cannot meet the needs indicated by the designer under each of these headings, they are rejected. The designer can ask why a particular question is being asked, and the system will respond by explaining the particular reasoning path it is using at that time. It will ultimately provide a list of coatings or treatments that will satisfy all of the indicated demands – or if none exist it will respond accordingly. The system has relative cost information encoded and information about economic quantities and availability which is also available to the designer. A case history file of similar applications is also incorporated, which the computer will display if it meets an application which it has seen before. Following the development of TRIBSEL, a more advanced system known as PRECEPT was devised by Robinson *et al.* (1993) and Franklin and Dijkman (1995). Subsequently, a proprietary system called ISIS was developed, which incorporated several of the methodologies within the earlier systems.

TRIBEXSY. The Philips TRIBEXSY tribological consultation system is designed to remove the need for an interview with the tribology expert when a wear problem arises. TRIBEXSY's dialogue-based approach guides the user through an analysis of the symptoms of the wear problem, by asking questions about the materials involved, their operating environment, the nature of the contact regime and the appearance of the damage to the components. This leads to an identification of the type of wear taking place. The program then searches for possible ways of overcoming the wear problem, by suggesting the use of different materials, coatings and lubrication regimes. As before, the user is asked questions about operating conditions and component geometry, in addition to being questioned as to the possible use of alternative materials and slight design changes.

TRIBEXSY may also be used for consultation in a design situation before a problem arises, although as with all expert systems, the more often an unknown is entered in response to the system's questioning, the less reliable will be the solutions proposed. TRIBEXSY is text based, and runs via remote terminals.

Experience with expert systems has shown that the essence of such systems, when developed to a satisfactory performance level, is basically a simple structure of rules. Those expert systems that have been developed rely first on identifying the correct rules and then on populating those rules with reliable knowledge, which mostly is coating-related data. The benefits gained in developing an expert system are then manifested in a systematic methodology, which has validity as either a paper-based or a computer-based system. A robust selection system having the desirable attributes needed for the computer expert system invariably creates a selection system which can also be implemented as a structured sequence, such as that presented within this chapter.

Often, if a design engineer or other person selecting coatings follows this sequence it will allow all aspects of the design, such as geometry, bulk material, coating, pre- and post-treatments as well as the manufacturing route, to be considered in an integrated way. Nevertheless, it should be pointed out that, in all cases where such systems have been used, their purpose has been to determine whether the coating can provide the basic functionality required, i.e. to operate within the pertaining contact conditions. They are not used to calculate the expected lifetime. When it comes to determining the lifetime of the coating then in-service and model tests are needed (Franklin, 2008). This is evident in the next chapter, which discusses applications-related performance enhancement.

6.7 Closing Knowledge Gaps

The authors, in their roles as coatings researchers, are often asked to advise on the selection of a suitable coating to fulfil a given operating need, often as a means of solving a wear problem which has arisen due to poor tribological design and practice.

Engineers, materials scientists and others who seek from coatings the same kind of easily specified performance and reliability parameters they achieve with a material chosen for its bulk strength are often frustrated by the apparent variability associated with selecting a surface coating. The reasons for this are twofold. One is the difficulty in understanding exactly what the contact conditions are, e.g. in terms of reaction products, operating pressures and temperatures. The second is the lack of comparative data about specific coatings. The successful consultant in this field is the one with the broadest experience of different coatings and specifically the conditions under which they will and will not operate.

In the same way, the value of the computer-based systems lies in the amount of accurate data which it contains. We stated earlier that there is no comprehensive coatings database available, but there are a number of tribology databases (e.g. Woydt, 2000) that are evolving, and as they become populated with coatings data then the ultimate goal of an integrated tribological property database can be reached.

While advanced computer and programming techniques can help in the coating selection problem, they cannot be the only solution – otherwise new untried coatings would never be specified. There is the clear need to bring together the separate skills of the coatings process developer, the coatings evaluator and the tribological modeller to design and optimize coatings to meet preset needs, with well-defined quality and reliability. In this way the potential of tribological coatings will be fully achieved, to the benefit of a wider range of application sectors. Some of those sectors currently benefiting are reviewed in the following chapter.

CHAPTER 7

Applications

Contents

7.1 General

In the previous chapters we have seen that due to the development of existing and new deposition techniques, designers are now, more than ever before, in a position to change the properties of surfaces and to produce the desired friction and wear properties precisely where they are needed.

The use to coated surfaces to control tribology in a variety of applications is of special interest because of the huge effects of friction and wear in relation to energy use, environment and economy. Oil as lubricant was the first significant solution for friction and wear control but there is a trend to limit its use due to both pollution and cost. Today, surface coatings offer a second major solution for friction and wear control in our society.

Examples of successful industrial applications of thin hard or soft coatings can be found in engineering components, tools for cutting and forming, electronic equipments, consumer goods and in special instruments. Examples of coated tribological products in mass production are PVD coated shaving blades, magnetic storage reader heads, tool inserts, drills, rolling bearings, diesel fuel injectors, piston rings, CD/DVD disk moulds, knives and optical lenses.

In engineering components such as sliding bearings, rolling bearings, gears, seals, cams and tappets, pistons, cylinders, valves, injectors, plungers, rotors, pumps and transmissions, both hard and soft coatings are used to reduce wear, increase component lifetimes and reduce friction. Industrial sectors in which such improvements are utilized include the automotive, aircraft and other transportation industries, the electronics, consumer goods and metals manufacturing industries as well as the chemical, food, pharmaceutical, medical and packing industries.

In the very severe surface conditions prevailing at the surface of metal cutting and forming tools, hard coatings are very successfully used to protect the substrate from wear and to increase the tool lifetime by several orders of magnitude. Several machinery parts in the pulp and paper industry such as cutter blades, rollers and grinders, as well as high-speed moving mechanical assemblies in the textile industry have improved properties due to hard coatings.

On components exposed to particle erosion, such as turbine blades and compressor components, or to scratching, such as optical lenses, hard coatings are used to protect against wear. In instruments where the wear of a critical component like a joint or slider influences accuracy or functionality, both hard and soft coatings are used to extend lifetimes and retain accuracy. These kinds of applications are found, for example, in computers and aerospace instruments.

Hard low-friction coatings are increasingly used on components in the automotive industry. Friction and adhesion are important in microdevices such as micro-electro-mechanical-systems (MEMS) and here thin film deposition is the solution with the most potential. A new and rapidly growing area for coating applications is biomedical components, where improved non-corrosive, non-toxic and biocompatible features are combined with high mechanical strength and fracture toughness.

In this chapter we will discuss experiences with the use of thin surface coatings in a number of different applications (HMSO, 1986; Matthews *et al.*, 1992a, 1998; Hilton and Fleischauer, 1992; Cartier, 2003; Lampe *et al.*, 2003; Merlo, 2003; Field *et al.*, 2004; Hauert, 2004, 2008; Tung and McMillan, 2004; Erdemir and Donnet, 2005; Vetter *et al.*, 2005; Neuville and Matthews, 2007; Héau, 2008; Van der Kolk, 2008; Matthews, 2008).

7.2 Sliding Bearings

7.2.1 Description of the application

A sliding or plain bearing is basically a very simple machine element where a moving shaft is supported against a stationary housing as shown in Fig. 7.1. Important functional design parameters are the dimensioning of the bearing and the choice of bearing material. It is common to use such bearings with oil lubrication. With an adequate rotational speed a hydrodynamic pressure will be created within the oil in the loaded contact zone between the shaft and the bushing. This oil film is then thick enough to keep the surfaces apart and inhibit wear and provide low friction (see section 3.1.5).

The desirable properties for bearing surfaces may be listed as follows:

1. Seizure resistance; i.e. the ability to resist asperity adhesion or welding. Materials, such as white metals based on lead and tin, which are soft and have low shear strength, are generally good in this respect.

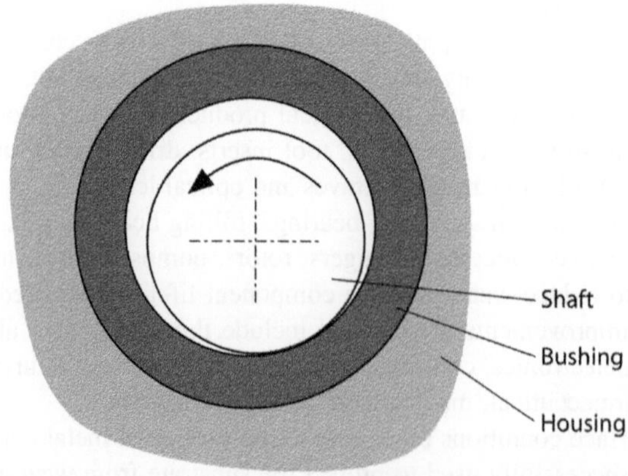

Fig. 7.1. In a sliding bearing the moving shaft is supported by a bearing bushing fixed to the stationary housing.

2. Compressive strength; i.e. the ability to cope with distortions and deflections of bearings or shafts. This is clearly a property which soft films may lack.
3. Conformability; i.e. the ability to cope with distortions and deflection of bearings or shafts. Also to cope with misalignment during manufacture.
4. Embeddability; i.e. the ability to embed hard particles and therefore prevent them from causing damage.
5. Fatigue strength. Since most metals have low fatigue strength if they have good deformability, a compromise must usually be made. High operating temperatures can sharply reduce fatigue life, especially of white metals.
6. Corrosion resistance; e.g. high temperatures can cause acid formation in mineral oils. Copper–lead, cadmium and silver alloys are especially vulnerable to attack.

The challenges in the design of hydrodynamic sliding bearings are to calculate and optimize the hydrodynamic conditions and the oil supply grooves. In advanced mechanical systems such as high-speed spindles, computer hard disks with arm sliders and videotape reading heads the interposing medium is air and they require very precise design and manufacturing. For more detailed information on sliding bearing design see Lang and Steinhilpher (1978), Barwell (1979), Welsh (1983), Constantinescu *et al.* (1985) or Brewe (2001).

7.2.2 Improvements by surface coatings

At start-up, slow-down, low-speed, very high load or in dry running conditions there will be direct contact between the shaft and the bushing. The wear and friction can then be minimized by selection of a suitable material combination and the tribological behaviour can again be manipulated by using surface coatings. This sliding situation is in many respects similar to the sliding contacts that we have referred to quite extensively in Chapter 4. Those sliding contact results and analyses are more applicable to sliding bearings than, for example, to rolling bearings or cutting tools.

In the earlier chapters we have frequently referred to friction and wear results of coated surfaces from pin-on-disk machine tests or similar. When one tries to apply these results to a certain sliding bearing application it is important to keep in mind the difference in heat generation and heat flow in a sliding bearing with conformal surfaces and a large contact area compared to the small contact spot and the possibility of surface cooling in a pin-on-disk geometry.

Historically, typical coatings for sliding bearings have been lead-based alloys of thicknesses from 10 to 40 μm and a microhardness of less than 20 H_V. The coatings may have additions of tin, indium or copper, although a number of other additions have also been used experimentally. Typical compositions are PbSn10, PbSn10Cu2 and PbIn7. An interlayer of nickel- or copper-based materials is often necessary to facilitate adhesion on an aluminium bearing and to retard tin diffusion on lead bronze materials.

A comparison of the electrodeposited layers shows that PbIn7, although the softest of the three, has the highest fatigue strength, but the lowest wear resistance, while lead–tin–copper has the best wear resistance. The tin coatings require a nickel interlayer to give the best corrosion resistance while the indium alloys do not. Thus, a decision on the overlay will inevitably be a compromise, with world-wide production generally favouring the nickel barriered lead–tin–copper (Eastham and Matthews, 1992). Additional information about bushing materials and thick layers or multilayers used in sliding bearings is found in, e.g., Welsh (1983), Constantinescu *et al.* (1985) and Brewe (2001).

Thin coatings are increasingly used today in sliding bearing applications. Soft coatings will decrease the shear strength in the contact between the shaft and bushing and can thus be used for friction reduction. Coatings that form a transfer layer on the countersurface are often favourable because the wear products will then remain in the contact and not cause material loss. Thin MoS_2, PTFE, graphite and polyimide coatings with thicknesses of 7 to 50 μm have successfully been used in foil or other types of gas bearing applications. Ion-plated soft metal coatings, like lead, gold and silver as well as sputtered MoS_2, with thicknesses of 50 to 500 μm, have proved to be good choices for high vacuum applications (Bhushan and Gupta, 1991).

Different kinds of sputtered thin coatings have been introduced in internal combustion engine bearings. One interesting solution is a bimetal coating containing both soft and hard components. The soft component gives good operating properties under mixed friction, such as during running-in, as well as a minimal tendency to seize and low vulnerability to foreign particles. The hard component provides wear resistance and fatigue strength. Figure 7.2a shows schematically how soft Sn-embedded particles of less than 3 μm in diameter are distributed in a 30 μm thick Al sputtered layer (Koroschetz and Gärtner, 1990). These AlSn20 sputtered coatings have excellent wear resistance in sliding bearing applications. The microhardness of the AlSn20 coating depends on the average grain size of the Sn-embedded particles, as shown in Fig. 7.2b.

In very low-load sliding conditions, such as a mechanical face seal contact, a PVD-deposited TiN coating on an Inconel 625 nickel-based alloy substrate, with coating thicknesses of 5 to 30 μm, has proved to be effective in combination with graphitic carbon as a counterface (Kennedy and Tang, 1990). The coefficient of friction in dry sliding conditions was about 0.1 and independent of coating thickness within the range tested of 5 to 30 μm. The wear rate of uncoated Inconel 625 under the same sliding conditions was seven times higher and the coefficient of friction was almost double and had a value of 0.18. Low friction and wear-resistant coatings like DLC and MoS_2 are candidates with great potential for low-load sliding bearings and for use as safety layers to give wear protection in situations when the lubricant does not carry the load.

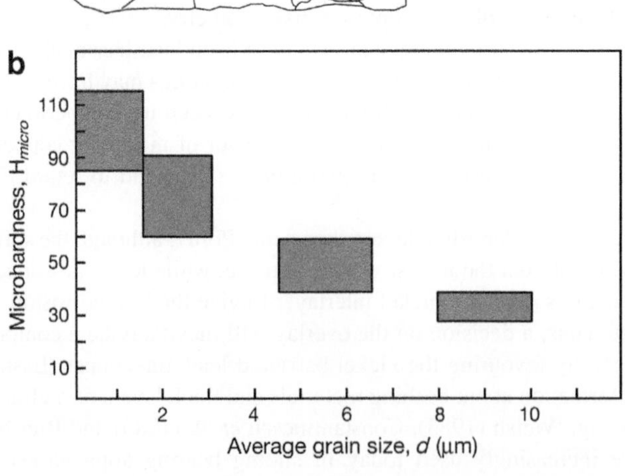

Fig. 7.2. Schematic representation of (a) soft Sn-embedded particles in the structure of a AlSn20 sputtered layer matrix and (b) microhardness of the sputtered layer in relation to the average grain size of the Sn embedments (after Koroschetz and Gärtner, 1990).

7.3 Rolling Contact Bearings

7.3.1 Description of the application

Rolling contact bearings are characterized by extremely low friction. In the most favourable situations the coefficient of friction may approach values as low as 0.0001 and in general machinery applications it is normally in the range of 0.001 to 0.005. Rolling contact bearings are used in both high- and low-load as well as in high- and low-speed applications. There are a great variety of different types. Typical geometries for ball bearings and for roller bearings are shown in Fig. 7.3.

In rolling contact bearings the contacting surfaces between both the inner race and the rolling elements, and the rolling elements and the outer race, are counterformal, resulting in high contact pressures. The commonly termed 'point contact' area in a ball bearing has the shape of an ellipse and the 'line contact' area in a roller bearing has a more rectangular shape. In heavily loaded applications the contact pressure in a ball bearing may be as high as 3 GPa while in common machinery applications it is often no more than 0.5 GPa.

The surface roughness and topography influence, on a microcontact level, the stress field at the top surface. It has been very clearly demonstrated that bearings with smoother surfaces exhibit 2 to 4.5 times longer endurance life, especially in boundary-lubricated conditions with small lambda values (Zhou, 1993). The general mechanism of rolling is described in Halling (1975) and Johnson (1985) and the contact conditions in rolling contact bearings are analysed in Hamrock and Dowson (1981), Stolarski (1990) and Ai and Moyer (2001).

A thin, about 0.5 to 5 μm, lubricant film is formed between the rolling surfaces according to the elastohydrodynamic lubrication theory (Dowson and Higginson, 1966; Hamrock and Dowson, 1981). This prevents adhesive wear when rolling even at moderate speeds. If, in addition, abrasive wear particles can be kept out of the contact with proper sealing, the bearing will run for a long time until it fails by rolling contact fatigue at one of the contacting surfaces. The rolling contact fatigue wear mechanism, also called pitting or spalling, is discussed in section 3.1.4. For design purposes, fatigue failure is commonly treated with well-established statistical methods where the probability for failure

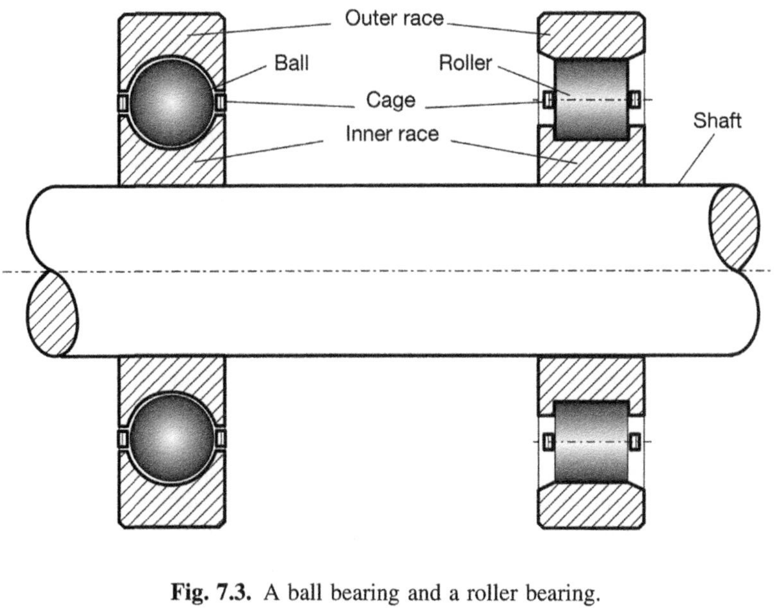

Fig. 7.3. A ball bearing and a roller bearing.

is analysed based on the Weibull distribution of experimental data, and the fatigue life that 90% of the bearing population will endure is estimated (Hamrock and Dowson, 1981; Shimizu, 2005).

Materials commonly used for rolling contact bearings are high-carbon chromium steel (SAE52100), molybdenum air-hardening steels (high-speed steels) for use at temperatures up to 430°C, stainless steel (AISI 440C) for corrosive environments and carburized or case-hardened steels. For high-performance applications several ceramic materials have been investigated. Silicon nitride has proved to be a suitable material for some applications because it has two to three times the hardness of bearing steels and gives low friction. In addition, it has a good fracture toughness compared with other ceramic materials and maintains its strength and oxidation resistance up to 1200°C. The limited application of ceramic materials is due to lower toughness and fracture resistance and high sensitiveness to shock loads (Bhushan and Gupta, 1991; Ai and Moyer, 2001).

7.3.2 Improvements by surface coatings

The bearing lifetime can be extended and the torque decreased by the deposition of thin soft or hard coatings on the raceways or the balls or both. Soft coatings like gold, lead, copper, chromium and MoS_2 and hard coatings like TiN, TiC, diamond-like carbon and multilayer coatings have been successfully deposited on bearing surfaces by CVD, ion plating or sputtering. Coated rolling contact bearings have mainly been developed for special requirement applications where the additional cost from the coating process can be justified. One considerable benefit is that with coated surfaces the bearing may run even for years without lubrication, which is a vital factor, in space applications, for example. Surface engineering approaches to rolling contact fatigue have been reviewed by Erdemir (1992).

It was previously believed that thin surface coatings will have little effect on the fatigue life because the maximum shear stresses often occur as deep as 10 to 150 μm below the surface and the coating is typically only about 1 μm thick. However, experimental results discussed earlier and below have shown that there are several situations where a soft or a hard thin coating may considerably increase the fatigue lifetime of rolling element bearings.

7.3.2.1 Soft coatings

Soft low shear strength surface coatings may influence the performance of rolling contact bearings by several mechanisms. The soft coating can inhibit adhesion between the substrate and the counterface, resulting in decreased adhesive wear. It can decrease the torque by offering a low shear strength layer for the shear to take place in contact areas where sliding occurs. Filling of surface roughness valleys by the coating material will smooth the surface and decrease the stresses on the asperity tips as the coating carries part of the load (Sainsot *et al.*, 1990). The soft coating may also make surface defects less harmful by moving crack initiation sites from the surface down to subsurface regions.

However, if the coating is too thick the rolling action may result in plastic deformation or grooving in the coating, which increases the torque. Soft and thin films, such as lead, behave elastically and the elastic hysteresis losses can be calculated and the film thickness optimized according to calculations presented by Halling *et al.* (1985) and Halling (1986). On the other hand, if the load is too high the rolling element may go through the coating by plastic deformation and interact with the substrate which increases wear. Thus, soft coatings are mainly suitable for rolling element bearings used at low loads and the coating should be very thin.

Rolling contact ball-on-rod fatigue tests, resulting in pitting or spall formation, have shown that a 0.5 μm thick ion-plated copper film can extend the fatigue life by almost one order of magnitude when deposited on martensitic stainless steel (440C), as shown in Fig. 7.4 (Hochman *et al.*, 1985). However, when the same coating was deposited on another stainless steel substrate (AMS5749) no beneficial effect was found.

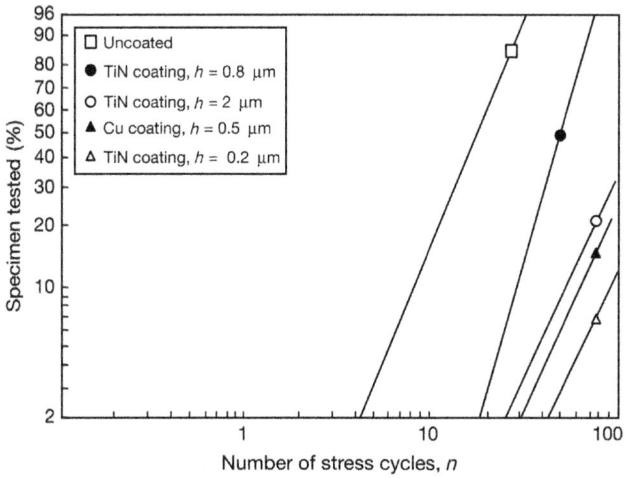

Fig. 7.4. Distribution of lifetimes given in number of stress cycles in rolling contact fatigue ball-on-rod tests for uncoated, copper coated and titanium nitride coated stainless steel (440C) specimens at a contact pressure of 4 GPa, rotational speed of 3600 rpm and oil lubrication (data from Hochman *et al.*, 1985).

Ion-plated gold and lead coatings with thicknesses of about 0.5 μm have satisfactorily been used in space bearings at low and medium speeds and low loads (Thomas *et al.*, 1980; Roberts, 1990b; Bhushan and Gupta, 1991). A major drawback in the use of lead coatings in high-precision applications is the debris generation from the cage matrix and the resultant torque noise. High torque noise has also been observed with gold coatings, probably originating from the generation of gold wear particles.

Plasma-polymer modified 1.4 μm thick silver coatings deposited on bearing races by thermal evaporation were tested in a thrust ball-bearing test rig at 147 N normal load and 31 m/s rotating speed in different environments by Yang *et al.* (2001). The rolling resistance was greatly affected by humidity. A hydrophobic silver film had minimum rolling resistance at a relative humidity of 80% and a hydrophilic silver film at relative humidity of 10%. It seems that the differences in rolling resistance is due to the effect of adsorbed water vapour on wear debris agglomeration.

Sputtered MoS_2 has been shown to provide excellent low-friction properties in sliding contacts, especially in vacuum. Even though the short lifetime in ball-bearing applications has been a problem, MoS_2 is used increasingly for space applications, in particular where there is a need for low torque and low-torque noise. The lowest torques have been obtained when only the ball-bearing races were coated with MoS_2. Coated balls increased the torque but, at the same time, also increased the lifetime fivefold (Roberts, 1990a, b). A comparison of the mean torque and the peak-to-peak torque for gold, lead and MoS_2 coatings in vacuum is shown in Fig. 7.5.

Bearings in rocket engines have been lubricated with PTFE transfer films with the bearing cage acting as the reservoir for the PTFE lubricant. The bearing cages are typically made from composites of PTFE and a matrix material such as bronze. Dareing (1991) developed a mathematical model to predict the effects of soft metal coatings on traction coefficients when placed in contact with PTFE films. The model takes into account the elastic deflections in coating, film and substrate, and predicts that the lowest traction coefficients can be achieved with a lead coating sliding against the PTFE transfer film. Miyakawa and Noudomi (1990) studied ball bearings in vacuum by using various PTFE composite materials in the retainer and MoS_2 sputtered as a coating on the balls and the inner and outer rings. The addition of MoS_2, Ag, graphite or W did not improve the lubricity but good results

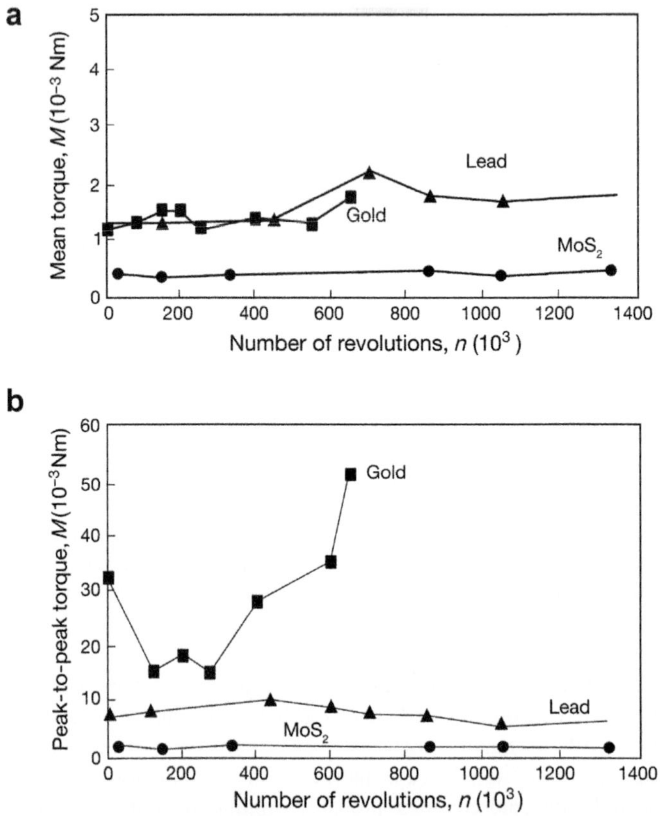

Fig. 7.5. (a) Comparison of mean and (b) peak-to-peak torques of ball bearings lubricated with thin films of gold, lead and MoS_2 in vacuum (data from Roberts, 1990b).

were achieved by a retainer composite material consisting of 50% PTFE, 25% glass fibre and 25% Mo. The addition of Mo to PTFE may improve the lubrication in vacuum and facilitate the formation of a transfer film of PTFE on the counterfaces.

7.3.2.2 Hard coatings

Wear protection and wear mechanisms. In dry or boundary-lubricated rolling conditions a thin hard coating on one or both of the surfaces can be beneficial because it inhibits adhesion between the substrate and the counterface. A hard coating can also inhibit ploughing, as explained in section 3.4.2 (Fig. 3.77). Even if the shear strength of the coating itself is high, low friction can be achieved if a soft microfilm is formed on top of the hard coating. Some coating deposition methods will result in a smoothing of the surface, which can have a beneficial effect because of lower stress levels at the asperities in contact. As with soft coatings, thin hard coatings may also make surface defects less harmful by moving crack initiation sites further into the substrate and suppressing the initiation of microcracks.

Many rolling bearings, especially at high speeds, work in elastohydrodynamically lubricated conditions, where adhesive and abrasive wear have been eliminated. Thus rolling contact fatigue will be the failure mechanism which determines the limit of the bearing lifetime.

There is clear experimental evidence that an increase in fatigue life can be achieved by the use of thin hard coatings. A theoretical explanation for the mechanisms behind this arises from calculations of the stress field in a surface coated by a hard thin coating and loaded by a rough counterface, carried out by Sainsot *et al.* (1990) and shown in Fig. 3.44. One conclusion is that for rough surfaces and thin hard coatings of thicknesses less than 15 μm the maximum shear stress will always occur in the interface between the coating and the substrate. This is in agreement with several experimental observations (Erdemir and Hochman, 1988; Chang *et al.*, 1990; Erdemir, 1992; Carvalho, 2001).

The advantage of thin hard coatings for improving rolling contact fatigue lifetime was well demonstrated by Thom *et al.* (1993) in a ball-on-rod rolling contact fatigue test where they measured lifetimes of eight different magnetron sputtered nitride coatings on hardened 440C stainless steel rollers at 4 and 5.4 GPa Hertzian stress and lubricated with a synthetic oil. The coatings tested were TiN, ZrN, HfN, CrN, Mo_2N, $Ti_{0.5}Al_{0.5}N$, $Ti_{0.5}Zr_{0.5}N$ and (Ti–Al–V)N with thicknesses in the range of 0.25 to 1 μm. Six of the eight coatings gave significant improvements in rolling contact fatigue life, whereas only CrN and $Ti_{0.5}Al_{0.5}N$ showed little or no improvement over an uncoated specimen. The best improvement was for a 0.5 μm thick HfN coating, which gave an increase in lifetime of 12 to 13 times that of an uncoated specimen. The coating thickness turned out to have a major influence on fatigue life, which is in agreement with the tribological wear mechanisms described earlier in Chapter 3. In general, for these coatings, the optimum thickness was in the range of 0.5 to 0.75 μm. Some reports suggest that CrN is not a good coating material for rolling contact applications; poor rolling fatigue life results have been reported for CrN and chromium-based coatings by both Remigiusz *et al.* (2004) and Yonekura *et al.* (2005). However, it is known that some proprietary rolling element bearings incorporate Cr–N compound coatings with good results.

An inhibiting effect of thin hard coatings on crack formation near the surface was observed by Chang *et al.* (1991) in microscopical analyses of worn surfaces of 1 μm thick TiN coatings on steel (AISI 4118) after a long period of rolling. In uncoated surfaces many near-surface microcracks were observed after 10 million rolling cycles in the region of 0 to 5 μm from the surface and parallel to it but inclined to the sliding direction. Chang *et al.* believe that the microcracks are initiated by asperity interaction. However, when TiN coated surfaces were used, the coating was still intact after 33 million rolling cycles and cracks were barely observable in the near-surface layers. XTEM showed a coherent interface structure for the coated specimens. No clear boundary could be seen at the interface compared to that of the as-deposited coating. This indicates that the interfacial activity at the elevated asperity contact temperature and the cyclic loading probably enhances the coating adhesion to the substrate instead of producing interfacial debonding.

If fatigue failure takes place in the substrate beneath a hard coating, the result is typically a large pitting crater, which rapidly results in catastrophic failure. However, fatigue failure at the interface or in the coating can result in several minor craters called fatigue spalls which slowly increase in number and area over a long period before they result in catastrophic failure, as shown in Fig. 7.6 (Erdemir and Hochman, 1988; Chang *et al.*, 1990, 1991; Erdemir, 1992).

It seems possible to improve the fatigue life in rolling contacts also by using diamond-like carbon coatings but the results are somewhat contradictory. In a twin-disk paraffin oil lubricated test with 10% slip and a Hertzian contact pressure of 1 to 1.5 GPa Klaffke *et al.* (2005b) tested seven different 2 to 3 μm thick DLC coatings in boundary/mixed lubricated conditions and found a large scatter in the results. The large scatter in lifetime of coatings did not only occur between samples of different coaters but also between the samples from one single deposition batch. They identified that the lifetime of the coatings was influenced not only by mechanical properties but also by coating thickness, interlayer type, topography and counterbody roughness.

In ball-on-rod rolling contact fatigue tests Rosado *et al.* (1997) found a 50% improvement in fatigue life with 33 nm thick DLC coated steel rods rolling with synthetic oil lubricant at 4.8 GPa Hertzian contact pressure against steel balls at a test temperature of 177°C while there was no significant improvement in room temperature when compared with uncoated steel rods. The wear

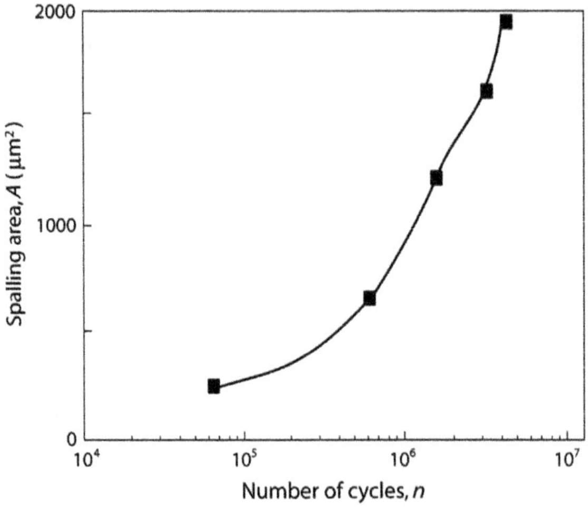

Fig. 7.6. Spalling area in a 2.5 μm thick TiN coated steel substrate (AISI4118) as a function of rolling contact cycles showing the extent of spalling propagation during rolling contact (data from Chang *et al.*, 1990).

volume was measured by profilometry and it was lower for DLC coated rods at both test temperatures. In roller bearing tests Olofsson *et al.* (2000) also recorded improved wear resistance and reduction in wear particle generation for metallic-doped DLC coated (Me–C:H) roller bearings. Based on Raman spectroscopic measurements Wei *et al.* (1993) suggest that increased lubricity and improved wear resistance of diamond-like amorphous-silicon–hydrocarbon (a-SiHC) coatings can be achieved because the cyclic stresses in rolling contacts transform the coating material from amorphous carbon to what they claim to be a more stable graphite phase.

DLC coated dry ball bearings can potentially be used in space applications (Doll, 2003). Ball-on-disk tests were carried out in air, vacuum and dry nitrogen with 1.1 μm thick highly hydrogenated DLC coatings (~50 at.% hydrogen) deposited on steel disks rotating against steel balls by Vanhulsel *et al.* (2007). The coefficient of friction was relatively high in air, 0.22 to 0.27; in dry nitrogen it was much lower, 0.02 to 0.03, and in vacuum very low, 0.007 to 0.013. It is interesting to note that the low-torque life, that is the lifetime when the coefficient of friction is less than 0.3, exceeded in similar conditions that of sputtered MoS_2 coated disks by a factor of two, representing the established technique in use.

Influence of coating thickness. It has been shown by several experimental investigations that the coating thickness is a key parameter for achieving improved lifetime in rolling contacts by the use of hard coatings. The optimum coating thickness for hard coatings like TiN or TiC seems to be in the range of 0.1 to 1 μm. A thin hard coating has the ability to deform elastically and follow the substrate deformation without coating debonding (Boving and Hintermann, 1987; Chang *et al.*, 1991). Because of the increased stiffness of the surface layer, thicker coatings do not have the same ability to easily follow substrate deformations. They become heavily fractured, have a greater tendency to spalling with larger initial spalls and can be partially removed from the rolling tracks at high Hertzian pressures (Erdemir and Hochman, 1988; Chang *et al.*, 1990; Erdemir, 1992). It has been reported (Kennedy and Tang, 1990) that thinner coatings contain higher compressive residual stresses which may have a beneficial compensating effect when tension is taking place due to substrate deformation.

Chang *et al.* (1990) studied lifetime improvements of reactive magnetron sputtered titanium nitride coatings on steel (AISI 4118) rollers within the coating thickness range of 0.25 to 5 μm at maximum

Hertzian stresses of 2.3 GPa. The tests were in mixed lubricated conditions. They found the best fatigue resistance with the thinnest, 0.25 μm, coatings and a decreasing fatigue resistance with increasing coating thickness, as shown in Fig. 3.90. The 0.25 μm thick coating underwent practically no spalling for the $60 \cdot 10^6$ cycles tested.

Results from tests with a ball pressing on a disk with oscillating load performed by Maillat *et al.* (1979) show that 2 μm thick CVD titanium carbide coatings have less fatigue damage than 5 μm thick coatings. In rolling contact fatigue tests with a ball on rod geometry at 5.5 GPa maximum Hertzian stresses rolling in boundary or mixed lubricated conditions with lambda values of 0.7, Wei *et al.* (1992) found that the optimum coating thickness for diamond-like hydrocarbon ion beam-deposited coatings on steel substrates is about 0.5 μm. Diamond-like carbon coatings in the thickness range of 0.2 to 3.4 μm were tested.

Lifetime improvements. Four to ten times lifetime improvements due to decreased fatigue wear for thin PVD-deposited titanium nitride coatings on steel substrates have been reported by several authors (Erdemir and Hochman, 1988; Chang *et al.*, 1990, 1991; Middleton *et al.*, 1991; Erdemir, 1992). Similarly, lifetime improvements of several times have been reported for CVD-deposited titanium carbide coatings (Boving *et al.*, 1987; Boving and Hintermann, 1987; Maillat *et al.*, 1979; Hanson, 1989) and up to tenfold increases by ion beam-deposited diamond-like hydrocarbon coatings on steel (Wei *et al.*, 1992, 1993), but also large scatter in lifetime results for different DLC coatings has been reported (Rosado *et al.*, 1997; Klafke *et al.*, 2005b; Manier *et al.*, 2008a, b). Considerable improvements in lifetime have been achieved by the use of duplex or multilayer hard TiC/CrC or soft MoS_2/TiN and MoS_2/Cr_2O_3 surface coatings on steel substrates (Dill *et al.*, 1984; Bhushan and Gupta, 1991).

7.4 Gears

7.4.1 Description of the application

Gears are the most commonly used elements for torque transmission because of their accurate transmission rate without any slip, high torque and high-speed capability, high efficiency and the availability of advanced design methods and production techniques.

One limitation in their use may be noise. The lifetime of a gear transmission is generally limited by tooth fracture, rolling contact fatigue failure and also scuffing or scoring failure, especially at low speeds.

The contact conditions for two gears are a combination of rolling and sliding. Only at one place on the gear tooth will pure rolling conditions prevail, along the pitch line, but elsewhere there is decreasing or increasing slip as shown in Fig. 7.7. The slide–roll ratio at the contact varies typically from about -15% to $+15\%$. A common place for fatigue failures, also frequently called pitting, to occur is at a slide–roll ratio of about -5%.

Gears are most commonly used with oil lubrication as, at high or medium speeds, a good elastohydrodynamic lubrication film is formed between the teeth and inhibits direct tooth contact. If the gears are correctly designed, no fracture, scuffing or scoring will take place, but the fatigue of the surfaces will limit their lifetime. The typical design level for the maximum Hertzian contact pressure in a gear contact of an industrial unit is about 1 GPa.

At low speed a complete EHD lubricant film cannot be formed and boundary lubrication is then the dominating lubrication mechanism. In such a situation scuffing or scoring failure may occur. Gear scuffing is mostly related to the flash temperature in a sliding gear contact. When it exceeds a critical temperature of the lubricant, the effective lubrication film will break down and scuffing is initiated. This is very typical, e.g., for worm gears, which, due to the gear geometry, have no pure rolling at all and sliding conditions completely dominate. Tribological aspects in gear design are presented by Stolarski (1990) and Cheng (2001).

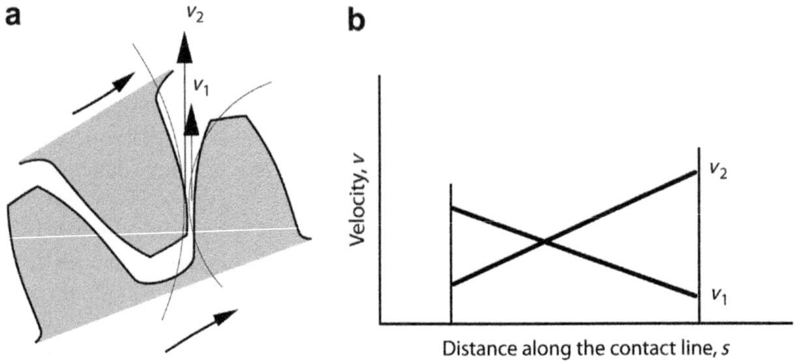

Fig. 7.7. The sliding velocities along the line of contact in a gear transmission.

7.4.2 Improvements by surface coatings

The potential advantages of using surface coatings on gears are:

- control of the running-in process,
- prevention of scuffing or scoring with wear-resistant low-friction coatings,
- increasing rolling contact fatigue life by surface stress level reduction or surface strengthening, or
- decreasing the noise level using a soft coating.

The use of hard coatings on gear teeth surfaces has been investigated to some extent, with promising results. The tungsten–carbon–carbide layer (WC/C) is a metallic-hydrocarbon surface coating (Me-C:H), which is increasingly used in automotive and other mechanical engineering products to reduce wear. The tribological behaviour of lamellar structured WC/C coatings as well as boron carbide coatings in gear transmissions has been studied extensively by Joachim *et al.* (2004). The coatings were applied by a PVD process permitting deposition temperatures less than 200°C and thus allowing the coating of case-hardened gears. They describe the boron–carbide coatings as amorphous and give them the designation B4C. Coating thickness was 1 to 4 μm, and the coatings had high hardness and high elasticity. In sliding against steel the coated surfaces showed low friction both in dry conditions, $\mu = 0.25$, and in boundary oil lubricated conditions, $\mu = 0.04$, while the friction coefficients for similar steel against steel contacts were 0.6 and 0.15, respectively.

FZG gear testing, where coated and uncoated gears were run with stepwise increasing load in lubricated conditions, showed a considerable improvement in scuffing load capacity. The improvement of the WC/C and B4C coated gears was about two damage load stages in the FZG rig, corresponding to an increase in torque capacity of about 50%, as shown in Fig. 7.8. Chemically accelerated super-finishing of the uncoated gear surfaces to a surface roughness value of $R_a = 0.09$ μm from the unfinished value of $R_a = 0.22$ μm resulted in a slight increase in scuffing load capacity due to improved EHD lubrication conditions. A reduction in oil temperature was recorded for coated gears compared to uncoated, indicating decreased friction. Transmission efficiency measurements showed a reduction in the coefficient of friction of 30% both for WC/C coated and super-finished gears compared to uncoated gears. The result shows very convincingly that the coating material itself, regardless of the formation of a lubricant film, is less prone to scuffing than the base material.

Pitting surface fatigue life testing in the same FZG test rig resulted in an increase in lifetime by a factor of 2 to 3, but only if both gears were coated, as shown in Fig. 7.9. The poor relative performance of the super-finished gears is not understood. Slow-speed wear testing, with a pitch-line velocity of $v = 0.03$ m/s, showed a surface material removal due to wear that was five times less for the

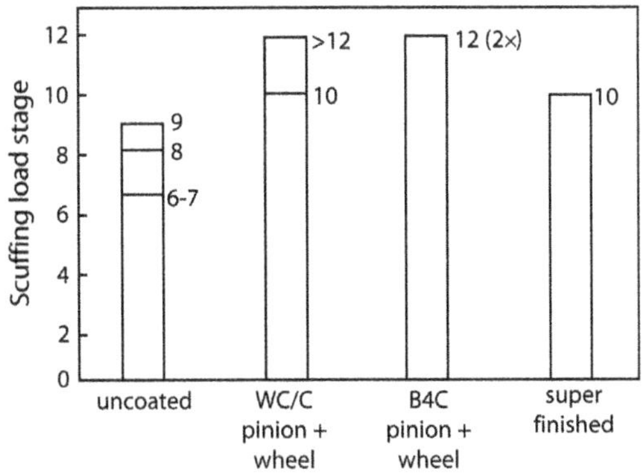

Fig. 7.8. Scuffing load capacity shown as scuffing load stage of uncoated, WC/C coated, B4C coated and super-finished gears in FZG scuffing test with oil lubrication (data from Joachim *et al.*, 2004).

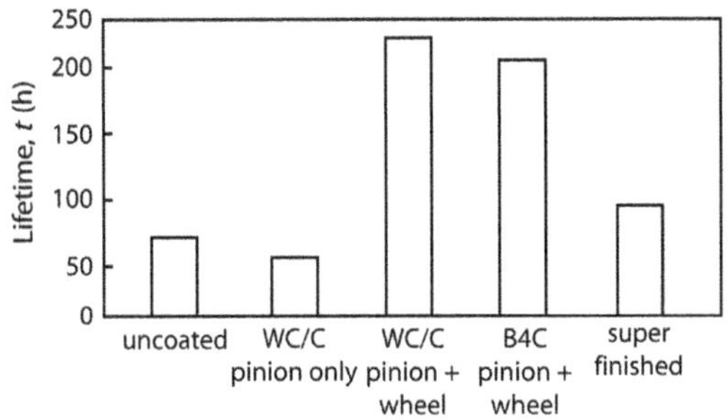

Fig. 7.9. Macropitting lifetime of uncoated, WC/C coated, B4C coated and super-finished gears in FZG pitting test with oil lubrication (data from Joachim *et al.*, 2004).

coated gears compared to the uncoated. Interestingly, simply coating one of the contacting gears was sufficient for the wear reduction and coating both gears did not result in any further improvement. The wide range investigation by Joachim *et al.* (2004) shows that hard coatings can improve the scuffing, pitting and slow-speed wear resistance as well as reduce the friction in gear contacts and transmissions considerably. However, the tribological contact conditions in various transmission applications vary and thus are expected to result in a fairly large scatter in the tribological effect of coatings on gears.

In gear tests metal-containing carbon-based coatings on spur gears resulted in a sixfold increase in wear life, reported by Krantz *et al.* (2003).

A marked increase in scuffing wear resistance and a higher load-carrying capacity were also observed by Douglas *et al.* (1987) for cathodic arc PVD titanium nitride coated gears compared to uncoated standard gears (DIN 17210), tested in an FZG power circuit and a Ryder gear testing machine. The TiN coating was approximately 3 μm thick at the tips of the teeth and tapered down to 1 μm at the tooth roots. Light scratches occurred on the standard gear surfaces at the third load stage and scuffing failure at the seventh load stage in the FZG test with a pinion speed of 1500 rpm. In the same conditions the TiN coated gears succeeded without scuffing failure in all 12 load steps with the coating intact and only minimal subsurface plastic deformation.

In dry running back-to-back gear testing Mao *et al.* (2007) found that TiN coated gears reduced the gear wear weight loss to less than half compared with uncoated gears while plasma-nitrided gears reduced it to one-tenth and duplex coated gears, with TiN on a plasma-nitrided surface, reduced it 50 times. The gears were run at 500 rev/min at a load of 40 Nm for periods of 3000 cycles. The duplex treatment of gears is certainly expensive but can considerably improve the wear performance in dry running conditions.

In a high-speed power circuit gear test machine, Terauchi *et al.* (1987) recorded improved scoring resistance of gears with a titanium nitride and a titanium carbide coated gear wheel running against an uncoated pinion of case-hardened chromium–molybdenum steel (SCM415H). The TiC was deposited by the CVD process and the TiN by PVD and the coating thickness was approximately 2 μm. In these tests the rotational speed of the pinion was first increased stepwise from 2000 to 10,372 rpm (from start-up to 45 MJ/mm in Fig. 7.10), with a constant load of 266 N/mm. The load was then increased stepwise from 266 to 390 N/mm (from 45 MJ/mm forward in Fig. 7.10) with a constant rotational speed of 10,372 rpm. Figure 7.10 shows that the best compromise between scoring resistance and low noise level was achieved with a TiN coated wheel against an uncoated pinion (UN). Improved scoring resistance, but at higher noise levels, was recorded for a TiC coated wheel against an uncoated pinion (UC) compared to the uncoated gear pair (U). Early failure by abrasive wear was observed for a TiN coated pinion against an uncoated wheel (NU) and for TiN coated gear pairs (N). This is not consistent with the good results from gear tests with TiN coated gears by Douglas *et al.* (1987) referred to earlier. It may have been due to unsatisfactory running-in of the gear surfaces.

Significant improvements in the scoring resistance of spur gears can be achieved by using copper plating or molybdenum disulphide coatings on the gears. Also graphite coatings have been found to be effective in increasing scoring and wear resistance (Terauchi *et al.*, 1986a; Field *et al.*, 2004). In an experimental study with a power circuit gear test machine, Terauchi *et al.* (1984) found that a 20 μm thick copper coating on the gears decreased the sound level by 7 dB and the best scoring resistance was achieved when both pinion and wheel were copper plated followed by the combination of unplated pinion and plated wheel. The test gears were made from chrome–molybdenum steel (SCM420H), had a roughness of $R_{max} = 2.4$ μm before copper plating and were lubricated by a straight mineral oil without additives.

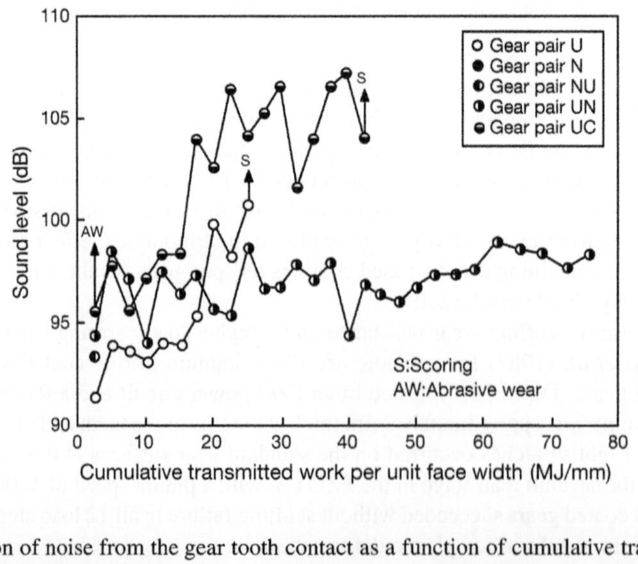

Fig. 7.10. Variation of noise from the gear tooth contact as a function of cumulative transmitted work per unit face width with the failure indicated by a vertical arrow. U stands for uncoated, N for TiN coated and C for TiC coated (data from Terauchi *et al.*, 1987).

Scuffing FZG gear tests carried out in additive-free oil lubricated conditions by Amaro *et al.* (2005) of MoS_2/titanium coated gears showed a very significant increase of the load-carrying capacity against the scuffing load correlating to 40% increased power transmission capacity at 1500 rev/min and 77% at 3000 rev/min rotational speed. The gearbox efficiency was improved 0.5% at high torque and low speed. The MoS_2/Ti coating was deposited by DC magnetron sputtering; it has a multilatered structure and a thickness of only 1.2 μm.

Gear tests with a 10 μm thick MoS_2 coating on the gears showed a significant increase in the scoring resistance but the effect of the MoS_2 film on seizure load was small (Terauchi *et al.*, 1986b). A significant improvement in scuffing load capacity has been reported when using multilayer composite MoS_2/Ti and C/Cr coatings on C-type gears in the FZG gear test rig (Martins *et al.*, 2008). The MoS_2/Ti coated gears increased the transfer gearbox's efficiency by 0.5%.

7.5 Tools for Cutting

7.5.1 Description of the application

Cutting of metals and other engineering materials has been and still is the major shaping process used in the production of engineering components. Even though new advanced manufacturing methods are being developed all the time, up to now no serious challenger has emerged that could replace cutting as the main shaping process within the next few decades.

This application is discussed in detail for two reasons. The first is that the tribological contact in cutting differs considerably from the general sliding or rolling contacts discussed earlier. The second is that cutting tools were the first successful commercial tribological application of thin surface coatings.

Even if the cost of consumable cutting tools is relatively low they have a remarkable influence on productivity and manufacturing costs. The material removal rate is very much determined by the tribological factors which dictate how fast a cutting tool can be moved through the material rather than other factors such as machine limitations. When a cutting tool wears out the operation is interrupted

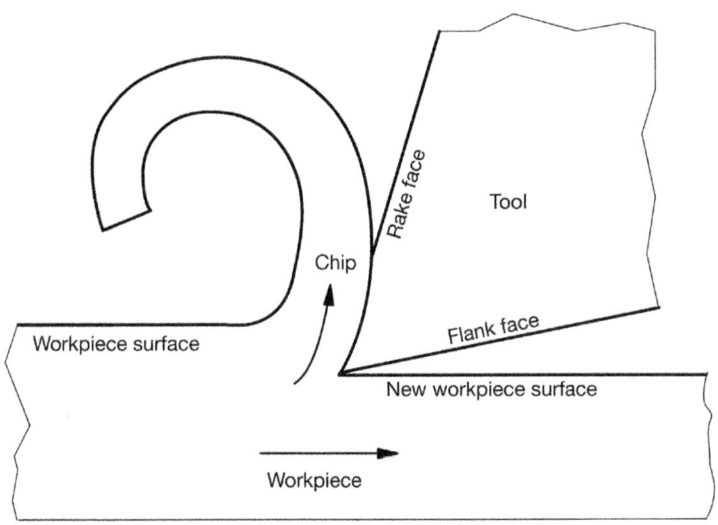

Fig. 7.11. Schematic illustration of the cutting process.

and the tool must be replaced or resharpened. The costs for machine downtime during replacement and the costs for regrinding and resetting worn tools are a significant part of the total machining cost (Trent, 1979; Tipins, 1980).

In cutting, the tool is forced to move through the workpiece so that a chip of material is removed from the surface, as shown in Fig. 7.11. The tribological contact conditions between the two moving surfaces, the tool and the workpiece, are very severe. Recently, both finite element and molecular dynamics models have been used to simulate the cutting process (Ikawa *et al.*, 1991; Shimada *et al.*, 1992; Belak *et al.*, 1993; Bouzakis *et al.*, 2003). The models considerably simplify the real cutting process but they are likely in the future to give valuable information on the different interactions in the extreme conditions at the tip/workpiece contact.

The surface of the tool continuously comes into contact with new virgin workpiece material which is unaffected by the environment, e.g. by oxidation. The temperature in the contact may be as high as 900 to 1300°C when cutting steel and it is distributed as shown in Fig. 7.12 (Boothroyd, 1963; Opitz, 1975; Suh, 1986; Rech *et al.*, 2004). In general, the temperature rise is sensitive to speed. Of the heat produced in the contact, 80% has been estimated to originate from heat generated by the mechanical deformation of the chip, 18% by friction between the chip and the upper tool surface and only 2% is created on the tool tip. Of the heat removed 75% is taken by the chip, 5% is absorbed by the workpiece and 20% is conducted through the tool (Lux *et al.*, 1986).

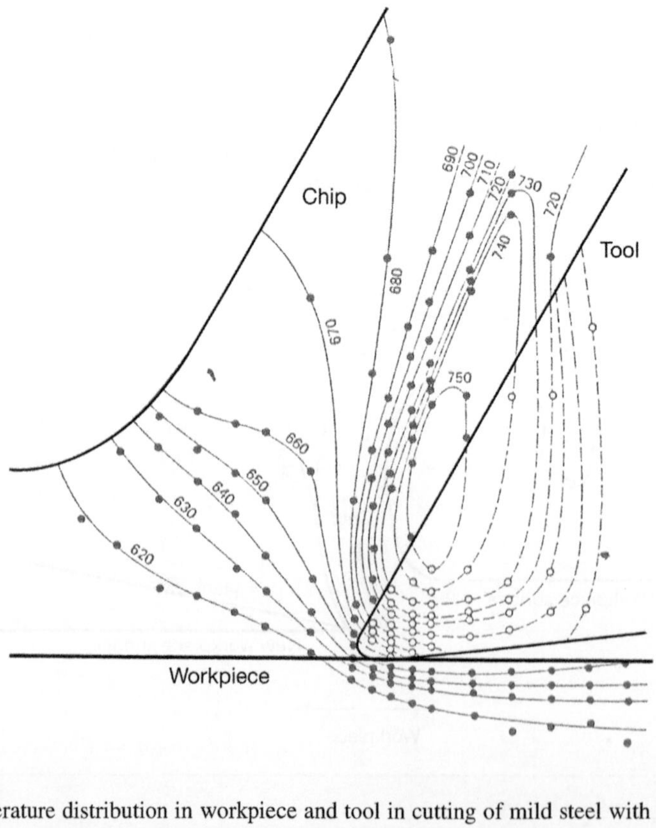

Fig. 7.12. Temperature distribution in workpiece and tool in cutting of mild steel with a cutting speed of 23 m/min, width of cut 6.3 mm, rake angle 30° and relief angle 7° (after Boothroyd, 1963).

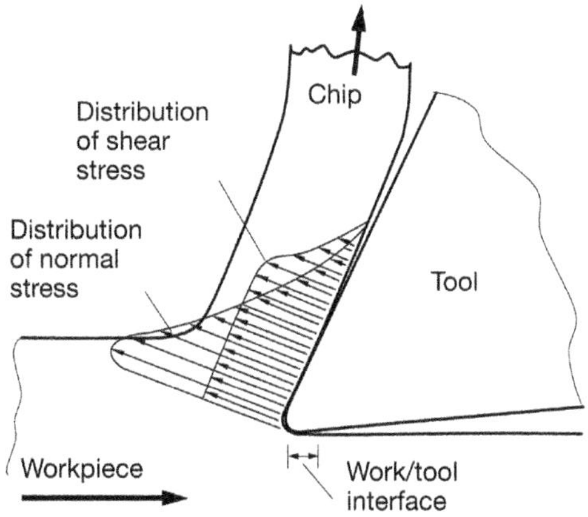

Fig. 7.13. Tool stress distributions in the chip/tool interface (after Moore, 1975).

The normal pressure on the rake face is very high and can be up to 1200 MPa when cutting steel. For tools of the same geometry the compressive stress on the tool is related to the yield stress of the work material (Trent, 1979). The distribution of the normal and the shear stress is shown in Fig. 7.13.

The coefficient of friction between the chip and the tool is influenced by factors such as cutting speed, feed rate, rake angle, etc. It has been observed to vary considerably, mainly because of the very high normal pressure at the surface. This can cause the real area of contact to approach the apparent area of contact over part of the tool/chip contact (Moore, 1975; Grzesik, 2000; Childs, 2006). The tool/workpiece contact friction is higher than the tool/chip friction and the frictional behaviour has been shown to be adhesion related (Chandrasekaran, 1976).

7.5.2 Tool wear

Four main mechanisms of tool wear have been identified, namely adhesive wear, abrasive wear, delamination wear and wear due to chemical instability, including diffusion, solution and electrochemical wear as shown in Fig. 7.14.

Adhesive wear is caused by the formation of welded asperity junctions between the chip and the tool faces and the fracture of the junctions by the shearing force so that tiny fragments of the tool material are torn out and adhere to the chip or the workpiece (Moore, 1975). This kind of wear may occur at the flank face in low-speed cutting when the contact temperatures are not so high. It may involve oxidation of the tool surface, or other chemical interaction with the surrounding atmosphere, followed by mechanical removal of the products of the reaction (Trent, 1979; Suh, 1986; Thangaraj and Weinmann, 1992).

Abrasive wear on the tool surfaces is caused by hard particles in the work material. Their effect on the tool wear can be explained theoretically by three abrasion models, these are microploughing, microchipping and microcracking (Knotek *et al.*, 1993). Many work materials, e.g. cast iron and steel, contain particles of phases which have a hardness that is much higher than that of the workpiece. The particles can typically be carbides or oxides, particularly Al_2O_3, but also silica and some silicates (Trent, 1979).

The particles may also be highly strain-hardened fragments from an unstable built-up edge on the tool (Moore, 1975). In particular, the wear at the flank face may be attributed to abrasive wear. In abrasive wear the influence of tool surface hardness is considerable (Budinski, 1980; Kramer, 1986; König and Kammermeier, 1992).

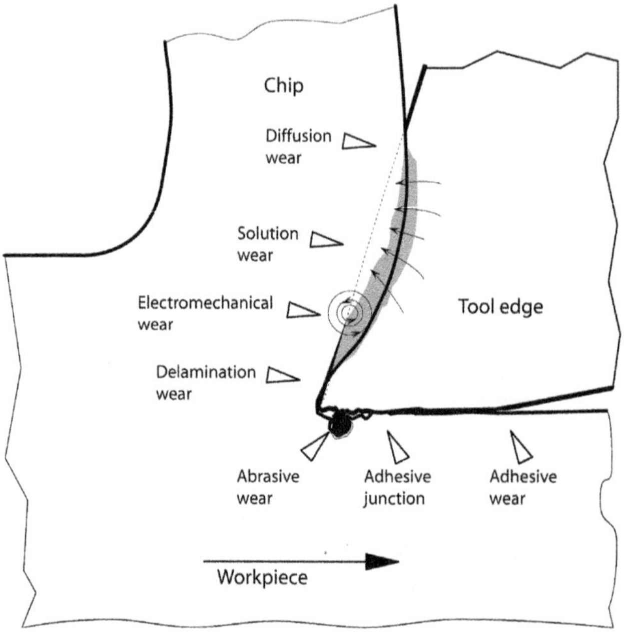

Fig. 7.14. The different mechanisms of tool wear.

Delamination wear is due to plastic deformation of the surface leading to subsurface crack nucleation, propagation and the liberation of tiny flakes from the tool surface. This has been observed when high-speed steel tools soften due to annealing during machining (Suh, 1986).

Wear due to chemical instability is very important in metal cutting because of the prevailing high temperatures at the contacting surfaces; it includes diffusion wear, solution wear and electrochemical wear. The dissolved material can be transported away by diffusion from the interface, convective transport by the chip material flowing over the interface and in the form of delaminated loose wear particles caused by subsurface failure below the dissolved layer.

Diffusion wear characterizes the material loss due to diffusion of atoms of the tool material into the workpiece moving over it. Requirements for diffusion wear are metallurgical bonding of the two surfaces so that atoms can move freely across the interface, a temperature high enough to make rapid diffusion possible and some solubility of the tool material phases in the work material (Trent, 1979; Subramanian *et al.*, 1993). Temperature and the rate of flow immediately adjacent to the surface are two major parameters influencing diffusion wear.

In a theoretical model of diffusion wear proposed by Kramer and Suh (1980) and Kramer (1986) the wear rate is considered to be controlled by the mass diffusion rate. However, tool wear prediction according to the model is not in agreement with experimental results reported by Sproul (1987). Ono and Takeyama (1992) have shown that in diffusion wear at the minor flank (relief face) of the tool the chemical reaction taking place at the interface has a major effect on wear. They showed that oxygen gas accelerates the formation of oxide layers that are continuously torn off resulting in increased wear, while wear is decreased by an environment of argon gas.

Solution wear describes the wear mechanism taking place when the wear rate is controlled by the dissolution rate rather than by convective transport (Kramer,1979; Kramer and Suh, 1980).

Electrochemical wear is due to a thermoelectric electromagnetic field generated at the chip/tool junction causing electric currents to circulate, resulting in the passage of ions from the tool surface to the workpiece material. This may result in breakdown of the tool material and scaling of the tool surface at the tool/chip interface (Moore, 1975). The thermoelectric current is formed at the chip/crater contact and at the flank/workpiece contact and its intensity has been found to be as high as 5 A with the tool being of negative polarity (Opitz, 1975; Shan and Pandey, 1980; Uehara *et al.*, 1992).

As shown in Fig. 7.15, for a given workpiece material, abrasive wear occurs under all cutting conditions to approximately the same extent while adhesive wear is found mainly at low cutting temperatures, corresponding to low cutting speeds. Wear due to chemical instability, including effects like diffusion and oxidation appears at high cutting speeds (König, 1984; Knotek *et al.*, 1993).

Flank wear at the front edge of the tool flank and *crater wear* at the tool face, as shown in Fig. 7.16, are the most typical modes of tool wear in metal cutting. Flank wear is believed to be caused mainly by abrasion of the tool by hard particles but there may also be adhesive effects. It is the dominating wear mode at low cutting speeds (Thangaraj and Weinmann, 1992; Schmid and Wilson, 2001). Mapping of wear data for flank wear of high-speed steel cutting tools in turning, as a function of feed rate and cutting speed, shows that by suitable choice of cutting parameters a safety zone with low wear can be reached (Lim *et al.*, 1993b).

Crater wear is the formation of a groove or a crater on the tool face, typically some 0.2 to 0.5 mm from the cutting edge, at the place where the chip moves over the tool surface, as shown in Fig. 7.16b. It is very commonly observed when cutting high melting point metals like steel at relatively high cutting speeds. Crater wear is caused primarily by the dissolution of tool material by diffusion or solution wear since it occurs in the region of maximum temperature rise, as shown in Fig. 7.12 (Subramanian *et al.*, 1993).

Two other mechanisms for tool failure, which do not include wear, are *fracture* of the tool and *plastic deformation* of the tool edge. Because of the high hardness and brittleness, as well as the sharp edges of the tool, fracture occasionally occurs. The high pressure and temperature near the tool edge may cause plastic deformation and initiate the wear process (Trent, 1979; Suh, 1986). A wear map for uncoated high-speed steel cutting tools with feed rate and cutting speed as co-ordinate axes and showing the transition from one wear mechanism to another has been published by Lim *et al.* (1993a).

A *built-up edge* of piled-up work material near the tip of the cutting edge is frequently formed at intermediate cutting speeds, as shown in Fig. 7.16a. This typically occurs when cutting alloys that

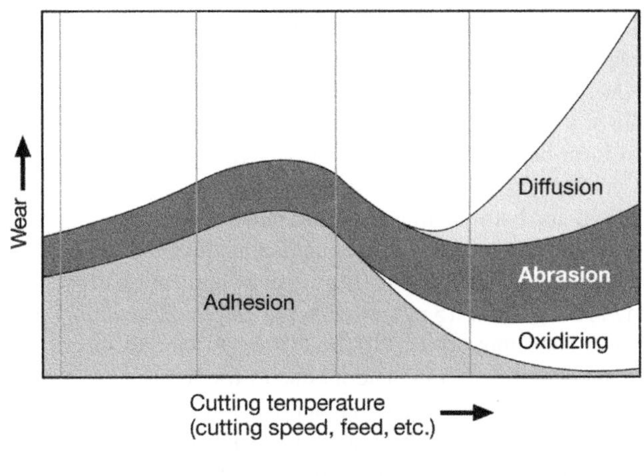

Fig. 7.15. Schematic diagram of tool wear mechanisms appearing at different cutting temperatures corresponding to cutting speed and feed (after König, 1984).

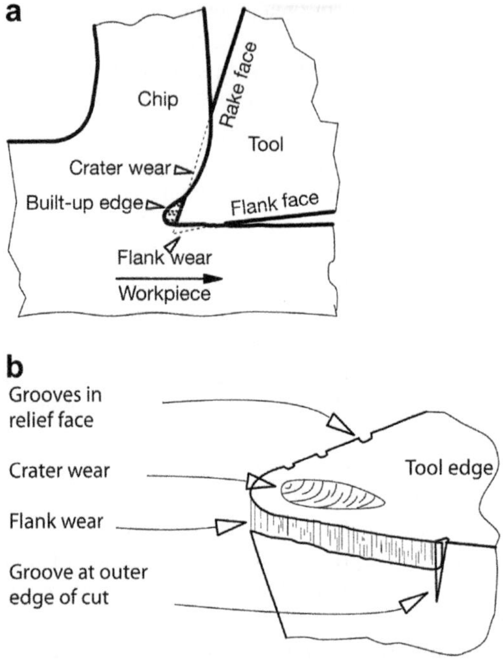

Fig. 7.16. Typical location of wear failure and built-up edge on the tool.

contain more than one phase, like carbon steel, cast iron and alpha–beta brass. As the first layer of work material that adheres to the tool tip is considerably strengthened by the shear strain, yield will occur in less-strained layers further from the tool surface. A continuous repetition of this process results in agglomerates that form relatively large layers of work-hardened material that accumulate to form a built-up edge (Trent, 1979; Suh, 1986).

The built-up edge is often unstable; it breaks away intermittently and is formed again. Depending on the materials and the cutting conditions the influence of the built-up edge may sometimes decrease or at other times increase the tool life. A stable built-up edge can be very beneficial and protect the tool surface from wear. On the other hand, loose highly strain-hardened fragments of the built-up edge may adhere to the chip or workpiece and cause abrasion of the tool. At higher cutting speeds a built-up edge is less likely to form because the temperature increases and the built-up edge can no longer support the stress of cutting and will thus be replaced by a flow zone.

Cutting fluids are used for both cooling and lubrication. Cooling is important at cutting speeds above 60 m/min, for the reduction of chip/tool temperatures thereby reducing crater wear, which is strongly temperature sensitive (Tipins, 1980). The effect of lubrication between chip and tool is most pronounced at cutting speeds below 15 m/min reducing chip compression, built-up edge formation, cutting forces and power consumption (De Chiffre, 1990). At cutting speeds greater than 10 m/min, for steels, local thermal softening of chip material due to friction heating results in a degree of self-lubrication. Some steels, such as free-machining and calcium treated steels, contain solid lubricant inclusions that can segregate on the rake face to reduce friction (Childs, 2006).

The mechanism of lubrication between chip and tool at the rake face has long been a contentious issue. A considerable effect of the cutting fluid is easily observed but it is difficult to explain how the lubricant can get into the contact area, characterized by high-temperature flow between chip and tool

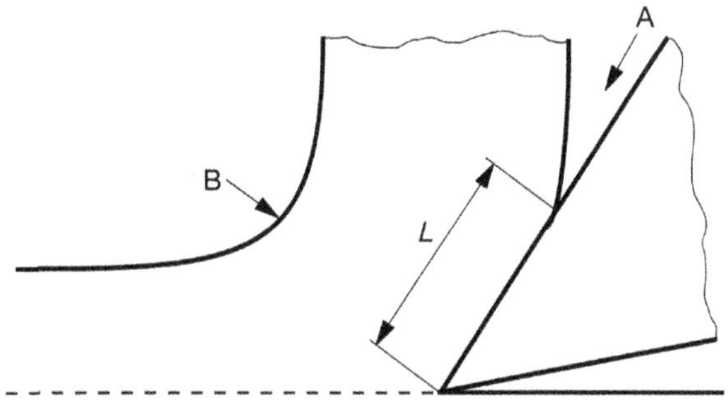

Fig. 7.17. Illustration of the contact length reduction theory by De Chiffre (1990). L is length of contact between chip and tool face, A and B are referred to in the text.

which follows immediately after the virgin chip surface has been produced. De Chiffre (1977, 1990) explains the lubricant action on the basis of a contact length approach which does not require any fluid access to the chip/tool interface. Over a distance from the cutting edge, intimate contact and full adhesion exist between chip material and tool. The cutting fluid, however, prevents further growth of the adhesion area, with the result of reducing the length of the contact as compared to dry cutting. The shorter contact length then reduces the frictional drag, and this in turn reduces the deformation work. Because the apparent area of contact is almost the same as the real area of contact, the contact length will be directly related to the friction force.

The reduction of the chip/tool contact length L is expected to occur due to the synergistic effect of the following three mechanisms (De Chiffre, 1988a, b, 1990), as shown in Fig. 7.17:

1. Contamination, or rather boundary lubrication, of the rake face at A. Adhesion of the virgin chip surface on to the tool is inhibited in the region far away from the cutting edge.
2. Chip curl promotion at B, due to the Rehbinder effect at low speeds and cooling at high speeds.
3. Overall reduction of the cutting temperature by cooling, lowering the adhesion tendency and promoting contact area restriction.

A similar reduction in the contact area on the rake face with TiN coated tools compared to uncoated has been observed by Posti and Nieminen (1989a). Here the solid coating acts as a lubricating surface by inhibiting adhesion.

Wear is almost always reduced by the use of cutting fluids, with some exceptions when it may promote thermal cracking, hardening of work material through quenching and embrittlement of tools by chemical or physical attack (Tipins, 1980).

7.5.3 Improvements by surface coatings

Improved performance of cutting tools to give lower wear, lower friction and extended tool lifetime can be achieved by depositing hard coatings on the tool surface. The main favourable effects from using coatings come from:

- the reduced friction resulting in lower heat generation and lower cutting forces,
- reduced adhesion to the workpiece resulting in less material transfer from the tool surface,
- increased hardness of the surface resulting in reduced abrasive wear, and
- reduced diffusion due to the improved diffusion barrier and chemical stability properties of the coating.

Thus to be successful the coating material should exhibit the following characteristics:

- low adhesion to the workpiece material,
- high adhesion to the tool material,
- good abrasive wear resistance,
- high chemical stability, and
- high toughness.

Thin hard coatings of nitrides or carbides match these requirements well and have been used successfully resulting in increased tool lifetime of 30% to up to 200-fold. Their small thickness gives them the high degree of elasticity needed for absorbing impacts and the high tool edge stresses. At the same time they also influence the basic cutting parameters like chip flow, chip form, cutting edge temperature and optimum working range. The tool thus needs to be reoptimized with respect to cutting edge angles, clearance angles, steel quality, hardness, roughness, etc. Successful applications of the use of hard coatings on tools has been reported for machining operations like turning, profiling, grooving, threading, drilling, tapping, reaming, broaching, milling and hobbing; and with workpiece materials like carbon steels, alloy steels, cast iron, stainless steels, superalloys, aluminium alloys and titanium alloys (Sudarshan, 1992; Günther *et al.*, 1989; Vogel, 1989; Quinto, 1996, 2007; Dearnley, 2006; Veprek and Veprek-Heijman, 2008).

The better tribological performance of coated cutting tools results in advantages like:

- reduced tool costs per component due to the extended tool lifetime,
- reduced downtime costs,
- increased productivity because faster cutting speeds can be used due to lower friction, which also reduces machine power consumption, and
- improved workpiece quality in the form of better surface finish due to improved lubricity and other effects such as improved dimensional accuracy in cutting.

7.5.3.1 Substrate materials

High-speed steel and cemented carbide are commonly used substrate materials for thin and hard PVD coatings (Hintermann, 1983a; Larsson, 1996; Grzesik, 2000). The powder metallurgy grades of high-speed steel are particularly useful, since their good carbide distribution ensures consistent quality and can even enhance adhesion due to the good lattice match between the carbides and TiN.

7.5.3.2 Wear mechanisms

Depending on the cutting conditions one or more of the wear mechanisms shown in Fig. 7.14 will dominate the wear process and thus be the limitation for the tool lifetime. If abrasive wear dominates, the hardness of the coating is important; if adhesive wear dominates, low adhesion to the work material is important; and if wear due to chemical instability dominates diffusion isolation or low solution properties of the coating are important.

Coating fracture and failure can result if the coating is too hard and thick in relation to the elasticity of the substrate or if the substrate beneath the coating is softened by the high temperatures. The deflection creates high stresses within the coating that may lead to fracture (Billgren, 1984). This mechanism has been observed to be the main cause of failure for nitride and carbide coatings on high-speed steel cutting tool inserts by Fenske *et al.* (1988) and is illustrated in Fig. 7.18. This is in line with the scratch test study by Hedenqvist *et al.* (1990b) where they described in detail 12 different mechanisms of coating fracture and damage partly due to substrate deformation. The very favourable properties of TiN and TiC coated HSS inserts in relation to the adhesive, abrasive and solution wear mechanisms have been shown by Billgren (1984) and Hedenqvist *et al.* (1990b).

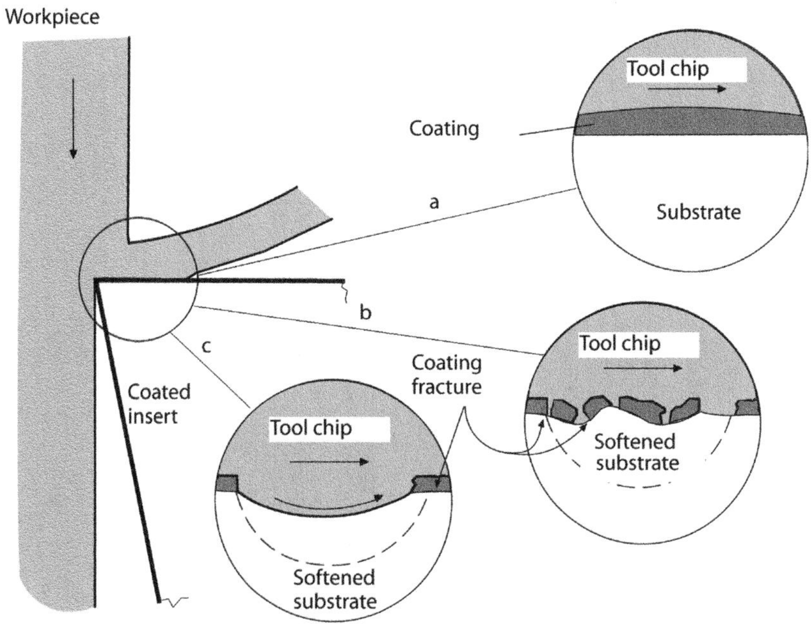

Fig. 7.18. Schematic diagram of the coating wear process at (a) the initial stage, (b) the substrate softening and coating fracture stage and (c) the failure stage, observed for nitride and carbide coated high-speed steel cutting tool inserts during dry cutting of 1045 steel at surface speeds of 100 m/min (after Fenske *et al.*, 1988).

7.5.4 Cutting test results

7.5.4.1 Turning tools

Considerable increases in tool lifetime have been attained with high-speed steel inserts coated with nitride or carbide coatings. The lifetime has been reported to increase by 2 to 10 times compared to uncoated tools (Bruno *et al.*, 1988; Kowstubhan and Philip, 1991). In lifetime comparisons Sproul (1987) found that TiN sputter coated inserts, with a coating thickness of 5 μm, have a longer lifetime than TiC, ZrC, HfC, ZrN or HfN coated inserts, as shown in Fig. 7.19. The results from turning tests by Harju *et al.* (1990) show that ion-plated TiN coated inserts, with a coating thickness of 1.5 to 3.5 μm, are superior to ZrN and HfN coated inserts. For TiN coatings longer lifetimes have been observed for coatings with thicknesses in the range of 5 to 6 μm compared to thinner coatings in the range 0.5 to 3.5 μm, as shown in Fig. 7.20.

Ronkainen *et al.* (1991a, b) and Knotek *et al.* (1992c, f) have shown that in turning tests ion-plated (Ti,Al)N and (Ti,V)N coatings improve the wear resistance considerably and give lifetimes two to three times better than TiN coatings; however, no major improvement was observed with Ti(B,N). Inserts coated by TiHfN, 2.5 μm thick, and TiNbN, 4.9 μm thick, have shown an improved crater wear resistance in cutting tests, compared to TiN coatings, but the flank wear increased (Boelens and Veltrop, 1987). A reduction in machining costs when using cBN–TiN composite carbide coated inserts for hard turning applications has been shown by cost analysis (More *et al.*, 2006). CVD coated Al_2O_3 inserts have a very similar wear behaviour compared to TiN coated inserts and no significant differences have been observed (Schintlmeister *et al.*, 1978).

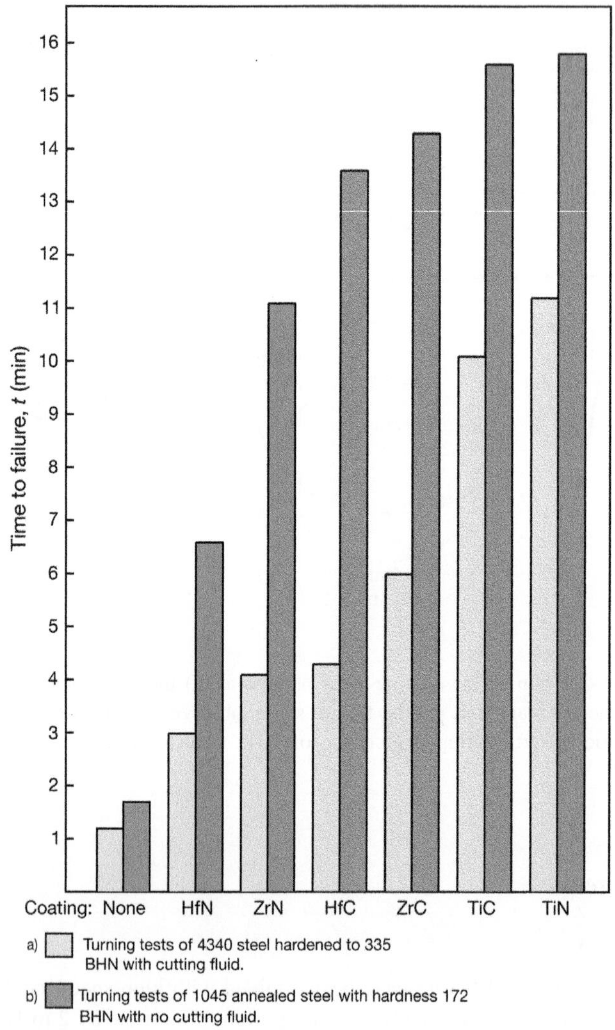

Fig. 7.19. Time to failure of coated and uncoated high-speed steel tool inserts, in turning tests with (a) cutting fluid and hard work material and (b) with no cutting fluid and soft work material according to data from Sproul (1987).

The use of hard coatings on cutting tools often gives a better surface finish on the workpiece but the performance of the tool may decrease after resharpening, depending of the type on wear and reground face (Vogel, 1989).

Diamond coatings have been successfully used for machining of high silicone content aluminium alloys for automotive applications. The major applications for Al–Si alloys are pistons, engine blocks and heads, wheels and chassis, and some transmission parts (Erdemir and Donnet, 2001). The increased use of lightweight materials in the automotive industry is likely to continue and the need for diamond coated tools will increase.

In the machining of aluminium, aluminium–silicon alloys, copper, graphite and other non-ferrous materials, CVD diamond coated inserts have given lifetimes 3 to 20 times longer than cemented

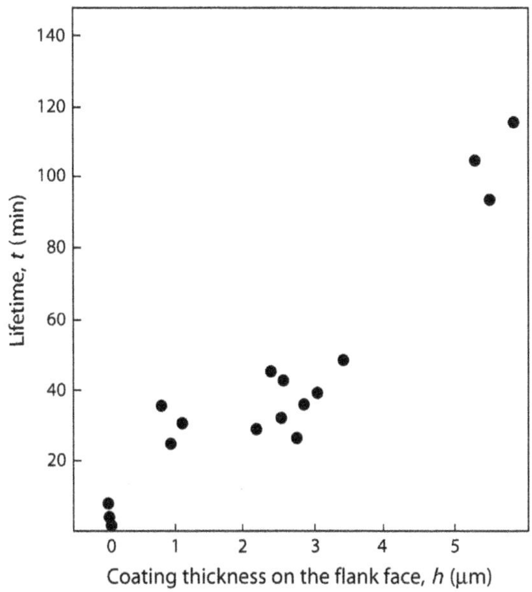

Fig. 7.20. Tool lifetime in turning tests of TiN coated ASP 30 tools as a function of the coating thickness on the flank face (data from Posti and Nieminen, 1989b).

carbide, silicon nitride or sintered diamond inserts (Kikuchi *et al.*, 1991; Söderberg *et al.*, 1991; Yazu and Nakai, 1991; Lux and Haubner, 1992; Inspector *et al.*, 1994; Quinto, 1996). Other materials suitable for machining by diamond tools are magnesium, manganese, lead, as well as various composites, bulk graphite, plastics, epoxy resins, green ceramic and concrete. However, tool wear of diamond coated inserts is extremely rapid when machining ferrous materials because of the high chemical reactivity and the solubility of carbon in ferrous materials at elevated temperatures (Evans, 1991; Söderberg, 1991).

7.5.4.2 Twist drills

Considerable lifetime improvements have been achieved by using twist drills coated with thin hard coatings. Titanium nitride has again proved to be an excellent coating material on high-speed steel drills (HMSO, 1986; Matthews, 1989; Monaghan *et al.*, 1993; Goller *et al.*, 1999; Cosemans *et al.*, 2003). In a large drilling test comparison, twist drills with 2 to 3 μm thick PACVD coated TiN layers had on average an eightfold lifetime increase compared to uncoated drills (Shizhi *et al.*, 1990). The performance of coated drills depends on the work material. Vogel (1989) has reported fourfold lifetime improvements of TiN coated twist drills when drilling an unalloyed carbon, case-hardened and tool steel, fivefold increases when drilling cast iron, non-ferrous metals and free-cutting steel, and ninefold increases when drilling stainless steel. He also reports a 17% drop in the feed force and improved cutting performance after resharpening of the drills. This can be explained by the main wear not occurring on the primary flank but more on the lip and on the margin. Further lifetime improvements by 15 to 100% in drilling tests have been reported by Goller *et al.* (1999) and Cosemans *et al.* (2003) when incorporating a soft MoS_x phase in the TiN matrix.

Drills coated with (Ti,Al)N perform, in general, even better than TiN coated ones, as shown in Fig. 7.21. This has been attributed to the formation of a dense, highly adhesive, protective Al_2O_3

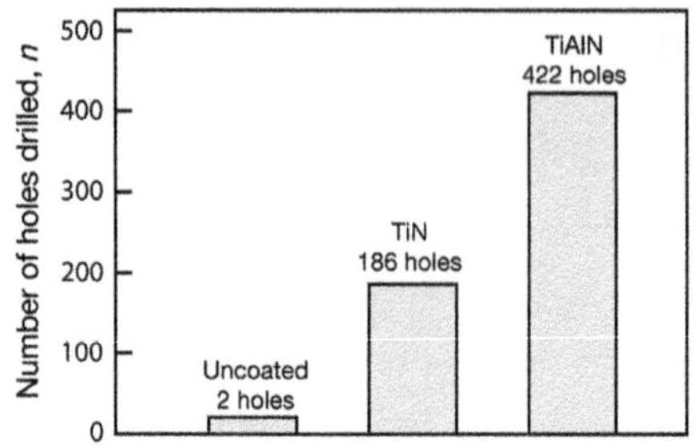

Fig. 7.21. Comparison of average drilling performance of coated and uncoated drills when drilling 4150 alloy steel with a 6.35 mm diameter drill (data from Randhawa *et al.*, 1988).

microfilm at its surface when it is heated, thus preventing diffusion of oxygen to the coating (Leyendecker *et al.*, 1989). Double lifetimes for (Ti,Al)N coated drills compared to TiN coated ones have been reported from drilling tests with different steel work materials (Randhawa *et al.*, 1988; Leyendecker *et al.*, 1989; Ronkainen *et al.*, 1991b; Monaghan *et al.*, 1993; Santos *et al.*, 2004). The biggest lifetime improvements were observed when drilling cast iron. In that case (Ti,Al)N coated drills had lifetime increases of two to ten times those of TiN coated drills (Leyendecker *et al.*, 1989; Roos *et al.*, 1990). Further lifetime improvements in drilling tests have again been reported for TiAlN coatings with MoS_2 incorporated in the coating (Pflueger *et al.*, 2000).

Co-deposition of molybdenum disulphide and chromium for 1 μm thick MoS_2–Crx% coatings on high speed drills showed that an addition of chromium gave the best result in drilling tests with twofold lifetime improvements compared to uncoated drills (Kao, 2005). Up to sevenfold lifetime improvements were achieved with graphitic hydrogen-free, amorphous carbon–chromium coatings (Cr:a–C) on drills when drilling aluminium alloy without lubricant at feed rates of 0.127 to 0.191 mm/rev (Field *et al.*, 2004).

TiN coatings give good improvement in machining most ferrous materials, but they do not offer much improvement with stainless steel as work material. DLC is normally unsuitable for machining ferrous alloys since the coating will react with carbon in the alloy when the temperature exceeds 350°C. However, since the machining of stainless steel has to be carried out at low feeds and speeds, the machining temperature will be relatively low. In drilling tests with DLC with 5% titanium (Ti:C–H) coated drills of stainless steel the average lifetime of the drills was increased more than fourfold compared to TiN coated and uncoated drills (Monaghan *et al.*, 1993).

Some of the most marked improvements with coated drills have been achieved when the drill geometry has been modified to take maximum benefit from the coating, i.e. in terms of increased penetration rates and speeds (Matthews, 1988c, 1989).

7.5.4.3 Milling cutters

In milling the cutting process is discontinuous and the cutting speeds are normally relatively low. The interrupted cutting results in a considerable variation of the maximum tool temperature of several hundred degrees centigrade during single milling cycles. Tönshoff and Denkena (1991) calculated that the rake face temperature varies during one milling cycle from 200 to 900°C when cutting cast iron with ceramic inserts at a cutting speed of 17 m/s.

Lifetime improvements by factors of 3 to 10 have been reported when using TiN coated tools compared to uncoated (Amberg, 1984; Vogel, 1989; Spur *et al.*, 1990). Significant cost benefits through greater productivity can be achieved because the cutting speed at a given feed rate per tooth can be doubled by using TiN coated tools. Full performance is achieved even after regrinding because no crater wear occurs at the cutting speeds normally used. The main wear mechanism is a continuous growth of flank wear by adhesion. As the cutting time increases, fracture of the coating at the rake face of the tool may take place, as shown in Fig. 7.22. This has been observed by Spur *et al.* (1990) when milling ductile materials by PVD TiN coated HSS indexible inserts. The result is a kind of self-sharpening effect.

Interestingly, similar phenomena have been observed for cutting blades such as knives by Matthews and Leyland (1990) and for metal stamping tools by Franklin (1990).

The lifetime of milling tools is greatly improved by TiN coated tools both in dry cutting and with a cutting fluid, as shown in Fig. 7.23. In steel milling tests König *et al.* (1992) found that the improvement of tool lifetime is slightly less with (Ti,Al)N coated tools compared to TiN coated ones, while Ti(C,N) coated tools showed the best wear resistance and had a lifetime three times longer than the TiN coated tools. Coating thicknesses in the range of 2 to 10 μm were investigated by Bouzakis *et al.* (2003) and they found the best performance in milling tests for relatively thick ($Ti_{46}Al_{54}$)N coatings with coating thicknesses of over 6 μm. Nordin (2000) found an up to 30% improvement in lifetime with TiN/TaN mulilayer coated tools compared to TiN coated.

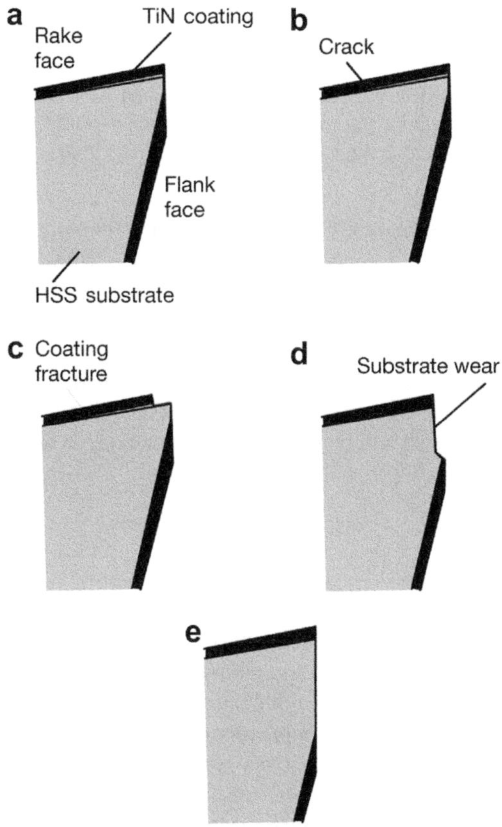

Fig. 7.22. Mechanisms of tool wear that results in self-resharpening when milling ductile materials with TiN coated inserts (after Spur *et al.*, 1990).

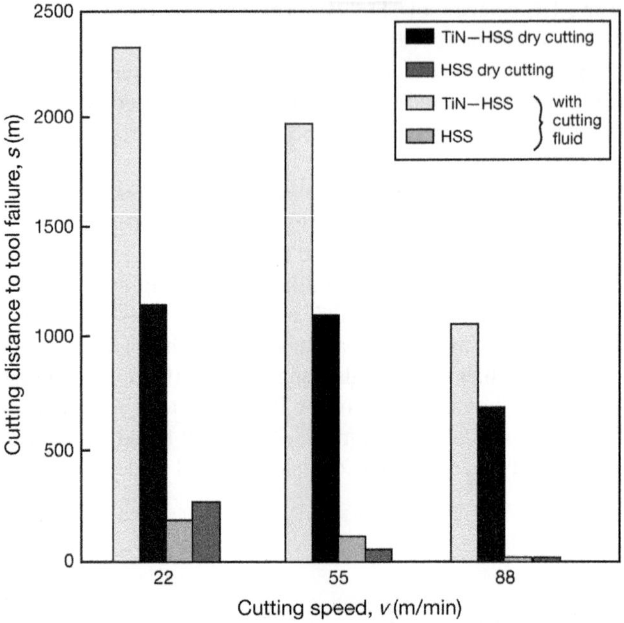

Fig. 7.23. A comparison of tool life obtained with uncoated and TiN coated HSS inserts in milling of a soft magnetic highly ductile material (Mumetal). Cutting distance to tool failure as a function of cutting speed for dry cutting and with cutting fluid (data from Spur *et al.*, 1990).

For machining of non-ferrous materials such as aluminium, aluminium–silicon and copper alloys, graphite and calcium silicate, tool lifetime improvements of two- to 18-fold have been reported when using CVD diamond coated tools compared to cemented carbide inserts (McCune *et al.*, 1989; Söderberg, 1991; Lux and Haubner, 1992; Quinto, 1996). The highest level of lifetime improvement was observed by Kikuchi *et al.* (1991) in face milling of Al–12%Si cylinder heads of automobile engines, when the CVD diamond coated cemented carbide tools had an 18-fold lifetime compared to uncoated. Quinto (2007) has presented the graphical representation shown in Fig. 7.24 illustrating how coatings for milling cutters should be selected for the workpiece material, taking account of the cutting speed and force.

7.5.4.4 Gear cutting tools

Titanium nitride coatings have been very successful for gear cutting tools. This is, to a great extent, because higher cutting speeds, over 120 m/s, can be used with coated tools. Lifetime improvements by factors of 3 to 8 for TiN coated tools (Vogel, 1989) and factors of 5 to 8 at double productivity, in terms of cutting speed and feed, have been reported for (Ti,Al)N (Zlatanovic and Munz, 1990). The use of increased cutting speeds changes the dominating wear mechanism from flank wear to crater wear. This results in a decreased lifetime improvement after resharpening of the tools. When the initial lifetime improvement for TiN coated tools has been observed to be 6 to 8 times, the improvement was on average only three times after resharpening of the tools (Vogel, 1989). A review of gear hob improvements by coatings was given in Matthews (1989), and subsequent reports have indicated steadily increasing performance, especially as the coating and gear producers work together on optimizing geometries and coatings for particular workpiece materials, in parallel with milling cutters.

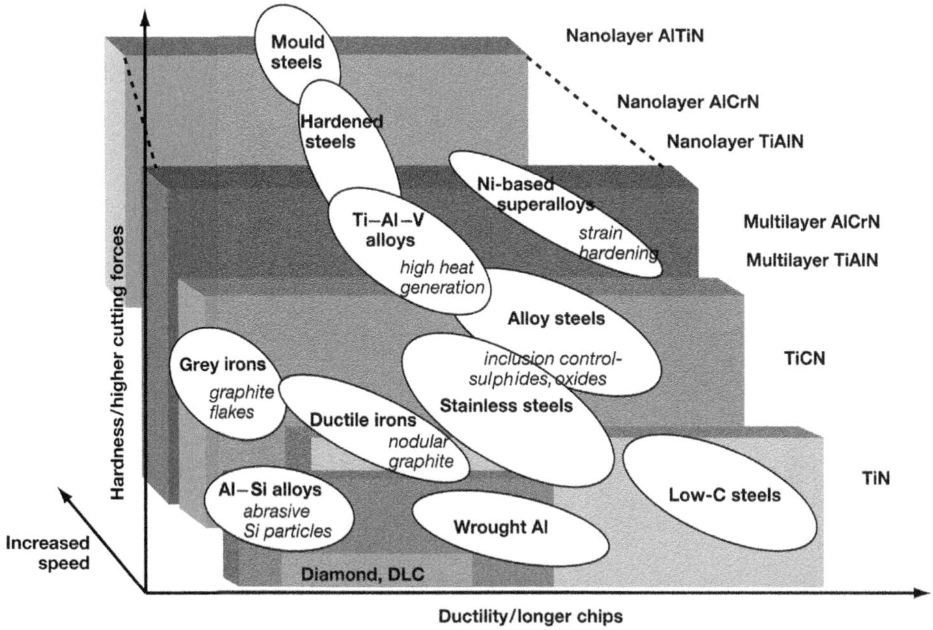

Fig. 7.24. Coating selection guide for milling cutters based on field performance and commercial acceptance. The multilayer versions are preferred in continuous cutting (Quinto, 2007, with permission).

7.5.4.5 Taps and reamers

Lifetime improvements by a factor of 1.5 to 6 have been observed when using TiN coated taps or reamers. The cutting cannot be carried out at high speeds so no additional improvement in productivity is likely to be achieved. However, especially in tapping of mild steels, the coating reduces cold welding, and in tapping of stainless steel the coating lowers the friction because the cutting forces are better distributed over the cutting edge (Bruno *et al.*, 1988; Vogel, 1989).

7.5.4.6 Wood cutting tools

The cutting conditions in wood and pulp product cutting are abrasive due to hard debris in the work material. Early experiments showed that TiN on tungsten carbide tipped circular saws can reduce wear on the rake face by 50% when sawing hardboard (Osenius *et al.*, 1987). No influence on wear was recorded when sawing particleboard, paperboard and plywood. Chrome nitride coating turned out to perform better in wood cutting compared to TiN, (Ti,Zr), Cr and W–C:H (DLC) (Djouadi *et al.*, 1999). Very promising investigations have related to the use of different compositions of chrome nitride often deposited on nitrided surfaces to form a duplex coating (Djouadi *et al.*, 2000; Chekour *et al.*, 2003; Nouveau *et al.*, 2001, 2003; Labidi *et al.*, 2005).

Chromium nitride (Cr_xN) coating on nitrided low alloy steel knife surfaces resulting in a duplex-type surface gave an increase in lifetime of more than 2.5 times in accelerated wear tests cutting medium density fireboard (MDF), which is an abrasive and relatively homogeneous wood derivative material (Beer *et al.*, 2003). One to 2 μm thick chromium and chromium nitride coatings have been used in wood cutting in milling-type cutting tests (Noveau *et al.*, 2005). A twofold lifetime increase in tool life has been reported for ZrBN coated tools (Chala *et al.*, 2005) and 2.5-fold lifetime improvements for ta-C and TiC/a-C DLC coated tools (Endler *et al.*, 1999; Kaminski *et al.*, 2005).

Table 7.1. Cutting tool improvements by hard coatings (mainly TiN). The data is collected from Hinterman (1983a), HMSO (1986), Fenske *et al.* (1988), Leyendecker *et al.* (1989), Posti and Nieminen (1989a, b), Vogel (1989), Matthews and Leyland (1990), Spur *et al.* (1990), Zlatanovic and Munz (1990), König *et al.* (1992), Monaghan *et al.* (1993), Goller *et al.* (1999), Pflueger *et al.* (2000), Cosemans *et al.* (2003), Field *et al.* (2004) and Kao (2005).

	Extended lifetime (times)	Surface quality	Improved cutting performance	Performance after resharpening	Work material/ Comments
1. Turning tools	2–10	Improved	Increased cutting speed and feed	Decreased	Steel. Aluminium: life-time twofold for (Ti, Al)N compared to TiN. Non-ferrous: lifetime 3–20-fold for diamond compared to CC and SiN.
2. Twist drills	4–20	Significantly improved	Drop in feed force. Increased cutting speed	Slight decrease	Carbon, case-hardened, tool and stainless steels. Cast iron
3. Milling cutters	3–10	Improved	Doubled cutting speeds	Slight decrease	Ductile magnetic materials. Threefold longer lifetime for Ti(C, N) compared to TiN
4. Gear cutting tools	3–8	Improved	Increased cutting speeds	Decreased	Carburizing steel
5. Taps	4–10	Improved	Increased cutting speeds	Retained	Carbon and stainless steel. Cast iron
6. Reamers	3–6	Improved	Increased speed	Improved	Aluminium. Carbon, stainless and high alloy tool steels. Cast iron
7. Counter bores, countersinks	2–10	Improved	Increased speed	Decreased	Carbon and cast steel. Ni–Cr castings
8. Broaching tools	3	Significantly improved	Increased speed	Retained	Steel
9. Punching tools	3–100	Significantly improved	Burring dramatically reduced	Decreased	Carbon steel. Spring steel. Nickel, silver and Cu–Be bronze. Soft iron for magnets
10. Saws	20	Improved	Reduced force	Decreased	Stainless steel. Ni alloys. Metal/plastic compounds
11. Knives, blades	2–500	Improved	Improved	Retained	Fabrics, foods, wood

Significant improvements in wear resistance was reported for CVD diamond coated tools in milling tests and machining fibreboard and particleboard (Endler *et al.*, 1999; Sheikh-Ahmad *et al.*, 2003).

7.5.4.7 Other cutting tools

Improved lifetimes, typically by factors of 2 to 5, have been achieved with other tools, such as boring tools, die-sinks, broaches and punching tools. Extremely high lifetime improvements with TiN coated tools have been reported for circular saws in cutting metal–plastic compounds with an extended lifetime of 20-fold and for punches for battery case production of 100-fold (Hinterman, 1983a; Vogel, 1989).

Knives are another important cutting application area for hard coatings. Matthews and Eskildsen (1993) report a 500-fold improvement in life between sharpening for a knife used for cutting multiple layers of synthetic fabric, when coated with diamond-like carbon.

Matthews and Leyland (1990) achieved considerable improvements on knives used in the packaging industry by coating them on one side with hard metal carbide, nitride and boride compounds, such that the coating had a dense columnar structure. This morphology is beneficial with respect to sharpening since it encourages a clean coating fracture between columns as the blade wears or is reground, analogous to the effect shown in Fig. 7.22 for milling cutters.

This structure also results in the effective adhesion being enhanced by ensuring that the adhesive bond to the substrate is greater than the lateral cohesive strength within the coating. A similar effect has been reported for turbine blade coatings (Fancey and Matthews, 1986; James and Matthews, 1990).

A summary of improvements in cutting tools by the use of thin hard coatings is presented in Table 7.1.

7.6 Tools for Forming

7.6.1 Description of the application

Forming of a solid work material is achieved through major plastic deformation along the forming tool surface. The quality of forming and metal-working from tools, such as forming dies, punches and rolls, is considerably influenced by the wear behaviour and the friction at the tool surface. The tribological situation is typically characterized by high surface pressures, low deformation velocities, sometimes high temperatures, and, depending on the lubrication, a variety of different friction conditions. Important criteria for tool material selection are a low tendency of adhesion to the workpiece in order to avoid adhesive tool wear and high hardness and sufficient toughness to avoid abrasive wear (HMSO, 1986; Matthes *et al.*, 1990b; Schmid and Wilson, 2001). Unlike metal cutting, texturing of the tool surface is frequently used to minimize wear, control friction and stabilize the forming process (Groche and Köhler, 2006). Furthermore, coatings on the product, such as paints on sheet metal, can be used to beneficially influence the tribological behaviour in the forming process.

7.6.2 Improvements by surface coatings

There are several published reports concerning major improvements in the lifetime of forming tools, with increases of up to 100 to 200 times, by the use of hard and thin surface coatings. However, very few of these articles describe and analyse in detail the tribological mechanisms involved, though it is obvious that the ability of the coating to decrease adhesion to the workpiece and to resist abrasive wear and galling are key parameters.

The selection of a proper tool coating to be used in metal forming depends greatly on the type of material to be formed. It is especially the resistance of the surface to galling that is critical. Podgornik *et al.* (2006b) studied PVD-deposited TiN, TiB_2, VN, TaC and DLC coatings on cold work tool steel in a cold forming simulating load-scanning test against austenitic stainless steel and alloys of aluminium and titanium. They found the following:

1. In the case of austenitic stainless steel, low friction materials, i.e. carbon-based coatings, give low and stable friction and excellent protection against galling. At the same time polishing of the surface reduces the probability of stainless steel transfer and improves the surface quality of the workpiece.
2. For very soft metals with a high degree of plastic deformation during forming, such as aluminium and aluminium alloys, nitrided forming tool steels and nitride-type coatings show a good resistance against galling and a relatively stable friction. Similar results were observed for DLC coated surfaces.
3. Materials that show a high tendency for galling and transfer layer reception, i.e. titanium and titanium alloys, started to adhere to the tool surface at very low loads, regardless of the coating used. Only plasma nitriding and VN coatings gave some improvement in galling performance as compared to untreated steel. In this case, and in the case of aluminium alloys, polishing of the steel surface, coated or uncoated, had practically no influence on the galling performance.

Drawing tools, dies and punches PVD coated with TiN, (Ti,Al)N, Ti(B,N) or TiB_2 films with thicknesses of about 3 to 5 μm have been reported to achieve lifetime improvements by factors of 2 to 10 (Sundquist *et al.*, 1983; Straede, 1989; Vogel, 1989; Matthes *et al.*, 1990b; Bhushan and Gupta, 1991; Björk *et al.*, 2001; Klimek *et al.*, 2003). For dies and punches, as well as for deep drawing tools, the use of 7 to 10 μm thick CVD-deposited TiC or TiN/TiC coatings has resulted in high lifetime improvement factors of 100 to 200 (Hintermann, 1979, 1983a; Bhushan and Gupta, 1991). Significant improvements in tool life with DLC and MoS_2:Ti coated tools in cold forming has also been reported (Field *et al.*, 2004; Amaro *et al.*, 2005).

It has been claimed to be more difficult to achieve lifetime improvements for forming tools operating in hot working conditions, according to work by Wiiala *et al.* (1987). In their simulating wear test experiments the coefficient of friction at 20°C, 500°C and 900°C was slightly higher for both TiN and TiAlN ion-plated tool material samples, compared to uncoated in both dry and lubricated conditions. Damage to the tool was initiated by cracking of the coatings, which immediately resulted in micro-pickup of work material along the cracks and the formation of large strongly adherent friction welded regions.

Groche and Köhler (2006) cite four classes of tool coatings which can be beneficial for unlubricated sheet forming: (1) sulphide/selenide films (e.g. MoS_2), (2) carbon films (e.g. Me–C:H), (3) boron films (e.g. hexagonal BN) and (4) oxide films (e.g. WO_x, Al_2O_3).

7.7 Erosion and Scratch Resistant Surfaces

7.7.1 Description of the application

There are many applications where surfaces are exposed to hard particles moving against and alongside the surface, resulting in plastic deformation and fracture of the surface seen as scratches, cracks, surface destruction and material removal. The particles may be carried by a gas or a liquid. The particle dimensions have a very large range, from large rocks in earth moving equipments to microsize particles bombarding optical instruments in space. In erosive wear, particles moving in gas or liquid bombard the surface and result in material removal, Fig. 3.1d. If the particles are pressed against the surface by another solid, three-body abrasive wear (see Fig. 3.11b) limits the endurance life of, for example, excavators, transporters, transfer tubes for bulk transportation, pipelines, valves, fluidized bed combustion systems, aeroplane turbine engine components, optical lenses, spectacles, displays, car chassis, etc.

The basic tribological mechanisms of three-body abrasion and erosion are in many respects alike and similar principal surface design solutions can often be applied. A commonly used relationship for erosion is that the erosion rate increases with increasing modulus of elasticity but decreases with increasing hardness and increasing fracture toughness. The angle of impact has, however, a considerable influence. Ductile materials exhibit their maximum erosion rate at impact angles between 15° and 30°, whereas brittle materials exhibit a continuously increasing erosion rate with increasing impact angle and the maximum occurs at 90° (Hillery, 1986; Olsson *et al.*, 1989; Hutchings, 1992; Davis, 2001).

7.7.2 Improvements by surface coatings

Thin hard coatings can be used on components for erosion and scratch protection. The impact angle as well as the shape of the particles have an influence on what kind of surface layer provides good erosion resistance. In general, harder and thicker coatings are more favourable at small impact angles where the wear mechanism is more abrasive.

Fatigue wear and brittle fracture dominate at impact angles close to 90°. If the impact mass is small a thicker hard coating can give good protection, but if the impacts cause material deflection a brittle coating may deflect, crack and fail. A thinner coating may, on the other hand, deflect with the substrate and provide scratch protection on the surface while the elastic substrate absorbs the impact energy.

The shape of the particles may also influence wear properties. Burnett and Rickerby (1988) showed that 10 μm thick PVD TiN coatings are more resistant to angular particle erosion while 3 μm thick TiN coatings have longer lifetimes when exposed to blunt particles.

7.7.2.1 Coated steel and ceramic surfaces

Hard surface coatings can offer good abrasive particle wear protection of steel surfaces. However, a direct correlation between the coating hardness and wear rate was not found in three-body TiO_2 particle abrasive wheel testing carried out by Van Acker and Vercammen (2004), except for very similar diamond-like carbon coatings, as seen in Fig. 4.86. The hard TiN, CrN, TiCN and DLC coatings had 25 to 200 times lower wear compared to the uncoated steel substrate and a hard chromium surface, while the lowest wear was measured for the hard DLC coating. A soft DLC coating showed very high wear, three times more than the uncoated steel surface.

It has been predicted and experimentally demonstrated that the optimum solution to abrasive and erosive wear problems might be in multilayered coatings such as a PVD titanium/titanium nitride (Matthews, 1980; Leyland and Matthews, 1993) or CVD tungsten/tungsten carbide (Garg and Dyer, 1993), combining hardness and toughness. This seems to depend on the particular case of application and surface design. Iwai *et al.* (2001) carried out a slurry jet type of erosive testing of several coated surfaces, both single and multilayers. All tested coatings, TiN, TiN/TaN, TiN/CrN, TiN/NbN and TiN/TiAlN, proved to have 2 to 10 times lower erosion wear rate than the uncoated high-speed steel surface. The single layer homogeneous TiN and the multilayered coating TiN/TiAlN with coarse lamellae of 0.1/0.1 μm layer thickness proved to have higher erosion resistance than the TiN/NbN, TiN/TaN and TiN/CrN multilayers with thin lamellae of 0.01/0.005 μm layer thickness.

Good erosion resistance has been obtained in laboratory experiments with single and multilayer CVD-deposited TiN, TiC, Al_2O_3 and $B_{13}C_2$ coatings on cemented carbide substrates. Of these coating materials $B_{13}C_2$ seems to be superior compared to the others, as shown in Fig. 7.25. The coating thicknesses were in the range of 2 to 11 μm (Olsson *et al.*, 1989; Shanov *et al.*, 1992). In a comparison of different 12 to 56 μm thick single and multilayer tungsten-based coatings on a titanium substrate Gachon *et al.* (1999) measured with a sand particle blast erosion device that the single layer W–N

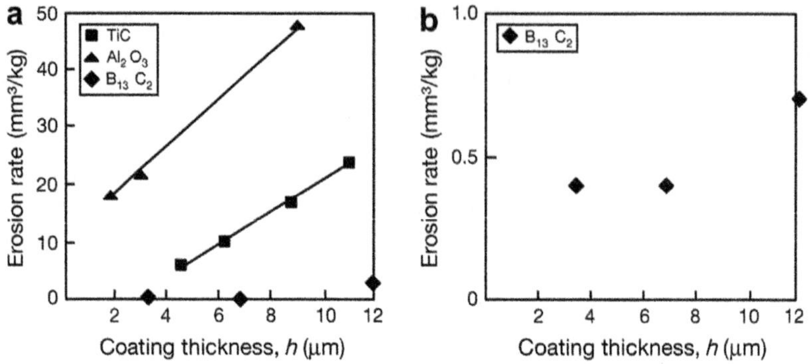

Fig. 7.25. Variation of erosion rate as a function of coating thickness for (a) CVD-deposited titanium carbide, aluminium oxide and boron carbide coatings; with (b) a close-up of the boron carbide part of the curve (data from Olsson *et al.*, 1989).

coated surfaces had as low an erosion rate as the multilayer surfaces when 250 μm grain size was used but they lost their wear resistance efficiency when the grain size was increased to 600 μm.

At elevated temperatures Sue and Troue (1991) observed that the erosion rate at 90° impact angle of PVD arc evaporated, 26 μm thick TiN, and 22 μm thick ZrN, increased markedly with increasing temperature above 316°C and 538°C, respectively. They suggest that this increase in erosion with temperature is related to changes in coating hardness, residual compressive stress and toughness.

Severe erosion of gas turbine components, such as compressor blades, can occur when sand and dust particles are ingested by aircraft turbine engines. CVD-deposited TiC–N hard coatings deposited on stainless steel blades have been found to give erosion resistance improved by factors of 70 to 80, when compared to those plasma sprayed with tungsten carbide (Yaws and Wakefield, 1973). Other coatings applied to compressor components are CVD-deposited TiB_2 and SiC, PVD-deposited MoB, B_4C and TiB_2 and sputtered $NiCr_3C_2$, TiB_2 and $NiTiB_2$ (Ramalingam, 1980; Hillery, 1986; Bhushan and Gupta, 1991). Other investigations show that turbine blades can also be successfully protected against erosive wear by 75 to 100 μm thick nickel aluminide diffusion coatings, 200 to 300 μm thick plasma-sprayed Y_2O_3-stabilized ZrO_2 coatings and 20 to 25 μm thick air–plasma-sprayed MCrAlY-type overlay coatings (Sivakumar and Mordike, 1989).

The erosion and corrosion resistance of tools for moulding reinforced plastics, especially screws used in injection moulding, have been improved using PVD and CVD coatings of TiN, TiC and TiC/TiN layers. A typical life improvement was a factor of 2.5, with improved anti-stick properties (Franklin, 1990).

CVD diamond coatings in the thickness range of 15 to 47 μm gave erosive wear protection to the W and SiC substrate surfaces giving erosive wear rates six to several hundred times lower than uncoated surfaces (Wheeler and Wood, 1999). The typical wear process was described as a three-stage process: initially microchipping, followed by nucleation of pin-holes and debonding areas at the coating/substrate interface, and finally failure of the coating.

In some applications, such as in the scratches produced by car wash brushes on car clearcoats, that is 40 to 50 μm thick thermoset resins which form the outer layer of automotive body coatings, the crucial wear is not measured as wear volume but rather estimated from the visible damage caused on the clearcoats. Bertrand-Lambotte *et al.* (2002) found that the scratch resistance in this respect can be improved by a decrease of the brittle scratches and favouring healing of ductile scratches. PVD coatings are also increasingly used for decorative purposes, such as spectacle frames, pens, watches, door hardware, bathroom taps, etc., where their visual appearance is crucial (Hurkmans *et al.*, 1997; Van der Kolk *et al.*, 1998).

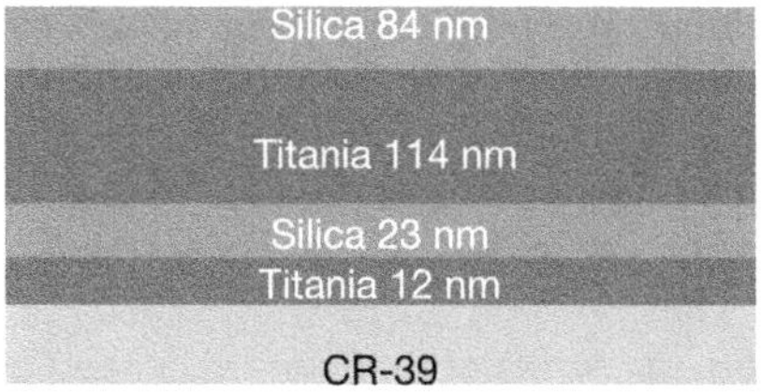

Fig. 7.26. A typical four-layer SiO_2/TiO_2 anti-reflection coating design on plastic (thermoset cast CR-39) eyeglass lenses (after Colton, 1997).

7.7.2.2 Coated glass and polymer surfaces

In optical applications there is often a requirement for scratch and erosion resistance of the lenses. These are traditionally of glass (Pulker, 1984, 1999; Hill and Nadel, 1999; Klafke and Beck, 2000) but especially in eyewear the majority of the lenses today are made of polymer composition, due to light weight and toughness requirements. Injection moulded polycarbonate and thermoset cast (CR-39) are widely used for a good combination of transparency, light weight and toughness. They are often used with anti-reflection (AR) coatings of four layer design with SiO_2, as a high index layer, and TiO_2 or alternatively ZrO_2 or Nb_2O_2, as a low index layer. The deposition is by electron beam or sputtering. The scratch resistance of CR-39 has been further improved by applying in addition a 3 μm thick layer of silica on the lens by using plasma-enhanced CVD (Colton, 1997). A typical multilayer coating design is shown in Fig. 7.26.

The scratch resistance of polycarbonate has been improved by applying a few micrometres thick sol-gel coatings (Charitidis *et al.*, 2004) and by ion implantation by high-energy ion irradiation, resulting in one order of magnitude better abrasion resistance (Bhattacharya, 1998).

DLC is infra-red transparent and can thus be used as an anti-reflective, scratch-resistant and wear protective coating for optical lenses made of Ge, ZnS and ZnSe (Erdemir and Donnet, 2001). DLC can be used to protect polycarbonate sunglass lenses from abrasion because it has a low deposited temperature and thus is suitable also for plastic products.

Improved erosion resistance of polyimide has been achieved by covering it with a sputtered 20 to 30 μm thick ZrB_2 surface coating deposited on a chromium interface layer (Chambers *et al.*, 1990).

7.8 Oscillating Contacts

7.8.1 Description of the application

A mechanical assembly that is connected to machinery, transmissions, transportation or other dynamic devices is often exposed to vibration from oscillating forces. The contacts that experience oscillating tangential movements need special attention. When the oscillating movement in the contact has larger amplitude, typically more than several hundred micrometres, it is tribologically classified as a sliding reciprocating contact and considered as a sliding contact.

If the amplitude is smaller, in the range of a few tens of nanometres to a 100 micrometres, the tribological mechanism is classified as *fretting*. In this section we will concentrate on the special case of fretting contacts. Examples of components where fretting typically occurs are shrink fits, bolted parts and splines. Hubs, flexible couplings, bearing houses on loaded rotating shafts, gas turbine blade root connections and fuel nozzles are particularly prone to fretting damage (Bhushan, 1999b; Kapsa *et al.*, 2005).

Fretting is a complex combination of adhesive, abrasive, tribochemical and fatigue wear. The normal load causes adhesion between asperities and the oscillatory movement generates surface fracture and wear debris. Because of the small amplitude, the wear debris stays in the contact for some time and thus has a role in separating the surfaces from each other, as shown in Fig. 3.11d. The new virgin surface produced by wear is exposed to the environment and chemical reactions take place.

Often, fretting is combined with corrosion whereby it is called *fretting corrosion*. When two steel surfaces slide against each other in oscillatory motion in air the freshly worn surfaces oxidize to Fe_2O_3 and a characteristic reddish-brown powder is produced. The oxide particles are hard and abrasive. The oscillatory movement may be so small that the contact experiences it as cyclic stress, which results in initiation of fatigue cracks and a damage called *fretting fatigue*. Fretting wear can be reduced by design changes which reduce oscillatory movement and stresses, or eliminate the moving contacts.

The limit between fretting and reciprocating sliding wear in a ball-on-flat contact is illustrated by Fig. 7.27a. The two cases of fretting, partial and gross slip, are situated in a domain with displacement amplitude/contact radius $= e < 1$, corresponding to a situation where a part of the wear scar of the flat is always in contact with the counterface. Reciprocating sliding wear corresponds to a situation where $e > 1$ and the entire wear scar is regularly out of contact and totally exposed to the environment. Four types of fretting contacts based on the motion type can be distinguished, as shown in Fig. 7.27b.

The fretting contact mechanisms have been analysed in detail using a contact energy approach by Fouvry and Kapsa (2001), Kapsa *et al.* (2005) and Fouvry *et al.* (2003, 2004a), and FEM modelling and simulation of the contact conditions has been calculated (Fouvry *et al.*, 2003; McColl *et al.*, 2004; Dick and Cailletaud, 2006b). The energy density approach has been extended to include coated surfaces by Fouvry *et al.* (2006).

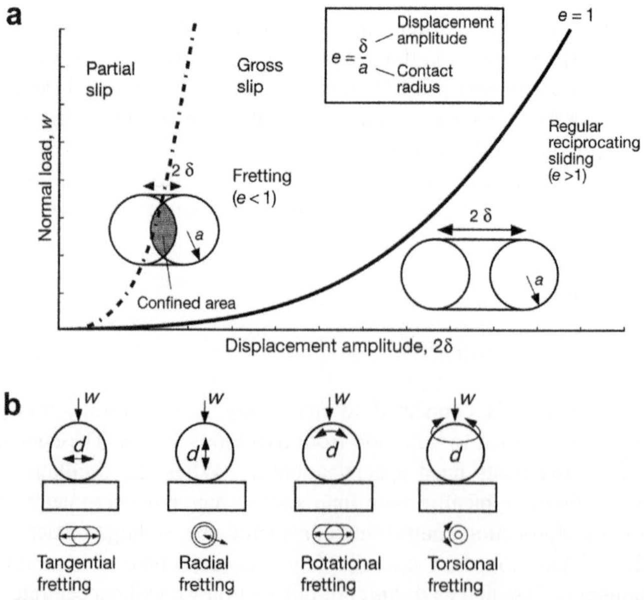

Fig. 7.27. (a) The regions for fretting and reciprocating sliding shown as a function of normal force and displacement amplitude in a ball-on-flat contact. (b) The four fretting modes in a ball-on-flat contact (after Kapsa *et al.*, 2005).

7.8.2 Improvements by surface coatings

One way of controlling fretting is to apply surface coatings on one or both of the surfaces in contact. With a coating it is possible to reduce the friction and thus the surface stresses to change the mechanical properties of the top layer resulting in changed stresses and surface toughness and to modify the chemical interaction conditions between the surfaces and the environment. Coatings can bring beneficial residual stresses to the top surface that inhibit crack propagation and wear debris generation.

In a fretting contact with a coated surface the surface wear process often proceeds in two phases. First, there is mild wear of the top part of the coating but as the coating gets thinner the surface loading is influencing the whole coating/substrate system and at some critical stage the coating is destroyed and delaminated. This process has been elegantly described by Liskiewicz and Fouvry (2005) and demonstrated by a 12.7 mm radius alumina ball sliding against a 1.6 μm TiC coated high-speed steel surface at 100 N normal load and ±50 and ±100 μm displacement gross slip amplitude with constant 5 Hz frequency and 50% relative humidity. Based on experimental observations and theoretical contact and contact energy analysis they conclude that the lifetime is not only related to the progressive wear mechanisms. It is also a function of more severe coating destruction mechanisms that ultimately drastically reduce the coating lifetime. They use a friction dissipation parameter, which is a unique parameter that takes into account the major loading variables, which are the pressure, sliding distance and friction coefficient.

As wear of the coating top surface progresses and the coating becomes thinner, the cyclic stresses imposed through the coating/substrate interface are increased. After reaching a threshold stress value, i.e. below a critical residual coating thickness, a severe decohesion is activated resulting in coating spalling and general surface failure. According to this hypothesis, damage evolves as progressive wear controlled by friction energy, followed by a quasi-instantaneous decohesion controlled by a critical overstressing of the coating/substrate interface, as shown in Fig. 7.28.

Liskiewitz and Fouvry (2005) further, very interestingly, calculate at different fretting cycles the ratio between the experimental coating lifetime and the predicted endurance extracted from the wear depth modelling and receive a fairly constant value of 0.8. This indicates that the decohesion phenomenon seems to be activated, in the studied case, for a constant residual coating thickness of 0.3 μm, which is related to a constant threshold of interface stress. The model enables the calculation of the endurance life of a coated surface in a fretting contact.

There are a few coating material fretting performance comparison studies published. Liskiewicz *et al.* (2005a) measured wear for a set of 1.6 to 4 μm thick hard PVD mainly titanium-based coated

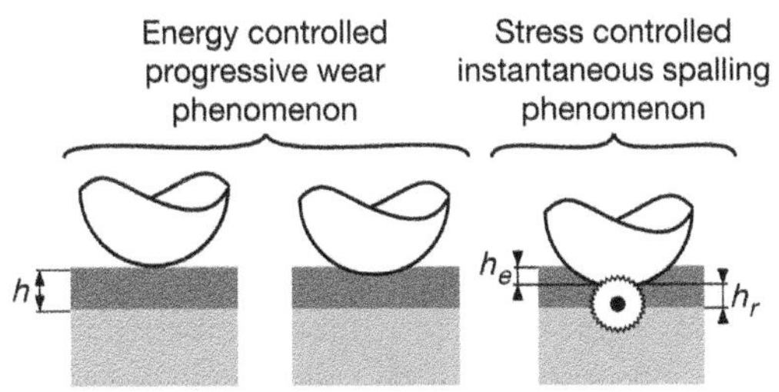

Fig. 7.28. Illustration of a coating fretting wear scenario involving first a progressive energy controlled wear mechanism followed by a stress controlled instantaneous decohesion of the coating. h_e = effective worn coating thickness; h_r = residual coating thickness at instant of spalling (after Liskiewitz and Fouvry, 2005).

high-speed steel surfaces sliding against an alumina ball at 15 to 150 N normal loads and with 50 to 200 μm sliding amplitude. The best fretting wear resistance was measured for VC followed by Ti/VC, $(TiC/VC)_{x2}$, TiC, TiCN and TiN, as shown in Fig. 7.29. Interestingly, the ranking does not correlate with the coating hardness but it directly correlates with their Young's modulus, the lowest modulus value giving best wear resistance.

The influence of friction in the fretting wear process is important. This is well illustrated by the study of Klafke (2004c) where he measured both the friction and the wear in fretting contacts with TiN, Ti(C,N), TiAlN and NiP coated high-speed steel surfaces, with and without a 2 μm thick DLC (a:C:H) top layer, in oscillating sliding against a 5 mm radius Al_2O_3 ball. The normal load was 10 N, the amplitude 200 μm and the frequency 20 Hz in non-lubricated conditions at 3, 50 and 100% relative humidity. The results in Fig. 7.30 show the very clear reduction of friction with the DLC coated surfaces; the coefficient of friction is 2 to 35 times higher without the DLC coating. Similarly, there is a reduction in wear in most cases from 20% to more than one order of magnitude. The friction and fretting wear performance of the four coatings without a DLC top layer is very similar. A reduction in friction and wear with higher relative humidity is also clearly observed.

A very large variety of different coating materials, all from hard diamond, DLC and ceramic coatings to soft and elastic PFTE and rubber coatings, have been studied and their tribological behaviour reported. In most cases some beneficial effects in fretting resistance are reported. However, the tests are typically carried out with different contact and test parameters and comparisons to uncoated or some reference surfaces is in most cases, unfortunately, not reported. This makes it impossible to rank the suitability of the coating materials for fretting resistance. Below is some of the reported coated surface fretting studies listed:

- *Diamond coating* (HF-CVD, $h = 0.9$ to 2.4 μm) on Ti–6Al–4V sliding against steel and alumina. Considerable reduction in friction and wear (Zeiler *et al.*, 2000).
- *DLC coating* (a-C:H, PA-CVD, $h = 2$–3 μm) on steel and on Ti-based coated steel surfaces sliding against steel and alumina. Significant reduction in friction and wear compared to Ti-based coated surfaces. Friction coefficients of 0.01 to 0.1 and wear rates of 0.1 mm^3/Nm have been reported. Friction and wear decreases with increasing humidity (Klafke, 2004c; Klafke *et al.*, 2005a; Wäsche and Klafke, 2008).
- *DLC coating lubricated* (a-C:H, PVD/CVD, $h = 2.7$ μm) in oil coated on steel and sliding against steel and DLC coated balls. DLC coatings reduced friction by more than a factor of 2 and wear by 3 to 10 times compared to steel/steel contacts (Kalin and Vižintin, 2006).

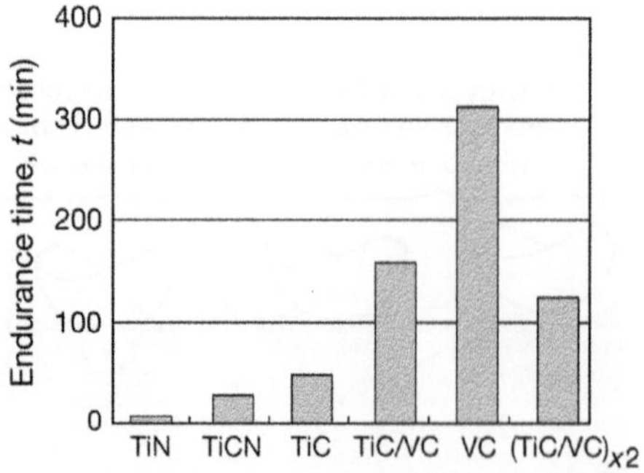

Fig. 7.29. Coated surface durability under fretting wear for a given coating thickness in oscillating sliding against a 12.7 mm radius alumina ball with 100 N normal load, 100 μm sliding amplitude and 5 Hz frequency (after Liskiewicz *et al.*, 2004).

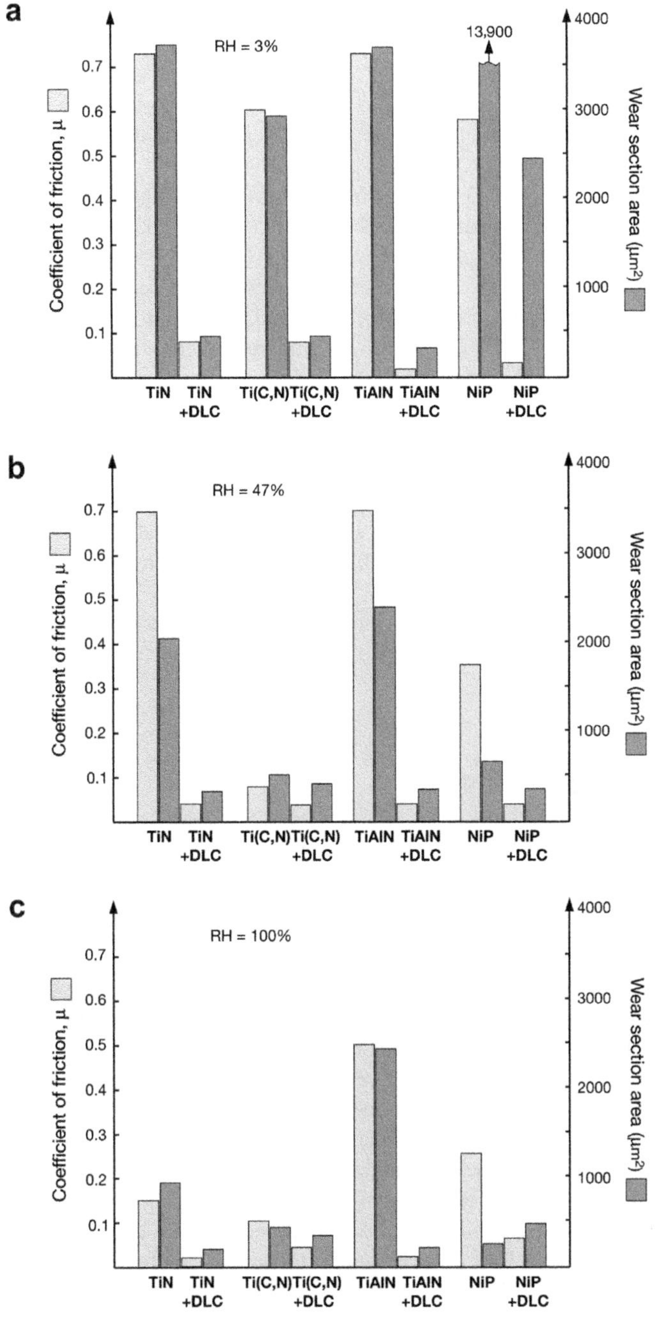

Fig. 7.30. Coefficient of friction and wear after 100,000 oscillating sliding cycles for TiN, Ti(C,N), TiAlN and NiP coated high-speed steel surface, with and without a DLC (a:C:H) top layer sliding against a Al_2O_3 ball in (a) 3%, (b) 44–51% and (c) 100% relative humidity. The wear is given as the section area between the worn wear groove and the unworn surface (data from Klafke, 2004c).

- *TiN coating* (PVD, $h = 1$ to 4 μm) on steel surfaces sliding against alumina. Decrease in friction and wear with increasing humidity (Cosemans, 2003; Zhang *et al.*, 2003a; Klafke, 2004c; Liskiewicz *et al.*, 2004; Caron *et al.*, 2006).
- *TiAlN, TiCN, TiC, VC, TiC/VC, NiP, a-*TiB_2 *coatings* (PVD, $h = 1$ to 4 μm) on steel sliding against alumina (Zhang *et al.*, 2003a; Klafke, 2004c; Liskiewicz and Fouvry, 2005; Liskiewicz *et al.*, 2004, 2005).
- *TiN–*MoS_x *coating* (PVD, $h = 3.3$ to 4 μm) on steel surface sliding against steel and alumina. Five hundred times lower wear compared to TiN (Cosemans *et al.*, 2003).
- *Multilayer TiAlCN/TiAlN/TiAl and TiC/VC coatings* (PVD, $h = 1.6$ and 4 μm) on steel sliding against alumina. Wear reduced with a factor of 3 compared to uncoated surface for the first mentioned multilayer (Fouvry *et al.*, 2004b; Larbi and Tlili, 2006).
- Si_3N_4 *and* SiO_2 *coatings* (PECVD, $h = 0.1$ to 1 μm) on silica and glass sliding against alumina (Klafke and Beck, 2000) and Si_3N_4 *coating* (PECVD, $h = 0.3$ μm) on titanium alloys sliding against titanium alloy balls (Yanito *et al.*, 2006).
- *Cu–Ni–In coating* (plasma spraying, $h = 150$ to 200 μm) on Ti–6Al–4V surface sliding against Ti–6Al–4V (Fridrici *et al.*, 2003).
- MoS_x *coating* (PVD, $h = 0.1$ to 1 μm) on steel sliding against alumina (Zhang *et al.*, 2003a).
- *PTFE coatings* ($h = 6$ to 30 μm) on steel surface sliding against steel. No correlation between coating thickness and endurance life but friction normally decreases with increasing thickness (Langlade *et al.*, 2005).
- *Rubber fluoropolymer coating* (screen-printing, $h = 500$ μm) on steel sliding against steel. Coefficient of friction decreases with increasing load and increases with increasing velocity (Baek and Khonsari, 2005).

7.9 Magnetic Recording Devices

7.9.1 Description of the application

The magnetic recording devices used in computers and other electronic instruments for audio, video and data-processing are tape drives, floppy disk (diskette) drives and rigid disk drives, as shown in Fig. 7.31. The magnetic read–write process takes place simultaneously with relative motion between the magnetic media and a stationary or rotating magnetic head. Under steady operating conditions, the magnetic head flies over the surface of the magnetic media supported by a hydrodynamic air film. At the moments of starting and stopping there is direct contact between the head and the media.

In rigid disk drives and computer tape drives the thickness of the air film is in the range of 10 to 50 nm and the RMS roughness of the head and magnetic medium surfaces ranges from 1 to 10 nm. The areal data densities of commercial tape and rigid disk drives are in the range of 10 to 100 Gbit/in^2. The higher value corresponds to track densities of 6000 to 8000 tracks/mm. Thus the size of a bit dimension for these densities is of the order of 160 nm × 40 nm. This small dimension forms a hard requirement on the defect size limit in the head and medium surfaces. The goal of the immediate future is a storage capacity of 1 Tbit/in^2, which would require flying heights to be reduced to 1 to 3 nm (Bhushan, 1988, 1990, 1996, 2001a, b; Talke, 1991, 2000; Menon, 2000; Carnes, 2004; Gellman and Spenser, 2005).

The need for higher recording densities requires that the flying heights are as low as possible and the surfaces are as smooth as possible. Closer flying heights result in occasional rubbing of high asperities and increased wear. Smoother surfaces will increase the friction and the interface temperatures. High interface reliability through the whole life of the magnetic storage is most important, since interface failure, even on a microscopic scale, can result in loss of valuable data recorded on the magnetic media.

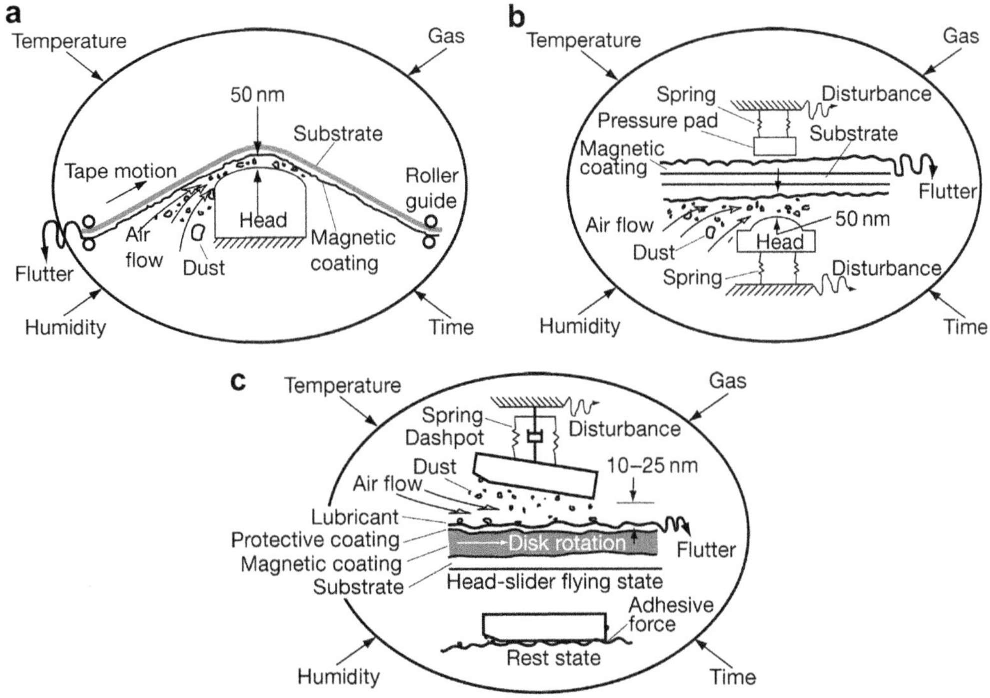

Fig. 7.31. Schematic illustrations, not drawn to scale, of (a) a magnetic head/tape interface, (b) a head/floppy disk interface (c) and a head/rigid disk interface (after Bhushan, 1996).

In a rigid disk drive the recording head is in contact with the disk when the disk is at rest. Thus most of the tribological problems are related to the transition of the head from sliding to flying. At low speeds there is boundary lubrication between the head and the disk. As speed is increased, the air bearing begins to take form and the suspension load is reduced by the air bearing forces, resulting in reduced friction forces between the head and the disk. Finally, when the head has been completely lifted by the air forces no contact between the surfaces occurs. The rotating speed is up to 10,000 rpm when the head flies over the disk surface. When stopping the drive the contact proceeds through the same steps in the reverse order. The coefficient of friction between a slider head and a disk is generally found to increase with the number of start/stop cycles. In order to prevent stiction and failure of the head/disk and head/tape interface surface texturing of either the disk or the head surface has been implemented (Talke, 2000; Zhou *et al.*, 2000; Bhushan, 2001b; Raeymaekers *et al.*, 2007).

The surfaces of most air bearing heads are made of hard materials such as Ni–Zn ferrite, Mn–Zn ferrite, SiC, calcium titanate or Al_2O_3–TiC. Plasma-sprayed coatings such as Al_2O_3–TiO_2 and yttria-stabilized zirconia have conventionally been used for tape drives.

7.9.2 Improvements by surface coatings

Surfaces of the magnetic media in common use are of two types. In *particulate media* magnetic particles are dispersed in a polymeric matrix and coated on to the polymeric substrate for flexible media, such as tapes and floppy disks, or on to the rigid substrate of rigid disks generally of aluminium, glass or glass ceramics, as shown in Fig. 7.32a and b. In *thin film media* continuous films of magnetic material are deposited on to the substrate by vacuum deposition techniques.

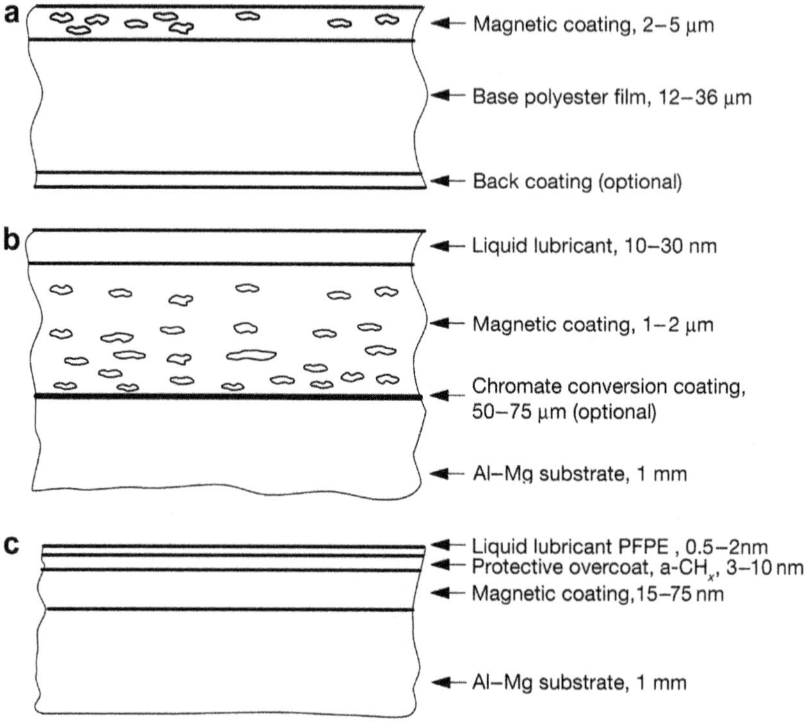

Fig. 7.32. Sectional views of (a) a particulate magnetic tape, (b) a particulate rigid disk and (c) a thin film rigid disk (data from Bhushan, 1996, 2001a; Gellman and Spenser, 2005).

In addition to the magnetic particles, hard ceramic particles of Al_2O_3 are generally added into the polymeric matrix, as shown in Fig. 7.33. The thickness of the magnetic disk coating may be less than 1 μm. Needle-like γ-Fe_2O_3 particles, about 1 μm long and 0.05 μm in diameter are longitudinally oriented in the magnetic coating. The embedded hard Al_2O_3 or Cr_2O_3 particles have a diameter of about 0.2 to 0.3 μm (Bhushan, 2001b).

The hard particles will inhibit strong sticking of the very smooth head and disk surfaces in contact situations and protect against wear. The concentration of particles necessary to provide good tribological properties is, however, critical. As the Al_2O_3 content increases the wear mechanism will change from adhesive wear to more abrasive wear. The optimum content of particles to provide good endurance of the contact has been estimated to be about 8% (Kita, 1991). The influence of the aluminium oxide particle content in the layer on disk wear, counterface wear and friction is shown in Fig. 7.34. The drastic drop of the coefficient of friction in the low particle content region of Fig. 7.34c is believed to be due to decreased adhesion between the surfaces.

A typical coating binder layer for a *recording tape* is 2 to 5 μm polyester–polyurethane on a 23 μm thick polyethylene terephthalate (PET) tape substrate (Bhushan, 1988, 1990, 2001b). A sectional view of a particulate magnetic tape is shown in Fig. 7.32a. The magnetic particles are acicular particles such as γ-Fe_2O_3, CrO_2, Fe or hexagonal platelets of barium ferrite.

In a large screening study performed with an accelerated wear test apparatus, Bhushan (1987) compared the wear resistance and friction of 24 different head material candidates from stainless steel and tool steel to ceramics, cermets and glass as well as coatings deposited by plasma spraying, sputtering and anodizing sliding against magnetic tape. Coating materials were Al_2O_3, Al_2O_3–TiC,

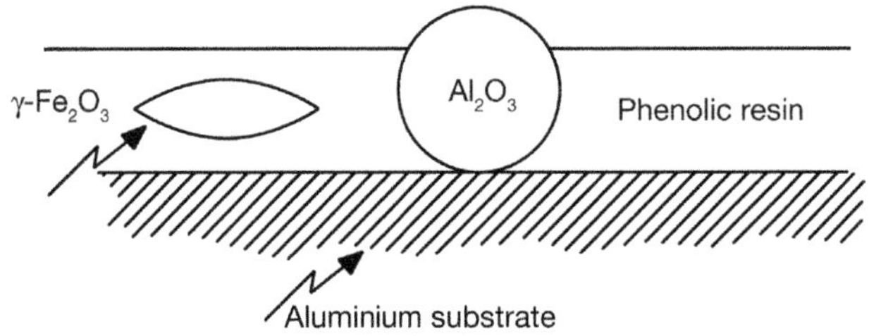

Fig. 7.33. Sectional view of a particulate magnetic disk with needle-like magnetic particles and hard aluminium oxide particles (after Kita, 1991).

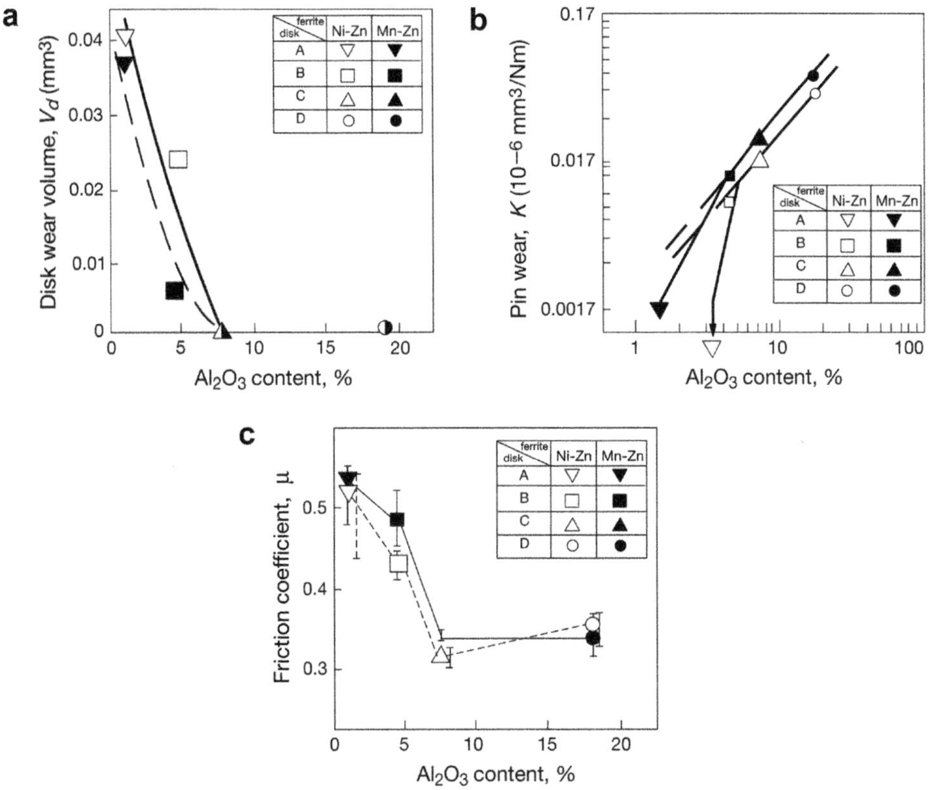

Fig. 7.34. Disk wear volume, pin counterface wear rate and coefficient of friction as functions of the aluminium oxide content in the magnetic layer of rigid disks. Data from experiments with a load of 60 mN, a sliding distance of 5.4 km and a speed of 3 m/s in (a) and (b), and 0.05 m/s in (c) (data from Kita, 1991).

Al_2O_3–TiO_2, WC–Co, TiC, Cr_2O_3, TiN, Triballoy 800 and hard-coat anodized aluminium. The lowest wear was found for solid ceramics and cermets.

Floppy disks have a similar construction and composition to tapes, except that the substrate is thicker and it has a magnetic coating on both sides. A typical coating binder layer is 2 to 5 μm polyester–polyurethane and epoxy on the 75 μm PET floppy disk substrate (Bhushan, 1988, 1990,

2001b). While the floppy disks operate only in occasional physical contact with the head, load-bearing Al_2O_3 or Cr_2O_3 particles are sometimes added to increase the wear resistance.

The *hard disk* in a magnetic recording disk drive is made of aluminium or glass which is coated by sputtering with a thin metallic magnetic layer. The magnetic layer is protected by a thin carbon coating, which is protected by a very thin layer of liquid lubricant, typically perfluoropolyalkyl ethers (PFPEs), as shown in Fig. 7.32c.

Today, sputtered hydrogenated diamond-like carbon (DLC) is almost exclusively used as the protective *overcoat material* for rigid disks. There are two types in use: amorphous hydrogenated carbon (a-C:H_x) and amorphous nitrogenated carbon (a-CN_x). These overcoats are quite complex materials and the details of their structure at the atomic level are not completely known. As they are exposed to air immediately after deposition their surfaces are partially oxidized (Gellman and Spenser, 2005).

In high-speed tests using a pin-on-disk apparatus Li *et al.* (1991) showed that glass-based thin film disks are more wear resistant than aluminium-based ones, and thin film disks are considerably more wear resistant than aluminium oxide ones. On the top of the glass substrate there was a sputtered chromium film with a thickness of 200 nm, which was used to enhance the magnetic properties of the medium. The magnetic medium was made of a cobalt-based alloy, which was sputtered over the chromium film and had a thickness of about 80 nm. A carbon overcoat was finally sputtered on top of the magnetic medium with a thickness in the range of 40 to 45 nm. Advantages of using glass as a substrate are its hardness and the fact that it can easily be polished to extreme smoothness.

A variety of different DLC coatings have been investigated for use on magnetic disks (Tsui *et al.*, 1995; Bhushan, 2001b; Schlatter, 2002). Several attempts have been made to try to correlate the coating structure to the tribological properties. Since the tribological properties of DLC coated rigid disks are affected by various factors the friction and wear characteristics can vary significantly during sliding.

The friction and wear mechanisms involved in the sliding contact of a commercially available thin film rigid disk with sputtered DLC overcoat sliding against an Al_2O_3–TiC magnetic recording head were studied by Wei and Komvopoulos (1997). They found that the friction behaviour of the head/disk interface is mainly due to the wear process occurring on the carbon film surface. The initial friction force was mainly due to asperity deformation and ploughing, while adhesional friction was secondary. At this stage the wear was mainly due to asperity nanofracture generating ultra-fine wear debris which caused an increase in the coefficient of friction. The steady-state friction was observed to be due to the concurrent effects of different wear processes on the DLC coated surface. The graphitization of amorphous carbon wear debris changes the interfacial energy and shear strength, thereby affecting the steady-state tribological behaviour. The topographies of unworn and worn carbon coatings have differences only in submicrometre-sized surface features, indicating that knowledge on nanoscale wear processes is of primary importance in magnetic recording.

Another approach has been to deposit a DLC coating on the sliding head. Kohira *et al.* (2000) investigated different slider designs with surfaces sputter deposited with a 1.5 nm silicon underlayer and a 5.5 nm thick DLC overcoat. The wear of the DLC coated sliders was proportional to the calculated contact force and the wear did level off after 10 km of sliding distance regardless of slider design.

Surface texturing has been investigated to reduce the stiction force caused by the contact of extremely smooth surface head sliders and rigid disks. Zhou *et al.* (2000) compared four kinds of hard coating materials on textured slider surfaces with a drag test. They found that the a-CN_x coating had a higher wear resistance than sputtered carbon, TiN_x and BN coatings.

The DLC overcoat is normally covered by a disk *lubricant layer*, most commonly perfluoro polyalkyl ethers (PFPE). The main function of the lubricant is to reduce the wear of the disk surface and to ensure that the friction remains low throughout the lifetime of the device. The lubricant should have the ability to adhere to the disk, have a good load-carrying capacity and be mobile in order to repair contact displacements. It should not evaporate from the disk during the lifetime of the disk, which is 5 to

Table 7.2. Microtribological applications in the information industry (Miyake and Kaneko, 1992). SP = sputtering, IP = ion plating and IM = ion mixing.

Application	Requirement	Material/Method
File memory	Ultrathin, durable	C, SiO_2/SP PFPE/spin coating
LSI manufacture	Dust free, maintenance free	MoS_2, Au, C PTFE/SP, IP PFPE/grease
Satellite communications	Reliable, long life, maintenance free	MoS_2, Au, Ag/SP, IP, IM, PTFE/transfer
Micromachines	Low torque, zero wear	Si_3N_4, C/SP, CVD

7 years, and it should not react with anything. The PFPE layers are liquid high-molecular weight polymers and have a very low vapour pressure, which keeps them from evaporating from the disk surface. They are of the most chemically inert base fluids available. The lubricant film is 2 to 3 monolayers thick even when the molecules lie down sideways. The lubricant layer is applied, after the deposition of the DLC coating, by dip coating the disks into a solution of the lubricant in a fluorinated solvent (Talke, 2000; Bhushan, 2001b; Tanaka *et al.*, 2003; Carnes, 2004; Gellman and Spenser, 2005).

In another application, which is not entirely unrelated to the above, *thermal printer heads* were coated with DLC at a thickness of 2 to 4 μm (Krokoszinski, 1991). These components are thin film resistors which are screen printed at the edge of a ceramic substrate and usually protected by an additional layer of thick film glass. The minimum necessary thickness of printed glass layers, of around 12 μm, introduces a thermally inert mass which reduces the heating and cooling speeds. In tests which replaced the 10 to 12 μm glass coating with 2 μm thick DLC the latter gave significantly increased lifetime and improved operating performance. Enke *et al.* (1987) also refer to the benefits of DLC on printer heads.

Miyake and Kaneko (1992) have outlined a number of applications for solid lubricant films, including DLC, in the information technology sector. They refer to these as microtribological applications, and details are given in Table 7.2. They investigated the microtribological properties of top surface films with an ultra-low load wear-testing rig and found that a silicon-containing carbon film can be improved tribologically by fluorination. This decreases the surface energy and reduces the wear on an atomic scale. The adhesion to the substrate and the strength of the carbon films were greatly improved by adding small quantities of silicon, which resulted in longer lifetimes (Miyake *et al.*, 1991; Miyake and Kaneko, 1991).

Grill *et al.* (1991) cite more applications of DLC in computer technology. An example is the use of DLC to prevent scratching of copper used in the conduction path of a high power chip, while also providing electrical isolation. DLC has been used to improve the quality and performance of photolithographic mask plates.

7.10 Microcomponents

7.10.1 Description of the application

There has been an increasing interest in studying mechanical systems with dimensions down to micrometre level, and sometimes even down to nanometre level. This has been triggered by the development in the electronics industry and especially the photolithographic process technology that made it possible to produce microcomponents and microdevices, known as microelectromechanical systems (MEMS). Today the lithographic processes have been complemented by a number of micromanufacturing methods like microcutting, microdrilling, micromilling, microsparc erosion, focused ion beam, laser ablation, laser machining and laser polymerization.

A variety of miniaturized components and devices have been fabricated for demonstration purposes: acceleration, pressure and chemical sensors, linear and rotary actuators, electric motors, gears, transmissions, gas turbines, pumps, valves, nozzles, cranks, sliders, springs, switches, grippers, pin joints, ratchets, tweezers, optoelectronic devices and magnetic recording (Muller, 1990; Komvopoulos, 1996; Bhushan, 1998b, 1999b, 2001c, d; Sundararajan and Bhushan, 2001a, b). The dimensions of the devices are in the range of one micrometer to some millimetres. One of the first large-scale commercial products was the acceleration MEMS sensor in cars for triggering the safety air-bags in collisions. Another is digital micromirror positioning device in laser printers and digital high-resolution projection devices. The fabrication of MEMS is described elsewhere (e.g. Komvopoulos, 1996; Sundarajan and Bhushan, 2001b).

The first used lithographic process was limited to silicon materials, which today are still dominating the MEMS applications due to the available machining technology. More recent micromachining processes extend the material choice to metals, ceramics and polymers. However, there are several design problems to overcome in downscaling a mechanical device from, e.g., 1 millimetre to 1 micrometre. The length decreases with three orders of magnitude, the related area with six orders of magnitude and the related volumes with nine orders of magnitude. Thus the forces that are proportional to the area, such as friction, adhesion, viscous drag and surface tension, will increase 1000 times in comparison with the forces related to volume, such as inertial and electromagnetic forces. This gives a more critical role to friction and wear in the functionality of the microdevices, see Fig. 7.35. In lubrication the continuum theory breaks down in the analyses of, e.g., fluid flow in microdevices. Another challenge is the material choice. The bulk silicon and polysilicon films frequently used in microdevices normally have high friction and wear in sliding applications. Their static coefficient of friction may vary in the range of 0.2 to 4.9 (Suzuki *et al.*, 1991; Deng and Ko, 1992; Komvopoulos, 1996; Krauss *et al.*, 2001; Sundararajan and Bhushan, 2001b; Rha *et al.*, 2005).

7.10.2 Improvements by surface coatings

Silicon-based MEMS devices are made from single crystal silicon, polycrystalline silicon (polysilicon) films deposited by low-pressure CVD and certain ceramic films on silicon. For high-temperature applications SiC films are used. Bhushan (2001d) studied the tribological performance of seven bulk and treated silicon samples, and polysilicon coated and SiC coated silicon wafers with 3 μm film thickness by surface characterization and tribological testing on both the macro- and microscale. The coefficient of friction was in general about one order of magnitude higher in macroscale reciprocating ball-on-flat testing with a 1.5 mm radius sapphire ball compared to microscale testing with a modified AFM/FFM with a 50 nm radius Si_3N_4 ball.

The macroscale coefficient of friction for the tested surfaces was in the range of 0.2 to 0.46 while on a microscale it was in the range of 0.02 to 0.07, as shown in Fig. 7.36. The microscale friction is lower because there is less ploughing friction contribution in the microscale measurements. Doping of the polysilicon coating did not significantly affect its tribological properties that were very similar to those of undoped single-crystal silicon. The SiC coated surfaces were superior with regards to their friction and wear performance. This is believed to be due to their higher hardness and fracture toughness. Polishing the surface reduced the coefficient of friction by 20 to 70% in microscale friction tests, as shown in Fig. 7.37.

There have been attempts to use diamond or DLC coatings to reduce friction and improve wear resistance in MEMS devices. Difficulties with using diamond coatings are related to fabricating surfaces with a morphology suitable for MEMS. Diamond coatings produced by conventional CVD

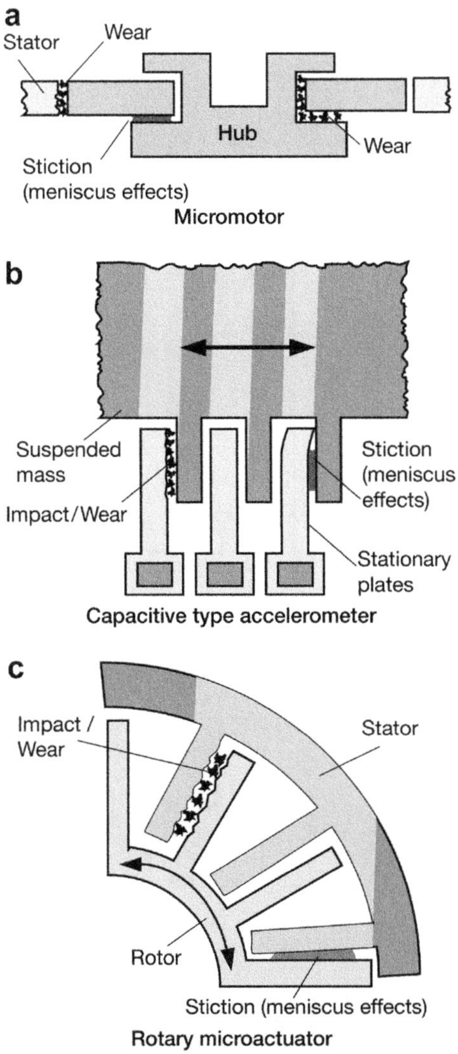

Fig. 7.35. Typical tribological issues in three MEMS devices in operation: (a) micromotor, (b) capacitive type accelerometer and (c) rotary microactuator (after Sundararajan and Bhushan, 2001b).

deposition techniques are unsuitable for MEMS components due to their high intrinsic stress, rough surface and inappropriate grain morphology. The nanocrystalline diamond films appear to be better suited because their grain size is very small, only 10 to 30 nm. They have been fabricated as free-standing MEMS structures and as very thin protective layers, with film thicknesses down to 50 nm on silicon MEMS components (Erdemir and Donnet, 2005). Krauss *et al.* (2001) developed a new microwave plasma technique for deposition of phase-pure ultra-nanocrystalline diamond films with mechanical and tribological properties that seem to match well for MEMS applications.

There are very promising attempts to use thin DLC layers on silicon MEMS components. The excellent tribological properties of DLC coatings can reduce the coefficient of static friction from the high values mentioned in silicon–silicon contacts to only 0.2 to 0.3 in silicon–DLC contacts and to

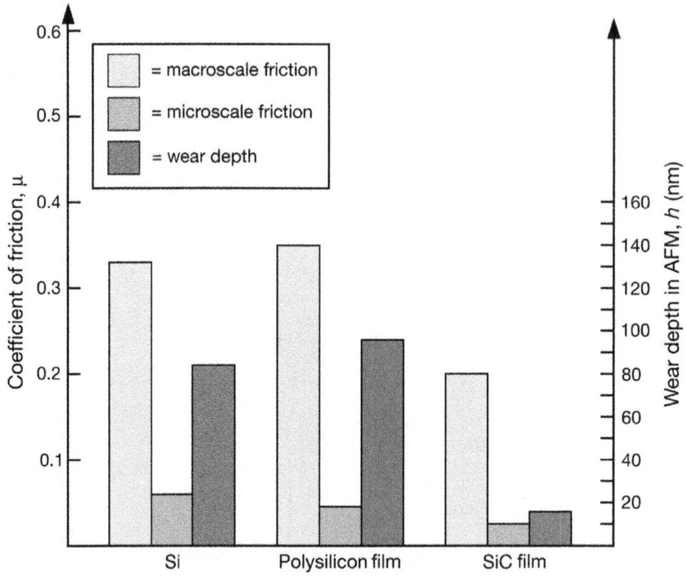

Fig. 7.36. Comparison of tribological scale effects: (a) macroscale coefficient of friction in reciprocating tribotester, (b) microscale coefficient of friction measured by AFM/FFM and (c) microscale wear measured by AFM (data from Bhushan, 2001d).

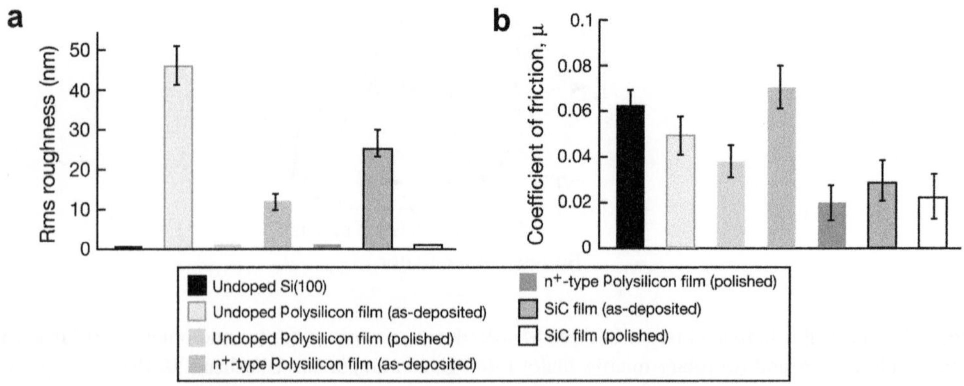

Fig. 7.37. Comparison of (a) rms surface roughness and (b) coefficients of microscale friction for various samples (data from Bhushan, 2001d).

0.04 to 0.06 in DLC–DLC contacts when both surfaces are coated (Suzuki *et al.*, 1991; Deng and Ko, 1992; Komvopoulos, 1996; Rha *et al.*, 2005; Smallwood *et al.*, 2006). High intrinsic stresses in ta-C DLC films may restrict their use.

The tribological properties of a coated or uncoated MEMS surface can be further improved by a suitable lubricant film. Again, the scaling down to microlevel brings new features into lubrication technology and the rules we used to apply on the macrolevel do not always apply. The continuum theory on which the classical approach to lubrication is based, with freely supported multimolecular layers of liquid lubricants, does not apply when approaching monolayer lubricant films. In traditional lubrication these can physically or chemically adsorb onto polar surfaces by their functional groups to form close-packed arrangements of lubricant chains oriented almost perpendicular to the substrate surface. The use of fluid lubricants on the microscale may introduce capillary and viscous shear mechanisms which result in energy dissipation. Consequently, the ideal lubricants for MEMS devices are ultra-thin layers covalently bonded to micromachined surfaces.

Grafting of densely packed organic molecules to obtain nanometre thick monolayers is especially advantageous because the typical surface separation distances are in the range of 0.5 to 2 μm. The monolayers must exhibit high hydrophobicity and low surface energies to minimize capillary forces, low frictional shear to yield minimal energy losses and high wear resistance and stability over long time periods. Such lubricant layers are typically long-chain hydrocarbon and fluorocarbon molecules and they are referred to as self-assembled monolayers (SAM). The PFPE lubricant films successfully used in magnetic hard disk drives may also be suitable for MEMS applications (Komvopoulos, 1996; Bhushan, 2001d; Sundararajan and Bhushan, 2001b; Hsu, 2004a, b).

7.11 Biomedical Applications

7.11.1 Description of the application

In the biological world, nature has created a huge number of fascinating and efficient tribological solutions. Tribology is related to mobility in humans and animals. High friction is required for acceleration, deceleration and manoeuvring while low friction is needed in joints for economic energy expenditure and low wear. A healthy human joint has a remarkable lifetime of typically 75 years with more than 1 million loading cycles per year (Fisher, 2001; Scherge and Grob, 2001; Dowson and Neville, 2006).

Biomedical tribological systems involve a large range of synthetic materials and natural tissues, which often operate in complex interaction with the biological environment. Their performance requirements are complex and the lifetime requirement often considerably exceeds that of engineering systems.

The field of medical implants in the human body is a growing area with several tribological aspects. This application sector has its own specific characteristics, dominated by stringent quality requirements, due to the human suffering and sometimes life-threatening consequences of a surface failing to fulfil its required function. Rejection of body implants is caused by poor biocompatibility of the surface of the implanted material. Surgical tools, which must be sharp, can corrode rapidly in physiological fluids. Other problems include fretting and wear in orthopaedics and dentistry, such as brackets, pins and plates for bridges, and haemocompatibility of cardiovascular applications such as avoidance of blood clots.

Total hip joint replacements, shown in Fig. 7.38b, are the largest field of tribological implants. Hip replacements are used in the case of either osteoarthritis or hip fracture. In the 1980s there were 120,000 total hip joint replacement surgeries per year and in 2006 the number had increased to about one million. Osteoarthritis of the hip joint is manifested by the degradation of the cartilage layer, leading to a breakdown of the low-friction contact between the femoral head and the acetabulum in the pelvis. In the absence of a viable drug therapy, complete or partial replacement of the degenerated joints is the main available treatment. The lifetime of hip joint replacements is about 10 to 15 years, and some 12% of patients require second replacements. There is currently an urgent need to extend the treatment to patients under 50 years who would require replacements to last at least 20 to 50 years. There are three main factors which limit this: (a) insufficient wear resistance, (b) loosening of the stem to hip caused by wear particles or (c) cement-to-bone interface failure or implant-to-bone interface failure.

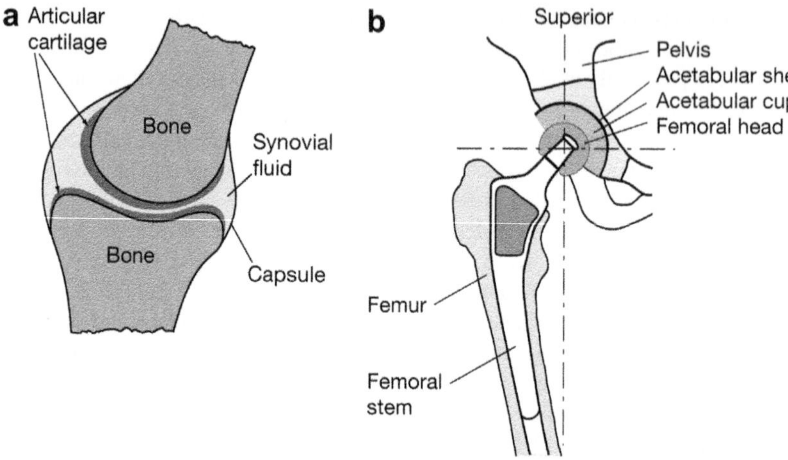

Fig. 7.38. Schematic illustration of (a) a natural joint in the lower limb and (b) a total replacement hip joint.

Current total hip joint implant material combinations are mainly:

- metal head against polymer cup, e.g. CoCrMo/UHMWPE, stainless steel/UHMWPE,
- ceramic head against polymer cup, e.g. alumina/UHMWPE , zirconia/UHMWPE,
- metal head against metal cup, e.g. CoCrMo/CoCrMo, and
- ceramic head against ceramic cup, e.g. alumina/alumina.

The metal/UHMWPE material combination has dominated the market in the last decade. Recent investigations shows that both metal–metal and ceramic–ceramic joint implants reduce the wear considerably and are very promising future long lifetime solutions (Rieker, 1998; Ajayi *et al.*, 2005; Dowson and Jin, 2006; Unsworth, 2006). However, there are concerns about the toxicity of metal wear particles and elevated metal ion levels in the human body (Fisher *et al.*, 2004). Total knee implants are predominately metal against polymer.

The lubrication mechanism in a natural hip joint is a complex combination of weeping, elastohydrodynamic fluid film, boosted, boundary and biphasic lubrication. The lubrication mechanism between the synthetic surfaces of hip implants is simpler and the lubrication conditions can be classified as boundary and mixed lubrication with occasional full surface separation. The lambda value (film thickness–surface roughness) is for metal–polymer joints typically <1 and for metal–metal joints typically <3. During a walking cycle a load of 4 to 5 times the body weight is transmitted and a typical load variation is 0.3 to 3 kN, the sliding velocity is up to 50 mm/s and the film thickness in a total hip joint implant is about 0.05 to 0.25 μm (Fisher, 2001; Unsworth, 2006; Dowson and Jin, 2006).

The coefficient of friction in a hip joint implant is typically in the range of 0.02 to 0.15, while in a healthy human joint it is as low as 0.002 to 0.025, which, indeed, is remarkably low at such low speeds. It appears, however, that the major tribological problem is not the change in surfaces due to wear but the large amount of wear debris produced in the contact and that plays a significant role in promoting loosening of both femoral stems and acetabular cups. Wear particles in the size range of 0.1 to 7 μm cause a macrophage biological reaction that causes bone resorption (Dowson, 1989; Davidson, 1992; Saikko, 1993; Hills, 2000; Fisher, 2001; Unsworth, 2006; Ahlroos, 2001; Calonius, 2002; Santavirta, 2003; Ajayi *et al.*, 2005; Briscoe and Sinha, 2005; Hutchings, 2007).

7.11.2 Improvements by surface coatings

Coatings are increasingly being used as enhancement, rather than protective, layers, e.g. on orthopaedic implants to facilitate bonding direct to the bone. One example of coatings to enhance properties are the bioactive calcium phosphate (hydroxyapatite) coatings. These accelerate bone fixation and reduce metal ion release (Habibovic *et al.*, 2002).

A potentially very promising coating in the medical sector is diamond-like carbon. Studies on its biocompatibility have shown that DLC is an inert, impervious hydrocarbon with properties suitable for use in the biomedical field, particularly in orthopaedic implants.

DLC adheres strongly to a range of materials used in bio-engineering including metals such as titanium and cobalt chromium, and various plastics including polyethylene and polyurethane. It can provide a chemically protective, wear-resistant and low-friction coating on such materials in a biological environment. Laboratory studies with mouse macrophage cells and fibroblast cells showed enzyme levels indicating no inflammatory response or loss of cell integrity from contact with DLC. Morphological examination confirmed the biochemical results that no cellular damage had occurred (Aisenberg, 1977; Franks *et al.*, 1990; Evans *et al.*, 1991; Thomson *et al.*, 1991; Cui and Li, 2000; Chu *et al.*, 2002; Reuter *et al.*, 2006).

In a reciprocating pin-on-disk tribotest study Ajayi *et al.* (2005) compared the tribological properties of typical in-use hip joint material combinations, stainless steel, alumina, zirconia and UHMWPE balls with a PACVD 2 μm thick DLC coating on steel ball, all sliding in commercial bovine blood serum against a polished stainless steel flat surface. The DLC coated surface had almost as low a coefficient of friction, 0.055, as the conventionally used UHMWPE surface, 0.03, while the other material combinations had considerably higher coefficients of friction, in the range of 0.09 to 0.115, as shown in Fig. 7.39. The DLC coated surface had the lowest wear, almost four orders of magnitude lower than the UHMWPE and several times lower than the other tested ball materials. Furthermore there was no wear of the uncoated steel surface.

There is some contradiction in the reported results of the tribological properties of DLC coatings used in synovial hip joints and that may be due to the quality of the deposited coatings and the countersurface material used. The 3 μm thick DLC coatings deposited by a hybrid PVD/CVD process on CoCr heads did not differ markedly in wear behaviour when sliding against UHMWPE in a hip joint simulator compared to alumina and uncoated CoCr heads (Saikko *et al.*, 2001).

On the other hand, remarkably low wear, low friction in the range of $\mu = 0.03$ to 0.06 and a negligible amount of wear debris was recorded when Santavirta *et al.* (1999) and Lappalainen *et al.* (2003) carried out both pin-on-disk and hip joint simulator tests with high-quality DLC coated metal heads. The DLC coatings were deposited by a filtered pulsed arc discharge method resulting in very smooth and pure DLC (ta-C) coatings. The hydrogen content was less than 0.1 at.% and the overall purity was greater than 99 at.%. It is interesting that they report tests with thin DLC coatings in the range of a few μm but also tests with comparably thick DLC coatings, in the range of 40 to 100 μm. They report wear rates of 0.0001 to $0.001 \cdot 10^{-6}\,mm^3/Nm$ for DLC/DLC pairs, 0.001 to 0.1 for alumina/alumina pairs, 1 to 30 for CoCrMo/CoCrMo pairs and 5 to 100 for UHMWPE/CoCrMo pairs, as shown in Fig. 7.40.

Promising results of the use of DLC coatings in hip joint implants have further been reported by Field *et al.* (2004) and reviewed by Hauert *et al.* (2004). In the investigation by Fisher *et al.* (2004) they compared in a hip joint simulator the wear properties of a conventional CoCrMo/CoCrMo hip joint to three 8 to 12 μm thick coatings, arc evaporation PVD TiN, CrN and CrCN, and one thin 2 μm PACVD DLC coating deposited on CoCrMo. They report that thin 1 to 2 μm TiN and DLC coatings can wear through while the most promising results were achieved for thicker 8 to 12 μm CrN and CrCN coated surfaces. The TiN coated surface may be improved by chlorine ion implantation of the surface (Gispert *et al.*, 2007). In a reciprocating ball-on-flat screening test with distilled water the

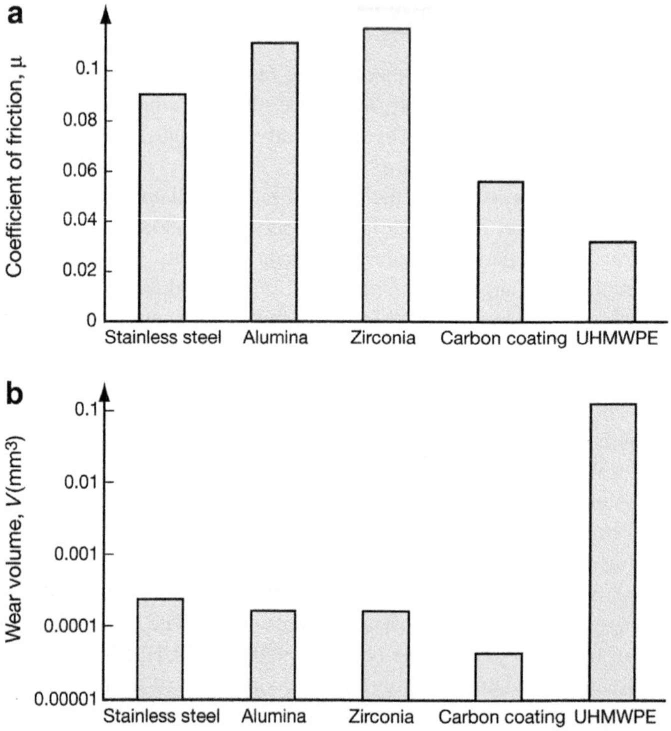

Fig. 7.39. (a) Average coefficient of friction and (b) wear volume of various in-use and potential hip joint ball materials sliding on a stainless steel flat surface in a reciprocating pin-on-disk test (after Ajayi *et al.*, 2005).

tested DLC coatings outperformed, in terms of friction and wear behaviour, standard TiN, TiC, Ti(C,N), CN, CrN coatings as well as more complicated Ti-based multilayer systems (Österle *et al.*, 2007). The test conditions were a reciprocating 10 mm diameter alumina ball sliding against the coatings with 2 mm stroke length, 2 Hz frequency and 10 N normal load resulting in a mean Hertzian pressure of 600 MPa.

The deposition of PVD diamond-like carbon coatings on prosthetic artificial heart valves of NiCr18 stainless steel or TC4 alloy seems to be a potentially useful application. Zheng *et al.* (1991) have shown that they satisfy both the mechanical and biological requirements of artificial heart valves.

The use of DLC coatings to improve surgical needles used in corneal surgery, such as corneal transplant and cataract operations, has been investigated by Grant *et al.* (1992). They carried out two types of test. One involved inserting 0.8 mm diameter AISI 430 stainless steel needles in pigs' eyes to determine the frictional resistance to penetration. The second involved a ball against flat fretting test with a 5 mm diameter ball, a 14 N normal force, a 110 μm amplitude at 10 Hz and the same materials. The penetration tests showed that a DLC coating would typically reduce the necessary penetration force by 30%. The fretting test showed, on the one hand, that deposition process optimization was still required but, on the other hand, that the driving force would be significantly reduced using the DLC coating and that excellent wear resistance can be achieved.

PVD titanium nitride coatings have been shown to offer 300 times reduction in anodic dissolution in physiological solutions compared to the uncoated stainless steel normally used for surgical knives (Gordashnik *et al.*, 1983).

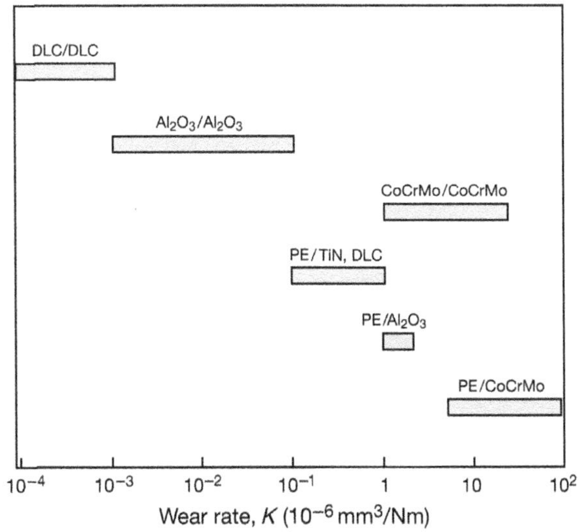

Fig. 7.40. Experimental wear rates of some material combinations used in knee and hip joint implants. Some of the values are based on data reported in the literature and other values are estimates using data from clinical surveys or hip joint simulator studies (data from Santavirta *et al.*, 1999).

Dental prostheses are required to possess very high strength and hardness, but also extremely good biocompatibility with the adjacent human tissue and teeth. Two competing material types are in use. Titanium materials are extremely compatible but their low strength means that the dentures have to be relatively thick. Chromium–cobalt alloys combine high hardness and strength with the greatest degree of denture-making know-how. Knotek *et al.* (1992e) have shown that it is possible to combine the good properties of these materials by developing a cobalt–chromium–molybdenum alloy with a multilayer PVD titanium nitride coating. Laboratory corrosion tests showed that a two-layer titanium nitride coating had excellent corrosion resistance. The special advantage of a two-layer structure is that intermediate etching reduces the number of defects remaining in the coating and hence increases its impermeability to corrosive attack, as shown in Fig. 7.41.

7.12 Future Issues

The applications described within this chapter, and those mentioned elsewhere in the book, show that thin coatings deposited by various, mainly PVD- and CVD-based, techniques have been a major success when applied on components and tools in many industrial sectors. The improvements have been a considerable reduction in friction, decreased wear and increased component lifetime, sometimes by even several orders of magnitude, resulting in improved production efficiency, productivity, reliability and cost efficiency. A summary of ten representative areas of applications for thin hard coatings including typical coating materials, thicknesses, substrates, counterfaces and achieved tribological improvements based on the examples referred to in this chapter are presented in Table 7.3.

The requirements from coatings have changed over the years, depending on the types of substrates in prevalent use and the demands of some key application sectors, such as automotive and aerospace. Two decades ago there was a perceived need to widen the applicability of advanced tribological coatings to low-cost substrate materials, for example by developing low-temperature coating processes and also devising duplex and hybrid processes which could ensure that those substrates could

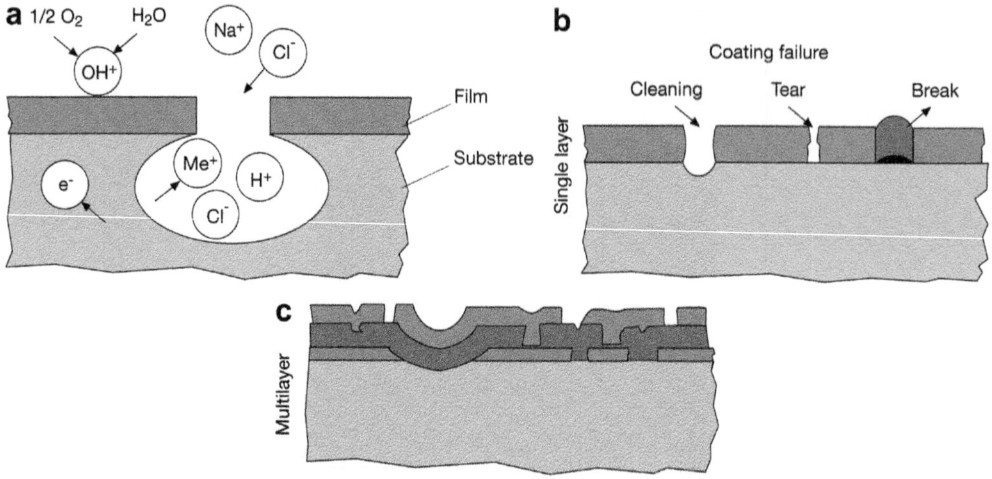

Fig. 7.41. (a) Pitting corrosion in chloride-containing electrolytes, (b) coating failure at a single layer and (c) the multilayer principle of coating density improvement (after Knotek *et al.*, 1992e).

provide the load support needed for PVD coatings. These objectives have been achieved to some extent.

However, what has also happened is that there has been a refinement in the way that the PVD coatings themselves have evolved, to match the requirements of the substrates and the applications. We now see many tailored coatings designed for specific needs. There is an increasing emphasis on frictional performance, which is not surprising, given the current interest in energy efficiency and resource conservation. We have seen in the book how frictional control and adaptability to operating requirements can be achieved through process and materials innovations such as nanocomposite, amorphous and glassy metal coatings.

We can expect these developments to continue – especially in the areas of adaptive and gradient coatings designed specifically for a given set of contact conditions. Given the excellent tribological performance largely confirmed in a wide range of operating conditions of coatings based on DLC and MoS_2, we can expect those materials in various forms and combinations to be used increasingly, often in conjunction with other materials.

The optimization of mechanical properties will be aided by the use of multiscale modelling techniques such as FEM, and other computer-based methods, including molecular dynamic simulation, which will provide a better understanding of atomic and molecular level tribomechanics and scaling effects.

We envisage that recent interesting findings relating to the influence of surface texture will lead to a widening of the use of surface topography control and enhancement methods including, for example, laser texturing.

A further important benefit will be the development of lubricant additives tailored to work together with specific coatings to achieve optimized surface interactions for improved tribological performance. This will be one example where controlled tribochemistry will enhance performance. Another will be the study and optimization of transfer layer development and third-body generation. We also anticipate improved control of coated surface fracture behaviour by the adoption of fracture mechanics analyses for hard coatings

We can see the need in all safety-critical parts to have means to monitor the condition of tribosurfaces. We can therefore expect to see intelligent coatings developed, which provide information on

Table 7.3. Typical coating applications and level of improvements.

Application	Coating	Thickness, μm	Substrate	Countersurface	Improvements
1. Sliding bearings	DLC, PTFE, MoS_2, graphite	5–50	Steel	Steel	Up to 7 times reduced wear rate. Decrease in friction.
2. Rolling contact bearings	DLC, TiN, TiC, HfN, MoS_2, Pb, Co, Cr, Au, PTFE, ML, DP	0.05–3 optimum typically 0.2–1	Steel, SS	Steel, SS	Increased fatigue life by up to 10 times. Large scatter in results. Increased fatigue limit.
3. Gears	WC/C, B4C, TiN, DP, TiC, Co, graphite, ML	1–10	Steel, case-hardened steel	Same as substrate or coated	Reduced friction and wear. Up to 70% torque capacity increase. Increased lifetime by 3 times.
4. Tools for cutting	TiC, Al_2O_3, TiN, ZrN, TiAlN, TiCN, TiVN, TiBN, MoS_2–Cr, diamond, Cr:a–C, Ti:C–H, ML, CrN, DP, DLC	1–6	HSS, cemented carbide, WC	Steel, cast iron, SS and super-, Al and Ti alloys, plastics, epoxy resins, ceramics, wood	Very commonly 2 to 20 times increased lifetime. Improved surface finish of workpiece. Improved cutting efficiency and productivity.
5. Tools for forming	TiN, TiB_2, VN, TaC, DLC, TiAlN, TiBN, TiC, MoS_2:Ti	1–10	Steel	Steel, Al and Ti alloys	Up to 200 times lifetime improvements. Enhanced surface quality of work piece and improved productivity.
6. Erosion and scratch resistance	TiN, CrN, TiCN, DLC, ML, WC, TiC, Al_2O_3, $B_{13}C_2$, WN, TiB_2, SiC, MoB, B_4C, $NiCr_3C2$, $NiTiB_2$, SiO_2, TiO_2, ZrO_2, Nb_2O_2, ZrB_2	2–50	Steel, SS, ceramics, glass, plastics,	Metal and ceramic debris, sand, dust, rocks, TiO_2	Up to 200 times decreased wear.

(Continued)

Table 7.3. Continued

Application	Coating	Thickness, μm	Substrate	Countersurface	Improvements
7. Oscillating contacts	TiC, VC, TiCN, TiN, ML, TiAlN, NiP, DLC, VC, TiB_2, Si_3N_4, SiO_2, MoS_x, diamond, PTFE, rubber	1–30	Steel, ceramics, glass	Steel, ceramics	Up to 10 times reduced wear. Reduction in friction.
8. Magnetic recording hard disk	DLC: a-CH_x or a-CN_x, ML	0.003–0.01, ML: 0.1	Magnetic coating on Al–Mg or glass	Ni–Zn ferrite, Mn–Zn ferrite, SiC, Al_2O_3–TiC, calcium titanate	DLC covered with 0.5–30 nm PEPF liquid layer. Reduced wear and increased lifetime.
9. Microcomponents	DLC, diamond	0.05–0.25	Si, SiC, ceramic films on silicon	Same as substrate	Reduced friction with a factor of 2 for Si–DLC and 10 for DLC–DLC contacts.
10. Biomedical applications	DLC, TiN, CrN, CrCN	1–100	Steel, CoCr, CoCrMo	SS, UHMWPE, DLC on steel	Low friction, reduced wear debris generation and reduced wear by several orders of magnitude.

ML = Multilayer Coating; DP = Duplex Coating

their operating performance level, and may even self-repair. In the absence of such smart coatings, it will be necessary to have advanced sensor technologies installed in tribological systems, to ensure that they are functioning effectively and to avoid catastrophic failure.

In the future a significant driver for coating development and the use of coated surfaces will be the trend towards sustainable industrial production and a sustainable society. This includes requirements for improved environmental protection, better safety and reliability in combination with economic efficiency, including end-of-life disposal and reclamation.

The move away from toxic materials and environmentally damaging processes will therefore continue. This process of environmental and energy conservation and protection requires that the full lifecycle impact of coatings and processes should be properly evaluated and acted upon. For example, a carburizing process to harden a gear might, during the hardening process, involve a significant CO_2 emission, but if that allows a car to operate more efficiently throughout its life then the emissions at the start might well be justified over the lifecycle.

Similarly, one might compare a modern high-strength case-hardened gear with one of equivalent power transmission capacity but made from cast iron. The weight saved by using the smaller and lighter high-strength gear material, made possible by surface treatment, will more than compensate for the energy which would have been consumed in carrying and moving the heavy large cast iron gear. So, the message is that tribological treatments and coatings have the capability to have a major impact on energy and resource usage – but this requires the need to educate not only tribological process users but also politicians and others to avoid judgements, e.g., relating to legislation, which are not in the overall interest of society.

We have mentioned in this book many application examples in which tribological coatings can provide gains in efficiency and performance. Modern high-economy diesel engines could not function reliably without carbon-based coatings applied to injector parts and advanced low-fuel consumption aero-engines could not operate without effective thermal- and erosion-barrier coatings. The move towards higher power densities for internal combustion engines, meaning smaller engines with higher power outputs and increased efficiency, can only be achieved by the extensive use of high-performance tribological coatings.

While these comments are made primarily with mechanical parts in mind, there are other product sectors in which tribological coatings are having an increasing impact. In particular, the medical and healthcare sector has a need to improve the lifetime and performance of implants and devices that come into contact with the body. In that case, reliability and safety are of absolute priority. Many advanced medical implant devices would fail without reliable biocompatible coatings which impart the appropriate triboresponse with the body.

Coatings tribology is at the forefront of developments in most mission-critical components, regardless of the sector in which they have to operate. When the technological demands are coupled with the economic imperative to reduce production costs and to improve reliability, it is clear that the challenges facing tribologists and coating process developers are enormous. Nevertheless the payoff from success in meeting these goals is huge with great societal impact, and it is our hope that this book will assist those who seek to achieve new solutions and breakthroughs.

APPENDIX A

Tribology data sheet for coatings: Polymers and elastomers

Reference			Coating					Counterface	
No.	Year	Authors	Film	Deposition method	Substrate	Thickness μm	Roughn./Hardn. R_a, μm/H_v	Material	Roughn./ Hardn. R_a, μm/H_v
1.	1988	Gladstone	Silicon rubber Silicon rubber Silicon rubber	Silicon adhesive	Steel	300 300 300	0.25	Glass Silicon coating Metal	
2.	1990	Adams	Rubbery copolymers		Glass	4 500	Hard rub. → Soft. rub. →	Glass	
3.	1977	Fusaro	Polyimide (PI)	Spraying + heating	Stainless steel 440C	25	0.01–1.3	Stainless steel 440C	
4.	1978	Fusaro	Polyimide (PI)	Spraying + heating	Stainless steel 440C	25	0.9–1.3	Stainless steel 440C	
5.	1981	Fusaro	Polyimide (PI-4701)	Spraying + heating	Stainless steel 440C	10–62	0.9–1.2	Stainless steel 440C	
6.	1987a	Fusaro	Polyim. PMDA " 4701 " 4701 + add. Polyim. PMDA " 4701 " 4701 + add.	Spraying + heating	Stainless steel 440C	25	0.9–1.2	Stainless steel 440C	
7.	1988	Sugimoto and Miyake	PTFE PTFE-PFPE PFPE PTFE PTFE-PFPE PFPE	Rf sputtering	Stainless steel	0.3		Stainless steel 440C	
8.	1990	Yamada *et al.*	PTFE PI	Sputtering	Glass	0.05–1		Sapphire	
9.	2004	Lee and Lim	PTFE PTFE + nano-diamond fillers	Spraying	Aluminium	30		Ball bearing steel ball	
10.	2005	Langlade *et al.*	PTFE + thermoset PTFE + fluorinated resin	Unknown commercial method	32CrMoVa13 steel	15–22 20–30		100Cr6 steel	

Experiments				Results		
Geometry	Speed m/s	Load N	Lubrication environment	Friction	Wear, lifetime, debris wear rate, $10^{-6}\cdot$ mm^3/Nm	Comments/ Conclusions
Cyl.-on-plate Reciprocating	0.01–0.15	40–80	Oil lubr.	0.006–0.06 0.02–0.06 0.02–0.03		Elastomer
Ball-on-plate Cone-on-plate	0.15	0.01–0.5 0.01 0.01	Dry	0.31–0.45 0.5–1 1.5–4		Elastomer μ increases with sharpness of cone for thick (500 μm) coatings
Ball-on-disk	1.6	10	Dry argon $T \approx 100°C$	0.03–0.23		$\mu < 0.03$ at $T > 80°C$ Friction transition independent of speed
Ball-on-disk	2.6	10	Dry air → Dry argon → Moist air →	0.25–0.39 0.19–0.25 0.08–0.14		$T = 25°C$ Transfer film on ball Increase in T gives decrease in μ in dry environment
Ball-on-disk	2.6	10	50% RH	0.08–0.22		Wear life proportional to film thickness Description of wear mechanisms No influence of film thickness on μ
Ball-on-disk ∅ 4.76 mm	0.26	10	50% RH Vacuum	0.41–0.48 0.11–0.18 0.13–0.18 0.35–0.45 0.05 0.05–0.1	90 50 3 100 0.5 9	Polyimid PMDA has high friction and low wear in vacuum
Ball-on-disk ∅ 6.35 mm Reciprocal	0.0015	5	Air Vacuum	0.18–0.38 0.15–0.23 0.15–0.2 0.18–0.45 0.15–0.45 0.2–0.6		
Pin-on-disk	0.0004	0.5–10	Air	0.3 0.2	Film-like debris Flake-like debris	μ independent of load and film thickness PI had longer wear life than PTFE
Ball-on-plate Reciprocating	0.01	0.8	Air Air at 150°C	0.16 0.08		Hard nanoparticle fillers in PTFE reduce friction and wear
Cyl.-on-plate Reciprocating	5 Hz with ±10 μm amplitude		Air	0.1–0.15 0.07–0.15	Wear depth: 1–2 μm after 10^4 cycles Wear depth: 2–25 μm after 10^4 cycles	Fretting test

Tribology data sheet for coatings: Molybdenum disulphide

Reference			Coating					Counterface	
No.	Year	Authors	Film	Deposition method	Substrate	Thickness µm	Roughn./ Hardn. R_a, µm/H_v	Material	Roughn./ Hardn. R_a, µm/H_v
1.	1990	Nabot *et al.*	MoS_x	Dc reactive magnetron sputtering	Stainless steel AISI 420	0.5–28	H_{knoop} = 0.6–15 GPa	Steel 100C6	
2.	1989	Roberts and Price	MoS_2	Magnetron sputtering	SiN Steel Ti alloy	1	0.05/19.9 0.05/8 0.05/3.2	SiN Steel Ti alloy	
3.	1972	Spalvins	MoS_2	Rf diode sputtering	Steel 440C	0.2–0.65	0.05	Steel 440C	
4.	1990	Fayeulle *et al.*	MoS_2	Sputtering	Steel 52100	1		Steel WC Sapphire	
5.	1990	Aubert *et al.*	MoS_x	Dc reactive magnetron sputtering	Stainless steel AISI 420	0.7–28		Steel 100C6	
6.	1985	Dimigen *et al.*	MoS_x	Rf diode sputtering	Steel 100Cr6	0.2–0.8		Steel	
7.	1989	Müller *et al.*	MoS_2	Dc magnetron sputtering	Al_2O_3 Ti–Al alloy Fibreglass	2.2 2.5 2		Steel AISI 52100	
8.	1988	Fleischauer and Bauer	MoS_2	Rf sputtering	Steel 440C			Steel 440C	
9.	1988	Müller *et al.*	MoS_2	Dc and rf magnetron sputtering	Steel AISI 440C	0.1–5	0.07	Steel 52100	
10.	1978	Fusaro	MoS_2	Mechanical rubbing, powder ∅ 10 µm	Steel 440C	1–2	0.09	Steel 440C	
11.	1983	Niederhänser and Hintermann	MoS_2 + add	Sputtering	Steel AISI D3			Steel Ruby TiC, Au	
12.	1990	Fleischauer *et al.*	MoS_2	Rf sputtering Dc " rf magn. sputtering	Steel 440C	0.5–1		Steel	
13.	1990	Didziulis and Fleischauer	MoS_2	Rf sputtering	HPSiC Steel 17-4PH	0.2		Al_2O_3 Single crystal 90°	
14.	1990	Singer *et al.*	MoS_2	Sputtering	Steel 440C	1		Steel 52100 → WC–Co → Sapphire →	Bearing grade

Experiments				Results		
Geometry	Speed m/s	Load N	Lubrication environment	Friction	Wear, lifetime, debris wear rate, $10^{-6} \cdot mm^3/Nm$	Comments/Conclusions
Ball-on-disk	0.01	5	Dry air Liquid nitrogen →	0.025–0.06 0.04	Low wear	$\mu < 0.05$ for $MoS_{1.5}$–$MoS_{2.1}$ Increased humidity → increased μ Good tribological properties in liquid nitrogen
Ball-on-disk	1.2	50	Vacuum	0.01 0.02 0.04	Life = 100,000 rev " = 10,000" " = 500 "	μ and life better for substrate SiN > steel > Ti alloy Higher roughness → lower friction Coating method → tribological properties
Pin-on-disk	0.12–0.25	10→ 2.5 →	Vacuum 10^{-11} torr	0.02 0.04	Life = 38,000 rev " = 250,000 " Debris agglomeration	No speed influence Same composition as bulk MoS_2 – no pinholes
Four ball ∅ 12.3 mm	0.2	50 1.15–1.49 GPa	Air and Ar	0.02–0.04	2–3 times longer life in Ar than in air	Longer lifetime for harder sliders Structure description of the sliding surfaces
Ball-on-disk	0.01	5	Dry air RH < 5%	0.02–0.06		Increased μ with smooth surfaces Increased μ with higher humidity $\mu = 0.025$ for $H_k = 5$ GPa and SiMo = 1.5
Ball-on-disk	4–200 rpm		Vacuum Dry nitrogen	0.012–0.04 0.016–0.035		Smallest μ in dry nitrogen with sulphur-deficient layers
Ball-on-disk	0.1	5	Humid air 98% RH		Life: 100–400,000 rev. ": 11–12,000 rev. " : 13–28,000 rev.	Little influence of h on lifetime Best lifetime with Al_2O_3 substrate
Pin-on-disk	0.033	32	Dry nitrogen			Lubrication by intercrystallite slip Optimum crystallite size → max wear life
Pin-on-disk	0.1	5	Humid air 98% RH	0.05	Life: 2000–1,000,000 rev.	Lifetime increases with thickness at $h = 0.1$–6 μm
Pin-on-disk	2.6	10	Dry air Dry argon Moist air	0.01–0.02 0.01–0.02 0.08		Transfer film formed
Pin-on-disk	0.0003–0.12	0.2–60	RH < 20% 20 < RH < 97% 97% < RH	0.02–0.3	Life: 0–4,005,000 rev.	Best results with MoS_2–PTFE on chemically active R_h interlayer Load and speed and humidity influence
Pin-on-disk		0.7 GPa	Air		Life: 45,000 rev. " : 156,000 rev. " : 200,000 rev.	
Pin-on-flat Oscillating 3 mm	0.004	0.6 1.75 GPa	Vacuum 10^{-5} torr	0.02–0.11 0.02–0.08	Life on SiC: 150–450 cycles Life on steel: 1000 cycles	Better wear life with MoS_2 on steel than with MoS_2 on SiC
Ball-on-disk ∅ 1.6–12.7 mm	0.02	1–50 0.2–1.5 GPa	Dry air RH <1%	0.03–0.14 0.02–0.12 0.03–0.18		Friction coefficient is inversally proportional to load

(Continued)

Tribology data sheet for coatings: Molybdenum disulphide (*Continued*)

Reference			Coating					Counterface	
No.	Year	Authors	ilm	Deposition method	Substrate	Thickness μm	Roughn./ Hardn. R_a, μm/H_v	Material	Roughn./ Hardn. R_a, μm/H_v
15.	1990	Kuwano	MoS_2 $B + MoS_2$ $BN + MoS_2$ $TiB_2 + MoS_2$	Fast atom beam sputtering	Steel 440C				
16.	1993	Miyoshi *et al.*	MoS_x	Magnetron sputtering	Steel 440C	0.11	0.009/6.8 GPa	Steel 440C	0.007/8.5 GPa
17.	1993	Wheeler	MoS_x	Sputtering	Steel 440C	0.11	0.016/6 GPa	Steel 440C	0.008/8.5 GPa
18.	1993	Ruff *et al.*	MoS_2	Dc sputtering Rf sputtering Magnetron sputtering Multilayer sputtering	Steel 440C	1–5	0.15/58 HRC	Au, Ni, Mo, Ti, Cu, bronze, PTFE, polyimide, different compositions	0.5
19.	1993	Matrin *et al.*	MoS_2 S:Mo-ratio = 2.04	Rf magnetron sputtering	Steel AISI 52100	0.12	$H = 7.8$ GPA $E = 165$–170 GPa	Steel	
20.	1998	Gilmore *et al.*	TiN–MoS_2	Dc magnetron co-sputtering	Stainless steel	0.5–3		Steel	
21.	1999	Wahl *et al.*	Pb–Mo–S amorphous	Dual ion-beam deposition (IBD)	Steel	0.2–0.51		Steel 52100	
22.	2003	Zhang *et al.*	$MoS_{1.3}$ basal plane orientation $MoS_{1.8}$ random orientation	Planar magnetron sputtering	Stainless steel 440C		$R_a = 0.05$ μm	Corundum	$R_a = 0.05$ μm
23.	2004	Steinmann *et al.*	MoS_2–C–TiB_2	Non-reactive dc magnetron sputtering	Steel		$R_z = 0.08$ μm	Steel 100Cr6	
24.	2005	Amaro *et al.*	MoS_2 –Ti	Dc magnetron sputtering	1200 SiC steel	1.22	M42 polished	Steel	

Experiments				Results		
Geometry	Speed m/s	Load N	Lubrication environment	Friction	Wear, lifetime, debris wear rate, $10^{-6} \cdot mm^3/Nm$	Comments/Conclusions
				0.025 0.005 0.005 0.01	Life: 2000 cycles " 10–20,000 cycles " 10–20,000 cycles	The friction coefficient is decreased when using a hard interlayer
Ball-on-disk Ball ∅ 6 mm	0.031 –0.107	0.5–3.6 0.33–0.69 GPa	Vacuum 10^{-7} Pa	0.004–0.1	0.8	Wear rate independent of load Friction coefficient decreases with increasing load
Ball-on-disk Ball ∅ 6 mm	0.03 –0.043	1 0.44 GPa	Vacuum 10^{-7}–10^{-4} Pa	0.01–0.05	High friction is associated with thick transfer films	Friction coefficient was 0.01–0.02 at oxygen pressures above $2 \cdot 10^{-5}$ Pa and 0.04–0.05 when it was below 10^{-5} Pa
Pin-on-ring	0.14	33 1.7 GPa	Dry argon	0.03–0.29	Pin: 10^{-7}–10^{-3} mm^3/m	
Pin-on-flat Reciprocating Ball ∅ 8 mm	0.0005	1.2 0.4 GPa	50 nPa vacuum	0.001 and below		Lowest friction ever measured between two solid surfaces in dry sliding!
Pin-on-disk Ball ∅ 6 mm	0.05	1	$T = 24C$ RH = 40%	0.1		$TiN_{1.0}$ $(MoS_{1.6})_{0.3}$ films turned out to give the best tribological performance
Ball-on-flat Reciprocating Ball ∅ 6.35 mm	0.004	9.8 0.92 GPa	RH < 2%	0.05	Wear depth < 25% of coating thickness after 10,000 sliding cycles	Pb content = 4–26%
Ball-on-flat Reciprocating Ball ∅ 10 mm		1–2	RH = 10% 50% 90 and 10% 50% 90%	0.06 0.08 0.15 0.25 0.40 0.43	4 5 20 10 60 80	Fretting wear test with 100 μm displacement oscillating at 10 Hz
Ball-on-flat Ball ∅ 4 mm	0.1	10 1.86 GPa	RH = 10–90%	0.12	Coated flat = 0.1–1.6 Ball = 0.00006 –0.00012	Coefficient of friction independent of RH
Pin-on-disk	0.2	80	RH = 50%	0.04–0.045		

Tribology data sheet for coatings: Soft metals

{Private} Reference			Coating					Counterface	
No.	Year	Authors	Film	Deposition method	Substrate	Thickness μm	Roughn./ Hardn. R_a, μm/H_v	Material	Roughn./ Hardn. R_a, μm/H_v
1.	1950	Bowden and Tabor	Lead		Copper	0.03–30		Steel	
2.	1985	Gerkema	Lead	Sputtered	Steel	0.1–0.2	0.05 → 0.28 → 0.46 →	Pb-coated steel	0.05
3.	1991	Liu *et al.*	Lead	Brush plated	Steel			Steel	
4.	1977	Sherbiney and Halling	Lead → Silver → Indium →	Ion plating	Steel ball (EN31)	0.01–10	0.75	Steel disk	
5.	1978	Arnell and Soliman	Lead → Silver → Indium →	Ion plating	Steel disk (EN31)	0.1–10	0.025–0.7	Steel	
6.	1976	Jahanmir *et al.*	Cadmium → Silver → Gold → Nickel →		Steel (AISI 1018, 1095 and 4140)	1 (0.05–10)	0.4/1700 MPa (840–4600 MPa)	Steel	
7.	1990	Erdemir *et al.*	Silver	Ion-assisted deposition	Al_2O_3	0.5–2	0.3	Al_2O_3	0.03
8.	1991	Tian *et al.*	Gold (Ni interlayer)	Electroplated	Brass	1.5 Ni = 2.5		Gold coating on Ni–Brass	
9.	1950	Bowden and Tabor	Indium		Tool steel	0.01–30		Steel	
10.	1989	Gawne and Ma	Chrom. → Nickel → Chrom. → Nickel →	Electroplated	Steel	40	0.25 0.5–0.6	Steel – " – Diamond – " –	0.25
11.	1991	Ajayi *et al.*	Silver	IBAD	Si_3N_4	1	0.072 GPa	Si_3N_4	/17.41 GPa
12.	1991a	Erdemir *et al.*	Silver	IBAD	ZrO_2 + 5%CaO	2	0.2	ZrO_2 + 5%CaO	0.03/15H_k

	{Private} Experiments			Results		
Geometry	Speed m/s	Load N	Lubrication environment	Friction	Wear, lifetime, debris wear rate, $10^{-6} \cdot mm^3/Nm$	Comments/Conclusions
Ball-on-flat ∅ 6 mm			Dry	0.3		$h = 30$ μm → minimum μ = 0.3 Softer substrate needs thicker film
Pin-on-disk Ball ∅ 5 mm	0.025	5 (2.5, 10)	Vacuum $5 \cdot 10^{-7}$ torr $T = 25$–300°C	0.08 0.10 0.16	Filmlife: 3–1300 m 9–10,000 m 4–11,000 m	Increase in μ with surface roughness Load (2.5–10 N) influence → μ very small Increased load → decreased endurance
Ball-on-disk	0.15 –3.75	10–3000	Lubricated Paraffin oil	0.05– 0.09		Higher hardness → lower μ and improved V No thickness influence
Pin-on-disk	0.04–1.2	5–40	Dry	0.1–0.6 0.15–0.5 0.05–0.5	Life: 500 rev. 150 " 150 "	Best lifetime for Pb Load, speed and thickness effects Film thickness μ min ~ $h = 0.5$–1 μm
Pin-on-disk ∅ 3 mm	0.01	1	Vacuum 10^{-8} torr	0.1–0.2 0.1–0.2 0.2–0.3		Increase in μ with h and R_a Increase in μ with speed
Cylinder-on-cylinder ∅ 6.3 mm	0.03	22.5	Argon	0.25–0.35 0.33 0.85–0.9 –	$1.8 \cdot 10^{-8}$ mg/cm – " – – " – Imediate failure	Optimum thickness = 1 μm
Pin-on-disk ∅ 9.5 mm	1	5 0.8 GPa	Dry air	0.4–0.9		Reduce wear rate and friction in Al_2O_3–Al_2O_3 contacts
Ball-on-flat Reciprocating ∅ 3.2 mm	0.004	1	Dry air → Polyphenyl ether lubrication →	0.3–1.2 0.1		Wear mechanisms described
Ball-on-flat ∅ 6 mm		40	Air	0.08–0.35		Minimum μ at $h = 0.1$–1 μm
Pin-on-cylind. Falex Scratch-test Reciprocating	0.1 0.04	225 20	Air	0.68–0.88 0.90–1.2 0.03–0.5 0.06–0.3		No correlation between coating hardness and wear Conventional chromium has lowest wear
Pin-on-disk Reciprocating	0.05	3–10	Air, $T = 23$°C RH = 40–45%	0.15–0.75	0.1	Friction coefficient reduced by 50% and wear rate by 1–2 orders of magnitude compared to uncoated contacts
Pin-on-disk Ball ∅ 9.5 mm	0.1–2	5 0.46 GPa	Air, $T = 23$°C RH <1%	0.2–0.45	Ball: 0.38 Disc: unmeasurable	Friction coefficient reduced by 50% and wear rate by several orders of magnitude compared to uncoated contacts

(Continued)

Tribology data sheet for coatings: Soft metals (*Continued*)

{Private} Reference			Coating					Counterface	
No.	Year	Authors	Film	Deposition method	Substrate	Thickness μm	Roughn./ Hardn. R_a, μm/H_v	Material	Roughn./ Hardn. R_a, μm/H_v
13.	1991b	Erdemir *et al.*	Silver	IBAD	MgO–PSZ	1.5	0.2/0.25 GPa	MgO–PSZ	0.03/11
14.	1992	DellaCorte *et al.*	Silver	Sputtered	Al_2O_3	1.5	0.1	Al_2O_3	/20
15.	1986	El-Sherbiny and Salem	Silver	Ion plating	Steel En-31 balls	1.5	0.75	Steel EN-31 plate	
16.	1992	Sander and Petersohn	Silver → Indium → Copper → Lead →	PVD	TiN coated steel (100Cr6)	<5		Steel (100Cr6)	
17.	2003	Yang *et al.*	Silver with hardultra-thin silverbond layer	IBAD and thermal evaporation	AISI 52100 steel	2	0.02/57–61 HRC	AISI 52100 steel	55–60 HRC
18.	2004	Lince	Gold $Au + MoS_2$	Sputtering	440C steel	0.26 0.11	0.03	440C steel	0.01–0.03

{Private} Experiments				Results		
Geometry	Speed m/s	Load N	Lubrication environment	Friction	Wear, lifetime, debris wear rate, $10^{-6} \cdot mm^3/Nm$	Comments/Conclusions
Pin-on-disk Ball ∅ 9.5 mm	0.05– 7.3	5 0.54 GPa	Air, $T = 22°C$ RH = 45%	0.3–0.4	0.005–0.5	Reduced friction and reduced wear rate by 2–3 orders of magnitude compared to uncoated contacts
Pin-on-disk Ball ∅ 24.5 mm	1	4.9	Air, $T = 25°C$ RH = 40–60%	0.25–0.4	20	Friction coefficient reduced by 50%, pin wear by a factor of 140 and disk wear by a factor of 2.5 Ti interlayer improved adhesion
Pin-on-disk Ball ∅ 6.35 mm	0.16	150	Air	0.2–0.6		Wear initially by microcutting, followed by wear debris abrasion and finally by fatigue wear process
Ball-on-flat Reciprocating Ball ∅ 10 mm	0.002	5 0.77	Air RH = 1–20%	0.17–0.19 0.15–0.24 0.27 0.28–0.32		Friction coefficient reduced by 50–80% compared to uncoated contact
Ball-on-disk Ball ∅ 12.7 mm	0.02–1	$p = 0.1–1$ MPa	Air	0.35–0.9	Wear coefficient = : 1.5×10^{-6}– 6.5×10^{-3}	Increased wear and friction with increased contact pressure and sliding speed
Pin-on-disk Ball ∅ 8 mm Pin ∅ 6.4 mm	0.08	5	Dry nitrogen RH 1%	Failed 0.013– 0.04	Failed after 10 m sliding 2000 m enurance life	15% MoS_2 content gives good endurance life at 0.1 MPa contact pressure and 41% at 730 MPa contact pressure

Tribology data sheet for coatings: Titanium nitride coatings

{Private} Reference			Coating					Counterface	
No.	Year	Authors	Film	Deposition method	Substrate	Thickness µm	Roughn./ Hardn. R_a, µm/H_v	Material	Roughn./ Hardn. R_a, µm/H_v
1.	1983	Sundquist *et al.*	TiN	Ion plating	Steel 40CrMoVo51	4		Mild steel	
2.	1987	Matthews *et al.*	TiN TiN TiN	Ion plating	HSS	1	0.06	Tool steel TiC Copper	
3.	1980	Jamal *et al.*	TiN	ARE	Stainless steel 304C	4–8	0.3	Steel TiC/440C TiN/440C Al	0.1
4.	1980	Habig *et al.*	TiN	CVD	Steel 205CrWMoV	3	0.9	Steel	1.7
5.	1989	Peebles and Pope	TiN	Reactive evaporation	Steel 304SS			Steel 440CSS	
6.	1990	Knight *et al.*	TiN	PVD sputtering evaporative CVD	Steel 835M30 Steel 835M30	2 2–5	0.14–0.88 0.65–0.88	Diamond	
7.	1987	Shimazki and Winer	TiN	Sputtering rf/reactive PVD	Tool steel	0.1–2.7	0.28–0.9	Steel	
8.	1990	Sue and Troue	TiN	PVD arc evaporation	Steel ring AISI 4620	10	0.1	Steel AISI 01 Block	0.25–0.3
9.	1990	Kennedy and Tang	TiN +Ti	PVD arc evaporation	Steel Inconel 625	25		Carbon graphite	
10.	1990	Bull *et al.*	TiN	PVD sputtering	Stainless steel	5		Steel TiN	
11.	1981	Hintermann	TiN	CVD	Steel		0.1	Steel TiC TiN	
12.	1989	Gardos	TiN					TiN	
13.	1985	Mäkelä and Valli	TiN	PVD	Steel HSS Pin	0.6–0.8	0.05–0.09	Tool steel TiN TiC	0.05–0.09
14.	1979	Suri *et al.*	TiN	Bias activated reactive evaporation	Stainless steel → Aluminium → Titanium →	4–6		Stainless steel 440C	
15.	1990	Habig	TiN	CVD	X205CrWMo 121	3	$R_z = 1$ µm	TiN on	$R_z = 1$ µm

	{Private} Experiments			Results		
Geometry	Speed m/s	Load N	Lubrication environment	Friction	Wear, lifetime, debris wear rate, $10^{-6} \cdot mm^3/Nm$	Comments/Conclusions
Pin-on-disk	0.2	0.5–10	Dry	0.2 at 0.5 N 0.4 at 1 N 0.7 at 10 N	2–10 times increased lifetime	Increase of friction with load Decreased wear even after failure Friction for St ~ St $\mu = 0.7$
Pin-on-disk	0.2	5	Dry	0.15–0.2 0.7	2.8 2.1	Influence of counterface material
Pin-on-disk	0.1	40	Dry	0.75 0.2 0.45		Low μ and *w* with hard wear couples Al does not support hard coatings Lubrication decreased μ
Pin-on-disk	0.1	2.5 5 50–500	Dry Dry Lubr.	0.17–0.21 0.1–0.38 0.1–0.06	0.76 0.2 –	$\mu = 0.38$ after coating failure
Pin-on-plate Pin ∅ 0.8, 15 and 54 mm	0.025	64 MPa 150 MPa 1060 MPa 0.33 N		0.17–0.24 0.15		No influence of pressure on μ or *w* Ti underlayer improves *w* but μ is the same Vacuum → rapid failure
Scratch test Stylus ∅ 100–200 μm	0.0005	0.25–10	Dry	0.2 0.2–0.35		μ increases with load
Pin-on-flat Reciprocating Pin ∅ 0.25 in	0.0025 Amplitude 2 cm	0.67 ~ 550 MPa 4.6 ~ 1000 MPa	Dry	0.6–0.9 Uncoated = 0.45–0.85		Formation of transfer layer TiN μ higher than for uncoated Influence of *h* on μ
Block-on-ring Reciprocating	0.07	1090	Dry	0.62–0.67	Reduced wear by 5 times Aggomer. α-Fe_2O_3 debris Little Ti_2O	Surface oxidation Abrasive and oxidative wear Transfer layer on steel
Ring-on-ring 'Drill press converted'	4.7	50 ~ 125 kPa 100 ~ 250 kPa	Dry		Spalling	Transfer film of carbon No influence of *h* on *w*
Pin-on-disk Ball ∅ 3 mm	0.075	2.5		0.3–0.7 0.1–0.7	1–10,000	Wear mechanisms: adhesion, oxidation, abrasion Infl. *v* → *w*:<0.11 m/s no influence; >0.11 increase Influence on microstructure → *w*
Pin-on-disk	0.01	5	Dry	0.49 0.16–0.31 0.19	10 1 0.5	Counterface influence Formation of TiO_2 layer
Pin-on-disk	0.2	5	Dry	0.6–0.9 0.1 0.15–0.7	0.5–5 0.2 2	Both pin and disk coated and uncoated → Influence of geometry Friction curves
Pin-on-disk 120° cone Tip ∅ 0.125 mm	0.2–0.23	4	Dry	1.15 0.92 0.92		Lubrication decreased μ to 0.175–0.225
Pin-on-disk	0.1	2.5	Air, 50% RH Vacuum	0.2 0.5		Friction depends strongly on the oxygen pressure of the environment

(Continued)

Tribology data sheet for coatings: Titanium nitride coatings (*Continued*)

{Private} Reference			Coating					Counterface	
No.	Year	Authors	Film	Deposition method	Substrate	Thickness μm	Roughn./ Hardn. R_a, μm/H_v	Material	Roughn./ Hardn. R_a, μm/H_v
16.	1991	Malliet *et al.*	TiN	Ion plating	HSS ASP23	1.5	0.24	Chromium steel 1.2080	0.02/63–66HRC
17.	1991	Singer *et al.*	TiN	Reactive magnetron sputtering	Tool steel M2	0.2–5	0.06–0.1 → 0.004 →	Steel 52100 → Steel 52100 → Sapphire →	
18.	1992	Stott *et al.*	TiN	PVD CVD	Stainless steel 321	3–5	0.35	Stainless steel 321	0.35
19.	1997	Huq and Celis	TiN	Magnetron sputtering ion plating	M2 tool steel	5	0.26/33 GPa	Alumina	0.2
20.	1999	Wiklund *et al.*	TiN	Electron beam evaporation + dc magnetron sputtering	HSS ASP 2030	4	0.015/2500	Steel AIAI 52100 Alumina	
21.	2004	Wiklund *et al.*	TiN	Reactive electron beam evaporation	HSS	2	–/2000	Steel Chromium Nickel Stainless steel	

{Private} Experiments				Results		
Geometry	Speed m/s	Load N	Lubrication environment	Friction	Wear, lifetime, debris wear rate, $10^{-6} \cdot mm^3/Nm$	Comments/Conclusions
Pin-on-disk Ball ∅ 10 mm	0.163	0.5–13	Air, $T = 22°C$ RH = 52%	0.55–1.15	0.1–0.8	
Ball-on-flat Unidirectional Ball ∅ 3.2–12.7 mm	0.0001	10	Air RH = 30%	0.5–0.7 0.15–0.2 0.05–0.15		Higher friction coefficients correspond to rougher surfaces and more debris in the wear track
Pin-on-disk Ball ∅ 50 mm Reciprocating	0.001	16.3	Carbon dioxide 20°C → 300°C → 500°C →	 0.2 0.4 0.6		Increased friction coefficient with increased temperature in carbon dioxide environment
Pin-on-disk Unidirectional Ball ∅ 10 mm	0.1 m/s	3–24	Dry	0.9–1.2	5–20	
Ball-on-disk Unidirectional Ball ∅ 5 mm .	0.05	5	Dry	0.5 1.1	Disc: −25; Ball: 40 Disc: 3; Ball: 0.1	
Ball-on-disk Unidirectional Ball ∅ 6 mm	0.01	30	Dry	On St: 0.6 On Cr: 0.7 On Ni: 0.35 On SS: 0.65		Initial friction 0.2 up to 20 revolutions Initial friction 0.2 up to 4 revolutions Initial friction immediately increasing from 0.2 Initial friction immediately increasing from 0.2

Tribology data sheet for coatings: Nitrides (−TiN excluded)

{Private} Reference			Coating					Counterface	
No.	Year	Authors	Film	Deposition method	Substrate	Thickness μm	Roughn./ Hardn. R_a, μm/H_v	Material	Roughn./ Hardn. R_a, μm/H_v
1.	1990a	Ronkainen *et al.*	(Ti,Al)N	Ion plating	Steel ASP23	2–8	0.04	Steel M50	
2.	1990	Roos *et al.*	(Ti,Al)N	PVD	Steel HSS	3.5		Steel Cr5020	
3.	1990a	Bull and Rickerby	CrN	Ion plating	Steel	2–9		Steel	
4.	1987	Miyoshi *et al.*	BN	Ion beam dep. PVD	Silicon, SiO_2, GaAs, InP	0.1		Diamond Tip ∅ 0.4 mm	
5.	1989	Miyoshi	BN	Ion beam dep. PVD				Stainless steel 304 and 440C Ti	
6.	1988	Elena *et al.*	BN	Rf magnetron sputtering	WC–Co	1		Diamond Tip ∅ 0.4 mm	
7.	1989	Inagawa *et al.*	C–BN	Reactive evaporation	Silicon	0.4		Stainless steel	
8.	1990b	Ronkainen *et al.*	Ti(B,N)	EB PVD rf magn. sput.	Steel ASP23	5 8	0.04	Steel M50	
9.	1990	Roos *et al.*	(Ti,Nb)N	PVD	Steel HSS	5.5		Steel Cr5020	
10.	1992	Coll *et al.*	(Ti,Al)N	Arc evaporation	Steel HSS M2 Ball	4		Steel disk AISI 4135	
11.	1993a	Vancoille *et al.*	(Ti,Al)N → (Ti,Nb)N → Ti(C,N) →	Ion plating	Steel HSS M2	2–2.5	0.01	Corrundum (α-Al_2O_3)	0.2
12.	1991	Watanabe *et al.*	C–BN → a-BN → h-BN →	Ion plating	Silicon wafer			SS 440C	0.1/4.3 GPa
13.	2001	Podgornik	TiAlN	PVD	Hardened AISI 4140 steel Nitrided AISI 4140 steel	4 4	0.4/1500 H_v 0.4/1900 H_v	AISI 52100 steel	–/700
14.	2003b	Tuszynski *et al.*	CrN Ti(C,N) Cr(C,N)	Vacuum arc PVD	Steel	2 2 2	0.05/2000 H_v 0.05/3100 H_v 0.05/2300 H_v	Al_2O_3	
15.	2004	Wiklund *et al.*	VN	Reactive electron beam evaporation	High-speed steel	1.1	–/1100 H_v	Steel Chromium Nickel Stainless steel	

	{Private} Experiments			Results		
Geometry	Speed m/s	Load N	Lubrication environment	Friction	Wear, lifetime, debris wear rate, $10^{-6} \cdot mm^3/Nm$	Comments/Conclusions
Pin-on-disk Ball ∅ 10 mm	0.1	10	Dry	0.7–0.9	Pin: 3 Disc: 25–60	
Pin-on-disk Ball ∅ 10 mm	0.4	4	Dry		Pin: 2 × TiN	
Pin-on-disk Ball ∅ 3 mm	0.1	2.75	Dry	0.6–0.8	Disc: 10 (0.1)	Good abrasive wear resistance Transfer layer on ball
Scratch test Single pass Dist. 10 mm	0.0002	1–15 5–25 GPa	Dry	0.05–0.2		Elastic and plastic deformation BN film is amorphous
Scratch test Single pass Dist. 10 mm	0.0002	1–15 5–25 GPa	Dry	0.08– 0.12		
Scratch test Single pass		2–20	Dry	0.05–0.3		
Ring-on-disk		10		0.35		μ = 0.05 in vacuum and 400°C
Pin-on-disk Ball ∅ 10 mm	0.1	10	Dry	0.68– 0.78 0.6–0.9	Pin: 0.8 Disc: 10–20	
Pin-on-disk Ball ∅ 10 mm	0.4	4	Dry		Pin = 3 × TiN ≈TiAlN	
Pin-on-disk Ball ∅ 30 mm	0.5–5	2	Dry, RH = 50%	0.37– 0.74	Pin: 0.37–1.24	
Pin-on-disk Ball ∅ 10 mm	0.1	5	Dry, $T = 25°C$ RH = 50–70%	1.0–1.4 0.9–1.4 0.1–0.4		Wear decreases with increasing speed
Ball-on-flat Reciprocal Ball ∅ 6.35 mm	0.0017	5	Dry RH = 50–60%	0.3 0.8–1.0 0.8–1.0	C–BN has less wear than a-BN and h-BN	
Pin-on-disk	1	60	Dry		Pin: 1 Pin: 0.5–0.8	Coated pin sliding against steel disk
Ball-on-flat Ball ∅ 10 mm	0.1	10	Dry RH = 50%	0.83 0.33 0.82	Disc: 0.03 Disc: 200 Disc: 1	The wear debris was removed from the contact continuously by a draught of dry argon
Ball-on-disk Unidirectional Ball ∅ 6 mm	0.01	30	Dry RH = 40%	St: 0.6 Cr: 0.6 Ni: 0.27 SS: 0.65		Initial friction 0.2 up to 20 revolutions Initial friction 0.2 up to 4 revolutions Initial friction immediately increasing from 0.2 Initial friction immediately increasing from 0.3

Tribology data sheet for coatings: Carbides

{Private} Reference			Coating					Counterface	
No.	Year	Authors	Film	Deposition method	Substrate	Thickness µm	Roughn./ Hardn. R_a, µm/H_v	Material	Roughn./ Hardn. R_a, µm/H_v
1.	1981	Hintermann	TiC	CVD	Steel		0.1	Steel TiC TiN AlO SiC	
			SiC	CVD	Steel		0.1	SiC Steel	
			Cr_7C_3	CVD	Steel		0.1	Steel	
2.	1960	Mordike			WC TiC TaC TiC NbC BC			WC WC TaC Graphite NbC BC	
3.	1987	Boving and Hintermann	TiC	CVD	Steel	4		Steel 52100	
4.	1987	Matthews *et al.*	TiC (ball)	Ion plating PVD	Steel			Tool steel TiN on steel	$58H_{RC}$ $0.02–0.04R_a$
5.	1990	Knight *et al.*	TiC	CVD	Steel	5	1700–2000	Diamond stylus	
6.	1980	Jamal *et al.*	TiC	Activated reactive evaporation = ARE	Steel 440C ball	4–8	0.3/ 2800–3000	Steel TiC TiN TiC/Ti TietAl $TiN + Ti_2N/Ti$ Ti + TiC/Al $TiN + Ti_2N/Al$	0.1/300 2800–3000 1800–2000
7.	1980	Habig *et al.*	TiC CrxCy	CVD "	Steel 62HRC	5 10	1/2000 1/1500	Steel – " –	1.7 1.7
8.	1979	Suri *et al.*	TiC	Bias activated reactive evaporation	Stainless steel Aluminium Titanium	4–6	0.1/3000	Stainless steel	0.1/300
9.	1989	Farges *et al.*	WC + Cr subl.	PACVD	Tool steel	16		Steel TiN + steel	52HRC
10.	1988	Srivastava *et al.*	WC	Rf reactive magnetron sputtering	Stainless steel	7	3200	Stainless steel	
11.	1987	Lindgren *et al.*	CrC	Detonation gun plasmagun		~20–100			
12.	1988	Rey	BC	CVD	Graphite			BC Steel 42CD4 SiC AlO Si_3N_4 ZrO	
13.	1990	Habig	TiC	CVD	Steel X205Cr WMo121	5	2.7	TiC coated steel X205Cr WMo121	62HRC

{Private} Experiments				Results		
Geometry	Speed m/s	Load N	Lubrication environment	Friction	Wear, lifetime, debris wear rate, $10^{-6} \cdot mm^3/Nm$	Comments/ Conclusions
Pin-on-disk	0.01	5	Dry	0.25 0.22–0.32 0.25–0.31 0.19 0.25–0.35 0.27–0.47 0.23 0.79	Disc: 1.5 Disc: 3 Disc: 5 Disc: 2 Disc: 2–7 Disc: 4–8 Disc: 0.5–0.7 Disc: 15	
Ring-on-ring	0.03	0.0025 –0.3	Dry	0.18 0.2 0.2 0.15 0.2 0.12		All carbides similar μ Increase in μ after ~1000°C Bulk carbides!
Pin-on-disk			Dry Humid Vacuum	0.3 0.2 0.3		No influence of vacuum Formation of TiO_2 improves friction
Pin-on-disk Ball ∅ 50 mm	0.2	5		0.3–0.4 0.15–0.2	Coated pin 0.4 " " 0.9	TiC sliding against TiN ⇒ low friction
Scratch Tip ∅ 0.1–0.25 mm	0.0005	0.25–10		0.2–0.25		μ increased with load μ depends on plastic flow
Pin-on-disk	0.1	40	Dry	0.5 0.2 0.2 0.175 0.8–0.85 0.2 0.3 0.35	High Low Medium Low High Low High High	Influence of counterface Lubrication decreased μ
Pin-on-disk "	0.1 "	2.5–5 "	Dry (lubr.) "	0.11–0.16 0.24–0.51	0.6 3–4	No running-in wear/ Lubr. $\mu = 0.09–0.105$ Higher wear/Lubr. $\mu = 0.06–0.1$
Pin-on-disk	0.2	4	Dry (lubr.)	0.7 0.9 1.07	Low Medium Medium	High friction and low wear
Pin-on-disk Ball ∅ 20 mm	0.015	1	Dry	0.3–0.4 0.25–0.5	Low Low	Pin wear reduced by polishing but not friction
Pin-on-disk	0.158	100	Lubr. SAE30 Mobil oil	0.09–0.25	Minimum with $h = 1–2$ μm	Coating thickness influence on wear
Pin-on-disk Reciprocating	0.007–0.7	0.7–7 MPa	$T = 20–800$°C	0.4–0.6		Different coating compositions and atmospheres
Pin-on-disk Reciprocating	0.0016	10	Dry	0.15 0.2–0.4 0.2–0.3 0.2–0.4 0.25–0.35 0.2–0.35		Not major influence of counterface BC/BC best tribocombination
Pin-on-disk	0.1	2.5	Air $R_h = 50\%$ → Vacuum →	0.2 0.5		Main wear mechanisms are polishing, tribooxidation and adhesion

Tribology data sheet for coatings: Oxides and borides

{Private} Reference			Coating					Counterface	
No.	Year	Authors	Film	Deposition method	Substrate	Thickness μm	Roughn./Hardn. R_a, μm/H_v	Material	Roughn./Hardn. R_a, μm/H_v
1.	1981	Hintermann	Al_2O_3	CVD	Cemented carbide tool			Steel 100Cr6 TiC	
2.	1987 1988	Roth *et al.*	Al_2O_3	Rf magnetron sputtering	Cast iron	4	0.36/300H_V 0.05	Steel 155CrVMo 121	0.26/1000H_v
3.	1987	Matthews *et al.*	Ti–O	Ion plating	HSS ball	2	0.06/2000–2400H_k	Steel Copper	58HRC 80H_v
4.	1987b	Fusaro	CdO	Sputtering	Steel	0.05–0.1		Steel	
5.	1988	Levy	TiB_2	CVD	WC–Co	30–40			
6.	1981	Hintermann	Fe_xB	CVD	Cemented carbide tools			Steel Cr6 pin	
7.	1980	Habig *et al.*	Fe_2B	CVD	Cemented carbide tools			Fe_2B	
8.	1990	Habig	TiB_2	Thermochemical treatment	Ti–6% Al–4%V	30	2.8	TiB_2 coated Ti–6% Al–4%V	

{Private} Experiments				Results		
Geometry	Speed m/s	Load N	Lubrication environment	Friction	Wear, lifetime, debris wear rate, $10^{-6} \cdot mm^3/Nm$	Comments/Conclusions
Pin-on-disk	0.01	20	Dry	0.45	Disc: 75 Pin: 0.5	Al_2O_3 sliding against TiC ⇒ low wear and friction
	0.01	20	"	0.19	Disc: 0.5 Pin: 0.005	
	0.01	5	"	0.62	Disc: 0.36 Pin: 0.66	
Plate-on-cylinder	0.04	100 Pa	Dry		Low wear compared to uncoated	
Pin-on-disk	0.2	5	Dry	0.2–0.35 0.5–0.65	Pin: 0.8	Coating wear ~TiN and TiC
Pin-on-disk	0.26	10	Dry	0.17–0.23	0.1	
		0.17 MPa	$T = 19–36°C$ $T = 540°C$	0.62–0.82 1.06	Extremely low compared to plasma sprayed	Very low wear
Pin-on-disk	0.01	5		0.76	Disc: 0 Pin: 12	
Pin-on-disk	0.01	5		0.40	0.06	Extremely low wear <TiN and TiC
Pin-on-disk	0.1	2.5	Air RH = 50% → Vacuum →	0.4–0.6 0.75		Wear volume was six times higher than for TiN

Tribology data sheet for coatings: Diamond and diamond coatings

{Private} Reference			Coating					Counterface	
No.	Year	Authors	Film	Deposition method	Substrate	Thickness μm	Roughn./ Hardn. R_a, μm/H_v	Material	Roughn./ Hardn. R_a, μm/H_v
1.	1973	Casey and Wilks	–	–	Natural diamond	–	Oriented	Natural diamond	Oriented
2.	1981	Mehan and Hayden	PCD	High pressure	Cemented WC	500	0.1–0.2 0.3–0.73	St.4620 St.Ni12Cr	0.2 0.76
3.	1988	Yust *et al.*	TiB_2	Ion implantation				Diamond	
4.	1990	Hayward and Singer	DC	μwave CVD	Single crystal silicon	1	0.001 –0.35	SCD Sapphire	
5.	1989	Jahanmir *et al.*	DC	HFCVD	SiC	2.5–4.3		SiC	
6.	1990	Blau *et al.*	DC	HFCVD	Silicon wafer		Pyramidal, μ-crystal plate	St52100 sapphire	
7.	1989	Wong *et al.*	DC	μwave CVD	Si, Mo...		0.047 –0.094	Steel	
8.	1990	Gardos and Soriano	DC	PACVD	α-SiC	1.5	0.1	DC on α-SiC → α-SiC →	0.1
9.	1990	Kohzaki *et al.*	DC	Hot filament CVD	SiC	10	0.02/ 80 GPa	SiC	
10.	1992	Miyoshi *et al.*	– – DC DC	– – μwave CVD μwave CVD	Natural diamond Natural diamond SiC SiC	– – 0.5–1.5	 0.013	Diamond → Si_3N_4 → Diamond → Si_3N_4 →	
11.	1992	Feng and Field	–	–	Diamond	–		Diamond	
12.	1993 a and b	Mohrbacher *et al.*	DC	Hot filament CVD	WC–Co (KIO)	1	0.97	DC → α-Al_2O_3 → Cr-steel → WC–Co →	

{Private} Experiments				Results		
Geometry	Speed m/s	Load N	Lubrication environment	Friction	Wear, lifetime, debris wear rate, $10^{-6} \cdot mm^3/Nm$	Comments/Conclusions
Stylus plane Oscillating	1–100 $\cdot 10^{-6}$	0.03–3	Dry and oil	0.04–0.11		No lubricant speed, load influence Orientation influence
Block ring Rotating	0.1	223 341 MPa	Dry → Oil →	0.1–0.7 0.05–0.1		
Stylus disk Rotating	$2.5 \cdot 10^{-3}$		Dry	0.01–0.04		Formation of a carbon layer
Stylus plane Oscillating – " –	$0.3 \cdot 10^{-3}$ – " –	2–5 – " –	Dry – " –	0.05–0.5 0.1–0.6		Strong topography influence DC = diamond coating SCD = single crystal diamond
Ball-on-3-flat	$38 \cdot 10^{-3}$	~40 Increasing	Dry	0.08	Much debris	Graphitization
Scratch 5 strokes			Dry	0.2–0.6	Much debris	
Block ring Rotating	1	54	Mineral oil	0.035–0.04		
Pin-on-flat Oscillating	0.0047	0.49 2900 MPa	Vacuum Air, 50% RH Vacuum	0.1 0.15 0.1	0.4 1 30	Friction measured at room temperature Test temperature cycle: 25°C, 850°C, 25°C
Ball-on-disk Ball ∅ 2.5 mm	0.04–0.08	4.9		0.09	Ball: 0.002 Disc: non-detectable	Smooth surfaces gives low friction
Ball-on-flat Reciprocating Ball ∅ 1.3–1.6 mm	0.0014	1	Air, 40% RH	0.01 0.03 0.09 0.1	0.2 0.2 0.05 0.05	No humidity effect on friction and wear in diamond–diamond or diamond–DC contacts Moisture increased friction in Si_3N_4–diamond and Si_3N_4–DC contacts
Stylus plane Oscillating	0.0001– 0.0002	1	Air	0.08		Most wear debris is hydrocarbon There is little graphite but no diamond
Ball-on-flat Ball ∅ 10 mm	0.001	1–6	Air, 22°C 50–60% RH	0.05 0.05 0.5–0.6 0.5		Transfer layers and high friction in DC–WC–C and DC–Cr steel contacts

(Continued)

Tribology data sheet for coatings: Diamond and diamond coatings (*Continued*)

{Private} Reference			Coating					Counterface	
No.	Year	Authors	Film	Deposition method	Substrate	Thickness μm	Roughn./ Hardn. R_a, μm/H_v	Material	Roughn./ Hardn. R_a, μm/H_v
13.	1994	Hollman *et al.*	DC	Hot flame CVD	Cemented carbide	1.5–5	0.53	Steel → Al–17Si → Al →	
14.	1997	Hollman	NDC	Biased hot flame CVD	Cemented carbide		0.03/1360	CC → Steel → Stainless steel Ti → Al →	/1360 /815 /165 /320 /110
15.	1998	Miyoshi	DC	Microwave plasma CVD	Silicon			CVD diamond	
16.	2003	Grillo *et al.*			Natural diam. → CVD diamond →			Natural diamond	
17.	2005	Abreu *et al.*	DC	Microwave plasma CVD	Si_3N_4	3	0.155	DC on Si_3N_4	0.155

{Private} Experiments				Results		
Geometry	Speed m/s	Load N	Lubrication environment	Friction	Wear, lifetime, debris wear rate, $10^{-6} \cdot mm^3/Nm$	Comments/Conclusions
Pin-on cylinder	0.062	50–1000	Dry	0.06–0.1 0.08–0.1 0.16–0.3	High wear against steel Very low wear against Al–Si	Experiments with ethanol lubrication also reported. Friction decreases with surface roughness
Ball-on-flat Ball ∅ 1.3–1.6 mm	0.1	2	Dry	0.08 0.095 0.07 0.5 0.53	0.00019 in sliding against stainless steel	Experiments with water and PAO oil lubrication also reported. Full transfer layer coverage with Ti and Al. Minor coverage with SS and no transfer layer with CC and BBC
Ball-on-flat Ball ∅ 1.6 mm	0.031–0.107	0.49/2 GPa	Air → Nitrogen → Vacuum → Water →	0.08 0.05 0.3 0.02	5 3 1000 2	
Ball-on-flat Ball ∅ 0.08/ Nd and 0.025/ CVDd mm	0.05	0.5 N/11 GPa and 24 GPa	Air	0.003–.008 0.002–.003		Friction coefficient increased from 0.002 to 0.035 with load increase from 0.5–2 N for CVD diamond sliding against diamond
Ball-on-flat Reciprocting Ball ∅ 5 mm	1 Hz and 6 mm stroke	20/6 GPa	Air, RH = 50–60%	0.03	Ball: 0.01	

Tribology data sheet for coatings: Diamond-like coatings

{Private} Reference			Coating					Counterface	
No.	Year	Authors	Film	Deposition method	Substrate	Thickness μm	Roughn./ Hardn. R_a, μm/H_v	Material	Roughn./ Hardn. R_a, μm/H_v
1.	1989 1990	Miyoshi *et al.* Miyoshi	a-C:H	Plasma CVD	Si_3N_4 Flat	0.06		Si_3N_4	
2.	1989 1989a 1990b	Anttila Hirvonen *et al.* – " –	a-C	Plasma arc discharge pulse PVD	WC–Co Flat	0.5		Steel Si_3N_4	
3.	1989	Matthews *et al.*	a-C:H	HFPVD		0.5–2		St52100	
4.	1988	Imamura *et al.*	a-C:H	Dc plasma CVD	Cemented carbide Flat	2		StMi–Mo	
5.	1990	Aksenov and Strel'nitskij	a-H	Plasma arc discharge		3–4		Steel	
6.	1991	Oguri and Arai	a-CiSi:H a-C:H	Dc discharge PACVD		2–3 1–5	0.1–0.3 0.1–0.2	St52100	Mirror
7.	1977	Ohja and Holland	a-C:H	Rf plasma PVD		0.2		Steel a-C:H	
8.	1990	Gropinchenko *et al.*	a-C:H	Pulse PVD		0.2–0.5		Quartz (St, germanium)	
9.	1991a	Yang *et al.*	a-C	Plasma arc-discharge PVD	Steel AISI316	1	0.03	PTFE	
10.	1987	Miyake *et al.*	a-C:H	Hot cathode glow discharge	Si(100)	0.5		Steel SUJ-1	
11.	1990a	Hirvonen *et al.*	a-C	Plasma arc-discharge PVD	Steel	0.6	0.03	St52100	/8.3 GPa
12.	1992	Koskinen *et al.*	a-C	Pulsed arc-discharge PVD	WC–Co → Si(100) →	1		Si_3N_4	
13.	1991	Itoh *et al.*	a-C:H	Silicone oil IBAD	St52100	0.15	0.1	St52100	
14.	1991c	Erdemir *et al.*	a-C:H	Ion beam deposition	SiC → Si_3N_4 → ZrO_2 →	2 2 2		Si_3N_4 → Si_3N_4 → Sapphire →	0.05/ 17 GPa 0.05/ 17 GPa 0.05/ 22 GPa
15.	1992	Wu *et al.*	a-C:H	Ion beam deposition	Si	0.3		Si_3N_4	
16.	1992 a and b	Ronkainen *et al.*	a-C	Pulsed arc-discharge PVD	Si	0.5		St52100 → StM50 → Al_2O_3	

{Private} Experiments				Results		
Geometry	Speed m/s	Load N	Lubrication environment	Friction	Wear, lifetime, debris wear rate, $10^{-6} \cdot mm^3/Nm$	Comments/Conclusions
Reciprocal Ball ∅ 1.6 mm	50–133 $\cdot 10^{-6}$	1–1.7 910 MPa	Air → Nitrogen → Vacuum → + high temp.	0.2–0.3 0.01– 0.15 0.1–1.3		a-C:H behaved like SCD Water vapor → increase of μ Annealed a-C:H → graphitic behaviour
Pin-on-disk Ball ∅ 6 mm	0.013	1.5 951– 1120 MPa	Dry	0.14 0.06	Small wear	Transfer of carbon → steel
Ball ∅ 10 mm		5–30	Dry	0.08–0.12		
Block-on- disk	0. 4	50	Dry	0.13–0.25		
	0.8	10–80	Vacuum → Air →	0.05–0.6 0.1		Contact → graphite on graphite Load influence → μ
Ball-on-disk Ball ∅ 6.35 mm	0.2–2	1.4–45	Dry RH = 50–70%	0.04–0.06 0.15–0.2	0.01–0.1 0.001–0.1	Optimized effect of carbon content No load effect Material transfer → steel surface → SiO_2
				0.28 0.2		
	0.05	0.2–1	Water	0.13–0.17 0.17		
Pin-on-disk Pin ∅ 5 mm	0.006	3.05	Dry	0.035		No transfer layer on DLC Low adhesion PTFE vs DLC
Ball-on-flat Oscillating Ball ∅ 1.5 mm	0.12	4.9	Air, 50% RH Vacuum	0.2–0.5 0.01–0.35		
Pin-on-disk Ball ∅ 6 mm	0.13	1.55 980 MPa	Air, 45% RH	0.12–0.18	0.07	Transfer layer on the pin
Pin-on-disk Ball ∅ 5.4 mm	0.1	1.8–2.2	Air, 72% RH	0.04–0.08 0.05–0.07	0.09–0.17 0.05	
Pin-on-disk Ball ∅ 5 mm	0.04	2.2	Air, 50–70% RH N_2 gas	0.04 0.02		The a-C film contained Si and O
Ball-on-disk Ball ∅ 9.5 mm Ball ∅ 6.3 mm	0.026	2	Dry air, 1% RH	0.07 0.07 0.07	0.02	Paper includes also comparison of tribological properties in humid, normal and dry air, argon and nitrogen
Pin-on-disk Reciprocating Ball ∅ 1.6 mm	0.0014	1	Air, 30–40% RH Dry nitrogen	0.15–0.20 0.01–0.03	0.1–1 0.01–0.1	
Pin-on-disk Ball ∅ 10 mm	0.1	5 5 10	Air RH = 50%	0.19–0.53 0.22–0.3 0.17–0.25	Pin: 0.045 Disc: 0.01 Pin: 0.015 Disc: 0.022 Pin: 0.0055 Disc: 0.042	Reaction layer was formed on the steel pins

(Continued)

Tribology data sheet for coatings: Diamond-like coatings (*Continued*)

{Private} Reference			Coating					Counterface	
No.	Year	Authors	Film	Deposition method	Substrate	Thickness μm	Roughn./ Hardn. R_a, μm/H_v	Material	Roughn./ Hardn. R_a, μm/H_v
17.	1992	Miyoshi *et al.*	Low-density DLC Low-density DLC High-density DLC High-density DLC	Glow discharge PVD	SiC	0.25–0.28		Diamond → Si_3N_4 Diamond → Si_3N_4	
18.	1992	Miyake	a-C:H DLC with mixed diamond and graphite	Plasma CVD				Stainless steel SUS440C	
19.	1992	Holiday *et al.*	a-C:H	FAB → Rf EBPVD → Rf CVD →	Tool steel ASP23	0.5	0.015–0.017/64R_c	St52100	
20.	1993	Blanpain *et al.*	a-C:H	Rf plasma → Arc ion plating → Laser ablation →	Stainless steel Cem. carbide Si(100)	2 1 1	0.17/5 GPa 0.25/13 GPa 0.06/25 GPa	Corundum	
21.	1996 1999c	Ronkainen *et al.* Ronkainen *et al.*	a-C:H	Rf plasma	Stainless steel AISI 440B	0.5–1	0.03/620	Steel → Alumina →	0.02 0.04
22.	1998	Ronkainen *et al.*	a-C → a-C:H → a-C:H (Ti) →	Rf plasma Pulsed vac arc Rf plasma	Stainless steel AISI 440B	0.5 0.5 0.5	0.03/620	Steel	0.02
23.	1998	Miyoshi	DLC	Ion beam rf	Silicon	0.52–0.66		CVD diamond	
24.	2000	Erdemir *et al.*	DLC	Plasma enhanced CVD	Steel → Sapphire →	1	0.05/8 GPa mirror/35 GPa	DLC on steel DLC on sapphire	0.05/ 8 GPa mirror/ 35 GPa
25.	2000	Jiang *et al.*	DLC	CFUBMF-PACVD	M42 tool steel	1.3	0.08 → 0.93 → 1.15 →	Tungsten carbide	

	{Private} Experiments			Results		
Geometry	Speed m/s	Load N	Lubrication environment	Friction	Wear, lifetime, debris wear rate, $10^{-6} \cdot mm^3/Nm$	Comments/Conclusions
Ball-on-flat Reciprocating Ball ∅ 1.3–1.6 mm	0.0014	1	Air, 40% RH	0.07 0.17 0.05 0.19	8 30 2 0.5	DLC more sensitive to moisture than DC and natural diamond
Ball-on-flat Reciprocating Ball ∅ 6.35 mm	0.120	4.9	Air Vacuum Air Vacuum	0.22–0.5 0.01–0.3 0.1–0.3 0.3–0.4		Low friction due to the formation of oriented hydrocarbons on the ball surface
Pin-on-disk Ball ∅ 10 mm	0.1		Air RH = 50%	0.07–0.11 0.18–0.45 0.38–0.48		
Ball-on-disk Reciprocating Ball ∅ 10 mm	0.0016	1	Air, 23°C RH = 70–80%	0.1–0.15 0.05–0.1 0.05–0.08	0.16 0.071 0.18	
Ball-on-disk Ball ∅ 10 mm	0.02– 3.0	5–40/ 0.78– 1.56 GPa 5–40/0.92 –1.84 GPa	Air RH = 50%	0.08–0.43 0.02–0.14	Disc: 0.05–0.18 Steel pin: 0.002–0.011 Disc: 0.04–0.16 AlO pin: 0.0001–0.00035	Transfer layer formed on pin surface at high load and high speed
Ball-on-disk Reciprocating Ball ∅ 10 mm	0.004	10/ 0.98 GPa	Air RH = 50%	0.15 0.22 0.1	Pin: 1 and disc: smoothing Pin: 0.06 and disc: 0.14 Pin: 0.03 and disc: 0.02	Results from oil with and without additives and water lubrication also reported
Ball-on-flat Ball ∅ 1.6 mm	0.031– 0.107	0.49/2 GPa	Air → Nitrogen → Vacuum → Water →	0.03 0.025 0.06 0.01	0.8 0.6 2 1	
Ball-on-flat, Ball ∅ 9.5/ St et 6.4/ Sa mm	0.2–0.5	10/1 GPa and 2.2 GPa	Dry nitrogen and argon	0.003 0.001	0.0005–0.01 0.0003	
Ball-on-flat Ball ∅ 6.3 mm	0.05	20 and 40	Dry air RH = 7%		0.1–0.3 0.15–0.55 1–3.5	No influence of normal load. Wear increases with surface roughness

(Continued)

Tribology data sheet for coatings: Diamond-like coatings (*Continued*)

{Private} Reference			Coating					Counterface	
No.	Year	Authors	Film	Deposition method	Substrate	Thickness µm	Roughn./ Hardn. R_a, µm/H_v	Material	Roughn./ Hardn. R_a, µm/H_v
26.	2003	Svahn *et al.*	a-C:W a-C:Cr	Magnetron sputtering	Ball bearing steel 100Cr6	2 3.5	0.007–0.39/ 10/W and 7/ Cr GPa	Steel (100Cr6)	
27.	2003b	Zhang *et al.*	DLC	Dc plasma CVD	Silicon	1		Stainless steel 440C	
28.	2004	Field *et al.*	DLC	Magnetron sputtering	M42 tool steel		/12–18 GPa	WC–6%Co	
29.	2005	Konca *et al.*	a-C	Magnetron sputtering	M42 tool steel	2.0–2.3	0.034/9.5 GPa	319 Al alloy (Al–7%Si)	
30.	2005a	Klaffke *et al.*	a-C:H (15% H)	RF PA-CVD	Steel			Steel Ceramic	/180–850 /1200–2700

{Private} Experiments				Results		
Geometry	Speed m/s	Load N	Lubrication environment	Friction	Wear, lifetime, debris wear rate, $10^{-6} \cdot mm^3/Nm$	Comments/Conclusions
Ball-on-flat Ball ∅ 9 mm	0.07	4/ 0.8 GPa	Air RH = 30–40%	0.1–0.3 0.15–0.34	0.2–0.4 0.18–1.2	Friction and wear increased with surface roughness
Ball-on-flat Ball ∅ 4.76 mm	0.025	1–3/ 1.2 – 1.8 GPa	Air RH = 20% → RH = 90% →	0.05–0.22 0.12–0.17	0.022–0.023 0.014–0.018	Formation of transfer layer was shown to control friction. No influence of speed on wear
Ball-on-flat Ball ∅ 5 mm	0.025	20/1 GPa 100/ 3.5 GPa	Air	0.07 0.02	0.025 0.01	Results from water lubrication also reported
Ball-on-flat Ball ∅ 4 mm	0.04– 0.15	4.9	Air RH = 52% Vacuum 0.01 Pa	0.16 0.52	$1.25 \cdot 10^{-4}$ mm^3/m $4.05 \cdot 10^{-4}$ mm^3/m	Very low friction coefficient of 0.006–0.02 was observed in vacuum when first run tribopair in ambient air. Presence of water vapour essential for low wear
Ball-on-flat Reciprocating Ball ∅ 10 mm	0.08 20 Hz with 200 μm stroke	10	Air RH = 2% RH = 50% RH = 100% Air RH = 2% RH = 50% RH = 100%	0.06–0.16 0.07–0.14 0.05–0.12 0.02–0.04 0.04–0.14 0.03–0.09	0.1–0.2 0.02–0.05 0.02–0.03 0.09–0.3 0.1–0.5 0.05–0.03	

APPENDIX B

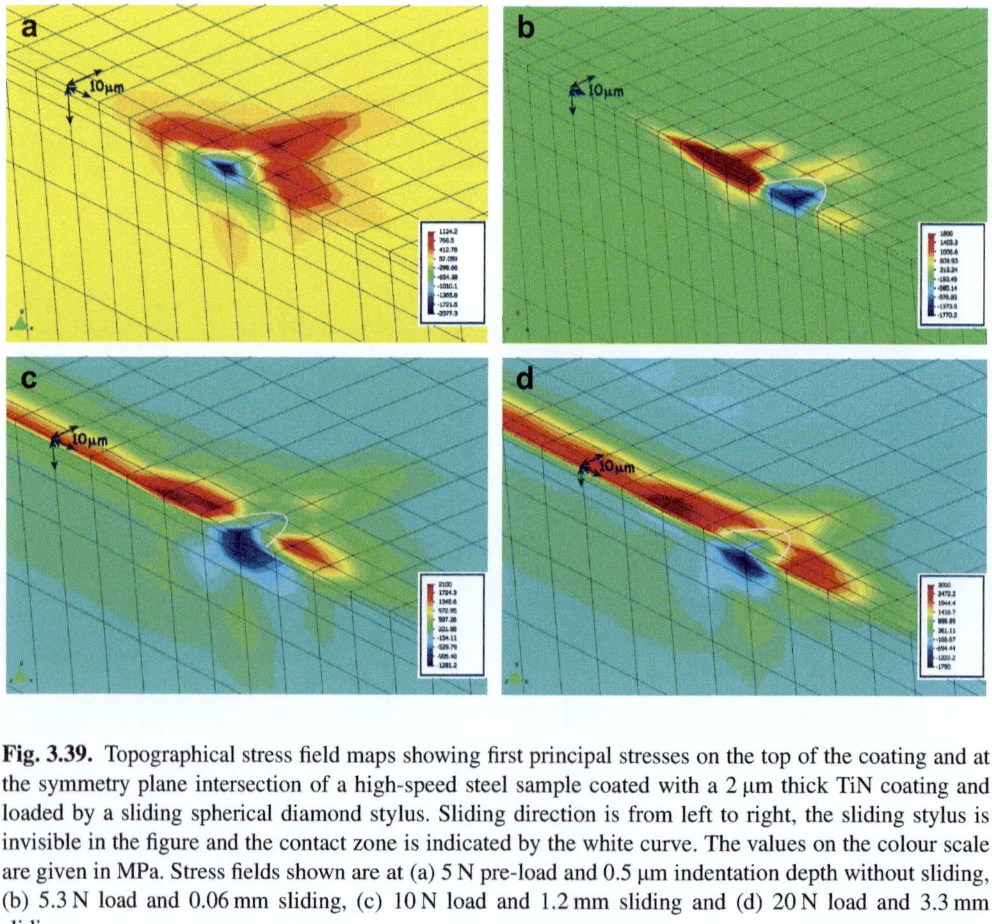

Fig. 3.39. Topographical stress field maps showing first principal stresses on the top of the coating and at the symmetry plane intersection of a high-speed steel sample coated with a 2 µm thick TiN coating and loaded by a sliding spherical diamond stylus. Sliding direction is from left to right, the sliding stylus is invisible in the figure and the contact zone is indicated by the white curve. The values on the colour scale are given in MPa. Stress fields shown are at (a) 5 N pre-load and 0.5 µm indentation depth without sliding, (b) 5.3 N load and 0.06 mm sliding, (c) 10 N load and 1.2 mm sliding and (d) 20 N load and 3.3 mm sliding.

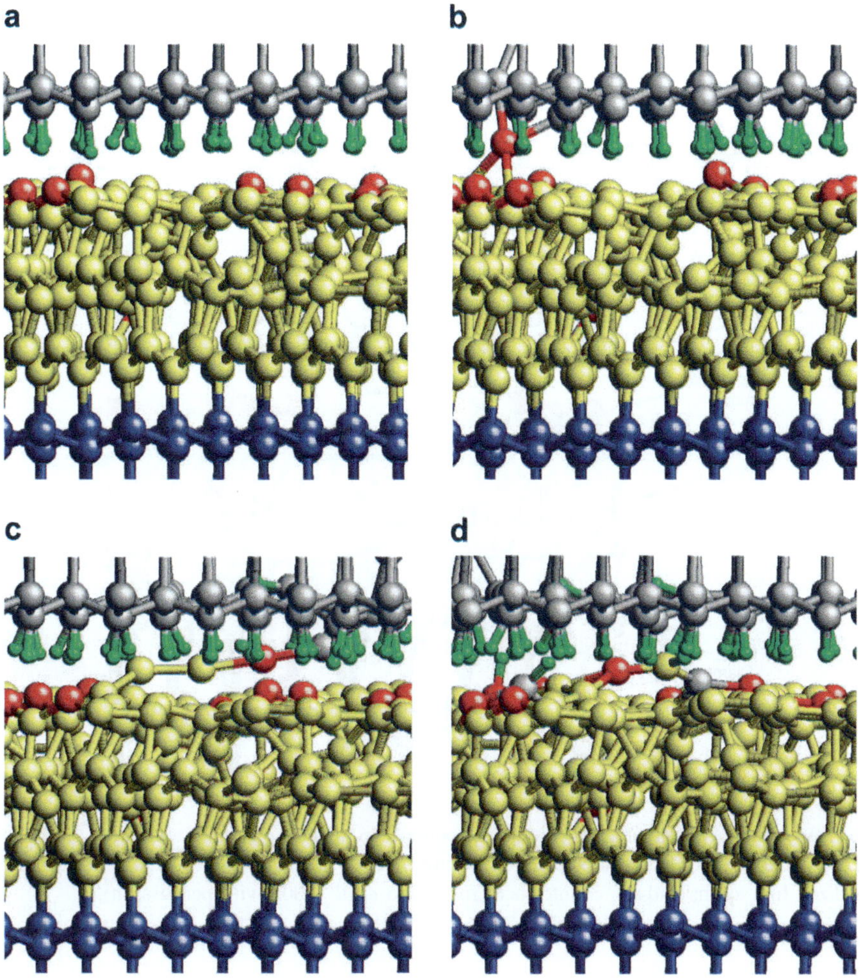

Fig. 3.102. Molecular dynamic simulation showing on an atomic level the interactions in a contact with a diamond surface sliding on a DLC coated diamond substrate. The diamond substrate is in blue, the DLC coating in yellow, the diamond contersurface in grey, hydrogen atoms in green and unsaturated sp-hybridized atoms in the DLC film in red (from Gao *et al.*, 2003, with permission).

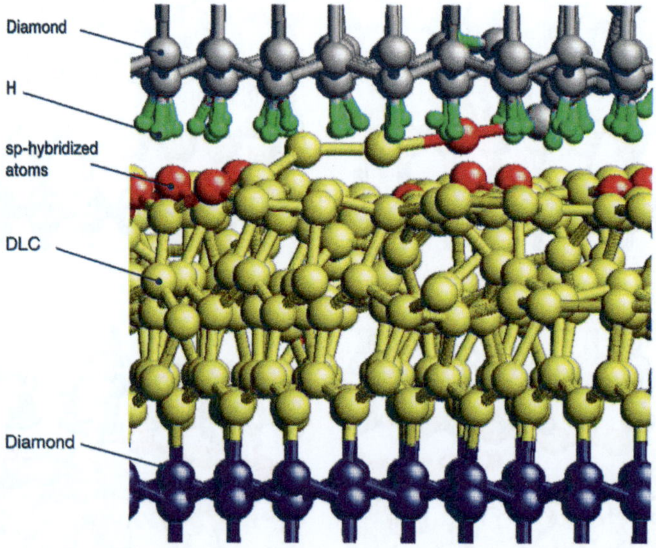

Fig. 4.63. A hydrogen-terminated (H) diamond counterface in sliding contact with an amorphous carbon layer (DLC) attached to a diamond substrate. Sliding-induced chemical reactions have occurred between sp-hybridized film atoms and the counterface (from Gao *et al.*, 2003 with permission).

REFERENCES

Abebe, M., Appl, F.C., Theoretical analysis of the basic mechanics of abrasive processes, Part I: General model. Wear, 126 (1988a) 251–266.

Abebe, M., Appl, F.C., Theoretical analysis of the basic mechanics of abrasive processes, Part II: Model of the ploughing process. Wear, 126 (1988b) 267–283.

Abel, P.B., Ferrante, J., Surface physics in tribology. In: Modern Tribology Handbook. Bhushan, B. (ed.), Vol. 1, CRC Press, New York, 2001, 5–47.

Abreu, C.S., Oliveira, F.J., Belmonte, M., Fernandes, A.J.S., Silva, R.F., Gomes, J.R., Grain size effect on self-mated CVD diamond dry tribosystem. Wear, 259 (2005) 771–778.

Achanta, S., Drees, D., Celis, J.P., Friction and nanowear of hard coatings in reciprocating sliding at milli-Newton loads. Wear, 259 (2005) 719–729.

Adachi, K., Hutchings, I.M., Wear-mode mapping for the micro-scale abrasion test. Wear, 255 (2003) 23–29.

Adams, F., Van Vaeck, L., Barrett, R., Advanced analytical techniques: platform for nano materials science. Spectrochimia Acta Part B – Atom. Spectrosc., 60 (2005) 1, 13–26.

Adams, M.J., Briscoe, B.J., Cartner, A.L., Tweedale, P.J., Assessing the durability of organic coatings. In: Mechanics of Coatings. Dowson, D. *et al.* (eds), Tribology series, 17, Elsevier, Amsterdam, 1990, 139–147.

Agrawal, D.C., Ras, R., Measurement of the ultimate shear strength of a metal–ceramic interface. Acta Metall., 37 (1984) 4, 1265–1270.

Ahmed, S.I.U., Bregliozzi, G., Haefke, H., Microfrictional properties of diamond-like carbon films sliding against silicon, sapphire and steel. Wear, 254 (2003) 1076–1083.

Ahn, H.S., A thermo-mechanical stress criterion for failure of surface-coated materials in concentrated sliding contacts. Tribology Int., 26 (1993) 1, 3–10.

Ahn, H.S., Roylance, B.J., Stress behaviour of surface-coated materials in concentrated sliding contact. Surf. Coat. Technol., 41 (1990) 1–15.

Ahn, H.S., Roylance, B.J., Thermal behaviour of surface-coated materials in concentrated sliding contacts. Tribology Int., 25 (1992) 4, 227–236.

Ahlroos, T., Effect of lubricant on the wear of prosthetic joint materials. Acta Polytechnica Scandinavica, Mech. Eng. Series No., 153, 2001, 26 pp.

Ai, X., Moyer, C.A., Rolling element bearings. In: Modern Tribology Handbook. Bhushan, B. (ed.), CRC Press, New York, 2001, 1041–1093.

Aindow, A.M., Dewhurst, R.J., Palmer, S.B., Laser-based non-destructive testing techniques for ultrasonic characterisation of subsurface flaws. In: Proc. Ultraconics International, Brighton, Dec. 1984, London, Butterworths, 1985, 329–335.

Aisenberg, S., Diamond-like carbon films – factors leading to improved biocompatilbility, Gulf and Western Co., USA, 1977, 87–88

Aisenberg, S., Chabot, R., Ion-beam deposition of thin films of diamond like carbon. J. Appl. Phys., 42 (1971) 2953–2958.

Aizawa, T., Mitsuo, A., Yamamoto, S., Sumitomo, T., Muraishi, S., Self-lubricated mechanism via the in situ formed lubricious oxide tribofilm. Wear, 259 (2005) 708–718.

Ajayi, O.O., Erdemir, A., Erck, R.A., Fenske, R.G., Nichols, F.A., The role of soft (metallic) films in the tribological behavior of ceramic materials. Wear, 149 (1991) 221–232.

Ajayi, O.O., Erdemir, A., Fenske, G.R., Nichols, F.A., Sproul, W.D., Graham, M., Rudnik, P.J., Tribological behavior of oil-lubricated TiN-coated steel. Surf. Coat. Technol., 54/55 (1992) 496–501.

Ajayi, O.O., Kovalchenko, A., Hersberger, J.G., Erdemir, A., Fenske, G.R., Surface damage and wear mechanisms of amorphous carbon coatings under boundary lubrication conditions. Surf. Eng., 19 (2003) 6, 447–453.

Ajayi, O.O., Soppet, M.J., Erdemir, A., Fenske, G.R., Shi, B., Liang, H., Assessment of amorphous carbon coating for artificial joints application. Trib. Lubr. Technol. (2005) Oct., 40–48.

Akagaki, T., Kato, K., Plastic flow process of surface layers in flow wear under boundary lubricated conditions. Wear, 117 (1987) 179–196.

Akarca, S.S., Altenhof, W.J., Alps, A.T., Subsurface deformation and damage accumulation in aluminum–silicon alloys subjected to sliding contact. Tribology Int., 40 (2007) 735–747.

Akhmatov, A.S., Molecular physics of boundary friction. Translation from: Molekularnaya fizica granichnogo treniya. Gosudarstvennoe Izdatel'stvo Fizico-Matematicheskoi Literatury. Moscow (1963) Israel Program for Scientific Translations, Jerusalem, 1966, 480 pp.

Aksenov, I.I., Strel'nitskij, V.B., Wear resistance of diamond-like carbon coatings. Surf. Coat. Technol., 47 (1991) 252–256.

Alahelisten, A., Mechanical and tribological properties of hot flame deposited diamond coatings. Acta Universitatis Upsaliensis. Dissertations from Faculty of Science and Technology 85, Uppsala, Sweden, 1994, 36 pp.

Alahelisten, A., Olsson, M., Hogmark, S., Mechanical and tribological properties of hot flame deposited diamond on cemented carbide substrates. Proc. 6th Int. Conf. on Tribology EUROTRIB'93, Budapest, Hungary, 3 (1993) 280–285.

Alliston-Greiner, A.F., Greenwood, J.A., Cameron, A., Thickness measurements and mechanical properties of reaction film formed by zinc dialkydithiophosphate during running. Proc. Inst. Mech. Engrs, C178/87 (1987) 565–572.

Almond, E.A., Aspects of various processes for coating and surface hardening. Vacuum, 34 (1984) 835.

Amaro, R.I., Martins, R.C., Sebara, J.O., Renevier, N.M., Teer, D.G., Molybdenum disulphide/titanium low friction coating for gears application. Tribology Int., 38 (2005) 423–434.

Amberg, S., Skiktbeläggning med PVD-teknik/Coating deposition with PVD technology (in Swedish). Verkstäderna, 30 (1984) 15, 44–48.

Anders, A. (ed.), Handbook of plasma immersion ion implantation and deposition. John Wiley and Sons, New York, USA, 2000, 736 pp.

Anderson, J.R. (ed.), Chemisorption and Reactions in Metallic Films. Academic Press, London, 1971, 555.

Anderson, T.L., Fracture Mechanics. Fundamentals and Applications, 2nd edition. CRC Press LLC, London, UK, 1995, 688 pp.

Andersson, J., Microengineered CVD diamond surfaces, triblogy and applications. Acta Universitatis Upsaliensis. Dissertations from Faculty of Science and Technology 977, Uppsala, Sweden, 2004, 50 pp.

Andersson, J., Erck, R.A., Erdemir, A., Frictional behavior of diamondlike carbon films in vacuum and under varying water vapor pressure. Surf. Coat. Technol., 163–164 (2003a) 535–540.

Andersson, J., Erck, R.A., Erdemir, A., Friction of diamond-like carbon films in different atmospheres. Wear, 254 (2003b) 1070–1075.

Anderssson, P., Holmberg, K., Limitations on the use of ceramics in unlubricated sliding applications due to transfer layer formation. Wear, 175 (1994) 1, 1–8.

Andersson, P., Koskinen, J., Varjus, S., Kolehmainen, J., Tervakangas, S., Lubrication improvements and friction reduction of DLC-coated sliding surfaces by in-process structuring an post polishing. Tribotest, 14 (2008) 97–112.

Andersson, S., Söderberg, A., Björklund, S., Friction models for sliding dry, boundary and mixed lubricated contacts. Proc. of 11th Nordic Symposium on Tribology – NORDTRIB2004, 1-4.6.2004, Tromsö, Norway, 651–665.

Angus, J.C., Koidl, P., Domitz, S., Carbon thin films. In: Plasma Deposited Thin Films. Mart, J., Jansen, F. (eds), CRC, Florida, 1986.

Angus, J.C., Diamond and diamond-like films. Thin Solid Films, 216 (1992) 126–133.

Anon, Diamond tooling takes off. Diamond Dep., Sci. Technol., 1 (1990) 2, 4.

Anthony, T.R., Angus, J.C., Glass, J.T., Diamond and diamond-like films. An intensive one-day course, lecture notes, part one. In connection with the 1st European Conf. on Diamond and Diamond-like Carbon Coatings, 17–19.9.1991, Crans-Montana, Switzerland.

Anttila, A., Ion-beam induced diamond-like carbon coatings. In: Structure–Property Relationships in Surface-modified Ceramics. McHargue, C.J. *et al.* (eds), Kluwer Academic Publishers, 1989, 455–475.

Aoki, Y., Ohtake, N., Tribological properties of segment-structured diamond-like carbon films. Tribology Int., 37 (2004) 941–947.

Archard, J.F., Contact and rubbing of flat surfaces. J. Appl. Phys., 24 (1953) 981–988.

Archard, J.F., Wear theory and mechanisms. In: Wear Control Handbook. Peterson, M.B., Winer, W. (eds), ASME, New York, 1980, 35–80.

Archer, N.J., Yee, K.K., Chemical vapour deposited tungsten carbide wear-resistant coatings formed at low temperatures. Wear, 48 (1978) 237–250.

Argon, A.S., Mechanical properties of near-surface material in friction and wear. In: Fundamentals of Tribology. Suh, N.P., Saka, N. (eds), MIT Press, London, 1980, 103–125.

Arnell, R.D., The mechanisms of tribology of thin film systems. Surf. Coat. Technol., 43/44 (1990) 674–687.

Arnell, R.D., Soliman, F.A., The effect of speed, film thickness and substrate surface roughness on the friction and wear of soft metal films in ultrahigh vacuum. Thin Solid Films, 53 (1978) 333–341.

Ashby, M.F., Materials Selection in Mechanical Design, 2nd edition. Butterworth- Heinemann, Oxford, UK, 1999, 502 pp.

ASM Handbook, Vol. 5, Surface Engineering. ASM International, Materials Park, Ohio, USA, 1994.

ASTM, Definition of terms relating to adhesion. D907–70, ASTM, Philadelphia, USA, 1970.

ASTM, Standard test method for adhesion strength and mechanical failure modes of ceramic coatings by quantitative single point scratch testing. Designation: C 1624-05, 2005, 28 pp.

Atar, E., Kayali, E.S., Cimenoglu, H., Reciprocating wear behaviour of (Zr,Hf)N coatings. Wear, 257 (2004) 633–639.

Atkins, A.G., Liu, J.H., Toughness and the transition between cutting and rubbing in abrasive contacts. Wear, 262 (2007) 146–159.

Aubert, A., Nabot, J.P., Ernoult, J., Renaux, P., Preparation and properties of MoS_2 films grown by d.c. magnetron sputtering. Surf. Coat. Technol., 41 (1990) 127–134.

Avallone, E.A., Baumeister, T., Marks' Standard Handbook of Mechanical Engineers, 9th edition. McGraw-Hill, New York, 1987.

Bachmann, P.K., Gartner, G., Lydtin, H., Plasma assisted chemical vapour deposited processes. MRS Bulletin, 13 (1988) 52–59.

Bachmann, P.K., Messier, R., Emerging technology of diamond thin films. Chem. and Engr. News, May 15, 1989, 24–39.

Baek, D.K., Khonsari, M.M., Friction and wear of rubber coating in fretting. Wear, 258 (2005) 898–905.

Bahadur, S., Wear research and development. J. of Lubrication Technology, Trans. ASME, 100 (1978) 449–454.

Bahadur, S., Friction and wear of polymers and composites. In: Tribology of Mechanical Systems. Vizintin, J., Kalin, M., Dohda, K., Jahamir, S. (eds), ASME Press, New York, 2004, 239–265.

Bair, S., Winer, W.O., A rheological model for elastohydrodynamic contacts based on primary laboratory data. J. Lubr. Tech., Trans.ASME F, 101 (1979) 258–265.

Baker, M.A., Kench, P.J., Joseph, M.C., Tsotsos, C., Leyland, A., Matthews, A., The nanostructure and mechanical properties of PVD CrCu(N) coatings. Surf. Coat. Technol., 162 (2003) 222–227.

Baker, M.A., Kench, P.J., Tsotsos, C., Gibson, P.N., Leyland, A., Matthews, A., Investigation of the nanostructure and wear property of physical vapor deposited coatings. J. Vac. Sci. Technol., A23/3 (2005) 423–433.

Bamberger, E.N., Clark, J.C., Development and application of the rolling contact fatigue test rig. In: Rolling Contact Fatigue Testing of Bearing Steels. Hoo, J.J.C. (ed.), ASTM STP 771, American Society for Testing Materials, 1982, 85–106.

Bandorf, R., Lüthje, H., Wortmann, T., Staedler, T., Wittorf, R., Influence of substrate material and topography on the tribological behaviour of submicron coatings. Surf. Coat. Technol., 174–175 (2003) 461–464.

Barriga, J., Kalin, M., Van Acker, K., Vercammen, K., Ortega, A., Leiaristi, L., Tribological performance of titanium doped and pure DLC coatings combined with a synthetic bio-lubricant. Wear, 261 (2006) 9–14.

Barros Bouchet de, M.I., Martin, J.M., New trends in boundary lubrication of DLC coatings. In: Tribology of Diamond-like Carbon Films. Donnet, C., Erdemir, A. (eds), Springer, New York, USA, 2008, 591–619.

Barros'Bouchet, M.I., Martin, J.M., Le-Mognet, T., Vacher, B., Boundary lubrication mechanisms of carbon coatings by MoDTC and ZDDP additives. Tribology Int., 38 (2005) 257–264.

Bartz, W.J., Xu, J., Wear behavior and failure mechanism of bonded solid lubricants. ASLE Preprint no. 86-AM-6F-1, ASLE An. Meeting May 1986, 8 pp.

Bartz, W.J., Holinski, R., Xu, J. A study on the behavior of bonded lubricating films containing molybdenum disulfide (MoS_2), graphite and antimony-thioantimonate ($Sb(SbS_4)$). Proc. 3rd Int. Conf. on Solid Lubrication, ASLE SP-14, 1984, 88–97.

Bartz, W.J., Holinski, R., Xu, J., Wear life and friction behavior of bonded solid lubricants. Lubr. Enging, J. ASLE, 42 (1986) 12, 762–769.

Barwell, F.T., Bearing Systems – Principles and Practice, Oxford University Press, Oxford, 1979, 565 pp.

Batista, J.C.A., Godoy, C., Matthews, A., Impact testing of duplex and non-duplex (Ti,Al)N and Cr–N PVD coatings. Surf. Coat. Technol., 163–164 (2003) 353–361.

Batista, J.C.A., Joseph, M.C., Godoy, C., Matthews, A., Micro-abrasion wear testing of PVD TiN coatings on untreated and plasma nitrided AISI H13 steel. Wear, 249 (2002) 971–979.

Bayer, R.G. (ed.), Selection and use of wear tests for metals. ASTM STP 615. American Society for Testing Materials, Philadelphia, 1977.

Bayer, R.G. (ed.), Selection and use of wear tests for coatings. ASTM STP 769. American Society for Testing Materials, Philadelphia, USA, 1982, 179.

Bayer, R.G., Shalkey, A.T., Wayson, A.R., Designing for zero wear. Machine Design, 1969, Jan., 142–151

Bayer, R.G., Mechanical Wear Prediction and Prevention, Marcel Dekker, New York, 1994, 657 pp.

Bec, S., Tonk, A., Georges, J.M., Coy, R.C., Bell, J.C., Roper, G.W., Relationship between mechanical properties and structures of zinc dithiophosphate anti-wear films. Proc. R. Soc. Lond. A 455 (1999) 4181–4203.

Becker, E.P., Trends in tribological materials and engine technology. Tribology Int., 37 (2004) 569–575.

Bedford, G.M., Richards, P.J., The fricsurf process – a new coating technology. Proc. Eng. Surface Conf., London, May 1986, Inst. of Metals, 5/1–5/2.

Beer, P., Rudnicki, J., Ciupinski, L., Djouadi, M.A., Nouveau, C., Modification by composite coatings of knives made of low alloy steel for wood machining purposes. Surf. Coat. Technol., (2003). 174–175 434–439.

Beerschwinger, U., Albrecht, T., Mathieson, D., Reuben, R.L., Yang, S.J., Taghizadeh, M., Wear at microscopic scales and light loads for MEMS applications. Wear, 181–183 (1995) 426–435.

Begelinger, A., De Gee, A.W.J., Failure of the film lubrication – the effect of running-in on the load carrying capacity of thin film lubricated concentrated contacts. J. Lubr. Technol., Trans. ASME, 103 (1981) 203–210.

Belak, J., Boercker, D.B., Stowers, I.F., Simulation of nanometer-scale deformation of metallic and ceramic surfaces. MRS Bulletin, 18 (1993) 5, 55–59.

Bell, T., Dong, H., Sun, Y. Realising the potential of duplex surface engineering. Tribology Int., 31/1–3 (1998a) 1–3, 127–137.

Bell, T., Mao, K., Sun, Y., Surface engineering design: modeling surface engineering systems for improved tribological performance. Surf. Coat. Technol., 108–109 (1998b) 360–368.

Bell, T., Sun, Y., The load bearing capacity of plasma nitrided steel under rolling/sliding contact. Surface Engng, 6 (1990) 133–139.

Bely, V.A., Sviridenok, A.I., Petrokovets, M.I., Savkin, V.G., Friction and Wear in Polymer-based Materials, Pergamon Press, Oxford, 1982, 415 pp.

Bemporad, E., Sebastiani, M., Pecchio, C., De Rossi, S., High thickness Ti/TiN multilayer thin coatings for wear resistant applications. Surf. Coat. Technol., 201 (2006) 2155–2165.

Ben Tkaya, M., Zahouani, H., Mezlini, S., Kapsa, P., Zidi, M., Dogui, A., The effect of damage in the numerical simulation of a scratch test. Wear, 263 (2007) 1533–1539.

Benjamin, P., Weaver, C., Measurement of adhesion of thin film. Proc. R. Soc., A254 (1959) 163–176.

Benjamin, P., Weaver, C., The adhesion of evaporated metal films on glass. Proc. R. Soc., A261 (1961) 516–531.

Benjamin, P., Weaver, C., The adhesion of metals to crystal faces. Proc. R. Soc., A274 (1963) 267–273.

Bennett, S., Matthews, A., Valli, J., Perry, A.J., Bull, S.J., Sproul, W.D., Multi-pass scratch testing at sub-critical loads. 19th Int. Conf. Metallurgical Coatings and Thin Films (ICMCTF 92), San Diego, USA 6-10.4.1992, 6 pp. Submitted to Tribologia – Finnish J. Tribology.

Berger, M., Thick physical vapour deposited TiB_2 coatings. Surf. Engng, 18 (2002) 3, 219–223.

Berger, M., Larsson, M., Mechanical properties of multilayer PVD Ti/TiB_2 coatings. Surf. Engng. 16 (2000) 2, 122–126.

Berger, M., Wiklund, U., Eriksson, M., Engqvist, H., Jacobson, S., The multilayer effect in abrasion – optimising the combination of hard and tough phases. Surf. Coat. Technol., 116–119 (1999) 1138–1144.

Berghaus, B., Improvements in and related to the coating of articles by means of thermally vapourized material. UK Patent specification 510992 (1938).

Bergman, B., Industriell försöksplanering och robust konstruktion/Industrial experiment and robust design. In Swedish, Studentlitteratur, Lund, Sweden, 1992, 328 pp.

Berman, A., Drummond, C., Israelichvili, J., Amonton's law at the molecular level. Tribology Lett, 4 (1998) 95–101.

Bernardo, A.T., Araújo, J.A., Mamiya, E.N., Proposition of a finite element-based approach to compute the size effect in fretting fatigue. Tribology Int., 39 (2006) 1123–1130.

Berthier, Y., Third-body reality – consequences and use of the third-body concept to solve friction and wear problems. In: Wear – Materials, Mechanisms and Practice. Stachowiak, G. (ed.), Tribology in Practice Series, John Wiley, Sons, Chichester, UK, 2005, 291–316.

Berthier, Y., Godet, M., Brendle, M., Velocity accommodation in friction. Tribology Trans., 32 (1989) 4, 490–496.

Berthier, Y., Vincent, L., Godet, M., Velocity accommodation in fretting. Wear, 125 (1988) 25–38.

Bertrand-Lambotte, P., Loubet, J.L., Verpy, C., Pavan, S., Understanding of automotive clearcoats scratch resistance. Thin Solid Films, 420–421 (2002) 281–286.

Beuth, J.L., Cracking of thin bonded films in residual tension. Int. J. Solids Structures, 29, 13 (1992) 1657–1675.

Bhansali, K.J., Kattamis, T.Z., Quality evaluation of coatings by automatic scratch testing. Wear, 14 (1990) 59–71.

Bhattacharya, K., James, R.D., A theory of thin films of martensitic materials with applications to microactuators. J. Mech. Phys. Solids, 47 (1999) 531–576.

Bhattacharya, R.S., Evaluation of high energy ion-implanted polycarbonate for eyewear applications. Surf. Coat. Technol., 103–104 (1998) 151–155.

Bhushan, B., Development of a wear test apparatus for screening bearing-flange materials in computer tape drives. ASLE Transactions, 30 (1987) 2, 187–195.

Bhushan, B., Tribological design – information storage and retrieval. Leeds-Lyon Symp. on Tribology, 6–9.9.1988, Leeds, UK, 13 pp.

Bhushan, B., Tribology and Mechanics of Magnetic Storage Devices, Springer-Verlag, New York, 1990.

Bhushan, B. (ed.), Handbook of Micro/Nanotechnology. CRC Press, London, 1995, 628 pp.

Bhushan, B., Tribology and Mechanisms of Magnetic Storage Devices, IEEE Press, Springer, New York, 1996, 1125 pp.

Bhushan, B., Contact mechanics of rough surfaces in tribology: multiple asperity contact. Tribology Lett., 4 (1998a) 1–35.

Bhushan, B., (ed.), Tribology issues and opportunities in MEMS. Proc.. of NSF/AFOSR/ ASME Workshop on Tribology Issues and Opportunities, MEMS, 9–11 November 1997, Columbus, USA, Kluver Academic Publishers, London, 1998b.

Bhushan, B., Chemical, mechanical and tribological characterization of ultra-thin and hard amorphous carbon coatings as thin as 3.5 nm: recent developments. Diamond and Related Materials, 8 (1999a) 1985–2015.

Bhushan, B., Nanotribology of ultrathin and hard amorphous carbon films. In: Tribology of Diamond-like Carbon Films. Donnet, C., Erdemir, A. (eds), Springer, New York, USA, 2008, 510–570.

Bhushan, B., Principles and Applications in Tribology. John Wiley, Sons, New York, 1999b, 1020 pp.

Bhushan, B. (ed.), Fundamentals of tribology and bridging the gap between the macro- and micro/nanoscales. NATO Science Series, II. Mathematics, Physics and Chemistry, Vol. 10, Kluver Academic Publishers, Boston, 2001a, 963 pp.

Bhushan, B., Macro- and microtribology of magnetic storage devices. In: Modern Tribology Handbook. Bhushan, B. (ed.), CRC Press, New York, 2001b, 1413–1513.

Bhushan, B., Macro- and microtribology of MEMS materials. In: Modern Tribology Handbook. Bhushan, B. (ed.), CRC Press, New York, 2001c, 1515–1548.

Bhushan, B., Tribology on the macroscale of microelectromechanical system materials: a review. Proc. Instn. Mech. Engrs Part J 215 (2001d) 1–18.

Bhushan, B., Surface roughness analysis and measurement techniques. In: Modern Tribology Handbook. Bhushan, B. (ed.), CRC Press, New York, 2001e, 49–119.

Bhushan, B., Nanotribology and nanomechanisms. Wear, 259 (2005) 1507–1531.

Bhushan, B., Gupta, B.K., Handbook of Tribology – Materials, Coatings and Surface Treatments, McGraw-Hill Inc., New York, 1991, 1168 pp.

Biederman, H., The properties of films preparation by the sputtering of PTFE and plasma polymerization of some freons. Vacuum, 31 (1981) 7, 285–289.

Billgren, P., The use of nitride and carbide coatings on high speed steel tools. Speedsteel Technical Report ZSD 22/84 (1984).

Bin-Sudin, M., Leyland, A., James, A.S., Matthews, A., Housden, J., Garside, B.L., Substrate surface finish effects in duplex coatings of PAPVD TiN and CrN with electroless nickel–phosphorus interlayers. Surf. Coat. Technol., 81 (1996) 215–224.

Birkholz, M., Fewsrer, P.F., Gentzel, C., Thin Film Analysis by X-ray Scattering, Wiley, New York, USA, 2006.

Bishop, H.E., Auger electron spectroscopy. In: Methods of Surface Analysis. Walls, J.M. (ed.), Cambridge University Press, 1989, 87–126, 36 pp.

Björk, T., Berger, M., Westergård, R., Hogmark, S., Bergström, J., New physical vapour deposition coatings applied to extrusion dies. Surf. Coat. Technol., 146–147 (2001) 33–41.

Blanchet, T.A., Kennedy, F.E., Sliding wear mechanism of polytetrafluoroethylene (PTFE) and PTFE composites. Wear, 153 (1992) 229–243.

Blanpain, B., Celis, J.P., Roos, J.R., A comparative study of the fretting wear of hard carbon coatings. Thin Solid Films, 223 (1993) 65–71.

Blanpain, B., Franck, M., Mohrbacher, E., Vancoille, J.P., Celis, J.P., Roos, J.R., Micro-Raman spectroscopy for the characterization of wear induced surface modifications on hard coatings. 19th Leeds–Lyon Symp. on Tribology, Thin films in tribology, Leeds 8–11.9.1992, 9 pp.

Blau, P.J., Friction and Wear Transitions of Materials – Break-in, Run-in, Wear-in, Noyes Publications, Park Ridge, New Jersey, USA, 1989, 476 pp.

Blau, P., Simulative friction and wear testing. In: Modern Tribology Handbook. Bhushan, B. (ed.), CRC Press, New York, 2001, 511–522.

Blau, P., Tribology web page.https://www.ms.ornl.gov/researchgroups/SPM/methods/Tribology/Tribology.htm. 25.6, 2008.

Blau, P.J., Yust, C.S., Clausing, R.E., Morphological aspects of the friction of hot-filament-grown diamond thin films. In: Mechanics of Coatings. Dowson, D. *et al.* (eds), Tribology Series 17, Elsevier, Amsterdam, 1990, 399–407.

Blomberg, A., Friction and wear of ceramics. PhD thesis, Uppsala University, Material Science Division, 1993, 47 pp.

Blomberg, A., Erickson, L., Lu, J., Hogmark, S., Olsson, M. Improved wear resistance of ceramics through the use of coatings. Surf. Coat. Technol., in press, 1993, 16 pp.

Blug, B., Konrath, G., Hollstein, T., Low friction coasting – a new design element for engineers. Proc. 2nd World Tribiology Congress, 3–7.9.2001, Vienna, Austria, Austrian Tribiology Society, 281–284.

Boelens, S., Veltrop, H., Hard coatings of TiN, (TiHf)N and (TiNb)N deposited by random and steered arc evaporation. Surf. Coat. Technol., 33 (1987) 63–71.

Booser, E.R., Tribology Data Handbook, CRC press, New York, USA, 1997, 1099 pp.

Boothroyd, G., Temperatures in orthogonal metal cutting. Proc. Inst. Mech. Engrs, 177 (1963) 29, 789–810.

Borland, D.W., Residual stress measurement – methods, limitations and significance. In: Quality Control and Assurance in Advanced Surface Engineering. Strafford, K.N., Subramanian, C. (eds), The Institute of Materials, The University Press, Cambridge, UK, 1997, 113–125.

Borodich, F.M., Chung, Y.W., Keer, L.M., Environmental and surface chemical effects on tribological properties of carbon-based coatings. In: Tribology of Diamond-like Carbon Films. Fundamentals and Applications. Donnet, C., Erdemir, A. (eds), Springer, New York, USA, 2008, 282–290.

Borodich, F.M., Harris, S.J., Keer, L.M., Cooper, C.V., Wear and abrasiveness of hard carbon-containing coatings under variation of the load. Surf. Coat. Technol., 179 (2004) 78–82.

Borodich, F.M., Keer, L.M., Harris, S.J., Self-similarity in abrasiveness of hard carbon-containing coatings. J. Tribology, Trans. ASME, 125 (2003) 1–7.

Borruto, A., Crivellone, G., Marani, F., Influence of wettability on friction and wear tests. Wear, 222 (1998) 57–65.

Bose, K., Wood, R.J.K., Optimum test conditions for attaining uniform rolling abrasion in ball cratering tests on hard coatings. Wear, 258 (2005) 322–332.

Bouslykhane, K., Moine, P., Villain, J.-P., Grilhé, J., Mechanical properties and wear resistance of ion-beam-assisted sputter-deposited NiTi(N) coatings. Surf. Coat. Technol., 49 (1991) 457–461.

Bouzakis, K.D., Hadjiyiannis, S., Skordaris, G., Anastopoulos, J., Mirisidis, I., Michailidis, N., Efstathiou, K., Knotek, O., Erkens, G., Cremer, R., Rambadt, S., Wirth, I., The influence of the coating thickness on its strength properties and on the milling performance of PVD coated inserts. Surf. Coat. Technol., 174–175 (2003) 393–401.

Boving, H.J., Hintermann, H.E., Properties and performance of chemical-vapour-deposited TiC coated ball-bearing components. Thin Solid Films, 153 (1987) 253–266.

Boving, H., Hintermann, H.E., Begelinger, A., De Gee, A.W.J., Load-carrying capacity of lubricated steel point contacts coated by chemical vapour deposition. Wear, 88 (1983) 13–22.

Boving, H., Hintermann, H., Hänni, W., Ball bearing with CVD–TiC coated components. Proc. 3rd European Space Mechanisms and Tribology Symposium, Madrid, Spain, 30.9–2.10.1987, 155–159.

Bowden, F.P., Tabor, D., Friction and Lubrication of Solids. Part I. Oxford, Oxford UP, Oxford, UK, 1950, 321 pp.

Bowden, F.P., Tabor, D., The friction and Lubrication of Solids. Part II. Oxford, Oxford UP, UK, 1964, 515 pp.

Bower, A.F., Fleck, N.A., Brittle fracture under a sliding line contact. J. Mech. Phys. Solids, 42 (1994) 9, 1375–1396.

Box, G.E.P., Hunter, W.G., Hunter, J.S., Statistics for Experimenters, John Wiley, New York, USA, 1978, 653 pp.

Boxman, R.L., Haber, D., Martin, P.J., Handbook of Vacuum Arc Science and Technology. William Andrew, New York, USA, 1996, 773 pp.

Boxman, R.L., Zhitomirsky, V.N., Grimberg, I., Rapoport, L., Goldsmith, S., Beiss, B.Z., Structure and hardness of vacuum arc deposited multi-component nitride coatings of Ti, Zr and Nb. Surf. Coat. Technol., 125 (2000) 257–262.

Bragallini, G.M., Cavatorta, M.P., Sainsot, P., Coated contacts: a strain approach. Tribology Int., 36 (2003) 935–941.

Bray, R.C., Film adhesion studies with the acoustic microscope. Thin Solid Films, 74 (1980) 15, 295–302.

Breton, E., Dubourg, M.C., Adhesion for coatings. 18th Leeds–Lyon Symp. on Tribology, 3–6.9.1991a, Lyon, France, Elsevier Tribology Series No. 21, Amsterdam, Netherlands, 41–47.

Breton, E., Dubourg, M.C., Behaviour of cracked coating submitted to Hertzian moving contact. 18th Leeds–Lyon Symp. on Tribology, 3–6.9.1991b, Lyon, France, 8 pp.

Brewe, D.E., Slider bearings. In: Modern Tribology Handbook. Bhushan, B. (ed.), CRC Press, New York, 2001, 969–1039.

Briggs, D., Grant, J.T. (eds), Surface Analysis by Auger and X-ray Photoelectron Spectroscopy. Wiley, New York, USA, 2003.

Briscoe, B.J., Pooley, C.M., Tabor, D., Friction and transfer of some polymers in unlubricated sliding. In: Advances in Polymer Friction and Wear – Polymer Science and Technology. Lee, H. (ed.), Plenum Press, New York, 1974, 191–204.

Briscoe, B.J., Tabor, D., The sliding wear of polymers: a brief review. In: Fundamentals of Tribology. Suh, N.P., Saka, N. (eds), The MIT Press, London, 1980, 733–758.

Briscoe, B.J., Sinha, S.K., Wear of polymers. Proc. Instn Mech. Engrs, Part J, 216 (2002) 401–413.

Briscoe, B.J., Sinha, S.K., Tribology of polymeric solids and their composites. In: Wear – Materials, Mechanics and Practice. Stachowiak, G.W. (ed.), John Wiley, Sons, Chichester, UK, 2005, 223–267.

Broek, D., Elementary Engineering Fracture Mechanics. Kluwer Academic Publishers, Dordrecht, 1997.

Bromark, M., Larsson, M., Hedenqvist, P., Olsson, M., Hogmark, S., Influence of substrate and surface topography on the critical surface force in scratch adhesion testing of TiN-coated steels. Surf. Coat. Technol., 52 (1992) 195–203.

Bromark, M., Larsson, M., Hedenquist, P., Olsson, M., Hogmark, S., Bergmann, E., PVD coatings for tool applications: tribological evaluation. Surf. Engng, 10 (1994) 3, 205–214.

Bromark, M., Wear resistance of PVD coatings for tool applications. PhD thesis at Uppsala University, Sweden, Acta Universitatis Upsaliensis, Faculty of Science and Technology Dissertations Nr 192, 1996, 42 pp.

Brookes, C.A., Brookes, E.J., Diamond in perspective: a review of mechanical properties of natural diamond. Diamond and Related Materials, 1 (1991) 13–17.

Brossa, F., Lang, E., Plasma spraying – a versatile coating technique. In: Advanced Techniques for Surface Engineering. Gissler, W., Jehn, H.A. (eds), Kluwer Academic Publishers, Dordrecht, 1992, 199–234.

Brudnyi, A.I., Karmodonov, A.F., Structure of molybdenum disulphide lubricant film. Wear, 33 (1975) 243.

Bruno, M., Bugliosi, S., Chiara, R., The performance of titanium-nitride coated HSS tools. J. Engng Mat. Tech., 110 (1988) 274–277.

Brydson, J.A., Handbook for Plastics Processors, Heinemann Newnes, Oxford, UK, 1990, 208 pp.

Buckley, D., The use of analytical surface tools in the fundamental study of wear. Wear, 46 (1978) 19–53.

Buckley, D., Surface effects in adhesion, friction, wear, and lubrication. Elsevier, Amsterdam, 1981, 631 pp.

Buckley, D., Properties of surfaces. In: CRC Handbook of Lubrication – Theory and Practice of Tribology. Volume II Theory and Design. Booser, E. (ed.), CRC Press Inc., Boca Raton, Florida, 1984, 17–30.

Budinski, K.G., Tool materials. In: Wear Control Handbook. Peterson, M.B., Winer, W. (eds), American Society of Mechanical Engineers, New York, USA, 1980, 931–986.

Bull, S.J., Failure modes in scratch adhesion testing. Surf. Coat. Technol., 50 (1991) 25–32.

Bull, S.J., Interfaces and adhesion. In: Advanced Techniques for Surface Engineering. Gissler, W., Jehn, H.A. (eds), Kluwer Academic Publishers, Dordrecht, The Netherlands, 1992, 31–68.

Bull, S.J., Failure mode maps in the thin film scratch adhesion test. Tribology Int., 30 (1997) 7, 491–498.

Bull, S.J., Physical vapour deposition methods for protection against wear. In: Surface Coatings for Protection against Wear. Mellor, B.G. (ed.), CRC Press, Woodhead Publishing Ltd, Cambridge, UK, 2006, 146–183.

Bull, S.J., Berasetegui, E.G., An overview of the potential of quantitative coating adhesion measurement by scratch testing. Tribology Int., 39 (2006) 99–114.

Bull, S.J., Berasetegui, E.G., Page, T.F., Modelling of the indentation response of coatings and surface treatments. Wear, 256 (2004) 857–866.

Bull, S.J., Bhat, D.G., Staia, M.H., Properties and performance of commercial TiCN coatings. Part 1: Coating architecture and hardness modeling. Surf. Coat. Technol., 163–164 (2003a) 499–506.

Bull, S.J., Bhat, D.G., Staia, M.H., Properties and performance of commercial TiCN coatings. Part 2: Tribological performance. Surf. Coat. Technol., 163–164 (2003b) 507–514.

Bull, S.J., Chalker, P.R., Lubricated sliding wear of physically vapour deposited titanium nitride. Surf. Coat. Technol., 50 (1992) 117–126.

Bull, S.J., Rickerby, D.S., Compositional, microstructural and morphological effects on the mechanical and tribological properties of chromium nitrogen films. Surf. Coat. Technol., 43/44 (1990a) 732–744.

Bull, S.J., Rickerby, D.S., New developments in the modelling of the hardness and scratch adhesion of thin films. Surf. Coat. Technol., 42 (1990b) 149–164.

Bull, S.J., Rickerby, D.S., Evaluation of coatings. In: Advanced Surface Coatings – A Handbook of Surface Engineering. Rickerby, D.S., Matthews, A. (eds), Blackie, London, 1991, 315–342.

Bull, S.J., Rickerby, D.S., Gent, J.T., Quality assurance assessment of thin films. Surf. Engng, 7 (1991) 2, 145–153.

Bull, S.J., Rickerby, D.S., Jain, A., The sliding wear of titanium nitride coatings. Surf. Coat. Technol., 41 (1990) 269–283.

Bull, S.J., Rickerby, D.S., Matthews, A., Leyland, A., Pace, A.R., Valli, J., The use of scratch adhesion testing for the determination of interfacial adhesion: the importance of frictional drag. Surf. Coat. Technol., 36 (1988a) 503–517.

Bull, S.J., Rickerby, D.S., Robertson, T., Hendry, A., The abrasive wear resistance of sputter ion plated titanium nitride coatings. Surf. Coat. Technol., 36 (1988b) 743–754.

Bunshah, R.F., Selection and use of wear tests for coatings. In: Selection and Use of Wear Tests for Coatings. Bayer, R.G. (ed.), ASTM STP 796, American Society for Testing Materials, 1982, 3–15.

Bunshah, R.F. (ed.), Handbook of Deposition Technologies for Films and Coatings. Noyes Publications, New Jersey, USA, 1994, 861 pp.

Bunshah, R.F., Nimmagada, R., Doerr, H.J., Movchan, B.A., Grechanuk, N.I., Dabizha, E.V., Structure and property relationships in microlaminate Ni–Cu and Fe–Cu condensates. Thin Solid Films, 72 (1980) 261–275.

Bunshah, R.F., Raghuram, A.C., Activated reactive evaporation process for high rate deposition of compounds. J. Vac. Sci. Technol., 9 (1972) 1385–1388.

Burnett, P.J., Page, T.F., Surface softening in silicon by ion implantation. J. Mat. Sci., 19 (1984) 854–860.

Burnett, P.J., Rickerby, D.S., The mechanical properties of wear-resistant coatings I: Modelling of hardness behaviour. Thin Solid Films, 148 (1987a) 41–50.

Burnett, P.J., Rickerby, D.S., The relationship between hardness and scratch adhesion. Thin Solid Films, 154 (1987b) 403–416.

Burnett, P.J., Rickerby, D.S., The erosion behaviour of TiN coatings on steels. J. Mat. Sci., 23 (1988) 2429–2443.

Byeli, A.V., Minevich, A.A., Stepanenko, A.V., Gick, L.A., Kholodilov, O.V., Wear resistance and structure of (Ti,Al)N coatings. J. Phys. D: Appl. Phys., 25 (1992) A292–A296.

Cai, S., Bhushan, B., Three-dimensional sliding contact analysis of multilayered solids with rough surfaces. Trans. ASME, J. Tribology, 129 (2007) 40–59.

Calonius, O., Tribology of prosthetic joints – validation of wear simulation methods. Acta Polytechnica Scandinavica, Mech. Eng. Series No, 159, 2002, 62 pp.

Cameron, A., The Principles of Lubrication, Longmans, London, 1966, 519 pp.

Cameron, A., The role of surface chemistry in lubrication and scuffing. Trans. ASLE, 23 (1980) 4, 388–392.

Cammarata, R.C., Stresses in thin films. In: Surface Modification and Mechanisms – Friction, Stress, and Reaction Engineering. Totten, G.E., Liang, H. (eds), Marcel Dekker, Basel, Switzerland, 2004, 17–29.

Campbell, D.S., The deposition of thin films by chemical methods. In: Handbook of Thin Film Technology. Maissel, L.I., Glang, R. (eds), McGraw-Hill, New York, 1970, 5.1–5.24.

Canter, N., How does ZDDP function? Tribology Lubr. Technol., June (2005) 20–26.

Cantley, R.E., Predicting the effects of lubricant chemistry on bearing fatigue life. ASLE Trans., 26 (1983) 1, 80–86.

Carlsson, J.-O., Thermally activated chemical vapour deposition. In: Advanced Surface Coatings: A Handbook in Surface Engineering. Rickerby, D.S., Matthews, A. (eds), Blackie, Glasgow, UK, 1991, 162–193.

Carnes, K., Hard-driving lubrication. Tribology Lubr. Technol., (2004) Nov., 31–38.
Caron, I., Gras, R., Ganne, T., De Monicault, J.M., Coatings on Ti6Al4V as palliatives for fretting fatigue in cryogenic environment. Tribology Int.l, 39 (2006) 1045–1051.
Cartier, M. (ed.), Handbook of Surface Treatments and Coatings. Tribology in Practice Series. Professional Engineering Publishing, London, 2003, 412 pp.
Carvalho, N.J.M., Low friction and wear resistant coatings: microstructure and mechanical properties. PhD thesis, Groningen University, Netherlands, 2001, 142 pp.
Casey, M., Wilks, J., The friction of diamond sliding on polished cube faces of diamond. J. Phys. D: Appl. Phys., 6 (1973) 1772–1781.
Cavaleiro, A., De Hosson, J.T.M. (eds), Nanostructured Coatings. Springer, New York, 2006, 648 pp.
Cavdar, B., Ludema, K.C., Dynamics of dual film formation in boundary lubrication of steels. Part I. Functional nature and mechanical properties. Wear, 148 (1991) 305–327.
Celis, B., Theoretical analysis of dry friction in brittle and ductile materials. Wear, 116 (1987) 287–298.
Celis, J.P., Ponthiaux, P., Wenger, F., Tribo-corrosion of materials: interplay between chemical, electrochemical, and mechanical reactivity of surfaces. Wear, 261 (2006) 939–946.
Celis, J.P., Roos, J.R., De Bonte, M., Developments and applications of electroless coatings. In: Surface Modification Technologies. Sudarshan, T.S., Bhat, D.G. (eds), Pennsylvania, USA, The Metallurgical Society, 1988, 215–235.
Chala, A., Chekour, L., Nouveau, C., Saied, C., Aida, M.S., Djouadi, M.A., Study of the duplex treatment on 32CrMoV13 low alloy steel: application in wood machining. Surf. Coat. Technol., 200 (2005) 512–516.
Chalker, P.R., Characterisation of coatings and interfaces. In: Advanced Surface Coatings – A Handbook of Surface Engineering. Rickerby, D.S., Matthews, A. (eds), Blackie, London, 1991, 278–314.
Chambers, K.W., Arneson, M.C., Waggoner, C.A., An on-line ferromagnetic wear debris sensor for machinery condition monitoring and failure detection. Wear, 128 (1988) 325–327.
Chambers, D.L., Taylor, K.A., Wan, C.T., Emrick, A.J., Sputtering of polymeric composites. Surf. Coat. Technol., 41 (1990) 315–323.
Chandrasekaran, H., Friction in machining – comparison of rake and flank wear-land friction. Wear, 36 (1976) 133–145.
Chang, T.P., Cheng, H.S., Sproul, W.D., The influence of coating thickness on lubricated rolling contact fatigue life. Surf. Coat. Technol., 43/44 (1990) 699–708.
Chang, T.P., Cheng, H.S., Chiou, W.A., Sproul, W.D., A comparison of fatigue failure morphology between TiN coated and uncoated lubricated rollers. Tribology Trans., 34 (1991) 408–416.
Chang, F.K., Cominou, M., Barber, J.R., Slip between a layer and a substrate caused by a normal force moving steadily over the surface. Int. J. Mech. Sci., 25 (1983) 11, 803–809.
Chang, T.P., Graham, M.E., Sproul, W.D., Cheng, H.S., Scuffing behaviour of TiN-coated steel rollers under rolling and sliding conditions. Surf. Coat. Technol., 54/55 (1992) 490–495.
Chapman, B., Glow Discharge Process, Wiley, New York, 1980, 177–296.
Charitidis, C., Gioti, M., Logotheidis, S., Kassavetis, S., Laskarakis, A., Varsano, I., Comparison of the nanomechanical and nanoscratch performance of antiscratch layers on organic lenses. Surf. Coat. Technol., 180–181 (2004) 357–361.
Chekour, L., Nouveau, C., Chala, A., Djouadi, M.A., Duplex treatment of 32CrMoV13 steel by ionic nitriding and triode sputtering: application to wood machining. Wear, 255 (2003) 1438–1443.
Chen, W.T., Engel, P.A., Impact on contact stress analysis in multilayer media. Int. J. Solids Structures, 8 (1972) 1257–1281.
Cheng, C.C., Erdemir, A., Fenske, G.R., Correlation of interface structure with adhesive strength of ion-plated TiN hard coatings. Surf. Coat. Technol., 39/40 (1989) 365–376.
Cheng, H.S., On some aspects of microelastohydrodynamic lubrication. In: Surface Roughness Effects in Lubrication, 4th Leeds–Lyon Symp. on Tribology, Lyon, Sept. 1977, MEP, London, 1978, 71–79.
Cheng, H.S., Fundamentals of elastohydrodynamic contact phenomena. In: Fundamentals of Tribology. Suh, N.P., Saka, N. (eds), MIT Press, Cambridge, MA, USA, 1980, 1009–1048.

Cheng, H.S., Gears. In: Modern Tribology Handbook. Bhushan, B. (ed.), CRC Press, New York, 2001, 1095–1129.
Cheng, J., Bull, S.J., Indentation fracture and toughness assessment for thin optical coatings on glass. J. Phys. D: Appl. Phys., 40 (2007) 5401–5417.
Cheng, C.C., Erdemir, A., Fenske, G.R., Correlation of interface structure with adhesive strength of ion-plated TiN hard coatings. Surf. Coat. Technol., 39/40 (1989) 365–376.
Childs, T.H.C., Friction modeling in metal cutting. Wear, 260 (2006) 310–318.
Chiu, Y.P., Harnett, M.J., A numerical solution for layered solid contact problems with applications to bearings. J. Lubrication Technology, Trans. ASME, 105 (1983) 585–590.
Cho, C.W., Hong, B., Lee, Y.Z., Wear life evaluation of diamond-like carbon films deposited by microwave plasma-enhanced CVD and RF plasma-enhanced CVD method. Wear, 259 (2005) 789–794.
Chollet, L., Perry, A.J., The stress in ion-plated HfN and TiN coatings. Thin Solid Films, 123 (1985) 223–234.
Christensen, H., Some aspects of the functional influence of surface roughness in lubrication. Wear, 17 (1971) 149–162.
Christensen, H., A theory of mixed lubrication. Proc. Instn. Mech. Engrs 186/41 (1972) 421–430.
Christie, A.B., X-ray protection spectroscopy. In: Methods of Surface Analysis. Walls, J.M. (ed.), Cambridge University Press, 1989, 127–168.
Chu, P.K., Chen, J.Y., Wang, L.P., Huang, N., Plasma-surface modification of biomaterials. Materials Science and Engineering, R 36 (2002) 143–206.
Clemens, B.M., Eesley, G.L., Relationship between interfacial strain and the elastic response of multilayer metal films. Material Research Society, Mat. Res. Soc. Symp. Proc., 130 (1989) 307–314.
Coll, B.F., Sathrum, P., Aharonov, R., Peyre, J.P., Benmalek, M., Tribological properties and oxidation resistance of graded (Ti,Al)N coatings. 19th Int. Conf. Metallurgical Coatings and Thin Films (ICMCTF 92), San Diego, USA, 6-10.4.1992, 20 pp.
Colton, E., Coating ophthalmic lenses. CERAC Inc. Technical Publications, 7 (1997) 4 Oct.–Dec., 5 pp.
Conrad, J.R., Radtke, J.K., Dodd, R.A., Worzala, F.J., Tran, N.C., Plasma source ion implantation technique for surface modification of metals. J. Appl. Phys., 62 (1987) 4591.
Constantinescu, V.N., Nica, A., Pascovici, M., Ceptureanu, G., Nedelcu, S., Sliding Bearings. Allerton Press Inc., New York, (1985) 555 pp.
Cookson, J., Automation sharpens sawing's edge. Metalworking Production, 127 (1983) 12, 72–93.
Cooper, J.A., Dewhurst, R.J., Palmer, S.B., Laser generated ultrasound. In: Proc. Ultrasonics Int., Butterworths, London, 1985.
Corso, S., Adamo, R., The application of ferrography in monitoring motor oils during engine development. Lubrication Engineering, J. STLE, 45 (1989) 9, 557–564.
Cosemans, P., Zhu, X., Celis, J.P., Van Stappen, M., Development of low friction wear-resistant coatings. Surf. Coat. Technol., 174–175 (2003) 416–420.
Cowan, R.S., Winer, W.O., Application of the thermomechanical wear transition model to layered media. In: Thin Films in Tribology. Dowson, D. *et al.* (eds), Tribology Series 25, Elsevier, Amsterdam, 1993, 631–640.
Cui, F.Z., Li, D.J., A review of investigations on biocompatibility of diamond-like carbon and carbon nitride films. Surf. Coat. Technol., 131 (2000) 481–487.
Cudworth, C.J., Higginson, G.R., Friction of lubricated soft surface layers. Wear, 37 (1976) 299–312.
Czichos, H., Tribology – a systems approach to the science and technology of friction, lubrication and wear. Tribology Series, 1, Elsevier, Amsterdam, 1978, 400 pp.
Czichos, H., Becker, S., Lexow, J., Multilaboratory testing: results from the Versailles advanced materials and standards programme on wear test methods. Wear, 114 (1987) 109–130.
Czichos, H., Becker, S., Lexow, J., International multilaboratory sliding wear tests with ceramics and steels. Wear, 135 (1989) 171–191.
Czichos, H., Advanced materials in tribology. Tribologia – Finnish J. of Tribology, 11 (1992) 3, 3–27.
Czichos, H., Habig, K., Tribologie handbuch – Reibung und Verschleiss/Tribology Handbook – Friction and Wear (in German), Vieweg, Wiesbaden, Germany, 1992, 560 pp.

Dag, S., Ciraci, S., Atomic scale study of superlow friction between hydrogenated diamond surfaces. Phys. Rev. B, 70 (2004). 241401(R).

D'agostino, R., Favia, P., Fracassi, F., Lamendola, R., Plasma-enhanced chemical vapour deposition. In: Advanced Techniques for Surface Engineering. Gissler, W., Jehn, H.A. (eds), Kluwer Academic Publishers, Dordrecht, The Netherlands, 1992, 105–133.

Damasceno, J.C., Camargo, S.S., Freire, F.L., Carius, R., Deposition of Si-DLC films with high hardness, low stress and high deposition rates. Surf. Coat. Technol., 133–134 (2000) 247–252.

Dareing, D.W., Traction coefficients for coated bearing races lubricated with Teflon transfer films. Journal of Tribology, Trans. ASME, 113 (1991) 343–348.

Davidson, J.A., Analysis of clinical and laboratory wear factors and the tribological performance of orthopaedic implants. Japanese J. Tribology, 37 (1992) 4, 399–413.

Davis, J.R., Davis, Associates (eds), Surface Engineering for Corrosion and Wear Resistance. ASM International and IOM Communications, USA, 2001, 279 pp.

Dayson, C., The friction of very thin solid film lubricants on surfaces of finite roughness. ASLE Trans, 14 (1971) 105–115.

De Chiffre, L., Mechanics of metal cutting and cutting fluid action. Int. J. Mach. Tool Des. Res, 17 (1977) 225–234.

De Chiffre, L., Function of cutting fluids in machining. Lubr. Engng, J. STLE, 44 (1988a) 6, 514–518.

De Chiffre, L., Restricted contact tools and cutting fluids in machining. In: Proc. 6th Int. Colloquium on Industrial Lubricants – Properties, Applications, Disposal. Bartz, W.J. (ed.), 12–14.1.1988b, Esslingen, Germany, Vol. II, 11.5-1–11.5-8.

De Chiffre, L., Metal cutting – mechanics and applications. Doctoral thesis, Technical University of Denmark, Inst. of Manufacturing Engineering, Lyngby, Denmark., PI-Publ. Nr 88-08-A, 1990, 102 pp.

De Wit, E., Blanpain, B., Froyen, L., Celis, J.P., The tribological behaviour of TiN-coatings during fretting wear. Wear, 217 (1998) 215–224.

Dearnaley, G., Ion implantation and ion-assisted coating. In: Advanced Surface Coatings: A Handbook in Surface Engineering. Rickerby, D.S., Matthews, A. (eds), Blackie, Glasgow, UK, 1991, 41–65.

Dearnley, P.A., Tribological challenges for plasma surface engineered materials – a personal review. Soc. of Vacuum Coaters. 49th Annual Technical Conf. Proc. (2006) 507–517.

Dearnley, P.A., Aldrich-Smith, G., Corrosion-wear mechanisms of hard coated austenitic 316L stainless steels. Wear, 256 (2004) 491–499.

Dedkov, G.V., Experimental and theoretical aspects of the modern nanotribology. Phys.. stat. sol.(a), 179 (2000) 3, 3–75.

Dehbi-Alaoui, A., Matthews, A., Franks, J., The optical and mechanical properties of carbon films grown using a fast atom beam source. Surf. Coat. Technol., 47 (1991) 722–729.

Dehbi-Alaoui, A. Olliver, A., Matthews, A., Evaluation of diamond-like carbon films on PET. Le Vide, les Couches Minces, Supplement au No. 261, March–April, 1992, 221–226.

DellaCorte, C., Pepper, S.V., Honecy, F.S., Tribological properties of Ag/Ti films on Al_2O_3 ceramic substrates. Surf. Coat. Technol., 52 (1992) 31–37.

Deng, K., Ko, W.H., Application of diamond-like carbon film for microdynamic devices. IEEE, 0-7803-0456-X/92$3.00, 1992, 98–101.

Derjaguin, B.V., Fedoseev, D.B., The synthesis of diamond at low pressure. Scientific American, 233 (1975) 102–109.

Descartes, S., Berthier, Y., Competition between mechanical and physico-chemical actions: consequences on friction. The case of a contact lubricated by a 3rd body stemming from a coating of MoS1.6. In: Proc. of 2nd World Tribology Congress, 3-7.9.2001, Vienna, Austria, Austrian Tribology Society, 67-71.

Diao, D., Finite element analysis on local yield map and critical maximum contact pressure for yielding in hard coating with an interlayer under sliding contact. Tribology Int., 32 (1999) 25–32.

Diao, D., Kandori, A., Finite element analysis of the effect of interfacial roughness and adhesion strength on the local delamination of hard coating under sliding contact. Tribology Int., 39 (2006) 849–855.

Diao, D., Kato, K., Interface yield map of hard coating under sliding contact. Thin Solid Films, 245 (1994) 115–121.

Diao, D., Kato, K., Hayashi, K., The local yield map of hard coating under sliding contact. In: Thin Films in Tribology. Dowson, D. *et al.* (eds), Elsevier Science Publishers, Amsterdam, The Netherlands, 1993, 419–427.

Diao, D., Kato, K., Hayashi, K., The maximum tensile stress on a hard coating Ander sliding friction. Tribology Int., 27 (1994a) 4, 267–272.

Diao, D.F., Kato, K., Hokkirogawa, K., Fracture mechanics of ceramic coatings in indentation. Journal of Tribology, Trans. ASME, 116 (1994b) 860–869.

Dick, T., Cailletaud, G., Analytic and FE based estimations of the coefficient of friction of composite surfaces. Wear, 260 (2006a) 1306–1316.

Dick, T., Cailletaud, G., Fretting modelling with a crystal plasticity model of TiAl4V. Comp. Mat. Sci., 38 (2006b) 113–125.

Dick, T., Paulin, C., Cailletaud, G., Fouvry, S., Experimental and numerical analysis of the local and global plastic behaviour in fretting wear. Tribology Int., 39 (2006) 1036–1044.

Dickrell, P.L., Sawyer, W.R., Erdemir, A., Fractional coverage model for the adsorption and removal of gas species and application to superlow friction diamond-like carbon. J. Tribology, Trans. ASME, 126 (2004) 615–619.

Didziulis, S.V., Fleischauer, P.D., Soriano, B.L., Gardos, M.N., Chemical and tribological studies of MoS_2 films on SiC substrates. Surf. Coat. Technol., 43/44 (1990) 652–662.

Dienwiebel, M., Frenken, J.W.M., Superlubricity between graphite surfaces. In: Superlubricity. Erdemir, A., Martin, J.M. (eds), Elsevier, Amsterdam, Netherlands, 2007, 199–206.

Dieter, G.E., Mechanical Metallurgy. Material Science and Engineering Series, McGraw-Hill Book Company, London, 1986, 751 pp.

Dill, J.F., Gardos, M.N., Hinterman, H.E., Boving, H.J., Rolling contact fatigue evaluation of hardcoated bearing steels. Proc. 3rd ASLE Int. Conf. on Solid Lubr., 5–10.9.1984, Denver, Colorado, USA, 12 pp.

Dimigen, H., Hubsch, H., Willich, P., Reichelt, K., Stochiometry and friction properties of sputtered MoS_2 layers. Thin Solid Films, 129 (1985) 79–91.

Dimigen, H., Klages, C-P., Microstructure and wear behavior of metal-containing diamond-like coatings. Surf. Coat. Technol., 49 (1991) 543–547.

Ding, Y., Jones, R., Khunell, B.T., Elastic–plastic finite element analysis of spall formation in gears. Wear, 197 (1996) 197–205.

Djabella, H., Arnell, R.D., Finite element analysis of the contact stresses in an elastic coating on an elastic substrate. Thin Solid Films, 213 (1992) 205–219.

Djabella, H., Arnell, R.D., Finite element analysis of contact stresses in elastic double-layer systems under normal load. Thin Solid Films, 223 (1993a) 98–108.

Djabella, H., Arnell, R.D., Finite element comparative study of elastic stresses in single, double layer and multilayered coated systems. Thin Solid Films, 235 (1993b) 156–162.

Djabella, H., Arnell, R.D., Two-dimensional finite-element analysis of elastic stresses in double-layer systems under combined surface normal and tangential loads. Thin Solid Films, 226 (1993c) 65–73.

Djouadi, M.A., Nouveau, C., Beer, P., Lambertin, M., Cr_xN_y hard coatings deposited with PVD method on tools for wood machining. Surf. Coat. Technol., 133–134 (2000) 478–483.

Doll, G.L., Thin film coatings on rolling bearings for interrupted or reduced lubrication. In: Proc. of Aerospace Transmission Systems: Who Needs Oil?, 18.9.2003, Solihull, UK, Institution of Mechanical Engineers, 7 pp.

Doll, G.L., Evans, R.D., Ribaudo, C.R., Improving the performance of rolling contact bearings with tribological coatings. In: Surface Engineering in Materials Science III. A Agarwal *et al.* (eds), Warrendale, PA: TMS, 2005, 153–162.

Doll, G.L., Osborn, K.B., Engineering surfaces of precision steel components. Proc. 44th Annual SVC Technical Conf., Philadelphia, USA, 21–26.4.2001, Albuquerque, NM: SVC, 2001, 78–84.

Doll, G.L., Ribaudo, C.R., Evans, R.D., Engineered surfaces for steel rolling element bearings and gears. Proc. Conf. on Materials Science, Technology 2004, New Orleans, USA, 26–29.9.2004, Warrendale, PA: AIST/TMS, 2004, 367–376.

Dong, H., Bell, T., Designer surfaces for titanium components, Anti-Corrosion Methods and Materials, 46/5 (1999) 338–345.

Donley, M.S., Zabinski, J.S., Tribological Coatings in Pulsed Laser Deposition of Thin Films. In: Chrisey, D.B., Hubler, G.K. (eds), Wiley, New York, 1994.

Donnet, C., Tribology of solid lubricant coatings. Condensed Matter News, 4 (1995) 6, 9–24.

Donnet, C., Problem-solving methods in tribology with surface-specific techniques. In: Handbook of Surface and Interface Analysis. Riviére, J.C., Myhra, S. (eds), Marcel Dekker, Basel, Chapter 16, 1998a, 697–745.

Donnet, C., Recent progress on the tribology of doped diamond-like and carbon alloy coatings: a review. Surf. Coat. Technol. 100–101 (1998b) 180–186.

Donnet, C., Erdemir, A. (eds), Tribology of diamond-like carbon films – fundamentals and applications. Springer, New York, USA, 2008, 664 pp.

Donnet, C., Fontaine, J., Le Mogne, T., Belin, M., Héau, C., Terrat, J.P., Vaux, F., Pont, G., Diamond-like carbon-based functionally gradient coatings for space tribology. Surf. Coat. Technol., 120–121 (1999) 548–554.

Donnet, C., Le Mogne, T., Martin, J.M., Superlow friction of oxygen-free MoS_2 coatings in ultrahigh vacuum. Surf. Coat. Technol., 62 (1993) 406–411.

Douglas, A., Doyle, E.D., Jenkins, B.M., Surface modification for gear wear. Int. Tribology Conf., 2–4.12. 1987, Melbourne, Australia, (1987) 52–57.

Dowson, D., Elastohydrodynamics. Proc. Instn. Mech. Engrs., /182/Pt3A (1968) 151–167.

Dowson, D., History of Tribology, 2nd edition, Professional Engineering Publishing, London, UK, 1998, 768 pp.

Dowson, D., Are our joint replacement materials adequate? Proc. Int. Conf. on the Changing Role of Engineering in Orthopedics, Instn Mech. Engrs, C384/KN1 (1989) 1–5.

Dowson, D., Higginson, G.R., New roller-bearing lubrication formula. Engineering, 4 (Aug. 1961) 158–159.

Dowson, D., Higginson, G.R., Elasto-hydrodynamic Lubrication – The Fundamentals of Roller and Gear Lubrication, Pergamon Press, Oxford, 1966, 235 pp.

Dowson, D., Jin, Z.M., Metal-on-metal hip joint tribology. Proc. Instn Mech. Engrs, Part H: Engineering in Medicine, 220 (2006) 107–118.

Dowson, D., Neville, A., Bio-tribology and bio-mimetics in the operating environment. Proc. Instn Mech. Engng, Part J: J. Engng Tribology, 220 (2006) 109–123.

Dowson, D., Whomes, T.L., Effect on surface quality upon the traction characteristics of lubricated cylindrical contacts. Proc. Inst. Mech. Engrs, 182 (1967-68) 292-321.

Doyle, E.D., Jewsbury, P., The surface engineering revolution. Materials Australiasia (1986) June, 8–10.

Drory, M.D., Thouless, M.D., Evans, A.G., On the decohesion of residually stressed thin films. Acta Metall., 36 (1988) 8, 2019–2028.

Dubourg, M.C., Villechaise, B., Analysis of multiple cracks – Part I: Theory. J. Tribology, Trans. ASME, 114 (1992) 455–461.

Dubourg, M.C., Godet, M., Villechaise, B., Analysis of multiple fatigue cracks – Part II: Results. J. Tribology, Trans. ASME, 114 (1992) 462–468.

Dunaevsky, V.V., Friction and wear equations. In: Tribology Data Handbook. Booser, E.R. (ed.), CRC Press, New York, USA, 1997, 445–454.

Dunaevsky, V.V., Jeng, Y-R., Rudzitis, J.A., Surface texture. In: Tribology Data Handbook. Booser, E.R. (ed.), CRC Press, New York, USA, 1997, 415–434.

Dunlop, E., Haupt, J., Schmidt, K., Gissler, W., Hardness and Young's modulus of diamond-like carbon films prepared by ion beam methods. In: Proc. 2nd European Conf. on Diamond, Diamond-like and Related Coatings, Nice, France, 2–6.9.1991, 644–649.

Dunn, M.L., Cunningham, J., Thermo- and electromechanics of thin-film microstructures. Chapter 35. In: Handbook of Nanotechnology. Bhushan, B. (ed.), Springer, Berlin, Germany, 2004, 1039–1081.

Dvorak, S.D., Wahl, K.J., Singer, I.L., Friction behavior of boric acid and annealed boron carbide coatings studied by in situ Raman tribometry. Tribology Trans., 45 (2002) 354–362.

Dwyer-Joyce, R.S., Drinkwater, B.W., Donohoe, C.J., The measurement of lubricant film thickness using ultrasound. Proc. R. Soc. Series A, 459 (2002) 957–976.

Dwyer-Joyce, R.S., Reddyhoff, T., Drinkwater, B., Operating limits for acoustic measurement of rolling bearing oil film thickness. STLE Tribology Trans., 47 (2004) 366–375.

Eason, R. (ed.), Pulsed Laser Deposition of Thin Films: Applications-led Growth of Functional Materials. Wiley, New York, 2006. ISBN: 978-0-471-44709-2.

Eastham, D., Matthews, A., Surface engineering applied to bearings. 1st European Workshop on Surface Engineering Technologies and Applications for SMEs, Newcastle-upon-Tyne, UK, 11–15.5.1992, L14–L15. Newcastle Polytechnic, UK.

Ebdon, P.R., Composite coatings with lubricating properties. Trans. Inst. Met. Fin, 65 (1987) 80–82.

Eberhardt, A.W., Peri, S., Surface breaking cracks in layered Hertzian contact with friction. STLE Tribology Trans., 38 (1995) 2, 299–304.

Echigoya, T., Hyakubu, T., Liu, Z.T., Sasaki, K., Suto, H., Crystal structure and scratch testing of κ-Al_2O_3 on WC–Co substrates with a TiC interlayer. Surf. Coat. Technol., 47 (1994) 111–117.

Edwards, C.M., Halling, J., An analysis of the plastic interaction of surface asperities and its relevance to the value of the coefficient of friction. J. Mech. Engng Sci., 10 (1968) 2, 101–110.

El Haddad, M.H., Topper, T.H., Smith, K.N., Prediction of non propagating cracks. Engng Fract. Mech., 11 (1979) 573–584.

El Shabasy, M., Adhesion measurements of thin metallic films: comparison of the direct pull-off and the scratch methods. Period. Polytech. Electr. Engng, 25 (1981) 283.

El-Sherbiny, M., The friction of solid film lubricants: a theoretical approach. 3rd Int. Conf. on Solid Lubricants, ASLE, Denver, August 1984, 39–43.

El-Sherbiney, M.G.D., Halling, J., The Hertzian contact of surfaces covered with metallic films. Wear, 40 (1976) 325–337.

El-Sherbiny, M.G., Salem, F.B., Initial wear rates of soft metallic films. Wear, 54 (1979) 391–400.

El-Sherbiny, M.G., Salem, F.B., Fatigue wear: a contribution to the wear phenomena of ion-plated thin metallic films. Wear, 66 (1981) 101–110.

El-Sherbiny, M., Salem, F., A wear equation for solid lubricant films. 3rd Int. Conf. on Solid Lubricants, ASLE, Denver, August 1984, 44–49.

El-Sheriny, M., Salem, F., Tribological properties of PVD silver films. ASLE Trans, 29 (1986) 2, 223–228.

Elena, M., Guzman, L., Gialanella, S., Production and characterization of boron nitride films obtained by r.f. magnetron sputtering and reactive ion-beam-assisted deposition. Surf. Coat. Technol., 36 (1988) 199–206.

Elsharkwy, A.A., Hamrock, B.J., A numerical solution for dry sliding line contact of multilayered elastic bodies. J. Tribology, Trans. ASME, 115 (1993) 237–245.

Endler, I., Bartsch, K., Leonhardt, A., Scheibe, H.J., Ziegele, H., Fuchs, I., Raatz, C., Preparation and wear behaviour of woodworking tools coated with superhard layers. Diamond and Related Materials, 8 (1999) 834–839.

Engel, L., Klingele, H., An Atlas of Metal Damage. Wolfe Science Books, Carl Hanser Verlag, Munich, Germany, 1981, 271 pp.

Engel, P.A., Hsue, E.Y., Bayer, R.G., Hardness, friction and wear of multiplated electrical contacts. Wear, 162–164 (1993) 538–551.

Enke, K., Dimigen, H., Hübsch, H., Frictional properties of diamondlike carbon layers. Appl. Phys. Lett., 36 (1980) 291–292.

Enke, K., Geisler, M., Kieser, J., Munz, W.D., Plasma-activated CVD for the production of protective layers in optics, electronics and mechanical engineering. Materials Research Society. Mat. Res. Soc. Proc., 93 (1987) 301–310.

Equey, S., Hauert, R., Rossi, A., Spenser, N.D., Crockett, R., Investigation on the tribofilms formed on diamond-like carbon coatings in presence of zinc dialkyl dithiophosphate. 34th Leeds–Lyon Symposium on Tribology, Lyon, France, 4–7.9.2007.

Equey, S., Roos, S., Mueller, U., Hauert, R., Spencer, N.D., Crockett, R., Tribofilm formation from ZnDTP on diamond-like carbon. Wear, 264 (2008) 316–321.

Erck, R.A., Erdemir, A., Fenske, G.R., Effect of film adhesion on tribological properties of silver-coated alumina. Surf. Coat. Technol., 43/44 (1990) 577–587.

Erdemir, A., Rolling-contact fatigue and wear resistance of hard coatings on bearing-steel substrates. Surf. Coat. Technol., 54/55 (1992) 482–489.

Erdemir, A., Solid lubricants and self-lubricating films. In: Modern Tribology Handbook. Bhushan, B. (ed.), CRC Press, New York, 2001a, 787–825.

Erdemir, A., Tribology of diamond, diamond-like carbon and related films. In: Modern Tribology Handbook. Bhushan, B. (ed.), CRC Press, New York, 2001b, 871–908.

Erdemir, A., Friction and wear of diamond and diamond-like carbon films. Proc. Instn Mech. Engrs, Part J: Engineering Tribology, 216 (2002) 387–400.

Erdemir, A., Design criteria for superlubricity in carbon films and related microstructures. Tribology Int., 37 (2004a) 577–583.

Erdemir, A., Genesis of superlow friction and wear in diamondlike carbon films. Tribology Int., 37 (2004b) 1005–1012.

Erdemir, A., Review of engineered tribological interfaces for improved boundary lubrication. Tribology Int., 38 (2005) 249–256.

Erdemir, A., Donnet, C., Tribology of diamond, diamond-like carbon and related films. In: Modern Tribology Handbook. Bhushan, B. (ed.), CRC Press, New York, 2001, 871–908.

Erdemir, A., Donnet, C., Tribology of diamond and diamond-like carbon films: an overview. In: Wear – Materials, Mechanisms and Practice. Stachowiak, G.W. (ed.), Tribology in Practice Series, John Wiley & Sons, Chichester, UK, 2005, 191–222.

Erdemir, A., Donnet, C., Tribology of diamond-like carbon films: recent progress and future prospects. J. Phys. D: Appl. Phys., 39 (2006). R331–R327.

Erdemir, A., Eryilmaz, O.L., Superlubricity in diamondlike carbon films. In: Superlubricity. Erdemir, A., Donnet, C. (eds), Elsevier, Amsterdam, Netherlands, 2007, 253–271.

Erdemir, A., Fenske, G.R., Tribological performance of diamond and diamondlike carbon films at elevated temperatures. Tribology Trans., 39 (1996) 4, 787–794.

Erdemir, A., Fontaine, J., Donnet, C., An overview of superlubricity in diamond-like carbon films. In: Tribology of diamond-like carbon films. Donnet, C., Erdemir, A. (eds), Springer, New York, USA, 2008, 237–262.

Erdemir, A., Bindal, C., Pagan, J., Wilbur, P., Characterization of transfer layers on steel surfaces sliding against diamond-like hydrocarbon films in dry nitrogen. Surf. Coat. Technol., 76–77 (1995) 559–563.

Erdemir, A., Busch, D.E., Erck, R.A., Fenske, G.R., Lee, R., Ion-beam-assisted deposition of silver films on zirconia ceramics for improved tribological behaviour. Lubr. Engng, 47 (1991a) 10, 863–872.

Erdemir, A., Erilmaz, O., Fenske, G., Synthesis of diamond-like carbon films with superlow friction and wear properties. J. Vac. Sci. Technol. A, 18 (2000) 1987–1992.

Erdemir, A., Fenske, G.R., Kazmanli, M.K., Eryilmaz, O.L., Development of superlow friction carbon films from hydrocarbon plasmas. Proc. of 2nd World Tribology Congress, 3–7.9.2001, Vienna, Austria, Austrian Tribology Society, 265–289.

Erdemir, A., Fenske, G.R., Erck, R.A., Cheng, C.C., Ion-assisted deposition of silver films on ceramics for friction and wear control. Lubr. Engng, 46 (1990) 1, 23–30.

Erdemir, A., Halter, M., Fenske, G., Preparation of ultralow-friction surface films on vanadium diboride. Wear, 205 (1997) 236–239.

Erdemir, A., Halter, M., Fenske, G., New oxide-based lubricants for sliding friction and wear applications at extreme temperatures. World Ceramic Congress and Forum on New Materials–CIMTEC, 14-19.6. 1998, Florence, Italy, 16 pp.

Erdemir, A., Hochman, R.F., Surface metallurgical and tribological characteristics of TiN-coated bearing steels. Surf. Coat. Technol., 36 (1988) 755–763.

Erdemir, A., Martin, J.M. (eds), Superlubricity. Elsevier, Amsterdam, The Netherlands, 2007, 499 pp.

Erdemir, A., Nichols, F.A., Fenske, G.R., Hsieh, J.H., Sliding friction and wear of ceramics with and without soft metallic films. MRS Bulletin, Oct. (1991b) 49–53.

Erdemir, A., Switala, M., Wei, R., Wilbur, P., A tribological investigation of the graphite-to-diamond-like behaviour of amorphous carbon films ion beam deposited on ceramic substrates. Surf. Coat. Technol., 50 (1991c) 17–23.

Erdogan, F., Gupta, G.D., Layered composites with an interface flaw. Int. J. Solids Structures, 7 (1971a) 1089–1107.

Erdogan, F., Gupta, G., The stress analysis of multi-layered composites with a flaw. Int. J. Solids Structures, 7 (1971b) 39–61.

ESDU – A guide to the design and selection of dry rubbing bearings. Engineering Sciences Data Unit, ESDU 87007, 1987, 43 pp.

European Standard, 2000, Advanced technical ceramics – methods of tests for ceramic coatings – Part 3: Determination of adhesion and other mechanical failure modes by scratch test, prEN1071-3:2000:E, CEN Management Centre, Stassartstraat 36, B-1050 Brussels, Belgium, 42 pp.

Evans, C., Cyrogenic diamond turning of stainless steel. Annals of the CIRP, 40 (1991) 1, 571–575.

Evans, A.C., Franks, J., Revell, P.J., Diamond-like carbon applied to bioengineering materials. Surf. Coat. Technol., 47 (1991) 662–667.

Evans, D.C., Lancaster, J.K., The wear of polymers. In: Treatise on Materials Science and Technology, Vol. 13, WEAR. In: Scott, D. (ed.), Academic Press, New York, 1979, 498 pp.

Eyre, T.S., Wear characteristics of metals. Tribology Int., Oct. (1976) 203–212.

Eyre, T.S., Friction and wear control in industry. Metals and Materials, March 1992, 143.

Fancey, K.S., Matthews, A., Ionisation assisted physical vapour deposition of zirconia thermal barrier coatings. J. Vac. Sci. Technol., A4 (1986). 6 2656–2660.

Faraday, M., Experimental relations of a gold (and other metals) to light. Phil. Trans., 147 (1857) 145.

Farges, G., Bosch, J.P., Bergmann, E., Friction and adhesive wear behaviour of W–C coatings against steel and titanium nitride. Wear, 135 (1989) 1–14.

Farr, J.P.G., Molybdenum disulphide in lubrication. A review. Wear, 35 (1975) 1–22.

Farrow, M., Selecting wear resistance surfaces. Int. Conf. on Metallurgical Coatings (ICMC 86), San Diego, USA, April 1986.

Farrow, M., Gleave, C., Wear resistant coatings. Trans. Inst. Met. Fin., 62 (1984), 7 pp.

Fayeulle, S., Ehni, P.D., Singer, I.L., Role of transfer films in wear of MoS_2 coatings. In: Mechanics of Coatings. Dowson, D. *et al.* (eds), Tribology Series 17, Elsevier, Amsterdam, 1990, 129–138.

Fein, R.S., Chemistry in concentrated-conjunction lubrication. In: Interdisciplinary Approach to the Lubrication of Concentrated Contacts. Ku, P.M. (ed.), NASA SP-237, 1970, 489–527.

Felder, E., Bucaille, J., Mechanical analysis of the scratching of metals and polymers with conical indenters at moderate and large strains. Tribology Int., 39 (2006) 70–87.

Fella, R., Holleck, H., Schulz, H., Preparation and properties of WC–TiC–TiN gradient coatings. Surf. Coat. Technol., 36 (1988) 257–264.

Fellowes, F.C.J., Steen, W.M., Laser surface treatment. In: Advanced Surface Coatings: A Handbook in Surface Engineering. Rickerby, D.S., Matthews, A. (eds), Blackie, Glasgow, UK, 1991, 244–277.

Feng, Z., Field, J.E., Friction of diamond and chemical vapour deposition diamond coatings. Surf. Coat. Technol., 47 (1991) 631–645.

Feng, Z., Field, J.E., The friction and wear of diamond sliding on diamond. J. Phys. D: Appl. Phys., 25 (1992) A33–A37.

Fenske, G.R., Kaufherr, N., Lee, R.H., Kramer, B.M., Bunshah, R.F., Sproul, W.D., Characterization of coating wear phenomena in nitride- and carbide-coated tool inserts. Surf. Coat. Technol., 36 (1988) 791–800.

Fernández-Ramos, C., Sánchez-Lopez, J.C., Justo, A., Rojas, T.C., Papst, I., Hofer, F., Fernández, A., Microstructural characterization of Ti–TiN/CN_x gradient-multilayered coatings. Surf. Coat. Technol., (2004) 180–181, 526–532.

Ferrante, J., Practical applications of surface analytical tools in tribology. Lubr. Engng, J. ASLE, 38 (1982) 4, 223–236.

Ferrari, A.C., Non-destructive characterization of carbon films. In: Tribology of Diamond-like Carbon Films – Fundamentals and Applications. Donnet, C., Erdemir, A. (eds), Springer, New York, USA, 2008, 25–82.

Ferrari, A.C., Robertson, J., Interpretation of Raman spectra of disordered and amorphous carbon. Phys. Rev. B, 61 (2000) 20, 14095–14107.

Field, S.K., Jarratt, M., Teer, D.G., Tribological properties of graphite-like and diamond-like carbon coatings. Tribology Int., 37 (2004) 949–956.

Fillot, N., Iordanoff, I., Berthier, Y., Wear modeling and the third body concept. Wear, 262 (2007) 949–957.

Finkin, E.F., A theory for the effect of film thickness and normal load in the friction of thin films. J. Lubr. Tech., July (1969) 551–556.

Fischer, K., Oettel, H., Microstructural gradients in thin hard coatings – tailor-made. Surf. Coat. Technol., 91 (1997) 301–312.

Fischer, T.E., Chemical effects in friction. In: Fundamentals of Friction: Macroscopic and Microscopic Processes. In: Singer, I.L., Pollock, H.M. (eds), NATO ASI Series, Series E: Applied Sciences, Vol. 220, Kluver Academic Publishers, London, 1992, 299–312.

Fischer-Cripps, A.C., Critical review of analysis and interpretation of nanoindentation test data. Surf. Coat. Technol., 200 (2005) 4153–4165.

Fischer-Cripps, A.C., Karvánkova, P., Vepřek, S., On the measurement of hardness of super-hard coatings. Surf. Coat. Technol., 200 (2006) 5645–5654.

Fisher, J., Biomedical applications. In: Modern Tribology Handbook. Bhushan, B. (ed.), CRC Press, New York, 2001, 1593–1609.

Fisher, J., Hu, X.Q., Stewart, T.D., Williams, S., Tipper, J.L., Ingham, E., Stone, M.H., Davies, C., Hatto, P., Bolton, J., Riley, M., Hardaker, C., Isaac, G.H., Berry, G., Wear of surface engineered metal-on-metal hip prostheses. J. Mat.Sci.: Materials in Medicine, 15 (2004) 225–235.

Fleischauer, P.D., Hilton, M.R., Bauer, R., Effect of microstructure and adhesion on performance of sputter-deposition MoS_2 solid lubricant coatings. In: Mechanics of Coatings. Dowson, D. *et al.* (eds), Tribology Series 17, Elsevier, Amsterdam, 1990, 121–128.

Fleischauer, P.D., Bauer, R., Chemical and structural effects on the lubrication properties of sputtered MoS_2 films. Tribology Trans., 31 (1988) 2, 239–250.

Flewitt, P.E.J., Wild, R., Physical methods for materials characterization, 2nd edition, Taylor and Francis, 2003.

Fontaine, J., Belin, M., Le Mogne, T., Grill, A., How to restore superlow friction of DLC: the healing effect of hydrogen gas. Tribology Int., 37 (2004) 869–877.

Fontaine, J., Donnet, C., Superlow friction of a-C:H films: tribochemical and rheological effects. In: Superlubricity. Erdemir, A., Donnet, C. (eds), Elsevier, Amsterdam, Netherlands, 2007, 273–294.

Fontaine, J., Donnet, C., Erdemir, A., Fundamentals of the tribology of DLC coatings. In: Tribology of Diamond-like Carbon Films. Donnet, C., Erdemir, A. (eds), Springer, New York, USA, 2008, 139–154.

Fouvry, S., Duo, P., Perruchaut, P., A quantitative approach of Ti–6Al–4V fretting damages: friction, wear and crack nucleation. Wear, 257 (2004a) 916–929.

Fouvry, S., Fridrici, V., Langlade, C., Kapsa, P., Vincent, L., Palliatives in fretting: a dynamical approach. Tribology Int., 39 (2006) 1005–1015.

Fouvry, S., Kapsa, P., An energy description of hard coating wear mechanisms. Surf. Coat. Technol., 138 (2001) 141–148.

Fouvry, S., Liskiewicz, T., Kapsa, P., Hannel, S., Sauger, E., An energy description of wear mechanisms and its applications to oscillating sliding contacts. Wear, 255 (2003) 287–298.

Fouvry, S., Wendler, B., Liskiewicz, T., Dudek, M., Kolodziejczyk, L., Fretting wear analysis of TiC/VC multilayered hard coatings: experiments and modeling approaches. Wear, 257 (2004b) 641–653.

Fowles, P.E., Jackson, A., Murphy, W.R., Lubricant chemistry in rolling contact fatigue – the performance and mechanism of one antifatigue additive. ASLE Trans., 24 (1981) 1, 107–118.

Franklin, S.E., The wear behaviour of CVD- and PVD-coated tools in metal stamping and plastics injection moulding. In: Surface Engineering Practice – Processes, Fundamentals and Applications in Corrosion and Wear. Strafford, K.N. *et al.* (eds), Ellis Horwood, 1990, 60–70.

Franklin, S., The friction and wear characteristics of several wear-resistant surface coatings. Philips CFT Technology, CTR 545.91.0043 (1991) 24 pp.

Franklin, S.E., Coating selection and tribological testing for professional equipment applications. Tribotest, 14 (2008) 2, 63–80.

Franklin, S.E., Dijkman, J.A., The implementation of tribological principles in an expert system ('PRECEPT') for the selection of metallic materials, surface treatments and coatings in engineering design. Wear, 181 (1995) 3, 1–10.

Franklin, S.E., Beuger, J., A comparison of the tribological behaviour of several wear-resistant coatings. Surf. Coat. Technol., 54/55 (1992) 459–465.

Franks, J., Enke, K., Richardt, A., Diamond-like carbon – properties and applications. Metals and Materials, Nov. (1990) 695–700.

Freller, H., Haessler, H., Evaluation of existing ion-plating processes for the deposition of multicomponent hard coatings. Surf. Coat. Technol., 36 (1988) 219–232.

Freller, H., Hempel, A., Lilge, J., Lorenz, H.P., Influence of intermediate layers and base materials on adhesion of amorphous carbon and metal–carbon coatings. In: Proc. 2nd European Conf. on Diamond, Diamond-like and Related Coatings, Nice, France, 2–6.9.1991, 563–569.

Frenk, A., Kurz, W., Laser surface treatments: microstructural aspects. In: Advanced Techniques for Surface Engineering. Gissler, W., Jehn, H.A. (eds), Kluwer Academic Publishers, Dordrecht, The Netherlands, 1992, 235–252.

Fridrici, V., Fouvry, S., Kapsa, P., Fretting wear behaviour of a Cu–Ni–In plasma coating. Surf. Coat. Technol., 163–164 (2003) 429–434.

Friedrich, K., Reinicke, R., Zhang, Z., Wear of polymer composites. Proc. Instn Mech. Engrs, Part J, 216 (2002) 415–426.

Fu, Y., Du, H., Zhang, S., Sputtering deposited TiNi films: relationship among processing, stress evolution and phase transformation behaviours. Surf. Coat. Technol., 167 (2003) 120–128.

Fuchs, E., Oppolser, H., Rehme, H., Particle beam microanalysis, VCH, Weinheim, Germany, 1990, 507 pp.

Fujino, T., Iwamoto, K., Tanaka, K., Shima, M., Stress distribution of coated film with a range of coated film thickness and elastic properties under a single EHL operating condition. Tribology Int., 40 (2007) 1638–1648.

Fujisawa, N., Sawain, M.V., James, N.L., Woodard, J.C., Tarrant, R.N., McKenzie, D.R., Carbon coating of Ti–6Al–4V for reduced wear in combined impact and sliding applications. Tribology Int., 36 (2003) 873–882.

Fujisawa, N., McKenzie, D.R., James, N.L., Woodard, J.C., Swain, M.V., Combined influences of mechanical properties and surface roughness on the tribological properties of amorphous carbon coatings. Wear, 260 (2006) 62–74.

Fujita, H., Spikes, H.A., The formation of zinc dithiophosphate antiwear films. Proc. Instn Mech. Engrs. Pt J: J. Engineering Tribology, 218 (2004) 265–277.

Fukumasu, N.K., Souza, R.M., Numerical analysis of the contact stress development during the indentation of coated systems with substrates with orthotropic properties. Surf. Coat. Technol., 201 (2006) 4294–4299.

Furlong, O., Gao, F., Kotvis, P., Tysoe, W.T., Understanding the tribological chemistry of chlorine-, sulphur- and phosphorus-containing additives. Tribology Int., 40 (2007) 699–708.

Fürstner, R., Barthlott, W., Neinhuls, C., Walzel, P., Wetting and self-cleaning properties of artificial superhydrophobic surfaces. Langmuir, 21 (2005) 956–961.

Fusaro, R.L., Molecular relaxations, molecular orientation and the friction characteristics of polyimide films. ASLE Trans., 20 (1977) 1, 1–14.

Fusaro, R.L., Effect of atmosphere and temperature on wear, friction, and transfer of polyimide films. ASLE Trans., 21 (1978) 2, 125–133.

Fusaro, R.L., Mechanisms of lubrication and wear of a bonded solid-lubricant film. ASLE Trans., 24 (1981) 2, 191–204.

Fusaro, R.L., Tribological properties of polymer films and solid bodies in a vacuum environment. NASA Technical Memorandum 88966, Cleveland, Ohio, USA, 1987a, 30 pp.

Fusaro, R.L., Sputtered cadmium oxide as a surface pretreatment for graphite solid-lubricant films. Lubr. Engng, 43 (1987b) 10, 790–799.

Gabelli, A., Jacobson, B., Finite element analysis of EHD lubrication of rubber layers. In: Mechanics of Coatings. Dowson, D. *et al.* (eds), Tribology Series 17, Elsevier, Amsterdam, 1990, 103–110.

Gachon, Y., Vannes, A.B., Farges, G., Sainte Catherine, M.C., Caron, I., Inglebert, G., Study of sand particle erosion of magnetron sputtered multilayer coatings. Wear (1999) 233–235, 263–274.

Gåhlin R., Björkman, H., Rangsten, P., Jacobson, S., Designed abrasive diamond surfaces. Wear 387–394 (1999) 233–235.

Galvan, D., Nanocomposite coatings: processing, structure and tribological performance, PhD thesis, Goringen University, Netherlands Institute for Metals Research, Netherlands, 2007, 144 pp.

Gangopadhyay, A., Willermet, P.A., Vassell, W.C., Tamor, M.A., Amorphous hydrogenated carbon films for tribological applications II. Films deposited on aluminium alloys and steel. Tribology Int., 30 (1997) 1, 19–33.

Gangopadhyay, A., Mechanical and tribological properties of amorphous carbon films. Tribology Lett., 5 (1998) 25–39.

Gao, C., Bredell, L., Kuhlmann-Wilsdorf, D., Makel, D.D., Micromechanics of MoS_2 lubrication. Wear, (1993) 162–164, 480–491.

Gao, G.T., Mikulski, P.T., Chateauneuf, G.M., Harrison, J.A., The effects of film structure and surface hydrogen on the properties of amorphous carbon films. J. Phys. Chem. B, 107 (2003) 40, 11082–11090.

Gao, J., Luedtke, W.D., Gourdon, D., Ruths, M., Israelachvili, J.N., Landtman, U., Friction forces and Amonton's law: from the molecular to the macroscopic scale. J. Phys. Chem. B, 108 (2004) 3410–3425.

Gardos, M., The synergistic effects of graphite on the friction and wear of MoS_2 films in air. Tribology Trans., 31 (1988) 2, 214–227.

Gardos, M.N., The tribooxidative behavior of rutile-forming substrates. Mat. Res. Soc. Symp. Proc., 140 (1989) 325–338.

Gardos, M., On the elastic constants of thin solid lubricant films. In: Mechanics of Coatings. Dowson, D. *et al.* (eds), Tribology Series 17, Elsevier, Amsterdam, 1990, 3–14.

Gardos, M.N., Tribology and wear behavior of diamond. In: Synthetic Diamond: Emerging CVD Science and Technology. Spear, K.E., Dismukes, J.P. (eds), John Wiley & Sons, 1994, 419–502.

Gardos, M.N., Tribological behaviour of polycrystalline diamond films. NATO Advanced Research Workshop Protective Coatings and Thin Films, 30 May–5 June 1996, Portimao, Algarve, Portugal.

Gardos, M., Re(de)construction-induced friction signatures of polished polycrystalline diamond films in vacuum and hydrogen. Tribology Let, 4 (1998) 175–188.

Gardos, M.N., Ravi, K.V., Tribological behavior of CVD diamond films. 1st Symp. on Diamond and Diamond-like Films. The Electrochemical Society, Los Angeles, 7–12.5.1989, 19 pp.

Gardos, M.N., Soriano, B.L., The effect of environment on the tribological properties of polycrystalline diamond films. J. Mater. Res., 5 (1990) 11, 2599–2609.

Garg, D., Dyer, P.N., Erosive wear behavior of chemical vapor deposited multilayer tungsten carbide coating. Wear, 162–164 (1993) 552–227.

Gawne, D.T., Ma, U., Friction and wear of chromium and nickel coatings. Wear, 129 (1989) 123–142.

Gee, M.G., Gant, A., Hutchings, I., Bethke, R., Schiffman, K., Van Acker, K., Poulat, S., Gachon, Y., Von Stebut, J., Progress towards standardisation in ball cratering. Wear, 255 (2003) 1–13.

Gee, M., Jennet, N.M., High resolution characterization of tribochemical films on alumina. Wear, 193 (1995) 133–145.

Gee, M.G., McCormick, N.J., The application of confocal scanning microscopy to the examination of ceramic wear surfaces. J. Phys. D: Appl. Phys., 25 (1992) A230–A235.

Geike, T., Popov, V.L., Reduction description of mixed lubrication. Tribology Int., 41 (2008) 542–548.

Gellman, A.J., Spenser, N.D., Surface chemistry in tribology. Proc. Instn Mech. Engrs Pt J: J. Engineering Tribology, 216 (2002) 443–461.

Gellman, A.J., Spenser, N.D., Wear and chemistry of lubricants. In: Wear – Materials, Mechanisms and Practice. Stachowiak, G.W. (ed.), Tribology in Practice Series, John Wiley & Sons, Chichester, UK, 2005, 95–122.

Georges, J.M., Tonck, A., Meille, G., Belin, M., Chemical films and mixed lubrication, ASLE Trans., 26 (1983) 3, 293–305.

Gerkema, J., Lead thin film lubrication. Wear, 102 (1985) 241–252.

Gerth, J., Wiklund, U., The influence of metallic interlayers on the adhesion of PVD TiN coatings on high-speed steel. Wear, 264 (2008) 885–892.

Ghosh, S.K., Kohler, M.S., Study of the relative wear and abrasion resistance of Ti(C,N) and TiN coatings. Surf. Coat. Technol., 54/55 (1992) 466–469.

Gilmore, R., Baker, M.A., Gibson, P.N., Gissler, W., Stoiber, M., Losbichler, P., Mitterer, C., Low-friction TiN-MoS_2 coatings produced by dc magnetron co-deposition. Surface and Coatings Technology, 108–109 (1998) 345–351.

Gispert, M.P., Serro, A.P., Colaco, R., Saramago, B., Friction and wear mechanisms in hip prosthesis: comparison of joint materials behaviour in several lubricants. Wear, 260 (2006) 149–158.

Gispert, M.P., Serro, A.P., Colaco, R., Botelho de Rego, A.M., Alves, E., Da Silva, R.C., Brogueira, P., Pires, E., Saramago, B., Tribological behaviour of Cl-implanted TiN coatings for biomedical applications. Wear, 262 (2007) 1337–1345.

Gissler, W., Jehn, H.A. (eds), Advanced Techniques for Surface Engineering. Kluwer Academic Publishers, Dordrecht, The Netherlands, 1992, 397 pp.

Gladstone, J.R., The tribology of elastomeric layers for use in hemiarthroplasty, MSc thesis, University of Waterloo, Waterloo, Ontario, Canada, 1988, 124 pp.

Gnecco, E., Bennewitz, R., Gyalog, T., Loppacher, C., Bammerlin, M., Meyer, E., Güntherodt, H.J., Velocity dependence of atomic friction. Phys. Rev. Lett., 84 (2000) 6, 1172–1175.

Goddard, W.A., Theoretical chemistry comes alive. Engng Sci., Sept., (1985) 2–8.

Godet, M., Third-bodies in tribology. Proc. 5th Int. Congr. on Tribology, Vol. 1. Holmberg, K., Nieminen, I. (eds), Helsinki 12–15.6.1989, 1–15.

Godet, M., The third body approach: a mechanical view of wear. Wear, 100 (1984) 437–452.

Godet, M., Berthier, Y., Dubourg, M.C., Vincent, L., Contact mechanisms: needs for broader applications. J. Phys. D: Appl. Phys, 25 (1992) A273–A278.

Godet, M., Berthier, Y., Vincent, L., Flamand, L., Hard coatings for tribological applications: a pluridisciplinary approach. Surf. Coat. Technol., 45 (1991) 1–8.

Godfrey, D., Boundary lubrication. In: Interdisciplinary Approach to Friction and Wear. Ku, P.M. (ed.), 1968, 335–384.

Goller, R., Torri, P., Baker, M.A., Gilmore, R., Gissler, W., The deposition of low-friction TiN–MoS_x hard coatings by a combined arc evaporation and magnetron sputter process. Surf. Coat. Technol., 120–121. (1999) 453–457.

Gong, Z.Q., Komvopoulos, K., Effect of surface patterning on contact deformation of elastic–plastic layered media. Trans. ASME, J. Tribology, 125 (2003) 16–24.

Gong, Z.Q., Komvopoulos, K., Mechanical and thermomechanical elastic–plastic contact analysis of layered media with pattered surfaces. Trans. ASME, J. Tribology, 126 (2004a) 9–17.

Gong, Z.Q., Komvopoulos, K., Surface cracking in elastic–plastic multi-layered media due to repeated sliding contact. Trans. ASME, J. Tribology, 126 (2004b) 655–663.

Gong, Z.Q., Komvopoulos, K., Contact fatigue analysis of an elastic–plastic layered medium with a surface crack in sliding contact with a fractal structure. J. Tribology, Trans. ASME, 127 (2005) 503–512.

Goods, S.H., Brown, L.M., The nucleation of cavities by plastic deformation. Acta Metallurgica, 27 (1979) 1–15.

Gordashnik, K.Z., Gorokhovsky, V.I., Uryukov, B.A., Study of the corrosion resistance of titanium nitride coated surgical tools. Proc. Int. Ion Eng. Congr. (ISIAT'83 and IPAT'83), Kyoto, Japan, Sept. 1983, 1315–1318.

Grainger, S. (ed.), Engineering Coatings – Design and Application. Abington, Abington Publishing, 1989, 200 pp.

Grant, D.M., McColl, I.R., Golozar, M.A., Wood, J.V., Plasma assisted CVD for biomedical applications. Diamond and Related Materials, 1 (1992) 727–730.

Green, A.P., Friction between unlubricated metals, a theoretical analysis of the junction model. Proc. R. Soc., 228A (1955) 191.

Green, D.C., McKenzie, D.R., Lukins, P.B., The microstructure of carbon thin films. Mater. Sci. Forum, 52/53 (1989) 103.

Greenwood, J.A., Williamson, J.B.P., Contact of nominally flat surfaces. Proc. Royal Soc. London, A295, 1442 (1966) 300–319.

Greshham, R.M., Bonded solid film lubricants. In: Tribology Data Handbook. Booser, E.R. (ed.), CRC Press, New York, 1997, 600–607.

Grill, A., Patel, V., Meyerson, B.S., Optical and tribological properties of heat-treated diamond-like carbon.J. Mater. Res., 11 (5) (1990) 2531–2537.

Grill, A., Tribology of diamondlike carbon and related materials: an updated review. Surf. Coat. Technol. (1997) 94–95, 507–513.

Grill, A., Patel, V., Meyerson, Applications of diamond-like carbon in computer technology. In: Applications of Diamond Films and Related Materials. Tzeng, Y. *et al.* (eds), Elsevier Science Publishers, Amsterdam, 1991, 683–689.

Grillo, S.E., Field, J.E., The friction of natural and CVD diamond. Wear, 254 (2003) 945–949.

Groche, P., Köhler, M., Development and application of functional surfaces. Proc. Instn Mech. Engrs part B: J, 220 (2006) 19–26.

Grosjean, A., Rezrazi, M., Takadoum, J., Berçot, P., Hardness, friction and wear characteristics of nickel–SiC electroless composite deposits. Surf. Coat. Technol., 137 (2001) 92–96.

Grosseau-Poussard, J.L., Moine, P., Villain, J.P., Microstructural and tribological characterization of MoS_x coatings produced by high-energy ion-beam-assisted deposition. Thin Solid Films, 224 (1993) 52–57.

Grove, W.R., On the electrochemical polarity of gases. Phil. Trans. Roy. Soc., B142 (1852) p.87.

Grubin, A.N., Fundamentals of the hydrodynamic theory of lubrication of heavily loaded cylindrical surfaces. In: Investigation of the Contact of Machine Components. Ketova, Kh.F. (ed.), Central Scientific Research Institute for Technology and Mechanical Engineering (TsNITMASh), Moscow, Book No. 30, 1949, 115–166.

Grundwürmer, M., Nuyken, O., Meyer, M., Wehr, J., Schupp, N., Sol-gel derived erosion protection coatings against damage caused by liquid impact. Wear, 263 (2007) 318–329.

Grzesik, W., The influence of thin hard coatings on frictional behaviour in the orthogonal cutting process. Tribology Int., 33 (2000) 131–140.

Guilemany, J.M., Nin, J., Thermal spraying methods for protection against wear. In: Surface Coatings for Protection against Wear. Mellor, B.G. (ed.), CRC Press, Woodhead Publishing Ltd, Cambridge, UK, 2006, 249–301.

Gunnars, J., Alahelisten, A., Thermal stresses in diamond coatings and their influence on coating wear and failure. Surf. Coat. Technol., 80 (1996) 303–312.

Guo, H., Alam, M., Residual stress in CVD diamond films. In: Applications of Diamond Films and Related Materials. Tzeng, Y. *et al.* (eds), Elsevier Science Publishers BV, 1991, 149–154.

Gupta, P., Meletis, E.I., Tribological behavior of plasma-enhanced CVD a-C:H films. Part: II multinanolayers. Tribology Int., 37 (2004) 1031–1038.

Gupta, P.K., Walowit, J.A., Contact stresses between an elastic cylinder and a layered elastic solid. J. Lubr. Tech., Trans. ASME, April (1974) 250–257.

Gupta, P.K., Walowit, J.A., Finkin, E.F., Stress distributions in plane strain layered elastic solids subjected to arbitrary boundary loading. J. Lubr. Tech., Trans. ASME, OCTOBER (1973) 427–433.

Gupta, P., Meletis, E.I., Tribological behavior of plasma-enhanced CVD a-C:H films. Part II: multinanolayers. Tribology Int., 37 (2004) 1031–1038.

Gupta, P., Singh, V., Meletis, E.I., Tribological behavior of plasma-enhanced CVD a-C:H films. Part I: effect of processing parameters. Tribology Int., 37 (2004) 1019–1029.

Guu, Y.Y., Lin, J.F., Ai, C.F., The tribological characteristics of titanium nitride coatings, Part I: Coating thickness effects. Wear, 194 (1996a) 12–21.

Guu, Y.Y., Lin, J.F., Ai, C.F., The tribological characteristics of titanium nitride coatings, Part II: Comparison of two deposition processes. Wear, 194 (1996b) 22–29.

Günther, K.G., Freller, H., Hintermann, H.E., König, W., Kammermeier, D., Advanced coatings by vapour phase processes. Annals of the CIRP, Keynote – Paper C,E,S, 38 (1989) 2, 11 pp.

Habibovic, P., Barrere, F., Van Blitterswijk, C.A., De Groot, K., Layrolle, P., Biomimetic hydroxyapatite coating on metal implants. J. American Ceramic Society, 85 (2002) 3, 517–522.
Habig, K.H., Friction and wear of sliding couples coated with TiC, TiN or TiB_2. Surf. Coat. Technol., 42 (1990) 133–147.
Habig, K.H., Evers, W., Hintermann, H.E., Tribologische Prufungen an Oberflächenschichten aus Titancarbid, Titannitrid, Chromcarbid und Eisenborid. Z. Werkstofftech., 11 (1980) 182–190.
Habig, K.H., Meier Zu Köcker, G., Simulation of the tribological behaviour of tools for cutting ductile materials. J. Phys. D: Appl. Phys., 25 (1992) A307–A312.
Hainsworth, S.V., Soh, W.C., The effect of the substrate on the mechanical properties of TiN coatings. Surf. Coat. Technol. 163–164 (2003) 515–520.
Håkansson, G., Hultman, L., Sundgren, J.E., Greene, J.E., Munz, W.D., Microstructure of thin films grown by various PVD techniques. Surf. Coat. Technol., 48 (1991) 51–67.
Halling, J. (ed.), Principles of Tribology. MacMillan Press, London, 1975, 401.
Halling, J., Surface films in tribology – Part one: A view for the future. Tribologia – Finnish J. Tribology, 1 (1982) 2, 15–29.
Halling, J., Arnell, R.D., Ceramic coatings in the war on wear. Wear, 100 (1984) 367–380.
Halling, J., El Shafei, T.E.S., Arnell, R.D., The rolling resistance of surfaces covered by soft metal films. Proc. Instn Mech. Engrs, 199C1 (1985) 51–55.
Halling, J., The tribology of surface coatings, particularly ceramics. Proc. Instn. Mech. Engrs., 200C1 (1986) 31–40.
Halling, J., Surface Coating – Thick or Thin? Soft or Hard? Mechanics of Coatings. In: Dowson, D. *et al.* (eds), Tribology Series 17, Elsevier, Amsterdam, 1990, 477–480.
Halling, J., El-Sherbiny, M., The role of surface topography in the friction of soft metallic films. Tribology Convention, Institution of Mechanical Engineers. Swansea, 1978, 155–158.
Halling, J., Sherbiney, M.G.D., The role of surface topography in the friction of soft metallic films. Tribology Convention, Instn Mech. Engrs, Swansea, UK, Paper C46/78 (1978) 131–134.
Ham, M., Lou, A., Diamond-like carbon films grown by a large-scale direct current plasma chemical vapor deposition reactor: system design, film characteristics, and applications. J. Vac. Sci. Technol., A8 (1990) 3, 2143–2149.
Hamilton, G.M., Explicit equations for the stresses beneath a sliding spherical contact. Proc. Inst. Mech. Engrs, 197C (1983) 53–59.
Hamilton, G.M., Goodman, L.E., The stress field created by a circular sliding contact. J. Appl. Mechanics, Trans ASME, 33 (1966) 371–376.
Hammer, B., Perry, A.J., Laeng, P., Steinmann, P.A., The scratch test adhesion of TiC deposited industrially by chemical vapour deposition on steel. Thin Solid Films, 96 (1982) 45–51.
Hamrock, B.J., Dowson, D., Ball Bearing Lubrication. The Elastohydrodynamics of Elliptical Contacts, John Wiley & Sons, New York, 1981, 386 pp.
Hanson, R.A., The adhesion and deformation properties of CVD TiC coated bearing balls under heavy loads. In: New Materials Approaches to Tribology: Theory and Applications. Pope, L. *et al.* (eds), Material Research Society, Mat. Res. Soc. Symp. Proc. Vol. 140 (1989) 477–482.
Harju, E.J., Penttinen, I.M., Korhonen, A.S., Lappalainen, R., Optimization of wear and corrosion resistance of troide-ion-plated nitride coatings. Surf. Coat. Technol., 41 (1990) 157–166.
Harris, P.G., Trigg, A.D., Surface analysis techniques and their application to materials characterisation. Materials Design, 9 (1988) 127–134.
Harris, S.J., Krauss, G., Siniawski, M.T., Wang, Q., Liu, S., Ao, Y., Surface feature variations observed in 52100 steel sliding against a thin boron carbide coating. Wear, 249 (2002) 1004–1013.
Harris, S.J., Weiner, A.M., Scaling relationships for the abrasion of steel by diamondlike carbon coatings. Wear, 223 (1998) 31–36.
Harris, S.J., Weiner, A.M., Grischke, M., Effects of load on the abrasion of steel by metal-containing diamond-like carbon. Surf. Coat. Technol., 120–121 (1999) 561–564.
Harris, S.J., Weiner, A.M., Meng, W.J., Tribology of metal-containing diamond-like carbon coatings. Wear, 211 (1997) 208–217.

Harrison, J.A., White, C.T., Colton, R.J., Brenner, D.W., Molecular-dynamics simulations of atomic-scale friction of diamond surfaces. The American Physical Society, Physical Review B, 46 (1992) 15, 9700–9708.

Harrison, J.A., Colton, R.J., White, C.T., Brenner, D.W., Effect of atomic scale surface roughness on friction: a molecular dynamics study of diamond surfaces. Wear, 168 (1993) 127–133.

Harrison, J., White, J., Colton, R., Brenner, D., Investigation of the atomic-scale friction and energy dissipation in diamond using molecular dynamics. Thin Solid Films, 260 (1995) 205–211.

Harrison, J., Stuart, S., Perry, M., The tribology of hydrocarbon surfaces investigated using molecular dynamics. In: Tribology Issues and Opportunities in MEMS. Bhushan, B. (ed.), Kluwer Academic Publishers, The Netherlands, 1998, 285–299.

Harry, E., Rouzaud, A., Ignat, M., Juliet, P., Mechanical properties of W and W(C) thin films: Young's modulus, fracture toughness and adhesion. Thin Solid Films, 332 (1998) 195–201.

Hauert, R., An overview on the tribological behavior of diamond-like carbon in technical and medical applications. Tribology Int., 37 (2004) 991–1003.

Hauert, R., DLC films in biomedical applications. In: Tribology of Diamond-like Carbon Films – Fundamentals and Applications. Donnet, C., Erdemir, A. (eds), Springer, New York, USA, 2008, 494–510.

Hayward, I.P., Friction and wear properties of diamonds and diamond coatings. Surf. Coat. Technol., 49 (1991) 554–559.

Hayward, I.P., Field, J.E.., Friction and wear of diamond. Int. Conf. on Tribology 50 years on, IMechE, C159/87, 1987, 205–209

Hayward, I.P., Field, J.E., The solid particle erosion of diamond. J. Hard Metals, 1 (1990) 1, 53–64.

Hayward, I.P., Singer, I.L., The tribological behaviour of diamond coatings. 2nd Int. Conf. on the New Diamond Science and Technology, Sept. 1990, Washington DC (1990), 5 pp.

Hayward, I.P., Singer, I.L., Seitzman, L.E., Effect of roughness on the friction of diamond on CVD diamond coatings. Wear, 157 (1992) 215–227.

Héau, C., DLC films in mechanical and manufacturing industry. In: Tribology of Diamond-like Carbon Films – Fundamentals and Applications. Donnet, C., Erdemir, A. (eds), Springer, New York, USA, 2008, 469–483.

Heavens, O.S., Collins, L.E., L'epitaxie dans les lames polycristallines. J. Phys. Rad., 13 (1952) 658.

Hedenqvist, P., Evaluation of vapour-deposited coatings for improved wear resistance. Acta Universitatis Upsaliensis, Comprehensive Summaries of Uppsala Dissertations from Faculty of Sciences, No. 360, Uppsala University, Sweden, 1991, 46 pp.

Hedenquist, P., Olsson, M., Solid particle erosion of titanium nitride coated high speed steel. Tribology Int., 23 (1990) 3, 173–181.

Hedenquist, P., Olsson, M., Sliding wear testing of coated cutting tool materials. Tribology Int., 24 (1991) 143–150.

Hedenquist, P., Olsson, M., Jacobson, S., Söderberg, S., Failure mode analysis of TiN-coated high speed steel: in situ scratch testing in the scanning electron microscope. Surf. Coat. Technol., 41 (1990a) 31–49.

Hedenqvist, P., Olsson, M., Wallén, P., Kassman, Å., Hogmark, S., Jacobson, S., How TiN coatings improve the performance of high speed steel cutting tools. Surf. Coat. Technol., 41 (1990b) 243–256.

Heilmann, P., Rigney, D.A., An energy based-model of friction and its application to coated systems. Wear, 72 (1981) 195–217.

Helmersson, U., Todorova, S., Barnett, S.A., Sundgren, J.E., Markert, L.C., Greene, J.E., Growth of single-crystal TiN/VN strained-layer superlattices with extremely high mechanical hardness. J. Appl. Phys., 62 (1987) 481–484.

Hertz, H., Über die beröhrung fester elastische Körper und über die harte / On the contact of rigid elastic solids and on hardness. Verhandlungen des Vereins zur Beforderung des Gewerbefleisses, Leipzig, Germany, Nov., 1882 (for English translation see Miscellaneous Papers by H. Hertz, Jones and Schott (eds), MacMillan, London, 1896).

Hill, R.J., Physical vapour deposition., Airco Inc., Berkeley, 1986, 249 pp.

Hill, R.J., Nadel, S.J., Coated glass applications and markets, BOC Coatings Technology, Fairfield, CA, USA, 1999, 143 pp.

Hillery, R.V., Coatings for performance retention. J. Vac. Sci. Technol.. A4, 6 (1986) 2624–2628.

Hills, B.A., Boundary lubrication in vivo. Proc. Instn Mech. Engrs, Part H, 214 (2000) 83–94.

Hills, D.A., Nowell, D., Sackfield, A., A survey of cracks in layers propelled by contact loading. In: Mechanics of Coatings. Dowson, D. *et al.* (eds), Tribology Series 17, Elsevier, Amsterdam, 1990, 203–208.

Hilton, M.R., Bauer, R., Didziulis, S.V., Dugger, M.T., Keem, J.M., Scholhamer, J., Structural and tribological studies of MoS_2 solid lubricant films having tailored metal-multilayer nanostructures. Surf. Coat. Technol., 53 (1992) 13–23.

Hilton, M.R., Fleischauer, P.D., Applications of solid lubricant films in spacecraft. Surf. Coat. Technol., 54/55 (1992) 435–441.

Hintermann, H.E., Verschleiss- und Korrosionschutz durch CVD- und PVD-Uberzuge. VDI-Berichte, (1979) 333, 53–67.

Hintermann, H.E., Tribological and protective coatings by chemical vapour deposition. Thin Solid Films, 84 (1981) 215–243.

Hintermann, H.E., Tough-hard composite materials: substrate–interface–coating relationships. Annals of the CIRP, 31 (1982) 1, 435–440.

Hintermann, H.E., Commercial aspects on CVD overlay and diffusion coatings. Oberfläche-Surface, 24/ Heft 4 (1983a) 7 pp.

Hintermann, H.E., Surface treatments. In: Science of Hard Materials. Plenum Press, New York, 1983b, 357–391.

Hintermann, H.E., Adhesion, friction and wear of thin hard coatings. Wear, 100 (1984) 381–397.

Hintermann, H., Laeng, P., The adhesion of hard and wear resistant coatings. Proc. Int. Conf. Recent Developments in Specialty Steels and Hard Materials (Materials Development '82), Pretoria, South Africa, 8–12.11.1982, 407–414.

Hirata, A., Igarashi, M., Kaito, T., Study on solid lubricant properties of carbon onions produced by heat treatment of diamond clusters or particles. Tribology Int., 37 (2004) 899–905.

Hirvonen, J.-P. (ed.), Timantti- ja timantinkaltaiset pinnoitteet (Diamond and diamond-like coatings). Technical Research Centre of Finland, Research Notes No. 1293, 1991, 74.

Hirvonen, J-P., Koskinen, J., Lappalainen, R., Anttila, A., Preparation and properties of high density, hydrogen free hard carbon films with ion beam or arc discharge deposition. Materials Science Forum, 52/53 (1989a) 197–216.

Hirvonen, J-P., Nastasi, M., Jervis, T., Tesmer, J., Zocco, T., Unlubricated sliding properties of ion beam and excimer laser mixed Fe–Ti–C multilayered films. In: New Materials Approaches to Tribology: Theory and Applications. Pope, L. *et al.* (eds), Material Research Society, Mat. Res. Soc. Symp. Proc. Vol. 140 (1989b) 183–188.

Hirvonen, J-P., Koskinen, J., Anttila, A., Lappalainen, R., Toivanen, R.O., Arminen, E., Trkula, M., Characterization and unlubricated sliding of ion-beam-deposited hydrogen-free diamond-like carbon films. Wear, 141 (1990a) 45–58.

Hirvonen, J-P., Koskinen, J., Likonen, J., Koponen, I., Kattelus, H., Modification of interfacial characteristics between diamond-like carbon films and substrates using ion bombardment. 8th Int. Conf. on Ion Beam Modification of Materials, Heidelberg, Sept. 1992, 11 pp.

Hirvonen, J.P., Lappalainen, R., Koskinen, J., Anttila, A., Jervis, T.R., Trkula, M., Tribological characteristics of diamond-like films deposited with an arc-discharge method. J. Mater. Res., 5 (1990b) 11, 2524–2530.

Hisakado, T., The influence of surface roughness on abrasive wear. Wear, 4 (1976) 179–190.

HMSO, Wear resistant surfaces in engineering. Her Majesty's Stationery Office, Department of Trade and Industry, London, 1986, 171 pp.

Hochman, R.F., Erdemir, A., Dolan, F.J., Thom, R.L., Rolling contact fatigue behavior of Cu and TiN coatings on bearing steel substrates. J. Vac. Sci. Technol., A3 6, (1985) 2348–2353.

Hogmark, S., Mechanical and tribological requirements and evaluation of coating composites. In: Modern Tribology Handbook, Vol. II, Bhushan, B. (ed.), CRC Press, London, UK, 2001, 931–963.

Hogmark, S., Jacobson, S., Larsson, M., Design and evaluation of tribological coatings. Wear, 246 (2000) 20–33.

Hogmark, S., Jacobson, S., Larsson, M., Wiklund, U., Mechanical and tribological requirements and evaluation of coating composites. In: Modern Tribology Handbook. Bhushan, B. (ed.), CRC Press, New York, 2001, 931–963.

Hokkirigawa, K., Kato, K., An experimental and theoretical investigation of ploughing, cutting and wedge formation during abrasive wear. Tribology Int., 21 (1988a) 1, 51–58.

Hokkirigawa, K., Kato, K., The effect of hardness on the transition of the abrasive wear mechanism of steels. Wear, 123 (1988b) 241–251.

Holiday, P., Dehbi-Alaoui, A., Matthews, A., A comparison of techniques for producing diamond-like carbon coatings. Materials Science Forum, 102–104 (1992) 643–654.

Holinski, R., Gänsheimer, J., A study of the lubricating mechanism of molybdenum disulfide. Wear, 19 (1972) 329–342.

Holleck, H., Material selection for hard coatings. J. Vac. Sci. Technol., A4 (1986) 6, 2661–2669.

Holleck, H., Designing advanced coatings for wear protection. Surface Engng, 7 (1991) 2, 137–144.

Holleck, H., Lahres, M., Woll, P., Multilayer coatings – influence of fabrication parameters on constitution and properties. Surf. Coat. Technol., 41 (1990) 179–190.

Holleck, H., Schier, V., Multilayer PVD coatings for wear protection. Surf. Coat. Technol., 76/77 (1995) 328–336.

Hollman, P., Micro- and nanocrystalline diamond coatings with extreme wear resistance and ultra low friction/Paper VII: Friction properties of smooth nanocrystalline diamond coatings. PhD theses, Uppsala University, Comprehensive Summaries of Uppsala Dissertations from the Faculty of Science and Technology, Sweden, 1997a, 11 pp.

Hollman, P., Alahelisten, A., Björk, T., Hogmark, S., CVD-diamond coating in sliding contact with Al, Al-17Si and steel. Wear, 179 (1994) 11–16.

Hollman, P., Micro- and nanocrystalline diamond coatings with extreme wear resistance and ultra low friction. Acta Universitatis Upsaliensis. Dissertations from Faculty of Science and Technology 325, Uppsala, Sweden, 1997b, 40 pp.

Holm, R., Electric Contacts. Hugo Gebers Förlag, Stockholm, 1946.

Holmberg, K., Elastohydrodynamic lubrication theory and its applications on machine elements. Helsinki University of Technology, Instn. for Machine Design, Series A1, 1980, 230 pp.

Holmberg, K., A survey of applications of EHL on machine elements. Tribology Int., June (1982) 123–131.

Holmberg, K., Friction in low speed lubricated rolling and sliding contacts. Espoo, Technical Research Centre of Finland, Publications 16, 1984, 81 pp.

Holmberg, K., The mechanism of lubrication in low speed rolling contacts. J. Tribology, Trans. ASME, 111 (1989) 703–707.

Holmberg, K., Tribological bases for accelerated testing. In: Operational Reliability and Systematic Maintenance. Holmberg, K., Folkeson, A. (eds), Elsevier, London, 1991, 31–50.

Holmberg, K., A concept for friction mechanisms of coated surfaces. Surf. Coat. Technol., 56 (1992a) 1–10.

Holmberg, K., A suggestion for the classification of tribological mechanisms in coated surfaces. Finnish J. Tribology, 11 (1992b) 1, 1–17.

Holmberg, K., The basic material parameters that control friction and wear of loaded surfaces under sliding. Tribologia – Finnish J. Tribology, 19 (2000) 3, 3–18.

Holmberg, K., Reliability aspects of tribology. Tribology Int., 34 (2001) 12, 801–808.

Holmberg, K., Tribology in the past and in the future. In: Tribology of Mechanical Systems: A Guide to Present and Future Technologies. Vizintin, J., Kalin, M., Dohda, K., Jahanmir, S. (eds), ASME Press, New York, USA, 2004, 1–23.

Holmberg, K., Folkeson, A. (eds), Operational reliability and systematic maintenance. Elsevier, London, 1991, 331p.

Holmberg, K., Koskinen, J., Ronkainen, H., Vihersalo, J., Hirvonen, J.-P., Likonen, J., Tribological characteristics of hydrogenated and hydrogen free diamond-like carbon coatings. Diamond Films and Technology, accepted for publication, 1994, 25 pp.

Holmberg, K., Kuoppala, R., Vuoti, A., Expert system for wear failure prediction. In: Proc. 5th Int. Congr. on Tribology – EUROTRIB'89. Holmberg, K., Nieminen, I. (eds), Helsinki, Finland, 12–15.6.1989, 32–37.

Holmberg, K., Laukkanen, A., Ronkainen, H., Wallin, K., Surface stresses in coated steel surfaces – influence of a bond layer on surface fracture. Tribology Int., in press – available online 7.7.2008 at Science Direct (2008b) 27 pp.

Holmberg, K., Laukkanen, A., Ronkainen, H., Wallin, K., Varjus, S., A model for stresses, crack generation and fracture toughness calculation in scratched TiN-coated steel surfaces. Wear, 254 (2003) 278–291.

Holmberg, K., Laukkanen, A., Ronkainen, H., Wallin, K., Varjus, S., Koskinen, K., Tribological contact analysis of a rigid ball sliding on a hard coated surface – Part I: Modelling stresses and strains. Surf. Coat. Technol., 200 (2006a) 3793-3809.

Holmberg, K., Laukkanen, A., Ronkainen, H., Wallin, K., Varjus, S., Koskinen, K., Tribological contact analysis of a rigid ball sliding on a hard coated surface – Part II: Material deformations, influence of coating thickness and Young's modulus. Surf. Coat. Technol., 200 (2006b) 3810–3823.

Holmberg, K., Matthews, A., Coatings Tribology – Properties, Techniques and Applications in Surface Engineering, Tribology Series 28, Elsevier, Amsterdam, 1994, 442 pp.

Holmberg, K., Matthews, A., Tribological properties of metallic and ceramic coatings. In: Modern Tribology Handbook. Bhushan, B. (ed.), CRC Press, New York, 2001, 827–870.

Holmberg, K., Matthews, A., Tribology of engineered surfaces. In: Wear – Materials, Mechanisms and Practice. Stachowiak, G.W. (ed.), Tribology in Practice Series, John Wiley & Sons, Chichester, UK, 2005, 123–166.

Holmberg, K., Matthews, A., Ronkainen, H., Coatings tribology – contact mechanisms and surface design. Tribology Int., 31 (1998) 1–3, 107–120.

Holmberg, K., Ronkainen, H., Laukkanen, A., Wallin, K., Friction and wear of coated surfaces – scales, modelling and simulation of tribomechnisms. Surf. Coat. Technol., 202 (2007) 1034–1049.

Holmberg, K., Ronkainen, H., Laukkanen, A., Wallin, K., Erdemir, A., Eryilmaz, O., Tribological analysis of TiN and DLC coated contacts by 3D FEM modelling and simulation. Wear, 264 (2008c) 877–884.

Holmberg, K., Ronkainen, H., Laukkanen, A., Wallin, K., Hogmark, S., Jacobson, S., Wiklund, U., Souza, R., Ståhle, P., Residual stresses in TiN, DLC and MoS2 coated surfaces with regard to their tribological fracture behaviour. In: Proc. of 13th Nordic Symposium on Tribology NORDTRIB 2008, 10–13.6.2008, Tampere, Finland, 2008d, 37 pp, submitted to Wear Special Issue for NORDTRIB 2008.

Holmberg, K., Ronkainen, H., Matthews, A., Wear mechanisms of coated sliding surfaces. In: Thin Films in Tribology. Proc. 19th Leeds–Lyon Symp. on Tribology, 8–11.9.1992, Elsevier, Leeds, UK, 1992, 9 pp.

Hooke, C.J., The elastohydrodynamic lubrication of a cylinder on an elatomeric layer. Wear, 111 (1986) 83–99.

Hopple, G.B., Keem, J.E., Loewenthal, S.H., Development of fracture resistant, multilayer films for precision ball bearings. Wear, 162–164 (1993) 919–924.

Howard, S.S., Tsui, Y.C., Clyne, T.W., The effect of residual stresses on the debonding of coatings. Part I: A model for delamination at a bimaterial interface. Submitted for publication to Acta Metall. Mater (1993).

Howson, R.P., J'Afer, H.A., Spencer, A., Substrate effects from an unbalanced magnetron. Thin Solid Films, 193/194 (1990) 127.

Howson, R.P., J'Afer, H.A., Reactive sputtering with an unbalanced magnetron. J. Vac. Sci. Technol., A, 10 (1992) 4, 1784–1790.

Hsu, S.M., Boundary lubrication of materials. MRS Bulletin, October (1991) 54–58.

Hsu, S.M., Nano-lubrication: concept and design. Tribology Int., 37 (2004a) 537–545.

Hsu, S.M., Molecular basis for lubrication. Tribology Int., 37 (2004b) 553–559.

Hsu, S.M., Gates, R.S., Boundary lubrication and boundary lubricating films. In: Modern Tribology Handbook. Bhushan, B. (ed.), CRC Press, New York, 2001, 455–492.

Hsu, S.M., Gates, R.S., Boundary lubricating films: formation and lubrication mechanism. Tribology Int., 38 (2005) 305–312.

Hsu, S.M., Lim, D.S., Wang, Y.S., Munro, R.G., Ceramic wear maps: concept and method development. Lubr. Engng, J. STLE, 47 (1991) 1, 49–54.

Hsu, S.M., Munro, R.G., Shen, M.C., Gates, R.S., Boundary lubricated wear. In: Wear – Materials, Mechanisms and Practices. Stachowiak, G.W. (ed.), John Wiley & Sons, Chichester, UK, 2005, 37–70.

Hsu, S.M., Shen, M.C., Wear mapping of materials. In: Wear – Materials, Mechanisms and Practices. Stachowiak, G.W. (ed.), John Wiley & Sons, Chichester, UK, 2005, 369–424.

Hsu, S.M., Ying, Z.C. (eds), Nanotribology – Critical Assessment and Research Needs. Kluwer Academic Publishers, London, UK, 2003, 442 pp.

Huang, J.H., Yu, K.J., Sit, P., Yu, G.P., Heat treatment of nanocrystalline TiN films deposited by unbalanced magnetron sputtering. Surf. Coat. Technol., 200 (2006) 4291–4299.

Huang, Y., Xiao, H., Ma, Z., Wang, J., Gao, P., Effects of Cu and Cu/Ti interlayer on adhesion of diamond film. Surf. Coat. Technol., 202 (2007) 180–184.

Huang, Z.P., Sun, Y., Bell, T., Friction behaviour of TiN, CrN and (TiAl)N coatings. Wear, 173 (1994) 13–20.

Huang, Y.S., Zeng, X.T., Annergren, I., Liu, F.M., Development of electroless NiP–PTFE–SiC composite coating. Surf. Coat. Technol., 167 (2003) 207–211.

Huong, P.V., Structural studies of diamond films and ultrahardened materials by Raman and micro-Raman spectroscopie. Diamond and Related Materials, 1 (1991) 33–41.

Huq, M.Z., Celis, J.P., Reproducibility of friction and wear results in ball-on-disc unidirectional sliding tests of TiN–alumina pairings. Wear, 212 (1997) 151–159.

Hurkmans, T., Hauzer, F., Buil, B., Engel, K., Tietema, R., A new large volume PVD coating system using advanced controlled arc and combined arc/unbalanced magnetron (ABS™) deposition techniques. Surf. Coat. Technol., 92 (1997) 62–68.

Hutchings, I.M., Tribology: Friction and Wear of Engineering Materials. Edward Arnold, London, 1992, 273 pp.

Hutchings, I., Tribology in joint replacement. Proc. TRIBOS Symposium, 10–11.5.2007, Florence, Italy, Stryker Ltd, Switzerland, 26 pp.

Huu, T.L., Zaidi, H., Paulmier, D., Lubrication properties of diamond-like coating. Wear (1995) 181–183, 766–770.

Hwang, D-H., Zum Ghar, K-H., Transition from static to kinetic friction of unlubricated or oil lubricated steel/steel, steel/ceramic and ceramic/ceramic pairs. Wear, 255 (2003) 365–375.

Hwang, D.H., Kim, D.E., Lee, S.J., Influence of wear particle interaction in the sliding interface on friction of metals. Wear, 225–229 (1999) 427–439.

Hwang, D.H., Zum Gahr, K.H., Transition from static to kinetic friction of unlubricated or oil lubricated steel/steel, steel/ceramic and ceramic/ceramic pairs. Wear, 255 (2003) 365–375.

Ikawa, N., Donaldson, R.R., Komanduri, R., König, W., McKeown, P.A., Moriwaki, T., Stowers, I.F., Ultraprecision metal cutting – the past, the present and the future. Annals of the CIRP, 40 (1991) 2, 587–594.

Iliuc, I., Tribology of thin layers, Tribology Series, No. 4, Amsterdam, Elsevier, 1980, 225 pp.

Iliuc, I., Wear and micropitting of steel ball sliding against TiN coated steel plate in dry and lubricated conditions. Tribology Int., 39 (2006) 607–615.

Imamura, A., Tsukamoto, T., Shibuki, K., Takatsu, S., Friction and wear properties of hard carbon films formed on cemented carbides by d.c. plasma chemical vapour deposition. Surf. Coat. Technol., 36 (1988) 161–169.

Inagawa, K., Watanabe, K., Saitoh, K., Itoh, A., Structure and properties of c-BN film deposited by activated reactive evaporation with a gas activation nozzle. Surf. Coat. Technol., 39/40 (1989) 253–264.

Inoue, A., Bulk amorphous and nanocrystalline alloys with high functional properties. Mater. Sci. Engng, A304/306 (2001) 1–10.

Inspector, A., Bauer, C.E., Oles, E.J., Superhard coatings for metal cutting applications. Surf. Coat. Technol., 68/69 (1994) 354–368.

Ioannides, E., Bergling, G., Gabelli, A., An analytical formulation for the life of rolling bearings, Acta Polytechnica Scandinavica, Mechanical Engineering Series, The Finnish Academy of Technology, No. 137, 1999, 80 pp.

Israelachvili, J.N., Adhesion, friction and lubrication of molecularly smooth surfaces. In: Fundamentals of friction: Macroscopic and microscopic processes. Singer, I.L., Pollock, H.M. (eds), NATO ASI Series, Series E: Applied Sciences Vol. 220, Kluwer Academic Publishers, London, 1992, 351–385.

Israelachvili, J., Spikes, H., Discussion forum report: Bridging the gap between macro- and micro/nanoscale lubrication. In: Fundamentals of tribology and bridging the gap between the macro- and micro/nanoscales, Bhushan, B. (ed) NATO Science Series, II. Mathematics, Physics and Chemistry, Vol. 10, Kluwer Academic Publishers, Boston, 2001, 799–802.

Israelachvili, J.N., Tabor, D., The measurement of van der Waals dispersion forces in the range 1.5 to 130 nm. Proc. Royal Soc. A, 331 (1972) 19–38.

Itoh, H., Lee, S.S., Sugiyama, K., Iwahara, H., Tsutsumoto, T., Adhesion improvement of diamond coating on silicon nitride substrate. Surf. Coat. Technol., 112 (1999) 199–203.

Itoh, T., Hibi, S., Hioki, T., Kawamoto, J., Tribological properties of metals modified by ion-beam assisted deposition of silicone oil. J. Mat. Res., 6 (1991) 4, 871–874.

Ivkovic, B., Lazic, M., Using radioactive techniques for tribological research. Tribology Int., Oct (1975) 209–213.

Iwai, Y., Honda, T., Yamada, H., Matsubara, T., Larsson, M., Hogmark, S., Evaluation of wear resistance of TiN hard coatings by a new solid particle impact test. Wear, 251 (2001) 861–867.

Jacobson, B.O., Rheology and elastohydrodynamic lubrication, Tribology Series, 19, Elsevier, Amsterdam, 1991, 382 pp.

Jacobson, S., Hogmark, S., On the tribological character of boundary lubricated DLC coated components. 2nd World Tribology Congress, 3–7.9.2001, Vienna, Austria, 2001, 9 pp.

Jaeger, G., Endler, I., Heilmaier, M., Bartsch, K., Leonhardt, A., A new method for determining strength and fracture toughness of thin hard coatings. Thin Solid Films, 377–378 (2000) 382–388.

Jahanmir, S., On the wear mechanisms and the wear equations. In: Fundamentals of Tribology. Suh, N.P., Saka, N. (eds), The MIT Press, London, 1980, 455–467.

Jahanmir, S., Abrahamson, E.P., Suh, N.P., Sliding wear resistance of metallic coated surfaces. Wear, 40 (1976) 75–84.

Jahanmir, S., Deckman, D.E., Ives, L.K., Feldman, A., Farabaugh, E., Tribological characteristics of synthesized diamond films on silicon carbide. Wear, 133 (1989) 73–81.

Jamal, T., Nimmagadda, R., Bunshah, R.F., Friction and adhesive wear of titanium carbide and titanium nitride coatings. Thin Solid Films, 73 (1980) 245–254.

James, D.H., Component reclamation by plasma spraying. Proc. 8th Int. Thermal Spraying Conf., Miami, Florida, American Welding Society, 27.9–1.10.1976, 148–155.

James, D.H., Balancing resources and requirement in surfacing. Surfacing J, 9 (1978) 3.

James, A.S., Matthews, A., Thermal stability of partially-yttria-stabilised zirconia thermal barrier coatings deposited by r.f. plasma-assisted physical vapour deposition. Surf. Coat. Technol., 41 (1990) 305–313.

Jantunen, E., Miettinen, J., Ollila, A., Maintenance and downtime costs of centrifugal pumps in Finnish industry. 13th Int. Pump Technol.. Conf., Pumps for a Safer Future, London, 21–23.4. (1993) 11 pp.

Jarvis, M.R., Perez, F.M., Van Bouwelen, C.M., Payne, M.C., Microscopic mechanism for mechanical polishing of diamond (110) surfaces. Phys. Rev. Lett., 80 (1998) 3428–3431.

Jayachandran, R., Boyce, M.C., Argon, A.S., Mechanics of the indentation test and its use to assess the adhesion of polymeric coatings. In: Adhesion Measurements of Films and Coatings. Mittal, K.L. (ed.), VSP, Utrecht, The Netherlands, 1995, 189–215.

Je, J.H., Gyarmati, E., Naoumidis, A., Scratch adhesion test of reactively sputtered TiN coatings on a soft substrate. Thin Solid Films, 136 (1986) 57–67.

Jehn, H.A., Multicomponent and multiphase hard coatings for tribological applications. Surf. Coat. Technol., 131 (2000) 433–440.

Jehn, H.A. Nanostructured hard coatings- architecture, properties and deposition. Proc. 7th Symposium of European Vacuum Coaters, Anzio, Rome, Italy, 2–4. 10. 2006, paper 20, Published in Romana Film Sottili, Anzio, 2007.

Jeng, Y.R., Tan, C.M., Study of nanoindentation using FEM atomic model. Trans. ASME, J. Tribology, 126 (2004) 767–774.

Jennet, N.M., Gee, M.G., Mechanical testing of coatings. In: Surface Coatings for Protection against Wear. Mellor, B.G. (ed.), Woodhead Publishing Ltd, Cambridge, UK, 2006, 58–78.

Jennet, N.M., Jacobs, R., Meneve, J., Advances in adhesion measurement good practice: Use of a certified reference material for evaluating the performance of scratch test instrumentation. MST Conference: Adhesion Aspects of Thin Films, 15–16.12, 2003, 20 pp.

Jiang, J., Arnell, R.D., The effect of substrate surface roughness on the wear of DLC coatings. Wear, 239 (2000) 1–9.

Jiang, J., Arnell, R.D., Tong, J., Some special tribological features of DLC coatings deposited on soft substrates. Wear, 211 (1997a) 254–264.

Jiang, J., Stack, M.M., Modelling sliding wear: from dry to wet environments. Wear, 261 (2006) 954–965.

Jindal, P.C., Quinto, D.T., Wolfe, G.J., Adhesion measurements of chemically vapour deposited and physically vapour deposited hard coatings on WC–Co substrates. Thin Solid Films, 154 (1987) 361–375.

Joachim, F., Kurz, N., Glatthaar, B., Influence of coatings and surface improvements on the lifetime of gears. Gear Technology, (2004) July/August, 50–56.

Johnson, C.E., Hasting, M.A.S., Weimer, W.A., Thermogravimetric analysis of the oxidation of CVD diamond films. J. Mater. Res., 5 (1990) 11, 2320–2325.

Johnson, K.L., Contact Mechanics, Cambridge University Press, Cambridge, 1985, 452 pp.

Johnson, K.L., Contact mechanics and wear of metals. Wear, 190 (1995) 162–170.

Johnson, K.L., Greenwood, J.A., Poon, S.Y., A simple theory of asperity contact in elastohydrodynamic lubrication. Wear, 19 (1972) 91–108.

Johnson, R.N., Principles and applications of electro-spark deposition. In: Surface Modification Technologies. Sudershan, T.S., Bhat, D.G. (eds), The Metallurgical Society, Pennsylvania, USA, 1988, 189–213.

Johnson, W., Mellor, P.B., Engineering Plasticity. Ellis Horwood, Chichester, UK, 1983, 646 pp.

Jones, M.H., Scott, D., Industrial tribology. Tribology Series 8, Elsevier, Amsterdam, 1981, 516 pp.

Jones, W.R., Nagaraj, H.S., Winer, W.O., Ferrographic analysis of wear debris generated in a sliding elastohydrodynamic contact. ASLE Trans., 21 (1978) 3, 181–190.

Jönsson, B., Hogmark, S., Hardness measurements of thin films. Thin Solid Films, 114 (1984) 257–269.

Jönsson, B., Askre, L., Johansson, S., Hogmark, S., Evaluation of hard coatings on steel by particle erosion. Thin Solid Films, 137 (1986) 65–77.

Ju, F.D., Chen, T.Y., Thermomechanical cracking in layered media from moving friction load. J. Tribology, Trans. ASME, 106 (1984) 513–518.

Ju, F.D., Liu, J.C., Parameters affecting thermo-mechanical cracking in coated media due to high-speed friction load. J. Tribology, Trans. ASME, 110 (1988) 222–227.

Jungk, J.M., Michael, J.R., Prasad, S.V., The role of substrate plasticity on the tribological behavior of diamond-like nanocomposite coatings. Acta Materialia, 56 (2008) 1956–1966.

Jönsson, B., Akre, L., Johansson, S., Hogmark, S., Evaluation of hard coatings on steel by particle erosion. Thin Solid Films, 137 (1986) 65–77.

Kadlec, S., Musil, J., Valvoda, V., Munz, W-D., Petersein, H., TiN film grown by reactive magnetron sputtering with enhanced ionization at low discharge pressures. Vacuum, 41 (1990a) 7–9, 2233–2238.

Kadlec, S., Musil, J., Munz, W-D., Sputtering systems with magnetically enhanced ionization for ion plating of TiN films. J. Vac. Sci. Technol., A, 8 (1990b) 3, 1318–1324.

Kaestner, P., Olfe, J., He, J-W., Rie, K-T. Improvement in the load-bearing capacity and adhesion of TiC coatings on TiA16V4 by duplex treatment. Surf. Coat. Technol., 142/144 (2001b) 928–933.

Kaestner, P., Olfe, J., Rie, K.T. Plasma-assisted boriding of pure titanium and TiAl6V4. Surf. Coat. Technol., 142/144 (2001a) 248–252.

Kajdas, C., Tribochemistry. In: Tribology 2001 – Scientific Achievements, Industrial Applications and Future Challenges. Franek, F., Bartz, W.J., Pauschitz, A. (eds), Proc. 2nd World Tribology Congress, 3–7.9.2001, Vienna, Austria, The Austrian Tribology Society, 2001, 39–46.

Kalin, M., Vižintin, J., A comparison of the tribological behaviour of steel/steel, steel/DLC and DLC/DLC contacts when lubricated with mineral and biodegradable oils. Wear, 261 (2006a) 22–31.

Kalin, M., Vižintin, J., The tribological performance of DLC coatings under oil-lubricated fretting conditions. Tribology Int., 39 (2006b) 1060–1067.

Kalin, M., Vižintin, J., Vercammen, K., Barriga, J., Arnsek, A., The lubrication of DLC coatings with biodegradable synthetic and vegetable oils. Proc.. of 11th Nordic Symposium on Tribology – NORDTRIB2004, 1–4. 6 (2004) Tromsö, Norway, 535–547.

Kaminski, J., Rudnicki, J., Nouveau, C., Savan, A., Beer, P., Resistance to electrochemical corrosion of Cr_xN_y- and DLC-coated steel tools in the environment of wet wood. Surf. Coat. Technol., 200 (2005) 83–86.

Kamiya, S., Sato, M., Saka, M., Abe, H., Residual stress distribution in the direction of the film normal in thin diamond coatings. J. Appl. Phys., Am. Inst. of Physics, 86 (1999) 1, 224–229.

Kanakia, M.D., Peterson, M.B., Literature review of solid lubrication mechanisms. Interim Report BFLRF No. 213, US Army Belvoir Research Development and Engineering Center, Nov. 1986, AD A185010.

Kano, M., Super low friction of DLC applied to engine cam follower lubricated with ester-containing oil. Tribology Int., 39 (2006) 1682–1685.

Kano, M., Yasuda, Y., Okamoto, Y., Mabuchi, Y., Hamada, T., Ueno, T., Ye, J., Konishi, S., Takeshima, S., Martin, J.M., De Barros Bouchet, M.I., Le Mogne, T., Ultralow friction of DLC in presence of glycerol mono-oleate (GMO). Tribology Lett., 18 (2005) 2, 245–251.

Kao, W.H., Tribological properties and high speed drilling application of MoS_2 coatings. Wear, 258 (2005) 812–825.

Kapoor, A., Johnsson, K.L., Williams, J.A., A model for the mild ratchetting of metals. Wear, 200 (1996) 38–44.

Kapoor, A., Wear by plastic ratchetting. Wear, 212 (1997) 119–130.

Kapsa, P., Fouvry, S., Vincent, L., Basic principles of fretting. In: Wear – Materials, Mechanisms and Practice. Stachowiak, G.W. (ed.), John Wiley & Sons, Chichester, UK, 2005, 317–338.

Karim, M.Z., Cameron, D.C., Murphy, M.J., Hashmi, S.J., Vapour deposition of boron nitride thin films: a review. European Engineering Research and Technology Transfer Congress (EUROTECH DIRECT'91). Instn Mech. Engnrs, Conf. Paper C412/057, 1991, 9 pp.

Kato, K., Micro-mechanics of wear – wear modes. Wear, 153 (1992) 277–295.

Kato, K., Classification of wear mechanisms/models. Proc. Instn Mech. Engrs, Part J: Engineering Tribology, 216 (2002) 349–355.

Kato, K., Diao, D.F., Tsutsumi, M., The wear mechanism of ceramic coating film in repeated sliding friction. Proc. Int. Conf. on Wear of Materials, 7–11.4.1991, Orlando, USA, American Soc. Mechanical Engineers, New York, 243–248.

Kato, K., Adachi, K., Metals and ceramics. In: Modern Tribology Handbook, Vol. 2. Bhushan, B. (ed.), CRC Press, New York, 2001a, 771–785.

Kato, K., Adachi, K., Wear mechanisms. In: Modern Tribology Handbook, Vol. 1. Bhushan, B. (ed.), CRC Press, New York, 2001b, 273–300.

Kato, K., Umehara, N., Adachi, K., Friction, wear and N_2-lubrication of carbon nitride coatings: a review. Wear, 254 (2003) 1062–1069.

Kelly, P.J., Liu, Z., Arnell, R.D., Tribological studies of titanium nitride coatings deposited by pulsed magnetron sputtering. Proc. 49th Annual Technical Conf., Washington DC, USA, Society of Vacuum Coaters, Albuquerque, USA, 2006, 544–547.

Kelly, P.J., Vom Braucke, T., Liu, Z., Arnell, R.D., Doyle, E.D., Pulsed DC titanium nitride coatings for improved tribological performance and tool life. Surf. Coat. Technol., 202 (2007) 774–780.

Kennedy, F.E., Espinoza, B.M., Pepper, S.M., Thermocracking and wear of ceramic-coated face seals for salt water applications. Lubr. Engng, 46 (1990) 10, 663–671.

Kennedy, F.E., Gibson, U.J., Tribological surface treatments and coatings. In: Tribology Data Handbook. Booser, E.R. (ed.), CRC Press, New York, 1997, 581–593.

Kennedy, F.E., Hussain, S.Z., Thermo-mechanical analysis of dry sliding systems. Computers and Structures, 26 (1987) 1/2, 345–355.

Kennedy, F.E., Tang, L., Factors affecting the sliding performance of titanium nitride coatings. In: Mechanics of Coatings. Dowson, D. *et al.* (eds), Tribology Series 17, Elsevier, Amsterdam, 1990, 409–415.

Khruschov, M.M., Principles of abrasive wear. Wear, 28 (1974) 69–88.

Kikuchi, N., Eto, H., Okamura, T., Yoshimura, H., Diamond coated inserts: Characteristics and performance. In: Applications of Diamond Films and Related Materials. Tzeng, Y. *et al.* (eds), Elsevier Scientific Publishers BV, 1991, 61–67.

Kim, D.S., Fischer, T.E., Gallois, B., The effects of oxygen and humidity on friction and wear of diamond-like carbon films. Surf. Coat. Technol., 49 (1991) 537–542.

Kim, D.E., Suh, N.P., On microscopic mechanisms of friction and wear. Wear, 149 (1991) 199–208.

Kim, G.S., Lee, S.Y., Hahn, J.H., Lee, B.Y., Han, J.G., Lee, J.H., Lee, S.Y., Effects of the thickness of Ti buffer layer on the mechanical properties of TiN coatings. Surf. Coat. Technol., 171 (2003) 83–90.

Kimura, Y., Updating the fracture theory of wear. J. Phys. D: Appl. Phys., 25 (1992) A177–A181.

King, R.B., O'Sullivan, T.C., Sliding contact stresses in a two-dimensional layered elastic half-space. Int. J. Solids Structures, 23 (1987) 5, 581–597.

Kita, T., The influence of aluminium oxide particles in magnetic coating on head/disk wear. Tribology Trans., STLE, 34 (1991) 1, 107–111.

Kitagawa, H., Takahashi, S., Applicability of fracture mechanics to very small cracks or the cracks in the early stage. Proc. 2nd Int. Conf. on Mechanical Behavior of Materials, Boston, USA, 627-631.

Kitsunai, H., Kato, K., Hokkirigawa, K., Inoue, H., The transition between microscopic wear modes during repeated sliding friction observed by a scanning electron microscope tribosystem. Wear, 135 (1990) 237–249.

Kitsunai, H., Tsumaki, N., Kato, K., Transitions of microscopic wear mechanism for Cr_2O_3 ceramic coatings during repeated sliding observed in an SEM-tribosystem. Wear, 151 (1991) 279–289.

Klaffke, D., Influence of normal force and humidity on the friction and wear of unlubricated reciprocating sliding steel/steel couples. Tribotest, 10 (2004a) 4, 311–325.

Klafke, D., Towards a tribological reference test – fretting test? 14th Int. Colloquium on Tribology, Esslingen, Germany, 13–15.1.2004b, 1839–1846.

Klaffke, D., Tribological characterisation of hard coatings with and without DLC top layer in fretting tests. 14th Int. Colloquium on Tribology, 13–15.1.2004, Esslingen, Germany, 2004c, 555–563.

Klafke, D., Beck, U., Tribological and mechanical investigations of SiO2 and Si3N4 coatings on fused silica and borosilicate glass. 12th Int. Colloquium on Tribology, 12–13.1.2000, Esslingen, Germany, 2000, 1953–1959.

Klaffke, D., Brand, J., Brand, C., Wittorf, R., Hartelt, M., Tribologische characterisierung metalldotierter DLC-schichten. Tribologie und Schmierungstechnik, 51 (2004a) 4, 15–19.

Klaffke, D., Brand, J., Brand, C., Wittorf, R., Tribological characterisation of a-C:H coatings at room temperature: effect of conterbody material. Tribotest, 11 (2005a) 213–232.

Klaffke, D., Sander, H., Hartel, M., Influence of temperature and relative humidity on the friction and wear of unlubricated reciprocating sliding steel/steel couples. Tribotest, 10 (2004b) 4, 327–346.

Klaffke, D., Santner, E., Spaltmann, D., Woydt, M., Influences on the tribological behaviour of slip-rolling DLC-coatings. Wear, 259 (2005b) 752–758.

Klaffke, D., Wäsche, R., Czichos, H., Wear behaviour of i-carbon coatings. Wear, 153 (1992) 149–162.

Klages, C., Deposition and properties of diamond thin films. 2nd Int. Conf. on Plasma Surface Engineering, 10–14.9.1990, Garmisch-Partenkirchen, Germany.

Kleer, G., Doell, W., Ceramic multilayer coatings for glass moulding applications. Surf. Coat. Technol., 94–95 (1997) 647–651.

Klimek, K.S., Ahn, H., Seebach, I., Wang, M., Rie, K.T., Duplex process applied for die-casting and forging tools. Surf. Coat. Technol., 174/175 (2003) 677–680.

Klimek, K.S., Gebauer-Teichmann, A., Kaestner, P., Rie, K.T., Duplex-PACVD coating of surfaces for die casting tools. Surface and Coatings Technology, 201 (2007) 5628–5632.

Kneissl, A.C., Kutschej, K., Wu, X., Production and characterisation of Cu–Al–Ni shape memory thin films. J. Physique IV, 112 (2003) 571–574.

Knight, J.C., Page, T.F., Techniques for investigating the mechanical response of interfaces in coated systems, Newcastle University, UK, 1992, 15 pp. (In press).

Knight, D.S., White, W.B., Characterization of diamond films by Raman spectroscopy. J. Material Res., 4 (1989) 385–393.

Knight, J.C., Page, T.F., Chandler, H.W., Ramsey, P.M., Techniques for investigating the mechanical responses of interfaces in coated systems. In: Designing Ceramic Interfaces II. Pexeves, S.D. (ed.), Proc. 2nd CEC Colloquium, Petten, November 1991, Official publication of the EC, EUR 15306, Luxemburg, 1993, 453–467.

Knight, J.C., Page, T.F., Hutchings, I.M., Surface deformation behaviour of TiC and TiN coated steels: 2 scratch response and friction. Surf. Engng, 6 (1990) 1, 55–63.

Knotek, O., Atzor, M., Prengel, H-G., On the reactively sputtered Ti–Al–V carbonitrides. Surf. Coat. Technol., 36 (1988) 265–273.

Knotek, O., Barimani, A., Bosserhoff, B., Löffler, F., Structure and properties of magnetron-sputtered Ti–V–N coatings. Thin Solid Films, 193/194 (1990a) 557–564.

Knotek, O., Breidenbach, R., Jungblut, F., Löffler, F., Superhard Ti–B–C–N coatings. Surf. Coat. Technol., 43/44 (1990b) 107–115.

Knotek, O., Elsing, R., Krämer, G., Jungblut, F., On the origin of compressive stress in PVD coatings – an explicative model. Surf. Coat. Technol., 46 (1991) 265–274.

Knotek, O., Bosserhoff, B., Schrey, A., Leyendecker, T., Lemmer, O., Esser, S., Thin film impact test. Surf. Coat. Technol., 54/55 (1992a) 102–107.

Knotek, O., Burgmer, W., Stössel, C., Arc-evaporated Ti–V–N thin films. 19th Int. Conf. Metallurgical Coatings and Thin Films (ICMCTF'92), San Diego, USA, 6–10.4.1992b, 18 pp.

Knotek, O., Löffler, F., Krämer, G., Multicomponent and multilayer PVD coatings for cutting tools. Surf. Coat. Technol., 54/55 (1992c) 241–248.

Knotek, O., Löffler, F., Krämer, G., Substrate and interface related influences on the performance of arc-PVD-coated cemented carbides in interrupted cut machining. Surf. Coat. Technol., 54/55 (1992d) 476–481.

Knotek, O., Löffner, F., Weitkamp, K., Physical vapour deposition coatings for dental prostheses. Surf. Coat. Technol., 54/55 (1992e) 536–540.

Knotek, O., Stössel, C., Burgmer, W., Arc-evaporated Ti–V–N thin films. Int. Conf. on Metallurgical Coatings and Thin Films (ICMCTF-1992), San Diego, USA, 6–10.4.1992f, 20 pp.

Knotek, O., Löffner, F., Beele, W., Krämer, G., Comparison of abrasion model test results and machining tests with PVD-coated indexable inserts. Wear. 162–164 (1993) 1033–1039.

Ko, P.L., Metallic wear – a review with special references to vibration induced wear in power plants components. Tribology Int., 20 (1987) 2, 66–78.

Ko, P.L., Robertson, M., Magel, E.E., Wear particle formation in lubricated sliding between a hardened sphere and a flat surface. In: Wear Particles: From the Cradle to the Grave. Dowson, D. *et al.* (eds), Elsevier Tribology Series 21, Amsterdam, 1992, 81–90.

Koboyashi, M., Doi, Y., TiN and TiC coating on cemented carbide by ion plating. Thin Solid Films, 54 (1978) 67–74.

Kodali, P., Walter, K.C., Nastasi, M., Investigation of mechanical and tribological properties of amorphous diamond-like carbon coatings. Tribology Int., 30 (1997) 8, 591–598.

Koehler, J.S., Attempt to design a strong solid. Phys. Rev. B, 2 (1970) 2, 547–551.

Kohira, H., Prabhakaran, V., Talke, F.E., Effect of air bearing design on wear of diamond-like carbon coated proximity recording sliders. Tribology Int., 33 (2000) 315–321.

Kohzaki, M., Higuchi, K., Noda, S., Frictional properties of chemical vapor deposited diamond thin films. Mat. Lett., 9 (1990) 2&3, 80–82.

Komvopoulos, K., Finite element analysis of a layered elastic solid in normal contact with a rigid surface. Trans. ASME, J. Tribology, 110 (1988) 477–485.

Komvopoulos, K., Elastic–plastic finite element analysis of indented layered media. Trans. ASME, J. Tribology, 111 (1989) 430–439.

Komvopoulos, K., Sliding friction mechanisms of boundary-lubricated layered surfaces: Part I – Basic mechanical aspects and experimental results. Tribology Trans., J. STLE, 34 (1991a) 2, 266–280.

Komvopoulos, K., Sliding friction mechanisms of boundary-lubricated layered surfaces: Part II – Theoretical analysis. Tribology Trans., STLE, 34 (1991b) 2, 281–291.

Komvopoulos, K., Surface engineering and microtribology for microelectromechanical systems. Wear, 200 (1996) 305–327.

Komvopoulos, K., Do, V., Yamaguchi, E.S., Ryason, P.R., Nanomechanical and nanotribological properties of an antiwear tribofilm produced from phosphorus-containing additives on boundary-lubricated steel surfaces. Trans. ASME. J. Tribology, 126 (2004) 775–779.

Komvopoulos, K., Saka, N., Suh, N.P., Plowing friction in dry and lubricated metal sliding. J. Tribology, Trans. ASME, 108 (1986a) 301–313.

Komvopoulos, K., Saka, N., Suh, N.P., The role of hard layers in lubricated and dry sliding. Trans. ASME, J. Tribology, 109 (1987) 223–231.

Komvopoulos, K., Suh, N.P., Saka, N., Wear of boundary-lubricated metal surfaces. Wear, 107 (1986b) 107–132.

Konca, E., Cheng, Y.T., Weiner, A.M., Dasch, J.M., Alpas, A.T., Vacuum tribological behavior of the non-hydrogenated diamond-like carbon coatings against aluminum: effect of running-in in ambient air. Wear, 259 (2005) 795–799.

König, W., Fertigungsverfaren/Manufacturing methods. Vol. 1, VDI-Verlag, Düsseldorf, Germany, 1984, 372 pp.

König, W., Fritsch, R., Performance and wear of carbide coated by physical vapour deposition in interrupted cutting. Surf. Coat. Technol., 54/55 (1992) 453–458.

König, W., Fritsch, R., Kammermeier, D., New approaches to characterizing the performance of coated cutting tools. Annals of the CIRP, 41 (1992) 1, 49–54.

König, W., Kammermeier, D., New ways towards better exploitation of physical vapour deposition coatings. Surf. Coat. Technol., 54/55 (1992) 470–475.

Koroschetz, F., Gärtner, W., New materials and methods for the manufacture of bearings, Miba Technical Information, Miba Gleitlager AG, Laakirchen, Austria, 1990, 15 pp.

Korunsky, A.M., McGurk, M.R., bull, S.J., page, T.F., On the hardness of coated systems. Surf. Coat. Technol., 99 (1998) 171–183.

Koskilinna, J., Quantum chemical studies on atomic-scale tribology of diamond and boron nitride, Doctoral thesis at University of Joensuu, Finland, 2007, 32 pp.

Koskinen, J., Abrasive wear resistance of ion-deposited hard-carbon films as a function of deposition energy. J. Appl. Phys., 63 (1988) 6, 2094–2097.

Koskinen, J., Hirvonen, J.P., Lappalainen, R., Anttila, A., Interfacial characteristics of arc discharge deposition diamond-like films on 19 different substrate materials. Diamond 1992 Conf. Heidelberg, Germany, 31.8.4.9.1992, 10 pp.

Kowstubhan, M.V., Philip, P.K., On the tool-life equation of TiN-coated high speed steel tools, Wear, 143 (1991) 267–275.

Kraghelsky, I., Calculation of dry friction forces. Conf. on Lubrication and Wear, Inst. Mech. Eng., 1–3. 10.1957, London, 1957, pp.

Kraghelsky, I.V., Calculation of wear rate. J. Basic Engng, Trans. ASME, Sept. (1965) 785–790.

Kragelsky, I.V., Dobychin, M.N., Kombalov, V.S., Friction and Wear Calculation Methods., Pergamon Press, Oxford, 1982, 464 pp.

Kragelskii, I.V., Marchenko, E.A., Wear of machine components. J. Lubr. Tech., Trans. ASME, 104 (1982) 1–8.

Kral, M.V., Davidson, J.L., Wert, J.J., Erosion resistance of diamond coatings. Wear, 166 (1993) 7–16.

Kral, E.R., Komvopoulos, K., Three-dimensional finite element analysis of subsurface stresses and shakedown due to repeated sliding on a layered medium. Trans. ASME, J. Appl. Mech., 63 (1996a) 967–973.

Kral, E.R., Komvopoulos, K., Three-dimensional finite element analysis of surface deformation and stresses in an elastic–plastic layered medium subjected to indentation and sliding contact loading. Trans. ASME, J. Appl. Mech., 63 (1996b) 365–375.

Kral, E.R., Komvopoulos, K., Three-dimensional finite element analysis of subsurface stresses and strain fields due to sliding contact on an elastic-plastic layered medium. Trans. ASME, J. Tribology, 119 (1997) 332–341.

Kral, E.R., Komvopoulos, K., Bogy, D.B., Finite element analysis of repeated indentation of an elastic–plastic layered medium by a rigid sphere, Part I: Surface results. Trans. ASME, J. Tribology, 62 (1995a) 20–28.

Kral, E.R., Komvopoulos, K., Bogy, D.B., Finite element analysis of repeated indentation of an elastic–plastic layered medium by a rigid sphere, Part II: Subsurface results. Trans. ASME, J. Tribology, 62 (1995b) 29–42.

Kramer, B.M., An analytical approach to tool wear prediction, PhD thesis, Department of Mechanical Engineering, MIT, Massachusetts, USA, June, 1979.

Kramer, B.M., Requirements for wear-resistant coatings. Thin Solid Films, 108 (1983) 117–125.

Kramer, B.M., Predicted wear resistances of binary carbide coatings. J. Vac. Sci. Technol., A4 (1986) 6, 2870–2873.

Kramer, B.M., Judd, P.K., Computational design of wear coatings. J. Vac. Sci. Technol., A3 (1985) 2439–2444.

Kramer, B.M., Suh, N.P., Tool wear by solution: a quantitative understanding. J. Engng Indust., Trans. ASME, 102 (1980) 303–309.

Krantz, T.L., Cooper, C.V., Townsend, D.P., Hansen, B.D., Increased surface fatigue lives of spur gears by application of a coating. NASA/TM-2003-212463 (ARL-TR-2971, DETC2003-48114) (2003) 9 pp.

Krauss, A.R., Auciello, O., Gruen, D.M., Jayatissa, A., Sumant, A., Tucek, J., Mancini, D.C., Moldovan, N., Erdemir, A., Ersoy, D., Gardos, M.N., Busmann, H.G., Meyer, E.M., Ding, M.Q., Ultrananocrystalline diamond thin films for MEMS and moving mechanical assembly devices. Diamond and Related Materials, 10 (2001) 1952–1961.

Krim, J., Friction at the atomic scale. Sci. Am., October (1996) 48–56.

Krim, J., Friction at macroscopic and microscopic length scales. Am. J. Phys. 70 (2002a) 9, 890–897.

Krim, J., Surface science and the atomistic-scale origins of friction: what once was old is new again. Surf. Sci., 500 (2002b) 741–758.

Krokoszinski, H.J., Abrasion resistance of diamond-like carbon coatings on thick film resistor materials. Surf. Coat. Technol., 47 (1991) 761–769.

Kullmer, R., Lugmair, C., Figueras, A., Bassas, J., Stoiber, M., Mitterer, C., Microstructure, mechanical and tribological properties of PACVD Ti(B,N) and TiB_2 coatings. Surf. Coat. Technol., 174–175 (2003) 1229–1233.

Kumar Wickramasinghe, H., Scanned-probe microscopes. Sci. Am., 261 (1989) 98–105.

Kuoppala, R., Scanning technique for wear particle analysis. 9th European Maintenance Congress. Espoo, Finland, 24–27.5 (1988) 10 pp.

Kurokawa, H., Mitani, J., Yonezawa, T., Applications of diamond-like carbon films to metallic thin film recording tape. New Diamond, 1 (1990) 107–109.

Kurokawa, H., Nakaue, H., Mitani, T., Yonezawa, T., Applications of diamond-like carbon films to electric components. In: Applications of Diamond Films and Related Materials. Tzeng, Y. *et al.* (eds), Elsevier Science Publishers, Amsterdam, 1991, 319–326.

Kuruppu, M.L., Negrea, G., Inanov, I.P., Rohde, S.L., Monolithic and multilayer Cr/CrN, Cr/Cr_2N, and Cr_2N/CrN coatings on hard and soft substrates. J. Vac. Sci. Technol., A 16 (1998) 3, 1949–1955.

Kustas, F.M., Misra, M.S., Wei, R., Wilbur, P.J., Diamondlike carbon coatings on Ti–6Al–4V. Tribology Trans., STLE, 36 (1993) 1, 113–119.

Kuwahara, K., Hirota, H., Umemoto, N., Adhesion measurement on thin evaporated films. In: Adhesion Measurements of Thin Films, Thick Films and Bulk Coatings. Mittal, K.L. (ed.), ASTM STP 640, American Society for Testing Materials, Philadelphia, 1978, 198–207.

Kuwano, H., Solid lubricant film formation using fast atom beam sputtering. Japanese J. Tribology, 35 (1990) 3, 291–300.

Kwietniewski, C., Dong, H., Strohaecker, T., Li, X.Y., Bell, T., Duplex surface treatment of high strength Timetal 550 alloy towards high load-bearing capacity. Surf. Coat. Technol., 139 (2001) 284–292.

Kwok, S.C.H, Wang, J., Chu, P.K., Surface energy, wettability, and blood compatibility phosphorous doped diamond-like carbon films. Diamond and Related Materials, 14 (2005) 78–85.

Labidi, C., Collet, R., Nouveau, C., Beer, P., Nicosia, S., Djouadi, M.A., Surface treatments of tools used in industrial wood machining. Surf. Coat. Technol., 200 (2005) 118–122.

Lackner, J.M., Stotter, C., Waldhauser, W., Ebner, R., Lenz, W., Beutl, M., Pulsed laser deposition of diamond-like carbon coatings for industrial tribological applications. Surf. Coat. Technol., 174–175 (2003) 402–407.

Lafaye, S., Gauthier, C., Schirrer, R., A surface flow line model of a scratching tip: apparent and true local friction coefficients. Tribology Int., 38 (2005) 113–127.

Lai, S.Q., Li, T.S., Wang, F.D., Li, X.J., Yue, L.I., The effect of silica size on the friction wear behaviors of polyimide/silica hybrids by sol-gel processing. Wear, 262 (2007) 1048–1055.

Lampe, T., Eisenberg, S., Rodriguez Cabeo, E., Plasma surface engineering in the automotive industry – trends and future prospectives. Surf. Coat. Technol., 174–175 (2003) 1–7.

Landheer, D., De Gee, A.W.J., Adhesion, friction and wear. MRS Bulletin, Oct (1991) 36–40.

Landheer, D., Zaat, J.H., The mechanism of metal transfer in sliding friction. Wear, 27 (1974) 129–145.

Landman, U., Luedtke, W.D., Ringer, E.M., Atomistic mechanisms of adhesive contact formation and interfacial processes. Wear, 153 (1992) 3–30.

Lang, O.R., Steinhilpher, W., Gleitlagern, Springer-Verlag, Berlin, 1978, 414 pp.

Langlade, C., Vannes, B., Taillandier, M., Pierantoni, M., Fretting behaviour of low-friction coatings. Tribotest J., 11 (2005) 193–206.

Lansdown, A.R., Use of the rotary particle depositor in wear debris analysis. Int. Conf. Friction, Wear and Lubricants, Tashkent, USSR, 22–26.5 (1985).

Lansdown, A.R., Molybdenum disulphide lubrication. Tribology Series No. 35, Elsevier, Amsterdam, The Netherlands, 1999, 380 pp.

Lappalainen, R., Selenius, M., Anttila, A., Konttinen, Y.T., Santavirta, S., Reduction of wear in total hip replacement prostheses by amorphous diamond coatings. J. Biomed. Mater. Res., 66B (2003) 410–413.

Larbi Ben Cheikh, A., Tlili, B., Fretting wear of multilayered PVD TiAlCN/TiAlN/TiAl on AISI 4140 steel. Surf. Coat. Technol., 201 (2006) 1511–1518.

Larsson, A., Microstructure and wear of CVD coatings used in metal cutting applications. PhD thesis No. 1588 of Chalmers University of Technology and Gothenburg University, Gothenburg, Sweden, 2000, 51 pp.

Larsson, M., Deposition and evaluation of thin hard PVD coatings. Acta Universitatis Upsaliensis, Comprehensive Summaries of Uppsala Dissertations from Faculty of Science and Technology, No. 191, Uppsala University, Sweden, 1989, 36 pp.

Larsson, M., Hedenqvist, P., Hogmark, S., Deflection measurements as method to determine residual stress in thin hard coatings on tool materials. Surf. Engng, 12 (1996) 1, 43–48.

Larsson, R., Höglund, E., Elastohydrodynamic lubrication, In: Tribology of Mechanical Systems: A Guide to Present and Future Technologies. Vizintin, J., Kalin, M., Dohda, K., Jahamir, S. (eds), ASME Press, New York, 2004, 25–40.

Lau, K.H., Li, K.Y., Mai, Y.W., Influence of hardness ratio on scratch failure of coatings. Int. J. Surf. Sci. Engng, 1 (2007) 1, 3–21.

Laugier, M.J., An energy approach to the adhesion of coatings using the scratch test, Thin Solid Films, 117 (1984) 234–249.

Laukkanen, A., Holmberg, K., Koskinen, J., Ronkainen, H., Wallin, K., Varjus, S., Tribological contact analysis of a rigid ball sliding on a hard coated surface – Part III: Fracture toughness calculation and influence of residual stresses. Surf. Coat. Technol., 200 (2006) 3824–3844.

Lawn, B.R., Evans, A.G., A model for crack initiation in elastic–plastic indentation field. J. Mater. Sci., 12 (1977) 2195–2199.

Lawn, B.R., Evans, A.G., Marshall, D.B., Elastic/plastic indentation damage in ceramics: the median/radial crack system. J. Am. Ceram. Soc., 63 (1980) 574.

Le Mogne, T., Donnet, C., Martin, J.M., Tonck, A., Millard-Pinard, N., Nature of super-libricating MoS_2 physical vapor deposition. J. Vac. Sci. Technol., A12 Jul./Aug. (1994) 4, 1998–2004.

Lee, E.J., Bayer, R.G., Tribological characteristics of titanium nitride thin coatings. Metal Finishing, July (1985) 39–42.

Lee, J., Lim, D., Tribological behavior of PTFE film with nanodiamond. Surface and Coatings Technology, 188–189 (2004) 534–538.

Lee, J.H., Xu, G.H., Liang, H., Experimental and numerical analysis of friction and wear behavior of polycarbonate. Wear, 251 (2001) 1541–1556.

Lee, S.C., Ren, N., The subsurface stress field created by three-dimensionally rough bodies in contact with traction. Tribology Trans., 37 (1994) 3, 615–621.

Lee, S.C., Ren, N., Behavior of elastic–plastic rough surface contacts as affected by surface topography, load, and material hardness. Tribology Trans., 39 (1996) 1, 67–74.

Legg, K.O., Cochran, J.K., Solnik-Legg, H., Mann, X.L., Modification of surface properties of yttria stabilized zirconia by ion implantation. Nucl. Instr. Meth. Res., 137/8 (1985) 535.

Lemoine, P., Quinn, J.P., Maguire, P.D., Laughlin, J.A., Mechanical characterisation and properties of DLC films. In: Tribology of Diamond-like Carbon Films – Fundamentals and Applications. Donnet, C., Erdemir, A. (eds), Springer, New York, USA, 2008, 83–101.

Leopold, J., Meisel, M., Wohlgemuth, R., Liebich, J., High performance computing of coating–substrate systems. Surf. Coat. Technol., 142–144 (2001) 916–922.

Leroy, J.M., Floquet, A., Villechaise, B., Thermomechanical behavior of multilayered media: theory. J. Tribology, Trans. ASME, 111 (1989) 538–544.

Leroy, J.M., Villechaise, B., Stress determination in elastic coatings and substrate under both normal and tangential loads. In: Mechanics of Coatings. Dowson, D. *et al.* (eds), Tribology Series 17, Elsevier, Amsterdam, 1990, 195–201.

Lesnjak, A., Tusek, J., Processes and properties of deposits in electrospark deposition. Science and Technology of Welding, Joining, 7 (2002) 391–396.

Levashov, E.A., Kudryashov, A.E., Vakaev, P.V., Shtansky, D.V., Malochkin, O.V., Gammet, F., Suchesstrunk, R., Moore, J.J., The prospects of nanodispersive powders application in surface engineering technologies. Surf. Coat. Technol., 180–181 (2004) 347–351.

Levashov, E.A., Vakaev, P.V., Zamulaeva, E.I., Kudryashov, A.E., Kurbatkina, V.V., Shtansky, D.V., Voevodin, A.A., Sanz, A., Dispersion-strengthening by nanoparticles advanced tribological coatings and electrode materials for their deposition. Surf. Coat. Technol., 201 (2007) 6176–6181.

Levy, A.V., Jee, N., Unlubricated sliding wear of ceramic materials. Wear, 121 (1988) 363–380.

Leyendecker, T., Lemmer, O., Esser, S., Ebberink, J., The development of the PVD coating TiAlN as a commercial coating for cutting tools. 16th ICMC, 7th IFTC Conferences, San Diego, USA, 2–6.4. 1989, 5 pp.

Leyland, A., Developments in plasma surface engineering, PhD thesis, Hull University, 1991, 121 pp.

Leyland, A., Bin-Sudin, M., Kalantary, M.R., Wells, P.B., Matthews, A., Housden, J., Garside, B., TiN and CrN PVD coatings on electroless nickel coated steel substrates. Surf. Coat. Technol., 60 (1993) 474–479.

Leyland, A., Fancey, K.S., James, A.S., Matthews, A., Enhanced plasma nitriding at low pressures: a comparative study of d.c. and r.f. techniques. Surf. Coat. Technol., 41 (1990) 295–304.

Leyland, A., Fancey, K.S., Matthews, A., Plasma nitriding in a low pressure triode discharge to provide improvements in adhesion and load support for wear resistant coatings. Surf. Engng, 7 (1991) 3, 207–215.

Leyland, A., Lewis, D.B., Stevenson, F.R., Matthews, A., Low temperature plasma diffusion treatment of stainless steels for improved wear resistance. Surf. Coat. Technol., 62 (1993a) 608–617.

Leyland, A., Matthews, A., Thick Ti/TiN multilayer coatings for abrasive and erosive wear resistance. Submitted for publication to Surf. Coat. Technol., (1993) 6.

Leyland, A., Matthews, A., Thick Ti/TiN multilayered coatings for abrasive and erosive wear resistance. Surf. Coat. Technol., 70 (1994) 19–25.

Leyland, A., Matthews, A., On the significance of the H/E ratio in wear control: a nanocomposite coating approach to optimized tribological behaviour. Wear, 246 (2000) 1–11.

Leyland, A., Matthews, A., Design criteria for wear-resistant nanostructured and glassy-metal coatings. Surf. Coat. Technol., 177/178 (2004) 317–324.

Leyland, A., Matthews, A., Optimization of nanostructured tribological coatings. In: Nanostructured Coatings. Cavaleiro, A., De Hosson, J.T.M. (eds), Springer, New York, 2006, 511–538.

Lhermitte-Sebire, I., Clomet, R., Naslain, R., The adhesion between physically vapour-deposited or chemically vapour-deposited alumina and TiC-coated cemented carbides as characterized by Auger electron spectroscopy and scratch testing. Thin Solid Films, 138 (1986) 221–233.

Li, C., Jordens, K., Wilkes, G.L., Abrasion-resistant coatings for plastic and soft metallic substrates by sol-gel reactions of a triethoxysilylated diethylenetriamine and tetramethoxysilane. Wear, 242 (2000) 152–159.

Li, H., Xu, T., Wang, C., Chen, J., Zhou, H., Liu, H., Tribochemical effects on the friction and wear behaviors of a-C:H and a-C films in different environment. Tribology Int., 40 (2007) 132–138.

Li, J., Beres, W., Three-dimensional finite element modelling of the scratch test for a TiN coated titanium alloy substrate. Wear, 260 (2006) 1232–1242.

Li, J.F., Huang, J.Q., Zhang, Y.F., Ding, C.X., Zhang, P.Y., Friction and wear behaviour of plasma-sprayed Cr_3C_2-NiCr against TiO_2 coating under water- and ethanol-lubricated sliding. Wear, 214 (1998) 202–206.

Li, X., Bhushan, B., Measurement of fracture toughness of ultra-thin amorphous carbon films. Thin Solid Films, 315 (1998) 214–221.

Li, X., Bhushan, B., Evaluation of fracture toughness of ultra-thin amorphous carbon coatings deposited by different deposition techniques. Thins Solid Films, 356–336 (1999) 330–355.

Li, X., Diao, D., Bhushan, B., Fracture mechanisms of thin amorphous carbon films in a nanoindentation. Acta Mater., 45 (1997) 11, 4453–4461.

Li, Z., Rabinowicz, E., Saka, N., Friction characteristics and wear resistance of carbon-coated glass-based rigid magnetic disks. Wear, 143 (1991) 241–253.

Ligot, J., Benayoun, S., Hantzpergue, J.J., Analysis of cracking induced by scratching of tungsten coatings on polyamide substrate. Wear, 243 (2000) 85–91.

Lim, S.C., Recent developments in wear-mechanisms maps. Tribology Int., 31 (1998) 1–3, 87–97.

Lim, S.C., Ashby, M.F., Wear-mechanism maps. Acta Metall., 35 (1987) 1, 1–24.

Lim, S.C., Ashby, M.F., Burton, J.H., The effects of sliding condition on the dry friction of metals. Acta Metall, 37 (1989) 3, 767–772.

Lim, S.C., Lee, S.H., Liu, Y.B., Seah, K.H.W., Wear maps for uncoated high-speed steel cutting tools. Wear, 170 (1993a) 137–144.

Lim, S.C., Liu, Y.B., Lee, S.H., Seah, K.H.W., Mapping the wear of some cutting-tool materials. Wear, 162–164 (1993b) 971–974.

Lin, H.C., Liao, H.M., He, J.L., Lin, K.M., Chen, K.C., Wear characteristics of ion-nitrided Ti50Ni50 shape memory alloys. Surf. Coat. Technol., 92 (1997) 178–189.

Lin, J.F., Liu, M.H., Wu, J.D., Analysis of the friction and wear mechanisms of structural ceramic coatings, Part 1: The effect of the processing conditions. Wear, 194 (1996a) 1–11.

Lin, J.F., Liu, M.H., Wu, J.D., Analysis of the friction and wear mechanisms of structural ceramic coatings, Part 2: The effect of operating conditions and substrate material. Wear, 198 (1996b) 7–14.

Lin, Y., Ovaert, T.C., A rough surface contact model for general anisotropic materials. Trans. ASME, J. Tribology, 126 (2004) 41–49.

Lince, J.R., Tribology of co-sputtered nanocomposite Au/MoS_2 solid lubricant films over a wide contact stress range. Tribology Lett., 17 (2005) 419–428.

Lindbauer, A., Haubner, R., Lux, B., Diamantbescheidung auf keramischen Materialen mit der Heissdrahtmetode. Wear, 159 (1992) 67–77.

Lindgren, J.R., Johnson, W.R., Friction and wear behavior of chromium carbide coatings. Surf. Coat. Technol., 32 (1987) 249–260.

Lindholm, P., Björklund, S., Svahn, F., Method and surface roughness aspects for the design of DLC coatings. Wear, 261 (2006) 107–111.

Lindholm, P., Svahn, F., Study of thickness dependence of sputtered-carbon coating for low friction valve lifters. Wear, 261 (2006) 241–250.

Lindquist, M., Self lubrication on the atomic scale. Doctoral thesis, Acta Universitatis Upsaliensis. Faculty of Science and Technology, Uppsala University, No. 391, 2008, 59 pp.

Liskiewicz, T., Fouvry, S., Development of a friction energy capacity approach to predict the surface coating endurance under complex oscillating sliding conditions. Tribology Int., 38 (2005) 69–79.

Liskiewicz, T., Fouvry, S., Wendler, B., Hard coatings durability under fretting wear. Leeds–Lyon Symposium on Tribology, 7–10.9.2004, Leeds, UK. In: Life Cycle Tribology. Dowson, D. (ed.), Tribology and Interface Series No. 48. 2004, 657–665.

Liskiewicz, T., Fouvry, S., Wendler, B., Development of a Wöhler-like approach to quantifying the $Ti(C_xN_y)$ coatings durability under oscillating sliding conditions. Wear, 259 (2005) 835–841.

Liu, C., Leyland, A., Bi, Q., Matthews, A., Corrosion resistance of multi-layered PAPVD TiN and CrN coatings. Surf. Coat. Technol., 141 (2001) 164–174.

Liu, D., Gao, W., Li, Z., Zhang, H., Hu, Z., Electro-spark deposition of Fe-based amorphous alloy coatings. Materials Lett., 61 (2007) 165–167.

Liu, E., Ding, Y.F., Li, L., Blanpain, B., Celis, J.P., Influence of humidity on the friction of diamond and diamond-like carbon materials. Tribology Int., 40 (2007) 216–219.

Liu, J., Zhang, X., Zhu, B., The effect of a soft metallic plated layer on the tribological behavior of steels under boundary lubrication. Tribology Trans., 34 (1991) 1, 17–22.

Liu, W., Chen, Y., Kou, G., Xu, T., Sun, D.C., Characterization and mechanical/tribological properties of nano Au-TiO_2 composite thin films prepared by a sol-gel process. Wear, 254 (2003) 994–1000.

Liu, Y., Singh, R., Poole, K., Diefendorf, R.J., Harriss, J., Cannon, K., Characterization of Al–Y alloy thin films deposited by direct current magnetron sputtering. J. Vac. Sci. Technol., B15/6 (1997) 1990–1994.

Löhr, M., Spaltmann, D., Binkowski, S., Santner, E., Woydt, M., In situ acoustic emission for wear life detection of DLC coatings during slip-rolling friction. Wear, 260 (2006) 469–478.

Longching, C., Qing, C., Eryu, S., Study on initiation and propagation angles of subsurface cracks in GCr15 bearing steel under rolling contact. Wear, 133 (1989) 205–218.

Louzguine, D.V., Inoue, A., Comparisons of the long-term thermal stability of various metallic glasses under continuous heating. Scripta Materialia, 47 (2002) 887–891.

Lowenheim, F.R., Deposition of inorganic films from solution. In: Thin Film Processes. Vossen, J.L., Kern, W. (eds), Academic Press, New York, USA, 1978, 209–256.

Lu, W., Komvopoulos, K., Nanomechanical and nanotribological properties of carbon, chromium and titanium carbide ultrathin films. J. Tribology, Trans. ASME, 123 (2001) 717–724.

Ludema, K., History of tribology and its industrial significance. In: Fundamentals of Tribology and Bridging the Gap Between the Macro- and Micro/Nanoscales. Bhushan, B. (ed.), NATO Science Series; II. Mathematics, Physics and Chemistry – Vol. 10, Kluwer Academic Publishers, London, UK, 2001, 1–11.

Ludema, K., Guest editorial. Tribology Int., 38 (2005) 207–210.

Lundberg, G., Palmgren, A., Dynamic capacity of rolling bearings. Acta Polytechnica Mechanical Engineering Series, Royal Swedish Academy of Engineering Sciences, 1 (1947) 3, 7.

Lundberg, G., Palmgren, A., Dynamic capacity of roller bearings. Acta Polytechnica Mechanical Engineering Series, Royal Swedish Academy of Engineering Sciences, 2 (1952) 4, 96.

Lux, B., Haubner, R., Low pressure synthesis of superhard coatings. Proc. Plansee-Seminar, Paper C2 (1989) 615–660.

Lux, B., Haubner, R., Diamond as a wear-resistant coating. To be published in Diamond Films and Coatings. In: Davis, R., Glass, J. (eds), Noyes Publications, New York, 1992. 46 pp.

Lux, B., Colombier, C., Altena, H., Stjernberg, K., Preparation of alumina coatings by chemical vapour deposition. Thin Solid Films, 138 (1986) 49–64.

Lyubimov, V.V., Voevodin, A.A., Yerokhin, A.L., Timofeev, Y.S., Arkhipov, I.K., Development and testing of multilayer physically vapour deposited coatings for piston rings. Surf. Coat. Technol., 52 (1992a) 145–151.

Lyubimov, V.V., Voevodin, A.A., Spassky, S.E., Yerokin, A.L., Stress analysis and failure possibility assessment of multilayer physically vapour deposited coatings. Thin Solid Films, 207 (1992b) 117–125.

Ma, X.G., Komvopoulos, K., Wan, D., Bogy, D.B., Kim, Y.S., Effects of film thickness and contact load on nanotribological properties of sputtered amorphous carbon thin films. Wear, 254 (2003) 1010–1018.

Ma, S., Procházka, J., Karvánkova, P., Ma, Q., Niu, X., Wang, X., Ma, D., Xu, K., Vepřek, S., Comparative study of the tribological behaviour of superhard nanocomposite coatings nc-TiN/a-Si_3N_4 with TiN. Surf. Coat. Technol., 194 (2005) 143–148.

Mabuchi, Y., Hamada, T., Kano, M., Yasuda, Y., Ariga, K., Diamond-like carbon coating for reducing valvetrain friction. In: Proc. FISITA 2004 World Automotive Congress, 23–27.5.2004, F2004F388, STA, Spanish Soc. for Automotive Engineers, 15 pp.

Maillat, M., Boving, H., Hintermann, H.E., Load-bearing capacities of TiC layers on steel and cemented carbides. Thin Solid Films, 64 (1979) 243–248.

Maillat, M., Hintermann, H.E., Coatings against fretting. Proc. of 2nd European Space Mechanisms, Tribology Symposium, 9–11.11.1985, Meersburg, FR, Germany, ESA SP-231, 161–166.

Maillet, B., Celis, J.P., Roos, J.R., Stals, L.M., Van Stappen, M., Wear phenomena in the system TiN-coated high speed steel disk against a chromium steel pin. Wear, 142 (1991) 151–170.

Mäkelä, U., Valli, J., Tribological properties of PVD TiN and TiC coatings. Finnish J. Tribology, 4 (1985) 2, 74–85.

Malzbender, J., The use of theories to determine mechanical and thermal stresses in monolithic, coated and multilayered materials with stress-dependent elastic modulus or gradient in elastic modulus exemplified for thermal barrier coatings. Surf. Coat. Technol., 186 (2004) 416–422.

Malzbender, J., De With, G., Sliding indentation, friction and fracture of a hybrid coating on glass. Wear, 236 (1999) 355–359.

Malzbender, J., De With, G., Cracking and residual stress in hybrid coatings on float glass. Thin Solid Films, 359 (2000a) 210–214.

Malzbender, J., De With, G., Modeling of the fracture of a coating under sliding indentation. Wear, 239 (2000c) 21–26.

Malzbender, J., De With, G., A model to determine the interfacial fracture toughness for chipped coatings. Surf. Coat. Technol., 154 (2002) 21–26.

Malzbender, J., Den Toonder, J.M.J., Balkenende, A.R., De With, G., Measuring mechanical properties of coatings: a methodology applied to nano-particle-filled sol-gel coatings on glass. Materials Science and Engineering, R36 (2002) 47–103.

Man, H.C., Laser surface treatment methods for protection against wear. In: Surface Coatings for Protection against Wear. Mellor, B.G. (ed.), CRC Press, Woodhead Publishing Ltd, Cambridge, UK, 2006, 377–391.

Manaila, R., Devenyi, A., Biro, D., David, L., Barna, P.B., Kovacs, A., Multilayer TiA1N coatings with composition gradient. Surf. Coat. Technol., 151–152 (2002) 21–25.

Manier, C.A., Spaltmann, D., Theiler, G., Woydt, M., Carboneous coatings by rolling with 10 % slip under mixed/boundary lubrication and higher initial Hertzian contact pressures. Diamond and Related Materials, in press, 2008a, 7 p.

Manier, C., Spaltmann, D., Woydt, M., Tribology of DLC films under slip-rolling conditions. In: Tribology of diamond-like carbon films, Donnet, C., Erdemir, A. (eds), Springer, New York, USA, 2008b, 383-409.

Mansour, H.A., Arnell, R.D., Salama, M.A., Mustfa, A.A.F., Finite elements studies of coated surfaces under normal and tangential loads. 6th Int. Conf. on Ion, Plasma Assisted Techniques, Brighton, UK, May 1987, 179–183.

Mao, K., Sun, Y., Bell, T., A numerical model for the dry sliding contact of layered elastic bodies with rough surfaces. Tribology Trans., 39 (1996) 2, 416–424.

Mao, K., Sun, Y., Bell, T., An initial approach to the link of multi-layer coatings contact stresses and the surface engineered gears. Surf. Coat. Technol., 201 (2007) 12, 5796–5803.

Martin, J.M., Superlubricity of molybdenum disulphide. In: Superlubricity. Erdemir, A., Martin, J.M. (eds), Elsevier, Amsterdam, Netherlands, 2007, 207–225.

Martin, J.M., Donnet, C., Le Mogne, T., Epicier, T., Superlubricity of molybdenum disulphide. Phys. Rev. B, 14 (1993) 48, 10583–10586.

Martin, J.M., Pascal, H., Donnet, C., Le Mogne, T., Loubet, J.L., Epicier, T., Superlubricity of MoS_2: crystal orientation mechanisms. Surf. Coat. Technol., 68/69 (1994) 427–432.

Martin, P.J., McKenzie, D.R., Nettesfield, R.P., Swift, P., Filipczuk, S.W., Müller, K.H., Pacey, C.G., James, B., Characteristics of titanium arc evaporation processes. Thin Solid Films, 153 (1987) 91–102.

Martin, P.J., Netterfield, R.P., Bendavid, A., Kinder, T.J., The deposition of thin films by filtered arc evaporation. Surf. Coat. Technol., 54/55 (1992) 136–142.

Martinez, E., Romero, J., Lousa, A., Esteve, J., Wear behavior of nanometric CrN/Cr multilayers. Surf. Coat. Technol., 163–164 (2003) 571–577.

Martins, R., Amaro, R., Seabra, J., Influence of low friction coatings on the scuffing load capacity and efficiency of gears. Tribology Int., 41 (2008) 234–243.

Matthes, B., Broszeit, E., Kloos, K.H., Fundamental and tribological properties of r.f. sputtered TiN coatings in plastic manufacturing model wear tests. Surf. Coat. Technol., 43/44 (1990a) 688–698.

Matthes, B., Broszeit, E., Kloos, K.H., Wear behaviour of r.f. sputtered TiN, TiB_2 and Ti(B,N) coatings on metal forming tools in a model wear test. Surf. Coat. Technol., 43/44 (1990b) 721–731.

Matthews, A., An investigation of adhesion of ion plated coatings. BSc dissertation, Salford University, 1974, 98 pp.

Matthews, A., The development of ion plated titanium nitride coatings for metal forming tools, PhD thesis, Salford University, 1980, 315 pp.

Matthews, A., The value of deposition processes for industrial tools. In: Proc. 1st Conf. on Materials Engineering, Leeds, UK, Inst. of Metallurgist, July 1984, 175–182.

Matthews, A., Developments in ionisation assisted processes. J. Vac. Sci. Technol., A3 (1985) 6, 2354–2363.

Matthews, A., Methods for assessing coating adhesion. Le Vide, les Couches Minces, No. Special, October, 1988a, 7–15.

Matthews, A., The role of physical vapour deposition as a manufacturing process. Advanced Materials and Manufacturing Processes, 3 (1988b) 1, 91–105.

Matthews, A., Industrial processes and applications of physical vapour deposition. In: Surface Modification Technologies. Sudarshan, T.S., Bhat, D.G. (eds), The Metallurgical Society, Warrendale, USA, 1988c, 23–34.

Matthews, A., Surface engineering in cutting technology. Autumn Meeting of the Inst. of Metals – 'A Cutting Edge for the 1990s', Sheffield, UK, 19–21.9.1989, Inst. of Metals, London, 1989, 51–61.

Matthews, A., Advances in surface processing technology. In: Surface Engineering Practice. Strafford, K. N., Datta, P.K., Gray, J.S. (eds), Ellis Horwood, Chichister, UK, 1990, 33–44.

Matthews, A., Plasma-based physical vapor deposition surface engineering processes. J. Vac. Sci. Technol., A, 21 (5) (2003) S224–S231.

Matthews, A., Artley, R., Holiday, P., 2005, Revisited: The UK surface engineering industry to 2010, NASURF/DERA, 1998, 67 pp., ISBN 09532872 03.

Matthews, A., Eskildsen, S.S., Engineering applications for diamond-like carbon. Diamond Films '93 Conf. Albufeira, Portugal, 20–24.9 1993, 6.

Matthews, A., Franklin, S., Holmberg, K., Tribological coatings: contact mechanics and selection. J. Phys. D: Appl. Phys., 40 (2007) 5463–5475.

Matthews, A., Holmberg, K., Mechanical properties II (Friction, Wear). In: Advanced Techniques for Surface Engineering. Gissler, W., Jehn, H.A. (eds), Kluwer Academic Publishers, Dordrecht, The Netherlands, 1992, 295–312.

Matthews, A., Leyland, A., Hybrid techniques in surface engineering, Surf. Coat. Technol., 71 (1995) 88–92.

Matthews, A., Leyland, A., Developments in PVD tribological coatings. 5th ASM Heat Treatment and Surface Engineering Conf. in Europe, 7–9.6.2000, Gothenburg, Sweden, 2000, 12 pp.

Matthews, A., Leyland, A., Hard tribological coatings: developments and applications. In: Total Tribology. Scherrington, I. *et al.* (eds), Wiley, West Sussex, UK, 2002, 39–63.

Matthews, A., Leyland, A., Dorn, B., Stevenson, P.R., Bin-Sudin, M., Rebholz, C., Voevodin, A., Schneider, J., Plasma-based surface engineering processes for wear and corrosion protection. J. Vac. Sci. Technol. A, 13/3 May/Jun., 1995, 1202–1207.

Matthews, A., Sundquist, H., Tribological studies of Ti-N films formed by reactive ion plating. Proc. Int. Ion Eng. Congr. (ISIAT'83 and IPAT'83). Takagi, T. (ed.), IEE, Kyoto, Japan, 1983, 1325-1330.

Matthews, A., Swift, K.G., Intelligent knowledge-based systems for tribological coatings selection. Thin Solid Films, 109 (1983) 305–311.

Matthews, A., Teer, D.G., Evaluation of coating wear resistance for bulk metal forming. Thin Solid Films, 73 (1980) 315–321.

Matthews, A., Artley, R., Holiday, P., Stevenson, P., The UK engineering coatings industry in 2005. Hull University, 1992a, 188 pp., ISBN 085958 7436.

Matthews, A., Fancey, K.S., James, A.S., Leyland, A., Ionisation in plasma-assisted physical vapour deposition systems. Surf. Coat. Technol., 61 (1993) 121–126.

Matthews, A., Holmberg, K., Franklin, S., A methodology for coating selection. 19th Leeds–Lyon Symp. on Tribology: Thin Films in Tribology, Leeds, UK, 8–11.9.1992b, 11 pp.

Matthews, A., James, A.S., Leyland, A., Fancey, K.S., Valli, J., Stainsby, J.A., Tribological characterization of some second generation ceramic coatings deposited by ionization-assisted physical vapour deposition. Inst. Mech. Eng. C163/87 (1987) 637–642.

Matthews, A., Jones, R., Dowey, S., Modelling the deformation behaviour of multilayer coatings. Tribology Lett., 11 (2001) 2, 103–106.

Matthews, A., Leyland, A., Ronkainen, H., Nieminen, I., Holmberg, K., Matthes, B., Broszeit, E., Standardization of test procedures in surface engineering. Instn Mech. Engrs, C412/038 (1991a) 213–220.

Matthews, A., Leyland, A., Stevenson, P., Widening the market for advanced PVD coatings. Proc. Int. Conf. on Advances in Materials and Processing Technologies. Hashmi, M.S.J. (ed.), Dublin City University, Dublin, 24–27.8.1993.

Matthews, A., Swift, K.G., Robinson, P., Expert systems in surface engineering. Proc. Surface Engineering and Heat Treatment: Past, Present and Future Conf., London, Inst. of Metals, Dec. 1991b, 363–384.

Matthews, A., Tither, D., Holiday, P., Diamond and carbon-based coatings. Proc. 5th Int. Conf. on Tribology – EUROTRIB 1989, 2 (1989), 96–101.

Matthews, D.T.A., Surface engineering: amorphous alloys and nanocrystallinity. PhD thesis, University of Groningen, The Netherlands, 2008, ISBN 978-90-367-3431-8.

Matthewson, M.J., Axi-symmetric contact on thin compliant coatings. J. Mech. Phys. Solids, 29 (1981) 2, 89–113.

Maugis, D., Adhesion of solids: mechanical aspects. In: Modern Tribology Handbook. Bhushan, B. (ed.), Vol. 1, CRC Press, New York, 2001, 163–203.

Maurin-Perrier, P., Heau, C., Jumaine, S., Lindic, M.H., Riand, R., Engelric, B., Sommer, A., A synergistic approach on DLC coatings for performance engines applications. Int. Conf. Metallurgical Coatings and Thin Films ICMCTF, 2–6.5. 2005, San Diego, USA.

Mayrhofer, P.H., Mitterer, C., Clemens, H., Self-organising nanostructures in hard ceramic coatings. Adv. Eng. Mater., 7/12 (2006) 1071–1082.

Mayrhofer, P.H., Mitterer, C., Musil, J., Structure-property relationships in single- and dual-phase nanocrystalline hard coatings. Surf. Coat. Technol., 174–175 (2003) 725–731.

McArthur, S.L., Applications of XPS in bioengineering. Surf. Interf. Anal., 38 (2006) 11, 1380–1385.

McClelland, G.M., Glosli, J.N., Friction at the atomic scale. In: Fundamentals of Friction: Macroscopic and Microscopic Processes. Singer, I.L., Polloc, H.M. (eds), NATO ASI Series, Series E: Applied Sciences, Vol. 220, Kluwer Academic Publishers, London, 1991, 405–422.

McColl, I.R., Ding, J., Leen, S.B., Finite element simulation and experimental validation of fretting wear. Wear, 256 (2004) 1114–1127.

McCune, R.C., Hoffman, D.W., Whalen, T.J., McHugh, C.O., Adherence of diamond films produced by microwave plasma deposition on SiAlON tool inserts. Mat. Res. Soc. Symp. Proc., 130 (1989) 261–266.

McGuire, G.E., Characterisation of thin films. In: Deposition Technologies for Films and Coatings. Bunshah, R.F. (ed.), Noyes Publications, New Jersey, USA, 1982. 584–569.

McKenzie, D.R., Muller, D., Pailthorpe, B.A., Compressive-stress-induced formation of thin-film tetrahedral amorphous carbon. Phys. Rev. Lett., 67 (1991) 773–776.

Meguid, S.A., Engineering Fracture Mechanics., Elsevier Applied Science, London, UK, 1989, 397 pp.

Mehan, R.L., Hayden, S.C., Friction and wear of diamond materials and other ceramics against metal. Wear, 74 (1981–82) 195–212.

Meijers, P., The contact problem of a rigid cylinder on an elastic layer. Appl. Sci. Res., 18 (1968) 353–383.

Melford, D.A., Quoted in: Future of Surface Engineering in the UK, CEST, London, 1991, 2.

Mellor, B.G. (ed.), Surface coatings for protection against wear. CRC Press, Woodhead Publishing Ltd, Cambridge, UK, 2006. 429 pp.

Memming, R., Tolle, H.J., Wierenga, P.E., Properties of polymeric layers of hydrogenated amorphous carbon produced by a plasma-activated chemical vapour deposition process, II: Tribological and mechanical properties. Thin Solid Films, 143 (1986) 31–41.

Menčik, J., Mechanics of Components with Treated or Coated Surfaces. Kluwer Academic Publishers, Solid Mechanics and its Applications, Vol. 42, London, 1996, 360 pp.

Meneve, J., Jacobs, R., Eersels, L., Smeets, J., Dekempeneer, E., Friction and wear behaviour of amorphous hydrogenated $Si_{1-x}C_x$ films. Surf. Coat. Technol., 62 (1993a) 577–582.

Meneve, J., Jacobs, R., Lostak, F., Eersels, L., Dekempeneer, E., Smeets, J., Micromechanical behaviour of amorphous hydrogenated silicon carbide films. MRS Spring Meeting, San Francisco, 1993b, 24 pp.

Meng, H.C., Ludema, K.C., Wear models and predictive equations: their form and content. Wear, 181–183 (1995) 443–457.

Menon, A.K., Interface tribology for 100 Gb/in^2. Tribology Int., 33 (2000) 299–308.

Menthe, E., Rie, K.T., Plasma nitriding and plasma nitrocarburising of electroplated hard chromium to increase the wear and the corrosion properties. Surf. Coat. Technol., 112 (1999) 217–220.

Merchant, M.E., Friction and adhesion. In: Interdisciplinary Approach to Friction and Wear. Ku, P.M. (ed.), NASA publication, 1968, 181–265.

Merlo, A.M., The contribution of surface engineering to the product performance in the automotive industry. Surf. Coat. Technol., 174–175 (2003) 21–26.

Messier, R., Giri, A.P., Roy, R.A., Revised structure zone model for thin film physical structure. J. Vac. Sci. Technol., A2 (1984) 500–503.

Meunier, C., Alers, P., Marot, L., Stauffer, J., Randall, N., Mikhailov, S., Friction properties of ta-C and a-C :H coatings under high vacuum. Surf. Coat. Technol., 200 (2005) 1976–1981.

Michalczewski, R., Piekoszewski, W., Wear and friction characteristics of low friction coatings in dry conditions. Tribologia – Finnish J. Tribology., 26 (2007) 3, 9–21.

Michler, T., Grische, M., Bewilogua, K., Hleke, A., Continuously deposited duplex coatings consisting of plasma nitriding and a-C:H:Si deposition. Surf. Coat. Technol., 111 (1999) 41–45.

Middleton, R.M., Huang, P.J., Wells, M.G.H., Kant, R.A., Effect of coatings on rolling contact fatigue behaviour of M50 bearing steel. Surf. Engng, 7 (1991) 4, 319–326.

Mihailidis, A., Bakolas, V., Drivakos, N., Subsurface stress field of a dry line contact. 12th Int. Colloquium on Tribology, Esslingen, Germany, (2000), 1117–1122.

Miller, K.J., The short crack problem. Fatigue of Engineering Materials and Structures, 5 (1982) 223–232.

Miller, K.J., De Los Rios, E.R., The Behaviour of Short Fatigue Cracks, Mechanical Engineering Applications, London 1986, 560 pp.

Misra, A., Finnie, I., A review of the abrasive wear of metals. J. Eng. Mat. Tech., Trans. ASME, 104 (1982) 94–101.

Mittal, K.L., Adhesion measurement of thin films. Electrocomponent Sci. Technol., 3 (1976) 21–42.

Mittal, K.L. (ed.), Adhesion Measurement of Films and Coatings. VSP, Utrecht, The Netherlands, 1995, 456 pp.

Mittal, K.L. (ed.), Adhesion Measurement of Films and Coatings, Vol. 2. VSP, Utrecht, The Netherlands, 2001, 351 pp.

Miura, K., Sasaki, N., Superlubricity of fullerene intercalated graphite composite. In: Superlubricity. Erdemir, A., Martin, J.M. (eds), Elsevier, Amsterdam, Netherlands, 2007, 161–177.

Miyakawa, Y., Noudomi, Y., Combined lubricating effect of PTFE composite materials and MoS_2 films (1st report) – effect of Mo addition to PTFE. Japanese J. Tribology, 35 (1990) 4, 459–473.

Miyakawa, Y., Noudomi, Y., Combined lubricating effect of PTFE composite materials and MoS_2 films (2nd report) – effects of adding MoS_2, Ag and graphite to PTFE. Japanese J. Tribology, 35 (1990) 4, 475–490.

Miyake, S., Tribological properties of hard carbon films: extremely low friction mechanism of amorphous hydrogenated carbon films and amorphous hydrogenated SiC films in vacuum. Surf. Coat. Technol., 54/55 (1992) 563–569.

Miyake, S., Kaneko, R., Micro-tribological properties and applications of super-hard and lubricating coatings. In: Applications of Diamond Films and Related Materials. Tzeng, Y. *et al.* (eds), Elsevier Scientific Publishers BV, Amsterdam, 1991, 691–698.

Miyake, S., Kaneko, R., Micro-tribological properties and potential applications of hard, lubricating coatings. Thin Solid Films, 212 (1992) 256–261.

Miyake, S., Kaneko, R., Kikuya, Y., Sugimoto, I., Micro-tribological studies on fluorinated carbon films. J. Tribology, Trans. ASME, 113 (1991) 384–389.

Miyake, S., Takahashi, S., Watanabe, I., Yoshihara, H., Friction and wear behavior of hard carbon films. ASLE Trans., 30 (1987) 1, 121–127.

Miyake, S., Watanabe, S., Murakawa, M., Kaneko, R., Miyamoto, T., Tribological study of cubic boron nitride film. Thin Solid Films, 212 (1992) 262–266.

Miyake, S., Saito, T., Yasuda, Y., Okamoto, Y., Kano, M., Improvement of boundary lubrication properties of diamond-like carbon (DLC) films due to metal addition. Tribology Int., 37 (2004) 751–761.

Miyoshi, K., Fundamental tribological properties of ion-beam-deposited boron nitride films. NASA Technical Memorandum 102088 (1989a) 23 pp.

Miyoshi, K., Plasma-deposited amorphous hydrogenated carbon films and their tribological properties. NASA Technical Memorandum 102379 (1989b) 12 pp.

Miyoshi, K., Studies of mechanochemical interactions in the tribological behavior of materials. Surf. Coat. Technol., 43/44 (1990) 799–812.

Miyoshi, K., Tribological studies of amorphous hydrogenated carbon films in a vacuum, spacelike environment. In: Applications of Diamond and Related Materials. Tzeng, Y. *et al.* (eds), Elsevier Scientific Publishers BV, 1991, 699–702.

Miyoshi, K., Lubrication by diamond and diamondlike carbon coatings. J. Tribology, Trans. ASME, 120 (1998) 379–384.

Miyoshi, K., Solid Lubrication – Fundamentals and Applications, Marcel Dekker, New York, 2001, 399 pp.

Miyoshi, K., Buckley, D., Pouch, J.J., Alterovitz, A., Sliney, H.E., Mechanical strength and tribological behavior of ion-beam-deposited boron nitride films on non-metallic substrates. Surf. Coat. Technol., 33 (1987) 221–233.

Miyoshi, K., Honecy, F.S., Abel, P.B., Pepper, S.V., Spalvins, T., Wheeler, D.R., A vacuum (10^{-9} torr) friction apparatus for determining friction and endurance life of MoS_x films. Tribology Trans., STLE, 36 (1993) 3, 351–358.

Miyoshi, K., Pouch, J.J., Alterovitz, A., Plasma-deposited amorphous hydrogenated carbon films and their tribological properties. NASA TM 102379 (1989) 11.

Miyoshi, K., Wu, R.L.C., Garscadden, A., Friction and wear of diamond and diamondlike carbon coatings. Surf. Coat. Technol., 54/55 (1992) 428–434.

Mohrbacher, H., Blanpain, B., Celis, J.P., Roos, J.R., Frictional behaviour of diamond-like carbon and diamond coatings in oscillating sliding. Surf. Coat. Technol., 62 (1993a) 583–588.

Mohrbacher, H., Blanpain, B., Celis, J.P., Roos, J.R., Low amplitude oscillating sliding wear on chemically vapour deposited diamond coatings. Diamond and Related Materials, 2 (1993b) 879–884.

Molarius, J.M., Korhonen, A.S., Harju, E., Lappalainen, R., Comparison of cutting performance of ion-plated NbN, ZrN, TiN and (Ti,Al)N coatings. Surf. Coat. Technol., 33 (1987) 117–132.

Moll, E., Daxinger, H., Method and apparatus for evaporating materials in a vacuum coating plant. UK Patent, 4147 (1980) 175.

Moll, E., Physical vapour deposition techniques II: ion plating, arc deposition and ion beam deposition. In: Advanced Techniques for Surface Engineering. Gissler, W., Jehn, H.A. (eds), Kluwer Academic Publishers, Dordrecht, The Netherlands, 1992, 181–197.

Monaghan, D.P., Laing, K.C., Logan, P.A., Teer, D.G., Advanced hard and soft coatings for high performance machining and forming. Finishing (1993) Nov., 7 pp.

Montgomery, D.C., Design and Analysis of Experiments, John Wiley & Sons, New York, 1991, 649 pp.

Moore, M.A., A review of two-body abrasive wear. Wear, 27 (1974) 1–17.

Moore, D.F., Principles and Applications of Tribology, Pergamon Press, Oxford, 1975, 388 pp.

Mordike, B.L., The frictional properties of carbides and borides at high temperature. Wear, 3 (1960) 374–387.

More, A.S., Jiang, W., Brown, W.D., Malshe, A.P., Tool wear and machining performance of cBN–TiN coated carbide inserts and PCBN compact inserts in turning AISI 4340 hardened steel. J. Mat. Proc. Technol., 180 (2006) 253–262.

Morina, A., Nevill, A., Tribofilms: aspects of formation, stability and removal. J. Phys. D: Appl. Phys., 40 (2007) 5476–5487.

Morley, J.R., Smith, H.R., High rate ion production for vacuum deposition. J. Vac. Sci. Technol., 9 (1972) 1377–1378.

Morrison, C.F., United States Patent No. 4351472 (1982).

Morrison, C.F., Welty, R.P., Anodic Plasma Generation in Magnetron Sputtering. Boulder Colorado, Vac-Tec Systems Inc., 1982.

Mounier, E., Pauleau, Y., Effects of energetic particles on the residual stresses in non-hydrogenated amorphous carbon films deposited on grounded surfaces by DC magnetron sputtering. J. Vac. Sci. Technol., A14 (1996) 2535–2543.

Mounier, E., Pauleau, Y., Mechanisms of intrinsic stress generation in amorphous carbon thin films prepared by magnetron sputtering. Diamond and Related Materials, 6 (1997) 1182–1191.

Movchan, B.A., Functionally graded EB PVD coatings. Surf. Coat. Technol., 149 (2002) 252–262.

Movchan, B.A., Demchishin, Study of the structure and properties of thick vacuum condensates of nickel titanium, tungsten, aluminium oxide and zirconium dioxide. Fizika Metal., 28 (1969) 653–660.

Movchan, B.A., Marinski, G.S., Gradient protective coatings of different application produced by EB-PVD. Surf. Coat. Technol., 100–101 (1998) 309–315.

Muller, C., Menoud, C., Maillat, M., Hintermann, H.E., Thick and compact MoS_2 coatings. 4th European Space Mechanics and Tribology Symp, Cannes, France, 1989, 20–22.9.1989 267–272.

Muller, C., Menoud, C., Maillat, M., Hintermann, H.E., Thick compact MoS_2 coatings. Surf. Coat. Technol., 36 1989 (1988) 351–359.

Müller, K.H., Model for ion-assisted thin-film densification. J. Appl. Phys., 59 (1986) 2803–2807.

Müller, K.-H., Stress and microstructure of sputter-deposited thin films, molecular dynamics investigations. J. Appl. Phys., 62 (1987) 1796–1799.

Muller, R.S., Microdynamics. Sensors and Actuators, A21–A23 (1990) 1–8.

Munz, W.D., Smith, I.J., Donohue, L.A., Deeming, A.P., Goodwin, R., TiAlN based PVD coatings tailored for dry cutting operations. Proc. 40th Annu. Technical Conf. Soc. Vacuum Coaters SVC, (1997) 89–93.

Murakami, Y., Sakae, C., Ichimaru, K., Morita, T., Experimental and fracture mechanics study of the pit formation mechanism under repeated lubricated rolling-sliding contact: effects of reversal of rotation and change of the driving roller. J. Tribology, Trans. ASME, 119 (1997) 788–796.

Murakawa, M., Watanabe, S., The possibility of coating cubic BN films on various substrates. Surf. Coat. Technol., 43/44 (1990) 145–153.

Murakawa, M., Takeuchi, S., Hirose, Y., An experiment in large area diamond coating using a combustion flame torch in its traversing mode. Surf. Coat. Technol., 43/44 (1990) 22–29.

Muratore, C., Voevodin, A.A., Hu, J.J., Zabinski, J.S., Tribology of adaptive nanocomposite yttria-stabilized zirconia coatings containing silver and molybdenum from 25 to 700°C. Wear, 261 (2006) 797–805.

Murray, M.J., Mutton, P.J., Watson, J.D., Abrasive wear mechanisms in steels. J. Lubr. Tech., Trans. ASME, 104 (1982) 9–16.

Musil, J., Physical and mechanical properties of hard nanocomposite films prepared by magnetron sputtering. In: Nanostructured Hard Coatings, Chapter 10. Kluwer Academic/Plenum Publishers, New York, USA, 2005, 48 pp.

Musil, J., Hard nanocomposite films prepared by reactive magnetron sputtering. In: Nanostructured Thin Films and Nanodispersion Strengthened Coatings. Voevodin, A.A. *et al.* (eds), Kluwer Academic Publishers, The Netherlands, 2004, 43–56.

Musil, J., Jirout, M., Toughness of hard nanostructured ceramic thin films. Surf. Coat. Technol., 201 (2007) 5148–5152.

Nabot, J., Aubert, A., Gillet, R., Renaux, P., Cathodic sputtering for preparation of lubrication films. Surf. Coat. Technol., 43/44 (1990) 629–639.

NAGl, M.M., Saunders, S.R.J., Evans, W.T., Hall, D.J., The tensile failure of nickel oxide scales at ambient and at growth temperature. Corrosion Science, 35 (1993) 971–977, 965–969.

Nakahigashi, T., Tanaka, Y., Miyake, K., Oohara, H., Properties of flexible DLC film deposited by amplitude-modulated RF P-CVD. Tribology Int., 37 (2004) 907–912.

Nakayama, K., Martin, J.M., Tribochemical reactions at and in the vicinity of a sliding contact. Wear, 261 (2006) 235–240.

Nakonechna, O., Cselle, T., Morstein, M., Karimi, A., On the behaviour on indentation fracture in TiAlSiN hard thin films. Thin Solid Films, 447–448 (2004) 406–412.

Nastasi, M., Kodali, P., Walter, K.C., Embury, J.D., Raj, R., Nakamura, Y., Fracture toughness of diamondlike carbon coatings. J. Mater. Res., 14 (1999) 5, 2173–2180.

Nélias, D., Dumont, M.L., Champiot, F., Vincent, A., Girodin, D., Fougères, R., Flamand, L., Role of inclusions, surface roughness and operating conditions on rolling contact fatigue. Trans. ASME. J. Tribology, 121 (1999) 240–251.

Neuville, S., Matthews, A., A perspective on the optimisation of hard carbon and related coatings for engineering applications. Thin Solid Films, 515 (2007) 6619–6653.

Neuville, S., Morina, A., Wear and chemistry of lubricants. In: Wear – Materials, Mechanisms and Practice. Stachowiak, G.W. (ed.), Tribology in Practice Series, John Wiley & Sons, Chichester, UK, 2005, 71–94.

Neville, A., Morina, A., Haque, T., Voong, M., Compatibility between tribological surfaces and lubricant additives – how friction and wear reduction can be controlled by surface/lube synergies. Tribology Int., 40 (2007) 1680–1895.

Ni, W., Cheng, Y.T., Grummon, D.S. Microscopic shape memory and superelastic effects under complex loading conditions. Surf. Coat. Technol., 177/178 (2004c) 512-517.

Ni, W., Cheng, Y.T., Lukitsch, M.J., Weiner, A.M., Lev, L.C., Grummon, D.S., Effects of the ratio of hardness to Young's modulus on the friction and wear behaviour of bilayer coatings. Appl. Phys. Lett., 85 (2004a) 18, 4028–4030.

Ni, W., Cheng, Y-T., Lukitsch, M.J., Weiner, A.M., Lev, L.C., Grummon, D.S. Novel layered tribological coatings using a superelastic NiTi interlayer. Wear, 259 (2004b) 842–848.

Nie, X., Wilson, A., Leyland, A., Matthews, A., Deposition of duplex Al_2O_3/DLC coatings on Al-alloys for tribological applications using a combined micro-arc oxidation and plasma-immersion ion implantation technique. Surface and Coating Technology, 131 (2000) 506–513.

Niemann, G., Machinenelemente II. Berlin, Springer Verlag, 1960.

Niederhäuser, P., Hintermann, H.E., Maillat, M., Moisture-resistant MoS_2-based composite lubricant films. Thin Solid Films, 108 (1983) 209–218.

Nieminen, I., Hammaspyörien vierintäväsymislujuuden arvioiminen, VTT – Technical Research Centre of Finland, Research Reports 695, 1990, 48 pp.

Nilsson, D., Svahn, F., Wiklund, U., Hogmark, S., Low-friction carbon-rich carbide coatings deposited by co-sputtering. Wear, 254 (2003) 1084–1091.

Nishimura, M., Watanabe, M., Yoshimurs, R., Friction and wear of sputtered PTFE films. Proc. Symp. on Tribochemistry, Chinese Academy of Science (1989) Lanzhou, P.R. China, 25–28.8.1989, 213–222.

Nordin, M., Larsson, M., Hogmark, S., Mechanical and tribological properties of multilayered PVD TiN/CrN, TiN/MoN, TiN/NbN and TiN/TaN coatings on cemented carbide. Surf. Coat. Technol., 106 (1998) 234–241.

Nordin, M., Larsson, M., Hogmark, S., Mechanical and tribological properties of multilayered PVD TiN/CrN. Wear, 232 (1999) 221–225.

Nordin, M., Design, synthesis and characterization of multilayered PVD coatings. Doctoral thesis, Uppsala University, Faculty of Sciences and Technology, No. 509, 2000, 55 pp.

Nosaka, M., Kikuchi, M., Oike, M., Kawai, N., Tribo-characteristics of cyrogenic hybride ceramic ball bearings for rocket turbopumps: bearing wear and transfer film. Tribology Trans., 42 (1999) 1, 106–115.

Nosonovsky, M., Bhushan, B., Scale effect in dry friction during multiple-asperity contact. Trans. ASME, J. Tribology, 127 (2005) 37–46.

Nouveau, C., Djouadi, M.A., Deèes-Petit, C., Beer, P., Lambertin, M., Influence of Cr_xN_y coatings deposited by magnetron sputtering on tool service life in wood processing. Surf. Coat. Technol., (2001) 142–144, 94–101.

Nouveau, C., Djouadi, M.A., Deèes-Petit, C., The influence of deposition parameters on the wear resistance of Cr_xN_y magnetron sputtering coatings in routing of oriented strand board. Surf. Coat. Technol., (2003) 174–175, 455–460.

Nouveau, C., Jorand, E., Decès-Petit, C., Labidi, C., Djouadi, M.A., Influence of carbide substrates on tribological properties of chromium and chromium nitride coatings: application to wood machining. Wear, 258 (2005) 157–165.

Noyan, I.C., Cohen, J.B., Residual stress – measurement by diffraction and interpretation, Springer-Verlag, Berlin, Germany, 1987.

Oberle, T.L., Wear of metals. J. Metals (1951) June, 438–439G.

O'Connor, R.F., Tomlinson, W.J., Blunt, L.A., Effect of waviness on the wear behaviour of 080 M40 (EN8) and 817 M40 (EN24A) steels. Tribology Int., 24 (1991) 5, 259–268.

Oguri, K., Arai, T., Low friction coatings of diamond-like carbon with silicon prepared by plasma-assisted chemical vapor deposition. J. Mater. Res., 5 (1990) 11, 1–5.

Oguri, K., Arai, T., Tribological properties and characterization of diamond-like carbon coatings with silicon prepared by plasma-assisted chemical vapor deposition. Surf. Coat. Technol., 47 (1991) 710–721.

Ogilvy, J.A., Numerical simulation of elastic–plastic contact between anisotropic rough surfaces. J. Phys. D: Appl. Phys., 25 (1992) 1798–1809.

Ogilvy, J.A., Predicting the friction and durability of MoS_2 coatings using a numerical contact model. Wear, 160 (1993) 171–180.

O'Hern, M.E., Parish, R.H., Oliver, W.L., Evaluation of mechanical properties of thin films by ultralow load indentation. Thin Solid Films, 181 (1989) 357–363.

Ohring, M., Materials Science of Thin Films, Academic Press, San Diego, USA, 2002, 794 pp.

Ojha, S.M., Holland, L., Some characteristics of hard carbonous films. Thin Solid Films, 40 (1977) L31–L32.

Ollendorf, H., Schulke, T., Schneider, D., Testing the adhesion of hard coatings including the non-destructive technique of surface acoustic waves. In: Adhesion Measurements of Films and Coatings, Vol. 2. Mittal, M. (ed.), VSP, Utrechnt, Germany, 2001, 49–77.

Ollivier, B., Matthews, A., Relationship between interlayer hardness and adhesion and pin-on-disc behaviour for FAB source DLC films. Accepted for publication to J. Adhes. Sci. Technol., 1993.

Oliveira, S.A.G., Bower, A.F., An analysis of fracture and delamination in thin coatings subjected to contact loading. Wear, 198 (1996) 15–32.

Oliver, W.C., Pharr, G.M., Am improved technique for determining hardness and elastic modulus using load and displacement sensing indentation experiments. J. Mater. Res., 7 (1992) 6, 1564–1583.

Oliver, W.C., Pharr, G.M., Measurement of hardness and elastic modulus by instrumented indentation: advances in understanding and refinements to methodology. J. Mater. Res., 19 (2004) 1, 3–20.

Olsson, M., Characterization of ceramic coatings – microstructure, adhesion and wear resistance. Acta Universitatis Upsaliensis, Comprehensive Summaries of Uppsala Dissertations from Faculty of Sciences, No. 200, Uppsala University, Sweden, 1989, 36 pp.

Olsson, M., Hedenqvist, P., Stridh, B., Söderberg, S., Solid particle erosion of hard chemically vapour-deposited coatings. Surf. Coat. Technol., 37 (1989) 321–337.

Olofsson, U., Sjöström, H., Sjödin, U., Increased wear resistance of roller bearing using Me-C:H coated rollers. J. Tribology, Trans. ASME, 122 (2000) 682–688.

Ono, T., Takeyama, H., A study for the cutting tool wear mechanism on the minor flank. Wear, 152 (1992) 47–56.

Onsöyen, E., Accelerated testing and mechanical evaluation of components exposed to wear. Norwegian Institute of Technology, Division of Machine Design. Trondheim, Norway, 1990, 178 pp.

Opitz, H., About wear on cutting tools. Proc. Conf. on Lubrication and Wear, Inst. Mech. Engrs, London, UK, 1–3.10.1975, 664–669.

Osenius, S., Korhonen, A.S., Sulonen, M.S., Performance of TiN-coated tools in wood cutting. Surf. Coat. Technol., 33 (1987) 141–151.

Österle, W., Klaffke, D., Griepentrog, M., Gross, U., Kranz, I., Knabe, C., Potential of wear resistant coatings on Ti–6Al–4V for artificial hip joint bearing surfaces. Wear, 264 (2007) 505–517.

Osuch-Slomka, E., Gradkowski, M., Influence of fillers on frictional properties of polytetrafluoroethylene (PTFE) composites, Proc. 14th Int. Colloquium on Tribology, 13–15.1.2004, Esslingen, Germany, 1431–1435.

O'Sullivan, T.C., King, R.B., Sliding contact stress field due to a spherical indenter on a layered elastic half-space. J. Tribology, Trans. ASME, 110 (1988) 235–240.

Overney, R., Meyer, E., Tribological investigations using friction force microscopy. MRS Bulletin, 18 (1993) 5, 26–34.

Öztürk, A., Ezirmik, K.V., Kazmanli, K., Ürgen, M., Eryilmaz, O.L., Erdemir, A., Comparative tribological behaviours of TiN-, CrN- and MoN–Cu nanocomposite coatings. Tribology Int., 41 (2008) 49–59.

Page, T.F., Knight, J.C., Factors affecting the tribological behaviour of thin hard TiN and TiC coatings. Surf. Coat. Technol., 39/40 (1989) 339–354.

Pahl, G., Beitz, W., Engineering Design, Springer-Verlag, Berlin, 1984, 413 pp.

Palmqvist, S., Brittish Iron and Steel Industry Translation No. 1865. Jernkorntorets Annaler, 141 (1957) 5, 300.

Palmqvist, S., Jernkontorets Annaler, 167 (1963) 4, 208.

Panich, N., Sun, Y., Mechanical characterization of nanostructured TiB_2 coatings using microscratch techniques. Tribology Int., 39 (2006) 138–145.

Panjan, P., Čekada, M., Navinšek, B., A new experimental method for studying the cracking behaviour of PVD multilayer coatings. Surf. Coat. Technol., 174–175 (2003) 55–62.

Pang, S., Zhang, T., Asami, K., Inoue, A., Formation of bulk glassy Fe75–x-yCrxMoyC15B10 alloys and their corrosion behaviour. J. Mater. Res., 17 (2002) 701–704.

Park, S.J., Lee, K.R., Ko, D.H., Tribochemical reaction of hydrogenated diamond-like carbon films: a clue to understand the environmental dependence. Tribology Int., 37 (2004) 913–921.

Parker, K., Hardness and wear resistance tests of electroless nickel deposits. Plating, 61 (1974) 834–841.

Patir, N., Cheng, H.S., An average flow model for determining effects of three-dimensional roughness on partial hydrodynamic lubrication. J. Lubr. Technol., Trans. ASME, F100 (1978) 12–17.

Patscheider, J., Zehnder, T., Diserens, M., Structure–performance relations in nanocomposite coatings. Surf. Coat. Technol., 146/147 (2001) 201–208.

Pauleau, Y., Physical vapour deposition techniques I: Evaporation and sputtering. In: Advanced Techniques for Surface Engineering. Gissler, W., Jehn, H.A. (eds), Kluwer Academic Publishers, Dordrecht, The Netherlands, 1992, 135–179.

Pauleau, Y., Residual stresses in DLC films and adhesion to various substrates. In: Tribology of Diamond-like carbon Films. Fundamentals and Applications. Donnet, C., Erdemir, A. (eds), Springer, New York, USA, 2008, 102–136.

Pawlowski, L., The science and engineering of thermal spray coatings, John Wiley & Sons, New York, USA, 2008, 656 pp.

Peebles, D.E., Pope, L.E., Reactive evaporation of thin titanium nitride films in ultrahigh vacuum and their friction and wear behavior as a function of contact stress. Thin Solid Films, 173 (1989) 19–37.

Pei, Y.T., Huizenga, P., Galvan, D., De Hosson, J.T.M., Breakdown of the Coloumb friction law in TiC/a-C:H nanocomposite coatings. J. Appl. Phys., 100 (2006) 114309, 9 pp.

Peker, A., Johnson, W.L., A highly processable metallic glass: $Zr_{41.2}Ti_{13.8}Cu_{12.5}Ni_{10.0}Be_{22.5}$. Appl. Phys. Lett., 63 (1993) 17, 2342–2344.

Pelletier, H., Durier, A., Gauthier, C., Schirrer, R., Viscoelastic and elastic–plastic behaviors of amorphous polymeric surfaces during scratch. Tribology Int., 41 (2008) 975–984.

Pepper, S.V., Effect of electronic structure of the diamond surface on the strength of the diamond–metal interface. J. Vac. Sci. Technol., 20 (1982) 3, 643–646.

Perry, A., The adhesion of chemically vapour deposited hard coatings on steel – the scratch test. Thin Solid Films, 78 (1981) 77–93.

Perry, A.J., Jagner, M., Sproul, W.D., Rudnik, P.J., The residual stress in TiN films deposited onto cemented carbide by high-rate reactive sputtering. Surf. Coat. Technol., 39/40 (1989) 387–395.

Perry, A.J., Valli, J., Steinmann, P.A., Adhesion scratch testing: a round-robin experiment. Surf. Coat. Technol., 36 (1988) 559–575.

Perry, M., Harrison, J., Molecular dynamics of the effects of debris molecules on the friction and wear of diamond. Thin Solid Films, 290–291 (1996) 211–215.

Persson, B.N.J., Albohr, O., Mancosu, F., Peveri, V., Samoilov, V.N., Sivebaek, I.M., On the nature of static friction, kinetic friction and creep. Wear, 254 (2003) 835–851.

Peterson, M.B., Winer, W.O. (eds), Wear Control Handbook. ASME, New York, 1980, 1358 pp.

Pethica, J.B., Hutchings, R., Oliver, W., Hardness measurements at penetration depths as small as 20 nm. Phil. Mag., A48 (1983) 593–606.

Petrov, I., Abidi, F., Greene, J.E., Sproul, W.D., Munz, W.-D., Use of an externally applied axial magnetic field to control ion/neutral flux ratios incident at the substrate during magnetron sputter deposition. J. Vac. Sci. Technol., 10/5 (1992) 3283–3287.

Petrov, I., Barna, P.B., Hultman, L., Greene, J.E., Microstructural evolution during film growth. J. Vac. Sci. Technol., A21/5 (2003) S117–S128.

Pettersson, U., Jacobson, S., Tribological texturing of steel surfaces with a novel diamond embossing tool technique. Tribology Int., 39 (2006) 695–700.

Pfluger, E., Schroer, A., Voumard, P., Donohoue, L., Munz, W.D., Influence of incorporation of Cr and Y on the wear performance of TiAlN coatings at elevated temperatures. Surf. Coat. Technol., 115 (1999) 17–23.

Pflueger, E., Savan, A., Haefke, H., Fleischer, W., MoS_2-based solid lubricants combined with wear resistant hard layers – materials selection, deposition process, tribological testing and applications results. 12th Int. Colloquium on Tribology. 12–13.1.2000, Esslingen, Germany, 2000, 993–1004.

Pfohl, C., Rie, K.T., Plasma duplex treatment of satellite. Surf. Coat. Technol., 142/144 (2001) 1116–1120.

Pharr, G.M., Measurement of mechanical properties by ultra-low load indentation. Mat. Sci. Engng, A253 (1998) 151–159.

Pippan, R., Cracks in composites: the concept of crack tip shielding. Proc. Int. School on Advanced Material Science and Technology. In: Baptiste, D., Rustichelli, F. (eds), Jesi, Italy, 1999, 97–114.

Podgornik, B., Influence of substrate pre-treatment on the tribological properties of hard coatings during sliding. Doctoral thesis, University of Ljubljana, Faculty of Mechanical Engineering, Slovenia, 2000, 104 pp.

Podgornik, B., Coated machine elements – fiction or reality? Surf. Coat. Technol., 146–147 (2001) 318–323.

Podgornik, B., Tribological behaviour of DLC films in various lubrication regimes. In: Tribology of Diamond-like Carbon Films. Donnet, C., Erdemir, A. (eds), Springer, New York, USA, 2008, 410–453.

Podgornik, B., Hren, D., Vižintin, J., Jacobson, S., Stavlid, N., Hogmark, S., Combination of DLC coatings and EP additives for improved tribological behaviour of boundary lubricated surfaces. Wear, 261 (2006a) 32–40.

Podgornik, B., Hogmark, S., Sandberg, O., Proper coating selection for improved galling performance of forming tool steel. Wear, 261 (2006b) 15–21.

Podgornik, B., Jacobson, S., Hogmark, S., Influence of EP and AW additives on the tribological behaviour of hard low friction coatings. Surf. Coat. Technol., 165 (2003) 168–175.

Podgornik, B., Sedlaček, M., Vižintin, J., Compatibility of DLC coatings with formulated oils. Tribology Int., 41 (2008) 564–570.

Podgornik, B., Vižintin, J., Tribological reactions between oil additives and DLC coatings for automotive applications. Surf. Coat. Technol., 200 (2005) 1982–1989.

Podgornik, B., Vižintin, J., Jacobson, S., Hogmark, S., Tribological behaviour of WC/C coatings operating under different lubrication regimes. Surf. Coat. Technol., 177–178 (2004) 558–565.

Podgornik, B., Wänstrand, O., An experimental method for coating–substrate interface investigation. Mat. Charact, 55 (2005) 173–178.

Podgornik, B., Zajel, B., Strnad, S., Stana-Kleinschek, K., Influence of surface energy on the interactions between hard coatings and lubricants. Wear, 262 (2007) 1199–1204.

Podsiadlo, P., Stachowiak, G.W., Multi-scale representation of tribological surfaces. Proc. Instn Mech. Engrs, Pt J: J. Engng Tribology, 216 (2002) 463–479.

Pollock, H., Nanoindentation. In: ASM Handbook, Vol. 18. Friction, Lubrication and Wear Technology, ASM Int., Ohio, 1992, 419–424.

Polonsky, I.A., Chang, T.P., Keer, L.M., Sproul, W.D., An analysis of the effect of hard coatings on near-surface rolling contact fatigue initiation induced by surface roughness. Wear, 208 (1997) 204–219.

Polonsky, I.A., Keer, L.M., Numerical analysis of the effect of coating microstructure on three-dimensional crack propagation in the coating under rolling contact fatigue conditions. J. Tribology, Trans. ASME, 124 (2002) 14–19.

Ponce De León, C., Kerr, C., Walsh, F.C., Electroless plating for protection against wear. In: Surface Coatings for Protection against Wear. Mellor, B.G. (ed.), CRC Press, Woodhead Publishing Ltd, Cambridge, UK, 2006, 184–225.

Popov, V.L., Psakhie, S.G., Numerical simulation methods in tribology. Tribology Int., 40 (2007) 916–923.

Posti, E., Nieminen, I., Coating thickness effects on the life of titanium nitride PVD coated tools. Mat. Manufact. Proc., 4 (1989a) 2, 239–252.

Posti, E., Nieminen, I., Influence of coating thickness on the life of TiN-coated high speed steel cutting tools. Wear, 129 (1989b) 273–283.

Prasad, S.V., Zabinski, J.S., Tribological behavior of nanocrystalline zinc oxide films. Wear, 203–204 (1997) 498–506.

Prasad, S.V., Zabinski, J.S., McDevitt, N.T., Friction behaviour of pulsed laser deposited tungsten disulphide films. STLE Tribology Trans., 38 (1995) 57–62.

Pulker, H.K., Coatings on glass. Thin Films Science and Technology 6, Elsevier, Amsterdam, 1984, 484 pp.

Pulker, H.K., Coatings on Glass, Elsevier, Amsterdam, The Netherlands, 1999, 394 pp.

Purandare, Y.P., Stack, M.M., Hovsepian, P.E., Velocity effects on erosion–corrosion of CrN/NbN 'superlattice' PVD coatings. Surf. Coat. Technol., 201 (2006) 361–370.

Qiao, X., Hou, Y., Wu, Y., Chen, J., Study on functionally gradient coatings of Ti–Al–N. Surf. Coat. Technol., 131 (2000) 462–464.

Quaeyhaegens, C., Van Strappen, M., Stals, L.M., Bodart, F., Terwagne, G., Vlaeminck, R., Interface study of physically vapour deposited TiN coatings on plasma-nitrided tool steel surfaces with Auger electron spectroscopy, resonant nuclear reaction analysis and Rutherford backscattering spectroscopy. Surf. Coat. Technol., 54/55 (1992) 279–286.

Quinn, T.F.J., Review of oxidational wear. Part I: The origins of oxidational wear. Tribology Int., 16 (1983a) 5, 257–271.

Quinn, T.F.J., Review of oxidational wear. Part II: Recent developments and future trends in oxidational wear research. Tribology Int., 16 (1983b) 6, 305–315.

Quinn, T.F.J., Winer, W.O., The thermal aspects of oxidational wear. Wear, 102 (1985) 67–80.

Quinto, D.T., Technology perspective on CVD and PVD coated metal-cutting tools. Int. J. Refractory Metals, Hard Materials, 14 (1996) 7–20.

Quinto, D.T., Twenty-five years of PVD coatings at the cutting edge. Oerlikon Balzers Bulletin, Fall (2007) 7–12.

Rabinowicz, E., Friction and Wear of Materials, John Wiley & Sons, New York, 1965, 244 pp.

Rabinowicz, E., Variation of friction and wear of solid lubricant films with film thickness. ASLE Trans., 10 (1967) 1–9.

Rabinowicz, E., Dry friction. In: Standard Handbook of Lubrication Engineering. McGraw-Hill, New York, 1968, Chapter 1, 26 pp.

Rabinowicz, E., The wear coefficient – magnitude, scatter and uses. ASEM Paper No. 80-C2/Lub-4, Century 2 ASME-ASLE Int. Lubrication Conf. San Francisco, USA, 18–21.8.1980, 6 pp.

Rabinowicz, E., Friction coefficients of noble metals over a range of loads. Wear, 159 (1992) 89–94.

Rabinovich, V.L., Sarin, V.K., Modelling of interfacial fracture. Mat. Sci. Engng, A209 (1996) 82–90.

Raeymaekers, B., Etsion, I., Talke, F.E., Enhancing tribological performance of the magnetic tape/guide interface by laser surface texturing. Tribology Lett., 27 (2007) 1, 89–95.

Ramalingam, S., New coating technologies for tribological applications. In: Wear Control Handbook. Peterson, M.B., Winer, W.O. (eds), American Society of Mechanical Engineers, New York, USA, 1980, 385–412.

Ramalingam, S., Tribological characteristics of thin films, applications of thin film technology for friction, wear reduction. Thin Solid Films, 118 (1984) 335–349.

Ramalingam, S., Zhen, L., Film–substrate interface stresses and their role in the tribological performance of surface coatings. Tribology Int., 28 (1995) 3, 145–161.

Ramsey, P.M., Chandler, H.W., Page, T.F., Modelling the contact response of a coated system. Surf. Coat. Technol., 49 (1991) 504–509.

Randhawa, H., Johnson, P.C., Cunningham, R., Deposition and characterization of ternary nitrides. J. Vac. Sci. Technol., A6 (1988) 3, 2136–2139.

Randhawa, H., Review of plasma-assisted deposition processes. 17th Int. Conf. on Metallurgical Coatings and Thin Films, San Diego, USA, 2-6.4.1990, 16 pp.

Rapoport, L., Nepomnyashchy, O., Lapsker, I., Verdyan, A., Moshkovich, A., Feldman, Y., Tenne, R., Behavior of fullerene-like WS_2 nanoparticles under severe contact conditions. Wear, 259 (2005) 703–707.

Raveh, A., Martinu, L., Hawthorne, H.M., Wertheimer, M.R., Mechanical and tribological properties of dual-trequency plasma-deposited diamond-like carbon. Surf. Coat. Technol., 58 (1993) 45–55.

Ravi, K.V., Koch, C.A., Hu, H.S., Joshi, A., The nucleation and morphology of diamond crystals and films synthesized by the combustion flame technique. J. Mater. Res., 5 (1990) 11, 2356–2366.

Rebholz, C., Ziegele, H., Leyland, A., Matthews, A., Structure, mechanical and tribological properties of Ti–B–N and Ti–Al–B–N thin films produced by electron beam evaporation. J. Vac. Sci. Technol., A16 (1998) 5, 2851–2857.

Rebholz, C., Leyland, A., Matthews, A., Deposition and characterisation of TiAlBN coatings produced by direct electron-beam evaporation of Ti and Ti–Al–B–N material from a twin crucible source. Thin Solid Films, 343/344 (1999a) 242–245.

Rebholz, C., Schneider, J.M., Voevodin, A.A., Steinebrunner, J., Charitidis, C., Logothetidis, S., Leyland, A., Matthews, A., Structure, mechanical and tribological properties of sputtered TiAlBN thin films. Surf. Coat. Technol., 113 (1999b) 126–133.

Rebholz, C., Ziegele, H., Leyland, A., Matthews, A., Structure, mechanical and tribological properties of nitrogen-containing chromium coatings prepared by reactive magnetron sputtering, Surf. Coat. Technol., 115 (1999c) 222–229.

Rebouta, L., Vaz, F., Andritschy, M., Da Silva, M.F., Oxidation resistance of (Ti,Al,Zr,Si)N coatings in air. Surf. Coat. Technol., 76–77 (1995) 70–74.

Rech, J., Kusiak, A., Battaglia, J.L., Tribological and thermal functions of cutting tool coatings. Surf. Coat. Technol., 186 (2004) 364–371.

Remigiusz, M., Witold, P., Szczerek, M., Waldemar, T., PVD coatings and environmentally friendly lubricants in highly-loaded contacts. Tribologia – Finnish J. Tribology, 23 (2004) 3–12.

Renji, Z., Ziwei, L., Zhouping, C., Qi, S., Studies on multilayer wear of CVD TiC–TiN multilayer composite coatings. Wear, 147 (1991) 227–251.

Renondeau, H., Taylor, R.I., Smith, G.C., Torrance, A.A., Friction and wear performance of DLC and Cr doped DLC coatings in contact with steel surfaces. 34th Leeds–Lyon Symposium on Tribology, Lyon, France, 4–7.9.2007, 15 pp.

Reuter, S., Wesskamp, B., Büscher, R., Fischer, A., Barden, B., Löer, F., Buck, V., Correlation of structural properties of commercial DLC-coatings to their tribological performance in biomedical applications. Wear, 261 (2006) 419–425.

Rey, J., Kapsa, P., Male, G., Dry friction and wear of chemically vapour deposited boron carbide coatings. Surf. Coat. Technol., 36 (1988) 375–386.

Reynolds, O., On the theory of lubrication and its application to Mr Beauchamp Tower's experiments, including experimental determination of the viscosity of olive oil. Phil. Trans. R. Soc., 177 (1886) 157–234.

Rezakhanlou, R., Von Stebut, J., Damage mechanisms of hard coatings on hard substrates: a critical analysis of failure in scratch and wear testing. In: Mechanics of Coatings. Dowson, D. *et al.* (eds), Tribology Series 17, Elsevier, Amsterdam, 1990, 183–208.

Rha, J.J., Kwon, S.C., Cho, J.R., Yim, S., Saka, N., Creation of ultra-low friction and wear surfaces for micro-devices using carbon films. Wear, 259 (2005) 765–770.

Rickerby, D.S., Internal stress and adherence of titanium nitride coatings. J. Vac. Sci. Technol. A, 4 (1986) 6, 2809–2814.

Rickerby, D.S., Bull, S.J., Robertson, T., Hendry, A., The role of titanium in the abrasive wear resistance of physically vapour-deposited TiN. Surf. Coat. Technol., 41 (1990) 63–74.

Rickerby, D.S., Burnett, P.J., Correlation of process and system parameters with structure and properties of physically vapour-deposited hard coatings. Thin Solid Films, 157 (1988) 195–222.

Rickerby, D.S., Burnett, P.J., The wear and corrosion resistance of hard PVD coatings. Surf. Coat. Technol., 33 (1987) 191–211.

Rickerby, D.S., Matthews, A. (eds), Advanced Surface Coatings: A Handbook of Surface Engineering, Glasgow, UK, Blackie, 1991a, 364 pp.

Rickerby, D.S., Matthews, A., Ceramic coatings by physical vapour deposition. Reviews on Powder Metallurgy and Physical Ceramics, 4 (1991b) 3–4, 155–295.

Rie, K.T., Recent advances in plasma diffusion processes. Surf. Coat. Technol., 112 (1999) 56–62.

Rie, K.T., Broszeit, E., Plasma diffusion treatment and duplex treatment – recent development and new applications. Surf. Coat. Technol., 76/77 (1995) 425–436.

Rieker, C., 20 times less wear. Sulzer Technical Review, 2 (1998) 12–15.

Rigney, D.A., Hirth, J.P., Plastic deformation and sliding friction of metals. Wear, 53 (1979) 2, 345–370.

Rigney, D.A., Chen, L.H., Naylor, M.G.S., Rosenfield, A.R., Wear processes in sliding systems. Wear, 100 (1984) 195–219.

Riviere, J.C., Myhra, S. (eds), Handbook of Surface and Interface Analysis. Marcel Dekker, New York, USA, 1998, 928pp.

Robbins, M., Krim, J., Energy dissipation in interfacial friction. MRS Bulletin, June (1998) 23–26.

Roberts, E.W., Towards an optimised sputtered MoS_2 lubricant film. Proc. 20th AMS, NASA Lewis, Cleveland, NASA Publications 2423, May 1986, 103–119.

Roberts, E.W., The tribology of sputtered molybdenum disulphide films. Proc. Int. Conf. on Tribology – Friction, Lubrication and Wear, Fifty Years On. 1–3.7.1987, London, Inst. Mech. Engrs, 1987, 503–510.

Roberts, E.W., Ultralow friction films of MoS_2 for space applications. Thin Solid Films, 181 (1989) 461–473.

Roberts, E.W., The advantages and limitations of sputtered molybdenum disulphide as a space lubricant. Proc. 4th European Symp. on 'Space Mechanisms and Tribology', Cannes, France, 20–22.9.1989 (ESA SP-299, March 1990a) 59–65.

Roberts, E.W., Thin solid lubricant films in space. Tribology Int., 23 (1990b) 2, 95–104.

Roberts, E.W., Price, W.B., In-vacuo, tribological properties of 'high-rate' sputtered MoS_2 applied to metal and ceramic substrates. Mat. Res. Soc. Symp. Proc., 140 (1989) 251–264.

Roberts, E.W., Williams, B.J., Ogilvy, J.A., The effect of substrate surface roughness on the friction and wear of sputtered MoS_2 films. J. Phys. D: Appl. Phys., 25 (1992) A65–A70.

Robertson, J., Mechanical properties and structure of diamond-like carbon. Diamond and Related Materials, 1 (1992a) 397–406.

Robertson, J., Properties of diamond-like carbon. Surf. Coat. Technol., 50 (1992b) 185–203.

Robinson, P.A., Matthews, A., Swift, K.G., Franklin, S., A computer knowledge-based system for surface coating and material selection. Surf. Coat. Technol., 62 (1993) 662–668.

Rogers, P.M., Hutchings, I.M., Little, J.A., TiN coatings for protection against combined wear and oxidation. Surf. Engng, 8 (1992) 1, 48–54.

Rohde, S.L., Munz, W.D., Sputter deposition. In: Advanced Surface Coatings: A Handbook in Surface Engineering. Rickerby, D.S., Matthews, A. (eds), Blackie, Glasgow, UK, 1991, 92–126.

Rohde, S.L., Petrov, I., Sproul, W.D., Barnett, S.A., Rudnik, P.J., Graham, M.E., Effects of unbalanced magnetron in a unique dual-cathode, high rate reactive sputtering system. Thin Solid Films, 193/194 (1990) 117–126.

Ronkainen, H., Tribological properties of hydrogenated and hydrogen-free diamond-like carbon coatings: VTT Publications 434, Technical Research Centre of Finland, Espoo, Finland, 2001, 52 pp.

Ronkainen, H., Holmberg, K., Environmental and thermal effects on the tribological performance of DLC coatings. In: Tribology of Diamond-like Carbon Films. Fundamentals and Applications. Donnet, C., Erdemir, A. (eds), Springer, New York, USA, 2008, 155–200.

Ronkainen, H., Holmberg, K., Fancey, K., Matthews, A., Matthes, B., Broszeit, E., Comparative tribological and adhesion studies of some titanium-based ceramic coatings. Surf. Coat. Technol., 43/44 (1990a) 888–897.

Ronkainen, H., Koskinen, J., Anttila, A., Holmberg, K., Hirvonen, J.P., Tribological characterization of hard carbon films produced by the pulsed vacuum arc discharge method. Diamond and Related Materials, 1 (1992a) 639–643.

Ronkainen, H., Koskinen, J., Likonen, J., Varjus, S., Vihersalo, J., Characterization of wear surfaces in dry sliding of steel and alumina on hydrogented and hydrogen free carbon films. Diamond Films '93 Conf. Albufeira, Portugal, 20–24.9.1993, 30 pp.

Ronkainen, H., Koskinen, J., Varjus, S., Holmberg, K., Experimental design and modelling in the investigation of process parameter effects on the tribological and mechanical properties of r.f. plasma deposited a-C:H films. Surf. Coat. Technol., 122 (1999a) 150–160.

Ronkainen, H., Koskinen, J., Varjus, S., Holmberg, K., Load-carrying capacity evaluation of coating/substrate systems for hydrogen-free and hydrogenated diamond-like carbon films. Tribology Lett., 6 (1999b) 63–73.

Ronkainen, H., Laukkanen, A., Holmberg, K., Three dimensional stress analysis of TiN, DLC and MoS2 coated contacts by 3D FEM modelling. In Proc. of ECOTRIB 2007 Joint European Conf. on Tribology, Vizintin *et al.* (eds), 12–15.6.2007, Ljubljana, Slovenia, submitted to Tribology Int. Special Issue on ECOTRIB 2007.

Ronkainen, H., Likonen, J., Koskinen, J., Tribological properties of hard carbon films produced by the pulsed vacuum arc discharge method. Surf. Coat. Technol., 54/55 (1992b) 570–757.

Ronkainen, H., Likonen, J., Koskinen, J., Varjus, S., Effect of tribofilm formation on the tribological performance of hydrogenated carbon coatings. Surf. Coat. Technol., 79 (1996) 87–94.

Ronkainen, H., Nieminen, I., Holmberg, K., Leyland, A., Fancey, K.S., Matthews, A., Matthes, B., Broszeit, E., Evaluation of some new titanium-based ceramic coatings in tribological model wear and metal-cutting tests. Mat. Sci. Engng, A140 (1991a) 602–608.

Ronkainen, H., Nieminen, I., Holmberg, K., Leyland, A., Matthews, A., Matthes, B., Broszteit, E., Evaluation of some titanium based ceramic coatings on high speed steel cutting tools. Surf. Coat. Technol., 49 (1991b) 468–473.

Ronkainen, H., Varjus, S., Holmberg, K., Fancey, K.S., Pace, A.R., Matthews, A., Matthes, B., Broszeit, E., Coating evaluation methods: a round robin study. In: Mechanics of Coatings. Dowson, D. *et al.* (eds), Tribology Series 17, Elsevier, Amsterdam, 1990b, 453–463.

Ronkainen, H., Varjus, S., Holmberg, K., Friction and wear properties in dry, water- and oil-lubricated DLC against alumina and DLC against steel contacts. Wear, 222 (1998) 120–128.

Ronkainen, H., Varjus, S., Koskinen, J., Holmberg, K., Friction and wear performance of a-C:H films in a wide normal load and sliding velocity range. Finnish J. Tribology, 18 (1999c) 1, 3–12.

Ronkainen, H., Varjus, S., Koskinen, J., Holmberg, K., Differentiating the tribological performance of hydrogenated and hydrogen-free DLC coatings. Wear, 249 (2001) 260–266.

Ronkainen, H., Vihersalo, J., Varjus, S., Zilliacus, R., Ehrnsten, U., Nenonen, P., Improvement of a-C:H film adhesion by intermediate layers and sputter cleaning procedures on stainless steel, alumina and cemented carbide. Surf. Coat. Technol., 90 (1997) 190–196.

Roos, J.R., Celis, J.P., Vancoille, E., Veltrop, H., Boelens, S., Jungblut, F., Ebberink, J., Homberg, H., Interrelationship between processing, coating properties and functional properties of steered arc physically vapour deposited (Ti,Al)N and (Ti,Nb)N coatings. Thin Solid Films, 193/194 (1990) 547–556.

Rosado, L., Jain, V.K., Trivedi, H.K., The effect of diamond-like carbon coatings on the rolling fatigue and wear of M50 steel. Wear, 212 (1997) 1–6.

Ross, R.B., Handbook of Metal Treatments and Testing, Chapman and Hall, London, 1988, 548 pp.

Ross, J.D.J., Pollock, H.M., Pivin, J.C., Takadoum, J., Limits of the hardness testing of films thinner than 1 μm. Thin Solid Films, 148 (1987) 171–180.

Rossi, F., Diamond and diamond-like carbon films. In: Advanced Technologies for Surface Engineering. Gissler, W., Jehn, H.A. (eds), Kluwer Academic Publishers, Dordrecht, The Netherlands, 1992, 371–397.

Rossi, S., Chini, F., Straffelini, G., Bonora, P.L., Moschini, R., Stampali, A., Corrosion protection properties of electroless nickel/PTFE, phosphate/MoS_2 and bronze/PTFE coatings applied to improve the wear resistance of carbon steel. Surf. Coat. Technol., 173 (2003) 235–242.

Roth, T., Broszeit, E., Kloos, K.H., Influence of elastohydrodynamic conditions in highly loaded lubricated contacts on the wear behaviour of TiN coatings prepared by r.f. bias sputtering. Surf. Coat. Technol., 36 (1988) 765–772.

Roth, T., Kloos, K.H., Broszeit, E., Structure, internal stresses, adhesion and wear resistance of sputtered alumina coatings. Thin Solid Films, 153 (1987) 123–133.

Rother, B., Siegel, J., Breuer, K., Muhling, K., Deutschmann, S., Vetter, S., Trommer, G., Rau, B., Heiser, C., Characterisation of composite carbon coatings deposited by d.c. cathodic arc technique. J. Mat. Res., 16 (1991) 1, 101–111.

Rowe, C.N., Armstrong, E.L., Lubricant effects in rolling contact fatigue. Lubr. Engng, 38 (1982) 1–23.

Roylance, B.J., Ferrography – then and now. Tribology Int., 38 (2005a) 857–862.

Roylance, B.J., Machine failure and its avoidance – tribology's contribution to effective maintenance of critical machinery. In: Wear – Materials, Mechanisms and Practice, Stachowiak, G.W. (ed.), John Wiley & Sons, Chichester, UK, 2005b, 425–452.

Ruff, A.W., Petersohn, M.B., Wear of selected materials and composites sliding against MoS_2 films. Wear, 162–164 (1993) 492–497.

Ruff, A.W., Shin, H., Evans, C.J., Damage processes in ceramics resulting from diamond tool indentation and scratching in various environments. Wear, 181–183 (1995) 551–562.

Rutherford, K.L., Hatto, P.W., Davies, C., Hutchings, I.M., Abrasive wear resistance of TiN/NbN multi-layers: measurement and neural network modelling. Surf. Coat. Technol., 86–87 (1996) 472–479.

Rutherford, K.L., Hutchings, I.M., A micro-abrasive wear test, with particular application to coated systems. Surf. Coat. Technol., 79 (1996) 231–239.

Rutherford, K.L., Hutchings, I.M., Theory and application of a micro-scale abrasive wear test. J. Test. Eval., 25 (1997) 2, 250–260.

Sabelkin, V., Mall, S., Elastic–plastic multi-asperity contact analysis of cylinder-on-flat configuration. J. Tribology, Trans. ASME, 129 (2007) 292–304.

Saijo, K. *et al.*, The improvement of the adhesion strength of diamond films. Surf. Coat. Technol., 43/44 (1990) 30–40.

Saikko, V., Tribology of total replacement hip joints studied with new hip joint simulators and a materials-screening apparatus. Acta Polyteknica Scandinavica, Mechanical Engineering Series, (1993) 110, 79.

Saikko, V., Ahlroos, T., Calonus, O., Keränen, J., Wear simulation of total hip prostheses with polyethylene against CoCr, alumina and diamond-like carbon. Biomaterials, 22 (2001) 1507–1514.

Sainsot, P., Leroy, J.M., Villechaise, B., Effect of surface coatings in a rough normally loaded contact. In: Mechanics of Coatings. Dowson, D. *et al.* (eds), Tribology Series 17, Elsevier, Amsterdam, 1990, 151–156.

Saka, N., Effect of microstructure on friction and wear of metals. In: Fundamentals of Tribology. Suh, N.P., Saka, N. (eds), The MIT Press, London, 1980, 135–170.

Salas, O., Kearns, K., Carrerra, S., Moore, J.J., Tribological behaviour of candidate coatings for Al die casting dies. Surf. Coat. Technol., 172 (2003) 117–127.

Salehi, M., Bell, T., Morton, P.H., Load bearing capacity of plasma nitrided and ion plated titanium alloys. Surface Modification Technologies IV, The Minerals, Metals, Materials Society (1991), 991–1002.

Salomon, G., De Gee, A.W.J., Zaat, J.H., Mechano-chemical factors in MoS_2-film lubrication. Wear, 7 (1964) 87–101.

Samuels, B., Wilks, J., The friction of diamond sliding on diamond. J. Mater. Sci., 23 (1988) 2846–2864.

Samyn, P., Schoukens, G., Quintelier, P., De Baets, P., Friction, wear and material transfer of sintered polyimides sliding against various steel and diamond-like carbon coated surfaces. Tribology Int., 39 (2006) 575–589.

Sanchette, F., Billard, A., Main features of magnetron sputtered aluminium-transition metal alloy coatings. Surf. Coat. Technol., 142/144 (2001) 218–224.

Sanchette, F., Billard, A., Frantz, C., Mechanically reinforced and corrosion-resistant sputtered amorphous aluminium alloy coatings. Surf. Coat. Technol., 98 (1998) 1162–1168.

Sanchette, F., Loi, T.H., Billard, A., Frantz, C., Structure–properties relationship of metastable Al–Cr and Al–Ti alloys deposited by r.f. magnetron sputtering: role of nitrogen. Surf. Coat. Technol., 74/75 (1995) 903–909.

Sánchez-López, J.C., Fernández, A., Doping and alloying effects on DLC coatings. In: Tribology of Diamond-like Carbon Films. Donnet, C., Erdemir, A. (eds), Springer, New York, USA, 2008, 311–338.

Sander, H., Petersohn, D., Friction and wear behaviour of PVD-coated tribosystems. 19th Leeds-Lyon Symposium on Tribology, Leeds, UK, 8–11.9.1992, 11 pp.

Sanders, D.M., Boercker, D.B., Falabella, S., Coating technology based on the vacuum arc – a review. IEE Trans. on Plasma Sci., 18 (1990) 883–894.

Sanders, D.M., Anders, A., Review of cathodic arc deposition technology at the start of the new millennium. Surf. Coat. Technol., 133–134 (2000) 78–90.

Santavirta, S., Compatibility of the totally replaced hip – reduction of wear by amorphous diamond coating, Kuopio University Publications C. Natural and Environmental Sciences No. 152, 2003, 35 pp.

Santavirta, S., Lappalainen, R., Pekko, P., Anttila, A., Konttinen, Y.T., The counterface, surface smoothness, tolerances, and coatings in total joint prostheses. Clinical Orthopaedics and Related Research (1999) 369, 92–102.

Santos, S.C., Sales, W.F., Da Silva, F.J., Franco, S.D., Da Silva, M.B., Tribological characteristics of PVD coatings for cutting tools. Surf. Coat. Technol., 184 (2004) 141–148.

Sargent, P.M., Factors effecting the microhardness of solids, PhD thesis, Cambridge University, UK, 1979.

Sargent, P.M., Page, T.F., The influence of microstructure on the hardness of ceramic materials. Pro. Brit. Ceram. Soc., 26 (1978) 209–224.

Sato, T., Ikeda, O., Hatsuzawa, T., Linzer, M., Real-time evaluation of wear particles using electromagnetic forced rotation and laser scattering. Wear, 115 (1987) 273–284.

Savan, A., Pflüger, E., Goller, R., Gissler, W., Use of nanoscaled multilayer and compound films to realize a soft lubrication phase within a hard, wear-resistant matrix. Surf. Coat. Technol., 126 (2000) 159–165.

Savvides, N., Window, B., Unbalanced magnetron ion-assisted deposition and property modification of thin films. J. Vac. Sci. Technol., A, 4(3) (1986) 504–508.

Schaefer, L., Gaebler, J., Mulcahy, S., Brand, J., Hieke, A., Wittorf, R., Tribological applications of amorphous carbon and crystalline diamond coatings. Proc. 43rd Annual Technical Conf., Society of Vacuum Coaters, 15–20.4.2000, Denver, USA, 311–315.

Schall, J.D., Mikulsky, P.T., Chateauneuf, G.M., Gao, G., Harrison, J.A., Molecular dynamics simulations of tribology. In: Superlubricity. Erdemir, A., Martin, J.M. (eds), Elsevier, Amsterdam, The Netherlands, 2007, 79–102.

Scharf, T.W., Singer, I.L., Monitoring transfer films and friction instabilities with *in situ* Raman tribometry. Tribology Lett., 14 (2003a) 1, 3–8.

Scharf, T.W., Singer, I.L., Thickness of diamond-like carbon coatings quantified with Raman spectroscopy. Thin Solid Films, 440 (2003b) 138–144.

Scharf, T.W., Singer, I.L., Quantification of the thickness of carbon transfer films using Raman tribometry. Tribology Lett., 14 (2003c) 2, 137–145.

Scharf, T.W., Singer, I.L., Third bodies and tribochemistry of DLC coatings. In: Tribology of Diamond-like Carbon Films. Fundamentals and applications. Donnet, C., Erdemir, A. (eds), Springer, New York, USA, 2008, 201–236.

Scherge, M., Grob, S., Biological Micro- and Nanotribology. Nature's Solutions. Nanoscience and Technology Series. Springer, Berlin, 2001, 304 pp.

Scherge, M., Pöhlmann, K., Gervé, A., Wear measurement using radionuclide-technique (RNT). Wear, 254 (2003a) 801–817.

Scherge, M., Shakhvorostov, D., Pöhlmann, K., Fundamental wear mechanism of metals. Wear, 255 (2003b) 395–400.

Schintlmeister, W., Pacher, O., Raine, T., Wear characteristics of hard material coatings produced by chemical vapour deposition with particular reference to machining. Wear, 48 (1978) 251–266.

Schiller, S., Heisig, U., Newmann, M., Beister, G., Processing, instrumentation in PVD techniques. Vakuum-Technik, 35 (1986) 35.

Schiller, S., Goedicke, K., Kirchoff, V., Kopte, T., Pulse technology – a new era of magnetron sputtering. Proc. of the 38th Annual Technical Conference of the Society of Vacuum Coaters, Chicago, April, 1995, SVC, Albuquerque.

Schiller, S., Goedicke, K., Reschke, J., Kirchoff, V., Schneider, S., Milde, F., Pulsed magnetron sputter technology. Surf. Coat. Technol., 61 (1993). 331–227.

Schlatter, M., DLC-based wear protection on magnetic storage media. Diamond and Related Materials, 11 (2002) 1781–1787.

Schmid, S.R., Wilson, W.R.D., Tribology in manufacturing. In: Modern Tribology Handbook. Bhushan, B. (ed.), CRC Press, New York, 2001, 1385–1411.

Schmidt, H., Inorganic–organic composites by sol-gel techniques. J. Sol-Gel Sci. Technol., 1 (1994) 217–231.

Schmidt, U., Der einfluss der Oberflächenrauheit auf die Schmierfilmausbildung in realen EHD-Wälzkontakten/The influence of surface roughness on lubricant film formation in cylindrical EHD contacts. VDI-Berichte, 549 (1985) 129–154.

Schmitt, M., Paulmier, D., Tribological behaviour of diamond coatings sliding against Al alloys. Tribology Int., 37 (2004) 317–325.

Schmutz, C., Jeanneret, J.P., Tranganida, S., Hintermann, H.E., Characterization of thin PVD coatings by microindentation. 7th Int. Conf. Ion and Plasma Assisted Techniques, Geneva, Switzerland, 31.5–2.6. 1989a, 6 pp.

Schmutz, C., Trangida, S., Jeanneret, J.P., Hintermann, H.E., New coating characterization tools. European Conf. on Advanced Materials and Processes (EUROMAT), Aachen, Germany, 22–24.11.1989b, 19 pp.

Schneider, D., Schultrich, B., Elastic modulus: a suitable quantity for characterization of thin films. Surf. Coat. Technol., 98 (1998) 962–970.

Schneider, D., Siemroth, P., Schülke, T., Berthold, J., Schultrich, B., Schneider, H.H., Ohr, R., Petereit, B., Hillgers, H., Quality control of ultra-thin and super-hard coatings by laser-acoustics. Surf. Coat. Technol., 153 (2002) 252–260.

Schneider, J.M., X-ray diffraction investigations of magnetron sputtered hard coatings. MSc thesis, Hull University, UK, 1994.

Scholl, R.A., Reactive PV deposition of insulators. Proc. of the 39th Annual Technical Conf. of the Soc. of Vacuum Coaters, Philadelphia, May, 1996, SVC, Albuquerque, 1996.

Schouten, M.J.W., Einfluss elastohydrodynamischer schmierung auf reibung, verschleiss und lebendsdauer von Getrieben/The influence of elastohydrodynamic lubrication on friction, wear and lifetime in gear transmissions/. Doctoral thesis, Eindhoven Technical University, 1973a. 388 p.

Schouten, M.J.W., Der einfluss elastohydrodynamischer schmierung und reibung, verschleiss und lebendsdauer von Getrieben/The influence of elastohydrodynamic lubrication on friction, wear and lifetime in gear transmissions. Schmiertechnik – Tribologie, 20 (1973b) 5, 147–151.

Scott, D. (ed.), Wear. Treatise on Materials Science and Technology, Vol. 13. Academic Press, New York, 1979, 498 pp.

Scott, K.T., Kingwell, R., Thermal spraying. In: Advanced Surface Coatings: A Handbook in Surface Engineering. Rickerby, D.S., Matthews, A. (eds), Blackie, Glasgow, UK, 1991, 217–243.

Seshan, K., Lacombe, R.H., Wagner, J.B., Effects of stress on the adhesion of metals to polyimide. Materials Research Society. Mat. Res. Soc. Symp. Proc. Vol. 130 (1989) 367–375.

Shan, H.S., Pandey, P.C., Influence of electrical characteristics of machine tools on cutting tool wear. Proc. 4th Int. Conf. on Production Engineering, Tokyo, 1980, 535–540.

Shanov, V., Tabakoff, W., Metwally, M., Erosive wear of CVD ceramic coatings exposed to particulate flow. Surf. Coat. Technol., 54/55 (1992) 25–31.

Sheikh-Ahmad, J.Y., Stewart, J.S., Feld, H., Failure characteristics of diamond-coated carbides in machining wood-based composites. Wear, 255 (2003) 1433–1437.

Sheldon, G.L., Wang, R., Clark, R.A., Characteristics of Ni–Ti surface alloys formed by electrospark deposition. Surf. Coat. Technol., 36 (1988) 445–454.

Shepard, S.R., Suh, N.P., The effects of ion implantation on friction and wear of metals. Trans. ASME,. J. Lubr. Tech., 104 (1982) 29–38.

Sherbiney, M.A., Halling, J., Friction and wear of ion-plated soft metallic films. Wear, 45 (1977) 211–220.

Shibuki, K. *et al.* Adhesion strength of diamond films on cemented carbide substrates. Surf. Coat. Technol., 36 (1988) 295–302.

Shimada, S., Ikawa, N., Ohmori, G., Tanaka, H., Uchikoshi, U., Molecular dynamics analysis as compared with experimental results of micromachining. Annals of the CIRP, 41 (1992) 1, 108–111.

Shimazaki, Y., Winer, W.O., Frictional behavior of sputtered TiN. Wear, 117 (1987) 161–177.

Shimizu, S., Fatigue limit concept and life prediction model for rolling contact machine elements. Tribology, Lubrication Technology (2005) May, 32–41.

Shinn, M., Hultman, L., Barnett, S.A., Growth, structure, and microhardness of epitaxial TiN/NbN superlattices. J. Mater. Res., 7 (1992) 901–911.

Shizhi, L., Cheng, Z., Xiang, X., Yulong, S., Hongshun, Y., Yan, X., Wu, H., The application of hard coatings produced by plasma-assisted chemical vapour deposition. Surf. Coat. Technol., 43/44 (1990) 1007–1014.

Shukla, A., Practical fracture mechanics in design, 2nd edition, Marcel Dekker, New York, USA, 2005, 525 pp.

Silva, F.J.G., Fernandes, A.J.S., Costa, F.M., Teixeira, V., Baptista, A.P.M., Pereira, E., Tribological behaviour of CVD diamond films on steel substrates. Wear, 255 (2003) 846–853.

Simmonds, M.C., Savan, A., Van Swygenhoven, H., Pflüger, E., Mikhailov, S., Structural, morphological, chemical and tribological investigations of sputter deposited MoS_x/metal multilayer coatings. Surf. Coat. Technol., 108–109 (1998) 340–344.

Simmonds, M.C., Savan, A., Pflüger, E., VAN Swygenhoven, H., Mechanical and tribological performance of MoS_2 co-sputtered composites. Surf. Coat. Technol., 126 (2000) 15–24.

Sin, H., Saka, N., Suh, N.P., Abrasive wear mechanisms and the grit size effect. Wear, 55 (1979) 163–190.

Singer, I.L., Solid lubricating films for extreme environments. Mat. Res. Soc. Symp. Proc., Materials Research Society, 140 (1989) 215–226.

Singer, I.L., A thermomechanical model for analyzing low wear-rate materials. Surf. Coat. Technol., 49 (1991) 474–481.

Singer, I., How third-body processes affect friction and wear. MRS Bulletin, June (1998) 37–40.

Singer, I.L., Bolster, R.N., Wegand, J., Fayelle, S., Hertzian stress contribution to low friction behavior of thin MoS_2 coatings. Appl. Phys. Lett., 57 (1990) 10, 995–997.

Singer, I.L., Dvorak, S.D., Wahl, K.J., Scharf, T.W., Role of third bodies in friction and wear of protective coatings. J. Vac. Sci. Technol., American Vacuum Society, A21 (2003) 5, 1–9.1.

Singer, I.L., Fayeulle, S., Ehni, P.D., Friction and wear behaviour of TiN in air: the chemistry of transfer films and debris formation. Wear, 149 (1991) 375–394.

Singer, I.L., Le Mogne, T., Donnet, C., Martin, J.M., Friction behavior and wear analysis of SiC sliding against Mo in SO2, O2 and H_2S at gas pressure between 4 and 40 Pa. Tribology Trans., 39 (1996a) 4, 950–956.

Singer, I.L., Fayeulle, S., Ehni, P.D., Wear behaviour of triode-sputtered MoS2 coatings in dry sliding contact with steel and ceramics, Wear, 195 (1996b) 7–20.

Singer, I.L., Le Mogne, T., Donnet, C., Martin, J.M., In situ analysis of the tribological films formed by SiC sliding against Mo in partial pressures of SO_2, O_2 and H_2S gases. J. Vac. Sci. Technol., A14 (1996c) 1, 38–45.

Singer, I., Pollock, H. (eds), Fundamentals of friction: macroscopic and microscopic processes. NATO ASI Series, Series E: Applied Sciences, Vol. 220, Kluwer Academic Publishers, London, 1992, 621 pp.

Sinha, S.K. (ed.), Scratching of materials and applications. Tribology and Interface Series No. 51. Elsevier, Amsterdam, Netherlands, 2006, 319 pp.

Siniawski, M.T., Harris, S.J., Wang, Q., A universal wear law for abrasion. Wear, 262 (2007) 883–888.

Siniawski, M.T., Harris, S.J., Wang, Q., Liu, S., Wear initiation of 52100 steel sliding against a thin boron carbide coating. Tribology Lett., 15 (2003) 1, 29–41.

Sirviö, E., Sulonen, M., Sundquist, H., Abrasive wear of ion-plated titanium nitride coatings on plasma-nitrided steel surfaces. Thin Solid Films, 96 (1982) 93.

Sivakumar, R., Mordike, B.L., High temperature coatings for gas turbine blades: a review. Surf. Coat. Technol., 37 (1989) 139–160.

Siwei, Z., Wear of Elastomers. Petroleum University Press, Shandong Province, China, 2000, 180 pp.

Sjöström, H., Wikström, V., Diamond-like carbon coatings in rolling contacts. Proc. Instn Mech. Engrs., Part J, 215 (2001) 545–561.

Smallwood, S.A., Eapen, K.C., Patton, S.T., Zabinski, J.S., Performance results of MEMS coated with a conformal DLC. Wear, 260 (2006) 1179–1189.

Smart, R.F., Selection of surface treatment. Tribology Int., April 1978, 97.

Söderberg, S., Westergren, K., Reineck, I., Ekholm, P.-E., Shahani, H., Properties and performance of diamond coated ceramic cutting tools. In: Applications of Diamond Films and Related Materials. Tzeng, Y. *et al.* (eds), Elsevier Scientific Publishers BV, 1991, 43–51.

Solecki, R., Ohgushi, Y., Contact stresses between layered elastic cylinders. J. Tribology, Trans. ASME, 106 (1984) 396–404.

Solomon, A.J., Antler, M., Mechanisms of wear of gold electrodeposits. Plating, Aug. (1970) 812–816.

Souza, R.M., Mustoe, G.G.W., Moore, J.J., Finite element modeling of the stresses, fracture and delamination during the indentation of hard elastic films on elastic–plastic soft substrates. Thin Solid Films, 392 (2001) 65–74.

Spalvins, T., Friction characteristics and lubrication with sputtered solid films. 2. Deutsche Sputtering-Schule: Tagung über Theorie und Anwendung der Kathodenzerstäuberung zur Herstellung dünner Schichten, 3-5.5.1971, Materials Research Corporation, Orangeburg, New York, 1971, 149–193.

Spalvins, T., Lubrication with sputtered MoS_2 films. ASLE Trans, 14 (1972) 4, 267–274.

Spalvins, T., Influence of various surface pretreatments on adherence of sputtered molybdenum disulphide to silver, gold, copper, and bronze. NASA Technical Note, NASA TN D–7169, 1973, 17 pp.

Spalvins, T., Morphological and frictional behavior of sputtered MoS_2 films. Thin Solid Films, 96 (1982) 17–24.

Spalvins, T., Sliney, H.E., Frictional behaviour and adhesion of Ag and Au films applied to aluminium oxide by oxygen-ion assisted screen cage ion plating. Surf. Coat. Technol.,, 68/69 (1994) 482–488.

Spear, K.E., Diamond – ceramic coating of the future. J. Am. Ceram. Soc., 72 (1989) 2, 171–191.

Spikes, H.A., Thin films in elastohydrodynamic lubrication: the contribution of experiment. Proc. Instn Mech. Engrs, 213 Pt J (1999) 335–352.

Spikes, H.A., Olver, A.V., MacPherson, P.B., Wear in rolling contacts. Wear, 112 (1986) 121–144.

Spitsyn, B.V., Bouilov, L.L., Derjaguin, B.V., Diamond and diamond-like films: deposition from the vapour phase, structure and properties. Prog. Crystal Growth Charact., 17 (1988) 79–170.

Springer, R.W., Hosford, C.D., Characterization of aluminium–aluminium nitride coatings sputter deposited using the pulsed gas process. J. Vac. Sci. Technol., 20 (1982) 3, 462–464.

Springer, R.W., Ott, N.L., Catlett, D.S., Effect of periodic chemical variation on the mechanical properties of Ta foils. J. Vac. Sci. Technol., 16 (1974) 3, 878–881.

Sproul, W.D., Very high reactive sputtering of TiN, ZrN and HfN. Thin Solid Films, 107 (1983) 141–147.

Sproul, W.D., Turning tests of high rate reactively sputter-coated T-15 HSS inserts. Surf. Coat. Technol., 33 (1987) 133–139.

Sproul, W.D., Multilayer, multicomponent, and multiphase physical vapour deposition coatings for enhanced performance. J. Vac. Sci. Technol., A12 (1994) 4, 1595–1601.

Sproul, W.D., Advances in reactive sputtering. Proc. of the 39th Annual Technical Conf. of the Soc. of Vacuum Coaters, Philadelphia, May, 1996, SVC, Albuquerque, 1996.

Sproul, W.D., Rudnik, P.J., Graham, M.E., Rohde, S.L., High rate reactive sputtering in an opposed cathode close-field unbalanced magnetron sputtering system. Surf. Coat. Technol., 43/44 (1990) 270–278.

Sproul, W.D., Tomashek, J.A., United States Patent No. 4 428 811 (1984).

Spur, G., Byrne, G., Bienia, B., The performance of high speed steel indexible inserts coated by physical vapour deposition in the milling of ductile materials. Surf. Coat. Technol., 43/44 (1990) 1074–1085.

Spurr, R.T., The abrasive wear of metals. Wear, 65 (1981) 315–324.

Srivastava, P.K., Vankar, V.D., Chopra, K.L., Mechanical properties of r.f. magnetron sputtered W–C films on stainless steel. Thin Solid Films, 161 (1988) 107–116.

Stachowiak, G.W., Batchelor, A.W., Engineering Tribology, 3rd edition, Elsevier, Amsterdam, The Netherlands, 2005, 801 pp.

Stachowiak, G.W., Batchelor, A.W., Stachowiak, G.W., Experimental Methods in Tribology. Elsevier, Amsterdam, The Netherlands, Elsevier Tribology Series No. 44, 2004, 372 pp.

Stachowiak, G.W., De Pellegrin, D.V., Podsiadlo, P., Characterization and classification of abrasive particles and surfaces. In: Wear – Materials, Mechanisms and Practice. Stachowiak, G.W. (ed.), Tribology in Practice Series, Wiley, Chichester, UK, 2005, 339–368.

Steinmann, P.A., Hintermann, H.E., A review of the mechanical tests for assessment of thin-film adhesion. J. Vac. Sci. Technol., A7 (1989) 3, 2267–2272.

Steinmann, P.A., Tardy, Y., Hintermann, H.E., Adhesion testing by the scratch test method: the influence of intrinsic and extrinsic parameters on the critical load. Thin Solid Films, 154 (1987) 333–349.

Steinmann, M., Müller, A., Meerkamm, H., A new type of tribological coating for machine elements based on carbon, molybdenum disulphide and titanium boride. Tribology Int., 37 (2004) 879–885.

Stephens, L.S., Liu, Y., Meletis, E.I., Finite element analysis of the initial yielding behavior of a hard coating/substrate system with functionally graded interface under indentation and friction. Trans. ASME. J. Tribology, 122 (2000) 381–387.

Stephenson, D.J., Nicolls, J.R., Hancock, P., Particle surface interaction during the erosion of alumina-coated MarM002. Wear, 111 (1986) 1, 31–39.

Stevenson, P., Leyland, A., Parkin, M., Matthews, A., The effects of process parameters on plasma carbon diffusion treatment of stainless steels at low pressures. 19th Int. Conf. Metallurgical Coatings and Thin Films (ICMCTH 92), San Diego, USA, 6–10.4.1992.

Stolarski, T.A., Tribology in Machine Design, Heinemann Newnes, London, UK, 1990, 298 pp.

Stoney, G.G., The tension of metallic films deposited by electrolysis. Proc. of the Royal Society of London. Series A, 82 (1909). 172–175, (DOI: 10.1098/rspa.1909.0021).

Stott, F.H., Mitchell, D.R.G., Wood, G.C., The influence of temperature on the friction and wear of thin ceramic coatings in carbon dioxide. J. Phys. D: Appl. Phys., 25 (1992) A189–A194.

Stout, K.J., Davis, E.J., Sullivan, P.J., Atlas of Machined Surfaces, Chapman and Hall, London, 1990, 247 pp.

Straede, C.A., Practical applications of ion implantation for tribological modification of surfaces. Wear, 130 (1989) 113–122.

Strafford, K.N., Subramanian, C. (eds), Quality control and assurance in advanced surface engineering. The Institute of Materials. The University Press, Cambridge, UK, 1997, 113–125.

Strondl, C., Van der Kolk, G.J., Hurkmans, T., Fleischer, W., Trinh, T., Carvalho, N.M., De Hosson, J.T.M., Properties and characteristics of multilayers of carbides and diamond-like carbon. Surf. Coat. Technol., 142–144 (2001) 707–713.

Stüber, M., Ulrich, S., Leiste, H., Kratzsh, A., Holleck, H., Graded layer design for stress-reduced and strongly adherent superhard amorphous carbon films. Surf. Coat. Technol., 116–119 (1999) 591–598.

Stupp, B.C., Synergistic effects of metal co-sputtered with MoS_2. Thin Solid Films, 84 (1981) 257–266.

Su, C., Lin, J.C., Thermal desorption of hydrogen from the diamond C(100) surface. Surface Science, 406 (1998) 149–166.

Subramanian, S.V., Ingle, S.S., Kay, D.A.R., Design of coatings to minimise tool crater wear. Surf. Coat. Technol., 61 (1993) 293–299.

Subramanian, C., Strafford, K.N., Review of multicomponent and multilayer coatings for tribological applications. Wear, 163 (1993) 85–95.

Sudarshan, T.S., Wear resistant coatings. TechTrends, International Reports on Advanced Technologies, Innovation 128, S.A. Paris, 1992, 195 pp.

Sue, J.A., X-ray elastic constants and residual stress of textured titanium nitride coating. Surf. Coat. Technol., 54/55 (1992) 154–159.

Sue, Y.L., Kao, W.H., Optimum multilayer TiN–TiCN coatings for wear resistance and actual application. Wear, 223 (1998) 119–130.
Sue, J.A., Troue, H.H., Friction and wear properties of titanium nitride coating in sliding contact with AISI 01 steel. Surf. Coat. Technol., 43/44 (1990) 709–720.
Sue, J.A., Troue, H.H., High temperature erosion behavior of titanium nitride and zirconium nitride coatings. Surf. Coat. Technol., 49 (1991) 31–39.
Sugimoto, I., Miyake, S., Solid lubricating fluorine-containing polymer film synthesized by perfluoropolyether sputtering. Thin Solid Films, 158 (1988) 51–60.
Suh, N.P., The delamination theory of wear. Wear, 25 (1973) 111–124.
Suh, N.P., Sin, H.C., The genesis of friction. Wear, 69 (1981) 91–114.
Suh, N.P., Tribophysics. Prentice-Hall, Englewood Cliffs, New Jersey, USA, 1986. 489 pp.
Suh, N.P., Jahanmir, S., Turner, A.P.L., Further investigation of the delamination theory of wear. J. Lubr. Tech., Trans. ASME, October (1974) 631–637.
Sullivan, J.L., The role of oxides in the protection of tribological surfaces – Part 1. Inst. Mech. Eng., C143/87 (1987a) 283–291.
Sullivan, J.L., The role of oxides in the protection of tribological surfaces – Part 2. Inst. Mech. Eng., C144/87 (1987b) 293–301.
Sullivan, J.L., Middleton, M.R., The mechanisms governing crack and pit formation in steel rolling sliding contact in aqueous lubricants. J. Synth. Lubr., 6 (1989) 1, 17–30.
Sundararajan, S., Bhushan, B., Micro/nanotribological studies of polysilicon and SiC films for MEMS applications. Wear, 217 (1998) 251–261.
Sundararajan, S., Bhushan, B., Micro/nanoscale tribology of MEMS materials, lubricants and devices. In Proc. 2nd World Tribology Congress, 3–7.9.2001a, Vienna, Austria, Austrian Tribology Society, 347–359.
Sundararajan, S., Bhushan, B., Micro/nanoscale tribology of MEMS materials, lubricants and devices. In: Fundamentals of Tribology and Bridging the Gap between the Macro- and Micro/Nanoscales. Bhushan, B. (ed.), Kluwer Academic Publishers, Dordrecht, The Netherlands, 2001b, 821–850.
Sundararajan, S., Bhushan, B., Micro/nanotribology of ultrathin hard amorphous carbon coatings using atomic force/friction force microscopy. Wear, 678–689 (1999) 225–229.
Sundershan, T.S., Wear resistant coatings. TechTrends, International Reports on Advanced Technologies, Innovation 128, Paris, 1992, 193 pp.
Sundquist, H., Rolling contact fatigue of case-hardened chromium steel. Wear, 66 (1981) 111–123.
Sundquist, H.A., Sirvio, E.H., Kurkinen, T.M., Wear of metal working tools ionplated with titanium nitride. Metals Technol., 10 (1983) 130–134.
Sunny, C.H.K., Wang, J., Chu, P.K., Surface energy, wettability and blood compatibility of phosphorus doped diamond-like carbon films. Diamond and Related Materials, 14 (2005) 78–85.
Suri, A.K., Nimmagadda, R., Bunshah, R.F., Influence of ion implantation and overlay coatings on various physico-mechanical and wear properties of stainless steel, titanium and aluminium. Thin Solid Films, 64 (1979) 191–203.
Suryanarayana, C., Koch, C.C., Nanocrystalline materials – current research and future directions. Hyperfine Interact., 130 (2000) 5–44.
Suzuki, M., Ludema, K.C., The wear process during the 'running-in' of steel in lubricating sliding. J. Tribology, Trans. ASME, 109 (1987) 587–593.
Suzuki, S., Matsuura, T., Uchizawa, M., Yura, S., Shibata, H., Friction and wear on lubricants and materials applicable to MEMS. IEEE, CH2957-9/91/0000-0143$01.00, 1991, 143–147.
Svahn, F., Kassman-Rudolphi, Å., Wallén, E., The influence of surface roughness on friction and wear of machine element coatings. Wear, 254 (2003) 1092–1098.
Swain, M.V., Mencik, J., Mechanical property characterization of thin films using spherical tipped indenters. Thin Solid Films, 253 (1994) 204–211.
Syan, C.S., Matthews, A., Swift, K.G., Knowledge-based expert systems in surface coating and treatment selection for wear reduction. Surf. Enging, 2 (1986) 249–255.

Syan, C.S., Matthews, A., Swift, K.G., Knowledge-based expert systems in surface coating and treatment selection for wear reduction. Surf. Coat. Technol., 33 (1987) 105.

Szczerek, M., Pytko, S., Wisniewski, M., Test methods – problems of repeatability and reproducibility. 2nd World Tribology Congress, Vienna, Austria, 3–7.9.2001, Proceedings, Austrian Tribology Society, 405–412.

Szeri, A.Z., Hydrodynamic and elastohydrodynamic lubrication. In: Modern Tribology Handbook. Bhushan, B. (ed.), CRC Press, New York, 2001, 383–453.

Tabor, D., Wear – a critical synoptic view. J. Lubr. Tech., Trans. ASME, Oct (1977) 387–395.

Tabor, D., Adhesion and friction. In: The Properties of Diamond. Field, J.E. (ed.), Academic Press, 1979, 325–350.

Tabor, D., Friction – the present state of our understanding. J. Lubr. Tech., Trans. ASME, 103 (1981) 169–179.

Takagi, R., Liu, T., The lubrication of steel by electroplated gold. ASLE Trans., 10 (1967) 115–123.

Takagi, R., Liu, T., Lubrication of bearing steels with electroplated gold under heavy loads. ASLE Trans., 11 (1968) 64–71.

Takagi, R., Tsuya, Y., Static friction between clean copper single crystal surfaces. Wear, 4 (1961) 216–217.

Takatsu, S., Saijo, K., Yagi, M., Shibuki, K., Echigoya, J., Microstructure of diamond films near the interface to WC substrate. In: Proc. 2nd Int. Conf. on Plasma Surface Engineering, Garmisch-Partenkirchen, Germany, 10–14.9 (1990) 747–752.

Talke, F.E., Tribologie in der Computertechnik. Tribologie und Schmierungstechnik, 38 (1991) 23–30.

Talke, F.E., Tribology in magnetic recording technology. 12th Int. Colloquium on Tribology, 12–13.1. 2000, Esslingen, Germany, (2000), 55–61.

Tallian, T.E., The theory of partial elastohydrodynamic contacts. Wear, 21 (1972) 49–101.

Tallian, T.E., A unified model for rolling contact life prediction. J. Lubr. Tech., Trans. ASME, 104 (1982) 336–346.

Tallian, T.E., Rolling contact failure control through lubrication. Conf. on Lubrication and Wear: Fundamentals and Application to Design, London, Proc. Inst. Mech. Eng., 182 Part 3A (1967–68) 205–236.

Tallian, T.E., Tribological design decisions using computerized databases. J. Tribology, Trans. ASME, 109 (1987) July, 381–387.

Tallian, T.E., McCool, J.I., An engineering model of spalling fatigue failure in rolling contact. Wear, 17 (1971) 447–461.

Tallian, T.E., McCool, J.E., Sibley, L.B., Partial elastohydrodynamic lubrication in rolling contact. Proc. Instn Mech. Engrs, 180 Pt 3B (1966) 169–184.

Tanaka, K., Mechanics and micromechanics of fatigue crack propagation. ASTM STP 1020, ASTM, Philadelphia, USA, 1989, 151–183.

Tanaka, K., Kato, T., Matsumoto, Y., Molecular dynamics simulation of vibrational friction force due to molecular deformation in confined lubricant film. Trans. ASME, J. Tribology, 125 (2003) 587–591.

Tanaka, A., Nishibori, T., Suzuki, M., Maekawa, K., Characteristics of friction surfaces with DLC films in low and high humidity air. Wear, 257 (2004a) 297–303.

Tanaka, A., Suzuki, M., Ohana, T., Friction and wear of various DLC films in water and air environments. Tribology Letters, 17 (2004b) 4, 917–924.

Tanaka, K., Kato, T., Matsumoto, Y., Molecular dynamics simulation of vibrational friction force due to molecular deformation in confined lubricant film. Trans. ASME. J. Tribology, 125 (2003) 587–591.

Tangena, A.G., Tribology of thin film systems. Doctoral thesis, Eindhoven Technical University, 1987, 130 pp.

Tangena, A.G., A low cycle fatigue wear model and its application to layered systems. 16th Leeds–Lyon Symposium on Tribology, 5–8.9.1989, Lyon, 1989, 6 pp.

Tangena, A.G., Hurkx, G.A.M., The determination of stress–strain curves of thin layers using indentation tests. J. Engng Mat. Technol., Trans. ASME, 108 (1986) 230–232.

Tangena, A.G., Van Wijngaarden, H., Fijnvandraat, J.G., A comment on an energy-based model of friction and its application to coated systems. Wear, 97 (1984) 303.

Tangena, A.G., Wijnhoven, P.J.M., The correlation between mechanical stresses and wear in layered system. Wear, 121 (1988) 27–35.

Tangena, A.G., Wijnhoven, P.J.M., Mijderman, E.A., The role of plastic deformation in wear of thin films. Trans ASME, J. Tribology, 110 (1988) 602–608.

Tankala, K., Debroy, T., Alam, M., Oxidation of diamond films synthesized by hot filament assisted chemical vapor deposition. J. Mater. Res., 5 (1990) 11, 2483–2489.

Taube, K.H., Measurement of mechanical properties of thin solid films I: Hardness, elasticity and stress. In: Advanced Techniques for Surface Engineering. Gissler, W., Jehn, H.A. (eds), Kluwer Academic Publishers, Dordrecht, The Netherlands, 1992, 275–294.

Tayebi, N., Polycarpou, A.A., Conry, T.F., Effects of substrate on determination of hardness of thin films by nanoscratch and nanoindentation techniques. J. Mater. Res., 19 (2004a) 6, 1791–1802.

Tayebi, N., Conry, T.F., Polycarpou, A.A., Reconciliation of nanoscratch hardness with nanoindentation hardness including the effects of interface shear stress. J. Mater. Res., 19 (2004b) 11, 3316–3323.

Teer, D.G., New solid lubricant coatings. Wear, 251 (2001) 1068–1074.

Tekmen, C., Ozdemir, I., Celik, E., Failure behaviour of functionally gradient materials under thermal cycling conditions. Surf. Coat. Technol., 174–175 (2003) 1101–1105.

Tendys, J., Conelly, I.L., Kenny, M.J., Pollock, J.T.A., Plasma immersion ion implantation using plasmas generated by radio frequency techniques. Appl. Phys. Lett., 53 (1988) 2143.

Terauchi, Y., Kohno, M., Nadano, H., Nakamoto, Y., Scoring resistance of spur gears with various coatings (1st report, Scoring tests under forced lubrication). Bulletin of JSME, 29 (1986a) 247, 235–240.

Terauchi, Y., Nadano, H., Kohno, M., Scoring resistance of copper-plated gears. Bulletin of JSME, 27 (1984) 232, 2287–2294.

Terauchi, Y., Nadano, H., Kohno, M., Effect of MoS2 films on scoring resistance of gears. J. Mechanisms, Transmissions and Automation in Design, Trans. ASME, 108 (1986b) 127–134.

Terauchi, Y., Nadano, H., Kohno, M., Nakamoto, Y., Scoring resistance of TiC- and TiN-coated gears. Tribology Int., 20 (1987) 5, 248–254.

Terheci, M., Wear by surface fatigue on a new foundation. Part: II Particle detachment mechanisms and quantitative aspects. Wear, 218 (1998) 191–202.

Thangaraj, A.R., Weinmann, K.J., On the wear mechanisms and cutting performance of silicon carbide whisker-reinforced alumina. J. Eng. Ind., Trans. ASME, 114 (1992) 301–308.

Thijsse, C.J., Tribology aids for designers. In: Tribological Design of Machine Elements. Dowson, D. *et al.* (eds), Elsevier Tribology Series 14, Amsterdam, 1989, 495–499.

Thom, R., Moore, L., Sproul, W., Chang, T.P., Rolling contact fatigue tests of reactively sputtered nitride coatings of Ti, Zr, Hf, Cr, Mo, Ti–Al, Ti–Zr and Ti–Al–V on 440C stainless steel substrates. Int. Conf. on Metallurgical Coatings and Thin Films (ICMCTF), San Diego, USA, 19–23.4.1993, 13 pp.

Thomas, A., Garnham, A.L., Todd, M.J., Current status of lead lubrication of ball bearings. Proc. Second Space Tribology Workshop, ESTL, Risley, U.K., 15–17.10.1980, ESA SP-158/Dec. 1980, 97–103.

Thomas, R., Acoustic micro imaging of cracks. Advanced Materials and Processes (2005) June, 49–51.

Thompson, C.V., Carel, R., Stress and grain growth in thin films. J. Mech. Phys. Solids, 44 (1996) 5, 657–673.

Thompson, P.A., Robbins, M.O., Simulation of contact-line motion: slip and the dynamic contact angle. The American Physical Society, Physical Review Letters, 63 (1989) 7, 766–769.

Thomson, L.A., Law, F.C., Ruchton, N., Franks, J., Biocompatibility of diamond-like carbon coating. Biomaterials, 12 (1991) Jan., 37–40.

Thornton, J.A., Influence of apparatus geometry and deposition conditions on structure and topography of thick sputtered coatings. J. Vac. Sci. Technol., 11 (1974) 666.

Thornton, J.A., High rate thick film growth. Ann. Rev. Mater. Sci., 7 (1977) 239–260.

Thorton, J.A., Coating Deposition by Sputtering. In: Deposition Technologies for Thin Films and Coatings. Bunshah, R.F. (ed.), Noyes Publications, New Jersey, USA, 1982, 170–243.

Thornton, J.A., Hoffman, D.W., Stress-related effects in thin films. Thin Solid Films, 171 (1989) 5–31.

Thouless, M.D., The role of fracture mechanics in adhesion. MRS Materials Research Society Symp., March 1988, Reno, Nevada, USA, Proc, 119 (1988) 51–62.

Thouless, M.D., Olsson, E., Gupta, A., Cracking of brittle films on elastic substrates. Acta Metall. Mater., 40 (1992) 6, 1287–1292.

Tian, H., Saka, N., Rabinowicz, E., Fretting failure of electroplated gold contacts. Wear, 142 (1991a) 265–289.

Tian, H., Saka, N., Rabinowicz, E., Friction and failure of electroplated sliding contacts. Wear, 142 (1991b) 57–85.

Tilbrook, T., Paton, D.J., Xie, Z., Hoffman, M., Microstructural effects on indentation failure mechanisms in TiN coatings: Finite element simulations. Acta Materialia, 55 (2007) 2489–2501.

Timoshenko, S., Strength of Materials, Van Nostrand, Toronto, 1941, 444 pp.

Timoshenko, S., Goodier, J.N., Theory of Elasticity, McGraw-Hill, New York, 1970.

Tipins, V.A., Cutting tool wear. In: Wear Control Handbook. Peterson, M.B., Winer, W. (eds), American Society of Mechanical Engineers, New York, USA, 1980, 891–930.

Tominaga, K., Preparation of AlN films by planar magnetron sputtering system with two facing targets. Vacuum, 41 (1990) 1154–1156.

Tönshoff, H.K., Denkena, B., Wear of ceramic tools in milling. Lubr. Engng, J. STLE, 47 (1991) 9, 772–778.

Topolovec Miklozic, K., Spikes, H.A., Application of atomic force microscopy to the study of lubricant additive films. Trans. ASME, J. Tribology, 127 (2005) 405–415.

Torrance, A.A., Debris detachment in plastic contacts. In: Wear Particles. Dowson, D. *et al.* (eds), Proc. 18th Leeds–Lyon Symp. Tribology, 3-6.9.1991, Lyon, France, Elsevier, London, 1992, 49–55.

Trent, E.M., Wear of metal-cutting tools. In: Treatise on material science and technology, Vol. 13, Wear. Scott, D. (ed), Academic Press, New York, USA, 1979, 443–490.

Tricoteaux, A., Jouan, P., Guerin, J., Martinez, J., Djouadi, A., Fretting wear properties of CrN and Cr_2N coatings. Surf. Coat. Technol., 174–175 (2003) 440–443.

Tsai, H., Bogy, D.B., Characterisation of diamond-like carbon films and their application as overcoats on thin film media for magnetic recording. J. Vac. Sci. Technol., A5 (1987) 6, 3287–3311.

Tsotsos, C., Kanakis, K., Davison, A., Baker, M.A., Matthews, A., Leyland, A., Mechanical and tribological properties of CrTiCu(B,N) glassy-metal coatings deposited by reactive magnetron sputtering. Surf. Coat. Technol., 200 (2006) 4601–4611.

Tsui, Y.C., Howard, S.J., Clyne, T.W., The effect of residual stresses on the debonding of coatings. Part II: An experimental study of a thermally sprayed system. Submitted to Acta Metall. Mater., 1993.

Tsui, T.Y., Joo, Y.C., A new technique to measure through film thickness fracture toughness. Thin Solid Films, 401 (2001) 203–210.

Tsui, T.Y., Pharr, G.M., Oliver, W.C., Bhatia, C.S., White, R.L., Anders, S., Anders, A., Brown, I.G., Nanoindentation and nanoscratching of hard carbon coatings for magnetic disks. Mat. Res. Soc. Symp. Proc., 383 (1995) 447–452.

Tsuya, Y., Takagi, R., Lubricating properties of lead films on copper. Wear, 7 (1964) 131–143.

Tucker, R.C., Considerations in the selection of coatings. Adv. Mat., Proc, 162 (2004) 3, 25–28.

Tung, S.C., McMillan, M.L., Automotive tribology overview of current advances and challenges for the future. Tribology Int., 37 (2004) 517–536.

Tuszynski, W., Szczerek, M., Michalczewski, R., Investigation of antiwear coatings deposited by the PVD process. Tribotest, 10 (2003) 3, 3–18.

Tysoe, W., Spencer, N., Why does Amonton's law work so well? Tribology, Lubrication Technology. June (2004) 64.

Uehara, K., Sakurai, M., Ikeda, T., On the problem of thermoelectric current in metal cutting. Annals of the CIRP, 41 (1992) 1, 75–78.

Uemura, M., Saito, K., Nakao, K., A mechanism of vapor effect on friction coefficient of molybdenum disulphide. STLE Preprint No. 89-TC-5D-4, 35th STLE/ASME Tribology Conf., Fort Lauderdale, USA, 16–19.10.1989, 5 pp.

Uetz, H., Sommer, K., Khosrawi, M.A., Übertragbarkeit von Versuchs- und Prüfergebnissen bei abrasiver Verschleissbeanspruchung auf Bauteile. VDI-Berichte No, 354, 1979, 107–124.

Uglov, V.V., Anishchik, V.M., Zlotski, S.V., Abadias, G., Dub, S.N., Stress and mechanical properties of Ti–Cr–N gradient coatings deposited by vacuum arc. Surf. Coat. Technol., 200 (2005) 178–181.

Uglov, V.V., Anishchik, V.M., Zlotski, S.V., Feranchuk, E.D., Alexeeva, T.A., Ulyanenkov, A., Brechbuehl, J., Lazar, A.P., Composition and phase stability upon annealing of gradient nitride coatings. Surf. Coat. Technol., 202 (2008) 2389–2393.

Unsworth, A., Tribology of artificial hip joints. Proc. Instn Mech. Engrs, Part J: Engineering Tribology, 220 (2006) 711–718.

Valli, J., A review of adhesion test methods for thin hard coatings. J. Vac. Sci. Technol., A4 (1986) 3007–3014.

Valli, J., Mäkelä, U., Matthews, A., Assessment of coating adhesion. Surf. Engng, 2 (1986) 1, 49–53.

Valli, J., Mäkelä, U., Matthews, A., Murawa, V., TiN coating adhesion studies using the scratch test method. J. Vac. Sci. Technol., A3 (1985a) 6, 2411–2414.

Valli, J., Palojärvi, J., Mäkelä, U., Pinnoitteen paksuuden mittaaminen kuulakokeella/ Coating thickness measurements using ball cratering (in Finnish). Technical Research Centre of Finland, Research Notes 435, 1985b, 22 pp.

Van Acker, K., Vercammen, K., Abrasive wear by TiO_2 particles on hard and on low friction coatings. Wear, 256 (2004) 353–361.

Van Ooij, W.J., Metal–polymer interfaces. In: Industrial Adhesion Problems. Brewis, D.M., Briggs, D. (eds), Orbital Press, Oxford, UK, 1985, 87–127.

Vancoille, M.J.S., Bogaerts, W.F., Prime – or the advent of second-generation expert systems. Corrosion Prevention and Control, (1986) 5.

Vancoille, E., Celis, J.P., Roos, J.R., Dry sliding wear of TiN based ternary PVD coatings. Wear, 165 (1993a) 41–49.

Vancoille, E., Celis, J.P., Roos, J.R., Tribological and structural characterization of a physical vapour deposited TiC/Ti(C,N)/TiN multilayer. Tribology Int., 26 (1993b) 2, 115–119.

Van der Kolk, G.J., Hurkmans, T., Trinh, T., Fleischer, W., Coating evaluations of decorative PVD finishes. Society of Vacuum Coaters. 41st Annual Technical Conf. Proc., (1998), 44–50.

Van der Kolk, G.J., Wear resistance of amorphous DLC and metal containing DLC in industrial applications. In: Tribology of Diamond-like Carbon Films – Fundamentals and Applications. Donnet, C., Erdemir, A. (eds), Springer, New York, USA, 2008, 484–493.

Vander Wal, R.L., Miyoshi, K., Street, K.W., Tomasek, A.J., Peng, H., Liu, Y., Margrave, J.L., Khabashesku, V.N., Friction properties of surface-florinated carbon nanotubes. Wear, 259 (2005) 738–743.

Vanhulsel, A., Velasco, F., Jacobs, R., Eersels, L., Havermans, D., Roberts, E.W., Sherrington, I., Anderson, M.J., Gaillard, L., DLC solid lubricant coatings on ball bearings for space applications. Tribology Int, 40 (2007) 1186–1194.

VDI-Richtlinien, Beschichten von Werkzeugen der Kaltmassivumformung: CVD- und PVD-Verfahren, Entwurf, No, 3198, September, 1991.

VDI-Richtlinien, Carbon films – basic knowledge, film types and properties. VDI 2840, November 2005, 44 pp.

Vepřek, S., The search for novel, superhard materials. J. Vac. Sci. Technol., A 17 (1999) 5, 2401–2420.

Vepřek, S., Argon, A.S., Towards the understanding of mechanical properties of super and ultra-hard nanocomposites. J.Vac. Sci. Technol., B20/2 (2002) 650–664.

Vepřek, S., Männling, H.D., Karvankova, P., Prochazka, J. The issue of the reproducibility of deposition of superhard nanocomposites with hardness of $\geq$50 GPa. Surf. Coat. Technol., 200 (2006) 3876–3885.

Vepřek, S., REIPRICH, S., A concept for the design of novel superhard coatings, Thin Solid Films, 268 (1995) 64–71.

Vepřek, S., Vepřek-Heijman, M.J.G., Industrial applications of superhard nanocomposite coatings. Surf. Coat. Technol., 202 (2008) 5063–5073.

Vercammen, K., Van Acker, K., Vanhulsel, A., Barriga, J., Arnsek, A., Kalin, M., Meneve, J., Tribological behaviour of DLC coatings in combination with biodegradable lubricants. Tribology Int., 37 (2004) 983–989.

Vetter, J., Barbezat, G., Crummenauer, J., Avissar, J., Surface treatment selections for automotive applications. Surf. Coat. Technol., 200 (2005) 1962–1968.

Vevrekova, J., Hainsworth, S.V., Effect of temperature and counterface on the tribological performance of W–DLC on a steel substrate. Wear, 264 (2008) 518–525.

Vickerman, J.C., Surface analysis: the principles techniques, John Wiley & Sons, New York, USA, 1997, 457 pp.

Vihersalo, J., Varjus, S., Ehrnstén, U., Zilliacus, R., Saarilahti, J., Nenonen, P., Ronkainen, H., Influence of substrate material, sputter cleaning and intermediate layers on the adhesion of rf-plasma a-C:H coatings. Proc. 2nd Int. Conf. on the Applications of Diamond Films and Related Materials (ADF'93). In: Yoshikawa, M. *et al.* (eds), MYU, Tokyo, 1993, 655–660.

Vingsbo, O., Wear and wear mechanisms. In: Wear of Materials. Ludema, K.C. *et al.* (eds), ASME, New York, 1979, 620–635.

Vingsbo, O., Hogmark, S., Jonsson, B., Ingemarsson, A., In: Microindentation Techniques in Material Science and Engineering, Blau, P.J., Lawn, B.R. (eds), ASTM STP889, Philadelphia, USA, 1986, 257–271.

Vingsbo, O., Odfalk, M., Shen, N., Fretting maps and fretting behaviour of some F.C.C. metal alloys. Wear, 138 (1990) 153–167.

Vizintin, J., Oil surface: additive reaction mechanisms. In: Surface Modification and Mechanisms – Friction, Stress, and Reaction Engineering. Totten, G.E., Liang, H. (eds), Marcel Dekker, Basel, Switzerland, 2004, 243–298.

Voevodin, A.A., Hard DLC growth and inclusion in nanostructured wear-protective coatings. In: Tribology of Diamond-like Carbon Films. Fundamentals and Applications. Donnet, C., Erdemir, A. (eds), Springer, New York, USA, 2008, 263–281.

Voevodin, A.A., Balbyshev, V.N., Khobaib, M., Donley, M.S., Nanostructured coatings approach for corrosion protection. Prog. Org. Coat, 47 (2003) 416–423.

Voevodin, A.A., Capano, M.A., Safriet, A.J., Donley, M.S., Zabinski, J.S., Combined magnetron sputtering and pulsed later deposition of carbides and diamond-like carbon films. Appl. Phys. Lett., 69 (1996) 188.

Voevodin, A.A., Donley, M.S., Preparation of amorphous diamond-like carbon by pulsed laser technology: A critical review. Surf. Coat. Technol., 82 (1996) 199–213.

Voevodin, A.A., Fitz, T.A., Hu, J.J., Zabinski, J.S., Nanocomposite tribological coatings with 'chameleon surface adaption'. J. Vac. Sci. Technol., A20/4 (2002) 1434–1444.

Voevodin, A.A., O'Neill, J.P., Zabinski, J.S., Tribological performance and tribochemistry of nanocrystalline WC/amorphous diamond-like carbon composites. Thin Solid Films, 342 (1999b) 194–200.

Voevodin, A.A., O'Neill, J.P., Zabinski, J.S., WC/DLC/WS2 nanocomposite coatings for aerospace tribology. Tribology Lett., 6 (1999a) 75–78.

Voevodin, A.A., Schneider, J.M., Stevenson, P., Matthews, A., Studies of atom beams produced by a saddle field source used for depositing diamond-like carbon films on glass. Vacuum, 46 (1995) 299–303.

Voevodin, A.A., Zabinski, J.S., Load-adaptive crystalline-amorphous nanocomposites. J. Mat. Sci., 33 (1998a) 319–327.

Voevodin, A.A., Zabinski, J.S., Superhard functionally gradient, nanolayered and nanocomposite diamond-like carbon coatings for wear protection. Diamond and Related Materials, 7 (1998b) 463–467.

Voevodin, A.A., Zabinski, J.S., Supertough wear-resistant coatings with 'chameleon' surface adaption. Thin Solid Films, 370 (2000) 223–231.

Vogel, J., Ion plating processes of wear-resistant coatings on tools. In: Wear and Corrosion Resistant Coatings by CVD and PVD. Pulkner, H.K. (ed.), Ellis Horwood Ltd, Chichester, UK, 1989, 165–196.

Von Stebut, J., Rezakhanlou, R., Anoun, K., Michel, H., Gantois, M., Major damage mechanisms during scratch and wear testing of hard coatings on hard substrates. Thin Solid Films, 181 (1989) 555–564.

Vossen, J.L., Measurements of film–substrate bond strengths by laser spallation. In: Adhesion Measurements of Thin Films, Thick Films and Bulk Coatings. Mittal, K.L. (ed.), ASTM STP 640, American Society for Testing Materials, Philadelphia, USA, 1978, 122–133.

Vossen, J.L., Cuomo, J.J., Glow discharge sputter deposition. In: Thin Film Processes. Vossen, J.L., Kern, J.L. (eds), Academic Press, 1978, 12–73.

Voumard, P., Savan, A., Pflüger, E., Advances in solid lubrication with MoS_2 multilayered coatings. Lubr. Sci., 13 (2001) 135–145.

Wahl, K.J., Dunn, D.N., Singer, I.L., Wear behaviour of Pb–Mo–S solid lubricating coatings. Wear, 230 (1999) 175–183.

Wahl, K.J., Dunn, D.N., Singer, I.L., Effects of ion implantation on microstructure, endurance and wear behavior of IBAD MoS_2. Wear, 237 (2000) 1–11.

Walck, S.D., Donley, M.S., Zabinski, J.S., Dyhouse, V.J., Characterisation of pulsed laser deposited PbO/MoS_2 by transmission electron microscopy. J. Mater. Sci., 9 (1994) 236–245.

Walck, S.D., Zabinski, J.S., McDevitt, N.T., Bultman, J.E., Characterisation of air-annealed pulse laser deposited ZnO–WS_2 solid film lubricants by transmission electron microscopy. Thin Solid Films, 305 (1997) 130–143.

Walls, J.M. (ed.), Methods of Surface Analysis. Cambridge University Press, Cambridge, 1989, 342 pp.

Walton, J., Fairley, N., Quantitative surface chemical-state microscopy by x-ray photoelectron spectroscopy. Surf. Interf. Anal., 36 (2004) 1, 89–91.

Wang, D.F., Kato, K., Coating thickness effects on initial wear of nitrogen-doped amorphous carbon in nano-scale sliding contact: Part I - in situ examination. Tribology Int., 36 (2003a) 649–658.

Wang, D.F., Kato, K., Nano-scale fatigue wear of carbon nitride coatings: Part I – Wear properties. J. Tribology, Trans. ASME, 125 (2003b) 430–436.

Wang, D.F., Kato, K., Nano-scale fatigue wear of carbon nitride coatings: Part II – Wear mechanisms. J. Tribology, Trans. ASME, 125 (2003c) 437–444.

Wang, D.F., Hu, N., Kato, K., Coating thickness effects on initial wear of nitrogen-doped amorphous carbon in nano-scale sliding contact: Part II – theoretical modelling. Tribology Int., 36 (2003) 649–658.

Wang, G.B., Wear mechanisms in vanadium carbide coated steels. Wear, 212 (1997) 25–32.

Wang, H., Akid, R., A room temperature cured sol-gel anticorrosion pre-treatment for Al2024–T3 alloys. Corrosion Science, 49 (2007) 4491–4503.

Wang, M., Miyake, S., Saito, T., Nanoindentation and nanowear of extremely thin protective layers of C–N and B–C–N. Tribology Int., 38 (2005) 657–664.

Wang, X., Xing, Y., Ma, S., Zhang, X., Xu, K., Teer, D., Microstructure and mechanical properties of MoS_2/titanium composite coatings with different titanium content. Surf. Coat. Technol., 201 (2007) 5290–5293.

Wang, Y.S., Hsu, S.M., Munro, R.G., Ceramics wear maps: alumina. Lubr. Engng, J. STLE. 47 (1991) 1, 63–69.

Wang, Y., Hsu, S.M., Jones, P., Evaluation of thermally-sprayed ceramic coatings using a novel ball-on-inclined plane scratch method. Wear, 218 (1998) 96–102.

Wang, Y., Lim, S., Tribological behavior of nanostructured WC particles/polymer coatings. Wear, 262 (2007) 1097–1101.

Wang, Y., Yan, F., Tribological properties of transfer films of PTFE-based composites. Wear, 261 (2006) 1359–1366.

Wang, M., Miyake, S., Saito, T., Nanoindentation and nanowear of extremely thin protective layers of C–N and B–C–N. Tribology Int., 38 (2005) 657–664.

Wäsche, R., Klafke, D., Tribology of DLC films under fretting conditions. In: Tribology of Diamond-like Carbon Films – Fundamentals and Applications. Donnet, C., Erdemir, A. (eds), Springer, New York, USA, 2008, 362–382.

Watanabe, S., Miyake, S., Murakawa, M., Tribological properties of cubic, amorphous and hexagonal boron nitride films. Surf. Coat. Technol., 49 (1991) 406–410.

Watanabe, S., Miyake, S., Jin, M., Murakawa, M., Frictional behavior of cubic BN films sliding against DLC. Tribology Int., 37 (2004) 923–927.

Watanabe, S., Noshiro, J., Miyake, S., Tribological characteristics of WS_2/MoS_2 solid lubricating multilayer films. Surf. Coat. Technol., 183 (2004) 347–351.

Watson, M., Byington, C., Edwards, D., Amin, S., Dynamic modeling and wear-based remaining useful life prediction of high power clutch systems. Tribology, Lubrication, 61 (2005) 12, 38–48.

Watts, J., Wolstenholme, J., An introduction to surface analysis by XPS and AES, John Wiley & Sons, New York, USA, 2005, 212 pp.

Wei, B., Komvopoulos, K., Friction and wear micromechnisms of amorphous carbon thin films. J. Tribology, Trans. ASME, 119 (1997) 823–829.

Wei, R., Wilbur, P.J., Liston, M., Effects of diamond-like hydrocarbon films on rolling contact fatigue of bearing steels. 3rd European Conf. on Diamond, Diamond-Like and Related Coatings – DIAMOND FILMS'92, 31.8–4.9.1992, Heidelberg, Germany, 11 pp.

Wei, R., Wilbur, P.J., Liston, M.-J., Lux, G., Rolling-contact-fatigue wear characteristics of diamond-like hydrocarbon coatings on steel. Wear, 162–164 (1993) 558–568.

Weiss, H., Characterization and quality assurance of surfaces and surface coatings. 1st Australian Int. Conf. on Surface Engineering – Practice and Prospects. Adeline, Australia, 12–14.3.1991, 24 pp.

Weissmantel, C., Applications of ion beams for the preparation of thin films. Proc. IX IVC-VICES, Madrid, Spain, 1983, 300.

Welsh, N.C., Surface hardening of non-ferrous metals by spark discharge. Nature, 181 (1958) 1005–1006.

Welsh, R.J., Plain Bearing Design Handbook. Butterworths, London, 1983, 162 pp.

Werner, H.W., Sims: from research to production control. Surf. Interf. Anal., 35 (2003) 11, 859–879.

Wheeler, D.R., Effect of oxygen pressure from 10^{-9} to 10^{-6} torr on the friction of sputtered MoS_x films. Thin Solid Films, 223 (1993) 78–86.

Wheeler, D.W., Chemical vapour deposition methods for protection against wear. In: Surface Coatings for Protection against Wear. Mellor, B.G. (ed.), CRC Press, Woodhead Publishing Ltd, Cambridge, UK, 2006, 101–145.

Wheeler, D.W., Wood, R.J.K., Solid particle erosion of CVD diamond coatings. Wear, 233–235 (1999) 306–318.

Whitehouse, D.J., The measurement and analysis of surfaces. Tribology Int., Dec. (1974) 249–259.

Wiiala, U.K., Kivivuori, S.O.J., Molarius, J.M., Sulonen, M., Wear of ion-plated hot-working tools. Surf. Coat. Technol., 33 (1987) 213–219.

Wiklund, U., Mechanics and tribology of micro- and nanolayered PVD coatings. Acta Universitatis Upsaliensis, Comprehensive Summaries of Uppsala Dissertations from Faculty of Sciences, No. 428, Uppsala University, Sweden, 1999, 47 pp.

Wiklund, U., Bromark, M., Larsson, M., Hedenqvist, P., Hogmark, S., Cracking resistance of thin hard coatings estimated by four-point bending. Surf. Coat. Technol., 91 (1997a) 57–63.

Wiklund, U., Hedenqvist, P., Hogmark, S., Multilayer cracking resistance in bending. Surf. Coat. Technol., 97 (1997b) 773–778.

Wiklund, U., Casas, B., Stavlid, N., Evaporated vanadium nitride as a friction material in dry sliding against stainless steel. Wear, 261 (2006) 2–8.

Wiklund, U., Gunnars, J., Hogmark, S., Influence of residual stresses on fracture and delamination of thin hard coatings. Wear, 232 (1999b) 262–269.

Wiklund, U., Nordin, M., Wänstrand, O., Larsson, M., Evaluation of flexible PVD TiC–C coating system. Article VIII in: Wiklund, U., Mechanics and tribology of micro- and nanolayered PVD coatings. Acta Universitatis Upsaliensis, Comprehensive Summaries of Uppsala Dissertations from Faculty of Sciences, No. 428, Uppsala University, Sweden, 1999c, 47 pp.

Wiklund, U., Wänstrand, O., Larsson, M., Hogmark, S., Evaluation of new multilayered physical vapour deposition coatings in sliding contact. Wear, 236 (1999a) 88–95.

Williams, B., Computer program for ferrous PM materials selection. Materials and Design, 8 (1987) 2, 89.

Williams, B.E., Glass, J.T., Characterisation of diamond thin films: diamond phase identification, surface morphology and defect structures. J. Mat. Res, 4 (1989) 2, 373–384.

Williams, J.A., Wear and wear particles – some fundamentals. Tribology Int., 38 (2005) 863–870.

Williamson, J.B.P., The shape of surfaces. In: CRC handbook of lubrication - Theory and practice of tribology. Volume II Theory and design. Booser, E. (ed), Boca Raton, Florida, 1984, 3–16

Wills, R.G.A., Walsh, F.C., Electroplating for protection against wear. In: Surface Coatings for Protection against Wear. Mellor, B.G. (ed.), CRC Press, Woodhead Publishing Ltd, Cambridge, UK, 2006, 226–248.

Wilson, S., Alpas, A.T., TiN coating wear mechanisms in dry sliding contact against high speed steel. Surf. Coat. Technol., 108–109 (1998) 369–376.

Wilson, A., Matthews, A., Housden, J., Turner, R., Garside, B., A comparison of the wear and fatigue properties of plasma-assisted physical vapour deposition TiN, CrN and duplex coatings on Ti–6Al–4V. Surf. Coat. Technol., 62 (1993) 600–607.

Window, B., Savvides, N., Charged particle fluxes from planar magnetron sputter sources. J. Vac. Sci. Technol., A, 4 (3) (1986) p.196–202.

Winer, W.O., Molybdenum disulfide as a lubricant: a review of the fundamental knowledge. Wear, 10 (1967) 422–452.

Wong, S.K., Kapoor, A., Effects of hard and stiff overlay coatings on the strength of surfaces in repeated sliding. Tribology Int., 29 (1996) 8, 695–702.

Wong, M.S. *et al.*, Thin diamond films for tribological applications. Mat. Res. Soc. Symp. Proc., 140 (1989) 483–488.

Wood, R.J.K., Tribo-corrosion of coatings: a review. J. Phys. D: Appl. Phys., 40 (2007) 5502–5521.

Woydt, M., Modern methods to retrieve innovative material solutions for tribosystems. Lubr. Engng (2000) May, 26–30.

Wriggers, P., Computational Contact Mechanics. John Wiley & Sons, Chichester, UK, 2002, 441 pp.

Wright, E.J., Westcott, C., Williams, D.E., Croall, I.F., Expert systems to aid design against corrosion. Materials and Design, 8 (1987) 3, 156.

Wu, R.L., Miyoshi, K., Vuppuladhadium, R., Jackson, H.E., Physical and tribological properties of rapid thermal annealed diamond-like carbon films. Surf. Coat. Technol., 54/55 (1992) 576–580.

Wu, J.H., Karthikeyan, S., Falk, M.L., Rigney, D.A., Tribological characteristics of diamond-like carbon (DLC) based nanocomposite coatings. Wear, 259 (2005) 744–751.

Xiaogang, L., Qing, C., Eryu, S., Initiation and propagation of case crushing cracks in rolling contact fatigue. Wear, 122 (1988) 33–43.

Xiaoyu, J., Lauke, B., Beckert, W., Schüller, T., Numerical simulation of micro-scratch tests for coating/substrate composites. Comp. Interf., 8 (2001) 1, 19–40.

Xiaoyu, J., Lauke, B., Schüller, T., Frictional contact analysis of scratch test for elastic and elastic–plastic thin-coating/substrate materials. Thin Solid Films, 414 (2002) 63–71.

Xie, Y., Hawthorne, H.M., The damage mechanisms of several plasma-sprayed ceramic coatings in controlled scratching. Wear (1999) 233–235, 293–305.

Xie, Y., Hawthorne, H.M., On the possibility of evaluating the resistance of materials to wear by ploughing using a scratch method. Wear, 240 (2000) 65–71.

Xie, Y., Hawthorne, H.M., Effect of contact geometry on the failure modes of thin coatings in the scratch adhesion test. Surf. Coat. Technol., 155 (2002) 121–129.

Yamada, Y., Tanaka, K., Saito, K., Friction and damage of coatings formed by sputtering polytetrafluoroethylene and polyimide. Surf. Coat. Technol., 43/44 (1990) 618–628.

Yan, W., Busso, E.P., O'Dowd, N.P., A micromechanics investigation of sliding wear in coated components. Proc. Royal Soc. London, A, 456 (2000) 2387–2407.

Yang, E.L., Hirvonen, J.P., Tribological transfer of polytetrafluoroethylene onto a diamond-like carbon film. Thin Solid Films, 226 (1993) 224–229.

Yang, E.L., Anttila, A., Toivanen, R.O., Hirvonen, J.P., Friction and wear of polytetrafluoroethylene on diamondlike carbon film. Thin Solid Films, 196 (1991a) L25–L29.

Yang, E.L., Hirvonen, J.P., Toivanen, R.O., Effect of temperature on the transfer film formation in sliding contact of PTFE with stainless steel. Wear, 146 (1991b) 367–376.

Yang, J., Komvopoulos, K., Dynamic indentation of an elastic–plastic multi-layered medium by rigid cylinder. J. Tribology, Trans. ASME, 126 (2004) 18–27.

Yang, J., Komvopoulos, K., A molecular dynamics analysis of surface interference and tip shape and size effects on atomic-scale friction. J. Tribology, Trans. ASME, 127 (2005) 513–521.

Yang, S.H., Kong, H., Choi, S., Kim, D.E., The effect of humidity on the rolling resistance of silver coatings modified by plasma surface treatments. Wear, 249 (2001) 780–787.

Yang, S.H., Kong, H., Yoon, E.S., Kim, D.E., An experimental study of the rolling resistance of bearings coated by pure silver. Wear, 225–229 (1999) 119–126.

Yang, S.H., Kong, H., Yoon, E., Kim, D.E., A wear map of bearing steel lubricated by silver films. Wear, 255 (2003) 883–892.

Yantio, G.R., Njankeu, J.Y., Paris, J., Denape, J., Pichon, L., Riviere, J.P., Study of seizure of coated and treated titanium alloy under fretting conditions. Tribology Int., 39 (2006) 1052–1059.

Yaws, C.L., Wakefield, G.F., Scale-up process for erosion resistant titanium carbide. Proc. 4th Int. Conf. on Chemical Vapor Deposition, Electrochemical Society, Princeton, NJ, USA, 1973, 577–588.

Yashar, P.C., Sproul, W.D., Nanometer scale multilayered hard coatings. Vacuum, 55 (1999) 179–190.

Yazu, S., Nakai, T., Tool application of diamond and CBN. In: Applications of Diamond Films and Related Materials. Tzeng, Y. *et al.* (eds), Elsevier Scientific Publishers BV, 1991, 37–41.

Ye, N., Komvopoulos, K., Effect of residual stress in surface layer on contact deformation of elastic-–plastic layered media. Trans. ASME, J. Tribology, 125 (2003a) 692–699.

Ye, N., Komvopoulos, K., Indentation analysis of elastic–plastic homogeneous and layered media: criteria for determining the real material hardness. Trans. ASME, J. Tribology, 125 (2003b) 685–691.

Ye, N., Komvopoulos, K., Three-dimensional finite element analysis of elastic–plastic layered media under thermomechanical surface loading. Trans. ASME, J. Tribology, 125 (2003c) 52–59.

Yerokhin, A.L., Nie, X., Leyland, A., Matthews, A., Dowey, S.J., Plasma electrolysis for surface engineering. Surf. Coat. Technol., 122 (1999) 73–93.

Yonekura, D., Chittenden, R.J., Dearnley, P.A., Wear mechanisms of steel roller bearings protected by thin hard and low friction coatings. Wear, 259 (2005) 779–788.

Yoshino, N., Shibuya, Y., Naoi, K., Nanya, T., Deposition of a diamond-like carbon film on a stainless steel substrate: studies of intermediate layers. Surf. Coat. Technol., 47 (1991) 84–88.

Yu, M.M.H., Bhushan, B., Contact analysis of three-dimensional rough surfaces under frictionless and frictional contact. Wear, 200 (1996) 265–280.

Yuan, F., Hayashi, K., Influence of the grain size of the aluminium coating on crack initiation in indentation. Wear 225–229 (1999) 83–89.

Yust, C.S. *et al.*, Friction and wear of ion-implanted TiB_2. Mat. Sci. Engng, A105/106 (1988) 489–496.

Zabinski, J.S., Donley, M.S., Lubricant Coatings. US Patent 5,282,985 (1994).

Zabinski, J.S., Donley, M.S., Dyhouse, V.J., McDevitt, N.T., Chemical and tribological characterisation of PbO–MoS_2 films grown by pulsed laser deposition. Thin Solid Films, 214 (1992) 156–163.

Zabinski, J.S., Donley, M.S. Walck, S.D., Schneider, T.R., McDevitt, N.T., The effects of dopants on the chemistry and tribology of sputter-deposited MoS2 films. Tribology Trans. 36, (1995a) 894–904.

Zabinski, J.S., Florkey, J.E., Walck, S.D., Bultman, J.E., McDevitt, N.T., Friction properties of WS2/ graphite fluoride thin films grown by pulsed layer deposition. Surf. Coat. Technol., 76–77 (1995b) 400–406.

Zabinski, J.S., Prasad, S.V., McDevitt, N.T., Advanced solid lubricant coatings for aerospace systems in tribology for aerospace systems. Proc.. of NATO AGARD SMP Conference 6–7 May 1996. Sesimbra, Portugal, AGARD-CP-589.

Zabinski, J.S., Voevodin, A., Ceramic and other hard coatings. In: Tribology of Mechanical Systems: A Guide to Present and Future Technologies. Vizintin, J., Kalin, M., Dohda, K., Jahanmir, S. (eds), ASME Press, New York, 2004, 157–181.

Zaretsky, E.V., A. Palmgren revisited – a basis for bearing life prediction. Lubr. Engng, J. STLE, Feb. (1997) 18–24.

Zeiler, E., Rosiwal, S.M., Singer, R.F., Klafke, D., Reducing friction and wear of titanium alloys by CVD-diamond coatings. 12th Int. Colloquium on Tribology, 12–13.1.2000, Esslingen, Germany (2000) 1861–1871.

Zhang, J., Moslehy, F.A., Rice, S.L., A model for friction in quasi-steady-state sliding. Part I. Derivation. Wear, 149 (1991a) 1–12.

Zhang, J., Moslehy, F.A., Rice, S.L., A model for friction in quasi-steady-state sliding. Part II. Numerical results and discussion. Wear, 149 (1991b) 13–25.

Zhang, L., Tanaka, H., Towards a deeper understanding of wear and friction on the atomic scale – a molecular dynamics analysis. Wear, 211 (1997) 44–53.

Zhang, S., Sun, D., Fu, Y., Du, H., Toughness measurement of thin films: a critical review. Surf. Coat. Technol., 198 (2005a) 74–84.

Zhang, S., Sun, D., Fu, Y., Du, H., Toughening of hard nanostructural thin films: a critical review. Surf. Coat. Technol., 198 (2005b) 2–8.

Zhang, S., Sun, D., Fu, Y., Pei, Y-T., De Hosson, J.T.M., Ni-toughened nc-TiN/a-SiNX nanocomposite thin films. Surf. Coat. Technol., 200 (2005c) 1530–1534.

Zhang, W., Liu, W., Wang, C., Tribological behavior of sol-gel TiO_2 films on glass. Wear, 253 (2002) 377–384.

Zhang, W., Tanaka, A., Wazumi, K., Koga, Y., Tribological properties of diamond-like carbon films in low and high moist air. Tribology Lett., 14 (2003b) 2, 123–130.

Zhang, W., Wang, C., Liu, W., Characterization and tribological investigation of sol-gel ceramic films on Ti-6Al–4V, Wear, 260 (2006a) 379–386.

Zhang, X., Prakash, B., Lauwerens, W., Zhu, X., He, J., Celis, J.P., Low-friction MoS_2 coatings resistant to wear in ambient air of low and high relative humidity. Tribology Lett., 14 (2003a) 131–135.

Zhang, Z.X., Dong, H., Bell, T., The load bearing capacity of hydrogen-free Cr-DLC coatings on deep-case oxygen hardened Ti6Al14V. Surf. Coat. Technol., 200 (2006a) 5237–5244.

Zhao, P., Hadfield, M., Wang, Y., Vieillard, C., Subsurface propagation of partial ring cracks under rolling contact Part I. Experimental studies. Wear, 261 (2006b) 382–389.

Zhao, P., Hadfield, M., Wang, Y., Vieillard, C., Subsurface propagation of partial ring cracks under rolling contact Part II. Fracture mechanics analysis. Wear, 261 (2006c) 390–397.

Zheng, C., Ran, J., Yin, G., Lei, W., Lei, W., Application of glow-discharge plasma deposited diamond like film on prosthetic artificial heart valves. In: Applications of Diamond Films and Related Materials. Tzeng, Y. *et al.* (eds), Elsevier Science Publishers BV, Amsterdam, 1991, 711–716.

Zhou, R.S., Surface topography and fatigue life of rolling contact bearings. Tribology Trans., STLE, 36 (1993) 3, 329–340.

Zhou, L., Kato, K., Umehara, N., Miyake, Y., Friction and wear properties of hard coating materials on textured hard disk sliders. Wear, 243 (2000) 133–139.

Zhou, Z., Rainforth, W.M., Lewis, D.B., Creasy, S., Forsyth, J.J., Clegg, F., Ehiasarian, A.P., Hovespian, P.E., Munz, W.D., Oxidation behaviour of nanoscale TiAlN/VN multilayer coatings. Surf. Coat. Technol., 177–178 (2004) 198–203.

Zhou, Z.R., Nakazawa, K., Zhu, M.H., Maruyama, N., Kapsa, P., Vincent, L., Progress in fretting maps. Tribology Int., 39 (2006) 1068–1073.

Zhitomirsky, V.N., Boxman, R.L., Goldsmith, S., Influence of external magnetic field on cathode spot motion and coating deposition using filtered vacuum arc evaporation. Surf. Coat. Technol., 68/69 (1994) 146–151.

Zlatanovic, M., Munz, W.D., Wear resistance of hard plasma-nitrided and sputter-ion-plated hobs. Surf. Coat. Technol., 41 (1990) 17–30.

Zoestbergen, E., De Hosson, J.T.M., Crack resistance of PVD coatings: influence of surface treatment prior to deposition. Surf. Engng, 18 (2002) 4, 283–288.

Zum Gahr, K.H., Abrasive wear. Proc. 5th Int. Congr. on Tribology, Vol. 2, Holmberg, K., Nieminen, I. (eds), Helsinki 12–15.6.1989, 34–39.

Index

Printed and bound by CPI Group (UK) Ltd, Croydon, CR0 4YY
27/08/2024
01029422-0003